Handbook of Optoelectronics
Second Edition

Series in Optics and Optoelectronics

Series Editors:
E. Roy Pike, Kings College, London, UK
Robert G. W. Brown, University of California, Irvine, USA

RECENT TITLES IN THE SERIES

Handbook of Optoelectronics

Second Edition
Enabling Technologies
Volume 2

Edited by
John P. Dakin
Robert G. W. Brown

CRC Press
Taylor & Francis Group
Boca Raton London New York

CRC Press is an imprint of the
Taylor & Francis Group, an **informa** business

CRC Press
Taylor & Francis Group
6000 Broken Sound Parkway NW, Suite 300
Boca Raton, FL 33487-2742

First issued in paperback 2019

ISBN-13: 978-1-4822-4180-8 (hbk)
ISBN-13: 978-0-367-87007-2 (pbk)

Library of Congress Cataloging-in-Publication Data

Names: Dakin, John, 1947- editor. | Brown, Robert G. W., editor.
Title: Handbook of optoelectronics / edited by John P. Dakin, Robert G. W. Brown.
Description: Second edition. | Boca Raton : Taylor & Francis, CRC Press, 2017. | Series: Series in optics and optoelectronics ; volumes 30-32 | Includes bibliographical references and index. Contents: volume 1. Concepts, devices, and techniques -- volume 2. Enabling technologies -- volume 3. Applied optical electronics.
Identifiers: LCCN 2017014570 | ISBN 9781482241808 (hardback : alk. paper)
Subjects: LCSH: Optoelectronic devices--Handbooks, manuals, etc.
Classification: LCC TK8320 .H36 2017 | DDC 621.381/045--dc23
LC record available at https://lccn.loc.gov/2017014570

Visit the Taylor & Francis Web site at
http://www.taylorandfrancis.com

and the CRC Press Web site at
http://www.crcpress.com

Contents

Series preface

This international series covers all aspects of theoretical and applied optics and optoelectronics. Active since 1986, eminent authors have long been choosing to publish with this series, and it is now established as a premier forum for high-impact monographs and textbooks. The editors are proud of the breadth and depth showcased by published works, with levels ranging from advanced undergraduate and graduate student texts to professional references. Topics addressed are both cutting edge and fundamental, basic science and applications-oriented, on subject matter that includes lasers, photonic devices, nonlinear optics, interferometry, waves, crystals, optical materials, biomedical optics, optical tweezers, optical metrology, solid-state lighting, nanophotonics, and silicon photonics. Readers of the series are students, scientists, and engineers working in optics, optoelectronics, and related fields in the industry.

Proposals for new volumes in the series may be directed to Lu Han, executive editor at CRC Press, Taylor & Francis Group (lu.han@taylorandfrancis.com).

Introduction to the Second Edition

There have been many detailed technological changes since the first edition of the *Handbook* in 2006, with the most dramatic changes can be seen from the far more widespread applications of the technology. To reflect this, our new revision has a completely new Volume III focused on applications and covering many case studies from an ever-increasing range of possible topics. Even as recently as 2006, the high cost or poorer performance of many optoelectronics components was still holding back many developments, but now the cost of many high-spec components, particularly ones such as light-emitting diodes (LEDs), lasers, solar cells, and other optical detectors, optoelectronic displays, optical fibers, and components, including optical amplifiers, has reduced to such an extent that they are now finding a place in all aspects of our lives. Solid-state optoelectronics now dominates lighting technology and is starting to dominate many other key areas such as power generation. It is revolutionizing our transport by helping to guide fully autonomous vehicles, and CCTV cameras and optoelectronic displays are seen everywhere we go.

In addition to the widespread applications now routinely using optoelectronic components, since 2006 we have witnessed growth of various fundamentally new directions of optoelectronics research and likely new component technologies for the near future. One of the most significant new areas of activity has been in nano-optoelectronics; the use of nanotechnology science, procedures, and processes to create ultraminiature devices across all of the optoelectronics domain: laser and LED sources, optical modulators, photon detectors, and solar cell technology. Two new chapters on silicon photonics and nanophotonics and graphene optoelectronics attempt to cover the wide range of nanotechnology developments in optoelectronics this past decade. It will, however, be a few years before the scale-up to volume-manufacturing of nano-based devices becomes an economically feasible reality, but there is much promise for new generations of optoelectronic technologies to come soon.

Original chapters of the first edition have been revised and brought up-to-date for the second edition, mostly by the original authors, but in some cases by new authors, to whom we are especially grateful.

Introduction to the First Edition

Optoelectronics is a remarkably broad scientific and technological field that supports a multibillion US dollar per annum global industry, employing tens of thousands of scientists and engineers. The optoelectronics industry is one of the great global businesses of our time.

In this *Handbook*, we have aimed to produce a book that is not just a text containing theoretically sound physics and electronics coverage, nor just a practical engineering handbook, but a text designed to be strong in both these areas. We believe that, with the combined assistance of many world experts, we have succeeded in achieving this very difficult aim. The structure and contents of this *Handbook* have proved fascinating to assemble, using this input from so many leading practitioners of the science, technology, and art of optoelectronics.

Today's optical telecommunications, display, and illumination technologies rely heavily on optoelectronic components: laser diodes, LEDs, liquid crystal, and plasma screen displays, etc. In today's world, it is virtually impossible to find a piece of electrical equipment that does not employ optoelectronic devices as a basic necessity—from CD and DVD players to televisions, from automobiles and aircraft to medical diagnostic facilities in hospitals and telephones, from satellites and space-borne missions to underwater exploration systems—the list is almost endless. Optoelectronics is in virtually every home and business office in the developed modern world, in telephones, fax machines, photocopiers, computers, and lighting.

"Optoelectronics" is not precisely defined in the literature. In this *Handbook*, we have covered not only optoelectronics as a subject concerning devices and systems that are essentially electronic in nature, yet involve light (such as the laser diode), but we have also covered closely related areas of electro-optics, involving devices that are essentially optical in nature but involve electronics (such as crystal light-modulators).

To provide firm foundations, this *Handbook* opens with a section covering "Basic Concepts." The "Introduction" is followed immediately by a chapter concerning "Materials," for it is through the development and application of new materials and their special properties that the whole business of optoelectronic science and technology now advances. Many optoelectronic systems still rely on conventional light sources rather than semiconductor sources, so we cover these in the third chapter, leaving semiconductor matters to a later section. The detection of light is fundamental to many optoelectronic systems, as are optical waveguides, amplifiers, and lasers, so we cover these in the remaining chapters of the Basic Concepts section.

The "Advanced Concepts" section focuses on three areas that will be useful to some of our intended audience, both now, in advanced optics and photometry, and now and increasingly in the future concerning nonlinear and short-pulse effects.

"Optoelectronics Devices and Techniques" is a core foundation section for this *Handbook*, as today's optoelectronics business relies heavily on such knowledge. We have attempted to cover all the main areas of semiconductor optoelectronics devices and materials in the eleven chapters in this section, from LEDs and lasers of great variety to fibers, modulators, and amplifiers. Ultrafast and integrated devices are increasingly important, as are organic electroluminescent devices and photonic bandgap and crystal fibers. Artificially engineered materials provide a rich source of possibility for next-generation optoelectronic devices.

At this point, the *Handbook* "changes gear"—and we move from the wealth of devices now available to us—to how they are used in some of the most important optoelectronic systems available today. We start with a section covering "Communication," for this is how the developed world talks and communicates by Internet and email today—we are all now heavily dependent on optoelectronics. Central to such optoelectronic systems are transmission, network architecture, switching, and multiplex architectures—the focus of our chapters here. In communication, we already have a multi-tens-of-billions-of-dollars-per-annum industry today.

"Imaging and displays" is the other industry measured in the tens of billions of dollars per annum range at the present time. We deal here with most if not all of the range of optoelectronic techniques used today from cameras, vacuum and plasma displays to liquid crystal displays and light modulators, from electroluminescent displays and exciting new three-dimensional display technologies just entering the market place in mobile telephone and laptop computer displays to the very different application areas of scanning and printing.

"Sensing and Data Processing" is a growing area of optoelectronics that is becoming increasingly important—from non-invasive patient measurements in hospitals to remote sensing in nuclear power stations and aircraft. At the heart of many of today's sensing capabilities is the business of optical fiber sensing, so we begin this section of the *Handbook* there, before delving into remote optical sensing and military systems (at an unclassified level—for here-in lies a problem for this *Handbook*—that much of the current development and capability in military optoelectronics is classified and unpublishable because of its strategic and operational importance). Optical information storage and recovery is already a huge global industry supporting the computer and media industries in particular; optical information processing shows promise but has yet to break into major global utilization. We cover all of these aspects in our chapters here.

"Industrial Medical and Commercial Applications" of optoelectronics abound, and we cannot possibly do justice to all the myriad inventive schemes and capabilities that have been developed to date. However, we have tried hard to give a broad overview within major classification areas, to give you a flavor of the sheer potential of optoelectronics for application to almost everything that can be measured. We start with the foundation areas of spectroscopy—and increasingly important surveillance, safety, and security possibilities. Actuation and control—the link from optoelectronics to mechanical systems is now pervading nearly all modern machines: cars, aircraft, ships, industrial production, etc.—a very long list is possible here. Solar power is and will continue to be of increasing importance—with potential for urgently needed breakthroughs in photon to electron conversion efficiency and cost of panels. Medical applications of optoelectronics are increasing all the time, with new learned journals and magazines regularly being started in this field.

Finally, we come to the art of practical optoelectronic systems—how do you put optoelectronic devices together into reliable and useful systems, and what are the "black art" experiences learned through painful experience and failure? This is what other optoelectronic books never tell you, and we are fortunate to have a chapter that addresses many of the questions we should be thinking about as we design and build systems—but often forget or neglect at our peril.

In years to come, optoelectronics will develop in many new directions. Some of the more likely directions to emerge by 2010 will include optical packet switching, quantum cryptographic communications, three-dimensional and large-area thin-film displays, high-efficiency solar-power generation, widespread biomedical and biophotonic disease analyses and treatments, and optoelectronic purification processes. Many new devices will be based on quantum dots, photonic crystals, and nano-optoelectronic components. A future edition of this *Handbook* is likely to report on these rapidly changing fields currently pursued in basic research laboratories.

We are confident you will enjoy using this *Handbook of Optoelectronics*, derive fascination and pleasure in this richly rewarding scientific and technological field, and apply your knowledge in either your research or your business.

Editors

John P. Dakin, PhD, is professor (Emeritus) at the Optoelectronics Research Centre, University of Southampton, UK. He earned a BSc and a PhD at the University of Southampton and remained there as a Research Fellow until 1973, where he supervised research and development of optical fiber sensors and other optical measurement instruments. He then spent 2 years in Germany at AEG Telefunken; 12 years at Plessey, research in Havant and then Romsey, UK; and 2 years with York Limited/York Biodynamics in Chandler's Ford, UK before returning to the University of Southampton.

He has authored more than 150 technical and scientific papers, and more than 120 patent applications. He was previously a visiting professor at the University of Strathclyde, Glasgow.

Dr. Dakin has won a number of awards, including "Inventor of the Year" for Plessey Electronic Systems Limited and the Electronics Divisional Board Premium of the Institute of Electrical and Electronics Engineers, UK. Earlier, he won open scholarships to both Southampton and Manchester Universities.

He has also been responsible for a number of key electro-optic developments. These include the sphere lens optical fiber connector, the first wavelength division multiplexing optical shaft encoder, the Raman optical fiber distributed temperature sensor, the first realization of a fiber optic passive hydrophone array sensor, and the Sagnac location method described here, plus a number of novel optical gas sensing methods. More recently, he was responsible for developing a new distributed acoustic and seismic optical fiber sensing system, which is finding major applications in oil and gas exploration, transport and security systems.

Robert G. W. Brown, PhD, is at the Beckman Laser Institute and Medical Clinic at the University of California, Irvine. He earned a PhD in engineering at the University of Surrey, Surrey, and a BS in physics at Royal Holloway College at the University of London, London. He was previously an applied physicist at Rockwell Collins, Cedar Rapids, IA, where he carried out research in photonic ultrafast computing, optical detectors, and optical materials. Previously, he was an advisor to the UK government, and international and editorial director of the Institute of Physics. He is an elected member of the European Academy of the Sciences and Arts (Academia Europaea) and special professor at the University of Nottingham, Nottingham. He also retains a position as adjunct full professor at the University of California, Irvine, in the Beckman Laser Institute and Medical Clinic, Irvine, California, and as visiting professor in the department of computer science. He has authored more than 120 articles in peer-reviewed journals and holds 34 patents, several of which have been successfully commercialized.

Dr. Brown has been recognized for his entrepreneurship with the UK Ministry of Defence Prize for Outstanding Technology Transfer, a prize from Sharp Corporation (Japan) for his novel laser-diode invention, and, together with his team at the UK Institute of Physics, a Queen's Award for Enterprise, the highest honor bestowed on a UK company. He has guest edited several special issues of *Applied Physics* and was consultant to many companies and government research centers in the United States and the United Kingdom. He is a series editor of the CRC Press "Series in Optics and Optoelectronics."

Contributors

Takao Ando
Research Institute of Electronics
Shizuoka University
Hamamatsu, Japan

Dominique Chiaroni
NOKIA Bell Labs
Paris-Saclay, France

John P. Dakin
Optoelectronics Research Centre
University of Southampton
Southampton, United Kingdom

Fernando Araujo de Castro
Materials Division
National Physical Laboratory
Middlesex, United Kingdom

Michel Digonnet
Stanford Photonics Research Center
Stanford University
Stanford, California

Uzi Efron
Holon Institute of Technology
Holon, Israel

Günter Gauglitz
Department of Analytical Chemistry
University of Tübingen
Tübingen, Germany

Ron Gibbs
Gibbs Associates
Dunstable, United Kingdom

Nicholas Holliman
University of Durham
Durham, United Kingdom

Kazuo Hotate
University of Tokyo
Tokyo, Japan

Michel Joindot
Laboratoire FOTON, UMR
CNRS
Lannion, France

J. Cliff Jones
School of Physics and Astronomy
University of Leeds
Leeds, United Kingdom

George K. Knopf
Department of Mechanical and Materials
    Engineering
University of Western Ontario
Ontario, Canada

Ton Koonen
Department of Electrical Engineering
Technische Universiteit Eindhoven
Eindhoven, the Netherlands

John N. Lee
Naval Research Laboratory
Washington, District of Columbia

Robert A. Lieberman
Lumoptix Inc
Redondo Beach, CA

Makoto Maeda
Home Network Company
SONY
Kanagawa, Japan

Michael A. Marcus
Lumetrics Inc.
Rochester, New York

Tom Markvart
University of Southampton
Southampton, United Kingdom

Susanna Orlic
Department of Optics
Technische Universität Berlin
Berlin, Germany

Tsutae Shinoda
Home Network Company
SONY
Kanagawa, Japan

Anthony E. Smart
Scattering Solutions, Inc.
Costa Mesa, California

Euan Smith
Light Blue Optics
Cambridge, United Kingdom

Masayuki Sugawara
Kochi University of Technology
Tokyo, Japan

Kenkichi Tanioka
Kochi University of Technology
Tokyo, Japan

Heiju Uchiike
Home Network Company
SONY
Kanagawa, Japan

J. Michael Vaughan
Worcestershire, United Kingdom

PART I

Enabling technologies for communications

Optical transmission

MICHEL JOINDOT
THE NATIONAL CENTRE FOR SCIENTIFIC RESEARCH

MICHEL DIGONNET
Stanford University

PROLOGUE

Optics has become the unique transmission technology in backbone networks, providing capacities absolutely unknown before, a very high transmission quality and a reduction of operational costs per transmitted bit. This is due to the development of wavelength division multiplexing (WDM) introduced in 1995 and allowing to transmit around 800 Gbit/s (80 and 10 Gbit/s channels) over one single mode fiber in 2005. This chapter describes this history, puts in evidence the basics of optical transmission and the development of WDM technology, and shows how the capacity of WDM systems can be increased by extending the used bandwidth, increasing the channel number or the bit rate per channel.

In 2005, coherent reception which had been explored by a many academic and industrial laboratories in the 1980s came back in foreground, but receivers were implemented quite differently from what had been proposed 30 years before. In fact, this very important technological step is closely related to the progress of electronics, allowing the implementation of complex and powerful digital signal processing (DSP) algorithms. The receiver consists now in a very "simple" optical front end (local oscillator and photodiodes), analogue to digital converters and a DSP unit compensating for all the transmission impairments due to the propagation over the fiber. Due to the fact that the optical demodulator just translates the channel transfer function into baseband, the baseband transfer function to be compensated for by the DSP unit is just the transfer function of the optical channel, which is not true with a quadratic detection. Coherent reception opens the way to complex and more than binary modulation schemes, which results into an increase of the transmitted bit rate within a given bandwidth.

Chromatic dispersion is no more in line compensated, which eliminates the dispersion-compensation fiber in each amplification site, polarization mode dispersion (PMD), which was a very serious problem can be compensated for very efficiently. Moreover, both polarizations can be used, which doubles the potential capacity of each channel by using polarization division multiplexing in conjunction with WDM. And DSP can cancel the interference between polarizations and then separate them without any problem.

Commercial coherent systems transmitting 80 100 Gbit/s channels in C band (8 Tbit/s over one single mode fiber) have been available since 2011 and are installed in backbone networks to face the always increasing traffic demand, while 200 and beyond 400 Gbit/s per channel are actively investigated in industrial laboratories.

1.1 INTRODUCTION

The enormous potential of optical waves for high-rate transmission of information was recognized as early as the 1960s. Because of their very high frequency, it was predicted that light waves could be ultimately modulated at extremely large bit rates, well in excess of 100 Gbit$\cdot$s^{-1} and orders of magnitudes faster than possible with standard microwave-based communication systems. The promise of optical waves for high-speed communication became a reality starting in the late 1980s and culminated with the telecommunication boom of the late 1990s, during which time a worldwide communication network involving many tens of millions of miles of fiber was deployed in many countries and across many oceans. In fact, much of the material covered in this handbook was generated to a large extent as a result of the extensive optoelectronics research that was carried out in support of this burgeoning industry. The purpose of this chapter is to provide a brief overview of the basic architectures and properties of the most widely used type of optical transmission line, which exploit the enormous bandwidth of optical fiber by a general technique called WDM. After a brief history of optical network development, this chapter examines the various physical mechanisms that limit the performance of WDM systems, in particular, their output power [which affects the output signal-to-noise ratio (SNR)], capacity (bit rate times number of channels), optical reach (maximum distance between electronic regeneration), and cost. The emphasis is placed on the main performance-limiting effects, namely fiber optical nonlinearities, fiber chromatic and group velocity dispersions (GVDs), optical amplifier noise and noise accumulation, and receiver noise. Means of reducing these effects, including fiber design, dispersion management, modulation schemes, and error-correcting codes, are also reviewed briefly. The text is abundantly illustrated with examples of both laboratory and commercial optical communication systems to give the reader a flavor of the kinds of system performance that are available. This chapter is not meant to be exhaustive, but to serve as a broad introduction and to supply background material for the following two chapters (optical network architecture and optical switching and multiplexed architectures), which dwell more deeply into details of system architectures. We also refer the reader to the abundant literature for a more in-depth description of these and many other aspects of optical communication systems (see, for example, [1,17,32,34]).

1.2 HISTORY OF THE INTRODUCTION OF OPTICS IN BACKBONE NETWORKS

Enabling the implementation of the optical communication concept required the development of a large number of key technologies. From the 1960s through the 1980s, many academic and industrial laboratories around the world carried out extensive research towards this goal. The three most difficult R&D tasks were the development of reliable laser sources and photodetectors to generate and detect the optical signals, of suitable optical fibers to carry the signals, and of the components needed to perform such basic functions as splitting, filtering, combining, polarizing, and amplifying light signals along the fiber network. Early silica-based fibers had a large core and consequently carried a large number of transverse modes, all of which travel at a different velocity, leading to unavoidable spreading of the short optical bits that carry the information and thus to unacceptably low bit rates over long distances. Perhaps, the most crucial technological breakthrough was the development of single-mode fibers, which first appeared in the mid-1970s and completely eliminated this problem. Over the following decade, progress in both material quality and manufacturing processes led to a dramatic reduction in the propagation loss of these fibers, from tens of decibels per kilometer in early

prototypes to the amazingly low typical current loss of 0.18 dB·km^{-1} around 1.5 μm used in submarine systems (or an attenuation of only 50% through a slab of glass about 17 km thick!). The typical attenuation of fibers used in long-distance terrestrial networks today is around 0.22 dB·km^{-1} at 1550 nm.

Fiber components were developed in the 1980s, including such fundamental devices as fiber couplers, fiber polarizers and polarization controllers, fiber wavelength division multiplexers [5,48], and rare-earth-doped fiber sources and amplifiers [17,18,41]. The descendants of these and several other components now form the building blocks of modern optical networks. Interestingly, the original basic research on almost all of these components was actually done not with communication systems in mind, but for fiber sensor applications, often under military sponsorship, in particular for development of the fiber optic gyroscope [5]. Parallel work on optoelectronic devices produced other cornerstone active devices, including high quantum efficiency, low-noise photodetectors, efficient and low-noise semiconductor laser diodes in the near infrared, in particular distributed-feedback (DFB) lasers, as well as semiconductor amplifiers, although these were eclipsed in the late 1980s by rare-earth-doped fiber amplifiers. The development of high-power laser diodes, began in the 1980s to pump high-power solid-state lasers, in particular for military applications, sped up substantially in the late 1980s in response for the growing demand for compact pump sources around 980 and 1480 nm for then-emerging erbium-doped fiber amplifiers. Another key element in the development of optical communication networks was the advent of a new information management concept called Synchronous Optical NETworks [31], especially matched to optical signals but also usable for other transmission technologies.

Up until the mid-1980s, long-distance communication network systems were based mostly on coaxial cable and radio frequency technologies. Although the maximum capacity of a single coaxial cable could be as high as 560 Mbit·s^{-1}, most installed systems operated at a bit rate of 140 Mbit·s^{-1}, while radio links could support typically eight 140 Mbit·s^{-1} radio channels. Intercontinental traffic was shared between satellite links and analogue coaxial undersea systems; digital undersea coaxial systems never existed. The switch to optical networks was motivated in part by the need for a much greater capacity, in part by the need for improved security and reliability of radio-based and cable-based systems. These systems were commonly affected by two different types of failures, namely signal fading and cable breaks due to civil engineering work, respectively. The first optical transmission systems were introduced in communication networks in the mid-1980s. Early prototypes were classical digital systems with a capacity that started at 34 Mbit·s^{-1} and rapidly grew to 140 Mbit·s^{-1}, i.e., comparable to established technologies. Optical communication immediately outperformed the coaxial technology in terms of regeneration span, which was tens of kilometers compared to less than 2 km for high-capacity coaxial-cable systems. However, there was no significant advantage compared with radio links, in terms of either capacity or regeneration span length, the latter being typically around 50 km. One could thus envision future long-distance networks based on a combination of secure radio links and optical fibers. Soon after optical devices became reliable enough for operation in a submerged environment, optical fiber links rapidly replaced coaxial-cable systems. The very first optical systems used multimode fibers and operated around 800 nm. This spectral window was changed to 1300 nm for the second generation of systems, when lasers around this wavelength first became available. In Japan, where optical communication links were installed early on, prior to the development of the 1550-nm systems, many systems operate in this window. Most of the current systems for backbone networks, especially in Europe and the United States, operate in the spectral region known as the C-band (1530–1565 nm). This has become the preferred window of operation because the attenuation of silica-based single-mode fiber is minimum around 1550 nm. The first transatlantic optical cable, TAT-8, was deployed in 1988. Containing two fiber pairs and a large number of repeaters, it spans a distance of about 6600 km under the Atlantic Ocean between Europe and the United States and carries 280 Mbits of information per second. In 1993, optical transmission systems carrying 2.5 Gbit·s^{-1} (16×155 Mbit·s^{-1}) over a single fiber with a typical regeneration span of 100 km began to be added to the growing worldwide optic–optic network. In terms of both capacity and transmission quality, radio-based systems could no longer compete, and optics became the unique and dominating technology in backbone networks.

The single most important component that made high-speed communication possible over great distances (≫100 km) without electronic regeneration is the optical amplifier. Although the loss of a communication optic around 1.5 μm is extremely small, after a few tens of kilometers, typically 50–100 km, the signal power has been so strongly attenuated that further propagation would cause the SNR of the signal at the receiver to degrade significantly, and thus the transmission quality, represented by the bit error rate (BER), to be seriously compromised. The SNR can be improved by increasing the input signal power, but the latter can only be increased so much before the onset of devastating nonlinear effects in the optic, in particular stimulated Raman scattering (SRS), stimulated Brillouin scattering (SBS), and four-wave mixing (FWM). Moreover, the gain in distance would be limited: a transmission over 200 km instead of 100 km of current optic would require the input power to be increased by roughly 20 dB!

This distance limitation was initially solved by placing optoelectronics repeaters along the optical line. Each repeater detects the incoming data stream, amplifies it electronically, and modulates the current of a new laser diode with the detected modulation. The modulated diode's output signal is then launched into the next segment in the optic link. This approach works well, but its cost is high and its bit rate is limited, on both counts by the repeaters' high-speed electronics. A much cheaper alternative, which requires high-speed electronics only at the two ends of the transmission line, is optical amplification. Each electronic repeater is now replaced by an in-line optical amplifier, which amplifies the low-power signals that have traveled through a long optic span before their SNR gets too low and then reinjects them into the next segment in the optic link. The advantage of this all-optical solution is clearly that the optical signal is never detected and turned into an electronic signal, until it reaches the end of the long-haul optical line, which can be thousands of kilometers long. Because the noise figure (NF) of optical amplifiers is low, typically 3–5 dB, the SNR can still be quite good even after the signals have traveled through dozens of amplifiers.

Starting as early as the 1960s, much research was devoted to several types of in-line optical amplifiers, first with semiconductor waveguide amplifiers [51], then with rare-earth-doped optic amplifiers [18], and more recently Raman optic amplifiers [28].

Semiconductor amplifiers turned out to have the highest wall-plug efficiency. However, at bit rates under about 1 Gbit·s⁻¹, in WDM systems they induce cross-talk between signal channels. Although solutions have been recently proposed, semiconductor amplifiers have not yet entered the market in any significant way, partly because of the resounding success of the erbium-doped optic amplifier (EDFA). First reported in 1987 [42], this device provides a high small-signal gain around 1.5 µm (up to ~50 dB) with a high saturation power and with an extremely high efficiency—the record is 11 dB of small-signal single-pass gain per milliwatt of pump power [52]. EDFAs used in telecommunication systems operate in saturation and have a lower gain, but it is still typically as high as 20–30 dB. The EDFAs can be pumped with a laser diode, at either 980 or 1480 nm, and they are thus very compact. Another key property is their wide gain spectrum, which stretches from ~1475 to ~1610 nm, or a total bandwidth of 135 nm (~16.4 THz!). For technical reasons, a single EDFA does not generally supply gain over this entire range, but rather over one of three smaller bands, called the S-band (for "short," ~1480–1520 nm), the C-band (for "conventional," ~1530–1565 nm), and the L-band (for "long," ~1565–1610 nm). Amplification in the S-band can also be accomplished with a thulium-doped fiber amplifier (TDFA) [49]. Gain has been obtained over the S- and C-band by combining an EDFA and a TDFA [50].

Perhaps more importantly, EDFAs induce negligible channel cross-talk at modulation frequencies above about 1 MHz. These unique features make it nearly ideally suited for optical communication systems around 1.5 µm. Since the mid-1990s, it has been the amplifier of choice in the overwhelming majority of deployed systems, thus eliminating the electrical regeneration bottleneck.

The very large gain bandwidth of EDFAs and other optical amplifiers also provided the opportunity of amplifying a large number of modulated optical carriers at different wavelengths distributed over the amplifier bandwidth. This concept of WDM had of course already been applied in radio links. One significant advantage of WDM optical systems is that the same amplifier amplifies many optical channels, in contrast with classical regenerated systems, which require one repeater per channel. Optical amplifiers thus reduce the installation

cost of networks in two major ways. First, the WDM technique results in an increase in capacity without laying new fibers, which reduces optic cost. Second, the cost of amplification is shared by a large number of channels, and because the use of a single optical amplifier is cheaper than implementing one regenerator per channel, the transmission cost is reduced proportionally. This critical economic advantage provided the final impetus needed to displace regenerated systems and launch the deployment of the worldwide optical WDM backbone networks that took place at the end of the 1990s.

1.3 GENERAL STRUCTURE OF OPTICAL TRANSMISSION SYSTEMS

1.3.1 Modulation and detection: RZ and NRZ codes

While radio systems use a wide variety of modulation formats in order to improve spectrum utilization, in optical systems data have been so far transmitted using binary intensity modulation. A logic 1 (resp. 0) is associated to the presence (resp. absence) of an optical pulse. Two types of line codes are mainly encountered: nonreturn to zero (NRZ), where the impulse duration is equal to the symbol duration (defined as the inverse of the data rate) and return to zero (RZ), where the impulse duration is significantly smaller than the symbol duration. This property explains why the name "return to zero" is used; if the impulse duration equals roughly one half of the symbol time, the modulation format is designed as RZ 50%. So at a bit rate of 10 Gbit·s⁻¹, NRZ uses impulses with a width of approximately 100 ps, while RZ 50% or RZ 25% will use 50 and 25 ps wide pulses, respectively. RZ has, for a given mean signal power, a higher signal peak power. This property can be used to exploit nonlinear propagation effects, which under certain conditions can improve system performance. Details will be provided further on.

Research is actively being conducted to investigate new modulation schemes for future high-bitrate systems. For instance, duobinary encoding, a well-known modulation scheme in radio systems, has been proposed because of its higher resistance to chromatic dispersion. Carrier-suppressed

RZ, an RZ modulation format with an additional binary phase modulation, is also extensively studied, as well as recently differential phase shift keying (DPSK); both provide a higher resistance to nonlinear effects. However, only NRZ and RZ are used in installed systems today.

Detection of the optical signals at the end of a transmission line is performed with a photodetector, which is typically a PIN or an avalanche diode. Photons are converted in the semiconductor in electron–holes pairs and collected in an electrical circuit. The generated current is then amplified and sent to a decision circuit, where the data stream is detected by comparing the signal to a decision threshold, as in any digital system. Detection errors can occur in particular because of the presence of noise on the signal and in the detector. The error probability is a measurement of the transmission quality. In practice, the error probability is estimated by the BER, defined as the ratio of error bits over the total number of transmitted bits.

Several sources of noise are typically present in the detection process of an optical wave. Shot noise, the most fundamental one, arises from the discontinuous nature of light. Thermal noise is generated in the electrical amplifiers that follow the photodetector. In PIN receivers, thermal noise is typically 15–20 dB larger than the quantum limit, and if the optical signal is low, thermal noise dominates shot noise. In the case of amplified systems under normal operating conditions, the amplified spontaneous emission (ASE) noise of the in-line amplifiers

is largely dominant compared to the receiver noise, which can thus be neglected.

1.3.2 Basic architecture of amplified WDM communication links

A typical amplified WDM optical link is illustrated in Figure 1.1. The emitter consists of N lasers of different wavelengths, each one representing a communication channel. The lasers are typically DFB semiconductor lasers with a frequency stabilized by a number of means, including temperature control and often Bragg gratings. Each laser is amplitude modulated by the data to be transmitted. This modulation is performed with an external modulator, such as an amplitude modulator based on lithium niobate waveguide technology. Direct modulation of the laser current would be simpler and less costly, but it introduces chirping of the laser frequency, which is unacceptable at high modulation frequencies over long distances [1,29]. The fiber-pigtailed laser outputs are combined onto the optical fiber bus using a wavelength division multiplexer, then generally amplified by a booster fiber amplifier.

The multiplexer can be based on concatenated WDM couplers (for low number of channels) or arrayed waveguide grating multiplexers. This is the technology of choice for high channel counts, in particular in the so-called dense wavelength division multiplexed systems: although this term has no precise definition, it applies usually to systems with a channel spacing less than 200 GHz. In-line

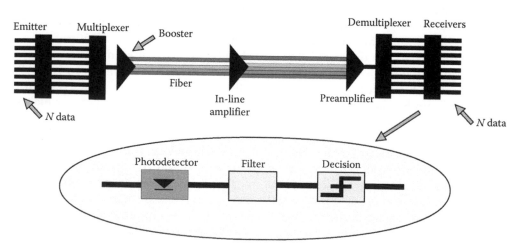

Figure 1.1 Structure of an amplified WDM optical system.

optical amplifiers are distributed along the fiber bus to periodically amplify the power in the signals, depleted by lossy propagation along the fiber. Ideally, each amplifier provides just enough gain in each channel to compensate for the loss in that channel, i.e., such that each channel experiences a net gain of unity. Because the gain of an optical amplifier and to a smaller extent the loss along the fiber are wavelength dependent, the net gain is different for different channels. If the difference in net gain between extreme channels is too large, after a few amplifier/bus spans the power in the strongest channel will grow excessively, thus robbing the gain for other channels and making their SNR at the receiver input unacceptably low. This major problem is typically resolved by flattening or otherwise shaping the amplifier spectral gain profile, or equivalently equalizing the power in the channels, using one of several possible techniques, either passive (for example, long-period fiber gratings) [58] or dynamic (e.g., with variable optical attenuators). While the first WDM systems that appeared in the mid-1980s used basic amplifiers without control system, the new very long reach systems include complex gain flattening devices that compensate for the accumulation of gain tilt. The gain of an in-line amplifier ranges approximately from 15 to 30 dB per channel. The distance between amplifiers is typically 30–100 km, depending on fiber loss, number of channels, and other system parameters. In deployed undersea systems, the amplifiers are equally spaced, whereas in terrestrial networks the amplifier location depends on geographical constraints, for instance, building location and amplifiers tend to be unevenly spaced. At the output of the transmission line, a wavelength division demultiplexer separates the N optical channels, which are then sent individually to a receiver, then electronically processed.

1.3.3 Basic architectures of repeaterless systems

A repeaterless communication system aims to accomplish a very long optical reach without in-line amplifiers. A common application is connecting two terrestrial points on each side of a straight or narrow arm of sea, in which case it is generally not worth incurring the cost of undersea amplifiers. Some deployed repeaterless systems are extremely long, as much as hundreds of kilometers,

and consequently they exhibit a high span attenuation, up to 50 or even 60 dB. The problem is then to ensure that the output power at the receiver is high enough, in spite of the high span loss, to achieve the required SNR at the end of the line.

This goal has been achieved with a number of architectures. A common solution involves using a preamplifier, i.e., an amplifier placed before the receiver to increase the detected power and reduce the receiver NF, as is also often done in classical WDM systems. Another one is to use a high-power amplifier, i.e., an amplifier placed between the emitter and the transmission fiber to boost the signal power launched into the fiber. Although such a booster amplifier is also present in some of the amplified WDM systems described before, the typical feature in repeaterless systems is the high power level, which can reach up to 30 dBm. In both cases, the amplifier can be either an EDFA or a Raman amplifier, or a combination of both. This solution, as we will see further on, is limited by nonlinear effects in the fiber, although they can be somewhat mitigated with proper dispersion management.

A third solution specific to repeaterless systems is to place an amplifier fiber in the transmission fiber itself and to pump it remotely with pump power launched into the transmission fiber from either end. The amplifier fiber can then be a length of Er-doped fiber; the entire transmission fiber can be lightly doped with erbium (the so-called distributed fiber amplifier); or the transmission fiber can be used as a Raman amplifier. The drawback of this general approach is that it requires a substantially higher pump power than a traditional EDFA, and it is therefore more costly. The reason is that the pump must propagate through a long length of transmission fiber before reaching the amplifier fiber, and because the transmission fiber is much more lossy at typical pump wavelengths than in the signal band, some of the pump power is lost. A fourth general solution, which is not specific to repeaterless systems, is to use powerful error-correcting schemes [12].

1.3.4 Optical reach and amplification span

Two important features of a WDM communication system are its total capacity, usually expressed as $N \times D$, where N is the number of optical channels and D the bit rate per channel, and the optical reach, which is the maximum distance over which the signal can be transmitted without regeneration. Even in amplified systems with a nominally unity net gain transmission, due to the accumulation of noise from the optical amplifiers and signal distortions, after a long enough transmission distance the bit error becomes unacceptably high, and the optical signals need to be regenerated. In practical deployed WDM systems in 2001, this limitation typically occurs after about seven amplifier spans with a loss of roughly 25 dB per span.

Another important parameter is the amplification span, i.e., the distance between adjacent amplifiers. The performance of an optical WDM system cannot be expressed only in terms of optical reach; the number of spans must also be introduced. As an example, the optical reach of commercially available terrestrial systems in 2002 was around 800 km, compared to 6500 km in transatlantic systems. A key difference between them is the amplification span, as will be explained in the next section. In the following, WDM systems with a bit rate per channel of 2.5, 10, and 40 Gbit·s⁻¹ will be designated as *WDM 2.5G*, *WDM 10G*, and *WDM 40G*, respectively.

1.4 LIMITATIONS OF OPTICAL TRANSMISSION SYSTEMS

1.4.1 Noise sources and bit error rate

1.4.1.1 AMPLIFIER NOISE

Amplification cannot be performed without adding noise to the amplified signals. In optical amplifiers, this noise originates from ASE [17], which is made of spontaneous emission photons emitted by the active ions (Er^{3+} in the case of EDFAs) via radiative relaxation subsequently amplified as they travel through the gain medium. The spectral power density of the ASE signal per polarization mode is given by

$$\gamma_{ASE} = n_{sp} h \nu (G - 1) \qquad (1.1)$$

where G is the amplifier gain, h Planck's constant, and ν the signal optical frequency. n_{sp} is a dimensionless parameter larger or equal to unity called the spontaneous noise factor. It depends on the amplifier's degree of inversion, and it approaches unity (lowest possible noise) for full inversion of

the active ion population. The ASE is a broadband noise generated at all frequencies where the amplifier supplies gain, and its bandwidth is nominally the same as that of the amplifier gain. The ASE power coming out of the amplifier, concomitantly with the amplified signals, is obtained by the integration of γ_{ASE} over the frequency bandwidth of the gain. As an example, in a particular C-band EDFA amplifying ten signals equally spaced between 1531 and 1558 nm and with a power of 1 μW each, and with a peak gain of 33 dB at 1531 nm, the total power in the amplified signals is 5.5 mW, whereas the total ASE output power is 0.75 mW, i.e., more than 10% of the signals' power.

The photodetectors used in receivers are the so-called quadratic detectors, i.e., they respond to the square of the optical field. Detection of an optical signal S corrupted by additive noise N (ASE noise in the case of amplified systems) in a photodetector thus gives a signal proportional to $|S+N|^2$. Expansion of this signal gives rise to the signal $|S|^2$ (the useful signal) plus two noise terms. The first term ($2SN$) is the beat noise between the signal and the ASE frequency component at the signal frequency; it is called the signal–ASE beat noise. The second term (N^2) is the beat noise between each frequency component of the ASE with itself (the ASE–ASE beat noise). The signal–ASE beat noise varies from channel to channel, but the ASE–ASE beat noise is the same for all channels. A third and fourth noise terms are of course the shot noise of the amplified signal and the shot noise of the ASE, and to this must be added a fifth term, namely the receiver noise discussed earlier. In high-gain amplifiers with low input signals, which are applicable to most in-line amplifiers in communication links, the dominant amplifier noise term is the signal–ASE beat noise. In the amplifier example given at the end of the previous paragraph, the SNR degradation (also known as the NF) is ~3.4 dB for all 10 signals, and it is due almost entirely to signal–ASE beat noise. This noise term is typically large compared to the receiver noise, which can usually be neglected. Note that the NF is defined as the SNR degradation of a *shot-noise-limited* input signal. The SNR degradation at the output of an amplifier is therefore equal to the NF only when the input signal is shot-noise limited. In a chain of amplifier, this is true for the first amplifier that the signal traverses. However, after traveling through several amplifiers, the signal is no longer shot-noise limited but dominated by signal–ASE beat noise, and the SNR degradation is smaller than the NF. Refer to "Accumulation of Noise" section for further detail on noise accumulation in amplifier chains.

1.4.1.2 PHOTORECEIVER THERMAL NOISE

As mentioned earlier, the photoreceiver thermal noise is generally fairly large compared to shot noise. However, it can become negligible when the signals are amplified with a preamplifier placed before the detector. To justify this statement, consider a receiver consisting of an optical preamplifier of gain G followed by a photodetector. The optical signal and ASE noise powers at the receiver input are proportional to G and $G-1$, respectively (in practice, G is very large and $G-1 \approx G$). Because the thermal noise does not depend on G, and because it is typically 15–20 dB worse than the quantum limit, it is clear that if the preamplifier gain is large enough, say 20 dB, the thermal noise is negligible compared to the signal–ASE beat noise. This is exactly the same phenomenon as in electronics, where the high-gain first stage of a receiver masks the noise of the following stages. This property illustrates another advantage brought by optical amplifiers: optical preamplifiers allow to get away from the relatively poor NF of electronic circuits and thus to achieve much better performance.

1.4.1.3 RELATIONSHIP BETWEEN BIT ERROR RATE AND NOISE

How does the error rate at the receiver depend on the noise level, or more exactly on the optical signal-to-noise ratio (OSNR), of the detected signal? To answer this question, we must make some assumptions regarding both the signal and the noise. First, because the signal–ASE beat noise depends on the signal power, it also depends on the state of the signal, i.e., on the transmitted data. If we assume an ideal on–off keying (OOK) modulation, signal–ASE beat noise is present only when the signal is on, whereas the ASE–ASE beat noise is present even in the absence of signal. Because the data can be assumed to be equally often on and off, the mean signal power is equal to half the peak power.

Second, to obtain an analytical expression for the bit error probability requires another assumption, common in communication theory, which is that the noise has a Gaussian statistics. This is true for signal–ASE beat noise, as a result of the linear processing of Gaussian processes, but it is not true for ASE–ASE beat noise. However, under normal operating conditions of amplified systems (i.e., with a sufficiently high OSNR), the influence of ASE–ASE beat noise remains relatively small. After a large number of optical amplifiers, however, the ASE–ASE beat noise component, which depends on the total ASE noise, can become significant. An effective way to reduce this noise component is then to place before the receiver an optical filter that cuts down the ASE power *between* the optical signals. This can be accomplished with a comb filter or with the demultiplexer that separates the channels. Such a filter reduces the ASE–ASE beat noise, but of course it does not attenuate either the signals or the ASE at the signals' frequencies, so it does not affect the signal–ASE beat noise. In the following, we assume that such a filter, with a rectangular transmission spectrum of optical bandwidth B_a, is placed before the receiver.

Third, because the noise variance is not the same conditionally to the transmitted data, the best decision threshold is not just at equal distance between the two signal levels associated with the two possible data values at the sampling time, but rather some other optimum threshold value that depends on signal power. Assuming that this optimum threshold value is used, the bit error probability (or BER) can be expressed as [30]:

$$P_{\text{exact}} = \frac{1}{2}\text{erfc}\left(\frac{\sqrt{2}R}{\sqrt{m}+\sqrt{m+4R}}\right) \qquad (1.2)$$

where erfc is the complementary error function and the SNR, R is the ratio of the mean signal power to the ASE power within the electrical bandwidth B, i.e., $\gamma_{\text{ASE}}B$. The electrical bandwidth B is the bandwidth of the electronic post-detection circuits. The parameter m is the normalized optical filter bandwidth, $m = B_a/B$. Equation 1.2 can be easily derived by computing the variances of the signal–ASE beat noise and ASE–ASE beat noise contributions. For the computation of the first term, the average power of the signal is used. An ideal rectangular optical filter is assumed, as well as a rectangular electrical filter

The SNR is usually measured not within the signal bandwidth, but over a much larger bandwidth B_0, corresponding generally to 0.1 nm in wavelength (or 12.5 GHz near 1550 nm). Calling this parameter the optical SNR R_0, the bit error probability can be rewritten as

$$P_{\text{exact}} = \frac{1}{2}\text{erfc}\left(\frac{Q}{\sqrt{2}}\right) \qquad (1.3)$$

where Q is the quality factor:

$$Q = \frac{2R_0}{\sqrt{\mu} + \sqrt{\mu + 4\beta R_0}} \qquad (1.4)$$

β is the ratio B/B_0 of the electrical to measurement bandwidths, $R_0 = \beta R$ and $\mu = m\beta^2$. An error probability of 10^{-9} (resp. 10^{-15}) requires, for example, $Q = 6$ (resp. 8). Neglecting the ASE–ASE beat noise contribution ($\mu \to 0$), the BER can be simply expressed as

$$P_{\text{exact}} = \frac{1}{2}\text{erfc}\left(\sqrt{\frac{R_0}{2\beta}}\right) \qquad (1.5)$$

From Equation 1.4, the value of R_0 needed to achieve a given quality factor Q_0 is given by

$$R_0 = \beta Q_0^2\left(1 + \frac{\sqrt{m}}{Q_0}\right) \qquad (1.6)$$

As an example, consider a 10 Gbit·s^{-1} system with an optical bandwidth B_a of 50 GHz, an electrical bandwidth B of ~7 GHz (as a rule of thumb, the electrical bandwidth is taken to be 70% of the bit rate), and a filter bandwidth B_0 of ~12.4 GHz (0.1 nm). Then $\beta = 0.56$ and $m = 7$, and a BER of 10^{-15} ($Q_0 = 8$) requires an OSNR R_0 of ~17 dB.

Based on the various degradation mechanisms that induce power penalty along the transmission system, equipment vendors specify a minimum SNR required by a given system. Typically, for *WDM 10G* systems, the minimum OSNR is in the range of 21–22 dB. It must be noted that the required OSNR increases with increasing electrical bandwidth, which is proportional to the bit rate. For instance, by going from 2.5 to 10 Gbit·s^{-1} the OSNR needs to be increased by about 6 dB (see Equation 1.6). This is a very important constraint, for system designers as well as operators.

1.4.1.4 ACCUMULATION OF NOISE

The accumulation of noise generated by successive amplifiers along an optical line degrades the OSNR. This accumulated noise limits the number of successive amplifiers that can be used, and thus the optical reach. Assuming a link with N equally spaced amplifiers of mean output power per channel P_0, an inversion parameter n_{sp}, and a gain G compensating exactly for the attenuation of the fiber span between them, the total noise power P_n (including both polarization modes) within a bandwidth B_0 at the last amplifier output is given by [29]:

$$P_n = 2Nn_{\text{sp}}(G-1)h\nu B_0 = n_{\text{sp}}\frac{L}{Za}(\exp(\alpha Z_a)-1)h\nu B_0$$

$$= 2n_{\text{sp}}\alpha L \frac{G-1}{\ln G}h\nu B_0 \qquad (1.7)$$

where L is the total length of the link, Z_a the length of the amplification span, and α the attenuation factor of the fiber. The SNR R is

$$R(N) = \frac{P_0}{2Nn_{\text{sp}}(G-1)h\nu B_0} \qquad (1.8)$$

or in dB:

$$\text{OSNR (dB)} = P_0(\text{dBm}) - P_n(\text{dBm}) \qquad (1.9)$$

These relationships show that for a given launched power and a given amplification span, the maximum transmission distance, represented by the maximum number of spans, is limited by the minimum required OSNR at the receiver input. For a given distance L, the OSNR increases when G decreases, i.e., when the amplification span becomes shorter. As an example, Equation 1.7 shows that for a link with a fixed length L and a fiber attenuation of 0.2 dB·km^{-1}, an OSNR improvement of 7 dB is achieved when the span length is reduced from $Z_a = 100$ km (a span loss of 20 dB) to $Z_a = 50$ km (a span loss of 10 dB). The link with the 50-km span length does require twice as many amplifiers, but the gain they each need to supply is reduced by 10 dB. So is their output noise power, and as a result the OSNR is improved. This illustrates how important a parameter the fiber attenuation is. For a given amplification span and a given number of spans, any reduction in this attenuation will result in a better OSNR simply because the amplifier gain G will be smaller. Equivalently, a lower fiber loss allows increasing the optical reach. For example, a fiber loss reduction as small as 0.02 dB·km^{-1} from 0.23 to 0.21 dB·km^{-1}, in 100-km spans, will allow to reduce the gain by 2 dB and thus to improve the

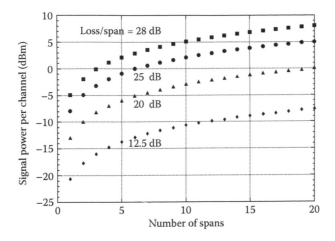

Figure 1.2 Output signal power per channel required to achieve an OSNR of 20 dB as a function of the number of spans for different span losses.

OSNR by as much as 2 dB. If with the 0.23 dB·km^{-1} fiber, the required OSNR was reached after eight spans; with the 0.21 dB·km^{-1} fiber, it will be possible to have 12 spans ($10\log(12/8) = 1.7$ dB), the same noise power being then produced by a larger number of less noisy amplifiers. This dependence of the OSNR on the span loss explains why the amplification span is significantly shorter in undersea lightwave systems compared to terrestrial ones, because transmission distances are much longer.

Figure 1.2 shows the output signal power per channel (calculated from Equation 1.8) needed to reach an OSNR of 20 dB as a function of the number of spans for different span losses. The inversion parameter of the optical amplifiers is taken to be $n_{sp} = 1.6$, and the output ASE is assumed to be filtered with a 0.1-nm narrowband filter ($B_0 = 12.5$ GHz). These curves also allow comparing the power required for achieving a given optical reach. For instance, for a link length $L = 1000$ km and a fiber attenuation $\alpha = 0.25$ dB km^{-1}, if there are $N = 20$ spans, each one of them will have a loss $\alpha L/N = 12.5$ dB, and Figure 1.2 shows that the required power per channel will be −7.5 dBm. When the number of spans is divided by two ($N = 10$), the span loss increases to 25 dB and the required signal power jumps nearly 10-fold, to 2 dBm.

1.4.2 Signal distortions induced by propagation

In addition to SNR degradation due to optical amplifiers, transmitted optical signals also suffer distortions induced by propagation along the fiber.

These effects become more and more important as the bit rate increases

1.4.2.1 CHROMATIC DISPERSION

Within a narrow bandwidth around the carrier angular frequency ω_0, a fiber of length L can be viewed as an all-pass linear filter, with an attenuation nearly independent of wavelength over the small signal bandwidth prevailing at the bit rates under consideration. Expanding the phase up to the second order in frequency ω about ω_0 allows to write the transfer function $H(\omega)$ of this filter as

$$H(\omega) = A \, \exp\left[i\left(\Phi(\omega_0) + \beta_1 L(\omega - \omega_0) + \frac{1}{2}\beta_2 L(\omega - \omega_0)^2 \right) \right]$$

(1.10)

Higher order terms in the expansion must be considered when, for instance, β_2 equals zero. The first and second terms of the exponent represent a constant phase shift and the delay of the impulse (also called group delay), respectively. The third term, proportional to the second derivative of the signal mode index with respect to the wavelength, originates physically from the dependence of the mode group velocity on wavelength. It is often referred to as GVD or chromatic dispersion. In an optical waveguide such as a fiber, GVD is approximately the sum of the material dispersion and the fiber dispersion. It is mathematically represented by β_2 usually expressed in ps^2·km^{-1} or by the so-called chromatic dispersion D, which is a more familiar parameter to system designers, expressed in ps nm^{-1}·km^{-1}. It is the group delay variation over a 1-nm bandwidth after propagation along a 1-km length of fiber. In a standard communication fiber, D is around 15 ps·nm^{-1}·km^{-1}. D (expressed in ps·nm^{-1}·km^{-1}) is related to β_2 (in ps^2·km^{-1}) and λ (in nm) by

$$D(\lambda) = -6\,\pi 10^5 \frac{\beta_2(\lambda)}{\lambda^2} \qquad (1.11)$$

Chromatic dispersion results in broadening of the signal as it propagates through the fiber. When the signal pulse amplitude is Gaussian, the pulse width evolution along the fiber can be computed analytically and pulse broadening can be expressed with simple expressions [1]. If a Gaussian pulse with a complex impulse envelope $u(t, 0)$ of temporal width θ_0,

$$u(t,0) = U_0 \exp\left(-\frac{t^2}{2\theta_0^2}\right) \qquad (1.12)$$

is launched into the fiber at $z=0$, the impulse at distance L is given by [1]:

$$u(t,L) = U \exp\left(-\frac{T^2}{2\theta^2(x)} + i\Psi(x,t)\right) \qquad (1.13)$$

where U is the amplitude taking into account the fiber attenuation, $T=it-\beta_1 L$ the time in local coordinates associated to the signal and $\Psi(x, T)$ a phase term. The parameter $\theta(x)$ is the temporal pulse width at distance L, given by

$$\theta(x) = \theta_0 \sqrt{1+x^2} \qquad (1.14)$$

where $x=L/L_D$ is the propagation distance normalized to the characteristic dispersion length $L_D = \theta_0^2/|\beta_2|$.

As expected from physical arguments, in the presence of chromatic dispersion the pulse width expands along the fiber in much the same way as a spatial beam expands in space due to diffraction (see Equation 1.14). Here the dispersion length L_D plays the same role as the Rayleigh range does in the diffraction of Gaussian beams. For a fiber of length L and dispersion coefficient β_2, there is an optimum value of the incident pulse width θ_0 that minimizes the pulse width at the fiber output. This optimum pulse width, obtained by setting the

derivative of Equation 1.14 with respect to θ_0 equal to zero, is given by

$$\theta_{0,\text{opt}} = \sqrt{|\beta_2|L} \qquad (1.15)$$

and the output pulse width is

$$\theta(L) = \sqrt{2}\theta_{0,\text{opt}} = \sqrt{2|\beta_2|L} \qquad (1.16)$$

Stated differently, Equation 1.15 shows that for a given input pulse width θ_0 the optimum fiber length that minimizes the output temporal pulse width is $L=\theta_0^2/|\beta_2|=L_D$, i.e., one dispersion length.

This analysis shows that the larger the chromatic dispersion is, the narrower the initial pulse needs to be, and the larger the pulse width will be at the output of the fiber. It implies that if the input pulse width is not properly selected, i.e., if it is either too narrow or too wide, chromatic dispersion will cause successive pulses to overlap, which creates what is known as intersymbol interference (ISI). This deleterious effect alters the decision process and thus increases the BER. It must be noted that a Gaussian pulse extends indefinitely in time and there is theoretically always a finite amount of ISI; the maximum distance is thus set by a "tolerable" level of ISI. This distance is exactly L_D if we define this acceptable ISI is reached when the initial width of the pulse has been multiplied by $\sqrt{2}$ (which is somewhat arbitrary). As an example, assuming Gaussian pulses with θ_0 equal to half the symbol duration, for a standard communication fiber ($D=17$ps·nm^{-1} km^{-1}, i.e., $\beta_2=20$ ps^2·km^{-1}) this distance is equal to 2000 km at 2.5 Gbit·s^{-1}, but only 125 km at 10 Gbit·s^{-1}.

This result demonstrates that propagation at bit rates of 10 Gbit·s^{-1} or greater in a standard fiber is not possible over distances longer than a few tens of kilometer without significant ISI. This problem is circumvented in practice by introducing along the transmission line components with a negative dispersion coefficient to compensate for chromatic dispersion, in much the same way as an optical lens is used to refocus a free-space beam after it has expanded as a result of diffraction. This method has been demonstrated in the laboratory with a number of optical filters, especially fiber Bragg gratings for dynamic compensation [19], or more simply and commonly with a length of dispersion-compensating fiber (DCF) designed to exhibit a strong negative dispersion coefficient D [17], in the range of -90 ps·nm^{-1}·km^{-1} for standard fiber. This last solution is the only one used in commercial systems today, and fiber suppliers try to develop the best compensation fiber matched to the fiber they sell. DCFs are typically more lossy than standard communication fibers, with attenuation coefficients around 0.5 dB·km^{-1}. To make up for this additional loss, the DCF is typically inserted near an amplifier. In order to reduce the impact of the DCF loss on the amplifier NF, in WDM systems the DCF is usually placed in the middle of a two-stage amplifier.

Chromatic dispersion depends on wavelength. This dependence is characterized by the dispersion slope (expressed in ps·nm^{-2}·km^{-1}). In order to completely compensate for the dispersion at any wavelength, the fiber and the associated DCF must exhibit the same D/S ratio, where S is the slope of the dispersion D. The existence of a perfectly slope-compensating DCF depends strongly on the type of fiber. For example, a DCF very well matched to standard single-mode fiber (SSMF) is available, but this is not true for all fibers. If the slope is not matched, some channels will exhibit a finite residual chromatic dispersion outside the "acceptance window," i.e., the interval within which the dispersion must lie to ensure a correct transmission quality. This window is typically 1000 ps·nm^{-1} wide for a *WDM 10G* system. In general, unless carefully designed a dispersion-compensation filter does not cancel dispersion perfectly for all channels. So even after correcting dispersion to first order, in long-haul WDM systems the residual dispersion can still limit the transmission length and/or the number of channels. To illustrate the magnitude of this effect, consider a link of length L carrying N channels spaced by $\Delta\lambda$ (i.e., $N\Delta\lambda$ is the total multiplexed width), with a dispersion slope after compensation S. The cumulated dispersion at the output of the link is then $SLN\Delta\lambda$. The receiver can be designed to tolerate a certain amount of residual cumulated dispersion within some spectral window, for instance, typically 1000 ps·nm^{-1} for a 10 Gbit·s^{-1} system, as stated earlier. For a typical dispersion slope $S=0.08$ ps·nm^{-2}·km^{-1}, $N=64$, and $\Delta\lambda=0.8$ nm (or a multiplexed width of 51.2 nm), the maximum possible fiber length for which the cumulated dispersion reaches 1000 ps·nm^{-1} is 244 km. This effect can of course be avoided by reducing the length or the number of channels, which impacts system performance. Better solutions include designing broadband dispersion-compensation filters with a dispersion curve matched to that of the fiber link. This is a key issue for WDM systems, which has received a lot of attention from system designers.

1.4.2.2 NONLINEAR EFFECTS

The maximum power that can be transmitted through an optical fiber, and thus the SNR of the signal at the fiber output are ultimately limited by a number of optical nonlinearities present in the optical fiber. These nonlinear effects are the Kerr effect (dependence of the fiber refractive index on the signal intensity), SBS (conversion of signal power into a frequency-shifted backward wave), SRS (conversion of signal power into a forward frequency-shifted wave), and FWM (optical mixing of a signal with itself or other signals and concomitant generation of spurious frequencies). The magnitude of these effects generally increases with increasing signal intensity, which can be relatively high in a single-mode fiber, even at low power, because of the fiber's large optical confinement. For example, in a typical single-mode fiber at 1.55 μm with an effective mode area of 80 μm^2, a 20-dBm signal has an intensity of ~1.2 kW·mm^{-2}! Because this intensity is sustained over very long lengths, and because the conversion efficiency of nonlinear effects generally increases with length, even the comparatively weak nonlinear effects present in silica-based fibers can have a substantial impact on system performance, even at low power. This section provides background on the magnitude of these nonlinear effects, describes their impact on system performance, and mentions typical means of reducing them.

1.4.2.3 SELF-PHASE MODULATION

When a signal propagates through an optical fiber, through the Kerr effect it causes a change Δn in the refractive index of the fiber material. In turn, this modification of the medium property reacts on the signal by changing its velocity and thus its phase. This nonlinear effect is known as self-phase modulation (SPM). For a signal of power P, the index perturbation Δn is expressed as

$$\Delta n = n_2 \frac{P}{A_{\text{eff}}} \tag{1.17}$$

where n_2 is the Kerr nonlinear constant of the fiber ($n_2 \approx 3.2 \times 10^{-20}$ m$^2\cdot$W^{-1} for silica) [35] and A_{eff} the signal mode effective area. The resulting change in the mode propagation constant β is

$$\Delta\beta = \frac{\omega\Delta n}{c} = \frac{2\,n_2}{\lambda} I \qquad (1.18)$$

where $I = P/A_{\text{eff}}$ is the signal intensity. In the case of a modulated signal, because the Kerr effect has an extremely fast response time ($\ll 1$ ps) each portion of the signal pulse modulates its own phase independently of other portions of the pulse. If $I_0(t)$ is the instantaneous intensity, or equivalently the intensity profile, of the signal launched into the fiber, and if α is the fiber loss at the signal wavelength, then the signal intensity at a point z along the fiber is $I(t, z) = I_0(t)\exp(-\alpha z)$. The amount of SPM experienced by the signal pulse after a propagation length L is simply [1]:

$$\Phi(t, L) = \int_0^L \Delta\beta\, dz = \frac{2\pi n_2}{\lambda} \int_0^L I_0 \exp(-\alpha z)\, dz = \frac{2\pi n_2 I_0}{\lambda} L_{\text{eff}}$$

$$(1.19)$$

where $L_{\text{eff}} = (1 - e^{-\alpha})/\alpha$ is the effective fiber length. In principle, because photodetectors are quadratic and thus phase insensitive, SPM should not cause any detrimental effect. This is true in AM systems provided the fiber is free of dispersion. However, in practice, the presence of chromatic dispersion converts SPM into amplitude fluctuations [1]. When β_2 is positive, SPM combines to chromatic dispersion to produce pulse broadening, just like chromatic dispersion alone does. When β_2 is negative, SPM combines to chromatic dispersion to produce pulse narrowing, i.e., they have opposite effects. In this case, SPM can be used to compensate for chromatic dispersion and thus improve the system performance. There is in fact a particular regime in which linear and nonlinear effects compensate mutually exactly at any moment in time. This particular solution of the nonlinear Schrödinger equation, valid only in a lossless fiber, is called an optical soliton. Provided that it has the proper shape and intensity, a soliton propagates without any temporal deformation. This phenomenon was extensively studied in the 1990s, and it continues to be an active research

topic, because fiber-optic solitons are very promising for ultralong distance transmission at high bit rates [26,43,44]. A soliton-based transmission encrypts the information in extremely short pulses that neither spread nor compress as they propagate along the fiber because the soliton has just the right peak power for the Kerr nonlinear phase shift to exactly compensate for chromatic dispersion. Soliton-based communication links are, however, not compatible with WDM-based links because in order to have a relatively low peak power, a soliton needs a low-dispersion fiber, which is not well suited for WDM (see "Four-Wave Mixing (FWM)" section). The WDM solution has obviously won so far, even for long-haul transmission. But the concept of soliton-based communication systems remains an interesting and promising approach that continues to stimulate a lot of research and development.

In parallel to these various schemes used to combat SPM, the most effective first-order solution to reduce SPM is to use a transmission fiber with a large mode effective area A_{eff}. This is of course also applicable to other undesirable fiber nonlinear effects, in particular cross-phase modulation (XPM), FWM, and stimulated scattering processes. The reason is that the magnitude of all of these processes increases as the reciprocal of A_{eff}, so a fiber with a higher A_{eff} can tolerate a higher signal power. Large mode effective areas are typically accomplished by designing fibers with a larger core and a concomitantly lower numerical aperture to ensure that the fiber carries a single mode. Communication-grade fibers have mode effective areas in the range of 50–100 μm^2, for example, 80 μm^2 for the so-called standard fiber. Substantially higher values are typically precluded for transmission fibers because they require such low numerical apertures that the fiber becomes overly susceptible to bending loss.

1.4.2.4 CROSS-PHASE MODULATION

XPM has the same physical origin as SPM, namely the Kerr effect, except that the phase modulation is not induced by a signal on itself, but by one or more different signals propagation through the fiber. A different signal means any signal with a different wavelength, a different polarization, and/or a different propagation direction. In a WDM system, the phase of a signal of wavelength λ_i is therefore modulated by itself (wavelength λ_i SPM) and by all the other channels (wavelengths $\lambda_{j \neq i}$, XPM).

The XPM affecting a particular channel i of a WDM system depends on the power (and therefore on the data) and wavelength of all other channels $j \neq i$. As in the case of SPM, XPM is converted into amplitude fluctuations through chromatic dispersion. However, the main detrimental effect of XPM is time jitter, due to the fact that the other signals also change the group delay of channel i. The position of the impulses is thus changed randomly around an average position, and sampling before decision does not occur always at the same instant within the pulse, which cause a BER penalty. If we consider the case of one interfering channel, interaction occurs when two pulses overlap. Because they propagate at different speeds (the group velocities at the wavelengths of the two channels), the interaction begins when the fastest impulse starts to overlap with the slowest pulse and ends when it has completely passed it. This phenomenon is called a collision. After one collision between symmetrical pulses, there is theoretically no memory on the perturbed pulse. The problem occurs in the case of an incomplete collision, for instance, when it begins just before an amplifier and then the powers change during the collision. In this case, the affected pulse keeps the memory through a shifted temporal position. A key parameter is to characterize this effect is the difference of group velocity between the two channels, which is equal to $D\Delta\lambda$, where D is the dispersion and $\Delta\lambda$ the channel spacing. If this parameter is high, the effect will be smaller, because collisions will be very rapid. Increasing the channel spacing will then reduce the interaction because the difference between group velocities is larger. The influence of chromatic dispersion is more complex. A higher dispersion reduces channel interaction and thus phase modulation, but as discussed earlier it also increases conversion into amplitude fluctuations. Further details can be found in Section 1.5.

1.4.2.5 FOUR-WAVE MIXING

FWM is another nonlinear process that results directly from the Kerr effect. Channels of a WDM system beat together in the receiver, giving rise to intermodulated sidebands at frequencies that are sums and differences of the channel's frequencies. Each of these sidebands is modulated with the

information encrypted on the channels that gave rise to it. When a sideband frequency happens to fall on or close to one of the channel's frequencies, this channel becomes modulated with unwanted information from other channels. This intermodulation has the same undesirable side effects as similar effects well known in radio systems.

As an illustration, consider a communication system utilizing channels that are equally spaced in frequency, which is usual in deployed systems, i.e., the channel frequencies are $f_0 + m\Delta f$, where m is an integer. The third-order beating between channels 0, 1, and 2 at respective frequencies f_0, $f_1 = f_0 + \Delta f$, and $f_2 = f_0 + 2\Delta f$ produces sideband signals at frequencies $pf_0 + qf_1 + rf_2$, where $|p| + |q| + |r| = 3$. In particular, an intermodulated sideband is generated at frequency $f_0 + \Delta f$ by interaction of three channels together ($p = 1$, $q = -1$, $r = -1$) but also by interaction of channels 0 and 1 only ($p = 1$, $q = 2$). This sideband has the same frequency as channel 1 and thus adds to channel 1 data modulation from channels 0 and 2. The same argument applied to other channels clearly shows that if the interaction is strong enough, every channel becomes contaminated with information from all other channels.

The magnitude of FWM effects can be characterized by the power in the intermodulation sideband P_{intermod}. This power can be calculated analytically for pure unmodulated waves, in which case it is given by [15,56]:

$$P_{\text{intermod}} = \eta_{\text{FWM}} d^2 \gamma^2 P^3 \exp(-2\alpha L) \qquad (1.20)$$

where $\gamma = 2\pi n_2 / \lambda A_{\text{eff}}$ represents the strength of the Kerr nonlinearity in the fiber, P is the power per channel, assumed the same for all channels, d is a constant equal to 6 if all channels are distinct and 9 if there are not. The factor η_{FWM} is the FWM efficiency, defined as

$$\eta_{\text{FWM}} = \frac{\alpha^2}{\alpha^2 + \Delta\beta_{\text{FWM}}^2} \left[1 + \frac{4\exp(-\alpha L)\sin^2(\Delta\beta_{\text{FWM}} L / 2)}{(1 - \exp(-\alpha L))^2} \right]$$

$$(1.21)$$

where $\Delta\beta_{\text{FWM}}$ is the phase mismatch between interacting waves, which depends on chromatic dispersion coefficient D, on its slope, and on the channel spacing Δf according to

$$\Delta\beta_{FWM} = \frac{2\pi\lambda^2}{c}\Delta f^2\left(D + \Delta f\frac{\lambda^2}{c}\frac{\partial D}{\partial\lambda}\right) \quad (1.22)$$

In the usual case where the total attenuation of the span is high enough ($\exp(-\alpha L) \ll 1$), the efficiency (Equation 1.21) is well approximated by

$$\eta_{FWM} = \frac{\alpha^2}{\alpha^2 + \Delta\beta^2_{FWM}} \quad (1.23)$$

FWM is a phase-matched process: for energy to flow effectively from one channel to another, the channels must remain in phase, i.e., the phase mismatch $\Delta\beta_{FWM}$ must be small. It means that the closer the channel frequencies are (small Δf), the more efficient FWM is, as indicated mathematically by Equations 1.21 and 1.22. This explains why the intermodulation power increases with decreasing channel spacing. Chromatic dispersion plays a beneficial role by increasing the phase mismatch between channels and thus reducing the FWM efficiency, as shown by Equation 1.22. The intermodulation power also increases with increasing channel power, and it does so rapidly (as the third power in P) because FWM is a nonlinear process.

Figure 1.3 shows the effect of both dispersion and channel spacing on the interference-to-carrier ratio, i.e., the difference between the channel power and the intermodulation product power. This quantity is plotted versus channel spacing for four values of the dispersion typical for channels located near the zero-dispersion wavelength. This figure simulates a fiber link with a length $L = 100$ km, a fiber attenuation of 0.2 dB·km^{-1}, a dispersion slope of 0.08 ps·nm^{-2}·km^{-1}, a nonlinear coefficient $\gamma = 3$W^{-1}·km^{-1}, and a launched power per channel of 4 dBm. It is clear that a higher dispersion reduces FWM and thus allows a better utilization of the available bandwidth. For example, if a ratio of -60 dB is required, Figure 1.3 shows that this can be accomplished with a 100-GHz channel spacing in a standard fiber ($D = 17$ ps·nm^{-1}·km^{-1}), but only 210 GHz or higher in a typical nonzero-dispersion-shifted fiber (NZDSF, family G.655) with a chromatic dispersion of 3 ps/(nm·km).

In single-channel transmission, a low dispersion is beneficial because it reduces the amount of pulse spreading induced by (1) dispersion and (2) SPM combined with dispersion, and thus it reduces the amount of dispersion compensation needed to correct for these effects. In multichannel transmission, the situation is not as simple because dispersion now brings protection against interchannel effects, XPM and FWM. But the situation depends strongly on the channel spacing: for *WDM 10G* systems with a typical channel spacing of 100 GHz or less, interchannel effects are dominant compared to intrachannel effects (SPM). This is the reason why a dispersion-shifted fiber (DSF) with zero dispersion around 1550 nm is much worse for WDM transmission than a standard G.652 fiber, and also why this fiber provides the smallest channel spacing at this bit rate (25 GHz). When higher bit rates are considered, the channel spacing cannot be reduced so much due to the spectral width of the modulated signals, and then intrachannel cannot be neglected compared to interchannel effects.

1.4.2.6 STIMULATED BRILLOUIN SCATTERING

SBS belongs to the family of parametric amplification processes. Through interaction between the

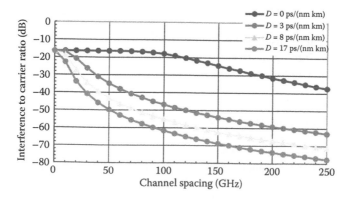

Figure 1.3 Dependence of the interference-to-carrier ratio due to FWM on the channel spacing.

optical signal and acoustic phonons, it causes power conversion from the signal into a counterpropagating signal shifted in frequency by the acoustic phonon frequency [1]. The power in the SBS signal grows as $\exp(g_B P_p - \alpha)z$, where g_B is the SBS gain, in $W^{-1} \cdot km^{-1}$, which depends on the wavelength separation between the two signals, and α is the attenuation of the medium. SBS is a narrowband process. In a silica fiber, the Brillouin frequency shift at 1.55 μm is $\nu_B \approx 11$ GHz and the gain bandwidth is only $\Delta \nu_B \approx 100$ MHz. It is customary to characterize SBS by its power threshold, i.e., the power required to compensate for the medium attenuation and thus just begin to provide a positive gain. For an unmodulated signal with a line width smaller or equal to the SBS gain bandwidth, the SBS threshold in a typical 1.5-μm fiber (mode effective area of 50 μm²) is around 21 mW·km, i.e., 21 mW in a 1-km fiber and 2.1 mW in a 10-km fiber [1]. For powers larger than the threshold, a fraction or all of the signal is converted in the backward SBS signal. It is therefore essential to keep the power in each signal below the SBS threshold, and because this threshold is fairly low, SBS limits the power of a narrowband signal that can be transported over a given distance. This is particularly critical in repeaterless systems [25].

Several solutions have been demonstrated and are routinely applied to increase the Brillouin threshold and thus increase the power and/or distance over which signal can be transported. When the signal amplitude is modulated at bit rates higher than ~100 MHz, as is the case in WDM systems, the signal bandwidth power exceeds the Brillouin line width and SBS is reduced. For this reason, SBS is generally not a concern in WDM systems. One caveat is that the carrier component of the modulated signal retains the original line width of the unmodulated signal, and it is still backscattered by SBS. Because the carrier component carries only half the signal average power, the SBS threshold is increased (by 3 dB) compared to an unmodulated narrowband signal, but in OOK schemes using high powers SBS acting on the carrier has been observed to induce signal distortion [33].

Because the SBS gain decreases with increasing carrier line width $\Delta \nu$ as $\Delta \nu_B / (\Delta \nu_B + \Delta \nu)$, another solution to further increase the SBS threshold is to use a larger carrier line width. This can be done with a directly modulated laser (direct modulation tends to chirp the laser frequency), by applying to the laser either a phase modulation [25,36] or a small amount of frequency modulation (at a frequency much lower than the bit rate) [45]. Other techniques include using a duobinary modulation scheme to suppress the carrier component [37], concatenating fibers with different Brillouin shifts to reduce the interaction length [40], and placing isolators along the fiber to periodically suppress the backward SBS signal [54].

1.4.2.7 STIMULATED RAMAN SCATTERING

Although caused by a different physical mechanism (interaction with vibrational modes of the medium structure instead of acoustic phonons), SRS can be modeled in a very similar manner, but its characteristics are quite different and so are its effects on transmission systems [1]. SRS is an optical process that causes power transfer between an optical pump and a co- or counterpropagating signal. Most solid media exhibit SRS, including silica-based fibers. *Spontaneous* Raman scattering occurs when a pump photon of frequency ω_p is scattered by a host phonon of frequency Ω, which results in the annihilation of the pump photon and the spontaneous emission of a signal photon at a frequency $\omega_s = \omega_p - \Omega$. This scattering process can also be stimulated when an incident signal photon of frequency ω_s interacts with a pump photon and a phonon, thus yielding the emission of a stimulated photon at frequency ω_s. This stimulated process thus provides what is known as Raman gain. The SRS gain spectrum is centered around a frequency downshifted from the pump frequency ω_p by the mean phonon frequency Ω of the material. The Raman gain spectrum and bandwidth are set by the finite-bandwidth phonon spectrum of the material. The Raman shift of silica is typically 13 THz (or ~100 nm at 1550 nm), which is much higher than for SBS. Similarly, the gain full width at half maximum is larger, around 8 THz (70 nm for a pump around 1.55 μm). However, the Raman gain coefficient for a silica fiber is much weaker than the SBS gain, by a factor of about 500, so the Raman threshold is typically much higher, for example, around 1.2 W in a 10-km length of 1.55-μm communication fiber with a 50-μm² effective mode area [1]. Although much weaker, SRS can still be deleterious in WDM systems because optical channels located at the highest gain frequencies act as pumps and can be depleted, while other channels can be amplified. In conventional systems using only the C-band (30 nm wide), SRS does not occur because the maximum separation between channels is much smaller than the Raman shift. But in systems using both the C- and L-bands, power transfer between channels of C- and L-bands can be induced by SRS and must be taken into account in system design.

It must be noted that SRS is also a useful mechanism: a pump signal injected in the fiber can transfer its power via SRS to one or more signals and thus provide amplification. This is the basic principle of fiber Raman amplifiers, which will be considered in Section 1.6.4.

1.4.2.8 POLARIZATION MODE DISPERSION

A standard single-mode optical fiber does not actually carry a single mode but two modes with orthogonal, nearly linear polarizations. Because the index difference between the fiber core and the cladding is small, these two polarization modes have nearly degenerated propagation constants. However, these propagation constants are not exactly the same. In a communication link, the signal launched into the fiber is typically linearly polarized. The fiber exhibits random linear and circular birefringence, and as the signal propagates through it the signal polarization evolves through many states. Because the two orthogonal polarization modes travel at slightly different velocity, one lags behind the other, and because the signal is temporally modulated into short pulses, after long-enough propagation each pulse is split into two pulses. This produces two electrical pulses with amplitudes that depend on the polarization of the optical signal at the receiver, and separated by a random delay called differential group delay (DGD). For fibers with strong coupling between polarization modes, DGD follows a Maxwell distribution. The mean value of DGD is called the PMD [3,16,20,22,23].

This multipath effect causes ISI and thus degrades the BER. Furthermore, random variations in the birefringence of the long fiber cause the DGD to be a random variable and thus the properties of the transmitted signals to be time dependent. Communication systems must then be characterized by their outage probability, or outage time, i.e., the probability that the BER exceeds the maximum tolerable value, above which transmission is no longer possible with the required quality. PMD is a linear effect, which, just like chromatic dispersion,

acts on each channel individually but does not cause coupling between them.

WDM systems are usually designed to tolerate a PMD approximately equal to one-tenth of the symbol duration or around 10–12 ps for a 10 Gbit·s^{-1} bit rate. When this value is exceeded by a small amount, transmission can still be sustained with fewer channels, which allows increasing the SNR of the remaining channels and provides a better resistance against PMD. When PMD is too high (for instance 20 ps or more for a *WDM 10G* system), distortion can cause closure of the eye diagram and increasing the power does not bring any improvement. Currently manufactured fibers allow 10 Gbit·s^{-1} transmission over several thousands of kilometers, and PMD is not a problem at this bit rate. Recent advances in manufacturing processes have led to fibers with low enough PMD values for 40 Gbit·s^{-1} transmission over more than 2000 km.

PMD compensation has been investigated in several laboratories, for example, using feedback equalizers in either the optical or the electrical domain [46,59]. The main application was the implementation of *WDM 10G* systems in existing fiber links that could originally not support this high bit rate because the fiber exhibited a high PMD. Although this method was successful, its economic viability has been questioned because it requires one equalizer per channel and its high cost cannot be shared. PMD compensation will certainly need to be implemented in the future in communication systems with higher bit rates over long distances, which have a reduced tolerance to PMD. For example, a *WDM 40G* system typically requires no more than 2 or 2.5 ps of PMD.

1.5 DESIGN OF AN OPTICAL WDM SYSTEM

1.5.1 Global performance of a system: BER and OSNR

As mentioned earlier, the performance of a WDM system is expressed in terms of its BER, which is obtained by measuring the number of error bits occurring over a given time interval. A minimum OSNR value is required in order to achieve the required transmission quality, i.e., the BER needs to be lower than a given threshold. Commercial equipment is typically specified in terms of OSNR: the maximum number of spans is specified for different losses per amplification span. For example, an OSNR of 22 dB will be guaranteed for seven spans of 25-dB loss (7×25 dB) or for ten spans of 23-dB loss (10×23 dB).

Another important feature is the system sensitivity to chromatic dispersion and dispersion-compensation strategy. The residual dispersion at the receiver input must remain within some interval. As an example, a 10 Gbit·s^{-1} receiver will only accept a cumulated dispersion between -600 and $+800$ ps·nm^{-1}. For a particular channel, it is always possible to bring the cumulated dispersion within this range with proper in-line compensation. However, due to the finite dispersion slope, the other channels will experience a different cumulated dispersion, and if the link is too long and/or the dispersion slope is too high, it will not be possible to meet this specification for all channels. This limitation could be lifted by adjusting the cumulative dispersion channel by channel, but this is not practical for economic reasons. As a result, chromatic dispersion generally imposes an upper limit on the bit rate and the optical reach.

1.5.2 Critical parameters and trade-offs for terrestrial, undersea, and repeaterless systems

As discussed earlier, the optical reach and amplification span are critical parameters in communication systems. For a given optical reach and a fixed launched power, a shorter amplification span improves the OSNR. Conversely, for a given required OSNR it increases the optical reach. However, a shorter amplification span also results in a more expensive system, more complex monitoring, and a higher operating cost. Moreover, in a terrestrial network the location of the amplification sites and the network topology in general are parameters that the operator does not want to change. The network infrastructure and fibers are long-term investments, and they are required to be compatible with several generations of systems. In particular, the attenuation per span is a constrained parameter. Its value is imposed by the characteristics of networks where systems have to be installed, and it is typically

in the range of 20–25 dB. Technical improvement goals in terrestrial WDM systems therefore consist in increasing the capacity and the optical reach within the framework of this attenuation per span.

Undersea systems benefit from an additional degree of freedom. Unlike in terrestrial networks, the fiber and the system are laid together and cannot be separated, and therefore the cable does not need to be designed for successive generations of systems. The amplification span can then be selected without any location constraint. The preferred solution is to space the amplifiers equally because it simplifies manufacturing, dispersion compensation, and cable maintenance. Because of the very long optical reach of undersea systems (for example, 6500 km between Europe and North America and 9000 km across the Pacific Ocean), the amplification span must be substantially reduced down to around 40 km.

In principle, the OSNR can always be improved by increasing the power launched into the fiber. However, the maximum *available* power is limited by the high cost of high-power amplifiers and also by safety rules. Quite independently, as discussed earlier, the maximum *usable* power is limited by nonlinear effects in the fiber (SPM, XPM, and FWM) to a level that insures that these effects are maintained at or below some acceptable level. Two strategies are possible to take into account this power limitation. The first strategy is to reduce the power level when the number of spans is increased, because impairments caused by nonlinear effects are cumulative along the transmission line. Systems using this type of approach are called linear, or also NRZ (because linear systems use NRZ pulses, i.e., pulses with a duration of roughly one symbol time). Another strategy is to exploit the beneficial effect of "soliton-like" propagation regimes, where linear and nonlinear effects cancel each other, which requires a precise compensation map. These systems are usually called nonlinear, or RZ (because they used pulses of the RZ type, i.e., significantly shorter than the symbol time and a higher peak power than NRZ pulses).

In summary, the amplifier gain is determined by the span loss that needs to be compensated, the output power is limited by the cost and technology of amplifiers and by fiber nonlinear effects, and the optical reach is then given by the minimum SNR that can be achieved.

1.6 STATE OF THE ART AND FUTURE OF THE WDM TECHNOLOGY

1.6.1 State-of-the-art WDM system capacity and distance

The first WDM systems appeared in 1995 and could transmit four 2.5 Gbit·s^{-1} channels. The number of channels was rapidly increased to 16, then 32, by reducing the frequency spacing between channels down to 100 GHz (0.8 nm). WDM systems were predicted to evolve toward even higher channel counts by reducing this spacing to 0.4 nm, but this change has taken place slowly because of the difficulty and higher cost of developing components, in particular multiplexers and demultiplexers, capable of handling signals so closely spaced in wavelength. Instead, in the next generation of systems the capacity was increased by increasing the bit rate to 10 Gbit·s^{-1}. WDM 10G systems provide a higher capacity, but they are also subject to more severe propagation impairments. Chromatic dispersion, which can be neglected at 2.5 Gbit·s^{-1} up to around 800 km, becomes a critical issue at higher bit rates, and dispersion-compensation units must be incorporated in all amplification sites. In 2001, the state of the art for typical engineering data provided to system operators by suppliers was 80×10 Gbit·s^{-1} channels with a 50-GHz spacing, or a total capacity of 800 Gbit·s^{-1} over a single fiber with an optical reach of around 700 km. Subsequent progress

in filtering and laser technology has allowed the achievement of 25-GHz channel spacing, offering a capacity of 1.6 Tbit·s^{-1} over the 30-nm-wide C-band of EDFAs. The capacity of commercially available fiber links has therefore been multiplied by a factor 160 in 7 years, and the symbolic barrier of 1 Tbit·s^{-1} has already been exceeded. Furthermore, much higher capacities have been demonstrated in laboratories, for example, up to 5 Tbit·s^{-1} over 12×100 km [7] and even 10.2 Tbit·s^{-1} over 100 km using polarization multiplexing [9].

Figure 1.4 depicts the state of the art of *WDM* technology at the time of this writing. Circles represent laboratory prototypes, and squares represent either commercially available or announced systems. The diagonal line represents the 1 Tbit·s^{-1} capacity boundary; systems that fall in the hatched quadrant above it have a capacity greater than 1 Tbit·s^{-1}. The highest commercial capacity is 3.2 Tbit·s^{-1} (80×40 Gbit·s^{-1}). Systems with a 1.6 Tbit·s^{-1} capacity have been proposed in two configurations, namely 160 channels at 10 Gbit·s^{-1} with a 25-GHz spacing and 40 channels at 40 Gbit·s^{-1} with a 100-GHz spacing [38]. Demonstrations have also been performed at 20 Gbit·s^{-1} per channel, but no commercial system has yet been developed at this bit rate.

In spite of these great advances in experimental communication links, most of the links deployed in the world operate at much lower bit rates. For example, in France most links operate at 2.5 Gbit·s^{-1}. The network is being upgraded to

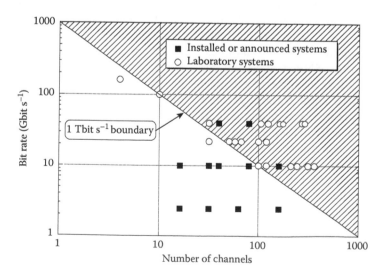

Figure 1.4 State of the art of WDM technology (see text for details).

10 Gbit·s^{-1}, but it will take several years before the conversion is complete [47]. Operational systems in the United States are further along; most of the deployed links run at 10 Gbit·s^{-1}. Very few commercial systems operate at 40 Gbit·s^{-1}. Again the main difficulty is that these systems require more precise correction for chromatic dispersion and dispersion slope. It means using accurately tailored and stable fiber Bragg gratings so that dispersion does not take over. The system is then less lenient on imperfections in dispersion-compensation circuits, and more difficult to develop and manage.

It is significant that all nodes in deployed terrestrial networks are opaque, i.e., the optical signals are detected, turned into electrical signals, amplified, switched (routed) electronically into the right direction, and turned back into a light signal with a local laser oscillator. Such optoelectronic nodes constitute a large fraction of the cost of the network. The reason why this function is not performed in a fully optical manner is the high cost, and to some extent loss, of optical components, in particular optical switches. Only a few companies have developed and are planning on deploying commercial transparent networks using all-optical nodes. The only market at the moment is secure military lines in the United States. Future WDM systems with extended optical reach (ultralong haul [ULH] and very long haul [VLH], see Section 1.6.3) are an efficient way to reduce the number of optoelectronic regeneration sites and thus reduce the network cost.

1.6.2 Forward error-correcting codes

As explained earlier, the compatibility of network topology with future system generations is a very strong requirement for terrestrial networks. For example, upgrading an existing link from 2.5 to 10 Gbit·s^{-1} requires an OSNR increase of 6 dB, which in turn should require a corresponding increase in power at the receiver. Instead of increasing the power, which again is not always possible for reasons covered earlier, a solution is the use of a forward error-correcting code [13], an approach widely used in satellite and radio systems. Redundancy is introduced in the input data (prior to transmission) and used by an electronic decoder to detect and eventually correct some of the errors affecting the data sequence through the detection process. A coding–decoding scheme is characterized by the proportion of errors it can detect

and correct, which is related to the redundancy. An error-correcting code therefore allows the receiver to accept a smaller OSNR, which results in a degraded BER that is corrected by the decoder. The price to pay is an increase in transmitted data resulting from the required redundancy.

A very simple example is the repeat code: binary bits of information are repeated an odd number of times, and the decoder corrects for errors by comparing the like bits of replica and retaining the bits that occur most often (a process known as majority decoding). This is not a very efficient code, because it requires a large amount of redundancy, but other, far more powerful codes exist, such as the well-known Reed–Solomon and BCH codes, that can reduce the BER substantially at the cost of surprisingly low redundancy [13]. Most of the *WDM 10G* systems today utilize a Reed–Solomon code, while other more powerful schemes have been extensively studied for the next generations of systems [2,27]. To illustrate the improvement brought by forward-error correction, a BER of 10^{-5} at the input of a Reed–Solomon decoder results into a BER of roughly 10^{-15} at its output. Assuming this last value is required by the user, it means that the transmission systems is only required to achieve a BER of 10^{-5} and can then operate at a lower OSNR. This approach allows in particular to gain the 6-dB difference in OSNR between *WDM 2.5G* and *10G* systems and thus to operate a 10 Gbit·s^{-1} system with the same OSNR as a 2.5 Gbit·s^{-1} system, at the cost of a modest redundancy (or data rate increase) of ~7%. Forward error-correcting codes can also clearly be used in 2.5 Gbit·s^{-1} systems to increase the optical reach or to operate with higher span losses. Forward error-correcting codes are likely to play a major role in the higher speed communication systems of the future.

1.6.3 Ultralong-haul technology: New problems arising

One key advantage of the WDM technology is that the high cost of amplifiers can be shared between all transmitted channels. Any reduction in the number of optoelectronic regeneration sites, or in other words any increase in the optical reach, is then very attractive. This is one major reason for the interest in ULH and VLH systems, which can operate today at 10 Gbit·s^{-1} per channel over respective distances of ~1500 and ~3000 km. Raman amplifiers are a key element in this technology: they allow to improve the OSNR for a given distance or equivalently to increase the optical reach for a given OSNR. The compatibility of ULH and VLH systems with existing infrastructures is also a difficult challenge because the OSNR decreases as the number of spans increases, and thus the accumulation of propagation impairments is much more critical. This limitation is of course mitigated in practice with error-correcting codes, as well as Raman amplifiers.

1.6.4 Raman amplification

Raman amplification relies on SRS, a nonlinear optical process that causes power transfer between an optical pump and a signal (see "Stimulated Raman Scattering" section). One of the greatest strengths of Raman amplification is that it can supply gain at any wavelength provided a suitable pump source is available. Since the Raman shift of silica is typically 13 THz, to obtain a gain peak at 1530 nm the pump wavelength must be ~1430 nm. With such a pump, gain will be available from about 1490–1546 nm. The Raman gain cross section of silica is unfortunately relatively small, so the power requirement is much higher than for an EDFA, but Raman fiber amplifiers present several important benefits that somewhat mitigate this disadvantages, including the flexibility of pump wavelength selection, the availability of gain anywhere where pump is available, and the fact that the gain medium is the transmission fiber itself.

Raman fiber amplifiers can be configured in a number of ways, each with its own benefits and applications. Forward-pumped Raman amplifiers, in which the pump and the signals to be amplified travel in the same direction, induce cross-talk between channels, which is undesirable in WDM systems. This is the reason why the backward-pumped Raman fiber amplifier is often preferred.

This configuration is illustrated in Figure 1.5. An obvious advantage of the Raman amplifier is that it can be easily implemented in any existing fiber link by simply adding a pump source at the proper wavelength. In particular, it does not require the insertion of a doped fiber, which keeps the cost down. Also, in the event of pump failure the fiber amplifier is still transparent, as opposed to an EDFA where an unpumped Er-doped fiber is essentially opaque at the signal wavelengths. We refer the reader to Chapter A1.6 for a more detailed description of Raman amplifiers.

To illustrate the system benefits of in-line Raman amplifiers, Figure 1.6 shows the calculated evolution of signal power with distance from the emitter with and without Raman amplification. This figure simulates a link with a fiber attenuation of $0.22\ \mathrm{dB\cdot km^{-1}}$ a total length of 100 km, and a Raman gain of 10 dB around 1.53 μm. In the absence of amplification, the signal power in dBm decreases linearly with distance. When pump is injected at the far end of the fiber, SRS adds gain at that end of the fiber. The gain decreases away from the pump, as a result of pump photons being either scattered by the fiber or converted into signal photons. There is practically no more incremental Raman gain after the pump has traveled about 60 km, because by then most of it has been consumed. The system advantage of Raman amplification is to increase the signal power at the receiver and thus to improve the SNR of the detected signal. As with any amplifier, the greatest benefit to the SNR occurs when the signal is amplified before its power becomes too low. To understand this basic principle, consider two configurations. Configuration (a) is a link of length L with an amplifier of gain G placed at the end, and configuration (b) is the same link with the same amplifier of gain G placed a distance d before the receiver. In both configurations, the signal power at the receiver is the same, but in (b) the noise power is reduced by a factor equal to the loss A of the fiber

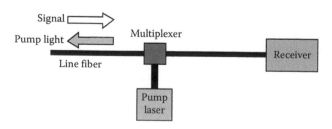

Figure 1.5 Schematic of a backward-pumped Raman fiber amplifier.

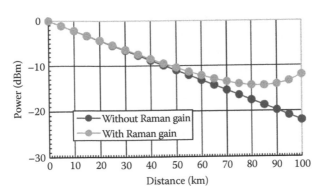

Figure 1.6 Typical variation of signal power versus distance with and without a fiber Raman amplifier.

length d. The in-line amplifier of configuration (b) is therefore equivalent to an amplifier located at the receiver with a NF reduced by A.

A backward-pumped Raman amplifier can be viewed as an in-line amplifier located before the receiver. The gain that it supplies allows (1) reducing the launched signal power and thus reducing nonlinear effects for a given topology (span loss and number of spans); (2) increasing the span loss for a given number of spans while maintaining the same OSNR; and (3) increasing the number of spans for a given span loss while maintaining again the same OSNR. Raman amplification is therefore a key technology to increase the optical reach and upgrade existing networks to higher bit rates.

1.6.5 Diversification of fibers and international telecommunications union fiber standards

We have seen how propagation phenomena become more and more critical as the bit rate increases. These phenomena depend on several system parameters, such as the channel power and channel spacing, as well as fiber parameters, especially scattering loss, chromatic dispersion, and effective mode area. It was therefore important, early on, to develop standard fibers with set ranges for all of their critical parameters. This task was accomplished by the International Telecommunications Union (ITU), which defined a number of other important standards and terminologies as well, such as the so-called ITU grid, i.e., the precise values of the discrete wavelengths used in optical communication systems around the world.

One of the most important single-mode standard fibers is the G.652 fiber (also known as SSMF), designed for early single-channel transmission in the 1550-nm band before the emergence of WDM and EDFAs.

Because the high dispersion of the G.652 fiber ($17\ \text{ps·nm}^{-1}\text{·km}^{-1}$) could be viewed as a drawback for single-channel transmission, the standard DSF, known as G.653, was designed to achieve zero dispersion at 1550 nm and thus drastically reduce signal distortion. As we saw earlier, a DSF is not well suited for WDM operation, because channels located near the zero-dispersion wavelength suffer severe nonlinear effects, especially FWM. G.653 fiber can be used in WDM systems but at the price of significant constraints, such as increased and eventually irregular channel spacing. This results in a poorer spectral efficiency and thus a higher cost, which is the reason why it is generally not used in this application.

A third family of fibers, the NZDSF, also known as G.655, became available somewhat later. Their main feature is a chromatic dispersion lower than that of the G.652 fiber. The initial idea was to choose a dispersion value high enough to keep deleterious nonlinear effects at a low level while requiring less dispersion compensation. Identifying the right trade-off between dispersion and dispersion compensation was not an easy task. The chromatic dispersion of commercially available G.655 fibers has increased progressively from $3\ \text{ps·nm}^{-1}\text{·km}^{-1}$ in early fibers to $8\ \text{ps·nm}^{-1}\text{·km}^{-1}$ for recent fibers. The dispersion of the NZDSF family of fibers therefore covers a wide range. Although G.652 is certainly the most widely used fiber in backbone networks, a considerable amount of G.655 fibers has been deployed, especially by new operators, in the second half of the 1990s. At a bit rate of 10 Gbit·s^{-1} per channel, the advantage over G.652 fibers is not obvious; excellent performance has been reached with G.652 fibers because their higher dispersion allows very efficient protection against interchannel effects, and thus a closer channel spacing. The question of fiber choice is more critical and certainly still an open issue at higher bit rates (40 Gbit·s^{-1} and above), although *WDM* 40G transmission has been successfully demonstrated over G.652 fibers in many laboratories.

1.6.6 Toward the future: WDM 40G systems and beyond

There is a definite economic advantage in operating transmission systems at higher bit rates, in part because integration reduces cost, at least when a certain level of production is reached and technology is stabilized. For instance, a 10 Gbit·s^{-1} emitter is less expensive than four 2.5 Gbit·s^{-1} emitters, yet both provide the same capacity. The same is true for receivers. A drawback of moving toward higher bit rates is that the narrower impulses required are more sensitive to propagation phenomena, and also that the implementation of electronic circuits is more difficult. Another consideration is the channel spacing. After its amplitude has been modulated, the optical signal has a much wider spectrum than the original signal produced by the emitter. The line width of the signal from a typical emitter is of the order of a few MHz, whereas after modulation this line width becomes comparable to the modulation frequency, i.e., many GHz or tens of GHz. Maintaining a tolerable level of cross-talk then requires a higher channel spacing: for instance, a bit rate of 40 Gbit·s^{-1} is certainly not compatible with a 25-GHz channel spacing. However, with a 100-GHz channel spacing a 40 Gbit·s^{-1} bit rate becomes possible. A sample of recent research in 40 Gbit·s^{-1} systems is given in "Selected Recent Results" section.

In response to the growing demand for higher speed communications, extensive research efforts have been expanded in industrial laboratories toward the development of *WDM* 40G systems, which are considered to be the next generation for optical data transmission. The main difficulties, as stressed earlier, are that this type of system has a much tighter tolerance to most design parameters,

including chromatic dispersion, FWM, and PMD, and more accurate forms of compensation will be required. In addition, advanced research is already being conducted on even higher bit rates per channel, 80 and even 160 Gbit·s^{-1} [53,61]. The objective of this research is primarily knowledge acquisition. At this point it is difficult to predict whether these systems will ever be viable and in what time frame, and whether they will be cost-effective compared to existing generations.

1.6.6.1 INCREASING THE NUMBER OF CHANNELS BY INCREASING THE AMPLIFICATION BANDWIDTH

For a given channel spacing, or when the minimum channel spacing imposed by the bit rate is reached, it is still possible to increase the capacity by increasing the number of channels. The channels used in practice fall in one of three bands, which are defined by the amplification bands of EDFAs, as opposed to other requirements. The C-band (1530–1565 nm) is the region where the EDFA is most efficient, i.e., where it provides the most gain per unit pump power. This is the band that is used first, and in many installed networks it is the only band that is used. The L-band (1565–1610 nm) is used when the C-band is full and additional channels need to be added, even though L-band EDFAs are less efficient and provide a lower gain than C-band EDFAs. Generally, an EDFA cannot be optimized to provide gain efficiently in both bands, so an in-line amplification site for both C- and L-bands typically consists of two separate EDFAs, one operating in the C-band and the other one in the L-band. The two amplifiers are often placed in a parallel configuration, as illustrated in Figure 1.7. A demultiplexer separates the incoming WDM signals into C-band signals, which are sent to the C-band EDFA, and L-band signals, which are sent to the L-band EDFA. A multiplexer placed at the output of the two EDFAs recombines the amplified signals onto the same fiber. Serial configurations are also possible. These configurations are not specific to the L- and C-bands: they can be used to multiplex any bands, in principle in any number. It should be pointed out that the economical advantage of using both the C- and L-bands in the same system is not obvious: amplification must be performed in two separate amplifiers, so that this technique allows only sharing

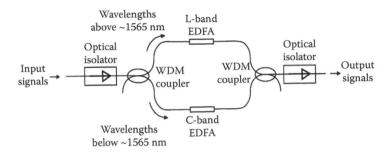

Figure 1.7 Diagram of a parallel arrangement of L-band and C-band EDFA.

the communication fiber. The benefits of using the L-band are also mitigated by other considerations. First, the fiber loss is higher in the L-band, and the C-band/L-band multiplexers introduce losses as well, in both bands, which degrade the power budget and the OSNR. The trade-off between these extraneous losses and OSNR degradation on one end, and the cost of deploying additional fiber on the other, needs to be carefully examined. Second, L-band amplification is more costly than C-band amplification. The reason is that it requires a different set of components than C-band amplifiers, and these components are more expensive because they do not benefit from the same economy of scale. As a result, very few systems in the world use the L-band, except in Japan where new technologies tend to be deployed earlier than elsewhere and where existing fibers happen to have the right amount of GVD for WDM communication in the L-band. Use of the L-band is often a good solution for G.653 fibers: chromatic dispersion is higher in the L-band than in the C-band, and it provides some protection against deleterious nonlinear effects such as FWM.

Very recently, EDFAs have also been designed to provide gain in the S-band (~1480–1530 nm) [4]. Although the gain and the gain efficiency in the S-band are even lower than in the L-band, they are both respectable, and the S-band is a possible future spectral window to extend the usable bandwidth of silica fibers. The S-band is in fact already used in metropolitan communication systems without amplification. However, amplification in the S-band is costly, even more than in the L-band. No deployed system uses amplification in the S-band, except perhaps isolated experimental systems. Because of the higher costs of L- and S-bands, communication companies generally find it economically preferable to use up the capacity of the C-band in a given link before starting to utilize other bands.

1.6.6.2 INCREASING THE NUMBER OF CHANNELS WITH A CLOSER CHANNEL SPACING

The above considerations show that filling the C-band is one of the most efficient solutions for increasing the bandwidth. The channel spacing has been normalized by ITU to 100 GHz, but sub-grids with a spacing of 50 GHz and even 25 GHz can be used. $N\times 10$ Gbit·s^{-1} systems available today have commonly 100- and 50-GHz channeling, and 25 GHz has been proposed [38]. As is well known in the radio domain, a convenient parameter to describe the fraction of the available bandwidth that is actually utilized is the spectral efficiency. It is defined as the total transmitted bit rate (number of channels times channel bit rate) divided by the occupied bandwidth. For example, a 16×10 Gbit·s^{-1} link with a channel spacing of 100 GHz has a total capacity of 160 Gbit·s^{-1} and occupies a bandwidth of 16×100 GHz, so the spectral efficiency is 0.1 bit s^{-1}·Hz^{-1}. The highest value obtained so far in commercially available optical systems based on OOK modulation is around 0.4 bit s^{-1}·Hz^{-1} for 10 Gbit·s^{-1} with a 25-GHz channel spacing in the C-band, and 0.8 bit s^{-1}·Hz^{-1} has been demonstrated in the laboratory. These values remain modest compared to the spectral efficiencies of 4 or 5 bit s^{-1}·Hz^{-1} currently achieved in radio systems. Such high values are made possible by multilevel modulation schemes. Similar schemes also exist in optics [6], but they have not yet been implemented in commercial systems.

Figure 1.8 depicts the evolution of the spectral efficiency of commercially available WDM systems

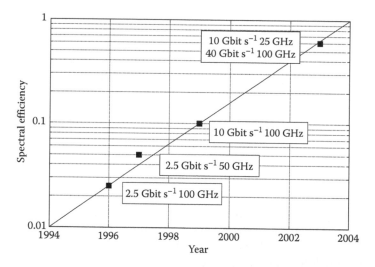

Figure 1.8 Historic evolution of the spectral efficiency (in bit s⁻¹·Hz⁻¹) of optical systems.

since 1994. Higher values have been achieved in the laboratory up to 1.28 bit·s⁻¹·Hz⁻¹ [9]. Reducing the channel spacing has required several technological improvements over the years, including increasing the stability of demultiplexing filters and of the channel wavelengths (to this end, lasers are now equipped with wavelength lockers). The ultimate achievable channel spacing depends strongly on the fiber. As mentioned earlier, G.652 fiber is a very good candidate because its high chromatic dispersion reduces the threshold of FWM.

1.6.6.3 MODULATION SCHEMES

Optical systems use currently an OOK intensity modulation format. The advantage of this modulation scheme is that it is rather simple, and that it does not require excessive signal power. However, its main drawback is that it does not allow to fully exploit the large available bandwidth of the fiber. As mentioned in the previous section, more sophisticated schemes can be used to improve the spectral efficiency, in particular multilevel modulation schemes. However, it is also well known that the modulation format has a strong influence on the penalties imposed by propagation phenomena. For example, to achieve the same BER, multilevel modulation schemes require a higher power than a binary modulation scheme, which means a higher sensitivity to propagation effects. Alternate schemes have been extensively investigated for long-haul systems, especially for undersea light wave transmission. The duobinary code has been

proposed by different laboratories as a good candidate, providing a better resistance to chromatic dispersion [60]; phase-shaped binary transmission is another technique, which is a modified version of duobinary code [14]. Vestigial sideband (VSB), a very well-known technique in radio communication systems, could also allow to reduce channel spacing and thus improve spectral efficiency [8,9]. More recently, DPSK has also been proposed [24]. Although OOK remains today practically the only one used, at least for terrestrial systems, it is clear that improving the modulation format to better utilize the capacity of existing systems is a key issue in research. This effort is likely to lead to major improvements in the future optical transmission systems.

1.6.6.4 SELECTED RECENT RESULTS

Recently, publications show that the field of communication systems is very prolific, and novel systems with very impressive performance continue to be demonstrated. Table 1.1 summarizes the characteristics of some of the most remarkable systems reported in recent years. The objective is not to be exhaustive—dozens of examples could be compiled—but to illustrate some of the concepts mentioned earlier and to give the reader a fair notion of the state of the art in research and development at the time of this writing. It has to be kept in mind that comparison of capacity is not necessarily meaningful because transmission distance, amplification span, and spectral efficiency are not identical.

Table 1.1 Selected examples of recent experimental high-capacity transmission lines

Bit rate (Gbit·s⁻¹)	Number of channels	Total capacity (Tbit s⁻¹)	Channel spacing (GHz)	Modulation scheme	Spectral efficiency (bit s⁻¹·Hz⁻¹)	Optical reach (km)	Amplification span (km)	Bands	References
Terrestrial Systems									
42.7	125	5		VSB		1200	100	C+L	[7]
40	80	3.2		Duobinary		300, G.655	100	C+Raman	[10]
40	273	10.9		VSB+Pol. div. mult	1.28	116	58	S+C+L	[21]
		10.9				100	NA	Raman	[8]
Transoceanic Systems									
365	10	3.65		NRZ+error correcting code		6850	~50		[57]
Repeaterless Systems									
160	10	1.6	25			380	NA	Raman+remote C-band	[39]
104	40	4.16	125		0.32	135	NA	EDFAS+C+L+Raman	[11]
25	40	1				306	NA		[55]

REFERENCES

1. Agrawal, G. P. 1995. *Nonlinear Fiber Optics.* 2nd ed. San Diego, CA: Academic.

2. Ait Sab, O. and Fang, J. 1999. Concatenated forward error correction schemes for long-haul DWDM optical transmission systems. *Proceedings of 25th European Conference on Optical Communication,* Nice, France, vol. 2, pp. 290–291.

3. Andresciani, D., Curti, F., Matera, F., and Daino, B. 1987. Measurement of the group delay difference between the principal states of polarization on a low birefringence terrestrial fibre cable. *Opt. Lett.* 12: 844–846.

4. Arbore, M. A., Zhou, Y., Keaton, G., and Kane, T. J. 2003. 30dB gain at 1500 nm in S-band erbium-doped silica fiber with distributed ASE suppression. *Proceedings of the SPIE—The International Society of Optical Engineering,* vol. 4989.

5. Bergh, R. A., Digonnet, M. J. F., Lefevre, H. C., Newton, S. A., and Shaw, H. J. 1982. Single mode fiber optic components. *SPIE Proceedings on Fiber Optics Technology'82,* vol. 326, pp. 137–142.

6. Betti, S., De Marchis, G., and Iannone, E. 1994. Toward an optimum use of the optical channel capacity. *Fiber Integr. Opt.* 13: 147–164.

7. Bigo, S. et al. 2001. Transmission of 125 WDM channels at 42.7 Gbit/s (5 Tbit/s capacity) over 12×100km of TeraLight Ultra fibre. *Proceedings of 27th European Conference on Optical Communication,* Amsterdam, Netherlands, vol. 6, pp. 2–3.

8. Bigo, S. 2002. Improving spectral efficiency by ultranarrow optical filtering to achieve multiterabit/s capacities. *Proceedings of Optical Fiber Communication Conference,* Anaheim, CA, vol. 1, pp. 362–364.

9. Bigo, S. et al. 2001. 10.2 Tbit/s (256×42.7 Gbit/s PDM/WDM) transmission over 100km TeraLight™ fiber with 1.28bit/s/Hz spectral efficiency. *Proceedings of Optical Fiber Communication Conference,* Anaheim, CA, vol. 4, Paper PD25-1-3.

10. Bissessur, H. et al. 2001. 3.2 Tbit/s (80×40 Gbit/s) C-band transmission over 3×100km with 0.8 bit/s/Hz efficiency. *Proceedings of 27th European Conference on Optical Communication,* Amsterdam, Netherlands, vol. 6, pp. 22–23.

11. Boubal, F. et al. 2001. 4.16 Tbit/s (104×40 Gbit/s) unrepeated transmission over 135km in S+C+L bands with 104 nm total bandwidth. *Proceedings of 27th European Conference on Optical Communication,* Amsterdam, Netherlands, vol. 1, pp. 58–59.

12. Brandon, E. and Blondel, J. P. 1998. Raman limited, truly unrepeated transmission at 2.5 Gbit/s over 453km with 30 dBm launch signal power. *Proceedings of 24th European Conference on Optical Communication,* Madrid, Spain, vol. 1, pp. 563–564.

13. Chan, V. W. S. 1997. Coding and error correction in optical fiber communications systems. *Optical Fiber Communications III,* vol. A, ed. I. P. Kaminow and T. L. Koch, pp. 40–62. New York: Academic.

14. Charlet, G. et al. 6.4Tb/s (159×42.7 Gb/s) capacity over 21×100km using bandwidth limited phase shaped binary transmission. *ECOC 2002 Copenhague Post Deadline.* Paper 4.1.

15. Chraplyvy, A. R. 1990. Limitations on lightwave communications imposed by optical-fiber nonlinearities. *J. Lightwave Technol.* 8: 1548–1557.

16. Ciprut, P., Gisin, B., Gisin, N., Passy, R., Von Der Weld, P., Prieto, F., and Zimmer, C. W. 1998. Second-order polarization mode dispersion: Impact on analog and digital transmissions. *J. Lightwave Technol.* 16: 757–771.

17. Desurvire, E. 1996. *Erbium-Doped Fiber Amplifiers: Principles and Applications.* New York: Wiley.

18. Digonnet, M. J. F. 2001. *Rare Earth Doped Fiber Lasers and Amplifiers.* 2nd ed. New York: Dekker.

19. Eggleton, B. J., Ahuja, A., Westbrook, P. S., Rogers, J. A., Kuo, P., Nielsen, T. N., and Mikkelsen, B. 2000. Integrated tunable fiber gratings for dispersion management in high-bit rate systems. *J. Lightwave Technol.* 18: 1418–1432.

20. Foschini, G. J. and Poole, C. D. 1991. Statistical theory of polarization mode dispersion in single mode fibers. *J. Lightwave Technol.* 9: 1439–1456.

21. Fukuchi, K. et al. 2001. 10.92 Tbit/s (273×40 Gbit/s) triple-band/ultra-dense WDM optical-repeated transmission experiment. *Proceedings of Optical Fiber Communication Conference*, Anaheim, CA, vol. 4, Paper PD 24-1-3.

22. Gisin, N., Passy, R., Bishoff, J. C., and Perny, B. 1993. Experimental investigations of the statistical properties of polarization mode dispersion in single mode fibers. *IEEE Photon. Technol. Lett.* 5: 819–821.

23. Gisin, N. et al. 1995. Definition of polarization mode dispersion and first results of the COST 241 Round Robin measurements. *Pure Appl. Opt.* 4: 511–522.

24. Griffin, R. A. and Carter, A. C. 2002. Optical differential quadrature phase-shift key (oDQPSK) for high capacity optical transmission. *Proceedings of Optical Fiber Communication Conference*, Anaheim, CA, vol. 1, pp. 367–368.

25. Hansen, P. B. et al. 1995. 529 km unrepeated transmission at 2.488 Gbit/s using dispersion compensation, forward error correction, and remote post-and pre-amplifiers pumped by diode-pumped Raman lasers. *Electron. Lett.* 31: 1460.

26. Hasegawa, A. 2000. An historical review of application of optical solitons for high speed communications. *Chaos.* 10: 475–485.

27. Helard, J.-F., Bougeard, S., and Citerne, J. 1999. Forward error correction coding schemes for optic fiber cable systems at 10 Gbit/s. *Proceedings of 25th European Conference on Optical Communication*, Nice, France, vol. 2, pp. 288–289.

28. Islam, M. N. 2002. Raman amplifiers for telecommunications. *IEEE J. Sel. Top. Quantum Electron.* 8: 548–559.

29. Joindot, I. and Joindot, M. 1996. Les Télécommunications par fibres optiques, *Collection Technique et Scientifique des Télécommunications*. Dunod, Paris, France.

30. Joindot, M. *Internal France Telecom R&D Publication*.

31. Kaiser, P. 1986. Network architecture and systems technology for future broadband ISDN systems. *Technical Digest of the 12th European Conference on Optical Communication, ECOC'86*, Barcelona, Spain.

32. Kaminow, I. P. and Koch, T. L. 1997. *Optical Fiber Communications III*, vols. A and B. New York: Academic.

33. Kawakami, H., Miyamoto, Y., Kataoka, T., and Hagimoto, K. 1994. Overmodulation of intensity modulated signals due to stimulated Brillouin scattering. *Electron. Lett.* 30: 1507.

34. Kazovsky, L., Benedetto, S., and Willner, A. 1996. *Optical Fiber Communication Systems*. Boston: Artech House.

35. Kim, K. S., Stolen, R. H., Reed, W. A., and Quoi, K. W. 1994. Measurement of the nonlinear index of silica-core and dispersion-shifted fibers. *Opt. Lett.* 19: 257–259.

36. Korotky, S. K., Hansen, P. B., Eskildsen, L., and Veselka, J. J. 1995. Efficient phase modulation scheme for suppressing stimulated Brillouin scattering. *Technical Digest of IOOC'95*, Hong Kong, Paper WD2-1.

37. Kuwano, S., Yonenaga, K., and Iwashita, K. 1995. 10 Gbit/s repeaterless transmission experiment of optical duobinary modulated signal. *Electron. Lett.* 31: 1359–1361.

38. Le Guen, D., Lobo, S., Merlaud, F., Billes, L., and Georges, T. 2001. 25 GHz spacing DDWDM soliton transmission over 2000 km of SMF with 25 dB/span. *ECOC 2001 Amsterdam Session WeF1*.

39. Le Roux, P. et al. 2001. 25 GHz spaced DWDM 160×10.66 Gbit/s (1.6 Tbit/s) unrepeated transmission over 380 km. *Proceedings of 27th European Conference on Optical Communication*, Amsterdam, Netherlands, vol. 6, pp. 10–11.

40. Mao, X. P., Tkach, R. W., Chraplyvy, A. R., Jopson, R. M., and Derosier, R. M. 1992. Stimulated Brillouin threshold dependence on fiber type and uniformity. *IEEE Photon. Technol. Lett.* 4: 66–69.

41. Mears, R. J., Reekie, L., Poole, S. B., and Payne, D. N. 1985. Neodymium-doped silica single-mode fibre lasers *Electron. Lett.* 21: 738–740.

42. Mears, R. J., Reekie, L., Jauncey, I. M., and Payne, D. N. 1987. Low-noise erbium-doped fibre amplifier operating at 1.54 μm. *Electron. Lett.* 23: 1026–1028.

43. Mollenauer, L. F., Evangelides, S. G. Jr., and Haus, H. A. 1991. Long-distance soliton propagation using lumped amplifiers and dispersion shifted fiber. *J. Lightwave Technol.* 9: 194–197.

44. Mollenauer, L. F. and Mamyshev, P. V. 1998. Massive wavelength-division multiplexing with solitons. *IEEE J. Quantum Electron.* 34: 2089–2102.

45. Park, Y. K. et al. 1993 A 5 Gb/s repeaterless transmission system using erbium-doped fiber amplifiers. *IEEE Photon. Technol. Lett.* 5: 79–82.

46. Penninckx, D. and Lanne, S. 2001. Reducing PMD impairments. *Proceedings of Optical Fiber Communication Conference*, Anaheim, CA, vol. 2, Paper TuP1-1-4.

47. Pureur, D. *Highwave Optical Technologies*. France, Private communication.

48. Ragdale, C. M., Payne, D. N., De Fornel, F., and Mears, R. J. 1983. Single-mode fused biconical taper fibre couplers. *Proceedings of the 1st International Conference on Optical Fibre Sensors*, London, pp. 75–78.

49. Sakamoto, T., Aozasa, S., and Shimizu, M. 2002. Recent progress on S-band amplifiers. *ECOC 2002 Copenhague Session Mo2.2*.

50. Segi, T., Aizawa, T., Sakai, T., and Wada, A. 2001. Silica-based composite fiber amplifier with 1480–1560 nm seamless gain band. *ECOC 2001 Amsterdam Session MoL3*.

51. Schicketanz, D. and Zeidler, G. 1975. GaAs-double-heterostructure lasers as optical amplifiers. *IEEE J. Quantum Electron.* 11: 65–69 (and references therein).

52. Shimizu, M., Yamada, M., Horiguchi, M., Takeshita, T., and Okayasu, M. 1990. Erbium-doped fibre amplifiers with an extremely high gain coefficient of 11.0 dB/mW. *Electron. Lett.* 26: 1641–1643.

53. Sunnerud, H., M. Westlund, Li, J., Hansryd, J., Karlsson, M., Hedekvist, P. -O., and Andrekson, P. A. 2001. Long-term 160 Gb/s-TDM, RZ transmission with automatic PMD compensation and system monitoring using an optical sampling system. *Proceedings of 27th European Conference on Optical Communication*, Amsterdam, Netherlands, vol. 6, pp. 18–19.

54. Takushima, Y. and Okoshi, T. 1992. Suppression of stimulated Brillouin scattering using optical isolators. *Electron. Lett.* 29: 1155.

55. Tanaka, K., Sakata, H., Miyakawa, T., Morita, I., Imai, K., and Edagawa, N. 2001. 40 Gbit/s×25 WDM 306 km unrepeated transmission using 175 µm²-Aeff fibre. *Electron. Lett.* 37: 1354–1356.

56. Tkach, R. W., Chraplyvy, A. R., Forghieri, F., Gnauck, A. H., and Derosier, R. M. 1995. Four-photon mixing and high-speed WDM systems. *J. Lightwave Technol.* 13: 841–849.

57. Vareille, G., Julien, B., Pitel, F., and Marcerou, J. F. 2001. 3.65 Tbit/s (365×11.6 Gbit/s) transmission experiment over 6850 km using 22.2 GHz channel spacing in NRZ format. *Proceedings of 27th European Conference on Optical Communication*, Amsterdam, Netherlands, vol. 6, pp. 14–15.

58. Vengsarkar, A. M., Lemaire, P. J., Judkins, J. B., Bhatia, V., Erdogan, T., and Sipe, J. E. 1996. Long-period fiber gratings as band-rejection filters. *J. Lightwave Technol.* 14: 58–64.

59. Wedding, B., Chiarotto, A., Kuebart, W., and Bulow, H. 2001. Fast adaptive control for electronic equalization of PMD. *Proceedings of Optical Fiber Communication Conference*, Anaheim, CA, vol. 2, Paper TuP4-1-3.

60. Yonenaga, K. and Kuwano, S. 1997. Dispersion-tolerant optical transmission system using duobinary transmitter and binary receiver. *J. Lightwave Technol.* 15: 1530–1537.

61. Yu, J., Kojima, K., Chand, N., Fischer, M. C., Espindola, R., and Mason, T. G. B. 2001. 160 Gb/s single-channel unrepeated transmission over 200 km of non-zero dispersion shifted fiber. *Proceedings of 27th European Conference on Optical Communication*, Amsterdam, Netherlands, vol. 6, pp. 20–21.

2

Optical network architectures

TON KOONEN
Technische Universiteit Eindhoven

2.1 INTRODUCTION

Telecommunication networks in all their various shapes are indispensable to bring information quickly anywhere and anytime, which is a vital need of our modern global society. Since the invention of the electrical telegraph by Samuel Morse in 1837, the variety of telecommunication services has grown at an increasing pace, as illustrated in Figure 2.1. In addition, the services are becoming ever more individualized, and along with the penetration of video-based services ("a picture says more than a thousand words") the request for information transport capacity has exploded and is continuing to do so. Since the early 1990s, the introduction of the worldwide Internet has drastically promoted this information transport explosion. The number of Internet hosts is still increasing exponentially; from January 1992 to

January 1997 to January 2002, it grew from 727 thousand to 19.5 million to 147 million worldwide. This is causing data traffic to take an ever-larger share of the telecommunication network capacity; since a few years, it has surpassed the volume of the traditional voice traffic (but not yet its revenues). Wireless mobile telecommunication is attracting ever more users, and enables a fast roll-out of services to the end users without the need to install extensive first-mile customer access networks. The telecommunication market liberalization has provided ample opportunities to the entry of more operators and service providers, and the resulting national and international competition is pressing for very efficient high-capacity telecommunication networks.

As a result, the volume of telecommunication traffic is ramping up at a compound annual growth rate of roughly 60%, which means an increase with

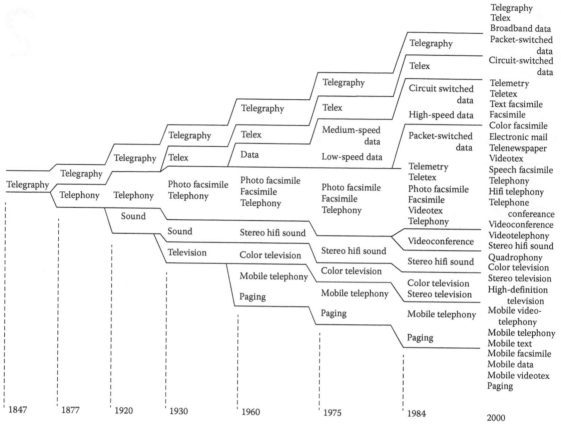

Figure 2.1 Evolution of telecommunication services. (From Consortium British Telconsult/Consultel/Detecon/Sofrecom et al.)

a factor of 10 in. no more than 5 years. This traffic is in vast majority carried along fixed-wired networks, due to their high reliability, security, and immunity for external disturbances. Wireless networks are coming up in customer access environments; but due to increasing microwave carrier frequencies and user densities, the wireless cells are shrinking and thus extensive fixed access network lines are still indispensable as the vessels to feed the wireless antenna stations. Traditional coaxial and twisted-pair copper cables have been the transport media of choice since the introduction of telecommunication networks. However, the advent of optical fiber with its extremely low losses and extremely large bandwidth as pioneered by Kao and Hockam in 1966, and the commercial introduction of optical fiber communication systems in the early 1980s has caused that single-mode optical fiber has become the transport medium uniquely used in long-distance fixed-wired core

transport networks, and that it is also conquering at increasing pace the area of metropolitan and access networks.

2.2 TELECOMMUNICATION NETWORKS HIERARCHY

Telecommunication networks are carrying traffic at various aggregation levels, as illustrated in Figure 2.2.

At the highest level, *long-haul core networks* are transporting huge data capacities in the tens of terabits/second over large distances, such as over transnational and transoceanic links up to 9000 km (transpacific). These global and wide area networks are transporting circuit-switched data following the SDH (or SONET) standards, where each fiber usually carries a number of wavelengths at bitrates of up to 10–40 Gbit s^{-1} each. They have to meet extremely high levels of reliability and availability,

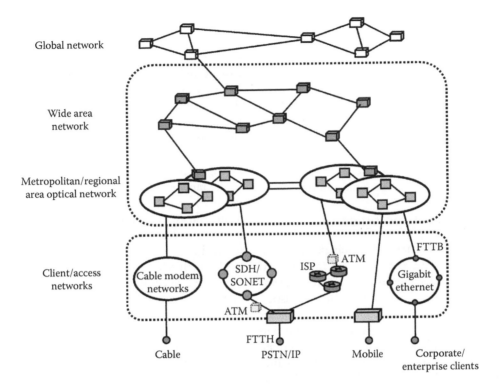

Figure 2.2 The hierarchy of telecommunication networks.

considering the huge volume of customers dependent on them. Taking into account the increasing dynamics in the traffic matrix describing the data flows between the various network nodes, packet switching techniques are being introduced, which can offer a more efficient utilization of the network's resources than circuit switching.

Metropolitan area networks (MANs) are covering large urban areas with a reach of up to 100 km and capacities of tens of gigabit/second, serving in particular business parks and residential customer access regions. High availability for large-volume fast file transfer is a major need for the business customers. Also these networks should be easily scalable for adding more network nodes, and flexible to accommodate new business needs. Storage area networks (SANs) are specifically employed for regularly moving large volumes of data between geographically separated sites, in order to safeguard vital business information.

Access networks are providing a wide variety of services to the end customers, and consist of fiber feeder networks followed by various first-mile networks. These networks are mostly optimized for a

particular set of services, and exploited by different operators. Coaxial cable network operators offer television and radio broadcast services, and since recently also data modem services and telephony, multiplexed in different frequency bands. The public switched telephone network (PSTN) uses twisted pair copper cables and is carrying voice telephony and data services, time-multiplexed according to the SDH/SONET or ATM standard; it is exploited by the incumbent telecom operators as well as new entrants. Mobile network operators are mainly providing wireless voice telephony, according to among others the GSM standard; also wireless data services using GPRS and UMTS are coming up. The statistics of traffic in access networks shows much higher dynamics than in metropolitan and core networks, due to the significantly lower traffic aggregation levels. Therefore, applying packet switching instead of circuit switching can improve remarkably the network utilization efficiency. Access networks have to be laid out very cost-effectively, as the factor with which network equipment is shared among customers is much lower than in metropolitan and in core networks.

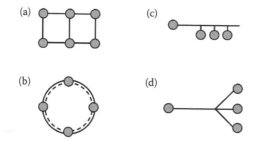

Figure 2.3 Network topologies. (a) Mesh, (b) Ring, (c) Bus, and (d) Tree -and- branch.

2.2.1 Network topologies

A number of general network topologies as shown in Figure 2.3 can be discerned for implementation of the various hierarchical network layers. Each of them has a specific set of characteristics, which makes it suited to match the requirements of a certain layer. Mesh networks, as exemplified in Figure 2.3a, provide a number of options to route traffic between two network nodes. This routing redundancy yields a large availability of the network services, which is a highly valued merit in long-distance core and metro-core networks. The entailed extra costs are of less concern due to the large resource-sharing factor among the huge customer base served. Ring networks, in particular when composed of both an inner and an outer ringlet as shown in Figure 2.3b, provide clockwise and counterclockwise traffic routing options; thus also network protection is established in order to yield a good network availability (albeit at a lower level than in mesh networks, but also at lower costs as less network resources are needed). The combination of good availability and moderate costs of resources makes ring topologies well suited to implement metropolitan networks. Bus networks, as shown in Figure 2.3c, use a common single linear medium along which signal power is tapped off to the various nodes (with power losses accumulating at each subsequent tap). This topology is quite cost-efficient, as a minimum of network resources is needed. The nodes may exchange information as peers in the network, which makes this topology suited for linking data processing equipment. However, no routing redundancy is provided, and therefore there is no guarantee for good network availability. In tree-and-branch networks, exemplified in Figure 2.3d, a relatively long single feeder line is running from a headend node

to a power splitting point, from where the signal power is distributed via short lines to a number of end nodes. This topology is most suited for broadcasting information from a single headend node to many customer end nodes. No routing redundancy is provided, so again network availability is limited. Also the topology is very cost-efficient, as the costs of the feeder line and the headend equipment is shared by all the end nodes. In addition, the signal power distribution is more efficient than in the bus network, as the power splitting loss (in decibels) to an end node increases only with the logarithm of the number of end nodes whereas it increases linearly with this number in the bus network. Every end node receives the same power level, which relaxes the dynamic range over which the end node equipment has to operate. The tree-and-branch topology is therefore well suited and popular for access networks. With a fully passive optical power splitter in the branching point, the topology is also widely known as the passive optical network (PON).

In the next sections of this chapter, architectural aspects and key functionalities needed at the subsequent hierarchical network layers (core networks, metropolitan networks, and access networks) will be discussed in more detail.

2.3 CORE NETWORKS

The main task of core networks is to transport huge amounts of telecommunication traffic over large distances in a highly reliable way. The network should not break down by failures in one or in a few links. Therefore, provisions have to be included for alternative routing of traffic, which are adequately offered by the mesh network topology with cross-connect functions in the nodes. The links between the nodes are usually very long (>100 km), and intermediate signal amplification and compensation of fiber dispersion effects is needed.

2.3.1 Optical signal multiplexing techniques

Using electrical time division multiplexing (ETDM), commercial systems support bitrates up to 10–40 Gbit s^{-1} and by direct laser diode modulation (or a laser diode followed by an external modulator) these bitrates can be carried through a fiber link via single wavelength channel. The transport capacity of the

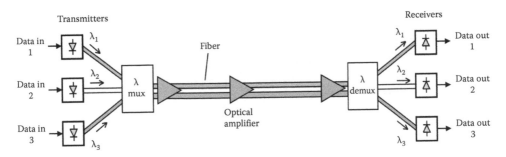

Figure 2.4 Wavelength division multiplexing.

fiber link can basically be increased further in two ways: by increasing the number of wavelength channels, and by increasing the bitrate per channel.

Multiple wavelength channels can be carried by a single fiber by combining them at the transmitting end by means of a wavelength multiplexing device, and separating them again at the receiving end by a wavelength demultiplexer. This wavelength division multiplexing (WDM) approach is shown in Figure 2.4. Standard single-mode fiber offers low dispersion in the wavelength window of 1285–1330 nm, which amounts to about 8 THz of bandwidth, and low attenuation in the 1500–1580 nm window, corresponding to about 10 THz. The popular 1530–1560 nm window (the C-band, the operation range of erbium-doped fiber optical amplifiers) represents about 3.8 THz of bandwidth. Wavelength channel spacings of 100 GHz according to ITU-T G.692 (or even down to 25 GHz) are being deployed, which enables hundreds of wavelength channels to be accommodated in a single fiber. Commercial systems are available which carry 160 wavelength channels at 10 Gbit s⁻¹ each, amounting to a total 1.6 Tbit s⁻¹ capacity. The record obtained in research stands at 273 wavelengths at 40 Gbit s⁻¹ each, amounting to 10.92 Tbit s⁻¹ [1].

Increasing the bitrate per wavelength channel beyond the limits of ETDM can be achieved with the so-called optical time division multiplexing (OTDM) technique. As illustrated in Figure 2.5, at the transmitter side a single laser diode generates a sequence of equidistant narrow optical pulses. After splitting and distribution to four fast optical gates, the pulse train is on/off switched in each gate by an electrical time-multiplexed data stream. The modulated pulse trains are delayed with respect to each other, and subsequently interleaved (like a "zipper"). The individual pulse trains may all have the same polarization (SP, single polarization), or alternating polarization (AP). The optical pulses need to be sufficiently narrow in order to avoid crosstalk. With the OTDM-AP scheme, somewhat broader optical pulses are allowed than with the OTDM-SP scheme. The resulting output signal is

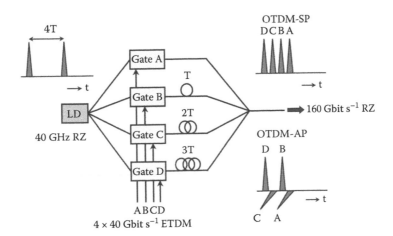

Figure 2.5 Optical time division multiplexing transmitter.

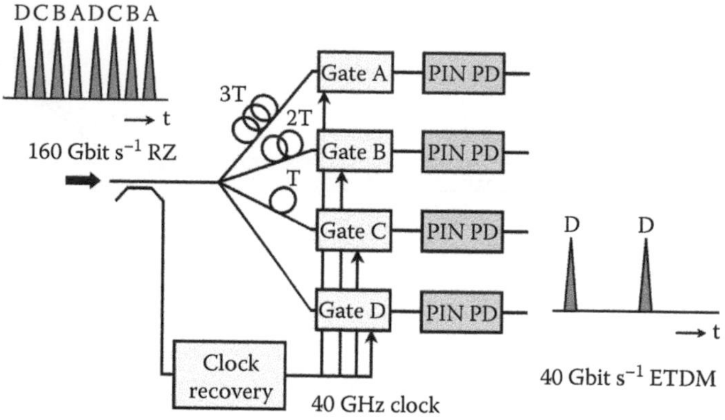

DCBADCBA

160 Gbit s⁻¹ RZ

160 Gbit s^{-1} RZ

Gate A — PIN PD
Gate B — PIN PD
Gate C — PIN PD
Gate D — PIN PD

Clock recovery

40 GHz clock

40 Gbit s^{-1} ETDM

Figure 2.6 Optical time division demultiplexing receiver.

a modulated optical pulse data stream at a speed that is the sum of the speeds of the electrical input data streams. At the receiver side, first the clock signal needs to be recovered from the high-speed data stream. Using this clock signal and appropriate time delays, fast optical gates followed by optical receivers with PIN photodiodes can demultiplex the original constituting pulse trains, as shown in Figure 2.6. Using these OTDM techniques, the next bitrate hierarchy of 160 Gbit s^{-1} can be realized with 40 GHz electronics. In research, the record has been set at 1.28 Tbit s^{-1} by polarization-multiplexing two OTDM-SP 640 Gbit s^{-1} streams [2].

The chart shown in Figure 2.7 indicates how by increasing the data rate per wavelength channel by means of advances in electrical and optical TDM at one hand, and by increasing the number of

wavelength channels at the other hand, the transport capacity of a single fiber has been enormously increased since the introduction of optical fiber communication systems in the early 1980s. The main leap forward was made with the introduction of wavelength multiplexing in the early 1990s. And opportunities for further capacity growth are still being created by opening new wavelength bands such as the S-band from 1450 to 1530 nm, and the L-band from 1560 to 1620 nm, supported by alternative optical amplifying processes due to fiber nonlinearities such as Raman gain. Optical gain in the 1300 nm window can be provided by fiber amplifiers using rare earth materials such as praseodymium and neodymium, and by semiconductor optical amplifiers. The present system capacity record stands at 10.92 Tbit s^{-1} deploying 273 wavelength channels at 40

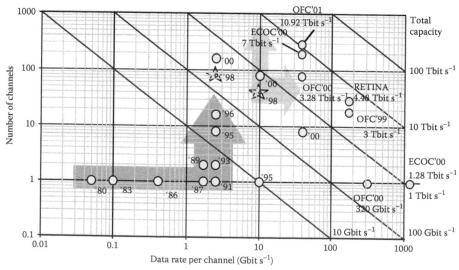

Figure 2.7 The evolution of transport capacity of a single fiber link.

Gbit s^{-1} each, as mentioned earlier. Applications for multi-terabits/second systems are found not only in transcontinental and transoceanic systems, but also for massive data processing systems such as huge synthesized antenna array systems for astronomical observations (e.g., the RETINA system [3]).

In addition to wavelength multiplexing, optical amplification and dispersion compensation are key techniques to enable high volume data transport over long fiber links. A single optical amplifier can handle many wavelength channels simultaneously; otherwise, an equivalent number of opto-electronic regenerators plus a wavelength multiplexer and demultiplexer would be needed, which is clearly more costly, requires more maintenance and powering, and requires a sizeable adaptation effort when the system needs to be upgraded with more wavelength channels.

2.3.2 Traffic routing

The deployment of multiple wavelength channels is not only beneficial for increasing the transport capacity of a single fiber link, but also provides more flexibility to route traffic streams in the network. Wavelength channels can constitute independent optical paths through the network, and each path may be laid out individually to establish an optimum connection between certain end nodes. As shown in Figure 2.8a, within each node the wavelength channels are routed by means of optical crossconnects which guide the incoming signals depending on their wavelength and the entrance port to a specific output port. The routing table is usually set by the network management system, and can be altered when needed by changing traffic conditions. When two optical paths do not touch each other, they may be established with the same wavelength. This wavelength re-use reduces the overall number of wavelengths needed in the network. In Figure 2.8a, for instance, wavelength Ai is used twice, but in different paths. As illustrated in Figure 2.8b, further reduction of the number of wavelengths needed can be obtained by using wavelength converters in the nodes; thus an optical path may be constituted by a sequence of different wavelengths in a series of links. By controlling the crossconnects in the nodes, the network management system can optimize the traffic flow routings to obtain a good balancing of the network load among the links, and thus to reduce the probability of congestion and to increase the network's efficiency. Also link failures can be circumvented by routing traffic through alternative links, which improves the reliability of the network.

The basic layout of an optical crossconnecting node is shown in Figure 2.9. The N wavelength channels carried by each of the M input fibers are firstly separated by a wavelength demultiplexer, and subsequently a large optical matrix switch with ($M \times N$) input ports and ($M \times N$) output ports is needed which can route any wavelength channel from any input fiber to any of the M output fibers. For larger numbers M and N, the internal architecture of the matrix switch becomes quite comprehensive. Composed of individual 2×2 optical switches, a Benes switch architecture as shown in Figure 2.10 would require $(MN/2)(2^2\log MN - 1)$ switches; and $2^2\log MN - 1$ switches have to be passed from any

(a)   Crossconnect node

(b)   Crossconnect node with λ-conversion

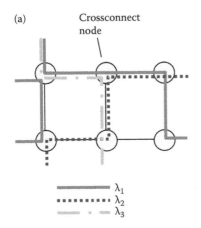

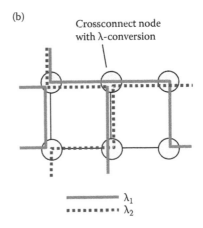

Figure 2.8 Establishing wavelength paths. (a) Without wavelength conversion and (b) with wavelength conversion.

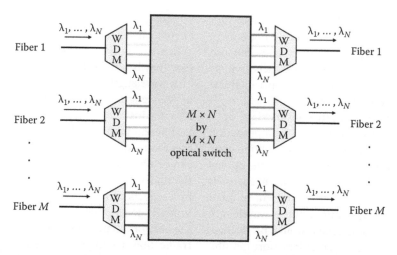

Figure 2.9 Optical crossconnect.

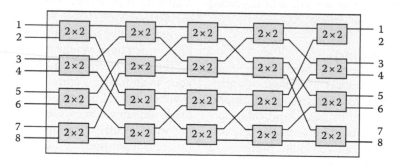

Figure 2.10 Re-arrangeable nonblocking Benes switching matrix.

input to any output causing losses which increase with the switch size. Three-dimensional free-space optical switches using micro-mechanical mirrors for beam steering between the array of input fibers and the array of output fibers exhibit losses which do not increase with the number of switch ports, and thus can outperform the Benes planar architecture [20, 21].

The amount of packet-based data traffic is growing fast in all telecommunication network layers, a.o. due to the steeply rising use of Internet services. IP packets are usually carried over SDH/SONET and/or ATM, which in their turn are carried in WDM channels; efficiency can be gained, however, by transporting the IP packets directly in the WDM channels, with some simple framing for basic synchronization functions. By using packet switching in the optical crossconnect nodes instead of circuit switching, the network capacity can be exploited much more dynamically in response to instantaneous traffic demands, and thus the network operation efficiency is improved. Following

the GMPLS protocol, wavelength-switched paths can be set up in the network in a similar way as label-switched paths in the MPLS protocol. The packets can be marked by assigning a wavelength to them, which acts as a label. Based on these labels, a path is established for each packet through the network; see Figure 2.11. The per packet label swapping needed for efficient forwarding and routing is achieved by wavelength converters in the nodes, which may be realized with fast tunable lasers and an all-optical wavelength converter (such as a Mach–Zehnder interferometer with semiconductor optical amplifiers in its branches). Another way to attach label information to a packet is to incorporate it in the packet data frame, or to modulate it on a subcarrier frequency outside the spectrum of the packet data. Even more comprehensive label information may be attached by frequency shift keying (FSK) modulation of the optical carrier (or differential phase shift keying, DPSK), orthogonally to the payload data that is intensity modulated on the carrier [4, 5]. Using wavelength converters and

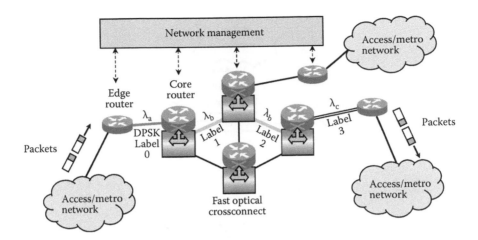

Figure 2.11 Optical packet routing.

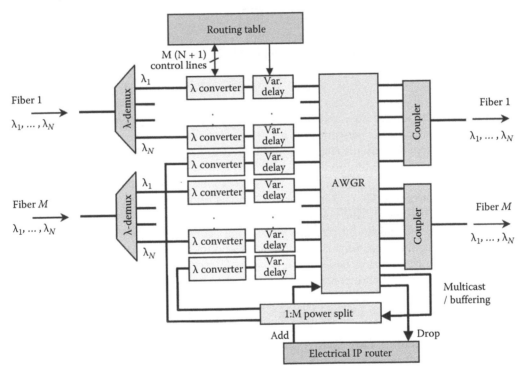

Figure 2.12 Fast optical crossconnect using wavelength conversion.

a wavelength-selective passive router (such as an arrayed waveguide grating router, AWGR), a fast optical crossconnect can be realized; e.g., in an architecture as shown in Figure 2.12. Variable delay lines are included to avoid collisions of packets that are heading for the same output fiber at the same wavelength. Congestion can also be avoided by temporary buffering in the recirculating loops; the loops plus power splitter enable optical multicasting as well.

2.4 METROPOLITAN AREA NETWORKS

MANs have to bring a variety of services to large urban areas, typically at bitrates up to 10 Gbit s^{-1} per wavelength over distances between nodes of less than 100 km. Major customers to be served are business parks, where customers mainly ask for fast large-volume file transfer, e.g., to interconnect their offices and for storage at safe locations

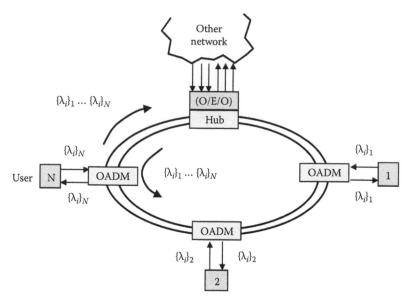

Figure 2.13 Multiwavelength MAN with fixed wavelength routing via hub.

of vital business data (storage area network). As high reliability is required while network costs per customer should be kept limited, the ring network topology is well suited for providing good network protection at moderate costs.

An example layout of a MAN ring network is shown in Figure 2.13. Typically, the ring is composed of two fibers, on which the traffic flows clockwise and counterclockwise, respectively. The fiber links between neighbouring nodes are less than 20 km, and the ring circumference is usually less than 100 km; therefore no in-line optical amplifiers are needed. Each node uses a specific set of wavelength channels, typically carrying bitrates up to 2.5 Gbit s^{-1} per channel. Each node is communicating with the hub node; in the hub, opto-electric-optical translation to another wavelength set is done in order to establish communication between ring nodes. So virtually in the network there are point-to-point node-to-hub connections. The MAN typically contains up to 16 nodes, and up to 40 wavelength channels. Per node, one or more out of the set of wavelength channels can be dropped and/or added by means of optical add/drop multiplexers (OADMs). Through the hub, also communication to other networks can be established.

2.4.1 Network protection

The two-fiber topology enables self-healing of the ring network in case of a cable break. As shown in Figure 2.14, a link failure may be circumvented by looping back the traffic at the nodes neighbouring the broken fiber cable. This procedure also allows a ring segment to be taken temporarily out of service without disrupting the traffic on the remainder of the ring network, e.g., for maintenance or for inserting a new node. In the SONET standard, the 2-fiber uni-directional path-switched ring (UPSR) is composed of an outer primary path ring, carrying clockwise the normal working traffic; see Figure 2.15. The inner ring provides the counterclockwise protection path. In the SDH standard, this concept is called sub-network connection protection (SNCP). Both rings are fed from transmitters at the nodes, and at the node receiver, the signal from the ring with the best quality is chosen. Bi-directional traffic between two nodes will span around the entire ring, and thus consumes resources on every link of the ring. The 4-fiber bi-directional line-switched ring (BLSR) in the SONET standard (called multiplex section shared protection ring, MS-SPRing, in the SDH standard) consists of an outer bi-directional 2-fiber primary

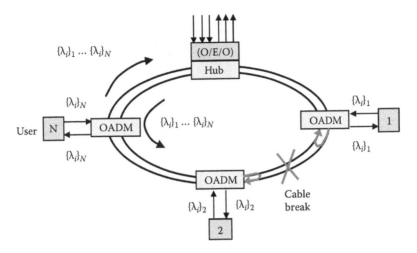

Figure 2.14 Self-healing by looping back at the OADMs.

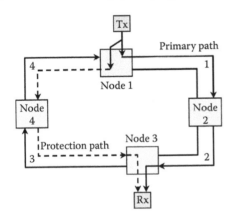

Figure 2.15 Protection in the 2-fiber SONET UPSR (SDH SNCP) network.

loop carrying the normal working traffic, and an inner bi-directional 2-fiber secondary loop for protection. Bi-directional traffic between two nodes is only sent along part of the ring, and does not involve resources on other parts of the ring. As shown in Figure 2.16, the system is well protected against cable breaks or node failures. In IEEE 802.17, the resilient packet ring (RPR) is being discussed for standardization [6]. It consists of a dual counter-propagating ring, with up to 256 node stations and spans up to 6000 km. The RPR is able to operate with any packet protocol (such as Ethernet), and uses frame-based transmission based on the standard gigabit Ethernet frame of variable length, but also supporting jumbo frames up to 9 kbytes.

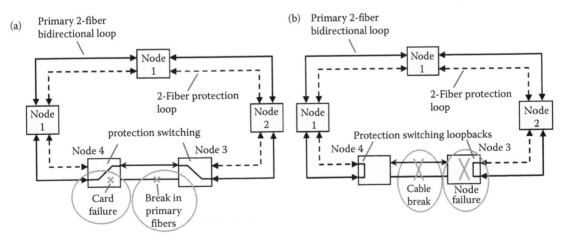

Figure 2.16 Protection in the 4-fiber SONET BLSR (SDH MS-SPRing) network. (a) Bypassing a fiber break or malfunctioning card and (b) reconfiguration in case of a cable break or node failure.

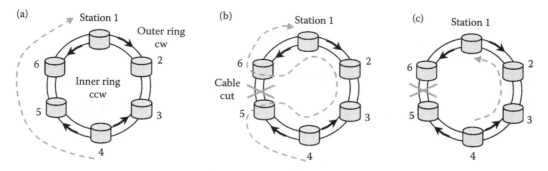

Figure 2.17 Protection in the resilient packet ring network. (a) Traffic flows along the outer ring before cable out, (b) wrapping of the data path after the cable cut and (c) traffic steering, yielding a shortest data path after discovery of the new topology.

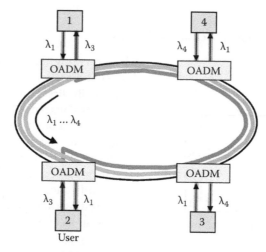

Figure 2.18 Multiwavelength MAN with dynamic wavelength routing between nodes.

It is optimized for providing multiple IP service classes: reserved traffic class A0 (guaranteed data rate, small bounded delay and jitter), high priority class A1 (committed rate, bounded delay and jitter), medium priority class B (committed rate, and excessive rate subject to fair access), and low priority class C (best effort, subject to fair access). The RPR management system is able to provide ring survivability within 50 ms. It selects the inner or outer ring to establish the shortest path. When a ring failure occurs, the traffic is first wrapped at the failure point, subsequently the damaged ring topology is discovered, and the source traffic is steered to establish the shortest path again taking the failure into account (see Figure 2.17).

The nodes in the MAN may also route the wavelength channels dynamically between themselves without relaying via a hub, as exemplified in Figure 2.18. By means of an appropriate algorithm, and by deploying tunable laser diodes and tunable OADMs at the nodes, the most appropriate wavelengths can be assigned to the communication paths needed between nodes. When paths do not overlap, the same wavelength may be used (wavelength re-use), which reduces the number of wavelengths needed.

2.4.2 Optical add/drop multiplexing

Essential network elements in the ring-shaped MANs are OADMs, which are able to extract data stream on a particular wavelength channel (or several channels) from the ring, and to insert new data streams on one or more wavelength channels into the ring. Signalling information controls whether a data stream will be dropped at a node or will pass through it. There are various ways to transfer this signalling information to the nodes, without interfering with the data streams. It may be modulated on a subcarrier frequency that is positioned above the data spectrum, and each wavelength channel may carry a unique subcarrier frequency. Thus, it is not needed to wavelength-demultiplex the channels first in order to detect and demodulate the signalling information on the various subcarriers. Another method to transfer the signalling information is to modulate it on a dedicated separate common wavelength channel, which is opto-electric-optically (O/E/O) converted at each node for inspection. And a third method is to put the signalling information in-band in a digital frame together with the data (e.g., the digital wrapper concept); then at each node all wavelength channels need to be demultiplexed and O/E/O converted for inspection.

The OADMs need to be wavelength-selective. When the pass-through wavelength channels experience some filtering, the passband of many OADMs put in cascade along the ring may narrow significantly. Thus the possibility to extend the number of nodes in the network scalability may be reduced. This scalability issue is avoided by applying notch-type OADMs, which do not exhibit pass-through filtering.

Another issue in the design of OADMs is crosstalk. Two types of crosstalk may be discerned: incoherent crosstalk and coherent crosstalk, as shown in Figure 2.19. Incoherent crosstalk may occur because part of the wavelength channel(s) $\lambda_i \neq \lambda_x$ to be passed through are dropped, and interfere with the intentionally dropped wavelength channel λ_x. Coherent crosstalk occurs because part of the dropped wavelength channel λ_x leaks through, and beats with the added wavelength channel λ'_x which has nominally the same wavelength. Thus the coherent crosstalk cannot be removed by subsequent optical bandpass filtering, and it may accumulate when cascading OADM nodes. The coherent crosstalk imposes the most stringent requirements on the device crosstalk characteristics. Mathematical analysis taking the specific statistical properties of the beat signal into account show that per node the crosstalk attenuation needs to be better than 32 dB to yield a bit error rate (BER) better than 10^{-12} [7]. For incoherent crosstalk in a four-channel OADM, the crosstalk attenuation needed to yield a BER < 10^{-12} needs to be better than 13 dB only.

An example of an OADM that can drop and add a fixed wavelength channel is shown in Figure 2.20. A fiber Bragg grating (FBG) reflects only wavelength λ_x; the other wavelength channels are passed

unaffected. This notch-type OADM does not put a limit to extending the number of nodes in the ring. The FBG is made by writing a grating with UV light into the fiber core. By putting thermal or mechanical stress on the FBG, by means of a local heater or a piezo-electric stretcher, respectively, the device may be slightly tuned at low speed to other wavelength channels. Thus circuit-switched connections may be set up. Another more widely tunable OADM is shown in Figure 2.21, which deploys a fiber Fabry–Perot (FFP) filter. The passband of the FFP can be tuned to any of the input wavelength channels, and thus the selected channel is dropped whereas the other channels are reflected without filtering and via

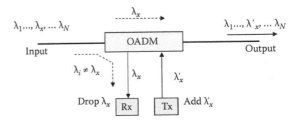

Figure 2.19 Crosstalk in an Optical Add/Drop Multiplexer (OADM)

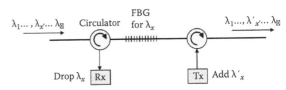

Figure 2.20 Fixed-channel OADM using fiber Bragg grating (FBG). (*Note:* the right-hand circulator may also be replaced by a power combiner.)

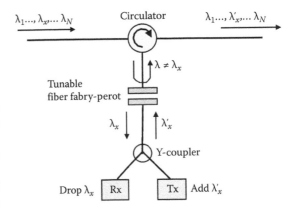

Figure 2.21 Tunable OADM using a fiber Fabry-Perot filter.

the optical circulator passed to the output port. The locally added channel will pass the FFP and join the other channels via the circulator. The residual reflection (a few percent) at the FFP of the added channel causes near-end crosstalk at the local receiver; this may be counteracted by echo cancelling, as the locally added signal is known. An FFP is usually tuned with piezo-electric means; tuning speed is therefore limited, and this OADM is suited for setting up circuit-switched connections. Its notch-type characteristic implies that this OADM architecture does not limit the extension of the number of nodes in the ring.

When the wavelength channels are arranged in groups of closely spaced channels, adding and dropping of a specific group per node may be accomplished by a cascaded architecture of a fine-grain demultiplexer/multiplexer and a coarse grain OADM. Figure 2.22 shows an example, where a silica-based AWGR demultiplexer with narrow

channel spacing and with a free spectral range of 500 GHz separates the 40 input wavelength channels spaced at 100 GHz into groups of eight wavelengths spaced at 500 GHz. Next, a compact OADM integrated in InP with a coarse channel spacing of 500 GHz separates the eight wavelength channels within a specific group.

For packet-switching applications, the OADM characteristics need to be fast tunable. Next to a wavelength demultiplexer stage and a wavelength multiplexer stage, the OADM architecture shown in Figure 2.23 is equipped with fast (lithium-niobate, or semiconductor-based) optical switches, and can drop and/or add multiple wavelength channels. The wavelength passband channels of the demultiplexing and the multiplexing stages need to be and to stay carefully aligned. By putting a 2×2 optical switch in the cross state (as indicated for channel λ_2), the corresponding wavelength channel can be dropped and added. The pass-through wavelength channels are filtered, and thus this design limits the cascadability of nodes. An OADM architecture with similar functionality but requiring only a single wavelength-selective element is shown in Figure 2.24 [8,9]. The wavelength-selective AWGR performs both the demultiplexing and the multiplexing of the wavelength channels, and thus avoids wavelength misalignment issues. Some crosstalk may occur due to direct leak-through from the input port to the output port of the AWGR. By looping back the wavelength-demultiplexed paths via the add/drop switch matrix not to the front side but to the back side of the AWGR, and positioning the output port also at the front side, this crosstalk is strongly reduced; this fold-back architecture requires,

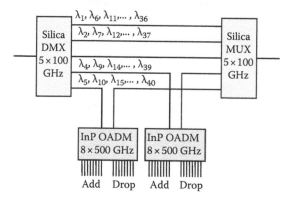

Figure 2.22 Two-stage OADM for handling wavelength groups.

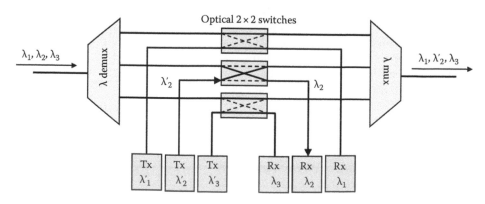

Figure 2.23 Fast wavelength-switchable OADM.

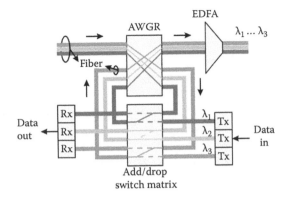

Figure 2.24 Fast wavelength-switchable OADM based on an arrayed waveguide grating router (AWGR).

however, a larger AWGR, with twice the number of ports at the back side.

2.5 ACCESS NETWORKS

Access networks carry a wide range of services to and from the residential end customers, ranging from voice-based services, audio-based ones, video-based ones, to Internet/data services. Also the capacities needed per service vary widely: from 64 kbit s^{-1} for traditional voice telephony to beyond 100 Mbit s^{-1} for high-speed Internet and data. The last link to connect to the end customer, the so-called first mile (or last mile, depending on the point of view), may be bridged with various types of transport media exploited by various network operators. Coaxial copper cable transports broadcast television and radio services, and increasingly also data services via cable modems. Twisted copper pair cables carry voice telephony, and data services via voice modems or high-speed ADSL and VDSL modems. Wireless systems bring mobile voice telephony via the GSM standard, and also data services via GPRS and UMTS. Optical fiber to the home/building is entering the market, but still has to surpass some cost barriers. It can offer the full set of integrated broadband services, from broadcast high bandwidth video services to gigabit Ethernet data services. A list of first-mile media with the bearer services, bitrates and reach is given in Table 2.1. The need for more bandwidth in the access network is growing continuously, due to the increasing amount of bandwidth required by each customer, mainly fuelled by

video-based services and high-speed Internet, the tailoring of services to individual customer needs, and the emergence of more competing operators due to liberalization. This spurs the introduction of optical fiber (mainly single mode fiber, being a future-proof solution with its virtually infinite bandwidth) into the access network. As the installation and equipment costs of fiber-to-the-home (FTTH) are still quite high in comparison to the traditional copper wired access lines, hybrid fiber access networks are the first step to introduce fiber. Fiber is used in the upper feeder part of the access network, where it runs from a local exchange (headend station) to a cabinet along the street (fiber-to-the-cabinet, FTTCab) or to the basement of a building (such as an apartment building with many living units; fiber-to-the-building, FTTB), where the optical signals are converted back into electrical ones which are then brought via copper-based first mile links or wirelessly to the end customers. In the following of this section, the attention will be focussed on the optical fiber part of the access network.

Basically, three architectures may be deployed for the fiber access network:

1. Point-to-point topology, where individual fibers run from the local exchange to each cabinet, home or building. Many fibers are needed, which entails high first installation costs, but also provides the ultimate capacity.
2. Active star topology, where a single fiber carries all traffic to an active node close to the end users, from where individual fibers run to each cabinet/home/building. Only a single feeder fiber is needed, and a number of short branching fibers to the end users, which reduces costs; but the active node needs powering and maintenance.
3. Passive star topology, in which the active node of the active star topology is replaced by a passive optical power splitter that feeds the individual short branching fibers to the end users. In addition to the reduced installation costs of a single fiber feeder link, the completely passive outside plant avoids the costs of powering and maintaining active equipment in the field. This topology has therefore become the most popular one for introduction of optical fiber into access networks, and is widely known as the passive optical network, PON.

Table 2.1 First-mile network technologies

Medium	Bearer service	Bitrate (down/up)	Reach (km)
Twisted pair	Analogue line	Rates up to 56 k/56 kbit s⁻¹	
Twisted pair	ISDN	144 k/144 k data incl. 64 k/64 k bit s⁻¹ voice or data circuits	<6
Twisted pair	SDSL	768 k/768 k bit s⁻¹	<4
Twisted pair	ADSL	1.5 M to 6 M/64 k to 640 k bit s⁻¹	<4–6
Twisted pair	VDSL	26 M to 52 M/13 M to 26 M bit s⁻¹	<0.3–1
Coaxial cable	CDMA/ OFDM+QAM/QPSK	<14 M/14 M (net 8.2 M) bit s⁻¹ in 6 MHz slot	
Fiber (single mode)	ATM	150 M to 622 M/150 M bit s⁻¹ shared up to 1:32 (FSAN ATM-PON); up 1.24 G/620 M bit s⁻¹ (FSAN/ITU B-PON)	< 20
Fiber (single mode)	Gigabit Ethernet	1 Gbit s⁻¹ (1.25 Gbit s⁻¹ 8B/10B coded)	<5
Fiber (multi mode)	Gigabit Ethernet	1 Gbit s⁻¹ (1.25 Gbit s⁻¹ 8B/10B coded)	<0.55
Wireless (mobile)	GSM	13 kbit s⁻¹ (at carrier frequency 900 and 1800 MHz, frequency duplex)	<16
Wireless (mobile)	GPRS	115 kbit s⁻¹	
Wireless (mobile)	UMTS	144 k to 2 Mbit s⁻¹ (at carrier frequency 2110–2200/1885–2025 MHz, frequency duplex)	
Wireless (fixed)	MMDS	6 Mbit s⁻¹ (at carrier frequency > 17 GHz)	
Wireless (fixed)	LMDS	45 Mbit s⁻¹ (at carrier frequency > 17 GHz)	

2.5.1 Multiple access PONs

The common fiber feeder part of the PON is shared by all the optical network units (ONUs) terminating the branching fibers. The traffic sent downstream from the optical line terminal (OLT) at the local exchange is simply broadcasted by means of the optical power splitter to every ONU. Sending traffic from the ONUs upstream to the local exchange, however, requires accurate multiple access techniques in order to multiplex collision-free the traffic streams generated by the ONUs onto the common feeder fiber. Four major categories of multiple access techniques for fiber access networks have been developed:

- Time division multiple access (TDMA)
- SubCarrier multiple access (SCMA)
- Wavelength division multiple access (WDMA)
- Optical code division multiple access (OCDMA)

In a TDMA system, as shown in Figure 2.25, the upstream packets from the ONUs are time-interleaved at the power splitting point, which requires careful synchronization of the packet transmission instants at the ONUs. This synchronization is achieved by means of grants sent from the local exchange, which instruct the ONU when to send a packet. At the local exchange in the OLT, a burst mode receiver is needed which can synchronize quickly to packets coming from different ONUs, and which also can handle the different amplitude levels of the packets due to differences in the path loss experienced.

In an SCMA system, illustrated in Figure 2.26, the various ONUs modulate their packet streams on different electrical carrier frequencies, which subsequently modulate the light intensity of the laser diode. The packet streams are thus put into different frequency bands, which are demultiplexed again at the local exchange. Each frequency band constitutes an independent communication channel from an ONU to the OLT in the local exchange and thus may carry a signal in a format different from that in an other channel (e.g., one channel may carry a high-speed digital data signal, and an other one an analogue video

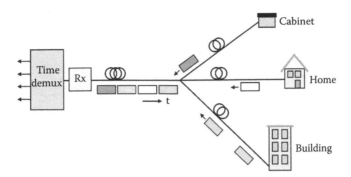

Figure 2.25 TDMA passive optical network.

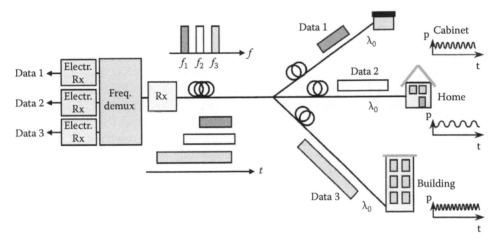

Figure 2.26 SCMA passive optical network.

signal). No time synchronization of the channels is needed. The laser diodes at the ONUs may have nominally the same wavelength. When the wavelengths of the lasers are very close to each other, the frequency difference between them may result in beat noise products due to optical beating at the photodetector in the receiver. These noise products may interfere with the packet data spectrum. The wavelengths of the laser diodes have to be adjusted slightly different (e.g., by thermal tuning) in order to avoid this optical beat noise interference.

In a WDMA system (see Figure 2.27), each ONU uses a different wavelength channel to send its packets to the OLT in the local exchange. These wavelength channels constitute independent communication channels and thus may carry different signal formats; also no time synchronization is needed. The same wavelength channel may be used for upstream communication as for downstream. The isolation requirements of the wavelength demultiplexer may be high to sufficiently suppress crosstalk, e.g., when high-speed digital data and analogue video are carried on two different wavelength channels. The channel routing by the wavelength multiplexer at the network splitting point prohibits broadcasting some channels to all ONUs, as needed for instance for CATV signal distribution. Every ONU needs a wavelength-specific laser diode, which increases costs, and complicates maintenance and stock inventory issues. An alternative is to use a light source with a broad spectrum at the ONU (e.g., a superluminescent LED), of which the in-field multiplexer cuts out the appropriate part of the spectrum. This "spectral slicing" approach reduces the inventory problems, but also yields a reduced effective optical power available from the ONU and thus limits the reach of the system. Another alternative is to use a reflective modulator at the ONU, which modulates the upstream data on a continuous light channel emitted at the appropriate wavelength by the OLT and returns it to the OLT [10]. Thus no light source is needed at the ONU, which eases maintenance; but again the power budget is limited.

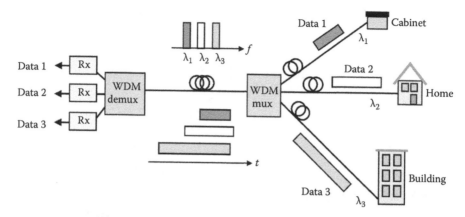

Figure 2.27 WDMA passive optical network.

In an OCDMA system, each ONU uses a different signature sequence of optical pulses, and this sequence is on-off modulated with the data to be transmitted. The duration of the sequence needs to be at least equal to that of a data bit, and thus a very high-speed signature sequence is needed to transmit moderate-speed data which limits the reach of the system due to the increased impact of dispersion and the decreasing power budget at high line rates. In the OLT at the local exchange, the received signals are correlated with the known signature sequences, in order to demultiplex the data coming from the different ONUs. As the signature codes may not be perfectly orthogonal, some cross-talk may occur.

TDMA systems have received the most attention for broadband access networks, as they are most suited for high-speed data transmission at relatively moderate complexity. Two types of TDMA passive optical networks have been addressed extensively in standardization bodies: the ATM PON (APON) carrying native ATM cells in the G.983 standard series of ITU-T SG15, and the Ethernet PON (EPON) carrying gigabit Ethernet packets in IEEE 802.3.

2.5.2 ATM PON

The full service access network (FSAN) group, a committee of presently 21 major telecommunication operators around the world, since 1995 is promoting the ATM PON for broadband access networks.

As laid down in the G.983.1 Recommendation of ITU-T [11], an ATM PON may have a downstream bitrate of 155 or 622 Mbit s^{-1} and an upstream one

of 155 Mbit s^{-1} The maximum optical splitting ratio is 32 (may grow to 64), and the maximum fiber length between the OLT in the local exchange and an ONU is 20 km. The range in which this length is allowed to vary is from 0 to 20 km. Standard single mode fiber (G.652) is foreseen. Coarse wavelength multiplexing is used for separating the bi-directional traffic: the downstream traffic is positioned in the 1.5 μm wavelength band, and the upstream traffic in the 1.3 μm band (using cheap Fabry–Perot laser diodes in the ONUs).

In the downstream direction of a 155 Mbit s^{-1} down/155 Mbit s^{-1} up system, 54 ATM cells of 53 bytes each are fitted together with two physical layer operation, administration, and maintenance (PLOAM) cells of 53 bytes in a frame [11]. The PLOAM cells contain each 53 upstream grants. A grant permits an ONU to send an ATM cell. By sending these grants, the OLT controls at each ONU the transmission of the upstream packets, and can therefore assign dynamically a portion of the upstream bandwidth to each ONU. In a 622 Mbit s^{-1} down/155 Mbit s^{-1} up system, a frame contains four times as many cells (i.e., 216 ATM cells and eight PLOAM cells). The downstream frame is broadcasted to all ONUs. An ONU only extracts those cells that are addressed to it.

In the upstream frame, both for the 155 Mbit s^{-1} down/155 Mbit s^{-1} up system and for the 622 Mbit s^{-1} down/155 Mbit sup system, 53 ATM cells are fitted of 53 bytes each plus an overhead of 3 bytes per cell. This overhead is used as guard time, as a delimiter and as preamble for supporting the burst mode receiver process in the local exchange.

The power budgets needed to bridge the fiber losses and the splitter losses are denoted by three

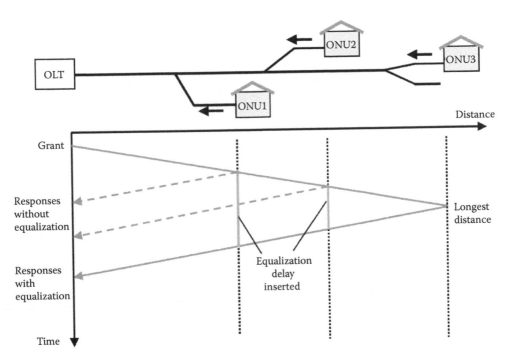

Figure 2.28 Time ranging in a TDMA PON.

classes of optical path losses: class A 5–20 dB, class B 10–25 dB, and class C 15–30 dB. At the ONU, a launched optical power of −4 to +2 dBm is specified for class B, and −2 to +4 dBm for class C [12]. The ONU receiver sensitivity at 155 Mbit s^{-1} should be better than −30 dBm for class B, and −33 dBm for class C.

The ONUs are usually positioned at different distances from the local exchange. Therefore, the upstream transmission of the packets from each ONU should be carefully timed, in such a way that the packets do not collide at the network splitter [11,14]. The OLT has to measure the distance to each ONU for this, and then instructs the ONU to insert an equalizing transmission delay such that all distances from the ONUs to the OLT are virtually equal to the longest allowable distance (i.e., 20 km); see Figure 2.28. To measure the distance to each ONU, the OLT emits a ranging grant to each ONU, and on receipt the ONU returns a ranging cell to the OLT. In this distance ranging process, the OLT can deduce the distance to each ONU from the round trip delay.

Each ONU sends an upstream cell upon the receipt of a grant. Because the path losses from each ONU to the OLT may be different, the power of the cells received by the OLT may vary considerably from cell to cell. The burst mode receiver

at the OLT should therefore have a wide dynamic range, and should be able to set its decision threshold quickly to the appropriate level to discriminate the logical ones from the zeros. Also the power of the ONU transmitter can be varied over a certain range to limit the requirements on the receiver dynamic range. In this amplitude ranging process, the overhead to each ATM cell is used for supporting the fast decision threshold setting at the OLT burst mode receiver and the power adaptation at the ONU burst mode transmitter.

Four types of network protection have been described in Recommendation G.983.1 [12], as shown in Figure 2.29. Type A protection involves protection of the feeder fiber only by a spare fiber over which the traffic can be rerouted by means of optical switches. After detection of a failure in the primary fiber and switch-over to the spare fiber, also re-ranging has to be done by the PON transmission convergence (TC) layer. Thus only limited protection of the system is realized. Mechanical optical switches are used up to now; when optical switching becomes cheaper, this protection scheme may become more attractive. Type B protection features duplication of both the feeder fiber and the OLT. The secondary OLT is on cold standby, and is activated when the primary one fails. Due to the high sharing factor of the duplicated resources

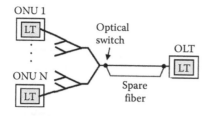

Type A: Feeder fiber protection

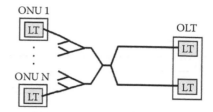

Type B: OLT and feeder fiber protection

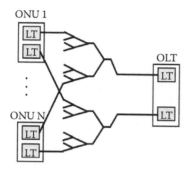

Type C: Full PON duplication

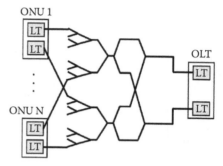

Type D: Independent duplication of feeder and branch fibers

Figure 2.29 PON protection schemes.

by the ONUs, this approach offers an economical yet limited protection. Type C protection implies full duplication of the PON, and all equipment is normally working which allows fast switch-over (within 50 ms) from the primary equipment to the secondary one. The branch fibers as well as the ONUs are protected; also a mix of protected and unprotected ONUs can be handled. Type D protection features independent duplication of the feeder fibers and the branch fibers. It cannot offer fast restoration. It is less attractive than C, as it requires more components but not a better functionality. In summary, types B and C are the most attractive schemes for a new recommendation.

To further increase the speeds laid down in Recommendation G.983.1, research is done into 622, 1244 and 2488 Mbit s^{-1} line rates, both for upstream and downstream. A key technical issue is the development of faster burst-mode circuitry to adequately retrieve the timing and set the decision threshold level, which becomes increasingly more difficult at higher line rates. Operation of 622 Mbit s^{-1} burst-mode circuitry has been achieved recently [12]. In January 2003, ITU has set standards for gigabit-capable PONs (G-PONs). These ITU-T Recommendations G.984.1 and G.984.2 cover downstream speeds of 1.25 and 2.5 Gbit s^{-1}

and upstream speeds of 155 and 622 Mbit s^{-1} and of 1.25 and 2.5 Gbit s^{-1}.

The G.983.1 ATM PON was initially mainly designed for high-speed data communication. However, in the residential access networks there is also a clear demand for economical delivery of CATV services, for which subcarrier multiplexing techniques are quite appropriate. In the enhanced Recommendation G.983.3 [12], room has been allocated in the optical spectrum to host video services or additional digital services next to the ATM PON services. As shown in Figure 2.30, the APON upstream services remain in the 1260–1360 nm band (as in G.983.1), but the band for downstream services is narrowed to 1480–1500 nm (1480–1580 nm in G.983.1). Next to those, an enhancement band for densely wavelength multiplexed bi-directional digital services (such as private wavelength services) is foreseen, or an enhancement band for an overlay of video delivery services. The latter is used in downstream direction only, and coincides with the C-band as thus economical erbium-doped fiber amplifiers (EDFAs) can be deployed for the power boosting required. When positioning an overlay of CATV distributive services in the C-band, stringent crosstalk requirements have to be put on the wavelength multiplexers and demultiplexers, to

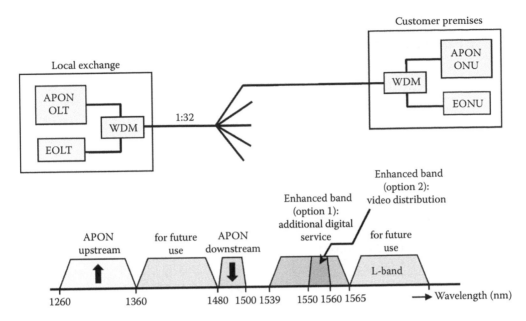

Figure 2.30 WDM enhancement G.983.3.

prevent noticeable interference of the CATV signals into the digital ATM signals, and vice versa [13].

In order to further improve the economics of ATM PON systems, an extended PON system with an increase of the network splitting factor to 128 and even 256 has been developed, while still maintaining a passive outside plant and compatibility with G.983.1 compliant ONUs [14]. This extended split is achieved by a larger optical power budget. In the downstream direction, at the OLT a high power laser diode or an EDFA is used to boost the power. In the upstream direction, the sensitivity of the burst-mode receiver is improved by applying an avalanche photo diode (APD). Also eight single-mode feeder fibers (each feeding a 1:16 or 1:32 power splitter in the field) are at the OLT coupled to a multimode fiber yielding a low-loss coupling to the receiver.

Even further extensions of the split factor and of the reach of an ATM PON have been realized in the SuperPON system [15]. An extension to a splitting factor of 1:2048 has been achieved; this needs, however, active equipment in the field. In the downstream direction exploiting the 1530–1560 nm wavelength window, EDFAs are used for overcoming the large path losses. In the upstream direction, gated semiconductor optical amplifiers (SOAs) are deployed. Each SOA gate is opened when upstream packets arrive, and is shut otherwise in order to avoid funneling of the amplified

spontaneous emission noise towards the OLT. This SuperPON approach is not compliant with present standards, and may be economically feasible only in the long term [14].

2.5.3 Ethernet PON

With the rapid penetration of Ethernet-based services, Ethernet PON (EPON) techniques are receiving increasing attention, and are promoted by the IEEE 802.3 Ethernet in the first mile (EFM) group. The major difference with ATM PONs is that an EPON carries variable-length packets up to 1518 bytes in length, whereas an ATM PON carries fixed-length 53 bytes cells. This ability yields a higher efficiency for handling IP traffic. The packets are transported at the gigabit Ethernet 1.25 Gbit s^{-1} speed using the IEEE 802.3 Ethernet protocol. However, ATM offers built-in quality of service for all traffic classes, whereas Ethernet does not. EPON thus cannot support voice services with quality of service as provided in the traditional public switched telephone network (PSTN), and also the support of real-time services still has issues due to latency and packet jitter.

The EPON features full-duplex transmission similarly as the ATM PON, with downstream traffic at 1490 or 1510 nm, and upstream traffic at around 1310 nm. As shown in Figure 2.31, standard IEEE 802.3 Ethernet packets are broadcasted

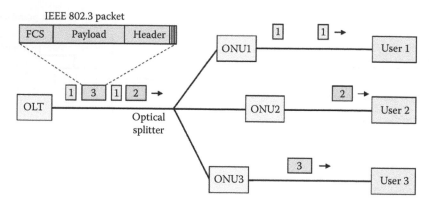

Figure 2.31 Downstream traffic in an EPON.

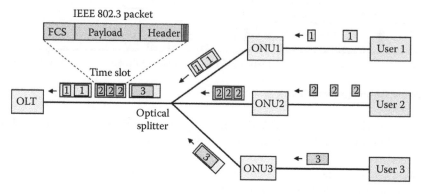

Figure 2.32 Upstream traffic in an EPON.

downstream by the OLT to all the ONUs. Each ONU inspects the headers, and extracts the packets that are addressed to it. Several variable-length packets are put into a fixed-length frame of 2 ms duration, and each frame begins with a 1-byte synchronization marker. In the upstream direction, also 2 ms frames are used. A frame contains time slots that each are assigned to one of the ONUs (see Figure 2.32). Each ONU puts one or more of its upstream variable-length IEEE 802.3 packet into a time slot; if it has no packets to send, the time slot may be filled with an idle signal. No packet fragmentation takes place. The time slot overhead consists of a guard band, and indicators for timing and signal power. The OLT thus allows only one ONU to send at a time, and no collisions occur. The time slot size is 125 or 250 µs.

2.5.4 Hybrid fiber coax networks

CATV networks usually are laid out over large geographical areas, and are mainly designed for downstream broadcasting of analogue TV channels that are frequency-division multiplexed in a carrier frequency grid extending up to 1 GHz. As shown in Figure 2.33, in a hybrid fiber coax (HFC) system a CATV headend station is collecting the CATV signals, remodulating them into a specific frequency grid, and sending them via single-mode fibers to fiber nodes. Each fiber node converts the composite optical signal into an electrical one, which is carried via a coaxial cable network including several RF amplifiers to the residential homes. A single headend may thus serve hundred thousands of customers, and a fiber node some thousands of customers. In particular during transmission in the coaxial cable network, the signal quality deteriorates due to the addition of noise from the electrical amplifiers and intermodulation products from nonlinearities in the system. On the fiber part of the network, the signals are carried with subcarrier multiplexing; see Figure 2.34. The TV channels each are amplitude-modulated on a separate frequency, and after summing all these modulated signals, a highly linear high power laser diode (or laser diode followed by a linearized external

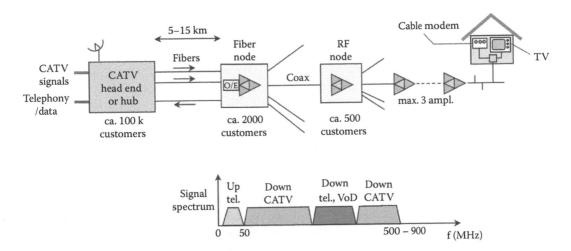

Figure 2.33 Hybrid fiber coax network.

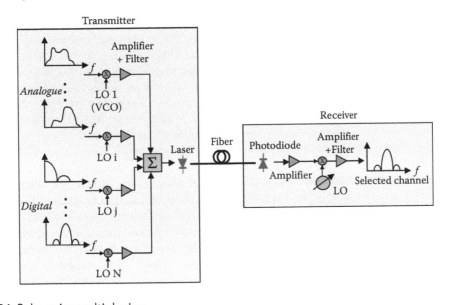

Figure 2.34 Subcarrier multiplexing.

modulator) generates an optical signal which is intensity-modulated with the composite CATV signal. At the receiver site, the optical signal is converted into the electrical CATV signal by means of a highly linear PIN photodiode, and subsequently the signal can be passed to the coaxial cable network or to a selective receiver. When using a laser diode with low relative intensity noise and high linearity (or a carefully linearized external modulator), the CATV signal can be transported with very little loss of quality. If a 1.5 μm wavelength laser diode is used, EDFAs may boost the power at the headend and compensate for the splitting losses; thus very extensive networks feeding

thousands of ONUs can be realized. In this wavelength region, however, with direct laser modulation second order intermodulation products may arise due to laser chirp in combination with fiber chromatic dispersion; with an external modulator, however, the chirp is small enough to avoid these intermodulation products.

The CATV signal quality that can be maintained in HFC networks is very high due to the fiber's low losses and high bandwidth in comparison with coaxial cable. Therefore, in HFC networks fiber is gradually brought deeper into the network, and fiber nodes have to serve fewer customers through a coaxial cable network of limited size (i.e., mini fiber nodes, each serving in the order of 40 customers).

At present, HFC networks are not only carrying CATV and FM radio broadcast services, but cable operators are also exploiting them for voice telephony and data transport using cable modems. For the upstream traffic involved with these interactive services, parts of the spectrum unused for CATV and FM radio broadcast can be used. In Europe, typically the 5–65 MHz band is used for this; in the US, the 5–42 MHz range. For downstream data, e.g., the 300–450 MHz range is used, taking into account that Internet traffic is usually highly asymmetric (much more downloading traffic than uploading). Downstream per 8 MHz CATV channel, 30–50 Mbit s^{-1} data can be accommodated deploying 64 or even 256 quadrature amplitude modulation (QAM). Upstream due to ingress noise less complicated modulation schemes are to be used; DQPSK offers about 3 Mbit s^{-1} per channel.

2.5.5 Dense wavelength multiplexing in access networks

In general, access networks have to meet a fast growth in capacity demand, due to several causes: customers are asking for second and more telephone lines; Internet data traffic is booming with higher data rates, more users and longer sessions on-line (even always on); an increasing amount of video-based services; fast growth in number of mobile phone users and session frequency; new operators entering asking to rent capacity on existing access networks; etc. This hunger for more capacity and the strive for convergence of services on a single network can most adequately be met by bringing fiber ever closer to the end users, from where only a short copper cable based (or wireless) link has to be bridged to the customer. Ultimately, when installation and equipment costs have come down sufficiently, the most powerful network is achieved when fiber runs all the way to the customer's home (fiber to the home, FTTH).

The upgradation of installed fiber plant to higher capacities while protecting the investments made is efficiently done by introducing wavelength multiplexing techniques. Wavelength channels may be allocated to specific sets of services (for service unbundling), and/or to separate service operators (leasing of network capacity).

2.5.6 Dynamic capacity allocation by flexible wavelength assignment

To cope with variation in service demand by the users and the sometimes quickly changing operator conditions, it is more efficient to flexibly allocate the augmented available network capacity across the access network. Dynamic wavelength routing techniques can be used for this, thus making more efficient use of the network's resources and generating more revenues. Figure 2.35 illustrates the principle: from the OLT in the headend station of the network, multiple wavelength channels are fed to the ONUs via a tree-and-branch PON. By wavelength-selective routing in the PON, or wavelength selection at the ONU, wavelength channels can be assigned to a number of specific ONUs. Thus capacity can be specifically shared between these ONUs. The ONUs subsequently transfer this capacity shares to their first-mile electrical network connecting the end users. The mapping of the network capacity resources to the first-mile networks can thus be changed by changing the wavelength channel assignment. Basically two approaches can be followed, as illustrated in Figure 2.36: a wavelength router in the field, or wavelength selection at the ONUs. As shown in Figure 2.36a, a tunable wavelength router directs the wavelength channels to specific output ports, and this routing can be dynamically adjusted by external control signals from the headend. In order to support in addition the delivery of broadcast services to all ONUs, extra provisions have to be made for enabling broadcast wavelength channel(s) to bypass the router. As the wavelength channels are routed to only those ONUs whose customers require the associated

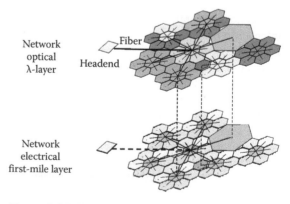

Network optical λ-layer   Headend   Fiber

Network electrical first-mile layer

Figure 2.35 Dynamic wavelength routing in hybrid access networks.

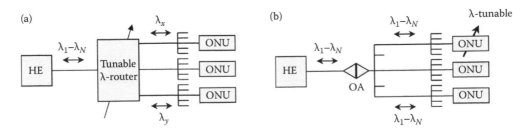

Figure 2.36 Dynamically allocating wavelength channels to ONUs. (a) Flexible wavelength routing and (b) broadcast-and-select.

services, no optical power is wasted. As shown in Figure 2.36b, another approach is to broadcast all wavelength channels to every ONU, and subsequently tune the ONU to the wavelength channel wanted. Clearly the power of the other wavelength channels is wasted by the ONU, and losses at the broadcasting power splitter are significant. An optical amplifier is usually needed to make up for these losses; the amplifier needs to operate bi-directionally to handle downstream as well as upstream traffic. No specific provisions in the network are needed for supporting broadcast services.

Figure 2.37 presents a multi-wavelength overlay of a number of ATM PON networks on a HFC network, following the wavelength channel selection approach [16]. Figure 2.37a shows a fiber-coax network for distribution of CATV services, operating at a wavelength λ_0 in the 1550–1560 nm window where EDFAs offer their best output power performance. Thus, using several EDFAs in cascade, an extensive optical network splitting factor can be realized and a large number of customers can be served. For example, with two optical amplifier stages and typical splitting factors of $N=4$ and $P=16$, and a mini-fiber node serving 40 users via its coaxial network, a total of 2560 users is served from a single headend fiber. For interactive services, the upstream frequency band in a standard HFC network (with a width of some 40–60 MHz) has to be shared among these users, thus allowing only limited bitrates per user for narrowband services such as voice telephony.

An upgrade of the system in order to provide broadband interactive services can be realized by overlaying the HFC network with a number of wavelength-multiplexed APON systems, as developed in the ACTS TOBASCO project [16] and shown in Figure 2.37b. Four APON OLTs at the headend site are providing each bi-directional 622 Mbit s^{-1} ATM signals on a specific downstream

and upstream wavelength. These eight wavelengths are positioned in the 1535–1541 nm window, where the up-and downstream wavelength channels are interleaved with 100 GHz spacing. The APON wavelengths are combined by a high-density wavelength division multiplexer (HDWDM), and subsequently multiplexed with the CATV signal by means of a simple coarse wavelength multiplexer (due to the wide spacing between the band of APON wavelengths and the CATV wavelength band). The system upgrade implies also replacement of the uni-directional optical EDFAs by bi-directional ones which feature low noise high-power operation for the downstream CATV signal, and for the bi-directional ATM signals a wavelength-flattened gain curve plus a nonsaturated behavior (to suppress crosstalk in burst-mode). At the ONU site, first the CATV signal is separated from the APON signals by means of a coarse wavelength multiplexer, and is subsequently converted to an electrical CATV signal by a highly linear receiver and distributed to the users via the coaxial network. The APON signals are fed to a wavelength-switched transceiver, of which the receiver can be switched to any of the four downstream wavelength channels, and the transmitter to any of the four upstream ones. The wavelength-switched transceiver may be implemented by an array of wavelength-specific transmitters and receivers, which can be individually switched on and off; this configuration allows to set up a new wavelength channel before breaking down the old one ("make-before-break"). Alternatively, it may use wavelength-tunable transmitters and receivers, which can in principle address any wavelength in a certain range; this eases further upgrading of the system by introducing more wavelength channels, but also implies a "break-before-make" channel switching. The network management and control system commands to which downstream

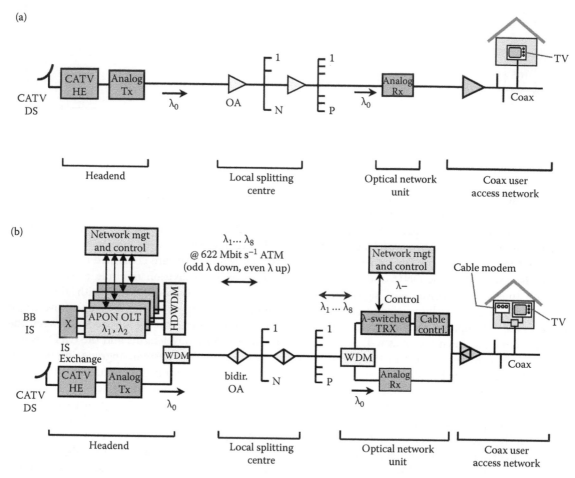

Figure 2.37 Flexible capacity assignment in a multi-wavelength fiber-coax network by wavelength selection at the ONUs. (a) fiber-coax network for distribution of CATV services and (b) upgrading of the fiber-coax network with multi-wavelength APON system for delivery of broadband interactive services.

and to which upstream wavelength channel each ONU transceiver is switched. By issuing these commands from the headend station, the network operator actually controls the virtual topology of the network, and thus is able to allocate the network's capacity resources in response to the traffic demands at the various ONU sites. The network management command signals are transported via an out-of-band wavelength channel in the 1.3 μm wavelength window. The APON signal channel selected by the ONU is converted into a bi-directional electrical broadband data signal by the transceiver, which is by a cable modem controller put in an appropriate frequency band for multiplexing with the electrical CATV signal. The upstream data signal is usually put below the lowest frequency CATV signal (so below 40–50 MHz),

and the downstream signal in empty frequency bands between the CATV broadcast channels. The signals are carried by the coaxial network (in which only the electrical amplifiers need to be adapted to handle the broadband data signals) to the customer homes, where the CATV signal is separated from the bi-directional data signals; the latter signals are processed by a cable modem, which interacts with the cable modem controller at the ONU site.

By remotely changing the wavelength selection at the ONUs, the network operator can adjust the system's capacity allocation in order to meet the local traffic demands at the ONU sites. As illustrated in Figure 2.39, the ONUs are allocated to the four upstream (and downstream) wavelength channels, which each have a maximum capacity of 622 Mbit s^{-1} for ATM data. As soon as the traffic

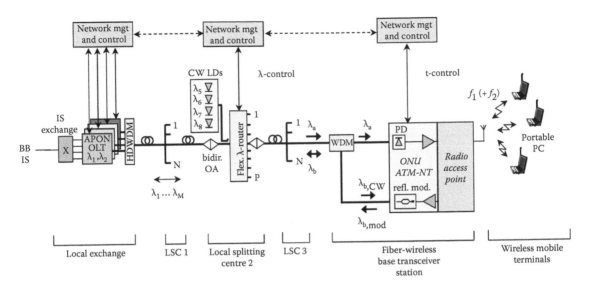

Figure 2.38 Flexible capacity assignment in a multi-wavelength fiber-wireless network by wavelength routing in the field.

to be sent upstream by an ONU grows and does not fit anymore within its wavelength channel, the network management system can command the ONU to be allocated to an other wavelength channel, in which still sufficient free capacity is available. Obviously, this dynamic wavelength re-allocation process reduces the system's blocking probability, i.e., it allows the system to handle more traffic without blocking and thus can increase the revenues of the operator.

Figure 2.38 presents the dynamic wavelength channel routing approach in a fiber-wireless network to allocate flexibly the capacity of a number of ATM PON systems among ONUs in a single fiber split network infrastructure [17]. The ONUs are each feeding a radio access point (RAP) of e.g., a wireless LAN, which wirelessly connects to a variable number of users with mobile terminals. These users move across the geographical area served by the network (e.g., a business park), and they may want to set up a broadband wireless connection to their laptop at any time anywhere in this area. When many users are within a wireless cell served by a certain RAP, this cell may have to handle much more traffic than the other cells; it has become a "hot spot" which has to be equipped with additional capacity. The corresponding RAP may switch on more microwave carriers to provide this additional capacity over the air, and also has to claim more capacity from the ONU. This local extra capacity can be provided

by reallocation of the wavelength channels over the ONUs, which is done by a flexible wavelength router positioned in the field. Similar to the architecture of the wavelength-reconfigurable fiber-coax network in Figure 2.37b, the architecture in Figure 2.38 developed in the ACTS PRISMA project has four 622 Mbit s^{-1} bi-directional APON OLTs with a specific downstream wavelength and an upstream one each. The four downstream wavelengths are located in the 1538–1541 nm range, with 100 GHz spacing, and the four upstream ones in the 1547–1550 nm range with the same spacing. The flexible wavelength router directs the downstream wavelength channels each to one or more of its output ports, and thus via a split network to a subset of ONUs. The RAPs could operate with up to five microwave carriers in the 5 GHz region, each carrying up to 20 Mbit s^{-1} ATM wireless LAN data in OFDM format. At the flexible router (or at the local exchange) a number of continuous-wave emitting laser diodes are located, which provide unmodulated light power at the upstream wavelengths. The flexible router can select one of these upstream wavelengths, and direct it to the ONUs that can modulate the signal with the upstream data and return it by means of a reflective modulator via the router to the local exchange. Thus no wavelength-specific source is needed at the ONU, the downstream light sources are shared by a number of ONUs, and all ONUs are identical, which reduces the system costs and the inventory issues.

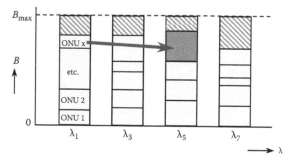

Figure 2.39 Re-allocating ONUs to wavelength channels.

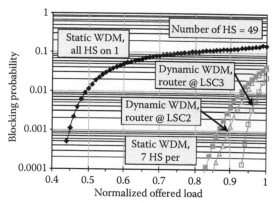

Figure 2.40 Improving the system performance by dynamic wavelength allocation.

The flexible wavelength router can be implemented with a wavelength demultiplexer separating the wavelength channels, followed by power splitters, optical switches and power couplers in order to guide the channels to the selected output port(s). Depending on the granularity of the wavelength allocation process, the flexible router may be positioned at desired different splitting levels in the network.

Using a similar strategy to assign wavelength channels to the ONUs as shown in Figure 2.39, a statistical performance analysis has been performed of the blocking probability of the system. It was assumed that the total network served 343 cells, of which 49 were "hot spots", i.e., generated a traffic load two times as large as a regular cell. It was also assumed that the system deployed seven wavelength channels, and that the calls arrived according to a Poisson process where the call duration and length were uniformly distributed. Figure 2.40 shows how the system blocking probability depends on the offered load (normalized on the total available capacity, which is 7 times 622 Mbit s⁻¹), using various system architecture options. When wavelength re-allocation would not be possible (i.e., static WDM) and all the 49 hot spots were positioned at cells served by ONUs assigned to the same wavelength channel, the blocking probability is obviously the worst case. On the other hand, in the static WDM case when the 49 hot spots were evenly spread over the seven wavelength channels, the blocking probability is much lower (i.e., best case). Unfortunately, a network operator cannot know beforehand where the hot spots will be positioned, so in this static WDM situation the system blocking probability will be anywhere between the best case and the worst case, and no guarantee for a certain

blocking performance can be given. When, however, dynamic re-allocation of the wavelength channels is possible, the system can adapt to the actual hot spot distribution. Figure 2.40 shows that when the flexible wavelength router is positioned at the second splitting point in the network, the blocking performance is better than the best-case static WDM performance; but more importantly, it is also stable against variations in the hot spot distribution, and thus would allow an operator to guarantee a certain system blocking performance while still optimizing the efficiency of his system's capacity resources. The blocking performance may even be better and stable when positioning the flexible router at the third splitting point; however, this implies that the costs of the router are shared by less ONUs. Locating the router at the second splitting point is a good compromise between adequate improvement of the system blocking performance and system costs per ONU.

2.5.7 Microwave signals over fiber

Fiber-wireless systems may also carry microwave signals directly over fiber. In wireless local area networks (WLANs), the evolution towards larger capacities necessitates higher microwave carrier frequencies. For example, the current IEEE 802.11b WLAN systems transport up to 11 Mbit s⁻¹ per carrier in the ISM 2.4GHz band. The upcoming IEEE 802.11a systems carry up to 54 Mbit s⁻¹ per carrier in the 5.2 GHz band, and 60 GHz systems are under study for providing more than 100 Mbit s⁻¹. With these increasing carrier frequencies, the microwave cells covered by the antenna of a radio access point (RAP) become smaller. Thus more RAPs are needed to serve e.g., all the rooms in an office building, and hence also a more extensive wired network to feed the RAPs. Instead of generating the microwave signals at each RAP individually, feeding the microwave signals from a central headend site to the RAPs enables to simplify the RAPs considerably. The signal processing functions can thus be consolidated at the headend site. Due to its broadband characteristics, optical fiber is an excellent medium to bring the microwave signals to the RAPs.

Carrying multi-gigahertz analogue signals over fiber requires very high frequency optical analog transmitters and receivers, including careful fiber dispersion compensation techniques. An attractive alternative avoiding the transport of multi-gigahertz intensity-modulated signals through the fiber is to apply heterodyning of two optical signals of which the difference in optical frequency (wavelength) corresponds to the microwave frequency. When one of these signals is intensity-modulated with the baseband data to be transported, and the other one is unmodulated, by optical heterodyning at the photodiode in the receiver the electrical microwave difference frequency signal is generated, amplitude-modulated with the data signal. This modulated microwave signal can via a simple amplifier be radiated by an antenna; thus a very simple low-cost radio access point can be realized, while the complicated signal processing is consolidated at the headend station. This approach, however, requires two light sources with narrow spectral linewidth and carefully stabilized difference in optical emission frequency. An alternative approach requiring only a single optical source is shown in Figure 2.41 [18]. The optical intensity-modulated signal from a laser diode is

subsequently intensity-modulated by an external Mach–Zehnder modulator (MZM) which is biased at its inflexion point of the modulation characteristic and driven by a sinusoidal signal at half the microwave frequency. At the MZM's output port a two-tone optical signal emerges, with a tone spacing equal to the microwave frequency. After heterodyning in a photodiode, the desired amplitude-modulated microwave signal is generated. The transmitter may also use multiple laser diodes, and thus a multi-wavelength radio-over-fiber system can be realized with a (tunable) WDM filter to select the desired wavelength radio channel at the antenna site. The system is tolerant to fiber dispersion, and also the laser linewidth is not critical as laser phase noise is largely eliminated in the two-tone detection process.

An alternative approach to generate microwave signals by means of a different kind of remote optical processing, named optical frequency multiplying, is shown in Figure 2.42 [19]. At the headend station the wavelength λ_0 of a tunable laser diode is swept periodically over a certain range $\Delta\lambda_{sw}$, with a sweep frequency f_{sw}. Alternatively, the wavelength-swept signal can be generated with a continuous-wave operating laser diode followed by an external phase modulator that is driven with the integral of the electrical sweep waveform. The intensity of the wavelength-swept signal is on/off modulated with low frequency chirp by the downstream data in a symmetrically driven MZM. After travelling through the fiber network, the signal transverses at the receiver an optical filter with a periodic bandpass characteristic. When the wavelength of the signal is swept back and forth over N filter transmission peaks, the light intensity impinging on the photodiode fluctuates at a frequency $2Nf_{sw}$. Thus the sweep frequency is multiplied, and a microwave

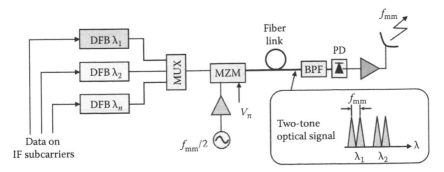

Figure 2.41 Generating microwave signals by heterodyning.

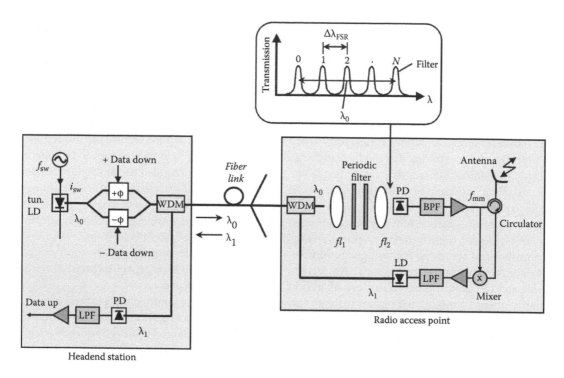

Figure 2.42 Generating microwave signals by optical frequency multiplying.

signal with carrier frequency $f_{mm} = 2Nf_{sw}$ plus higher harmonics is obtained. The intensity-modulated data is not affected by this multiplication process, and is maintained as the envelope of the microwave signal. The microwave signal is subsequently transversing an electrical bandpass filter (BPF) to reject the unwanted harmonics. Simulations have shown that the microwave signal is very pure; its linewidth is nearly independent of the linewidth of the tunable laser. The periodic optical bandpass filter can be advantageously implemented by a Fabry–Perot filter with a free spectral range $\Delta\lambda_{FSR}$ which is N times as small as the wavelength sweep range $\Delta\lambda_{sw}$. The microwave signal can also carry more advanced data modulation schemes; e.g., 16-level quadrature amplitude modulated (16-QAM) signals may be modulated on a subcarrier first, and then drive the MZM. The main advantage of this optical frequency multiplying method is that the fiber network is carrying only signals at moderate frequencies (up to the sweep frequency f_{sw}, e.g., up to 1 GHz), while at the antenna site microwave signals with carrier frequencies in the tens of gigahertz region are generated. Thus dispersion problems in the fiber network and other

bandwidth limitations in the transmitter are effectively circumvented. The system does not rely on heterodyning, and thus may also operate on multimode fiber networks (such as polymer optical fiber).

The system can also transport upstream data from the antenna station to the headend. When no data are sent downstream, the upstream microwave signal arriving at the antenna can be downconverted with the locally generated unmodulated microwave signal. Thus data can be conveyed bidirectionally in time-division duplex mode.

2.6 CONCLUDING REMARKS

Optical fiber is now generally recognized to be the most powerful medium for transporting information, due to its very low losses and extremely wide bandwidth. Next to space and time multiplexing, the wavelength dimension offers unprecedented opportunities to extend not only the data traffic transport capacity, but also the traffic routing possibilities in networks.

In core networks, optical amplifiers together with wavelength multiplexing techniques have allowed transoceanic links to be bridged with tera-bits/second total transport capacity. In terrestrial mesh-shaped core networks, wavelength routing also provides link protection and thus improves network availability. IP packet streams may be very efficiently transported through mesh-shaped networks by wavelength-selective optical bypassing of the electronic processing in the packet routers located in the nodes.

In MANs, wavelength multiplexing techniques enable simultaneous broadband data communication between many nodes. They also improve the network availability by offering alternative routing, which enables fast recovery from link or node failures. Through wavelength add-drop multiplexers, multiple connections can be set up between nodes via a hubbing node, or directly between them using an appropriate wavelength assignment strategy.

In access networks, fiber is penetrating steadily towards the end user. Infrastructure costs are the major nut to crack here. Shared-feeder concepts such as the passive optical network tree-and-branch one greatly reduce the installation costs of the fiber network, and can support various multiple access techniques (a.o. ATM and Ethernet for time multiplexed access). In the past, operators have invested a lot in various last-mile networks (e.g., twisted pair, coaxial cable) to reach their residential customers. In upgrading these networks to higher capacity and larger service variety, fiber can support a wide range of last-mile technologies by hybrid combinations such as fiber-twisted pair, fiber-coax, and fiber-wireless. By means of wavelength multiplexing techniques, the fiber feeder part of such hybrid networks can very flexibly host different operators and service categories. Augmented with wavelength routing, capacity-on-demand can be realized, e.g., for handling hot spots, while respecting quality of service requirements. Carrying microwave radio signals in analog form over the fiber offers clear advantages in the implementation of mobile communication networks; the antenna stations can be considerably simplified and thus reduced in costs, while also the mobility functions can be consolidated in the headend station yielding improved efficiency of the signal processing. By combining radio over fiber techniques with flexible wavelength routing, the mobility offered by wireless communication can be powerfully complemented with the broadband dynamics of optical networks.

2.7 FUTURE PROSPECTS

With the ongoing improvements in fiber characteristics and development of novel optical amplifier structures, the wavelength range available for communication will stretch from below 0.8 to beyond 1.6 μm, covering a bandwidth of some 200 THz. Further improvements in signal coding yielding a higher spectral efficiency, in ultra-dense WDM and in ultra-high speed OTDM will enable us to exploit this huge bandwidth, and will push the transport capacity of a single fiber beyond 100 Tbit s^{-1}.

Realizing these tremendous link transport capacities over fiber links is of little value if the network nodes cannot keep up with handling the data streams. Present-day electronic routing will be replaced by fast optical processing. All-optical packet routing will provide the ultimate in node throughput, by optically inspecting the header, and by making routing decisions at light speed by means of ultra-fast optical logic. Optical memories will provide the intermediate buffering in the nodes, and electronics may be expelled to the edges of the network, thus providing a fully-optical network path.

The end user is to benefit from all these ultra-broadband communication possibilities. Therefore, the optical fiber will not only reach up to his house, but will penetrate into it as well, benefiting of the low installation cost techniques made possible by deploying e.g., large-core multimode (polymer) optical fiber. It will reach close to his personal area network, which due to the ongoing miniatuarization may consist of a myriad of small wireless power-lean terminals, sensors and actuators. These wireless devices will be incorporated not only in his residential living environment, but also in his clothes, his car, etc. Next to the traditional wired terminals such as the TV, these wireless devices will be connected to the fixed in-home and access network by a myriad of small intelligent antennas. Fiber-radio techniques, augmented with optical routing to accommodate dynamically the hot spots, will provide the best match of the ultimate capacity of fiber with the user freedom of wireless. Even in the wireless domain, optics may penetrate by means of intelligently-steered free-space light beams providing the ultimate in wireless transport capacity.

Which in the far foreseeable future will make the communication world an end-to-end globally transparent one, with nearly unlimited communication capacity for anybody, anytime, anywhere, for any kind of service... the ultimate global crystal ball! Today we are now seeing such visionary statements (explained in previous paragraphs) being implemented as the 'Internet of Things,' though the final outcomes and potential are still not yet clear.

REFERENCES

1. Fukuchi, K., Kasamatsu, T., Morie, M., Ohhira, R., Ito, T., Sekiya, K., Ogasahara, D., and Ono, T. 2001. 10.92 Tb/s (273×40Gb/s) triple-band/ultra-dense WDM optical-repeatered transmission experiment. *Proceedings of OFC 2001* Post-Deadline Paper PD24 (Anaheim, CA, March 22, 2001).

2. Nakazawa, M., Yamamoto, T., and Tamura, K. R. 2000. 1.28 Tbit/s—70 km OTDM transmission using third-and fourth-order simultaneous dispersion compensation with a phase modulator. *Proceedings of ECOC 2000* Post-Deadline Paper 2.6 (Munich, September 3–7, 2000).

3. Koonen, T., de Waardt, H., Jennen, J., Verhoosel, J., Kant, D., de Vos, M., van Ardenne, A., and van Veldhuizen, E. J. 2001. A very high capacity optical fiber network for large-scale antenna constellations: the RETINA project. *Proceedings of NOC 2001* (Ipswich, June 25–29, 2001) pp. 165–172.

4. Koonen, T, Morthier, G., Jennen, J, de Waardt, H., and Demeester, P. 2001. Optical packet routing in IP-over-WDM networks deploying two-level optical labeling. *Proceedings of ECOC'OI* paper Th.L.2.1 (Amsterdam, September 30–October 4, 2001) pp. 608–609.

5. Koonen, T., Sulur, S., Monroy I. T., Jennen J., and de Waardt, H. 2002. Optical labeling of packets in IP-over-WDM networks. *Proceedings of ECOC'02* (Copenhagen, September 8–12, 2002) vol. 2, paper 5.5.2.

6. van As, H. R. 2002. Overview of the evolving standard IEEE 802.17 resilient packet ring. *Proceedings of NOC 2002* (Darmstadt, June 18–21, 2002) pp. 277–284.

7. Koonen, A. M. J. and van Veen, D. T. 1997. Estimating error probabilities caused by in-band and out-band crosstalk in multi-wavelength all-optical networks. *Proceedings*

of *IEEE/LEOS Symposium* Chapter Benelux (Eindhoven, November 26, 1997) pp. 169–172.

8. Tachikawa, Y., Inoue, Y., Kawachi, M., Takahashi, H., and Inoue, K. 1993. Arrayed-waveguide grating add-drop multiplexer with loopback optical paths. *Electr. Lett.* 29: 2133–2134.

9. van Veen, D. T., Krommendijk, F. N., Sikken, B. H., and Koonen, A. M. J. 1999. Impact of OADM architecture design on the performance of ring networks. *Proceeedings of ECOC '99* vol II (Nice, September 1999) pp. 48–49.

10. Frigo, N. J., Iannone, P. P., Reichmann, K. C., Walker, J. A., Goossen, K. W., Arney, S. C., Murphy, E. J., Ota, Y., and Swartz, R. G. 1995. Demonstration of performance-tiered modulators in a WDM PON with a single shared source. *Proceedings of ECOC '95* (Brussels, Seprtember 17–21, 1995) pp. 441–444.

11. Ueda, H., Okada, K., Ford, B., Mahony, G., Hornung, S., Faulkner, D., Abiven, J., Durel, S., Ballart, R., and Erickson, J. 2001. Deployment status and common technical specifications for a B-PON system. *EEE Commun. Mag.* 39: 134–141.

12. Effenberger, F. J., Ichibangase, H., and Yamashita, H. 2001. Advances in broadband passive optical networking technologies. *IEEE Commun. Mag.* 39: 118–124.

13. Schoop, R., Fredericx, F., Koonen, T., and Hardalov, C. 2002. WDM isolation requirements for CATV in BPON. *Proceedings of ECOC 2002* (Copenhagen, September 8–12, 2002).

14. Vetter, P., Fredricx, F., Ringoot, E., Janssens, N., De Vos, B., Bouchat, C., Duthilleul, F., Dessauvages, C., Stubbe, B., Gilon, E., and Tassent, M. 2002. Study and demonstration of extensions to the standard FSAN BPON. *Proceedings of ISSLS 2002* (*XIVth International Symposium on Services and Local Access*) (Seoul, April 14–18, 2002) paper 4–2, pp. 119–128.

15. van de Voorde, I., Martin, C., Ringoot, E., Slabbinck, H., Tassent, M., Bouchat, C., Goderis, D., and Vetter, P. 2000. The super PON demonstrator: A versatile platform to evaluate possible upgrades of the G.983

APON. *Proceedings of ISSLS 2000: XIIIth International Symposium on Services and Local Access* (Stockholm, June 18–23, 2000) paper 14–3, p. 10.

16. Koonen, T., Muys, T., van der Plaats, C., Heemstra de Groot, S. M., Kenter, H. J. H. N., Niemegeers, I. G. M. M., and Slothouber, F. N. C. 1997. TOBASCO: An innovative approach for upgrading CATV fiber-coax networks for broadband interactive services. *IEEE Commun. Mag.* 35: 76–81.

17. Koonen, T., Steenbergen, K., Janssen, F., and Wellen, J. 2001. Flexible reconfigurable fiber-wireless network using wavelength routing techniques: The ACTS project AC349 PRISMA. *Photon. Network Commun.* 3: 297–306.

18. Griffin, R. A., Lane, P. M., and O'Reilly, J. J. 1999. Radio-over-fiber distribution using an optical millimeter-wave/DWDM overlay. *Proceedings of OFC '99* (San Diego, February 22–25, 1999) paper WD6.

19. Koonen, T., Ng'oma, A., Smulder, P., van den Boom, H., Monroy, I. T., van Bennekom, P., and Khoe, G.-D. 2002.

In-house networks using polymer optical fiber for broadband wireless applications. *Proceedings of ISSLS 2002: XIVth International Symposium on Services and Local Access* (Seoul, Apr. 14–18, 2002) paper 9–3, pp. 285–294.

20. Koonen, T., Ng'oma, A., Smulders, P., van den Boom, H., Monroy, I. T., and Khoe, G.-D. 2003. In-house networks using multimode polymer optical fiber for broadband wireless services. *Photon-Network Commun.* 5: 177–187.

21. Ryf, R. 2002. Optical MEMS for optical networking. *Proceedings of ECOC'02* (Copenhagen, September 8–12, 2002) vol. 1, Tutorial 2.

22. Ryf, R., Neilson, D. T., Kolodner, P. R., Kim, J., Hickey, J. P., Carr, D., Aksyuk, V., Greywall, D.S., Pardo, F., Bolle, C., and Frahm, R. 2002. Multi-service optical node based on low-loss MEMS optical crossconnect switch. *Proceedings of OFC 2002* (Anaheim, CA, March 17–22, 2002) paper ThE3, pp. 410–411.

Optical switching and multiplexed architectures

DOMINIQUE CHIARONI
NOKIA Bell Labs

PROLOGUE

The optical packet switching technology has been identified as a promising technology to identify a new generation of systems and networks during the past few years. However, the lack of optical memory, and the complexity of optical packet switching systems offering functionalities comparable to full electronic systems, has motivated the constructors of equipment to envisage new directions: reconfigurable optical add/drop multiplexers, currently deployed in the field or hybrid packet switching systems. In the early 2000s, the hybrid approach combining the best of optics and electronics has been then identified as a new promising direction for optical packet switching. Several research laboratories have investigated new concepts limiting this technology to fast space switches [based on pure spatial technique or exploiting the combination of tunable lasers arrayed waveguide gratings (AWG)] and some processing functions. The capability of the optical technology to switch ultrahigh capacities has concentrated its efforts on systems for the metro core or backbones where the capacity was very challenging. But the recent reorientation of information communication technologies (ICT) toward its cloudification forces a repositioning of this technology in new network segments, closer to the users where new key performance indicators

(KPIs) are asked. The new challenges are then to find a technology offering ultralow latencies, with systems more simple, cheaper, and less power consuming than existing products. This new direction then creates a need for new low-cost systems, ecodesigned at the convergence of new network concepts and of the emergent technologies driven in particular by the GPON2 and the DATA COM communities.

3.1 INTRODUCTION

This chapter gives a positioning of optical switching technologies in the next generation of systems and networks. After giving two introduction scenarios, one for the metro part and the other for the backbone part, this chapter addresses feasibility issues. For the metro part, three introduction scenarios are presented. The first one exploits the well-known circuit switching techniques that can be introduced rapidly on the market. The second one proposes a packet switching technique to have a better bandwidth exploitation. Finally, with the emergence of new optical functions/devices, the third scenario describes how it is possible to propose a full flexible packet ring network that is really competitive with respect to other electronic alternatives. For the backbone part, the first scenario will be probably in the core of large routers, competing with current smart routers (router + cross-connects). The second scenario is for a new network concept, being disruptive with what exists but pushing toward a transparent compliant with a multiservice environment and fully evolutive in capacity. Finally, the last scenario describes how it is possible to go into an all-optical approach through the description of key optical functions required to make this concept realistic at a lower cost.

With the introduction of the Internet protocol in the network, the telecommunication domain has turned a new corner. The broadband access to this new technology, opening the way to many residential applications, creates a revolution for the next generation of switching systems. The first revolution is the traffic volume. Personal computers becoming more and more powerful are generating a traffic through files that could not be envisaged even 2 years before. The second revolution is probably the traffic profile evolution, moving from a constant bit rate to a variable bit rate, always driven by personal computer capabilities (video applications, high definition TV (HDTV), net shopping, net courses, games, etc.).

Optical technologies could appear in the next 4 years as an important technology to grow the capacity of systems while preserving the simplicity, reliability, and performance of the systems. But, more importantly, optical packet switching technologies could become efficient techniques to really fit with the statistical behavior of the traffic profile to preserve bandwidth utilization as much as possible. One of the key issues in such packet-switched networks is the identification of the best packet format (variable packets or fixed packets). Several European projects have concentrated their efforts on this important topic such as the RACE 2039 ATMOS project, the ACTS 043 KEOPS project, and more recently the IST DAVID project.

Thus, in this chapter, after a positioning of optical switching in the next generation of systems and networks, the benefits of multiplexed architectures will be presented. Solutions for a progressive introduction of this technology in the metro are described, highlighting the required technology and addressing physical feasibility as well as performance issues. Opportunities for the backbone are also presented with the objective of highlighting

the most promising approaches. For a pragmatic approach, criteria introducing this technology on the market are listed, but, more importantly, a basic cost approach leading to the winning solution is mentioned. Finally, a conclusion is drawn.

3.2 POSITIONING OPTICAL SWITCHING TECHNIQUES IN THE NEXT GENERATION OF SYSTEMS AND NETWORKS

In this section, we will present the advantage of optics with respect to electronics, but, more importantly, how optics can be exploited to complement electronic technology to really make the most of both technologies.

3.2.1 Why optical switching?

To give some arguments, we must list the advantages and drawbacks of optics.

The main advantages of optics are as follows:

- Low-power consumption: An example is a laser exploited in direct modulation. A laser operating at 622 Mbit s^{-1} or at 2.5 Gbit s^{-1} will be electrically modulated with the same electrical modulation amplitude.
- High reliability: For passive devices, it is evident that for active devices [lasers, EDFAs, semiconductor optical amplifiers (SOAs), etc.], the reliability is quite high because we exploit a carrier density dynamic, characteristic of the material used.
- Good mode adaptation: The default on the coupling simply introduces losses, and the reflection can be easily managed by exploiting tilts.
- Low-power dissipation: The photon–photon interaction dissipates less energy than the electron–electron interaction, mainly because of the mass.
- High bit rate compliant: A passive guide is *a priori* a high bandwidth medium capable of supporting several terabits of capacity.
- Management of large granularity: The switching can be done at the wavelength level but also at the waveband (group of wavelengths) level.

The main drawbacks of optics are as follows:

- Slow progress in integration: There is still a debate between monolithic and hybrid integration.
- Slow progress in low-cost packaging: The coupling between passive and active guides still remains a costly technique.
- Polarization sensitive: The characteristic of some materials often depends on the polarization state of the light.

In summary, optics is very interesting when the switching granularity is high, exploiting the WDM dimension to make simple structures. In electronics, we need to demultiplex at the wavelength level and then at the bit rate level; in optics, we simply need one device.

To switch at the WDM granularity, we have commercially available devices such as opto-mechanical switches, thermo-optical switches, electro-optical switches, and micro electro mechanical systems (MEMs) for slow switching applications and digital optical switches for fast switching SOA.

3.2.2 Granularities of switching

In optics, we can switch wavebands (group of wavelengths), wavelengths, or optical packets.

3.2.2.1 SWITCHING OF WAVEBANDS

The switching of wavebands is particularly interesting in the following cases:

- To reduce the number of switching elements inside systems. This is the case for a large number of optical cross-connects or switches.
- At the network level, when the traffic is aggregated enough to tolerate a waveband switching with a good bandwidth utilization. This is the case for pipes bridging networks and where the traffic matrix is quite stable.

The waveband switching really exploits the potential of optics because it reduces the complexity of the switching process with respect to electronic techniques. This technique has to be exploited as much as possible to make a system or a network concept really competitive to electronic techniques but only when the traffic matrix is stable enough not to penalize the average load of the waveband.

3.2.2.2 SWITCHING OF WAVELENGTHS

The switching of wavelengths is particularly interesting as follows:

- To switch at the line bit rate without time demultiplexing inside optical systems, which is an advantage with respect to electronic techniques, especially when the bit rate is high (≥ 10 Gbit s^{-1}). Generally, the wavelength switching is used to relax the power budget to offer higher switching capacity. In fact, with one important parameter being the optical signal-to-noise ratio, it is fundamental to preserve it to be able to address the higher throughputs. This is generally imposed when input powers launched into optical amplifiers cannot exceed a certain value. By this way, as the optical signal-to-noise ratio is always a function of the channel input power, if the total input power is the channel power, then we reach the maximum optical signal-to-noise ratio value in the architecture. This is the case for a major part of the optical cross-connects or switches.
- At the network level, when the traffic is aggregated enough and stable enough not to cope with traffic transient effects. This technique is still efficient when it is possible. The switching is done at the line bit rate. It is more efficient for backbones than for metro networks, for example, simply because of the traffic matrix characteristic.

3.2.2.3 SWITCHING OF OPTICAL PACKETS

The switching of optical packets can be processed at the wavelength level or at the waveband level. The only difference with respect to wavelength or waveband switching is that the ON state is relatively short, on the order of a few hundred nanoseconds or microseconds. The switching of optical packets is particularly interesting as follows:

- To create a datagram connection or a virtual circuit connection inside optical switching systems. The switching can be done at the wavelength level (classical optical packet switching) or at the waveband level (we define then a WDM packet) for a short time. Like for the previous cases, the WDM dimension is preferred where possible to reduce the number of active components in the architecture considered.
- At the network level, when the traffic profile is sporadic, we have to cope with time constants

that cannot fit with seconds or minutes, and where the packet connection is the only one realistic to exploit efficiently the available bandwidth. This is the case for metro networks, in particular, but also for backbones if the application bit rate is growing at bit rate.

3.2.3 Optical packet switching: An interesting approach context

Among the different switching techniques, the optical packet switching technique is probably the most promising technique for the next generation of networks. The main indicator is the natural evolution of the traffic profile versus packet techniques. Driven mainly by the Internet protocol, we need to cope more with a traffic profile than a traffic matrix as could have been the case in the past. The main reason is the drastic change of telecommunication applications moving from telephony to data. In addition, the rapid introduction of personal computers (PCs) at home as multimedia machines pushes telecom companies to find solutions to offer a higher quality of service at a lower cost. This new form of traffic imposes new infrastructures capable of handling the required capacity and to provide at the same time the required flexibility to offer low-cost connections.

3.2.3.1 WHY OPTICAL PACKET SWITCHING?

When analyzing the traffic profile at the output of a local area networks (LAN), the sporadic behavior of the traffic, often modeled with self-similar functions, clearly points to the problem. We then need to adapt the network concepts to the traffic nature coming from the access. And how do we adapt such a variability of the traffic profile with

circuit connections while having a good efficiency? The answer lies in the high aggregation level that requires a grouping of different LANs, and this is not always possible. Another solution is to cut the circuit into pieces (packets) trying to follow the traffic evolution at a scale comparable to the scale of the incoming traffic. Even if the technique is more complex to manage, it is undoubtedly the most efficient way to optimize the bandwidth utilization, more in the time domain today, than in the volume domain in the recent past.

3.2.3.2 WHAT KIND OF PACKETS: FIXED PACKET OR VARIABLE PACKETS?

There is still a debate on the choice of the packet size: for main arguments in favor of the variable packet, the optical packet size always follows the incoming packet size, whereas for the fixed packet, the optical packet format contributes to a better management of the performance. In both cases, arguments are acceptable, but the reality is more complex.

It is evident to say that where the contention can be managed easily (a case of small or simple topologies), the variable packet has to be envisaged. But in the case of a large topology (meshed or other), the resolution of the contention is then local in each node, and the control of the traffic profile inside the network becomes fundamental. In that case, the fixed packet format is the only reasonable format. Another alternative, probably the best, is the adoption of a concatenated packet. For best effort, the concatenation is created to really follow the incoming packet profile. For high priority traffic where the delay is fundamental, small packets will always experience the smallest delay in the network. In this way, the technique can be adapted to a multiservice environment (Figure 3.1).

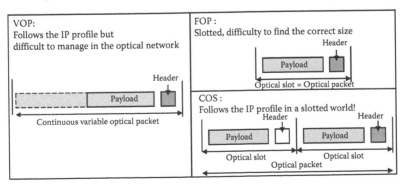

Figure 3.1 Different optical packet types that can be adopted.

Arguments	VOP	FOP	COS
Management in the ingress edges	☺	☺	☹
Management in the core nodes (Perf)	☺	☺	☺
Management in the egress edges	☺	☺	☺
High priority CoS	☹	☺	☹
Best effort	☺	☹	☺
Interesting solution		FOP + COS	

Figure 3.2 Comparison: variable optical packets versus fixed optical packets versus concatenated optical slots.

Figure 3.2 gives a comparison among fixed, variable, and concatenated packets.

3.3 BENEFITS OF MULTIPLEXED ARCHITECTURES

3.3.1 The WDM dimension, a dimension not only for transmission systems

The WDM technique is often assimilated to the transmission domain. But, in fact, this technique is very useful in optics for many purposes.

The WDM technique can be exploited for the following objectives:

- To increase the capacity of a link without increasing the TDM bit rate at values difficult to manage. This first application was found rapidly. But there was also the requirement to reduce the number of active components. As a good example, the EDFA amplifying a group of wavelengths has rapidly replaced the classical single-channel repeater.
- To facilitate the optical processing to avoid the interferometric noise when interleaving several pulses at high bit rate. The technique is used in optical time division multiplexing (OTDM), coding format techniques, multiplexing techniques (bit level or packet level), etc.
- To ease the contention resolution. In fact, the wavelength becomes a new support, with different colors, and the color is selected in real time to avoid collisions. This technique is particularly used in optical packet switching network concepts.
- To switch high granularities. It can be introduced to reduce the number of components in an optical architecture or in cases of low connectivity while addressing high throughputs.

3.3.2 Benefits of the WDM dimension in multiplexed architectures

In multiplexed architectures, the WDM dimension is exploited for different purposes.

In the following, we will describe where the WDM dimension can be exploited efficiently and what kind of benefit we can expect.

3.3.2.1 IN METROPOLITAN NETWORKS

In the case of optical rings, the WDM dimension can be exploited first to provide an upgradability of the network in terms of allocated resources, simply by attributing progressively bands of wavelengths. The advantage of the band is mainly to relax the filtering constraints during the cascade of several nodes. This advantage was raised many times in circuit switching platforms. If we want to exploit the same WDM infrastructure while introducing packet techniques, the notion of band can then be advantageously exploited to reduce the latency in the transmitting parts. In fact, we can exploit the statistical time multiplexing of packets over a group of wavelengths to increase the chance to insert a packet in the line when we can have access to a group of wavelengths instead of one.

In summary, we can see that for optical rings, the WDM dimension can relax physical constraints and, in addition, improve the performance in packet rings in terms of latency.

To illustrate the benefit of the WDM dimension, the IST DAVID project is proposing a multiring optical packet MAN exploiting the WDM dimension for these two main aspects.

3.3.2.2 IN BACKBONE NETWORKS

In backbone networks where the topology is generally meshed, the WDM dimension is exploited for four main reasons:

- *Cost and compactness reasons*: In optical cross-connect/packet switching architectures, the WDM dimension can be exploited to reduce the number of switching components to make the architecture compact and low cost. The objective is to switch in the WDM dimension as much as possible in the limit of the required power budget. In cross-connect architectures, the dynamic management of the wavelength dimension can, in addition, contribute to a drastic reduction of the number of interconnected fibers.
- *Power consumption reasons*: By exploiting the WDM dimension, the processing is done at the WDM dimension. In electronics, a double demultiplexing is required: at the wavelength level and at the bit rate level. This is, for example, the case for coarse synchronization stages where the WDM dimension can be exploited efficiently at the WDM level to reduce the complexity of many electronic structures.
- *Physical reasons*: The number of channels switched can also vary inside the architecture to preserve the optical signal-to-noise ratio at the output. For example, in the first stage of an optical architecture, the WDM switching can be done on a large number of channels (8 or 16), and then progressively reduced to a lower number, converging then to one channel switched.
- *Performance reasons*: In packet architectures, the WDM dimension can be exploited to reduce the packet loss rate and the latency at the same time. When the packets can be addressed onto different wavelengths, the statistical multiplexing on these wavelengths

let a freedom degree for the choice of the wavelength at the output of an optical packet switch. By this way, the contention can be avoided simply by reaffecting a new wavelength to a packet instead of putting it in a queue. The benefit is double because the packet loss rate can be improved, with the same amount of digital memory, and the latency is preserved because the packet will not experience a queue.

3.4 SPECIFICITY OF THE METRO AND PROPOSED SOLUTIONS

3.4.1 Introduction scenario

To take a pragmatic approach, it is fundamental to identify what could be the introduction scenario of a technology at the time. These paragraphs intend to position the optical switching technology for the metro in a timescale.

Before developing these scenarios, it is also important to list the specificity of the metro part.

In the metro, it is clear that the cost is probably the most important parameter together with the performance. The cost addresses the hardware part but also the means adopted to exploit the bandwidth in the best way. Due to the traffic profile coming from the access, the important point is really to preserve the bandwidth. Thus, packet techniques will be preferred in this part of the network.

3.4.1.1 WHAT ABOUT THE WDM DIMENSION?

The WDM dimension is expensive in the metro, but there are also some arguments in favor of the introduction of the WDM in this part of the network.

Due to the current traffic profile, not constraining the bandwidth utilization too much (because the native bit rate is still quite low), a circuit platform is interesting. To reduce the cost of the WDM dimension while having enough resources to cope with the traffic volume and profile, probably the waveband approach is the best one. It guarantees upgradability (sub-band per sub-band), physical performance (relaxing filtering constraints), and simplicity (circuit switching) with an existing infrastructure (fibers already installed). But if the native bit rate increases together with the variance, there will be a need to go into a lower granularity to have a better utilization of the optical bandwidth. Then the optical packet technique could be easily introduced making use of an existing infrastructure. The gap becomes natural.

Thus, in the following presentation, we will introduce the circuit switching technique as a first step with a progressive migration toward optical packet switching techniques, which lead to a really efficient platform.

3.4.1.2 SHORT-TERM INTRODUCTION: OPTICAL CIRCUIT SWITCHING TECHNOLOGY

Due to the current traffic profile, optical circuit switching could be rapidly introduced in the market for many reasons:

- The optical technology is mature enough and commercially available to envisage its utilization in optical platforms. It requires active devices such as integrated laser modulators (ILMs), receivers, and basic passive devices (demultiplexers, couplers, filters, etc.).
- It is a very simple technique that can be implemented rapidly.
- The optical technology can exploit advantageously an existing WDM infrastructure. Currently, all the fibers installed are not exploited.
- The management of such a network is mature enough to propose products. Management studies have been carried out leading to clear information models, protection scenario, and well-defined monitoring techniques.
- The traffic profile is not yet so critical to envisage using such a technique in the metro. The bit rate at the output of personal computer (PC) is sufficiently low to have enough aggregation level at the output of the LANs.
- This kind of platform can be easily upgraded with optical packet techniques to be fully compatible with the future traffic profile driven by a powerful PC and fiber to the home (FTTH). This is also a strong argument for operators who want to invest with the possibility to upgrade their platform according to traffic constraints.
  Examples of implementation are ring topologies, star topologies, and mesh topologies. In ring topologies, we can find the single ring or

multiring approach. Figure 3.3a and b illustrates these three interesting topologies.

The particularity of this approach is in the optical add/drop multiplexing structure. We can list mainly as follows:

- Passive optical add/drop multiplexers
- Dynamic optical add/drop multiplexers

The passive structure can be used when the traffic matrix is very stable, whereas the dynamic structure can allow some adaptation of the allocated resources to follow at least the envelope of the traffic matrix. This last type of the active structure is particularly interesting when the time constants are in the range of a few hours.

Figure 3.4a and b describes the two main structures of add/drop multiplexers.

In the metro part, the WDM granularity for the upgradation of the network capacity will depend on the cost of the intervention. And in some cases, the upgradation of one wavelength is not the most cost-saving solution. For that purpose, it is important to identify the minimum WDM granularity for the upgradation. This minimum granularity, which can be on the order of few wavelengths (2 or 4), can impact dramatically the network infrastructure. If we take into account the physical limitations in the cascade of filters, it is clear that the sub-band approach is an interesting approach. Figure 3.5 gives an overview of a ring MAN adopting a sub-band strategy.

3.4.1.3 MEDIUM-TERM INTRODUCTION: OPTICAL PACKET SWITCHING TECHNOLOGY

The optical packet switching technology could be introduced as a required second step simply to face the traffic profile evolution. In that case, a packet technique will be required on the basis of the infrastructure already installed. The upgradation is made simply by changing the optical ring access node. Two new sub-blocks are then mandatory: the opto-electronic part and the electronic interface compliant with different classes of services.

The more pragmatic approach is the adoption of a commercially available technology. Several concepts are proposed, all based on the adoption of ILMs and synchronous optical

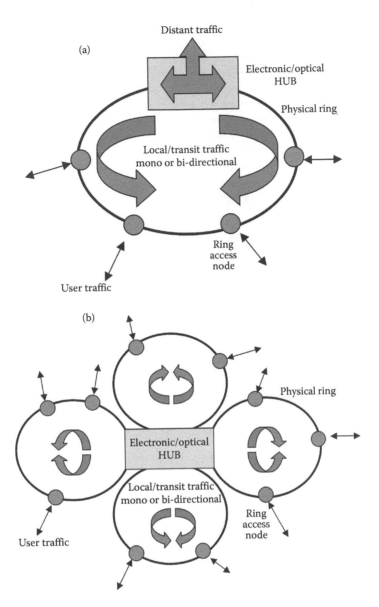

Figure 3.3 (a) Conventional ring topology. (b) Multiring topology.

NETwork (SONET)-like receivers. The resilient packet ring (RPR) concept is probably the most representative one.

3.4.1.4 MEDIUM-TERM INTRODUCTION: FULL FLEXIBLE OPTICAL PACKET SWITCHING TECHNOLOGY

In the case of long-term approaches, the adoption of an advanced technology is then mandatory.

In particular, two important components are a fast tunable source and a fast wavelength selector. These two components have already been demonstrated feasible (e.g., agility for the tunable sources, nippon telegraph and telephone (NTT) for the wavelength selectors).

These components are:

- *At the transmission side (tunable laser)*: limitation of the latency in the output queue by exploiting the WDM dimension
- *At the receiving side (wavelength selector)*: the guarantee of the wavelength transparency to receive any packet from the optical bandwidth

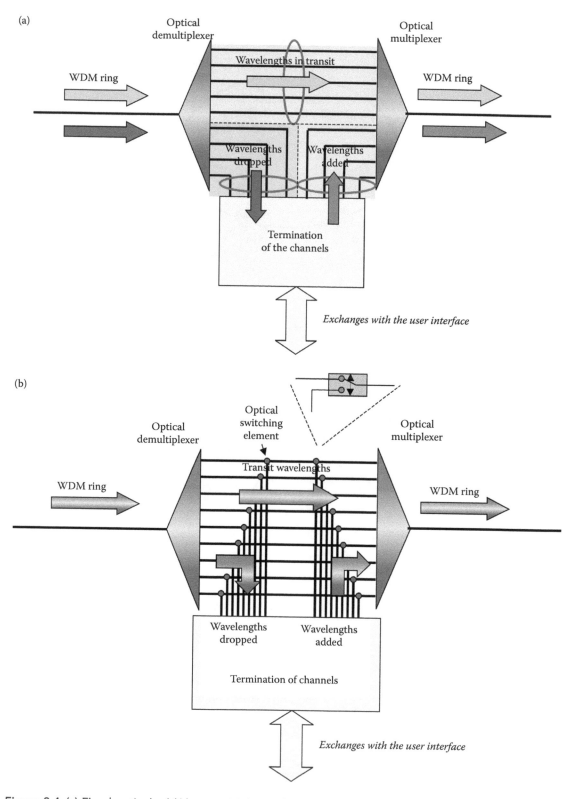

Figure 3.4 (a) Fixed optical add/drop multiplexer. (b) Dynamic optical add/drop multiplexer: an optical switching element guarantees the dynamic behavior.

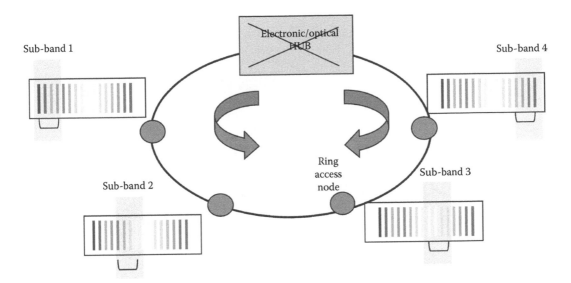

Figure 3.5 Sub-band introduction in an optical ring.

- *At the transit side (wavelength selector)*: the guarantee to introduce some fairness mechanisms when a high priority traffic with a minimum of latency is required
- The structure of the optical add/drop multiplexer is depicted in Figure 3.6a and b.

3.4.2 Technology identification and feasibility

The technology required to introduce these concepts can be split into two categories:

- Commercially available technology
- Advanced technology

In the following, we will list the required commercial devices, and we will describe in more detail the potential new advanced technologies that open the way to really attractive system functionalities.

3.4.2.1 COMMERCIALLY AVAILABLE TECHNOLOGY

There are two types of devices: passive devices and active devices

- For passive devices, we need couplers, attenuators, isolators, fixed filters, circulators, interconnection fibers, connectors, (de)multiplexers,

etc. For filters and (de)multiplexers, we can have different specifications if we are addressing channel filtering or band of channel filtering.

- In all the cases, these components are products and today fit all the required specifications for system applications.
- For active devices, we will need lasers, ILMs, slow tunable lasers, optical amplifiers, photodiodes, etc. Particular interest must be devoted to optical amplifiers because we can distinguish two types of optical amplifiers: fiber-based amplifiers and semiconductor amplifiers. For fiber-based amplifiers, largely introduced in point-to-point transmission systems, the main preoccupation is to find the best way to make these devices very cost-effective. For SOAs, even if they are on the market today, they currently suffer from a small market. However, the generic potential of such a component for different functions, such as optical gating or optical conversion, makes this component a promising one for future applications.

3.4.2.2 ADVANCED TECHNOLOGY AND FEASIBILITY ISSUES

This is the most promising technology to really propose something new and disruptive with respect to what exists today.

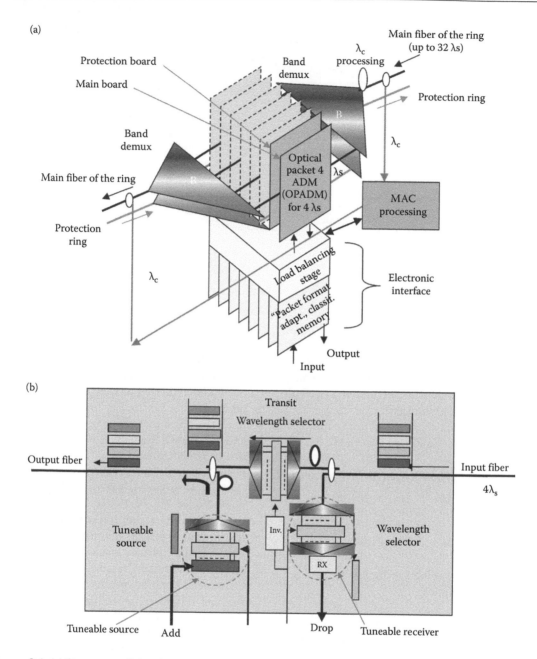

Figure 3.6 (a) Structure of the ring access node for an upgrade of the circuit switching platform into a packet switching platform. (b) Structure of the optical add/drop multiplexer including the required functionality to fully manage the transit traffic.

In the following, we will illustrate the four main components:

- The SOA for gating functions
- The tuneable source
- The wavelength selector, a strategic fast tuneable filter
- The packet mode receiver

3.4.2.2.1 SOA for gating

A SOA is basically a laser in which the facets have been treated in order to eliminate the resonant cavity. Only the amplification medium is exploited. To be used as an optical gate, the SOA requires a high-frequency driver interconnected to the SOA. The driver sends a control signal, which can be forced at the ON or OFF state. Because of the short carrier lifetime, a SOA can be switched with response times on the order of few tens of nanoseconds. At present, this component is exploited by many laboratories and has been demonstrated to be feasible for many applications.

In the metro area, and according to the two network model described previously, different SOA structure are particularly interesting.

There are roughly three types of SOA: SOA with high confinement factor or long active section to achieve wavelength conversion at high bit rate, SOA with low confinement factor or short active section to exploit more the linear characteristic of the gain, and SOA with an internal clamping to have a strictly linear response not to create distortion on the signal crossing the component.

As for fiber amplifiers, there are three classes of SOAs: preamplifiers, in-line amplifiers, and boosters.

3.4.2.2.2 SOA gate at the output of an ILM

A SOA interconnected at the output of a modulated source is the basic schematic we can envisage.

The advantage of this solution is that it is commercially available today. The main drawback is that there is a need to have control of the cross-gain modulation. One interesting solution is the use of a clamped-gain SOA to avoid any cross-gain modulation. The SOA used only in its linear characteristic will provide enough gain to guarantee a sufficiently high ON/OFF ratio with no degradation of the pulse shape. This solution is currently studied in different laboratories

to analyze network concepts based on a packet transmission.

3.4.2.2.3 Tunable source using a hybrid integration of a SOA gate array

The SOA gate array is located in front of a laser array. At the output, a multiplexer and an integrated modulator are interconnected. The SOAs see only continuous waves and can be switched, thus selecting the wavelength that must be transmitted. The SOAs are in a stable regime because the input power is a constant. Therefore, they do not experience any cross-gain modulation as could be the case in the previous use. The preservation of the signal quality (no degradation of the extinction ratio and no distortion of the bits) makes this SOA array an important device for the building of hybrid tunable sources.

Figure 3.7a illustrates the structure of a tunable source based on a gate array, and Figure 3.7b shows an integrated four gate-array (OPTO+ realization).

3.4.2.2.4 Tunable switching sources based on a sampled grating— Distributed Bragg reflector structure

Another solution to build a fast tunable source is to use an Simple Grating Distributed Bragg Reflector (SG-DBR) laser while integrating a SOA section and an electro-absorption section. The following schematic illustrates the structure of the source. As for the hybrid tunable source, the SOA sees only a continuous wave that again prevents cross-gain modulation. The structure is currently studied by agility (Figure 3.8).

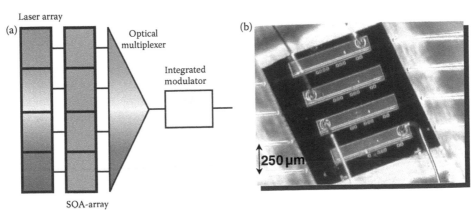

Figure 3.7 (a) Structure of a hybrid tunable source using a SOA gate array. (b) Photo of a four gate-array, key building block of the hybrid tunable source.

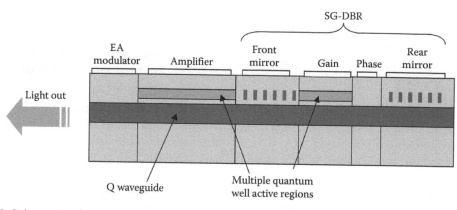

Figure 3.8 Schematic of an integrated tunable source integrating a SOA section for amplification of the signal or optical gating.

3.4.2.2.5 Wavelength selector

The wavelength selector is probably one of the other key devices because it can be comparable to a fast tunable filter. The principle of operation is very simple. A first demultiplexer demultiplexes the wavelengths, then each wavelength is selected or not, depending on the orders coming from the control part, and finally an output multiplexer regroups the wavelengths selected and contributes to reject the wideband amplified spontaneous emission coming from the SOAs. In principle, only one wavelength is selected among a group in a normal scenario. However, in the case of the third network scenario, the number of output wavelengths can vary from 1 to N (N being the total number of wavelengths at the input of the device).

Figure 3.9a shows the basic structure of a wavelength selector.

3.4.2.2.6 Packet mode receiver

In the case of the third scenario, another important technology is required, not at the optical level but more at the electronic level: the packet mode receiver.

The packet mode receiver has to experience a continuous or noncontinuous packet stream exhibiting different packet phases (when aligned to a common reference clock) and suffering from a packet power dispersion. Such kinds of receivers are currently studied in different laboratories. We can cite NTT, Nichiden (NEC), Lucent, Alcatel, etc.

Currently, this technology has been demonstrated to be feasible for use in different coding formats: Manchester but also return to zero (RZ) or non return to zero (NRZ).

Figure 3.10a and b shows the performance obtained with a packet mode receiver operating at 10 Gbit s^{-1}.

3.4.3 Performance expected for the packet approach

The performance will depend on the packet format adopted.

- In the case of variable packets, the contention resolution becomes a bidimensional problem: to be able to insert a packet in the ring, we have to check if we have a void and if the void is large enough to insert the packet.
- In the case of fixed packets, we need to also experience two dimensions, but in that case, they are not cojoined. The first problem to solve is the packet filling ratio, imposing some time out to have a good efficiency, whereas the second problem is in the add part: check if there is a free slot.

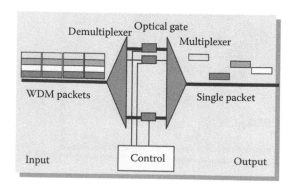

Figure 3.9 Structure of a wavelength selector.

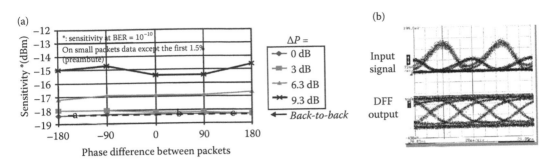

Figure 3.10 (a) Characteristics of a packet mode receiver. Large power dynamic ranges and fully transparent to the packet phase fluctuation. (b) Eye diagram recorded: before and after the 10 Gbit s^{-1} packet mode receiver. The phase is preserved and the amplitude completely equalized.

Therefore, the problem is not the same for both cases, and the impact on the performance is not the same.

- In the case of variable packets, the main problem to solve will be the management of the voids in the ring so that the last ring access node in cascade is not blocked.
- In the case of fixed packets, the main problem to solve is the choice of the good packet size to be compatible with the incoming traffic profile.

Therefore, concerning the performance, the variable packet format exhibits a poorer performance compared to the fixed packet. This is the reason why the concatenated approach is probably the optimum solution, providing performance and reliability (packet rhythm present in the ring to ease monitoring aspects).

3.5 OPPORTUNITIES FOR THE BACKBONE

3.5.1 Introduction scenario

In this part, the objective is to draw a progressive introduction of optical switching techniques for the backbone. For the short term, a cross-connect interconnected to a router could be envisaged to reach high-throughputs and high throughput routers. Both approaches are of prime importance because they correspond to two reality cases and two classes of products required to cover different network specificity. As a medium-term approach, we will present a multiservice Opto/Electro/Opto

(O/E/O) network concept based on new features. Finally, we will describe how an all-optical packet switching network could become a reality in the longer term because of attractive features.

Short-term introduction: smart router (router + cross-connect) versus multi-terabit class routers/ switches.

3.5.1.1 WHERE SMART ROUTERS AND WHERE HIGH-THROUGHPUT ROUTERS?

Smart routers are required for the dorsal network and where the aggregation is forced by the poor connectivity of the nodes and by the huge amount of traffic that must be transported. This type of product is particularly interesting when the traffic matrix is stable enough to make semipermanent connections realistic and efficient. This is particularly the case in the United States when connections have to be established between states. The router is mandatory to collect the traffic coming from regional networks or national networks.

High-throughput routers are required when it is not possible to process a part of the traffic with semi-permanent connections through optical cross-connects. It can be the case for metro core networks, where the dynamicity of the traffic could require a transfer mode at the packet level to increase the network efficiency while optimizing the resource cost.

3.5.1.1.1 Smart router

Figure 3.11 shows the global structure of a smart router. The router collects the traffic at the packet

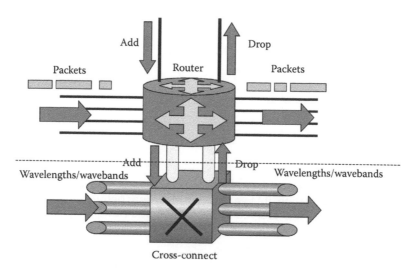

Figure 3.11 Smart router schematic.

granularity. Packets are put into queues and are sent on a specific wavelength. The optical cross-connect has to manage high throughputs. The structure can be based on a MEM technology. The approach is very interesting when the traffic is strong enough to open large pipes, thus enabling the establishment of a waveband. The cross-connection at a waveband level is the best guarantee of simplicity and reliability without creating breakthrough between the transmission system and the switching cross-connect.

3.5.1.1.2 High-throughput routers

Figure 3.12a shows the generic structure of a high-throughput router (multi-terabit-class router) exploiting an optical core in its center part, whereas Figure 3.12b and c shows a prototype realized and the bit error ratio (BER) curve. The optical matrix is basically a fast space switch, creating connections at the packet level. The burst card is responsible for the packet format adaptation, whereas the line card is used to manage the incoming traffic. Buffers are at the input and the output of the optical matrix, and in the burst and line cards. An internal speedup can be exploited to guarantee the full functionality even in the case of failure of one of the switching planes. This approach is currently adopted by many constructors.

Both approaches are really important and demonstrate the potential of optical switching for basic functions, mainly focused on space switching: slow or fast.

3.5.1.2 MEDIUM-TERM APPROACH: NETWORK CONCEPT BASED ON O/E/O NODES

In the previous cases, solutions are based on a traffic profile assumption, which enables the circuit switching technique for long-haul networks or enables the packet technique for small-scale backbones. However, the problem will occur when the application bit rate is increased together with the variance. This can rapidly happen if merging low bandwidth connections coming from mobile phones and high bandwidth coming from more and more powerful PCs reinforced by an optical connection giving access to very high bit rates per user. In this particular case, the variance can be increased leading to huge problems for efficient aggregation and forcing the telecommunications companies to think differently toward an optical packet platform for the backbone.

This means that, due to the traffic profile evolution, the packet will be used extensively. And the key question could be as follows: how to realize an efficient network capable of handling the required capacity while providing a mandatory flexibility?

3.5.1.2.1 The generic approach

In the network concept, we mainly exploit the edges to prepare the traffic in such a way that the traffic constraints inside the core of the network are relaxed. This means that all the complex functions are located in the edges such as aggregation,

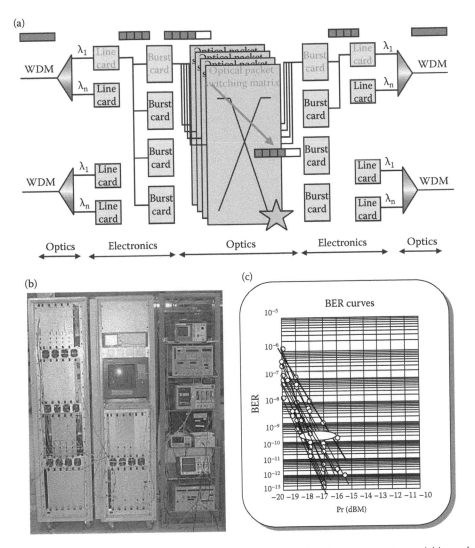

Figure 3.12 (a) High-throughput router schematic. A fast optical switching matrix could be adopted in the center of the architecture. (b) Photo of a 640 Gbit s^{-1} throughput optical matrix. (c) BER performance.

switching per destination, classification, packetization, traffic shaping, load balancing, and admission control. The traffic profile, having a better shape, is then sent to the network. The core nodes will be responsible for the synchronization, the contention resolution, and the switching. In this case, the packet being created in the edges only simplifies the structure of the core router.

3.5.1.2.2 What type of packet format?

As a first introduction, and if possible, it can be envisaged to introduce an existing packet format. The G709 framing is currently being investigated to identify the potential of the concept, but other packet formats could be considered.

For a second introduction, it seems clear that a more smaller packet size is required to relax the problems of aggregation. This also imposes a standard on a packet that does not exist today.

It must be noted that some universities and laboratories are currently studying the possibility of managing variable packets called bursts. The advantage of this approach is that the edge part is simplified in its functionality, but the core nodes are more complex to control, and the overall performance is affected by a highly sporadic traffic profile.

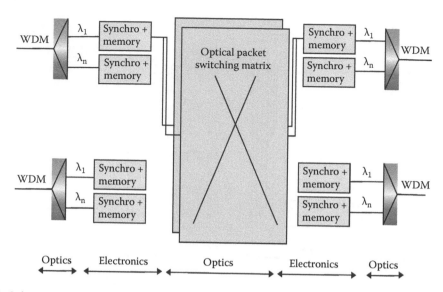

Figure 3.13 Schematic of a core router in the concept of a packet network.

3.5.1.2.3 Structure of the core node

The core node is strongly simplified with respect to the first approach because all the complex functions are located in the edge nodes.

Figure 3.13 describes a representative structure of such a core router.

It can be noticed that the particularity of this architecture is to have synchronization stages and memory stages before and after the optical matrix in the core of the high-throughput routers. This optical matrix has been introduced in the first introduction scenario, so the step is quite easy to cross. The challenge is greater at the management level than at the node level.

3.5.1.3 LONG-TERM SOLUTION: ALL-OPTICAL NETWORK

In the previous scenario, we still needed a lot of costly O/E conversion. One key question is as follows: can we efficiently reduce the number of O/E conversion stages in the core routers?

For this, we need to solve three key problems:

- The synchronization (to realign the packets before the switching)
- The regeneration (to enable the cascade of several optical core nodes).
- The contention resolution (to be able to offer the required packet loss rate and delay with respect to the Class of Services requirements)

3.5.1.3.1 Synchronization stage

In electronics, we have bit memories making the synchronization process very simple. The information is stored at the distant clock rhythm, and it is extracted at the local clock rhythm. However, the structure is quite complex because the process is done at low bit rate, thus imposing one stage of WDM demultiplexing and one stage of bit rate demultiplexing.

As in optics, we do not have bit memories, and therefore, it is important to think differently. By imposing in the packet format a sufficiently large guard band, we simply need to preserve the phase between consecutive packets in order not to have collisions. This means that we need simple synchronization structures capable of having a resolution so as to avoid the problem mentioned. Typically, the resolution that can be handled is in the range of few nanoseconds. It is exactly what we will adopt for the synchronization.

3.5.1.3.2 What kind of problem we need to face?

The first problem is the thermal effect in the fiber modifying the index and creating variable delays during the propagation of data depending on the average temperature of the fiber. This means that all the WDM multiplexes will be affected. A WDM structure could bring a solution to this problem.

The second problem is that we do not have control of the time jitter created in any optical switching fabric. This packet jitter can be a blocking point in the cascade of several nodes. We need to control the packet jitter packet per packet, which indicates that the control must be done at the wavelength level but not at the WDM level.

Therefore, in summary, we can easily solve the problem of the synchronization by using one or two stages and combining a processing at the WDM level and a processing at the wavelength level. In both cases, we operate at the line bit rate. The gain is in the simplicity of the synchronization process and in the complexity of the structure, making this synchronizer reliable.

3.5.1.3.3 Regeneration stage

The regeneration stage is mandatory if we want to exploit the maximum throughput of the optical switching matrix. The regeneration separates the two systems: the transmission and the switching in order to lead to a maximum throughput for the nodes. It is also the only way to cascade nodes when the line bit rate is high. Once again, as the processing is done at the line bit rate level, the structure of the optical regenerator is really simple, being a guarantee of simplicity and robustness.

3.5.1.3.4 Contention resolution

The contention resolution is still an issue in optical architectures because we do not have any efficient optical memory. To solve the problem, we will exploit the WDM dimension and more particularly the statistical multiplexing over the different available wavelengths. Therefore, the technique adopted is to avoid collision by reaffecting the wavelength to the packet at the output of the switching fabric. As the number of wavelengths per fiber can be limited, a recirculation buffer is then mandatory to solve the contention properly. By combining both techniques, the performance can easily reach the performance of a classical electronic switch but offering here all the switching capacity in one unique stage.

3.5.1.3.5 Optical matrix adapted to optical interfaces

If photodiodes have very low sensitivity, it is not the case when introducing all-optical interfaces.

Therefore, the optical switching fabric must be adapted to these optical interfaces by providing the required power. Once again, different techniques can be proposed to achieve this goal.

3.5.1.3.6 Generic structure of optical core nodes exploiting optical techniques

A generic structure for an all-optical packet core is described in Figure 3.14. The particularity of this architecture is that there is no O/E conversion except for the electronic control and the memory, making this architecture cost-effective. Based on the previous concept, this architecture is simply an evolution of the core node exploiting optical functions for a better efficiency, and a potential line bit rate increases at a lower cost. This approach is fully compatible with future point-to-point transmission systems at 40 Gbit s^{-1}.

3.5.2 Required technology available technology

For the short- and medium-term approaches, we need the following:

- An optical technology for space switching
  - For optical cross-connects, the MEM technology is probably the most promising technology.
  - For fast optical matrix, we need the following:
    - *For the optical matrix itself*: free space-like technologies (Chiaro-like), tunable source-based technologies (Lucent), and SOA-based technologies (NEC, Alcatel, etc.).
    - *For the receivers (still in the laboratory)*: packet mode receivers capable of being fully transparent to the packet phase and capable of absorbing packet power variations arriving at the packet rhythm. Several companies have proposed such a kind of receiver: Lucent, NTT, NEC, Alcatel, etc.
    - *Burst cards (still in the laboratory)*: fully electronic adaptation interfaces for the packet format used inside the fabric.

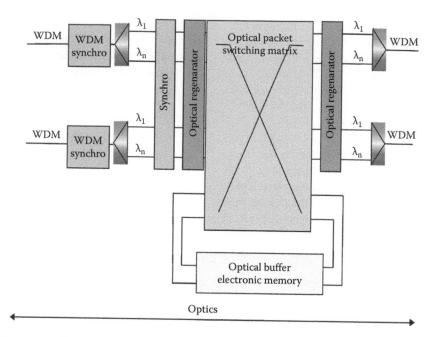

Figure 3.14 Structure of an optical core router exploiting the optical resources but including an electronic memory stage in recirculation to guarantee the performance.

3.5.2.1 ADVANCED TECHNOLOGY

To realize compact systems, there is first a need for an integrated technology.

To realize the key sub-blocks presented previously, we need the following:

- *For the fast optical switching matrix*: compact tunable lasers, SOA gate arrays, and integrated wavelength selectors
- *For the optical regenerators*: integrated Mach–Zehnder, self-pulsating lasers, etc.
- *For optical synchronization*: optical gates

3.5.3 Technological feasibility viability of advanced sub-blocks and feasibility issues

3.5.3.1 OPTICAL MATRIX

In the case of opto-electronic interfaces, the fast optical switching matrix (introduction scenario for the short and medium terms), the constraints are quite relaxed because the sensitivity of the receivers allows the design of switches with small output powers. Typically, reception powers as low as −10 dBm can be considered at the output of the switching fabrics. This also means that the amplification is limited in the core of the switch,

leading to very compact and less power-consuming architectures.

One typical matrix is the SOA-based matrix, requiring simply an amplification stage before the splitting stage is the broadcast-and-select architecture. Another one is using tunable lasers and a wavelength router in the center. Both are represented in Figure 3.15.

The first architecture (Figure 3.15a) takes advantage of broadcasting functionalities and exploits robust devices such as ILMS or SOAs. However, it is limited in capacity, mainly due to large losses that have to be compensated by amplification. The optical signal to noise ratio (OSNR) affected mainly limits the capacity.

The second architecture (Figure 3.15b) has *a priori* a larger potential in terms of capacity because the architecture simply includes a tunable source and a passive wavelength router. However, this architecture is not adapted to the broadcast of the packets, and the fast tunable laser is probably the most challenging switching element.

In the perspective of the long-term scenario, with optical interfaces, the constraint comes from the output power that must be high enough to be compatible with optical interfaces. In addition, the polarization is responsible for problems in the optical regenerative structures, it is then fundamental

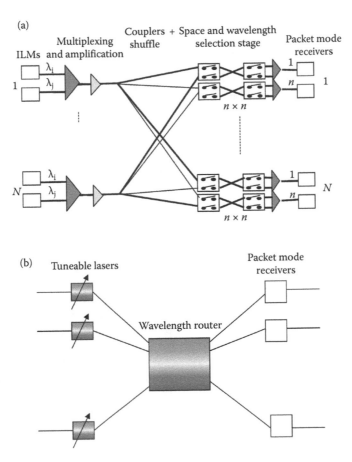

Figure 3.15 (a) Optical matrix-based on SOAs. (b) Optical matrix-based on tunable lasers.

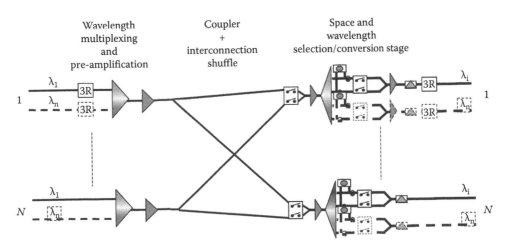

Figure 3.16 SOA-based matrix compatible for optical interfaces.

to transform a switched packet stream into a packet stream in a transmission-like configuration. This is the reason why an optical conversion is mandatory in the switching matrix.

Figure 3.16 shows an optical matrix based on a SOA technology but including a new element: the wavelength selection/conversion stage, as it is studied in the frame of the 1ST DAVID project.

Figure 3.17 A 32 SOA gate-array module.

3.5.3.2 SOA TECHNOLOGY FEASIBILITY

The SOAs have been used in different system applications for amplification but also for wavelength conversions or for optical gating. To realize large systems as described previously, there is a need for a large amount of components. In this case, the integration is then mandatory to make such a matrix very compact. OPTO+ has designed and realized 32 SOA gate array modules. The module shown in Figure 3.17 includes 32 SOAs and their respective drivers. It has been used to realize a 640 Gbit s^{-1} switching matrix.

3.5.3.3 TUNABLE SOURCE FEASIBILITY

The tunable source is a key component for many system applications.

In the case of slow switching, we can identify the tunable wavelength conversion to provide the required flexibility to achieve a best utilization of the wavelengths in a network. Another evident application is the replacement of ILMs with tunable sources. The advantage is mainly in the spare cost: instead of duplicating all the sources, the objective is to have only one source capable of emitting at any wavelength of the comb exploited in the WDM system. Finally, another application is the monitoring of optical switching systems. In this particular case, we need a compact structure capable of testing the different wavelengths and paths of a switching system. To be compatible with the system constraints, the requirements are switching times in the range of milliseconds or more (for monitoring or for sources), large tunability, high output power, and good extinction ratio.

In the case of fast switching, the main applications are for the metro and the backbone.

The tunability is fundamental in providing the required flexibility to exploit the WDM dimension in optical packet switching network concepts. The requirements are fast switching time in the range of a few nanoseconds, small tunability (four or eight channels), high output power, good extinction ratio, and high ON/OFF to guarantee no impact of the cross talk on the signal quality.

For slow structures, a DBR laser has been tested by different laboratories, and feasibility is not an issue.

For fast structures, the main problem is the stability of the wavelength. DBR can be considered if the tunability is small. These components have been demonstrated to be feasible, with switching times in the range of a few tens of nanoseconds. Another alternative is the selective source. Based on the cascade of a laser array, a SOA gate array, a phasar, and an integrated modulator, this structure has been demonstrated to be feasible.

3.5.3.4 OPTICAL SYNCHRONIZATION

The optical synchronization is probably the most challenging function. The objective is to process the signal, if possible, in the WDM regime or at the wavelength level. The second important point is probably the lack of digital memory that forces designers to think differently. In that context, the synchronization cannot be done at the bit level. We assume that the synchronization can be efficient in a resolution of a few nanoseconds. When we have said that the other point is to identify the source of desynchronization with respect to a reference clock.

The first source of loss of synchronization is the thermal effect in the transmission fiber. With a value between 40 and 200 ps km^{-1}, depending on the mechanical protection scheme adopted for the fiber, thermal effects can dramatically affect the phase of the packet streams. The WDM dimension can be advantageously exploited to make the synchronization stage compact and low cost.

The second source is the loss of synchronization in switching fabrics due to a nonideal path equalization. This occurs at the packet level, imposing a synchronization at the wavelength level.

The structure adopted is shown in Figure 3.18.

3.5.3.5 OPTICAL REGENERATION

The optical regeneration is one of the fundamental functions to make the approach realistic. To build all-optical networks while having optical switches

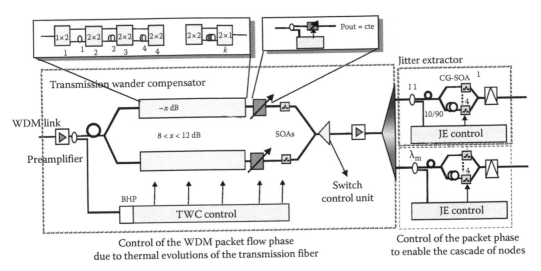

Figure 3.18 Optical synchronizer as proposed in the frame of the 1ST DAVID project.

capable of handling terabits/second throughputs, the optical regeneration is then mandatory at the periphery of switching architectures. The main functions are the total reshaping of the pulses in the amplitude and in the time domains. To achieve this reshaping, several techniques can be adopted. We will retain one, particularly adapted to the characteristic of switching fabrics creating strong impairments between pulses or between the groups of pulses. The technique adopted is a total reshuffle of the pulses adopting nonlinear elements such as Mach–Zehnder or Michelson structures.

The main distortions identified are the following: bits affected at the periphery of packets due to the switching regime, nonlinear effects such

as cross-gain modulation and four-wave mixing, cross talk (in-band and out-of-band), patterning effects by crossing active devices, noise accumulation, and jitter accumulation.

To overcome these effects, a structure has been proposed in the frame of the DAVID project. This structure, presented in Figure 3.19, has the following characteristics: by using a cascade of two nonlinear elements, the convoluted function creates a much more nonlinear transfer function, thus limiting the noise transferred in the first cascade. This has an important impact on the OSNR specification, which can be close to the back-to-back value (before the first stage) even in the case of a large number of cascades.

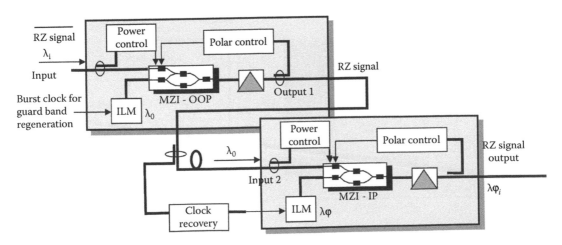

Figure 3.19 Structure of the Re-amplification–Reshaping–Retiming (3R) regenerator as it is studied in the frame of the DAVID project.

To really reshuffle the pulses, different techniques will be adopted. We can retain an amplitude and a phase modulation creating an amplitude modulation in interferometric structures to really enhance the extinction ratio and remove the noise. The second technique adopted is a sampling technique of each pulse with a clock to remove the jitter. The wavelength conversion technique will then be preferred to reallocate the wavelength in the correct wavelength comb of the new system.

3.5.3.6 FEASIBILITY OF NETWORK CONCEPTS

The feasibility of the approach was demonstrated for the first time in 1998, at the end of the ACTS KEOPS project.

In this project, we have cascaded 40 network sessions error free at 10 Gbit s^{-1} per wavelength demonstrating for the first time the possibility of building an all-optical network at a backbone scale.

Figure 3.20 gives the network session tested and put in a loop to demonstrate the concept.

3.5.4 Performance expected

The performance is probably one of the most important indicators for the feasibility of such concepts. When the physical aspects are verified, the challenge becomes the performance in a real traffic environment.

3.5.4.1 ENVIRONMENT AND SPECIFICITY OF THE BACKBONE

If, in the metro, the capacity is limited, in the backbone this is the major characteristic. To provide the capacity with a technology limited today to 10 Gbit s^{-1}, the only solution is in the exploitation of the WDM dimension.

Therefore, the WDM dimension will be fully exploited to provide the required capacity but also

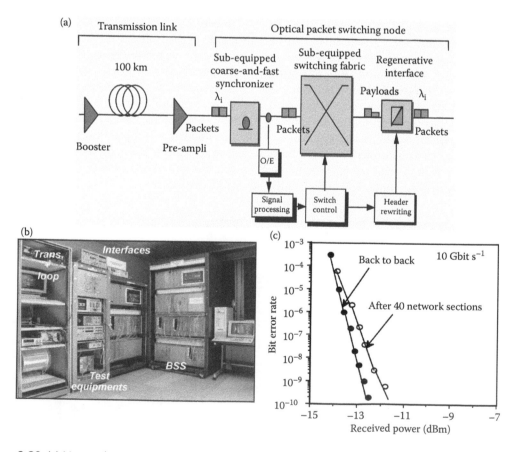

Figure 3.20 (a) Network session put in a loop to test the feasibility of an all-optical network. (b) Photo of the demonstrator. (c) BER curves giving the physical performance.

to avoid collisions due to the natural statistical multiplexing of packets on the wavelengths.

The second particularity is the aggregation. Depending on the traffic profile, circuit switching or packet switching will be preferred.

3.5.4.2 CIRCUIT SWITCHING TECHNIQUES FOR AN IMMEDIATE INTRODUCTION

Circuit switching techniques can be envisaged in the first scenario as a transport layer to provide the capacity of transport.

To be compatible with a DATA traffic, the coupling of a cross-connect with a packet router is, even today, the more pragmatic approach. This is a subject of strong interest for products that are used at present.

However, this solution is not really cost-effective because only two alternatives can be adopted:

- All the wavelengths are connected to the packet router, and in this case, the number of TX/RX dramatically increases the cost per port.
- Only a part of the wavelength is connected to the packet router, and in this case, the traffic matrix needs to be very stable. The deterministic approach for the number of connections becomes nonrealistic when the traffic profile becomes highly statistical.

3.5.4.3 OPTICAL PACKET SWITCHING TECHNIQUES TO COPE WITH A TRAFFIC PROFILE EVOLUTION

In this case, what could be the benefit of the concept proposed?

- The packetization at the edge level with an optimized size can reduce the latency in the creation of the packets. A second technique to accelerate the filling ratio is in the upgrade of the best effort in a premium class of service. Therefore, the advantage of this solution is that latency can be controlled to reduce the latency in the rest of the network. Calculations show that there is a global benefit in terms of end-to-end latency.
- The exploitation of the WDM dimension once again reduces the latency. The packets cross the architecture, and they see only a transmission path, even if it is switched. No buffers are crossed so the resultant latency is minimum.

Figure 3.21 shows a table summarizing the performance in terms of packet loss rate established

Total load	CoS 1 PLR	CoS 2 PLR	BE no recir. PLR	BE recirc 16λ, PLR
0.5	<10e−6	<10e−6	~10e−04	<10e−06
0.7	<10e−6	<10e−6	0.012	<10e−06
0.8	<10e−6	<10e−6	0.03	<10e−04
0.9	<10e−6	~10e−5	0.065	0.007

Figure 3.21 Performance of the all-optical packet switching network concept.

in the frame of the Reseau Optique Multiservice (ROM) project. It appears that on the three class of services considered, the end-to-end performance can be obtained. From dimensioning issues, it appears that for the WAN, a sum of 30% for Class of Service 1 (CoS 1) and CoS 2, and a best effort (BE) lower than 80% is tolerated.

This demonstrates the viability of an all-optical concept and as a consequence the viability of the opto-electronic scenario.

3.6 INTRODUCTION ON THE MARKET: CRITERIA

3.6.1 Criteria of selection for a new technology

To select a technology to task is not easy, but we can draw some conclusions:

- Bit rate evolution at the user part creating a convergence of the bit rate in all the layers of the network and forcing a transfer of the traffic profile even in the backbone. This will create a need for high-flexible networks to cope with a traffic profile and not with a traffic matrix. Packet technique is today the only pragmatic solution with a coexistence of circuit switching techniques.
- The key bit rate is 10 Gbit s^{-1}. All the companies are focusing on 10 Gbit s^{-1} that will develop a volume to make this technology compact and cost-effective. This also reinforces the packet technique because the bit rate is now totally independent of the physical bit rate; the granularity is offered by the packet size and not by the bit rate of the wavelength.
- If a circuit switching technique is adopted today, it must be compatible with a migration toward packet switching.

3.6.2 Cost approach

For the cost approach, everything will depend on the aggregation efficiency. In the following, we have computed the relative cost of different approaches, comparing mainly packet switching and circuit switching.

3.6.2.1 METRO PART

If the average load of a wavelength is high enough, due to an efficient aggregation process, then circuit switching is probably viable. However, if the load is low, below 20%, even if the cost of switches are more expensive, the gain in statistical multiplexing creates a real opportunity for packet techniques making them less expensive than circuit switching techniques.

The main reason for this gain is probably the high cost of the wavelength due to expensive infrastructure costs, pushing all telecommunications companies to prefer an increase of the bit rate rather than an exploitation of the WDM dimension.

Therefore, the tendency is probably packet techniques to decorrelate the bit rate from the granularity of switching and high-bit rates to adopt the most cheap technology while providing the required capacity.

Figure 3.22 shows the areas where optical packet switching is better than circuit switching.

From the figure, the numbers 2, 4, 6, and 8 indicate the ratio in terms of cost per port (wavelength) between an optical packet switch and a cross-connect targeting the same size (256 × 256) and the same technology.

The load of a wavelength is the average load.

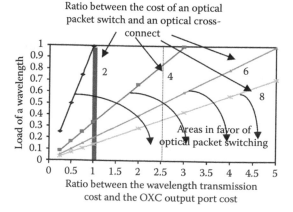

Figure 3.22 Need to exploit packet techniques in the near future.

The ratio on the horizontal axis is a ratio between the wavelength transmission cost (including the installation costs) and the cross-connect port.

For example, if the ratio between the cost of a wavelength in the transmission system and the cost of a cross-connect port is equal to 1 (red bar), optical packet switching techniques are interesting:

- 2: always, whatever the load is
- 4: if the average load of a wavelength in circuit switching is lower than 33%
- 6: if the average load of a wavelength in circuit switching is lower than 20%
- 8: if the average load of a wavelength in circuit switching is lower than 13%

Therefore, the tendency is the following: If the cost of a wavelength in a transmission system is high (the case of the metro where the installation cost is not negligible), or if the aggregation is not efficient enough forcing an average load very low (this is a serious tendency with the increase of the application bit rate and the sporadicity of the traffic profile), packet switching techniques always exhibit a better performance than circuit switching.

3.6.2.2 BACKBONE PART

As an example, we have computed, for two levels of aggregation, the load of a network with respect to the distance of the network (for the access to the backbone). It appears that in major cases, if the aggregation process is not enough, even in the backbone, the packet switching technique is the cost-effective solution.

Figure 3.23 shows the importance of packet switching techniques, also for the backbone. This is one of several curves that could be drawn. However, it once again shows a tendency.

The grooming or the aggregation efficiency depends on the traffic profile in large part. Therefore, we plot two indicative curves:

- One curve exhibits an efficient aggregation (a realistic case is when the constant bit rate (CBR) is higher than VBR or when the number of connection points is high to facilitate the grooming process).
- The other curve exhibits a less efficient aggregation (a realistic case is when the variable bit rate (VBR) becomes dominant). In that case, we cannot have a stable traffic matrix, and we

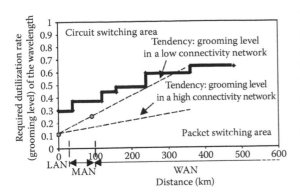

Figure 3.23 Curve showing strong interest to introduce packet switching techniques even in the backbone.

are addressing a sporadic traffic profile (a real-istic case if we have a bit rate convergence from the access to the backbone).

The vertical axis indicates that the required average load of a wavelength is cost-effective. The horizontal axis indicates the average distance of a transmission system with respect to an average network session representative of the network considered. Therefore, the WAN starts for transmissions higher than 100 km. The calculations show the importance of the time multiplexing. If the distance is long, there are a large number of clients sharing the same network infrastructure. Therefore, the cost per client is reduced. In addition, the cost of the installation of a wavelength is considered lower than that of the metro. The reason is that in the WAN, natural infrastructures are exploited to reduce the installation costs (such as highways or railways). If we are under the curve in bold, there is an advantage of introducing packet switching techniques. The grooming tendency gives values for the required load in circuit switching.

For example, in the case of a good grooming efficiency, if the average distance of propagation of a representative network session is lower than 300 km, you will have a cost gain by exploiting packet techniques. In the case of a low grooming efficiency, packet switching techniques are always more efficient.

3.7 CONCLUSIONS

In this chapter, we have suggested the optical switching technique as a potential technique for the next generation of systems or networks.

However, more importantly, an evolution scenario is given for the metro part and the backbone part describing what could be the most promising solutions. Optical packet switching techniques appear very attractive because they really offer a solution compliant with the traffic constraints.

Circuit switching techniques will be introduced as a first step, but we must not forget optical packet switching techniques that will improve the bandwidth utilization.

We have seen that there is no problem building any of the network concepts proposed, because all the functions have already been demonstrated to be feasible. The solution is now in the availability of the technology and in the cost. The progress on this integrated/low-cost optical technology will be fundamental for the future systems and could really provide new advantages with respect to classical solutions exploiting electronics only.

Today, we can imagine two scenarios:

The first one will consist of the introduction of a circuit switching platform to give a concrete answer to an immediate need at a lower cost. Circuit switching is probably the best today. However, we cannot forget the evolution of the traffic profile to increase the bit rate at the access part. Therefore, the migration scenario is an important argument to propose solutions that can be rapidly adapted to packet switching techniques with the best flexibility and upgradability.

The second scenario is in the adoption of packet switching techniques such as RPR for the metro or routers for the backbone. And we then need to think about competitive solutions with serious added values to justify the introduction of optical techniques in the network. Optical packet switching is probably one technique that can emerge. In the metro part, the benefit is mainly in the exploitation of the WDM dimension and in the very simple in-line processing (without any buffer) to reduce the latency and the number of Transceiver (TRX). In the backbone part, the benefit is probably in the adoption of large packets assimilated to containers in order to be able to exploit techniques to reshape and manage the traffic profile in the edge nodes and WDM techniques to reduce mainly the latency without constraining the capacity expansion in the core nodes.

However, to build subsystems, there is also a need for an advanced technology. Without any advanced technology such as tunable sources or tunable filters, there will be no chance to provide the functionality required to be really competitive on other aspects. Therefore, the development of this new technology (components and systems) is then fundamental and will position a constructor of equipment as a leader in the future market.

ACKNOWLEDGMENTS

The author acknowledges his colleagues from Alcatel, the European Commission, and the French ministry for funding for the following projects—RACE 2039 ATMOS, ACTS 043 KEOPS, REPEAT, IST DAVID, RNRT and ROM, particularly T. Atmaca from INT, M. Renaud from Opto+ who provided key results in terms of network performance and optical component illustration, and all the partners involved in these projects.

FURTHER READING

Blumenthal, D.J. et al. 1994. Photonic packet switches: architectures and experimental implementations. *Proc. IEEE*, 82, 1650–1667.

Bostica, B. et al. 1997. Synchronization issues in optical packet switched networks. In G. Prati (ed.), *Photonic Networks*. London: Springer, pp. 362–376.

Brackett, C.A. 1996. Is there an emerging consensus on WDM networking. *J. Light Technol.*, 14, 936–941.

Callegati, F. 1997a. Which packet length for a transparent optical network? *SPIE Symposium on Broadband Networking Technologies*, Dallas, TX, November 1997.

Callegati, F. 1997b. Efficiency of a novel transport format for transparent optical switching. *IEEE ICT 97*, Melbourne, Australia, April 1997.

Chiaroni, D. et al. 1997a. Feasibility assessment of a synchronization interface for photonic packet-switching systems. *ECOC'97*, Edinburgh, Scotland, September 1997.

Chiaroni, D. et al. 1997b. A 160 Gbit/s throughput photonic switch for fast packet switching systems. *Proceedings of Photonics in Switching'97*, paper PWB3. Stockholm, Sweden, April 2–4, pp. 37–40.

Chiaroni, D. et al. 1997c. Demonstration of full optical regeneration based on semiconductor optical amplifiers for large scale WDM networks. *Postdeadline ECOC'97*, Edinburgh, Scotland, September 1997.

Danielsen, S.L. et al. 1996. Bit error rate assessment of 40 Gbit/s all-optical polarization independent wavelength converter. *Electron. Lett.*, 32, 1688–1689.

Gabriagues, J.M. et al. 1995. Performance evaluation of a new photonic ATM switching architecture based on WDM. *Australian Telecommunication and Network Application Conference*, Sydney, Australia, December 1995.

Gambini, P. 1997. State of the art of photonic packet switched networks. In G. Prati (ed.), *Photonic Networks*. London: Springer, pp. 275–284.

Guillemot, C. et al. 1995. A two stage transparent packet switch architecture based on wavelength conversion. *ECOC'95*, vol. 2. Brussels, Belgium, September 17–21, 1995, pp. 765–768.

Guillemot, C. et al. 1998. Transparent optical packet switching: the European ACTS KEOPS project approach. *Special issue of J. Light Technol.*, vol. 16, December 1998, pp. 12.

Hansen, P.B. et al. 1997. 20 Gbit/s experimental demonstration of an all-optical WDM packet switch. *ECOC'97*, vol. 4. Edinburgh, Scotland, September 1997, pp. 13–16.

Hunziker, W. et al. 1995. Self-aligned flip chip packaging of tilted semiconductor optical amplifier arrays on Si motherboard. *Electron. Lett.*, 31, 488–490.

Janz, C. et al. 1998. Low-penalty 10Gbit/s operation of polarization-insensitive Mach–Zehnder wavelength converters based on bulk-tensile active material. *OFC'98* WB1, San Jose, CA, February 1998, pp. 101–102.

Mestric, R. et al. 1997. Up to 16 channel phased array wavelength demultiplexer on InP with -20 dB crosstalk. *ECIO '97*, paper EThE3, Stockholm, Sweden, 1997.

Misawa, A. et al. 1996. 40Gbit/s broadcast-and-select photonic ATM switch prototype with FDM output buffers. *22nd European Conference on Optical Communication, 1996. ECOC '96*, Oslo, Norway, September 19, 1996.

Renaud, M. et al. 1997. Network and system concepts for transparent optical packet switching. *IEEE Commun. Mag.*, 35, 96–102.

Zucchelli, L. et al. 1996. New solutions for optical packet delineation and synchronization in optical packet switched networks. *ECOC'96*, vol. 3. Oslo, Norway, September 15–19, 1996, pp. 301–304.

PART II

Enabling technologies for imaging and displays

4

Camera technology

KENKICHI TANIOKA
Kochi University of Technology

TAKAO ANDO
Shizuoka University

MASAYUKI SUGAWARA
Kochi University of Technology

4.1 THE CAMERA TUBE AND CAMERA

4.1.1 Introduction

The history of photoconductive camera tubes using the internal photoelectric effect began in 1950 with the Vidicon camera tube [1]. A photoconductive camera tube performs both photoelectric conversion and signal storage on a photoconductive target, which is a vapor-deposited film of Sb_2S_3 in Vidicon tubes. Although the Vidicon tube boasts a simple, small, and lightweight structure compared with the Image Orthicon tube that used the external photoelectric effect, it also suffers from several weak points such as large lag and dark current. The Vidicon, as a result, has not found much use in broadcasting-type television cameras that require high levels of picture quality. These weak points stem from the fact that excited carriers are easily trapped in Sb_2S_3-deposited film and that charge is injected into the target from external electrodes (injection-type target). However, the Plumbicon camera tube announced in 1963 features a target formed by a PbO film with a p–i–n structure that blocks the injection of charge from external electrodes (blocking-type target) [2]. With this type of target, the Plumbicon became the first photoconductive camera tube to feature low lag and low dark current among other superior features. In the 1970s, the Plumbicon rode the wave of change to color television broadcast facilities and became the leading tube for broadcast-class color television cameras replacing the Image Orthicon tube that used the external photoelectric effect. The research and development of photoconductive camera tubes was quite active. The 1970s, for example, saw the back-to-back development and commercialization of various blocking-type photoconductive camera tubes, including the Chalnicon using CdSe, $CdSeO_3$, and As_2S_3 as targets [3] and the Saticon using Se–As–Te [4,5].

In the 1990s, the solid-state imaging device such as the charge-coupled device (CCD) became the mainstream image sensor even for cameras used in the field of HDTV broadcasting. There is still a demand, though, for the camera tube, which was once the predominant type of image sensor in the form of the Plumbicon, Saticon, etc., for purposes of camera maintenance. As a consequence, modern camera tubes normally employ past technology.

An exception, however, is the high-gain avalanche rushing amorphous photoconductor (HARP) camera tube developed by the Japan Broadcasting Corporation (NHK) and Hitachi, Ltd, whose novel technology has become the focus of attention [6]. This camera tube achieves a level of sensitivity higher than that of CCDs and conventional pickup tubes by converting light to an electric signal in a photoconductive target and simultaneously amplifying that signal by an avalanche multiplication effect. It features higher quality pictures than past ultrahigh-sensitivity image sensors using image intensifiers. The HARP camera tube can be used in high-definition (HD) cameras and for a wide range of applications, including scientific and medical fields.

Section 4.1.2 describes the mechanism of the photoconductive camera tube covering past types and the ultrahigh-sensitivity HARP tube based on a new operating principle.

4.1.2 Basic configuration and operating principle of camera tubes

Camera tubes come in two main types: an image camera tube that uses an external photoelectric (photoemission) effect and a photoconductive camera tube that uses the photoconducting effect, a type of internal photoelectric effect. For the image camera tube, a typical example is the Image Orthicon developed during the monochrome television age. For photoconductive pickup tubes, there is the Vidicon, which can be called the original tube of this type, and tubes such as the Plumbicon and Saticon that played a great role in improving the performance of broadcast color cameras and achieving handheld video cameras. The HARP camera tube to be described here is also a photoconductive pickup tube.

The basic configuration of the photoconductive camera tube is shown in Figure 4.1. The tube consists of a photoconductive target that performs photoelectric conversion and charge storage, and a scanning electron beam system for reading stored charge. The operation of this tube is described later.

Referring to Figure 4.2, light incident on the target generates electron–hole pairs in the film (using, for example, a blocking-type target as described later). Here, for an ordinary pickup tube, the scanning electron beam irradiates the target at

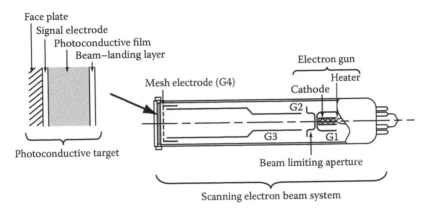

Figure 4.1 Basic structure of photoconductive camera tube.

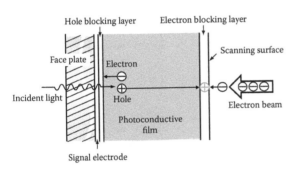

Figure 4.2 Behavior of charge in target.

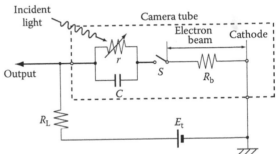

Figure 4.3 Equivalent circuit for one pixel in camera tube.

low velocity, and voltage is applied in such a way that the signal–electrode side takes on a positive potential with respect to the scanning surface. This causes electrons to move toward the transparent signal electrode and holes to move toward the target's scanning surface. The material used for the target's photoconductive film, however, generally has a high value of resistance with resistivity at 10^{12} Ω cm, and these charges accumulate at both ends of the target as a result. Meanwhile, the scanning electron beam system works to focus and deflect the electron beam emitted from the electron gun so as to make it incident on the target at low velocity (low-velocity beam landing). This causes stored holes to recombine and disappear and an equivalent amount of signal current to flow from the transparent signal electrode.

We can make this operation even easier to understand by focusing on a single pixel and using the equivalent circuit shown in Figure 4.3 consisting of resistors, a capacitor, and other elements. In the figure, the symbols r and C correspond to the resistor and capacitor making up the pixel. Here, the value of resistor r changes according to

the intensity of the incident light. In addition, R_b denotes the equivalent resistance of the scanning electron beam, S the scanning switch, R_L the load resistance provided externally, and E_t the power supply for applying voltage to the target. The arrival of the scanning electron beam on a certain pixel corresponds to the closing of switch S at which time C charges via R_b. Conversely, departure of the beam from that pixel corresponds to the opening of switch S at which time the charge accumulated in C discharges via r. The discharge period is determined by the number of frames per second and is 1/30 s (one frame's worth) in principle in the NTSC system. In actual camera tubes, however, the beam is broad compared to the scanning line interval and beam scanning overlaps the odd and even fields. As a consequence, the discharge period becomes 1/60 s, or one field's worth, despite interlaced scanning. Because the value of r changes according to the intensity of incident light, the amount of discharge is large for a bright subject and small for a dark one. This means that the charging current flowing to C from the power supply when S is closed

MM (electromagnetic–focusing/electromagnetic-deflection)

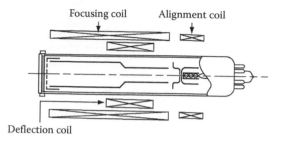

MS (electromagnetic-focusing/electrostatic-deflection)

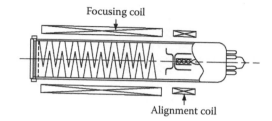

SM (electrostatic-focusing/electromagnetic-deflection)

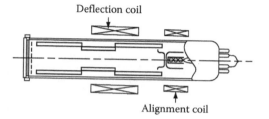

SS (electrostatic-focusing/electrostatic-deflection)

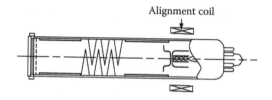

Figure 4.4 Focusing and deflection system of camera tube.

is equivalent to the current, i.e., signal current, modulated by the brightness of the subject.

4.1.2.1 TYPES OF SCANNING ELECTRON BEAM SYSTEMS AND THEIR FEATURES

The scanning electron beam system of the camera tube described here is classified in terms of electric-field/magnetic-field combinations used for focusing and deflecting the beam. There are the electromagnetic-focusing/electromagnetic-deflection (MM) type, the electrostatic-focusing/electromagnetic-deflection (SM) type, the electromagnetic-focusing/electrostatic-deflection (MS) type, and the electrostatic-focusing/electrostatic-deflection (SS) type as shown in Figure 4.4, respectively. Of these, the MS type features high resolution up to the corners of the screen in principle. The SS type, moreover, requires no coil for focusing and deflection and can therefore achieve a compact, light, and low-power configuration.

4.1.2.2 INJECTION AND BLOCKING TYPES OF TARGETS

Targets can be divided into injection type and blocking type. The structure of the injection-type target is such that charge comes to be injected into the photoconductive film from both the signal-electrode side and electron-beam-scanning side

or from either one of these sides. As a result, an amplification effect called "injection amplification" occurs within the target and high sensitivity with a quantum efficiency of 1 or greater can be obtained. (Quantum efficiency is defined here as the number of output electrons per number of unit incident photons in the target; it is denoted as η.) This injection amplification effect is described here using the target shown in Figure 4.5 in which electrons are injected from the beam scanning side. Now, one hole created by one incident photon will come to be stored on the beam scanning side as shown in the figure. Then, when the scanning electron beam comes to read this hole, the hole will not immediately recombine with an electron but will instead have to wait until N electrons are first injected into

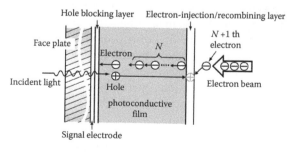

Figure 4.5 Operating principle of electron-injection-amplification target.

the target until it can recombine with the $(N+1)$ th electron. This means that $N+1$ electrons flow out of the target from the signal electrode. In other words, this operation provides an amplification effect with a gain of $N+1$ in which electrons greater than the number of incident photons can be read out to an external circuit. Before the invention of the HARP target, the injection-type target was researched as the only photoconductive target that could achieve high sensitivity of $\eta > 1$. This target, however, suffers from sharp increases in lag and dark currents under high-sensitivity operation and consequent drops in picture quality, and could not, as a result, be viewed as a new approach to camera tubes.

On the other hand, a blocking-type target has a structure in which both the signal–electrode side and electron-beam-scanning side block the injection of charge from the outside. As an example, Figure 4.6 shows a Saticon target whose main component is amorphous selenium (a-Se). Here, the injection of holes is blocked at the junction between the first layer consisting of Se + As (arsenic) film and the SnO_2 (tin oxide) signal electrode and CeO_2 (cerium oxide) film. The injection of electrons, meanwhile, is blocked by the fifth layer consisting of Sb_2S_3 (antimony trisulfide) on the beam scanning side. In addition, the second layer in the figure is a layer to increase sensitivity; layer 3 plays the role of conducting holes created in the first and second layers to the fourth layer; and the fourth layer acts to decrease storage capacitance in the target and reduce capacitive lag. A target such as this that blocks the injection of charge from external electrodes means that increase in dark current will be small even for an increase in applied voltage. A sufficient electric field (1.25×10^7 Vm^{-1} in

the Saticon) can therefore be given to the photoconductive film. As a result, most electron–hole pairs excited by incident light can be separated by a strong electric field to form a signal current, and a relatively high level of sensitivity can be obtained as a consequence. This kind of target also features little dark current and low lag. For the above reasons, targets for recently developed photoconductive camera tubes have been of the blocking type for which high picture quality can be obtained. In a target of this type, however, a scanning-beam electron immediately recombines with a hole stored on the scanning side and subsequently disappears, as shown in Figure 4.2. To put it another way, the number of scanning-beam electrons landing on the target per one hole is simply one, which in turn means that no more electrons than the number of incident photons can, in principle, be read to the outside, i.e., the limit of sensitivity is $\eta = 1$. In the past, this was referred to as the sensitivity barrier in blocking-type targets.

4.1.3 Ultrahigh sensitivity photoconductive camera tube

The more sensitive imaging devices are, the better they are able to produce clear pictures even in low lighting conditions. Consequently, achieving increased sensitivity has always been the most important theme throughout the more than 80-year history of research into imaging devices, and even today it is a matter of fierce competition between researchers.

From the 1960s to the 1980s, NHK researched and developed a variety of high-sensitivity imaging devices such as secondary electron conduction tubes and I-CCDs, which are made by combining image intensifiers with CCDs.

But because these conventional high sensitivity devices had problems associated with their picture quality, such as high noise levels and poor resolution, demand grew in the 1980s for the development of imaging devices that combine high sensitivity with high picture quality. During this period, HDTV camera using camera tubes such as Saticon began to be used in practical applications. However, their sensitivity was still rather poor, and as reporting breaking news programs and science programs began to attach increasing importance to camera sensitivity even for standard TV broadcasts, it became even more important to develop a

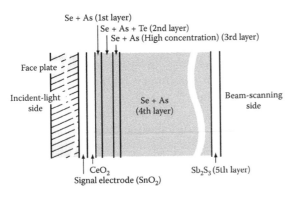

Figure 4.6 Structure of Saticon target.

TV camera with high sensitivity and high picture quality, capable of producing clear images even from poorly lit subjects.

Consequently, from about 1980, NHK began a fresh study with the aim of realizing a high sensitivity and high picture quality imaging device suitable for use in HDTV applications. This study focused on using the amplification effect of an a-Se to obtain a high level of sensitivity.

In 1985, it was found that when an a-Se target in the camera tube is operated in a strong electric field of about 10^8 Vm^{-1}, continuous and stable avalanche amplification takes place, allowing high sensitivity to be obtained with little picture degradation. Based on this discovery, NHK and Hitachi, Ltd, went on to develop a new kind of imaging device called HARP camera tube, which has been studied to this day to achieve further increases in sensitivity and a wider range of applications.

HARP camera tubes, which have achieved sensitivities roughly 100 times greater than CCDs, are used not only for standard TV broadcasts, but also in HDTV handheld cameras, and are used in the production of night-time news flashes and special programs such as imaging the aurora.

The following section summarizes the research conducted so far into HARP camera tubes and describes the features of ultrahigh sensitivity cameras that use them.

4.1.4 The development of HARP camera tubes

4.1.4.1 THE INVESTIGATION OF HIGH SENSITIVITY AND HIGH PICTURE QUALITY IMAGING DEVICES

The basic technique employed in conventional high-sensitivity imaging devices involves directing the incident light toward a photocathode and accelerating the photoelectrons emitted from this surface with a large voltage inside a vacuum. For example, in the I-CCD mentioned earlier, these accelerated electrons impinge on a fluorescent surface where they form a bright picture that is imaged using a CCD. Using a photocathode has the advantage that a high level of sensitivity can be obtained quite easily, so it has also been developed in various other types of high-sensitivity imaging devices, such as the silicon intensifier target tube [7].

But conventional high-sensitivity imaging devices that use a photocathode also suffer from drawbacks, such as the following:

1. Because they use the external photoelectric effect for photoelectric conversion, it is difficult to increase their conversion efficiency to values close to 100%. A low conversion efficiency results in increased picture quality degradation due to shot noise.
2. Picture quality can also be degraded by other forms of noise that are characteristic to the device, such as ion feedback noise that arises from residual gases inside the tube.
3. It is difficult to achieve the high resolution needed for HDTV cameras with a compact imaging device.

To address these problems, NHK decided to work on developing high sensitivity and high image quality imaging devices that do not rely on the use of a photocathode and to investigate the possibility of achieving substantial increases in the sensitivity of the photoconductive target in Saticon tubes that were also used in HDTV cameras.

In the mid-1980s when this investigation got underway, the mainstream of imaging devices had started to shift from the camera tubes to solid-state devices (CCDs). Not only are CCDs compact, lightweight, easy to use, and highly reliable, but it is also possible to suppress the noise from their internal amplifier circuits to a much lower level than can be achieved with the external amplifiers used with camera tubes. CCDs, therefore, seemed to have greater potential than camera tubes in terms of sensitivity.

Nevertheless, NHK decided to take a fresh look at photoconductive camera tubes because it was considered the targets in these camera tubes to have the best potential for meeting the conditions necessary for realizing the ultimate ultrahigh-sensitivity imaging device, i.e., the conditions for obtaining a high S/N ratio at the theoretical limit.

To achieve the ultimate ultrahigh-sensitivity imaging device, the following three conditions had to be met:

1. All of the incident photons must be guided to the photoelectric conversion part (100% fill factor).

2. All the photons must be converted into electrons in the photoelectric conversion part (100% photoelectric conversion efficiency).
3. It must be possible to amplify the converted electronic signal without adding any noise.

A camera tube has a fill factor of 100% and thus satisfies condition (1). Also, because the target uses the internal photoelectric effect, it is also easier to increase the photoelectric conversion efficiency than in imaging devices based on the external photoelectric effect, such as image devices using intensifiers. In other words, it can also satisfy condition (2). Accordingly, if a way can be found to satisfy condition (3), then it will be possible to obtain imaging devices with unparalleled sensitivity and picture quality. To achieve this, it is first necessary to bring about some form of amplification within the target. Based on this reasoning, NHK began to research targets with in-built amplification capabilities, as described later.

4.1.4.2 TARGET TYPES AND AMPLIFICATION EFFECTS

As described earlier, targets can be classified into two types: an injection type where electrical charge is injected into the film from outside and a blocking type where the injection of electrical charge is blocked. In the injection type, an amplification effect is obtained whereby the external circuitry extracts a greater number of electrons than the number of photons incident on the target. Although blocking types result in good picture quality with low lag and low dark current, it has not been possible to produce amplification effects in such targets. We, therefore, concentrated our studies on injection-type targets.

However—to cut a long story short—the HARP camera tube target is actually a blocking type, not an injection type. Injection-type targets suffer from drawbacks such as a susceptibility to increased dark current and the amplification of lag by the same gain. However, the reason why we applied ourselves to the study of injection-type targets is because at that time there were thought to be no other ways of conferring amplification properties to the target.

Although NHK's research initially focused on injection-type targets for these reasons, in 1985, an unusual experiment was conducted involving making a blocking-type target behave like an injection-type target by forcibly applying a large voltage. This led to the discovery of a phenomenon whereby the sensitivity is increased within the photoelectric conversion film in a manner that could not be explained in terms of charge injection. This was the starting point for the development of the HARP camera tube. The experiment is described below.

4.1.4.3 EXPERIMENTAL DETAILS

At the time, NHK was working on an injection-type target using a-Se, which can produce an amplification effect with a relatively weak electric field (5×10^6 Vm^{-1}). A weak electric field reduces the efficiency with which photons are converted into electrons, and thus gives rise to problems such as increased shot noise.

It was deduced that this noise could be reduced by subjecting the target to a suitably strong electric field before injecting the charge, so the target was initially designed for an experiment to inject electrons as shown in Figure 4.5 by forcibly applying a very high voltage to a blocking type target that has a structure in which the injection of charge is blocked. Figure 4.7 shows the structure of the prototype target we produced for this experiment. The photoconductive film in this target is a 2-μm-thick a-Se film formed by vacuum deposition (vacuum: 1.33×10^{-4} Pa). The target is of the blocking type. Like the Saticon, it blocks the injection of holes at the junction formed between the a-Se film and the transparent signal electrode (indium–tin oxide) and CeO$_2$ layers and blocks the injection of electrons through the use of an Sb$_2$S$_3$ layer. But, in contrast to the Saticon, it does not include high-concentration Te- and As-doped layers to concentrate an electric field near the signal–electrode interface, which means that even better hole-injection-blocking characteristics can be expected. Also, for the Sb$_2$S$_3$

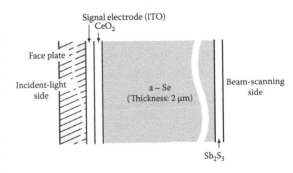

Figure 4.7 Structure of prototype target.

layer, inert-gas (Ar) pressure at the time of deposition was set to 31.9 Pa considering the porous-film fabrication conditions that would suppress the emission of secondary electrons even when target voltage is exceptionally high and promote stable low-velocity beam landing. The thicknesses of the CeO_2 and Sb_2S_3 films in the target are 20 and 100 nm, respectively, indicating that these two films are considerably thinner than the Se layer. Target thickness can, therefore, be regarded as essentially the same as that of the Se layer.

4.1.4.4 CURRENT–VOLTAGE CHARACTERISTICS

Figure 4.8 shows target current–voltage characteristics of the experimental tube with the prototype target. Blue light (center wavelength: 440 nm) is used here as incident light. From Figure 4.8, we see that signal current increases rapidly as target voltage increases from 0 V but comes to saturation, at least temporarily, starting at about 20 V. This saturation region is thought to correspond to the state where most electron–hole pairs excited in the a-Se film by incident light have come to separate under a strong electric field within the film becoming signal current as a result. As target voltage continues to increase; however, we see the phenomenon of signal current again rising dramatically beyond this saturated region.

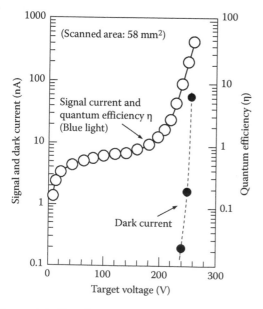

Figure 4.8 Signal current and dark current versus target voltage.

Quantum efficiency η with respect to blue light in a-Se film is estimated to be 0.9 for an operating electric field of 8×10^7 Vm^{-1} [8]. In the experimental tube, this electric-field strength corresponds to a target voltage of 160 V and this fact enables us to establish a scale for η on the right vertical axis in Figure 4.8.

This scale tells us that η exceeds 1 at a target voltage of 180 V and reaches 10 at 240 V. Furthermore, at a target voltage of 260 V, η is 40 and extremely high sensitivity occurs in the experimental tube. As for dark current, it also becomes large in the high-voltage region, but it is nevertheless quite small at 0.2 nA under target-voltage operating conditions of 240 V (η = 10). As described earlier, the phenomenon of increased sensitivity with η exceeding 1 has been observed when operating an a-Se photoconductive target with a blocking-type structure in a very strong electric field.

4.1.4.5 THE ORIGIN AND OPERATING PRINCIPLE OF THE HARP CAMERA TUBE

Because η was greater than 1 in the prototype target, it was thought that an amplifying action occurred due to the expected blocking type target behaving as an injection type. However, we found that this target exhibited hardly any dependence of lag on the applied voltage, which one would normally expect to see in an injection-type target. Specifically, according to the operating theory of injection-type targets, the effective storage capacitance of the film increases by an amount corresponding to the magnitude of the gain, so when the applied voltage exceeds 180 V the lag ought to increase steeply. However, such phenomenon was not seen in the prototype target. We, therefore, performed several new experiments. As a result, it became clear that the phenomena exhibited when η > 1 conform to the following properties:

1. The effective storage capacitance of the target is constant and does not increase even when η is greater than 1.
2. The amplification effect is dependent on the direction of the incident light, and compared with the face plate side, the degree of amplification is smaller when light is incident from the beam scanning side. That is, the target has a higher gain for hole transport than for electron transport.

3. When the electric field inside the target has a constant intensity, the amplification gain increases as the a-Se layer gets thicker.

Based on these findings, it was concluded that the amplification action obtained with the test targets is not due to the injection of charge but is due to an avalanche amplification effect that occurs stably and continuously in blocking-type targets for imaging devices. This marks the origin of HARP camera tubes that use the avalanche amplification phenomenon, and in this way HARP camera tubes were born out of research into completely different targets where charge injection activity is taken into consideration.

Figure 4.9 schematically illustrates the operating principle of this target. Electrons and holes produced by the incident light are accelerated inside the target, to which a strong electric field of about 10^8 Vm^{-1} is applied, and new electron–hole pairs are then generated successively by impact ionization. As a result, a large number of electrons are extracted from the signal electrode for each incident photon. The high sensitivity of HARP camera tubes is due to the avalanche multiplication effect in the a-Se target and the fact that this multiplication results in hardly any added noise. Furthermore, a HARP camera tube also has superior lag characteristics and resolution as mentioned later.

4.1.4.6 BASIC STRUCTURE OF TARGETS FOR PRACTICAL USE

Figure 4.10 shows the basic structure of a HARP camera tube target for practical use. Like the prototype target as shown in Figure 4.7, it uses layers of a-Se, CeO$_2$, and Sb$_2$S$_3$. However, the target for practical use also contains arsenic (As), lithium fluoride

(LiF), and tellurium (Te). The arsenic suppresses crystallization of the a-Se, thereby preventing the generation of defects. The lithium fluoride serves to control the electric field inside the target, and prevents the generation of defects by decreasing the electric field near the interface between the a-Se film and the CeO$_2$. The tellurium increases the target's sensitivity to red light and is added to the target for the red channel. The parts to which LiF and Te are added are exceedingly thin and are no more than a few per cent of the overall target film thickness.

4.1.5 The evolution of HARP camera tubes

When HARP camera tubes were first developed, they had a target film thickness of 2 μm and their sensitivity was about 10 times that of conventional Saticon camera tubes (Figure 4.11).

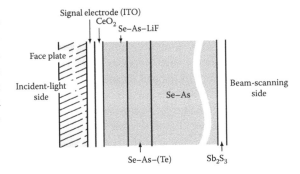

Figure 4.10 Structure of the HARP target.

Figure 4.11 Appearance of the ultrahigh-sensitivity HARP camera tube.

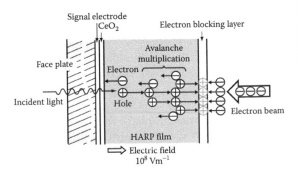

Figure 4.9 Operating principle of the HARP target.

Table 4.1 The evolution of HARP camera tubes

	Year in which developed		
	1985	1990	1995
Target film thickness (μm)	2	6–8	25
Sensitivity (relative to a Saticon tube)	About 10×60–80×	About 600×	
Lag (50 ms)	4.6%	1.5%–1.2%	Below measurable limit (theoretical value: 0.09%)

By taking advantage of the fact that these targets also have high resolution, HDTV camera tubes were developed, which were put to use at the Seoul Olympics [9]. But as they came to be used increasingly for TV programs, there was a demand for even higher sensitivity. Furthermore, because CCDs, which had by then become the most common imaging devices, have virtually no lag problems, there was also a demand for improving the lag characteristics of HARP camera tubes.

As can be seen from the operating principle shown in Figure 4.9, the avalanche multiplication factor of a HARP camera tube increases as the target gets thicker, resulting in greater sensitivity. The lag also decreases as the film thickness increases. This is due to reduction of the target storage capacitance, which dominates the lag characteristics. Consequently, by increasing the thickness of the layer consisting primarily of a-Se in the target to around 6–8 μm, it was able to develop a practical HARP camera tube with improved lag characteristics that had 60–80 times the sensitivity of a Saticon tube.

Furthermore, experiences such as news coverage of the Kobe earthquake disaster in Japan (1995) resulted in increased demand for the development of an ultrahigh-sensitivity imaging device capable of producing, for example, aerial nighttime shots of the stricken region, which had been plunged into darkness due to power failures. NHK, Hitachi, Ltd, and Hamamatsu Photonics K.K., therefore, studied ways of making the HARP camera tube even more sensitive, as described later, and developed an ultrahigh-sensitivity HARP camera tube with the target film thickness increased to 25 μm whose sensitivity is 600 times greater than that of a Saticon tube. The lag of this camera tube was reduced to a level below the measurable limit. Note that because modern CCDs are about six times as sensitive as Saticon tubes, the sensitivity of

this HARP camera tube is about 100 times greater than that of a CCD.

Table 4.1 shows how the target film thickness, sensitivity (relative to a Saticon tube), and lag (the value 50 ms after the incident light is cut off) of HARP camera tubes have changed over the years. In the following, characteristics of the HARP camera tube with a target film thickness of 25 μm, whose sensitivity exceeds that of the naked eye, will be mentioned.

4.1.6 Principal characteristics of the ultrahigh sensitivity HARP camera tube

This section describes the principal characteristics of the ultrahigh-sensitivity HARP camera tube with a target film thickness of 25 μm (2/3 in MM type, shown in Figure 4.4) [10,11].

4.1.6.1 SENSITIVITY

Figure 4.12 shows how the signal current (which represents the sensitivity) and dark current vary with the applied voltage. By way of comparison, this figure also shows the signal current measured using an ordinary (Saticon) camera tube subjected to the same amount of incident light. With an applied voltage of 2500 V, the HARP camera tube is over 600 times more sensitive than the Saticon tube. The dark current in this case is about 2 nA.

Note that since the HARP camera tube's sensitivity can be varied greatly by controlling the applied voltage, it can be adjusted to the sensitivity of ordinary imaging devices by reducing this voltage. In other words, it can also be used to take pictures in very bright situations such as day lit outdoor scenes.

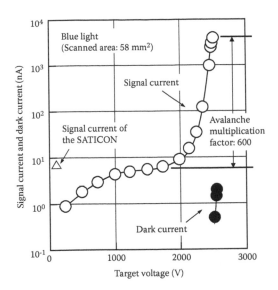

Figure 4.12 Signal current and dark current versus target voltage in the ultrahigh-sensitivity HARP camera tube.

4.1.6.2 SPECTRAL RESPONSE CHARACTERISTICS

Figure 4.13 shows the spectral response characteristics of an ordinary HARP target and a Te-doped target with increased sensitivity to red light (for use in the red channel). In the a-Se layer of the HARP

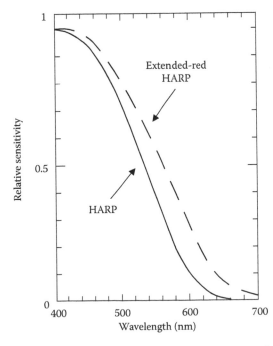

Figure 4.13 Spectral response characteristics of the ultrahigh-sensitivity HARP camera tube.

target, there is little cancellation of charge due to recombination, even close to the junction interface with the CeO_2, so a high photoelectric conversion efficiency is obtained for short-wavelength (blue) light that is absorbed in this part. On the other hand, because the band gap of a-Se is about 2.0 eV, the limit of sensitivity to light of longer wavelengths (red light) is the corresponding wavelength which is about 620 nm (standard type). Because the red channel of a color camera should exhibit sensitivity up to about 700 nm [12], the target for the red channel is made more sensitive to red light by doping it with Te.

4.1.6.3 LAG CHARACTERISTICS

The lag characteristics are determined by the storage capacitance of the film and the electron-beam temperature of the scanning electron beam. When the target film thickness is 25 μm, the theoretical value of the lag in the third field after turning off the incident light is 0.09%. This value is calculated from a target-layer storage capacitance of 130 pF and an electron-beam temperature of 3000 K. As shown in Table 4.1, the lag of the HARP camera tube is below the measurable limit when the target film thickness is 25 μm.

4.1.6.4 RESOLUTION CHARACTERISTICS

The HARP camera tube has a limiting resolution of more than 800 TV lines, and no degradation of resolution due to the avalanche multiplication action was observed. Because the resolution characteristics of the camera tube are controlled by the spot diameter of the scanning electron beam, even higher resolution can be obtained by combining with electron optics having a smaller beam spot diameter.

4.1.6.5 NOISE

The magnitude of the noise added as a result of avalanche multiplication is represented by the excess noise factor, but in the case of a HARP camera tube its value is approximately 1. In other words, the amplification achieved with this camera tube is almost noise free. The reason why this result is obtained is thought to be because of the large ratio of the respective ionization coefficients of holes and electrons in the a-Se film and because the beam scanning side of the photoelectric conversion film is in a floating state (its potential is not fixed), whereby a type of negative feedback action takes place which controls the noise.

In this way, the HARP camera tube combines the characteristics needed for high sensitivity with the characteristics needed for superior picture quality.

4.1.7 Ultrahigh sensitivity HARP cameras and their applications

A handheld HARP color camera has been produced using HARP tubes with a 25-μm-thick target. Its appearance is shown in Figure 4.14. Table 4.2 shows major specifications of the camera. This three-tube color camera with the target voltage of 2500 V has greater sensitivity than the naked human eye and can obtain clear color images even in lighting conditions equivalent to moonlight. Figure 4.15 shows an example of how the images taken with such a camera compare with the images taken by a CCD handheld camera under the same conditions (subject illumination: 0.3 lux, lens aperture: F1.7). The dark subject was difficult to view with the CCD camera even when its gain was boosted by 18 dB, whereas the HARP camera was

able to produce a clear image. It was confirmed by the color camera test that the HARP camera is about 100 times as sensitive as the CCD camera. It goes without saying that the sensitivity of the HARP camera can be decreased by decreasing the target voltage, so that the camera is capable of producing excellent picture quality over a wide range of shooting conditions from daylight to moonlight.

(a)

(b)

Figure 4.15 Monitor pictures produced by color cameras with the ultrahigh-sensitivity HARP tubes and CCDs. Illumination is 0.3 lux and lens irises are at F1.7. (a) Image taken with the HARP camera and (b) image taken with a CCD camera (+18 dB).

Figure 4.14 Appearance of the ultrahigh-sensitivity HARP color camera.

Table 4.2 Specifications of the ultrahigh-sensitive HARP camera

Maximum sensitivity	11 lux, F8
Minimum scence illumination	0.03 lux (F1.7, +24 dB)
Signal-to-noise ratio	60 dB
Resolution	700 TV lines
Amplifier gain selection	0, +9 dB, +24 dB
Weight	5 kg
Power consumption	25 W

In addition, the HARP tube offers other excellent features, such as insensitivity to burning, compared with the Saticon camera tube. It was also confirmed that the additional noise produced by the avalanche multiplication was negligibly small.

We have also developed a handheld HDTV HARP camera as shown in Figure 4.16. Figure 4.17

Figure 4.16 Appearance of the hand-held HDTV HARP camera.

(a)

(b)

Figure 4.17 Monitor pictures produced by HDTV cameras with HARP tubes and CCDs. (a) Image taken with the HDTV HARP (25-μm-thick) camera and (b) image taken with a HDTV CCD camera (+42 dB).

shows the results of comparing the HDTV HARP camera with an HDTV CCD handheld camera when viewing a night-time scene (lens aperture: F2). Here, the image from the CCD camera has been made brighter by boosting the gain by 42 dB. However, this has caused a corresponding increase in noise, and in the darker parts of the image (e.g., the trees in the park) the details are completely swamped by noise. On the other hand, the HARP camera produces a clear image with low noise. In this case, the difference in sensitivity manifests itself as a difference in the S/N ratio.

This camera was developed for use by the broadcast. Therefore, the HARP camera was not only used for emergency night-time broadcast, but was also used in the making of a wide variety of programs including subjects such as nocturnal animals, the aurora, and the rainbow at night in falls. However, the high sensitivity and superior picture quality of HARP cameras has also led to a considerable amount of interest from scientific and medical fields. This section describes how it is applied to deep-sea exploration and research into X-ray medical diagnosis.

In oceanic research involving studies of deep-sea organisms, mineral resources on the sea floor, and so on, underwater TV cameras are an indispensable means of gathering information. In particular, the TV cameras carried onboard unmanned deep-sea vehicles perform an essential role as the eyes of the remote control pilot. Consequently, the cameras used on such equipment must have superior characteristics such as highly sensitive and high picture quality. Artificial lighting equipment is essential for taking pictures deep underwater where no sunlight reaches. However, when developing an underwater camera it is important to bear in mind that light propagates very differently underwater because it attenuates much faster than in air. Red light attenuates particularly quickly, whereas blue light can penetrate somewhat further. This means that to obtain images of distant objects it is necessary to use a highly sensitive imaging device that can cope efficiently with blue light.

Because HARP camera tubes exhibit just such characteristics, in 2000, the Japan Agency for Marine-Earth Science and Technology developed an HDTV HARP camera for deep sea use, which was carried on the 3000 m class Hyper Dolphin unmanned deep-sea vehicle in conjunction with NHK. In test dives, this camera obtained clear

Figure 4.18 Deep-sea organism captured with the HARP camera.

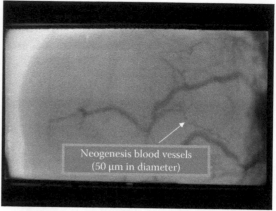

Neogenesis blood vessels
(50 μm in diameter)

Figure 4.19 Minute blood vessels (mouse cancer).

pictures of deep-sea organisms as shown in Figure 4.18 and was confirmed to be a highly sensitive HDTV camera.

We are also researching the use of HARP cameras in the medical applications. A notable example is the potential use of HARP cameras in next-generation X-ray medical diagnosis systems. This research was done in collaboration with other research institutions such as the National Cardiovascular Center Research Institute, the Tokai University School of Medicine, and the High Energy Accelerator Research Organization in Japan.

The X-ray equipment currently used in hospitals is only able to see large blood vessels with a diameter of at least 0.2–0.5 mm, but this study aims to make it possible to obtain clear images of blood vessels that are several times smaller. It has been said that if narrow blood vessels with a diameter of 0.1 mm or less can be imaged, then it should be possible to detect cancer earlier and make better diagnosis of conditions such as heart disease and cerebrovascular disorders.

For this purpose, it is necessary to have special X-rays that are absorbed well by a tiny quantity of contrast medium inside the narrow blood vessels to be imaged, and a TV camera that can clearly reproduce the image formed on a fluorescent screen (placed behind the subject being viewed) due to this absorption. For the special X-rays, we have been using monochromatic X-rays with a specific energy obtained from the synchrotron radiation. The TV camera is required to have superior sensitivity and resolution. This is because the image on the fluorescent plate is finely detailed and very dark (so as to restrict the exposure of the subject to X-rays).

We have, therefore, conducted experiments involving the use of an ultrahigh-sensitivity and high-resolution HDTV HARP camera in the imaging section of a next-generation X-ray medical analysis system. Figure 4.19 shows a photograph (obtained using this system) of tiny blood vessels called neogenesis blood vessels that developed in cancerous parts of a mouse. This image also shows narrow blood vessels of a characteristic shape with a diameter of about 50 μm, which it has not been possible to see hitherto. This technology is attracting interest as an X-ray diagnosis technique that can lead to the early detection of cancer.

In addition, because a HARP target can convert X-rays into electrons directly, it should be possible to exploit this capability to produce X-ray imaging devices with unparalleled levels of resolution and sensitivity. Consequently, this technology is attracting high levels of interest for applications such as the early detection of cancer and diagnosis of heart disease as a clinical diagnostic study.

4.1.8 Conclusion

This section has provided a description of the mechanism of the photoconductive camera tube covering past types and the ultrahigh-sensitivity HARP camera tube based on a new operating principle.

In general, ultrahigh-sensitivity imaging devices have so far been regarded as special-purpose devices, in a separate class to ordinary high picture quality imaging devices such as CCDs. But if we can develop the ultimate ultrahigh-sensitivity imaging device—with noise-free internal amplification and extremely high gain, and whose fill factor and photoelectric conversion efficiency are 100%—then we will have a device with ultrahigh picture quality and an S/N ratio close to the theoretical limit. Such a device will be able to take clear pictures with less noise than any other existing device under all lighting conditions. Although HARP camera tubes are coming very close to this ultimate goal, we would still like to make further improvements to the photoelectric conversion efficiency.

REFERENCES

1. Weimer, P. K., Forgue, S. V., and Goodrich, R. R. 1950. The Vidicon photoconductive camera tube. *Electronics* 23: 71–73.
2. De Hann, F. F., Vander Drift, A., and Schampers, P. P. M. 1963/64. The 'Plumbicon' a new television camera tube. *Philips Tech. Rev.* 25: 133–151.
3. Shimizu, K., Yoshida, O., Aihara, S., and Kiuchi, Y. 1971. Characteristics of experimental CdSe Vidicons. *IEEE Trans.* 18: 1058–1062.
4. Shidara, K., Goto, N., Maruyama, E., Hirai, T., and Nonaka, N. 1981. The advanced composition of SATICON photoconductive target. *IEEE Electron Device Lett.* 2: 101–102.
5. Maruyama, E. 1982. Amorphous build-in-field effect photoreceptors. *Japan. J. Appl. Phys.* 21: 213–223.
6. Tanioka, K., Yamazaki, J., Shidara, K., Taketoshi, K., Kawamura, T., Ishioka, S., and Takasaki, Y. 1987. An avalanche-mode amorphous selenium photoconductive layer for use as a camera tube target. *IEEE Electron Device Lett.* 8: 392–394.
7. Robinson, G. A. 1977. The silicon intensifier target tube. *SMPTE J.* 86: 414–418.
8. Pai, D. M. and Enck, R. C. 1975. Onsager mechanism of photogeneration in amorphous selenium. *Phys. Rev. B* 11: 5163–5174.
9. Okano, F., Kumada, J., and Tanioka, K. 1990. The HARP high-sensitivity handheld HDTV camera. *SMPTE J.* 99: 8.
10. Kubota, M., Kato, T., Suzuki, S., Maruyama, H., Shidara, K., Tanioka, K., Sameshima, K., Makishima, T., Tsuji, K., Hirai, T., and Yoshida, T. 1996. Ultrahigh-sensitivity new super-HARP camera. *IEEE Trans. Broadcast.* 42: 251–258.
11. Tanioka, K., Ohkawa, Y., Miyakawa, K., Suzuki, S., Takahata, T., Egami, N., Ogusu, K., Kobayashi, A., Hirai, T., and Kawai, T. 2001. Ultra-high-sensitivity new super-HARP pickup tube. IEEE Workshop on CCD and Advanced Image Sensors, June 7–9, pp. 216–219.
12. Neuhauser, R. G. 1987. Photoconductors utilized in TV camera tubes. SMPTE J. 96: 473–484.

5

Vacuum tube and plasma displays

MAKOTO MAEDA, TSUTAE SHINODA, AND HEIJU UCHIIKE
Home Network Company, SONY

5.1 VACUUM TUBE DEVICES

5.1.1 CRT structure and its operation

The cathode ray tube (CRT) is a luminescent display invented by K. F. Braun in 1897. The display, which is inexpensive but can show resolute pictures on its screen, has been the leading technology in the display field over more than 100 years. But the heavy weight and long depth are the CRT's weaknesses. The liquid crystal display (LCD) and other new flat panel displays that have no weaknesses as such are expanding their market share in recent years.

The CRT comes in several types—direct view, monochrome and projection. Based on the television broadcast specifications, the CRT with the screen's length-height ratio of 4:3 was the most common but the CRT with the ratio of 16:9 is now popularized. Even the 1:1 ratio CRT is produced for some special customers. The CRT screen was once round to intensify the glass strength but a series of recent technological innovations has made it possible to design more varieties of flat-screen-type CRT.

5.1.1.1 CRT OPERATION

Figure 5.1 shows the monochrome CRT structure. Video signal is fed to the cathode which is a part of the electron gun. The cathode generates free electron. The electron is focused by the electron gun like light is focused by a lens. Electrons travel freely in its evacuated inside, while glass is used to "envelope" the CRT. Electro-conductive substance is applied to the funnel's inside to form a film that keeps the inside's electric potential stable. The film builds as a high-voltage condenser in-between with another film made on the outside of the funnel. The condenser stabilizes the supply of anode voltage. Emitted by the electron gun and accelerated with the voltage of 20–30 kV, electrons travel fast in the form of a beam onto the panel coated with phosphor. The deflection yoke creates a magnetic field that bends the electron beams and makes them scan the entire panel. The electrons hit against the phosphor that emits light. Individual components and their function are described below.

5.1.1.2 ELECTRON GUN

Figure 5.2 shows the cross-section of the electron gun. This device is equipped with a cathode and operates in an electric lens system. The cathode is usually an oxide composition of barium (Ba), calcium (Ca) and strontium (Sr). Activated by heat at about 800°C, the cathode emits electrons. When varying picture signal voltage with an amplitude of about 100 V is applied to the cathode, the volume of electrons going through the grit 1 (G1) changes: the bigger the volume, the brighter the picture and vice versa. But to focus the dispersed electrons emitted from the cathode in large current and to create pictures on the panel, we must carefully design the structure and shape of the gun's electrode and the arrangement of applied voltage. Using 100 V signals, the electron gun can control electrons of 30 keV of energy. This function as a noise free amplifier is unique and not found in any other device of flat panel displays.

5.1.1.3 DEFLECTION

An electron beam put out from the gun travels straight toward the center of the CRT. To let it reach all over the screen, we usually use two methods—electrostatic deflection and electromagnetic deflection. The former method requires two flat-plate electrodes facing each other. The beam travels between the two. When electrical potential between them varies, the beam changes the direction. In spite of its lower deflection efficiency, the method is quite effective in deflecting high frequency. Meanwhile, the latter method of electromagnetic deflection characterized by its higher deflection efficiency is used for television and many other CRTs, and the deflection yoke performs electromagnetic deflection, using two pair of

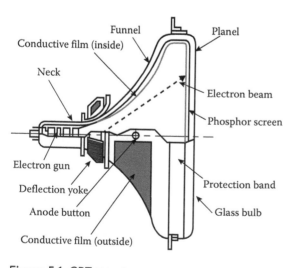

Figure 5.1 CRT structure.

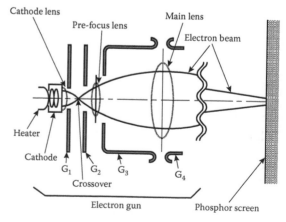

Figure 5.2 Electron gun cross-section.

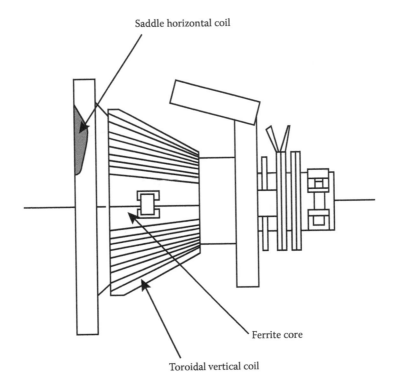

Saddle horizontal coil

Ferrite core

Toroidal vertical coil

Figure 5.3 Deflection yoke.

coils. As a pair of coils facing each other generates one magnetic field, the two pairs create two magnetic fields, directing the beam both horizontally and vertically. See the structure in Figure 5.3 for magnetic deflection yoke.

5.1.1.4 PHOSPHOR SCREEN

The inside of the CRT panel is coated with layers of phosphor particles. Each particle is 3–10 μm in diameter. The aluminium film covers and protects the phosphor. Gasses still remaining in the CRT would be hit by electrons and become ionized. Without the film, the ion would crash against the phosphor screen and damage the phosphor. This film is also effective in raising CRT luminance by reflecting light coming from the phosphor. It stabilizes electric potential around the screen as well.

5.1.2 Monochrome CRT

Monochrome rather than color CRTs are mainly used in the medical field where high resolution and high brightness pictures are required. The electromagnetic focusing method is applied to the electron gun to achieve high resolution. Electric current running through the coils attached around the CRT neck induces magnetic field and works as electro-magnetic lens. Being put outside the neck, the "lens" has a very small aberration, making the beam spot extremely tiny.

5.1.3 Projection CRT

The color picture projection CRT is a combination of three monochrome CRTs in red, green and blue, respectively. Pictures on each of the three single CRT panels are expanded through each optical lens and projected onto the outside screen where the pictures are combined. The system is depicted in Figure 5.4. The phosphor screen of the single CRT is usually 7–9 in. in diagonal. The size of projected pictures is as large as 40–60 in. in diagonal, and higher brightness (10,000 cd·m^{-2}) and higher resolution (the spot size is 0.2 mm in diameter) are requisite. If larger electric current is applied to illuminate phosphor, however, the phosphor becomes extremely hot; CRT brightness is saturated; and phosphor quality degrades. Then, the CRT fails to operate. To prevent this problem, the CRT panel is

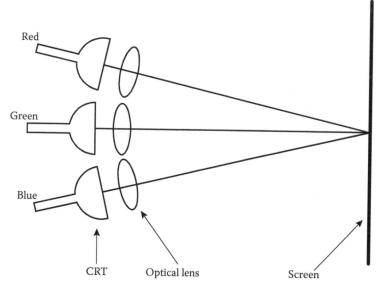

Figure 5.4 CRT projection system.

kept cool with coolant, while a kind of phosphor strong enough to bear such heavy load is applied.

5.1.4 Color CRT

5.1.4.1 PRINCIPLE

Phosphor in three different colors is applied to produce color pictures on the CRT screen. Figure 5.5 shows the mechanism for exciting phosphor. The color selection device (mask) is put in front of the phosphor-coated panel. Three pieces of cathode

in the electric gun put out electrons, which travel through the mask and hit onto red, green and blue phosphor and let it illuminate.

5.1.4.2 COLOR SELECTION MECHANISM

Among several types of color selection masks, the shadow mask and the aperture grille are widely applied to the CRT these days. The former is steel plate of 0.2 mm in thickness with round-or rectangular-shaped holes 0.2–1.0 mm apart. The mask's opening must be put in place to let a beam from one particular color cathode reach the same color phosphor. Hit by electron beams, the mask becomes hot and expands and the position of the opening moves. Therefore, the material Invar that has the very small thermal expansion coefficient is sometimes used for the mask. To match more varieties of flat-faced CRT designed these days, new shadow masks are being developed.

The aperture grille has a shape of a vertical reed screen. Both the shadow mask and the aperture grille have light transmittance of about 20%. This means 80% of the electrons out of the gun do not go through the mask. Figure 5.6 shows the aperture grille system and Figure 5.7 shows the shadow mask system.

5.1.4.3 PHOSPHOR SCREEN

Slurry that contains phosphor and photo sensitizer is applied to the panel to create the phosphor

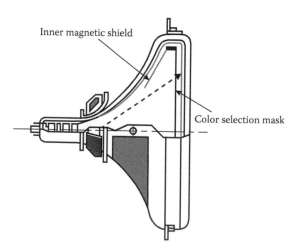

Figure 5.5 Color CRT structure.

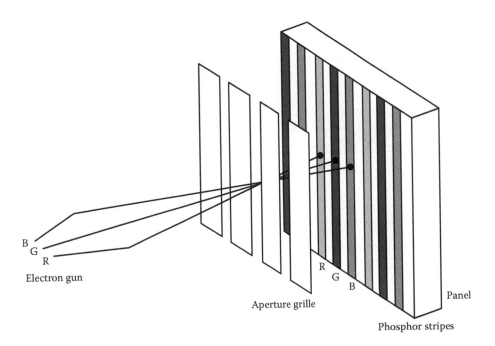

Figure 5.6 Aperture grille system.

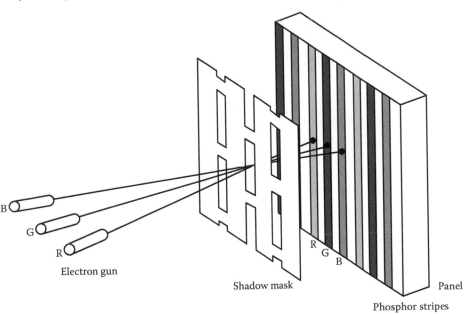

Figure 5.7 Shadow mask system.

screen. After being dried, the coated slurry is exposed to ultraviolet (UV) radiation light emitted from the lamp. When the panel is washed with water, the material except the light-exposed and hardened phosphor flows away. After this process is repeated three times, red, green and blue phosphor stripes appear; black stripes between each

different color phosphor are also created by a similar method. Figure 5.8 illustrates the cross-section of the phosphor screen.

5.1.4.4 ELECTRON GUN

The modern electron gun for a color picture tube is quite complicated compared to the basic structure

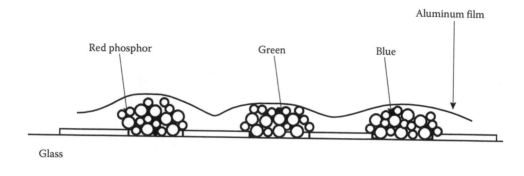

Red phosphor   Green   Blue   Aluminum film

Glass

Figure 5.8 Phosphor screen cross-section.

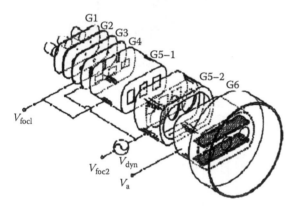

Figure 5.9 Electron gun for color CRT by Wada et al. [6].

shown in Figure 5.2. Figure 5.9 depicts an example of the electron gun for color CRT [6]. As one gun is equipped with one piece of cathode and its electrons are designated to illuminate the corresponding color phosphor, the color CRT generally requires three electron guns to illuminate three colors to produce color pictures.

The trinitron gun is equipped with one large electric lens for three electron beams. A large lens generally shows better performance than a small lens.

5.1.4.5 DEFLECTION YOKE

Unlike the monochrome CRT, the color CRT needs to deflect three electron beams at once. Those beams are required to converge into every spot spread all over the phosphor screen in order to realize quality pictures. The deflection yoke helps distribute needed magnetic field.

5.1.4.6 PURITY

The CRT is designed for electron beams to travel through the mask and hit the designated phosphor. But this is disturbed when the path is affected by terrestrial magnetism. To prevent this problem, the CRT is equipped with a magnetic shield. When the CRT power switch is turned on, attenuating alternating current runs through the coil installed around the CRT to produce attenuating magnetic field. This "degauss" process is designed to magnetize the shield in the intensity opposite to that of the outside magnetic field and to alleviate the influence.

5.1.5 Contrast

Several measures are taken to prevent ambient light reflection off the surface of the panel and the phosphor screen and to maintain good picture contrast. For instance, the swath between color stripes (dots) is blackened to halve the light reflection without blocking phosphor light.

Glass of lower transmittance is applied to the CRT panel. Incident rays come through the panel glass, reflect against the phosphor screen and go out through the glass again. Glass of low light transmittance can reduce light reflected through the two passages substantially. Phosphor light travels through the glass but the passage is only once. As the lower the glass transmittance, the better the picture contrast, the transmittance rate of about 50% is the most favourable (see Figure 5.10).

However, 4% of incoming light is still reflected at the surface of the panel. Computer display CRTs are equipped with glass whose surface is treated in the non-reflection process.

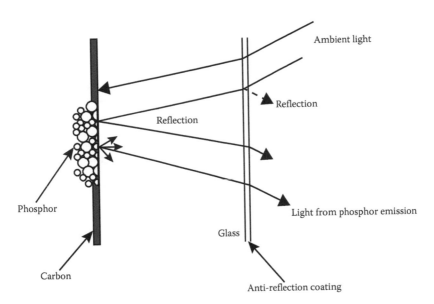

Figure 5.10 Light from phosphor emission and ambient light reflection.

5.1.6 Safety

Glass is used for the CRT and air in its inside is exhausted. Being exposed to the atmospheric pressure, even a small fault of the CRT could lead to a dangerous implosion, dispersing glass pieces all around. A metal band is applied around the CRT to prevent such accidents. The band that is a little smaller in circumference than the CRT reinforces the glass strength by canceling out glass stress caused by the atmospheric pressure. The band prevents damage even if the glass is broken.

5.1.7 Other CRTs

5.1.7.1 FLAT CRT

A big challenge for the CRT is how to shorten the long depth. As the deflection yoke bends electron beams, the CRT depth becomes shorter if the yoke's deflection angle is wider. Most TV CRTs these days have the yoke with the angle of 110° and new CRT models even wider 120°. But the larger deflection angle requires much more deflection power to bend the beam more sharply. The magnetic field gets distorted and the beams running through it end up producing poor pictures.

Maeda [7] designed a thin monochrome CRT (Figure 5.11). So far, no color CRT adopting this technology has been sent to the market.

Unlike the traditional CRT, the field emission display (FED) is shorter in depth. It has no deflection device. Now, engineers are intensifying R&D effort to produce this type of product too.

5.1.7.2 BEAM INDEX CRT

Only 20% of electrons are transmitted through the color selection mask and the rest are absorbed into the mask. Many years ago, the color CRT without the mask to raise power efficiency was developed and sold in the market. This CRT is designed to make one electron beam scan all over the phosphor screen. The beam should have a very small beam spot size not to strike more than one stripe of phosphor. That stripe stands next to another stripe of phosphor, which emits none of the three colors but UV as a signal when the beam moves onto it. Detecting that signal and finding its own location, the beam immediately changes its picture color information signal. The problem of this system is that the beam spot size becomes bigger when high beam current for high luminance is generated. CRT engineers want to solve this problem and succeed in developing this CRT some day. Their effort will continue until that day (see Figure 5.12).

5.1.8 Recent developments

Efforts to improve performance and cost reduction are continuing. Some of the recent developments are described below.

Okano developed a 21-in. CRT having very high resolution. Aperture grille pitch at the screen

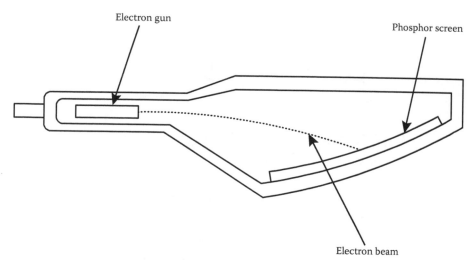

Figure 5.11 Flat monochrome CRT.

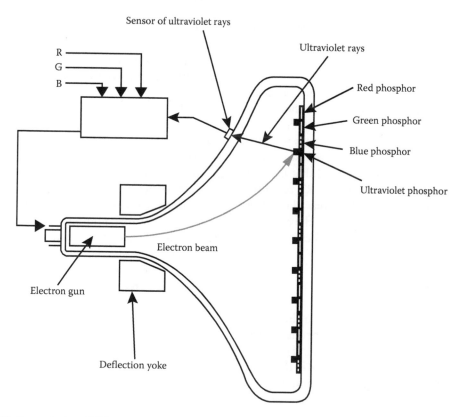

Figure 5.12 Beam index CRT.

center of CRT measures 0.126 mm. Horizontal resolution of 2800 dots was realized [8].

Beam index tube is a CRT which has no mask. A new idea was proposed by Bergman et al. [9]. Primary function of the mask, color selection, is taken over by an electronic control system. This CRT (called F!T tube) employs the system which has phosphor stripes parallel to electron beam scanning lines. Figure 5.13 shows the tracking principle of this new beam index tube.

One of the major problems of CRT is its weight. Most of the weight comes from the glass envelope.

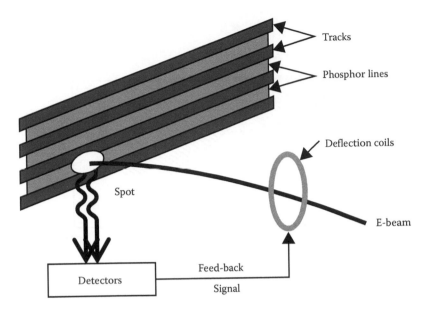

Figure 5.13 Tracking principle of F!T tube by Bergman et al. [9].

Sugawara [10] reduced funnel weight by redesigning the shape of the funnel shown in Figure 5.14.

The cathode is another component that needs to improve its performance. Oxide cathode is used for most CRTs. The current density from oxide cathode is limited. The new cathode called a hopping electron cathode (Figure 5.15) was proposed by van der Varrt et al. [11]. It is based on a self-regulation secondary emission process enabling transport of electrons over insulation surface. The cathode utilizes this mechanism to compress electrons coming from a large conventional cathode into a small funnel structure of insulating material. The exit of the funnel serves as a high-brightness electron source for a CRT and can be used to reduce the spot size.

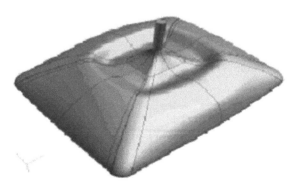

Figure 5.14 Novel funnel shape to reduce its weight by Sugawara et al. [10].

5.2 PLASMA DISPLAY

5.2.1 Introduction

Plasma displays have greatly advanced in the 1990s and are getting a position in the mainstream of the large area flat panel television and display. The road to the development was long and not peaceful in 30 years. Both successes of the color moving picture presentation on plasma display panels [12,13] and production of a 42-in. diagonal PDP [14] promise great business growth. The further developments of the interlaced and progressive displays from 32 to 60 in. HDTVs are making a new value-added market in addition to the market replacing the conventional TV of CRT because PDP has made a new large area and beautiful displays possible and is giving people a new impression.

5.2.2 Development of color plasma displays

There were two ways researched to present color image on PDPs. One was to use visible light caused by the discharge. The other was to use visible light from excited phosphors by UV rays or electrons in discharge. As a result of these researches, phosphor system with UV ray excitation has been applied to the recent color PDPs, because the system was superior to the other methods due to high luminance and color purification. In particular, the gas

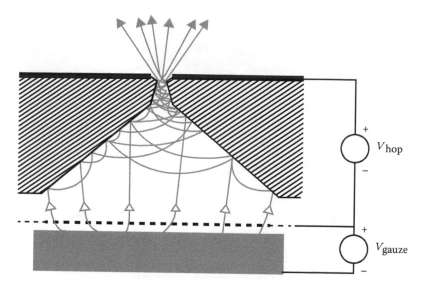

Figure 5.15 Hopping electron cathode by van der Vaart et al. [11].

system of Ne and Xe including 4–5% of Xe contents showed the excellent results to achieve high luminance and luminous efficiency. Vacuum UV rays of 147, 152 and 172 nm are radiated from Xe and Xe-dimers as shown in Figure 5.16.

At the early stage of color plasma display development, both AC and DC PDPs were carried out

to accomplish computer monitors and color television. Most of the color PDP researches to achieve color television were DC ones, because color AC PDPs had difficulty in attaining long lifetime due to the degradation of color phosphors caused by the ion sputtering [15]. Because phosphors in DC color PDPs were deposited around the anodes and

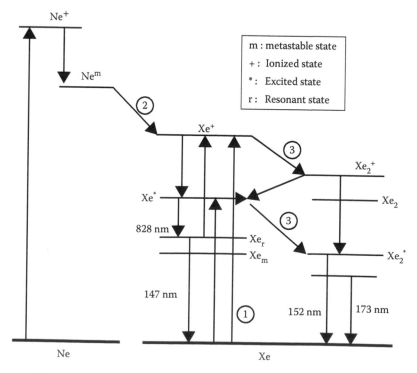

Figure 5.16 Energy level of transition for Ne+Xe mixture gas.

then no ion bombardment, color DC PDPs had an advantage of longer phosphor lifetime compared to that of AC ones at that stage.

Color DC PDPs for television were investigated by the application of negative glow and positive column region. Even though the color DC PDPs using positive column had higher luminous efficiency compared to those of negative glow region, the luminance and luminous efficiency were still not sufficient to be released into the market.

In 1983, NHK developed 16-in. diagonal color DC PDPs whose luminance and luminous efficacy were 21 cd·m^{-2} and 0.05 lm·W^{-1} respectively [16]. The performance of luminance and luminous efficacy by NHK was not able to cross the value of 150 cd·m^{-2} and 0.4 lm·W^{-1} respectively [17].

The practical panel structure for a color plasma display panel is called a three-electrode surface discharge belonging to a kind of AC-type plasma display [18]. AC PDP was invented by Bitzer and Slottow in 1966 [19] and the monochrome display has been put into practical use with opposed discharge technologies. The color PDP, however, did not succeed in practical use although there were many researches with the opposed discharge technologies. There were evolutional developments in structural and operational technologies from monochrome PDP to put the color AC PDP into practical use. The new direction to develop the color PDP was opened by introducing surface discharge technologies.

Table 5.1 shows the summary of the development of the color PDP technologies with surface discharge. Takashima [20] firstly reported the application of surface discharge technology on the color PDP for segment-type display in 1973. Dick [21] has also reported the surface discharge technology of a matrix-type monochrome PDP in 1974. The matrix-type color PDP with surface discharge technologies has been proposed and followed by authors from 1979 [22,23]. From the beginning of the research and development, the surface discharge color AC PDPs showed extremely excellent luminous efficacy performance of 0.75, 0.4 and 0.15 lm W^{-1} for green, red and blue phosphors, respectively [24]. Figure 5.17 shows the comparison between opposed discharge and surface discharge PDPs. The phosphors were deposited on the surface of the MgO layers on the opposed electrodes as shown in Figure 5.17a and alternated pulses were applied between the electrodes to ignite the discharge. Then the phosphors exposed to the discharge resulted in the rapid degradation due to the ion bombardment in the discharge. On the other hand, the phosphors were deposited on the front cover glass substrate placed away from the discharge area as shown in Figure 5.17b; therefore, the surface discharge color PDPs have got an advantage of long life. All of these surface discharge color PDPs employed two electrodes. The two-electrode surface discharge system, however, did not succeed in developing the practically available color PDPs.

The research has finally resulted in a new structure with the three electrodes as shown in Table 5.1 [17,25]. Shinoda invented both essential technologies such as the three-electrode structure and a new greyscale driving technology, or address-, display-period separation method (ADS method), which enabled the realization of practical color plasma displays. Figure 5.18 summarizes

Table 5.1 Development of color PDP technologies with surface discharges

Color technologies history							
Year	Researcher	Phosphors	Electrode	Electrode On	Type	Electrode Configuration	Ribs
1973	Takashima	Green	Two	Single substrate	Transmitting	Segment	Glass sheet
1974	Dick	Non	Two	Single	—	Matrix	Stripe
1979	Shinoda	RGB	Two	Single	Transmitting	Matrix	Non
1984	Shinoda	RGB	Three	Single	Transmitting	Matrix	Stripe+mesh
1985	Dick	Non	Three	Double	—	Matrix	
1987	Shinoda	RGB	Three	Double	Reflecting	Matrix	Stripe+mesh
1992	Shinoda	RGB	Three	Double	Reflecting	Matrix	Stripe

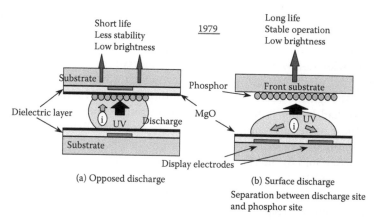

Figure 5.17 Comparison between surface discharge and opposed discharge. Phosphors are deposited on the front cover glass substrate placed away from the discharge area in the surface discharge.

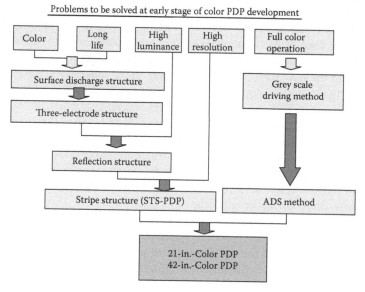

Figure 5.18 Summary of the technical issues to develop the surface discharge color PDP.

the technical issues to develop the color PDP, such as realizations of color PDP, long operating life, high luminance, high resolution and full color operation when the research for the surface discharge color PDP was started in 1979. Developing a three-electrode panel structure has solved these first four issues. And development of the new driving technology has solved the last one. Finally, the three-electrode PDP structure with stripe rib and phosphor structure has been completed and the practical 21-in. diagonal color PDP has been developed with these technologies in 1992 as shown in Figure 5.19. The larger 42-in. diagonal plasma display shown in Figure 5.20 was put onto the market in 1996, which started the era of plasma television.

5.2.3 Essentials of color AC PDP

5.2.3.1 OPERATING PRINCIPLE OF COLOR PDP

The color PDP is the display using the luminance from the combination of the phosphors and gas discharge. It is possible to think of a color PDP model as if some millions of miniature florescent lamps are arranged between the glass plates of an area of 1 m². Figure 5.21 shows the luminance model in the each lamp, which is usually called a cell. When the voltage is applied to the gas, discharge is ignited in each cell making ions and electrons from the atoms. The ions and electrons lose energy by emitting a UV ray. The plasma display is designed to

Item	Performance
Display area	422 mm × 316 mm
Aspect ratio	4:3
Number of pixels	640(R,G,B) × 480
Pixel pitch	0.66 mm × 0.66 mm
Number of colors	260,000
Luminance	180 cd m^{-2}
Viewing angle	>160˚
Power consumption	100 W$_{max}$
Weight	4.8 kg

(1992)

Figure 5.19 Specification of 21-in. diagonal color plasma display.

Item	Performance
Display area	920 mm × 518 mm
Aspect ratio	16:9
Number of pixels	852 (R,G,B) × 480
Pixel pitch	1.08 mm × 1.08 mm
Number of colors	16.7 million
Luminance	350 cd m^{-2}
Viewing angle	>160˚
Power consumption	300 W$_{max}$
Weight	18 Kg

(1996)

Figure 5.20 A 42-in. diagonal color plasma display.

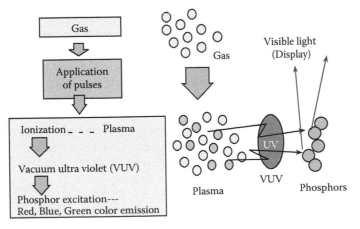

Figure 5.21 The luminance model of each cell in plasma display.

irradiate UV rays of Xe resonance emission (147 nm) and Xe molecular emission (173 nm). The irradiated UV rays stimulate the phosphors and visible lights are emitted. The three prime color phosphors are arranged in each of the discharge cells that are at the cross point of the electrodes.

To display a color image on PDP, the discharge needs to be controlled between the ON and OFF

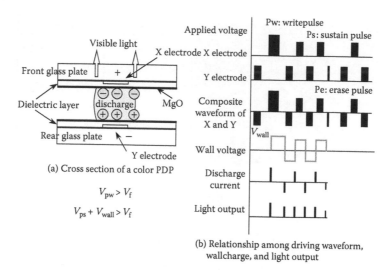

(a) Cross section of a color PDP

$$V_{pw} > V_f$$

$$V_{ps} + V_{wall} > V_f$$

(b) Relationship among driving waveform, wallcharge, and light output

Figure 5.22 The operating principle with the opposed discharge structure without phosphors.

states in each discharge cell (sub-pixel) that has the three prime color phosphors. The operating principle is simply explained with the opposed discharge structure without phosphors as shown in Figure 5.22. The electrodes are arranged orthogonally on each opposite glass plate and covered by a dielectric layer. The dielectric on the front substrate is also covered by an MgO protecting layer. The plates are assembled with a gap of about 100 μm and a Ne+Xe gas system is introduced between them.

The operating waveform is composed of the write pulses, the sustain pulses and the erase pulses. The write pulse whose voltage is higher than the firing voltage (V_f), i.e., a threshold voltage to ignite the discharge, is applied to the X electrode and then the discharge is ignited. The ions and electrons generated by the discharge are absorbed to the opposed dielectric surfaces by the electric fields. As a result, the internal electrical field of the cell is reduced rapidly by these absorbed charges and then the discharge is stopped. Therefore, the discharge has a pulse shape. The charge deposited on the dielectric is called a wall charge and the voltage through the capacitance between the surface of the dielectric and the electrode under the dielectric is called a dielectric voltage. The voltage due to the wall charge across the gas is called a wall voltage and the voltage across the gas due to both the wall voltage and externally applied voltage is called a cell voltage. The width of the write pulse is adjusted to accumulate a sufficient wall charge on the dielectrics. The sustain pulses are applied on Y electrodes successive to the write pulse. The

polarity of the voltage is reversed to the write pulse and the voltage is superimposed on the formerly accumulated wall voltage and then the cell voltage can exceed the firing voltage to ignite discharge, although the applied sustain voltage itself is lower than the firing voltage. The discharge also has a pulse shape similar to the discharge by the write pulse. In the same manner the successive sustain pulses sustain the discharge state and make display. The luminance of the display is proportional to the number of the sustain pulses and from 30,000 to 50,000 pulses per second are usually applied for sustaining. The narrower width than the sustain pulse is applied to erase the discharge. Although the discharge is ignited, the wall charge cannot be accumulated on the dielectric because of the narrow width of the erase pulse. The successive sustain pulses cannot sustain the discharge because the cell voltage is lower than the firing voltage.

In this manner the discharge state (ON state) and the non-discharge state (OFF state) are maintained. This phenomenon is called the memory effect of AC PDP.

5.2.3.2 FEATURES OF COLOR AC PLASMA DISPLAY PANEL

The developed color AC PDP has all the advantages of conventional monochrome AC PDP.

The following are the advantages: (1) good nonlinearity, (2) memory function, (3) high addressing speed, (4) wide viewing angle, (5) high luminance for large area and large display capacity, (6) high contrast ratio, (7) high greyscale, (8) full color,

(9) digital display, (10) flat panel, (11) simple structure and (12) large area.

The good nonlinearity means that the relationship between an applied voltage and the luminance has a clear nonlinearity. That is, when the applied voltage is increased, the discharge current is increased rapidly and then bright illumination begins at a certain voltage, and when the applied voltage is decreased, the discharge current decreases rapidly at a certain voltage and then illumination is eliminated. As the luminance levels are quite different between ON and OFF states, the high quality display with high contrast ratio is possible.

The memory effect is the function to maintain the ON and OFF states on the panel and is a great advantage for realizing a large area or a large display capacity. For example, in the case of the CRT without memory effect, the luminance decreases by increasing the display size or the display capacity because the electron beam stimulating the phosphors stays for a short duration at the phosphor surface of a display spot. In contrast, as all of the cells can be illuminated at the same time when the common sustain pulses are applied to the electrodes in the AC PDP, the high luminance level can be kept not depending on the display size or the display capacity.

The high speed addressing is due to the gas device. As the discharge is finished within 1 μs and the wall charge is accumulated within 2 μs after the pulse is applied, the data input is possible within 2 μs. There is a report that the data input is possible within 1.5 μs for one scan line. As a result, the PDP can display a beautiful moving image with 16.8 million colors.

The PDP is expected to play an important role in the future digital society because it is essentially a digital device. The digital ON and OFF states of the discharge cell are easily controlled by the digital signals of the computer and the network.

5.2.3.3 PRINCIPLE OF THREE-ELECTRODE PLASMA DISPLAY PANEL

Although the color PDP has been researched since the end of the 1960s when the AC PDP was invented, it was not successful with the two-electrode structure because of the phosphor and MgO degradation and a difficulty in the operation [26]. One of the key breakthroughs that solved the issues was the introduction of the surface discharge and three-electrode structure. Figure 5.23 shows the

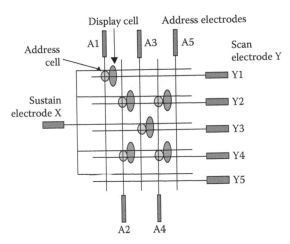

Figure 5.23 The principal electrode configuration of three electrode PDP.

principal electrode configuration and operation. There are three kinds of electrodes, such as two display electrodes (sustain and scan electrodes) and an address electrode. The sustain electrodes are connected to each other with a common electrode and the scan electrodes are independent. And then the one electrode terminal is added to the conventional two-electrode system. There are two kinds of operation method. One is write-in operation, which means the discharges are ignited to make wall charges between the selected address electrodes and a scan electrode for input display data and then sustain the discharge in the cells to be displayed by applying the sustain pulses between the display electrodes. Another one is erase operation, which means discharges are ignited to make the wall charges in all of the cells between display electrodes, scan and sustain electrodes, along a display line at once and then ignite discharges between the address electrodes and a scan electrode to erase the wall charges resulting in elimination of the discharges in the cells not to be displayed.

5.2.4 Practical panel structure and fabrication process

The first successful full color PDP has a diagonal size of 21 in. and the fabrication processes were completed while developing the panel. The new simple panel structure with a shape of stripe ribs and phosphors was developed to realize the fine pixel pitch (0.66×0.22 mm) for high resolution and was the most suited structure for mass production.

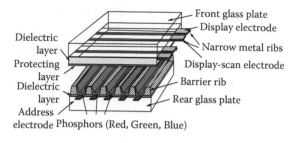

Figure 5.24 The practical panel structure of the three-electrode surface discharge PDP with the stripe structure.

Figure 5.24 shows the practical panel structure which is called the three-electrode surface discharge PDP with the stripe structure [27]. Paired parallel display electrodes, sustain electrode and scan electrode, are formed on the front glass substrate. Each display electrode is composed of a transparent ITO (indium–tin oxide) and a narrow bus electrode of multi-layered Cr, Cu and Cr, to emit a luminance effectively through the transparent electrode and reduce the electrode resistance. These electrodes are covered by a dielectric layer, which is made of low melting point glass materials. These are also covered with a thin MgO layer. On the other rear substrate, the striped address electrodes are arranged. Striped barrier ribs are on both sides of the address electrodes to separate the adjacent discharge cells and to eliminate the optical cross talk between them. Three primary color phosphor materials for red, blue and green colors are deposited in the neighboring channels made by the ribs to cover both on the side walls of the ribs and on the dielectric layer. The structure has realized good performances such as a high luminance, a high luminous efficacy and a wide viewing angle. Phosphor materials are $BaMgAl_{14}O_{23}$:Eu for blue, $(Y\times Ga)BO_3$:Eu for red and Zn_2SiO_4:Mn for green. The substrates are assembled to each other with about $150\,\mu m$ gap. A Ne+Xe gas mixture is introduced between the substrates. The panel structure developed for the 21-in. diagonal color PDP is the simplest one of conventionally researched color PDPs. And the fabrication process is also simple enough for mass production. So the PDP has advantages such as a low cost process, and an easiness in the manufacture of the large area panels and the high resolution panels.

The essential fabrication process as shown in Figure 5.25 is also completed to develop the 21-in. PDP. The transparent conductive ITO film is made on the front glass. The plural paired display electrodes are formed by a photolithography technology. The metal electrode film of a Cr/Cu/Cr multi-layer is sputtered on these transparent electrodes. The bus electrode is also formed by a photolithography technology. These electrodes are covered with a frit glass

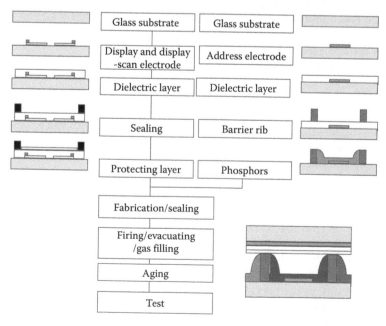

Figure 5.25 The essential fabrication process of color plasma display panel.

layer with a printing technology and then heated at about 600°C to make a transparent dielectric layer. The seal glass layer with a width of about 3 mm is made on the outside of the display area and then pre-heated. An MgO protecting layer is evaporated on the dielectric layer over the display area of inside of the seal layer. The front plate is completed with these processes.

A small hole of a diameter of about 1 mm is made on a corner of the rear plate. The Ag address electrodes are printed and heated. The frit glass is printed on the electrodes in the display area and then heated at about 600°C to make a dielectric layer. The barrier ribs are made by sandblasting the frit glass on both sides of the address electrode. The red, blue and green phosphors are printed inside the channel between the barrier ribs. Each color phosphor is printed at a same time and then the printing is repeated three times and then dried. The rear plate is completed with these processes.

Both plates are assembled and fixed with clips. The assembled plates are heated to melt the seal layer and the plates are glued resulting in a panel. At the next gas filling process, the panel is connected to an evacuation/gas-filling system through the evacuating tube. After the baking, the discharge gas is introduced. Finally, the PDP is completed after melting off the evacuating tube. The driving pulse is applied to the panel and discharges are ignited in every discharge cell to reduce and make the operating voltage stable.

5.2.5 Improvement of the cell structure

Fujitsu Hitachi Plasma Display (FHP) had reported on the alternate lighting of surfaces (ALIS) technology as shown in Figure 5.26 [28]. The ALIS method does not have a non-luminous area. Discharge takes place between adjacent display electrodes, instead of scan and sustain electrode pairs one by one. This system, however, does not permit line progressive scanning by upper and lower discharge cells with the use of shared cells. This interlaced scanning can be operated by the drive circuits, which are the same as those in existing PDPs with 480 scanning lines. It does not require special high speed addressing technology or dual scanning, which requires twice the usual number of ICs. It is possible to double its definition by using the same manufacturing and driving technology as that for the conventional method. In other words, they can apply conventional methods to turn a VGA color PDP into an SXGA PDP.

The ALIS system itself does not improve luminous efficacy. However, the adoption of the method makes a non-luminous area unnecessary and raises the aperture rate to 65%. As a result, ALIS improved the luminance by 150% of that of the conventional method. By adopting the ALIS system, FHP made a 32-in. color PDP with 852×1024 pixels capable of displaying high definition pictures.

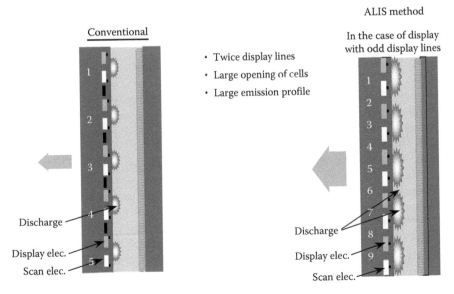

Figure 5.26 The alternate lighting of surfaces (ALIS) technology developed by FHP.

Method	Alternate	lightning surface method
Display size		922 mm (H) × 522 mm (V)
Number of cells		1024 (H) × 1024 (V)
Sub-pixel pitch		0.9 mm (H) × 0.51 mm (V)
Colors		16.77 million colours
Luminance		500 cd m^{-2}
Contrast ratio		250 : 1 (dark room)
Power consumption		250 W$_{max.}$

Figure 5.27 A 42-in. HDTV with ALIS technologies.

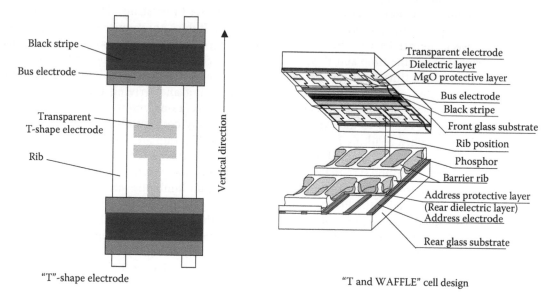

"T"-shape electrode

"T and WAFFLE" cell design

Figure 5.28 Three-electrode surface discharge plasma display with waffle rib and T-shaped electrode structure developed by Pioneer.

Figure 5.27 shows the 42-in. diagonal HDTV with ALIS technologies.

Pioneer devised PDPs with a high luminous efficacy and a high contrast ratio by adopting a T-shaped electrode structure as shown in Figure 5.28 [29]. The T-shaped electrode structure produces a favorable effect on luminous efficacy and contrast ratio.

In the conventional system, the ribs on the rear plate side, to which the RGB phosphors are applied, adopted a stripe structure. The waffle structure arranges these ribs in parallel crosses [30]. The waffle rib structure eliminates light leaks vertically, to reproduce sharp image contours. At the same time, the arrangement can widen the per cell area of applied phosphors. The adoption

of the T-shaped electrode structure and the waffle rib structure enables a luminance of 560 cd m^{-2}, which is 60% higher than that of the conventional system, raises luminous efficacy by 40% and produces a color PDP with optimum high resolution.

5.2.6 Gradation

The display is performed by controlling the two states, ON and OFF, in AC PDP. Changing the number of sustain pulses changes the luminance. The wall charge especially plays an important role to control the ON and OFF states in AC PDP and a sufficiently wide pulse width is needed to get a sufficiently wide operating margin. The ADS method is adopted for operating the AC PDP to meet the

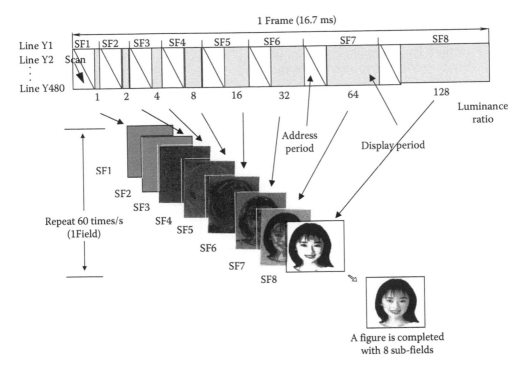

Figure 5.29 The ADS method, address-, display-period separation method, adopted for operating the AC PDP to realize an HDTV image with 256 greyscales.

requirement and realizes an HDTV image with 256 greyscales. Figure 5.29 shows the ADS method [31]. An image is constructed with 60 fields and each field is divided into eight sub-fields. Each of the eight sub-fields has different luminous level and the example of the luminance ratio for the eight sub-fields is 1:2:4:8:16:32:64:128 to realize the 256 greyscales. The luminance level is determined by setting the number of sustain pulses for each sub-field. The 256 greyscales are realized with the combination of the sub-fields in which the ON and OFF states are controlled by depending on the display data.

Each sub-field is divided into two periods, such as address period and display period, as shown in Figure 5.30. In the address period the discharge is ignited between selected address electrodes and a scan electrode sequentially depending on the display data. And then the wall charges are formed in the selected cells all over the display area and the successive sustain pulses are applied in the display period between all display electrodes. The width of the address pulse is less than 2 µs, and then the addressing speed is sufficiently high to realize the 256 greyscales for HDTV format.

The address period is further divided into the reset step and address step. The reset step is important for operating AC PDP. The wall charges accumulated in the previous display period are eliminated and then the pre-condition is made to ignite the discharge stable in the address step. Although the display period and address step are indispensably important, the reset step is

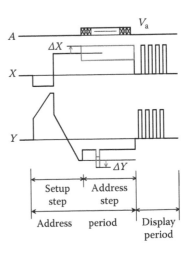

Figure 5.30 Detailed waveform for ADS method. Each sub-field is divided into two periods, such as address period and display period.

also important for stable addressing, high speed addressing and controlling the contrast.

The ADS method has advantages, such as an easiness in setting the width of sustain pulses, the number of sustain pulses, low power consumption, controllability of the greyscale and stable operation.

The ADS has been improving to realize a high quality display image and stable operation.

5.2.7 Improvement of drive systems

Characteristically, many color PDPs now adopt software-based improvements in addition to the hardware-based improvements such as the improvement of cell structures.

False contour issue was one of the essential issues to degrade the image quality when PDP adopts a driving method for greyscale using the luminous combination of different luminance sub-fields as shown in Figure 5.29. They have always been one of the most serious problems that degraded the display image on PDPs. To diminish the false contours, engineers previously resorted to methods that made false contours as invisible as possible. There are many methods to reduce the false contours, such as error diffusion, dithering, duplicated sub-field method, and so on [32]. With the combination of the methods, the false contour issue is improved as an acceptable level for the television application. Figure 5.31 shows the duplicated sub-field method.

Pioneer achieved a high contrast ratio of 560:1 and solved the problem of false contours by adopting a new ADS sub-field method called the Hi-Contrast and Low Energy Address and Reduction of False Contour Sequence (CLEAR) system, to achieve a Hi-Definition progressive display with excellent picture quality [33].

The CLEAR method resolved this problem by preventing, in principle, the generation of false contours by using the luminance accumulation of different luminance sub-fields as shown in Figure 5.32. Although one TV field is divided into sub-fields, which have different luminance in each the same as the conventional one, the reset period is only once. The discharges are ignited at once in all cells of the panel and form wall charges in the reset step. The wall charges are erased at the address period in the selected sub-fields according to the display data and then eliminate the sustaining discharge in the

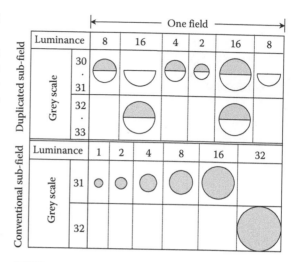

Figure 5.31 Suppression of the dynamic false contour by the duplicated sub-field method.

display period of the sub-field. So the luminance of unselected cells is accumulated from the first sub-field to the sub-field just advanced to the sub-field in which the erase pulses are applied. When the luminance is gradually varied, the light emission pattern of the sub-field does not change largely as in the case of the conventional sub-field method. Then the false contour issue is essentially solved. If the principle described above is applied simply, the grey levels are insufficient to display a beautiful image. To get more grey levels, the number of sustain pulses of the sub-field is changed in each TV field. When a TV field is composed of the m sub-fields and the sustain pulse number is changed between the n TV fields, an $[m \times n + 1]$-step greyscale is obtained. And dither method and error diffusion method are also applied; then, the CLEAR system yields a color PDP with the same gamma characteristic as that of CRTs. Consequently, a display capability of over 256-step grey levels for each RGB cell can be realized.

5.2.8 Future prospects of color PDP

One of the latest innovative developments is the delta arrangement PDP with meander barrier ribs reported by Fujitsu Labs [34]. This improves luminance and luminous efficacy by increasing the discharge cell size and also the area of phosphor application as shown in Figure 5.33. The luminance

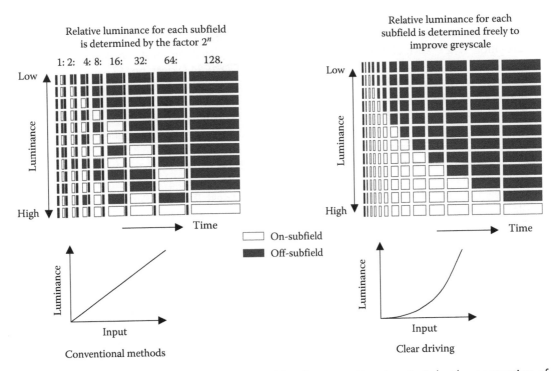

Conventional methods

Clear driving

Figure 5.32 The CLEAR method resolved the problem by preventing, in principle, the generation of false contours by using the luminance accumulation of different luminance sub-fields.

and luminous efficacy of this structure are about double those of the conventional method.

The next approach to improving performance was the idea of raising the concentration of Xe in the conventional rate of 4 or 5%. It is well known that this method did not attract much attention, because it did not solve the problem of the extra cost for the driver ICs that must handle the

additional driving voltage. Nevertheless, it certainly was effective in improving the luminous efficacy and luminance of PDPs.

REFERENCES

1. Piece, J. T. 1954. *Theory and Design of Electron Beams*. Princeton, NJ: van Nostrand.
2. Boekhorst, A., and Stolk, J. 1962. *Television Deflection Systems*. London: Cleaver-Hume.
3. Morrel, A. M., Law, H. B., Ramberg, E. G, and Herold, E. W. 1974. *Color Television Picture Tubes*. New York: Academic.
4. Tannas, L. E. Jr., 1985. *Flat-Panel Displays and CRTs*. New York: Van Nostrand Reinhold.
5. Hawkes, P. W. 2002. *Advances in Imaging and Electron Physics*, vol. 105. New York: Academic.
6. Wada, Y., and Daimon, T. 2001. An electron gun for 76 cm 120-degree 16.9 color TV tubes. *SID01 Digest* 32(1): 1116–1119.
7. Maeda, M., 2 inch flat CRT. *Japan Display 83, PD*.

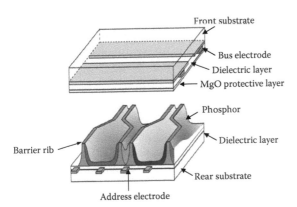

Figure 5.33 Delta arrangement PDP with meander barrier ribs. This improves luminance and luminous efficacy by increasing the discharge cell size and also the area of phosphor.

8. Okano, N., Maeda, M., Saita, K., and Horiuchi, Y. 1999. Development of ultra-high resolution 17/21" CRT. *SID99 Digest,* pp. 254–257.

9. Bergman, A. H., van den Brink, H. B., Budzelaar, F. P. M., Engelaar, P. J., Holtslag, A. H. M., Ijzerman, W. L., Krijn, M. P. C. M., van Lieshout, P. J. G., Notari, A., and Willemsen, O. H. The fast intelligent tracking (F!T) tube: A CRT without a shadow mask. *SID Digest* 12: 10–1213.

10. Sugawara, T., and Murakami, T. Status of glass bulb development for flat and thin CRTs. *SID02 Digest,* pp. 1218–1221

11. van der Vaart, N. C., van Gorkom, G. G. P., Hiddink, M. G. H., Niessen, E. M. J., Rademakerts, A. J. J., Rosink, J. J. W. M., Winters, R., and de Zwart, S. T. A nobel cathode for CRTs based on hopping electron transport. *SID02 Digest,* pp. 1392–1395.

12. Shinoda, T., Wakitani, M., and Yoshikawa, K. 1998. High level gray scale for AC plasma display panels using address-display-period-separated sub-field method. *Trans. IEICE* 100(3): 349–355 (in Japanese).

13. Yoshikawa, S., Kanazawa, Y., Wakitani, M., Shinoda, T., and Ohtsuka, A. 1992. Full color AC plasma display with 256 gray scale. *Japan Display 92:* 605–608.

14. Hirose, T., Kariya, K., Wakitani, M., Ohtsuka, A., and Shinoda, T. 1996. Performance features of a 42-in.-diagonal color plasma display. *SID 1996 Digest,* pp. 279–282.

15. Hoehn, H. J., and Martel, R. A. 1973. Recent developments on three-color plasma panels. *IEEE Trans. Electron Devices* 20: 1078–1081.

16. Kojima, T., Toyonaga, R., Sakai, T., Tajima, T., Sega, S., Kuriyama, T., Koike, J., and Murakami, H. 1979. Sixteen-inch gas-discharge display panel with 2-lines-at-a-time driving. *Proceedings of SID,* vol. 20, pp. 153–158.

17. Kurita, T., Yamamoto, T., Takano, Y., Ishii, K., Koura, T., Kokubun, H., Majima, K., Yamaguchi, K., and Murakami, H. 1998. Improvement of picture quality of 40-in.-diagonal HDTV plasma display. *Proceedings of IDW'96,* vol. 2, pp. 287–290.

18. Shinoda, T., and Niinuma, A. 1984. Logically addressable surface discharge ac plasma display panels with a new write electrode. *SID 1984 Digest,* pp. 172–175.

19. Bitzer, D. L., and Slottow, H. G. 1966. *AFIPS Conf. Proc.* 29: 541.

20. Takashima, K. et al. 1973. Surface discharge type plasma display panel. *SID 1973 Digest,* pp. 76–77.

21. Dick, G. W., 1974. Single substrate AC plasma display. *SID International Symposium, Digest Technical Papers,* pp. 124–125.

22. Shinoda, T. et al. 1980. Surface discharge color AC-plasma display panels. *Late News in Biennial Display Research Conference.*

23. Uchiike, H., Yamamoto, H., Niwa, A., Ohhashi, H., and Fukushima, Y. 1986. Mechanisms of 3-phase driving operation in surface-discharge AC-plasma display panels. *International Display Research Conference,* pp. 358–361

24. Shinoda, T., Miyashita, Y., Sugimoto, Y., and Yoshikawa, K. 1981. Characteristics of surface-discharge color AC-plasma display panels. *SID International Symposium Digest,* pp. 164–165.

25. Dick, G. W. 1985. Three-electrode per pel AC plasma display panel. *International Display Research Conference,* pp. 45–50.

26. Dedule, M. C., and Chodil, G. J. 1975. High-efficiency, high-luminance gas-discharge cells for TV display. *SID Symposium Digest,* pp. 56–57.

27. Shinoda, T., Wakitani, M., Nanto, T., Awaji, N., and Kanagu, S. 2000. Development panel structure for a high resolution 21-in.-diagonal color plasma display panel. *IEEE Trans. Electron Devices* 47(1): 77–81.

28. Kanazawa, Y., Ueda, T., Kuroki, S., Kariya, K., and Hirose, T. 1999. High-resolution interlaced addressing for plasma displays. *Proceedings of SID'99,* pp. 154–157.

29. Amemiya, K., and Nishio, T. 1997. Improvement of contrast ratio in co-planar structured AC-plasma display panels by confined discharge near the electrode gap. *Proceedings of IDW'97,* pp. 523–526.

30. Komaki, T., Taniguchi, H., and Amemiya, K. 1999. High luminance AC-PDPs with waffle-structured barrier ribs. *IDW'99 Digest,* pp. 587–590.

31. Shinoda, T. *United States Patent* 5, 541, 618.

32. Makino, T., Mochizuki, A., Tajima, M., Ueda, T., Ishida, K., and Kariya, K. 1995. Improvement of video image quality in plasma display panels by suppressing the unfavourable coloration effect with sufficient grey shades capability. *Proceedings of Asia Display*, pp. 381–384.

33. Tokunaga, T., Nakamura, H., Suzuki, M., and Saegusa, N. 1999. Development of new driving method for AC-PDPs *Proceedings of IDW'99*, pp. 787–790.

34. Toyoda, O. 1999. A high performance delta arrangement cell PDP with meander barrier ribs. *Proceedings of IDW'99 Digest*, pp. 599–602.

Liquid crystal displays

J. CLIFF JONES
University of Leeds

OBJECTIVE

Over the past half century, liquid crystal displays (LCDs) have grown to be one of the most successful optoelectronic technologies, becoming an integral part of communication devices and often an enabling technology. This success has required many adaptions to meet the requirements of ever-increasing complexity and performance. Indeed, it is the adaptability of liquid crystal devices that grounds their success. After outlining the basic physics of liquid crystals and device construction, the various modes used in commercial displays will be reviewed, both for mainstream and niche markets, together with a summary of the important complimentary technologies. Finally, liquid crystal devices that have promise for future applications in optoelectronics will be discussed.

6.1 INTRODUCTION

On May 28, 1968, the cinema audience at the Loews Capitol Movie Theatre on Broadway, New York, USA, would have been astounded by the vision of the future presented to them in Stanley Kubrick's masterpiece *2001: A Space Odyssey* [1]. Among the technological advances envisaged, from an International Space Station, video telephony to the omnipresence of computers, the viewers will have been no less captivated by the tablet computers that astronauts Dave Bowman (Keir Dullea) and Frank Poole (Gary Lockwood) apprise themselves with from the BBC12 podcast, after being awakened from hibernation. These devices had flat panel displays that are able to present full color, video information, something that would be as unfamiliar to the audience in 1968, as it is familiar to us now. Just one block away, at the Headquarters of *RCA* in the Rockefeller Plaza on 6th Avenue, George Heilmeier was at a press conference [2] to announce his recently published patent of the technology that would eventually lead to Kubrick's vision becoming reality: the first practical LCD [3,4].

Although the first device utilizing liquid crystal electro-optical effects dates back to 1934 [5], the era of LCD research had begun in earnest with

the groundbreaking inventions at *RCA* in the mid-1960s [2,6,7]. This work was driven by *RCA* CEO David Sarnoff's dream for "hang on the wall" television displays [6] and inspired by the work of George Gray [8] of Hull University in the UK, Glenn Brown at Kent State University in the US [9], and Richard Williams [10,11] at *RCA*'s Sarnoff Laboratories in Princeton. LCDs offered both reflective mode and backlit transmissive modes, in displays with a simple construction, light weight, and flat form factor. Display power consumption and operating voltages were attractively low and suited for being driven by Complementary metal-oxide-semiconductor (CMOS) circuitry.

Soon after Heilmeier's invention of the Dynamic Scattering Mode LCD [3,4], *RCA* terminated their LCD investment but the seed had been sown. Research groups from around the world, notably from Europe [12] and Japan [7,13], entered the field, stimulated by the *RCA* Press announcement. Having teamed up with Martin Schadt at Hoffman La Roche in Switzerland, ex-*RCA* researcher Wolfgang Helfrich invented the twisted nematic (TN) LCD in 1970 [14], a device with a more attractive optical appearance than its Dynamic Scattering Mode predecessor and a design that was to become the mainstay of flat panel displays for the following three decades. In the UK, Gray's group had formulated the first stable room temperature liquid crystal compounds [15] allowing mixtures with operating temperatures below zero degrees centigrade [16] to be formulated by Peter Raynes at the Royal Signal and Radar Establishment in Malvern, UK. In Japan, Tomio Wada at *Sharp* led a joint collaboration with Dainippon Ink and Chemicals that launched the world's first commercial LCD, incorporated in *Sharp*'s electronic calculator EL-805 in early 1973.

The promise of flat-screen, low-power, high-information content displays that met Sarnoff's goal for television displays seemed imminent [17]. Developments in the fabrication of LCD devices [7,12,13] led to early success in applications such as wristwatches and pocket calculators. However, it would take a further three decades of investment, invention, and development before the LCDs replaced the dominant cathode ray tube (CRT) displays used in televisions. Today's state-of-the-art ultrahigh-resolution LCDs for Ultra-high Definition Television (UHDTV) combines technologies such as thin-film transistors behind each color subpixel, optical compensation layers to give the widest angle of view, new alignment modes and addressing methods to achieve 120 Hz frame rates, liquid crystal mixtures that can operate from −20°C to +80°C to produce displays from several millimeters to over 100″ diagonal, at a price affordable for mass market adoption. Although it took this period to achieve Sarnoff's goal, the road map includes many consumer products that have been enabled by the emerging LCD: flat screen desktop monitors, laptop computers, mobile phones, and tablets. LCDs are the purveyors of the Internet age.

There is a plethora of different LCD types, each with properties that have been optimized for different markets. After reviewing some basics of liquid crystal science and display construction, a selection of typical LCD modes is described in this chapter, together with some of the more esoteric devices that meet the needs of niche markets or are yet to achieve commercial success.

6.2 LIQUID CRYSTALS AND THEIR DISPLAY RELEVANT PHYSICAL PROPERTIES

6.2.1 Orientational order

Although discovered by Friedrich Reinizer in 1888 [18], the true nature of liquid crystals, and the oxymoronic term *liquid crystal* itself, was described a year later by Otto Lehmann [19,20]. The type of liquid crystal used in all devices to date is formed from rigid rod-like (calamitic) organic compounds, such as those shown in Figure 6.1. There are a variety of liquid crystal phases that may appear between the

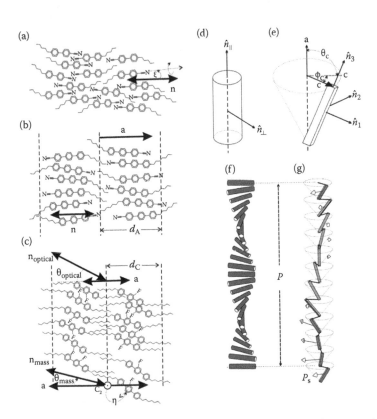

Figure 6.1 Some liquid crystals basics: (a) Schematic representation of (a) the nematic phase and the **n** director; (b) the smectic A phase; and (c) the smectic C phase. The principal axes for (d) the cylindrical nematic and smectic A phases and (e) the monoclinic smectic C. The spontaneous twist of the chiral nematic or cholesteric and definition of the helical pitch length P are shown in (f), and the spontaneous twist/bend of the chiral smectic C and ferroelectric spontaneous polarisation P_s are shown in (g).

isotropic liquid and solid crystal of certain organic compounds. Another, less common expression for liquid crystals is the term mesogenic, with molecules capable of forming liquid crystals being called mesogens, and the various phases termed mesophases. The simplest of the liquid crystal phase is the nematic (denoted N for shorthand, Figure 6.1a), which has no positional order, as with a conventional isotropic liquid, but the liquid crystal phase has long-range orientational order of the long molecular axes. The molecular axes tend to orientate in a common direction called the director, represented by the unit vector **n**. This orientational order leads to crystal-like anisotropy of physical properties such as the refractive indices, dielectric constants, and conductivities. However, the lack of positional order means that the nematic is fluid and can flow when perturbed. The statistical distribution of the molecular axes is quantified by order parameters. For a cylindrically symmetric phase, such as the nematic, which is composed of cylindrical molecules, the degree of order is defined using the S order parameter:

$$S = \frac{1}{2}\langle 3\cos^2\xi - 1\rangle, \qquad (6.1)$$

where the brackets <-> represent the statistical average over the coherence volume of the material and the deviation from the director of an individual long molecular axis is ξ. The anisotropic physical properties of the liquid crystal are related to S, which varies with temperature with the approximate form [21]:

$$S = \left(1 - \frac{T}{T_{NI}}\right)^{\nu}, \qquad (6.2)$$

where ν is a material constant, typically $0.15 \leq \nu \leq 0.2$, T_{NI} is the nematic to isotropic transition temperature, and T is the temperature below T_{NI}. Above T_{NI}, $S = 0$ by definition. This expression neglects the weakly first-order nature of the nematic to isotropic phase transition, which jumps immediately from 0 to about $S = 0.43$ at T_{NI} in actuality. However, Equation 6.2 is satisfactorily a couple of degree celsius below T_{NI}, predicting typical nematic materials to have $S \approx 0.60$–0.80 for typical nematic material at ambient temperatures.

A phial containing a nematic liquid crystal is milky in appearance. Unlike the common colloidal fluids with this type of behavior (such as milk itself), this appearance occurs for a pure, single compound. It arises because, although the liquid crystal molecules tend to point in the same direction locally, this direction changes over micron length scales due to thermal fluctuations, causing strong Tyndall scattering of the light. When heated to the isotropic liquid, the scattering disappears suddenly and the liquid is clear; for this reason, the temperature of the liquid crystal to isotropic transition (for example, T_{NI}) is often termed the *clearing point*.

Almost all LCDs sold are based on the simple nematic type of liquid crystal because it is the easiest to handle and most well understood. However, there are many other phases that can form that are intermediate between the liquid and crystal states, from those with higher degrees of order (layered smectics, helical cholesterics, and columnar phases), those using different shaped molecules (discs, banana-shaped, bowl-shaped, main-chain, and side-chain polymeric) to those systems where the liquid crystal nature is due to changes in concentration, rather than, or in addition to, temperature (lyotropic and chromonic phases). The interested reader is referred to Reference [22] for a comprehensive review on all aspects of liquid crystal science. For the present chapter, only the smectic and cholesteric phases are considered because of their (limited) application to displays. Smectic liquid crystal phases have some degree of positional order in addition to orientational order. The simplest smectics are termed smectic A and C (for which the shorthand SmA and SmC is used, respectively) and they exhibit one-dimensional positional order to form layers of nematic-like material. The layers of such phases are described by the unit vector **a**. SmA have **n** parallel to **a** (Figure 6.1b), whereas the director **n** is at an angle θ_C to **a** in the SmC phase (Figure 6.1c). Both the nematic and SmA phases have cylindrical symmetry about **n** (Figure 6.1d), whereas the SmC has monoclinic symmetry. Nematics comprised, at least in part, of chiral molecules undergo a spontaneous twist of the director to form a helical structure (Figure 6.1f). Chiral nematics, represented by N*, are often called cholesterics after the first liquid crystal behavior was observed for this phase in a compound derived from cholesteryl benzoate [18]. A full 2π rotation of the **n** director is defined as the pitch length, *P*, which may vary from submicron in some pure compounds, to several hundreds of microns in a nematic material doped with a small concentration of a cholesteric dopant. In the chiral smectic C (SmC*) phase (Figure 1g), the loss of mirror symmetry induced through the addition of chirality has two notable effects. First, the **n** director undergoes a spontaneous twist and bend deformation (so that the **c** director forms a twist in a similar way to the cholesteric), again with pitch *P*. Second, the combination of the loss of the mirror plane with the monoclinic symmetry of the SmC leads to a spontaneous ferroelectric polarization **P$_s$** in the plane of the layers and normal to the tilt plane [23]. As the **c**-director precesses about the layer normal in a helical fashion, the net ferroelectric polarization cancels out throughout the bulk of the phase because the **c**-director precesses about the layer normal in a helical fashion.

Liquid crystal phases are formed for molecules that comprise both a rigid core structure and pendant flexible hydrocarbon chains. The occurrence, temperatures, and phase sequences of liquid crystals depend on the relative proportion of the flexible chains and rigid core: the flexible groups are needed to space the cores sufficiently to prevent crystallization but sufficiently small to allow orientational ordering of the cores. Predicting the temperature range over which liquid crystal phases form is a complex subject, requiring skill and experience for the synthetic chemist. For the simple rod-shaped calamitic molecules used in commercial displays, the transition from isotropic to liquid crystal phase is related to the length of the molecule. However, viscosity is also dependent on molecular size and so most practical liquid crystal compounds are formed from two or three ring structures to keep viscosity low and switching speeds high. The ring structures in the core may be benzene, cyclohexane, pyrimidine, thiophene, etc., joined together with parasubstitution (e.g., 1,4 benzene) or with near-linear linking groups (such as esters, ethyl groups), and terminated on one or both sides with alkyl chains. In each case, the molecular core retains its near-cylindrical symmetry with each of any low-energy conformers allowed by the core, such as rotations about the ester or ethyl linkages. Table 6.1 lists a range of different compounds that are typical for nematic liquid crystal mixtures, together with some of the key physical parameters [24] related to typical device behavior.

Table 6.1 Typical nematic liquid crystal compounds and their physical properties

#	Compound Structure	Use	TNI (°C)	Δn (20°C, 589 nm)	Δε (20°C, 1 kHz)	K$_{11}$; K$_{22}$; K$_{33}$ (pN, 20°C)	γ$_1$ (Pa·s, 20°C)
1	C$_5$H$_{11}$—◯—◯—C≡N	TN	35 / 23	0.237	+21.6	7.4 / 4.3 / 10.4	0.112
2	C$_5$H$_{11}$—◯—(O)(F)—◯—C≡N	TN	(24) / 70	0.182	+50.2	k_{33}/k_{11} = 1.78	0.270
3	C$_3$H$_7$—◯—C≡C—◯—OC$_2$H$_5$	STN	95 / 88	0.300	+2.3	k_{33}/k_{11} = 1.38	0.054
4	C$_2$H$_7$—⬡—◯—C≡N	STN	74 / 68	0.17	+22.5	11.2 / 6.6 / 24.3	0.170
5	C$_3$H$_7$—⬡—⬡—◯(F,F,F)—F	TFT	94 / 65	0.073	+8.3	9 / 7 / 17	0.025
6	C$_3$H$_7$—⬡—◯—◯(F,F)—OC$_2$H$_5$	VA	145 / 67	0.156	−5.9		0.217

6.2.2 General anisotropic physical properties

All LCDs operate because of the combination of the anisotropic physical properties combined with fluidity exhibited by liquid crystals. Tensorial anisotropic physical properties such as the refractive indices and electric permittivities can be diagonalized to give principal values parallel $\left(\hat{n}_{\|}\right)$ and perpendicular $(\hat{n}_{\perp})$ to the director [25]. Most commercially available devices use reorientation of the director induced by an applied electric field E coupling to the dielectric tensor ε, to reduce the electrostatic free energy G_E, given by

$$G_E = -\frac{1}{2}D.E = -\frac{1}{2}\varepsilon_0\left(E.\varepsilon.E\right) = -\frac{1}{2}\varepsilon_0\Delta\varepsilon(n.E)^2,$$

(6.3)

where the dielectric anisotropy $\Delta\varepsilon$ is the difference of the principal components $\varepsilon_{\|}-\varepsilon_{\perp}$. Equation 6.3 predicts that the director tends to reorient parallel to the applied field if the material has a positive $\Delta\varepsilon$ and perpendicular to the field if $\Delta\varepsilon$ is negative. Similarly, many devices rely on changes in optical retardation from the material birefringence Δn $(= n_{\|}-n_{\perp} = n_e-n_o)$, where e and o refer to the extraordinary and ordinary rays, respectively, to give the perceived optical modulation. For example, if viewed between crossed polarizers, a device will appear isotropic, and therefore black, when viewed along the optic axis (parallel to n_e) and will exhibit some white light transmission when viewed in a direction at an angle to the optic axis due to the effect of the optical retardation.

The physical parameters depend on both chemical structure and temperature. Anisotropic properties, such as Δn and $\Delta\varepsilon$, are also related to the order parameter S, whereas the fluid properties, such as density, bulk viscosity, and the mean refractive indices and permittivities, are primarily related to absolute temperature. The uniaxial refractive indices are given by [25]

$$n_o^2 = 1 + \frac{\rho N_A hF}{M.\varepsilon_0}\left\{\bar{\alpha} - \frac{1}{3}\Delta\alpha S\right\},$$

$$n_e^2 = 1 + \frac{\rho N_A hF}{M.\varepsilon_0}\left\{\bar{\alpha} + \frac{2}{3}\Delta\alpha S\right\},$$

(6.4)

where ρ is the density, N_A is the Avogadro number, M is the molecular weight, $\bar{\alpha}$ is the mean molecular polarizability, and $\Delta\alpha$ is the difference between the molecular polarizability parallel to the long α_l and short α_t axes. The cavity field factor h and the Onsager reaction field F relate to the internal field experienced by an average molecule, approximated for a spherical cavity by

$$h = \frac{3\bar{\varepsilon}}{2\bar{\varepsilon}+1} \approx 1; \quad F = \left\{\frac{1}{1 - \frac{2}{3}\frac{\rho N_A\bar{\alpha}}{M.\varepsilon_0}\left(\frac{\bar{\varepsilon}-1}{2\bar{\varepsilon}+1}\right)}\right\} \approx 1,$$

(6.5)

where mean anisotropic properties are expressed using a bar above the symbol, such as for the mean polarizability on the molecular scale:

$$\bar{\alpha} = \frac{1}{3}\left(2\alpha_t + \alpha_l\right)$$

(6.6)

and permittivity and refractive index on the macroscopic scale:

$$\bar{\varepsilon} = \frac{1}{3}\left(2\varepsilon_{\perp} + \varepsilon_{\|}\right); \quad \bar{n} = \frac{1}{3}\left(2n_o + n_e\right)$$

(6.7)

The birefringence is given by

$$\Delta n = \frac{\rho N_A hF}{M.\varepsilon_0}\left\{\frac{\Delta\alpha}{3\bar{n} - n_o}\right\}S$$

(6.8)

and is largely dictated by the polarizability anisotropy $\Delta\alpha$ of the constituent molecules. The density ρ is usually close to that of water at 20°C, with a selection of nematic compounds $\rho = 990 \pm 50$ kgm^{-3} [26], and with a near linear temperature dependence through the liquid crystal (LC) phases that is typically 1 kgm^{-3}·K^{-1}. Thus, the temperature dependence of the birefringence is dominated by the order parameter S. A calamitic liquid crystal has a positive birefringence because $\Delta\alpha$ is positive for the cylindrical molecular core structures. If the rigid core is short and comprises weakly polarizable saturated moieties such as cyclohexanes, the birefringence will be low. It will be higher for longer, unsaturated groups such as phenyls and ethynes, reaching $\Delta n \approx 0.4$, but more typically being maximum at $\Delta n \approx 0.26$ for light stable compounds.

Similarly, the uniaxial electric permittivities are given by [25]

$$\varepsilon_\perp = n_o^2 + \frac{\rho N_A h F^2}{3M.\varepsilon_0 kT}\mu^2\left\{1+\frac{1}{2}(1-3\cos^2\beta)S\right\},$$

$$\varepsilon_\| = n_e^2 + \frac{\rho N_A h F^2}{3M.\varepsilon_0 kT}\mu^2\left\{1-(1-3\cos^2\beta)S\right\}, \tag{6.9}$$

where β is the angle between the molecular dipole μ and the long molecular axis. The dielectric anisotropy is

$$\Delta\varepsilon = n_e^2 - n_o^2 - \frac{\rho N_A h F^2}{6M.\varepsilon_0 kT}(1-3\cos^2\beta)\mu^2 S \tag{6.10}$$

and

$$\bar{\varepsilon} = 1 + \frac{\rho N_A h F}{M\varepsilon_0}\left(\bar{\alpha}+\frac{\mu^2}{3kT}\right). \tag{6.11}$$

There is more freedom for controlling the dielectric anisotropy than for birefringence by placement of strong dipole moments onto the core structure. Strongly positive materials result from polar moieties such as cyano-terminal groups, or 3, 4, 5 substitution of fluoro-groups onto a terminal phenyl group. Such placement ensures that β is kept low, with $\Delta\varepsilon \approx +50$ being readily achievable. It is harder to make strongly negative materials for a number of reasons. First, the $n_e^2 - n_o^2$ term of Equation 6.10 is always positive and leads to $\Delta\varepsilon \approx +2$ even if there is a negligible dipole moment. Equation 6.10 predicts that transverse dipoles, where $\beta \approx 90°$, are only half as efficient at contributing to a negative $\Delta\varepsilon$ due to the cylindrical symmetry. Moreover, bulky polar side groups detract from the rod-like shape of the molecule that leads to liquid crystallinity. Nevertheless, mixtures with $\Delta\varepsilon \approx -6$ have been achieved for modern LCD TV applications. Note, Equation 6.10 also predicts that a dipole moment at $\beta = 48.2°$ contributes equally to $\varepsilon_\|$ and $\varepsilon_\perp$; even a large dipole moment oriented at $\beta \approx 52°$ gives $\Delta\varepsilon \approx 0$.

Implicit in the definition of liquid crystal behavior is the concept of orientational elasticity and the energetic cost associated with deforming the director field. Whereas a crystal solid has elasticity associated with the positional translation of the constituent molecules, the liquid crystal has elasticity associated with changing director orientation. The curvature strain tensors of phases with cylindrical symmetry contain terms in splay, twist, and bend deformations [27] (as shown in Figure 6.2a). The elasticity of the chiral nematic is equivalent to that of the achiral nematic but includes $2\pi/P$ subtracted from the twist term to represent the spontaneous twist of director. The elastic bulk free-energy density for nematics and cholesterics G_K is given by

$$G_K = \frac{1}{2}\iiint\left[\begin{array}{c}k_{11}(\nabla.\mathbf{n})^2 + k_{22}\left(\mathbf{n}.\nabla\times\mathbf{n}-\frac{2\pi}{P}\right)^2 \\ + k_{33}(\mathbf{n}\times\nabla\times\mathbf{n})^2\end{array}\right], \tag{6.12}$$

where $P = \infty$ for the usual achiral nematic case and k_{ii} ($i = 1, 2, 3$) are the splay, twist, and bend elastic constants, respectively.

The elastic constants are important to the display engineer because they dictate the amount of deformation induced by the applied electric field. Both splay and bend elastic constants contribute to distortions in the plane of the director and deforming torque, whereas twist occurs where the distortion is perpendicular to the director and torque. There have been extensive studies of the relationship between molecular structure and the elastic constants and because of this important role they play in the operation of LCDs. To a first approximation, the temperature dependence of the elastic constants follows S^2, a relationship that holds well for k_{11} and k_{22}, where the ratio k_{22}/k_{11} is relatively insensitive to temperature and chemical structure with $0.5 \leq k_{22}/k_{11} \leq 0.8$ and usually $k_{22}/k_{11} \approx 0.5$. The behavior of the bend elastic constant k_{33} is more complex [28], with k_{33}/k_{11} being both temperature dependent and showing a strong dependence on structure, with values ranging from $1.0 \leq k_{33}/k_{11} \leq 2.2$. Approximating the constituent molecules of the liquid crystal to rigid, hard rods of length L and diameter W lead to the approximate relationship:

$$k_{33}/k_{11} \sim L^2/W^2, \tag{6.13}$$

which provides a useful rule of thumb for the LC mixture designer. However, there is a strong influence on this elastic ratio from short-range local

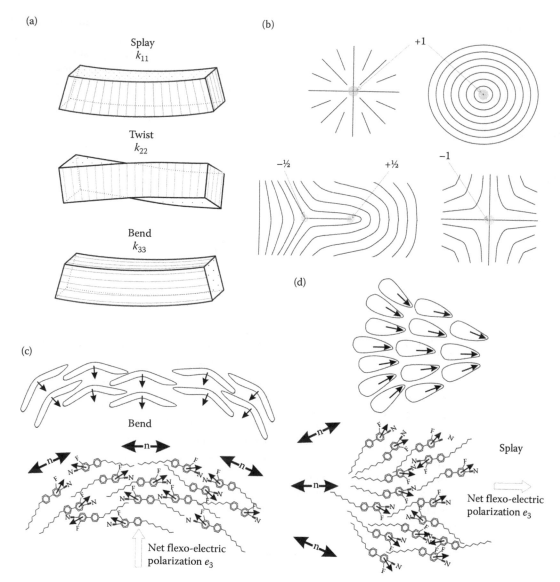

Figure 6.2 Nematic elasticity and disclinations: (a) Splay, twist, and bend deformations; (b) line disclinations of strength +1, +½, −½, and −1. In each case, the disclination continues normal to the page, and the director remains invariant in this direction (cylindrical symmetry). The defect core is indicated in gray; (c) flexoelectric behavior from banana-like molecules in a bend deformation; and (d) flexoelectric behavior from wedge-like molecules in a splay deformation.

ordering in the nematic phase due to intermolecular dipole correlations or local smectic ordering. For example, smectic local ordering of positive $\Delta\varepsilon$ materials leads to a higher k_{33}/k_{11}.

Liquid-like fluidity of a liquid crystal is dictated by the viscosity, which determines the switching speed of LCDs. How quickly the director reorients with respect to a deforming torque is determined by the ratio of the viscosity and the elastic constants. However, even for the simplest nematic case, there are five viscosity coefficients [29], describing flow of the director in directions parallel and perpendicular to the director and shear force directions. The effect of fluid flow can play an important part in device behavior, such as the optical bounce that occurs in twisted-nematic displays due to backflow [30]. Full understanding of these effects requires numerical modeling [31].

For simplicity, backflow is often ignored and the time dependence of director reorientation without coupling to mass flow is represented by the single twist viscosity γ_1. Viscosity has a strong temperature dependence related to the change in order as well as the Arrhenius dependence common to liquids [32]:

$$\gamma_1 = \left(a_1 S + a_2 S^2\right) e^{-\frac{U}{kT}}, \tag{6.14}$$

where a_1 and a_2 are material-dependent constants, U is the activation energy for molecular movement, k is the Boltzmann constant, and T is the absolute temperature. Most commercial liquid crystal mixtures are based on aromatic systems, where $a_1 \gg a_2$ for compounds with low birefringence, $a_2 \gg a_1$ for highly birefringent compounds, and $a_1 \approx a_2 \approx 10^{-14}$ s·m^{-2} and $U \approx 0.3$ eV for $\Delta n \approx$ 0.18. At low temperatures, the viscosity often deviates from this Arrhenius type of behavior as it diverges toward a low temperature glass transition. The standard Vogel–Fulcher–Tammann equation has been applied successfully to a range of nematic compounds [26]:

$$\log \gamma_1 = A + \frac{B}{T - T_0}, \tag{6.15}$$

where A and B are material-dependent constants and T_0 is a temperature that is typically about 20°C below the liquid crystal glass transition.

Continuum descriptions of the director field are not always satisfactory because the director field often also includes topological discontinuities, called defects or disclinations [27]. Indeed, it was the thread-like appearance of nematic samples containing such defects that originally led to the naming of the phase, from the Greek for thread (*nematos*). Both point defects and line defects are common; examples of nematic line disclinations with different strengths are shown in Figure 6.2b. At each discontinuity in the director field, there is local melting of the phase and the liquid crystal order is zero at the disclination core, which is typically of the order of 10 nm in diameter. For the nematic phase, disclinations of strengths +1, −1, +½, and −½ are common, with +½ and −½ appearing as pairs and usually forming defect loops or may terminate at interfaces such as the containing substrates of a device. Avoiding defects is a key part of the design of most LCDs, although recently there has been an insurgence of interest in the potential applications for liquid crystal devices with controllable formation of topological defects, such as the zenithal bistable display (ZBD) described in Section 5.3.

The constituent molecules for the liquid crystal do not exhibit perfect cylindrical symmetry but have more complex shapes. For example, they may not have a twofold rotational symmetry about either short or long axes, leading to "wedged" or "banana" molecular shapes, respectively. Normally, the nematic includes all possible orientations of the molecules that maintain the symmetry of the phase. However, with bend and splay deformations, the distribution of asymmetric molecules is shifted slightly as the molecules pack. For example, banana-like molecules will tend to orient so that molecular curvature follows that of the director field (Figure 6.2c). Similarly, the apex of wedge-like molecules points into the splay (Figure 6.2d). In either case, any molecular dipole will then contribute to a macroscopic polarization [33] and flexoelectric terms are required in the free energy. The degree of polarization for splay and bend deformations is represented by the flexoelectric coefficients e_1 and e_3, respectively, which are dependent on the material shape, dipole moments, and density.

6.2.3 Formulating liquid crystal mixtures

Table 6.1 includes some of the important display-related physical properties for a number of common nematic compounds, including temperatures for the nematic to isotropic transition T_{NI}, as well as the room temperature values for the optical, electrical, elastic, and viscous properties. Each of these compounds is stable at operating temperatures and when exposed to light, allowing device lifetimes of scores of years to be achieved. As shall be shown in the following sections, the properties of each of the compounds might be suited to different display technologies. For example, devices using optical scattering effects require a high birefringence to become strongly scattering, whereas polarized light displays using the retardation effect of the liquid crystal often require a lower birefringence to enable higher device spacing (and hence higher yields). Modes such as the supertwist nematic (STN) require k_{33}/k_{11} to be high, whereas the simpler TN display requires k_{33}/k_{11} to be kept low. Such requirements are rarely isolated, being combined with a range of target physical properties, such as appropriate temperature range of the required liquid crystal phase, low viscosity, to more complex needs, such as low temperature dependence of elastic constant ratios, etc. This necessitates mixtures to be used, frequently involving scores of components.

Properties such as the birefringence Δn, the isotropic to nematic transition temperature T_{NI}, and splay elastic constant k_{11} depend roughly linearly with concentration. The dielectric anisotropy $\Delta\varepsilon$, and twist and bend elastic constants k_{33} and k_{22}, are slightly more complex because of the effects of dipole correlations and smectic ordering. For example, the epoch-making compound pentyl-cyano-biphenyl (5CB, compound 1 in Table 6.1) has a measured $\Delta\varepsilon$ of about 15 close to room temperature and can be treated as $\Delta\varepsilon$ of about 24 when adding into mixtures. This is because the antiparallel dipole correlations of the pure compound that effectively reduce the parallel dipole moment are disrupted in a multicomponent mixture. Care still has to be taken when formulating mixtures because the breaking of the dipole correlations can lead to the unmasking of smectic behavior: it is quite common for a mixture of polar and apolar nematic compounds to exhibit an unwanted smectic phase despite neither component having smectic behavior. Such a phase is called an "injected" smectic and arises because the dipole correlations of the polar compound were preventing the formation of the smectic layers.

Ignoring the order parameter-related terms in Equation 6.14 and considering only the viscosity to follow the Arrhenius behavior of conventional liquids, the viscosity of an n component mixture is given by

$$\log(\gamma_1) = \sum_i^n C_i \cdot \log(\gamma_1)_i, \qquad (6.16)$$

where $i = 1, 2, 3, \dots n$, and C_i is the concentration of component i such that $\sum_i^n C_i = 1$. This logarithmic concentration dependence means that highly viscous additives can be used at low concentration without increasing the mixture viscosity significantly. For example, it is common practice to increase the clearing point through the addition of three- or four-ringed compound, where the linear increase in T_{NI} is accompanied with a disproportionately small viscosity increase.

None of the compounds in Table 6.1 show a room temperature liquid crystal phase, rather freezing to a crystal form. The liquid crystal phases usually supercool below the melting point because of the strong first-order nature of crystallization; this is particularly true in a thin container such as an LCD, where crystallization is suppressed by the surfaces. However, LCDs require operation typically between −20°C and +70°C and to be stored for months down to −40°C. Such temperature ranges require the formation of eutectic mixtures to suppress melting, at concentrations approximated by the Schroeder–Van Laar equation:

$$\ln(C_i) = -\frac{\Delta H_i}{R}\left(\frac{1}{T} - \frac{1}{T_i}\right), \qquad (6.17)$$

where R is the Rydberg constant; C_i, ΔH_i, and T_i are the molar concentrations, enthalpy of freezing and melting point of the ith component, respectively. This provides a guide for the material scientist to calculate the eutectic composition, given that, at the eutectic temperature, the concentrations sum to 100%. In practice, more thorough empirical methods [34] are required to formulate commercial mixtures. Table 6.2 includes the composition of three typical positive $\Delta\varepsilon$ eutectic mixtures, E7, ZLI2293, and ZLI 4792, together with their important physical properties.

6.2.4 Functional liquid crystal compounds

As shall be shown, it is not just the nematic phase that is used for LCDs. A variety of mesogenic compounds have been used to impart some new functionality to the system. Table 6.3 lists examples of mesogenic compounds, which are used for important display-related purposes, as summarized below.

1. The inclusion of one or more chiral centers (denoted *) in the flexible end chain of a mesogen imparts a tendency for spontaneous twist of the director. The material may exhibit inherent cholesteric (N*) or chiral SmC* phases (such as CE3 in Table 6.3) or

Table 6.2 Physical properties of selected nematic LCD mixtures

Mixture	E7	ZLI 1132	ZLI–4792
Typical use	Passive matrix TN	Passive matrix TN	Active matrix TN
Composition	C_5H_{11}—⬡—⬡—C≡N 51% C_7H_{13}—⬡—⬡—C≡N 25% C_8H_{17}O—⬡—⬡—C≡N 16% C_5H_{11}—⬡—⬡—⬡—C≡N 8%	C_3H_7—⬡—⬡—C≡N 24% C_3H_{11}—⬡—⬡—C≡N 36% C_7H_{15}—⬡—⬡—C≡N 25% C_5H_{11}—⬡—⬡—⬡—C≡N 15%	C_6H_{13}—⬡—⬡—F 7% C_7H_{15}—⬡—⬡—F 5% C_2H_5—⬡—⬡—⬡—OCF_3 7% C_3H_7—⬡—⬡—⬡—OCF_3 11% C_4H_9—⬡—⬡—⬡—OCF_3 8% C_5H_{11}—⬡—⬡—⬡—OCF_3 8% C_3H_7—⬡—⬡—F,F 11%
Nematic temperature range (°C)	S < –30 N 58 I	S < –6 N 71 I	S < –40 N 92 I
Δn; n_o (589 nm, 20°C)	0.2253; 1.5211	0.1396; 1.4830	0.0969; 1.4794
$\Delta\varepsilon$; $\varepsilon_\perp$ (1 kHz, 20°C)	13.8; 5.2	13.1; 4.6	5.2; 3.1
k_{11}; k_{22}; k_{33} (pN, 20°C)	11.7; 8.8; 19.5	1.95	13.2; 6.5; 18.3
Dynamic bulk viscosity η (cP, 20°C)	465	200	150
Twist viscosity γ_1 (mPa·s, 20°C)	180	250	109
Flexoelectric coefficients $e_1 + e_3$; $e_1 - e_3$ (pCcm^{-2}, 20°C)	+15; 12.2		|10|; –15

Source: Data collected from References [35–37].

Table 6.3 Examples of functional mesogenic compounds

#	Compound	Use	Key properties
1	(+)-4-*n*-Hexyloxyphenyl-4-(2-methylbutyl) biphenyl-4'-carboxylate (CE3) K 65°C S_C* 79°C N* 162°C I	N* ; Chiral Dopant; Ferroelectric compound	Helical twisting power; Induced Ferroelectric polarisation
2	2,3-Difluoro-1-ethyloxy-4-[*trans*-4-propyl cyclohexyl] biphenyl 	VAN S_C; FLC host	Low γ_1 High $\partial\varepsilon$ N phase; Wide S_C
3	Example anthraquinone dye 	Pleochroic dye	High order parameter; High dichroism; High miscibility
4	1,2-Bis(4-hex-5-enyloxyphenyl)diazene = C_4H_8O —⬡—N≫N—⬡—OC_4H_8 =	Optically induced reorientation	Solubility; S; Absorption efficiency
5	RM257: 1,4-Bis[4-(3-acryloxyproyl benzoxyl]-2-methylbenzene. K 64°C N 126°C I	Bifunctional reactive mesogen	Solubility; Photoreactivity
6	DB126: Triphenylene hexa-(2-methyl-4-*n*-decyloxy)benzoate K 109°C (Col 75°C) N_D 164°C	Discotic nematic. Acrylate version used for optical compensator films	Negative Δn

K 65°C S_C* 79°C N* 162°C I

K 109°C (Col 75°C) N_D 164°C I

may induce chirality through doping into a nematic or SmC host material. The important characteristic of the chiral compound as a dopant is its helical twisting power and handedness. These are related to molecular structure through various empirical rules, such as those of Reference [38].

2. Smectic phases occur with longer terminal chain groups at the mesogenic core. The formation of tilted phases, such as the SmC, occurs with suitable placement of transverse dipole moments, though precise control of phase transition temperatures and sequences remains somewhat an art for the chemist [e.g., Reference 39].

3. Pleochroic and fluorescent dyes can be added to liquid crystals to give appropriate optical functionality. Pleochroic dyes such as the

anthraquinone shown in Table 6.3 have reasonably good miscibility in the cyano-biphenyls, leading to anisotropic absorption that is much more parallel to the director than perpendicular to it [40]. This is used in Guest–Host displays, where the polarizer is replaced by the constituent dye mixed in the liquid crystal, allowing switching of the appropriate wave bands between absorbing and transmitting states. Performance is dictated by the combined order parameter of the system and the direction of the dye transition moment with respect to the liquid crystal director.

4. Including a central azo-moiety into the core of a mesogen allows photoinduced realignment of the director due to anisotropic absorption of the dye. Irradiating the molecule with polarized light causes reorientation of the dye molecule to the direction normal to the polarization, usually via transition between the *trans-* and *cis*-isomers. This may be used to effect photoreorientation of the nematic director, either reversibly [41] or irreversibly by combining the azo-mesogen into polymerizable groups [42].

5. Reactive mesogens have become a major tool for the displays field [43]. These compounds exhibit conventional liquid crystal behavior, either individually or when used in mixtures. The inclusion of a photoreactive group in one or both of the terminal end chains allows the material to be polymerized by exposure to UV light. Hence, the reactive mesogen may be aligned appropriately (through interaction with the surface, electric fields or polarized light) and the liquid crystal structure stabilized by polymerizing the reactive mesogen. The solubility of the reactive mesogens is high and so a variety of systems are possible, from linear elastomers to polymer-stabilized networks and gels, and liquid crystal polymers. The photoreactive groups are often acrylates and so may also be used in conjunction with standard acrylate cross-linkers and photoinitiators. Reactive mesogens have been used widely, from creating patterned optical retarders, functional alignment layers, to creating templates for extending the temperature range of narrow liquid crystal phases.

6. Discotic liquid crystals have a symmetry axis that is parallel to the short molecular axis [44].

This leads to distinctive physical properties, such as negative birefringence, one-dimensional conductivity, and semiconductivity, and the formation of columnar rather than smectic phases. Many of the other anisotropic physical properties, such as the order parameter, permittivities, and elastic constants are similar to those of their calamitic (i.e., rod-like) counterparts. A particularly important application for LCDs is the use of polymerizable discotic liquid crystals to form optical compensation plates.

6.3 BASICS OF LIQUID CRYSTAL DEVICES

6.3.1 Basic display construction

Much of the basic scientific understanding of liquid crystals was available in the first half of the 20th century. Mauguin [45] had found that, contrary to the optical rotation caused by a quarter-wave plate, polarized white light could be guided by a liquid crystal with a twisted structure, allowing light to be transmitted between crossed polarizers by samples with a twist of 90°. Fréedericksz had shown field-induced switching and the existence of a critical threshold for magnetic [46] and electric [47] fields applied to a uniform sample. Châtelain [48] achieved uniform alignment of the liquid crystal director using thinly spaced samples and rubbing to define the orientation. Such scientific advances prompted the first liquid crystal device to be patented by the Marconi Wireless Company in the UK in 1936 [49]. However, the first LCD commercialization had to wait until the early 1970s, and the availability of various other components, many of which were invented and developed independently for other applications. These associated technologies include the following:

- *Glass.* Typically 0.7 mm thick, polished glass is remarkably optically uniform ($n = 1.520 \pm 0.005$), transparent (91% transmission from 320 to 700 nm) and flat (± 0.1 μm). Together with its excellent mechanical, thermal, and economic properties it is hard to imagine a better substrate material. Today's 40″ TV displays are made on $2B production lines that handle Generation 10 glass (2850 mm × 3050 mm), though most small displays are still made using Generation 4 production lines (550 mm × 650 mm). Two types are common: soda lime glass for low-cost displays and borosilicate glass for high-end TFT (thin-film transistor) displays. The latter has a low alkali ion content and so adds far fewer ionic impurities to the contacting LC that would otherwise prevent the TFT maintaining its charge across the pixel. Recently, ion-implanted reinforced glass and ultrathin (<100 μm) glass have been used for high-end displays, particularly for use in portable display applications and curved screen HDTV.

- *Indium tin oxide.* Following the initial work done in the mid-1960s at Philips [50,51], and in the Far East at the Japanese Government Research Institute [52], indium tin oxide (ITO) has become the dominant transparent conductor for the display industry, as well as for many other applications. The layer is sputtered onto the glass surface between 50 and 200 nm, depending on the display requirements. The thin layer causes loss of light through interference and reflection, but these thicknesses usually lead to about 88% and 85% transmissivity of the glass overall and correspond to sheet resistances of 80 $\Omega/\square$ and 15 $\Omega/\square$, respectively. It is important to understand how sheet resistance works: a square of ITO gives a resistance of 15 Ω regardless of its area. Calculation of the resistance of any electrode requires the number of squares to be multiplied by the sheet resistance. So, for example, the resistance of a display electrode that is 10 cm long and 200 μm wide is typically about 15 $\Omega/\square \times L/W = 7.5$ kΩ.

- *CMOS circuitry.* CMOS integrated circuits were invented by Wanlass at Fairchild in 1963 [53], whereby p-type and n-type transistors are used in parallel to provide low-power logic circuitry (and hence suitable for use with battery operated equipment) and logic output voltages of between 3.5 and 5 V, which could be used directly to address the liquid crystal. The first LCD product was launched by *Sharp* in April 1973: the EL-805 electronic calculator. The CMOS provided both the logic for the calculator operation and the driving circuitry for the dynamic scattering mode LCD. This was followed in October 1973 by the 06LC digital watch from Seiko, which used a TN LCD.

- *Polarizing film.* Although prism and wire-grid polarizers were established technologies, Land's invention of iodine doped stretched polymer film polarizers in 1929 [54] and the subsequent use of polymerized dichroic nematics by Dreyer in 1946 [55] enabled low-cost flat-panel displays to operate by controlling the polarization state. Today's polarizers use aligned silver nanoparticles, and comprise protective films, adhesives, release liners, and other filters required for different display purposes [56].

- *Thin-film transistor* [57]. It was invented at *RCA* by Paul Weimar in the early 1960s. The original demonstration used tellurium as the semiconductor but other materials researched at that time included silicon, germanium, cadmium selenide, and cadmium sulfide. Most displays today use either amorphous or polycrystalline silicon.

There are various different types of LCD but many of the design principles and modes of operation are common throughout. Consider the common TN LCD [14] shown in Figure 6.3. The display is formed from two containing plates, the front one at least being transparent and coated with transparent conducting electrodes, such as ITO. The plates are spaced a few microns apart by glass beads, adhered together and the laminate filled with the liquid crystal material and sealed. The most common substrates in use are 0.5 or 0.7 mm glass, though optically isotropic plastic substrates may also be used for low weight and flexibility. Reflecting metal foil layer has been used as the rear substrate where flexibility and high temperature thermal processes are required for a reflective display. The ITO is etched to form the appropriate electrode pattern. This is often a series of rows and column electrodes on the opposing internal faces of the device. In such an arrangement, the electrodes form a parallel plate capacitor, wherein the individual pixels are formed in the regions of overlap of the rows and columns. Of course, other designs are possible, such as polar coordinates (formed from axial and radial electrodes), alphanumeric characters and icons, or interdigitated electrodes on one of the substrates to provide an in-plane electric field. The most sophisticated displays, used from mobile phones to HDTV, use a TFT on one of the plates, addressed using copper bus lines to provide the

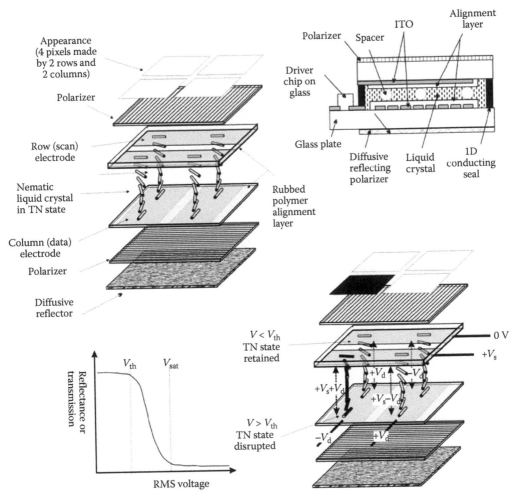

Figure 6.3 Construction and operation of a twisted nematic LCD: (a) Schematic diagram of a 4 pixel TN in the off-state; (b) cross section of a modern chip-on glass LCD; (c) typical electro-optic response for a monostable LCD, such as the TN; (d) schematic diagram of the 4 pixel TN with voltage V_s applied to one row, and data $\pm V_d$ applied to the columns, such that one pixel (bottom left) is switched dark.

signals to a pixel electrode. In these instances, the opposite plate is a single electrode, held at Earth.

Creating an electro-optic effect requires some optical property, such as the reflectance or transmittance for a backlit display, to be changed with an applied field. Today's commercial LCDs almost exclusively use reorientation of the director from some initial prealigned state, dictated by alignment layers on the inner surfaces of the display. Various alignment layers are possible, but most devices use a polymer coating to impart either homeotropic (i.e., normal to the surface), planar homogeneous, or tilted homogeneous alignment of the director (Figure 6.4). For example, the TN device includes polyimide coatings rubbed in a direction to impart

a homogeneous alignment with a surface pretilt θ_s of typically 1° to 2°, though an STN would require a somewhat higher pretilt of $2° < \theta_s < 6°$. For the simple TN mode, a 90° twist from the top to bottom surfaces is induced in the device by arranging the rubbing directions on the opposing surfaces to be normal to each other. The prevention of tilt and twist disclinations is ensured by including a small helical twisting power to the nematic through the addition of very low quantities of a chiral nematic and by matching the sign of the helicity to the two pretilts to minimize bend and splay distortion [58]. Ignoring biaxial surface terms, the orientation of the director at each surface is governed by the surface energy G_s:

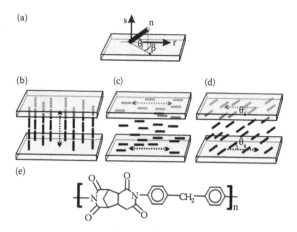

Figure 6.4 Nematic surface alignment leading to uniform director profile: (a) Definition of in-plane azimuthal angle β and out-of-plane zenithal tilt angle θ; (b) homeotropic; (c) planar homogeneous; and, (d) uniform tilted from antiparallel surfaces with pretilt θ_s; (e) basic structure of the dedicated polyimide AL 1051 from Japan Synthetic Rubber KK.

$$G_S = \frac{1}{2} \int_S \left[\begin{array}{c} W_\theta (\mathbf{n} \cdot \mathbf{s} - \theta_s)^2 + W_\beta (\mathbf{n} \cdot \mathbf{r})^2 \\ + W_p (\mathbf{P} \cdot \mathbf{s})^2 \end{array} \right] dS, \quad (6.18)$$

where **s** is the surface normal and **r** is the preferred alignment direction in the plane of the surface (Figure 6.4a). Changes to the surface tilt are related to the zenithal anchoring energy, W_θ. Typically, $10^{-7} \leq W_\theta \leq 10^{-3}$ Jm^{-2} and $10^{-9} \leq W_\beta \leq 10^{-5}$ Jm^{-2} [59] though for typical commercial devices the anchoring energies are greater than 10^{-5}·Jm^{-2} and are usually considered as fixed boundaries. The polar surface term W_p is insignificant for nematics but is important for ferroelectric liquid crystal (FLC) devices, where there is often a preferred orientation of the spontaneous polarization with respect to the surface normal.

Each device is constructed in two stages: the back-end process is done under strict clean room conditions, and the front-end, wherein the devices are filled with liquid crystal and the associated optical layer and driving electronics added and testing is done. The back-end processes are done in a clean room and consist of producing empty laminates each with the potential to form multiple displays. The factory equipment dictates the size of the plates used; it is important for high yield and efficiency that the glass is large enough to allow at least six or eight displays to be made on a single laminate. Even 47 in. diagonal television displays are produced eight per laminate using generation 8

sized glass (2200 × 2500 mm) and large-scale associated manufacturing equipment.

The plates are made on a production line preventing the use of many standard laboratory-based practices, such as oven baking and spin coating. Each step occurs within the TAKT time, which is the time that one plate moves from one step to the next; the TAKT time, by definition, must be the same for all of the steps on a given line. Following a brief inspection, the laminate is then sent to the front-end, where it is singulated, filled, and the remaining components attached.

The processes for constructing a passive matrix display are shown in Figure 6.5. Many of these processes are common to either passive matrix or TFT-driven active matrix devices. Common back-end processes include

1. *Glass cleaning*: Glass is supplied with the appropriate thickness of ITO deposited onto one side. The plate is washed with deionized water, ultrasound, soaps and sometimes solvents such as IPA (though this is not preferred due to flammability).
2. *Electrode patterning*: This is done using photolithography. A thin layer of positive photoresist is printed onto the substrate, soft baked (to harden) and a mask is placed in contact with the layer. Where exposed to UV, the photoresist is washed off by developer but the photoresist remains in contact with the ITO everywhere else. The exposed ITO is then

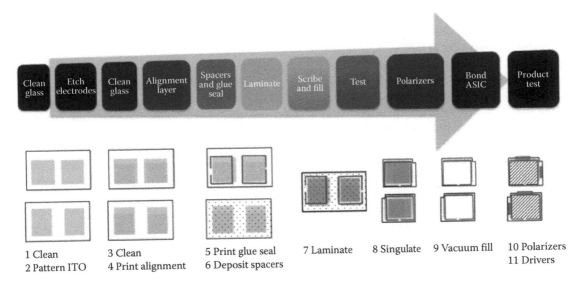

1 Clean
2 Pattern ITO

3 Clean
4 Print alignment

5 Print glue seal
6 Deposit spacers

7 Laminate   8 Singulate   9 Vacuum fill   10 Polarizers
11 Drivers

Figure 6.5 Construction steps for a passive matrix display.

removed by acid and finally the remaining photoresist is stripped from the patterned ITO using an aggressive solvent. It is common for the electrode patterning to be done separately from the standard production process.

3. *Alignment layer*: Following another clean step, the alignment layer is pad-printed onto the patterned glass in the areas that form the viewing area of each display. The alignment layer is patterned to avoid the part of the electrodes where bonding (electrical access) is required and where the glue seal will be deposited (to ensure a strong seal). The polyimide is then baked at high temperature (e.g., 180°C) to harden it and then rubbed using a rayon cloth mounted onto a roller. The resulting pretilt is not only dependent on the choice of the alignment polymer but also on the rubbing strength L:

$$L = N \cdot l \cdot \left(1 + \frac{2\pi r n}{60v}\right), \qquad (6.19)$$

where N is the number of passes of the roller, l is its contact length (mm), n is the roller rotation speed (rpm), r is the roller radius (mm), and v is the speed of the plate (mm·s^{-1}). Each of these steps needs to be controlled to minimize display variation.

4. *Spacers*: Polystyrene spheres of the required spacing are either deposited in air or solvent onto one of the glass plates that form

the laminate. The density of the spacers depends on the cell gap but is typically 50–100 per mm^2.

5. *Glue seal*: A thermal epoxy is syringe deposited onto the other substrate to form the boundaries for each individual display. The seal is designed to be as thin as can be reliable, with four edges comprising the sealant but with one edge having a ≈5 mm wide hole to allow filling. After deposition, the glue solvent is removed in a soft bake and the plate brought into contact with its opposing, spacer-coated plate. The laminate is then sealed under pressure using a heated press. Usually, the glue seal includes glass rod spacers to ensure the most uniform spacing to the edge of the panel. Often, the glue seal will also include a low density of gold beads, each with a diameter 20% or so higher than the spacers. These gold beads are distorted to near cylinders under the pressure used to seal the device. In this fashion, they form multitudinous one-dimensional conductors randomly distributed throughout the sealant, linking the top and bottom substrates. These allow connection from one plate to the other and thereby allow a single driver to be used to supply the row and column signals. Careful patterning of the ITO electrodes ensures that no unwanted shorts occur, within the alignment tolerance of the plates.

The completed laminate then leaves the high-end clean room for the front-end processes. Common front-end processes are as follows:

6. *Singulation*: Each display is cut from the laminate using a diamond scribe. Each display is cut to give at least one bonding ledge to allow electrical access to the electrodes. The panels are inspected for any nonuniform cell gaps and tested for shorts.

7. *Filling and sealing*: For small area LCDs, each device is filled in a vacuum chamber, where the cell is initially evacuated, the liquid crystal is brought into contact with the sealant's filling hole where it begins to fill by capillary action. Although still in contact, the vacuum is removed and the air pressure used to force the liquid crystal into the remainder of the display. Once complete, each display is sealed with a low-ion content UV glue while pressure is applied to the cell to ensure that the correct cell gap is obtained on the sealed cell. Large display panels are filled before singulation using a one-drop-fill method [60], which is combined with sealant deposition and vacuum assembly steps. One-drop filling is a major enabling technology that helped LCDs meet the cost requirements for large area applications.

8. *Polarizers*: After cleaning, each cell has polarizers and associated optical films mounted front and back. The films are supplied with an index-matching adhesive to reduce extra reflections from the layers. If a reflective display is required, the rear polarizer also comprises a diffusive mirror. Front polarizers too may include lightly scattering elements that act to remove unwanted reflections and compensation films for improving viewing angle.

9. *Driver bonding*: Connections are made from the exposed ITO on the bonding ledge to the driving circuitry by one of four methods:
   a. Surface mount technology: Connectors are made directly from the bonding ledge to the printed circuit board (PCB);
   b. Chip-on-board: The silicon die is mounted onto the PCB, connected to the panel using gold wires, and protected by adhesive;
   c. Tape automated bonding-mounted drivers. The driver manufacturers often supply tape-automated bonded chips. The chip is mounted onto a tape with gold I/O lines etched into it. The tape is adhered to the bonding ledge using a one-dimensional conducting adhesive;
   d. Chip-on-glass (COG). COG uses gold-bump soldering to attach the silicon chip directly to the bonding ledge. The chip must be protected from light and handling by embedding in a black epoxy adhesive.

10. *Inspection and test*: Obviously, the degree of testing is minimized to reduce costs, but manufacturers will operate a quality system to determine and minimize the occurrence of optical and electrical defects.

11. *Provision of color*: Full-color LCDs are made in the same way as described above but the glass plates are prepatterned with color filters (for the front plate) and TFT (for the rear plate). Usually, Red-Green-Blue filters are used and aligned over subpixels in the column electrodes. A black matrix surrounds each filter to maximize contrast, minimize color leakage, and optically isolate the TFT (thereby preventing problems with photo-generated charge). The resulting structure has a lower pixel fill factor, typically <50%. Given the losses associated with the low fill factor and absorption of 67% of the light by the color filters, reflective full-color LCDs are yet to be successful commercially: the images are too dim in all conditions but the brightest sunshine. Instead of using a reflective rear polarizer, color panels are combined with a backlight unit to provide illumination.

6.3.2 LCD polarization optics

Although liquid crystals can be used to emit light [61], all LCDs use the liquid crystal medium to modulate light incident on the panel, whether from ambient light or from a built-in source. The anisotropic nature of the liquid crystal presents a number of means through which contrast can be produced, whether by scattering, absorption, selective reflection, or changes of optical retardation and hence polarization state. The great majority of LCDs are sandwiched between film polarizers and use changes of optical retardation to give the required appearance. The state of polarization of light, as it travels through an optically transparent medium, can be linear, elliptical, or circular, depending on the relative magnitudes of two orthogonal polarization components. As the light goes from one transparent medium to another, light of each of the polarization components will be refracted, reflected, and the polarization reoriented. Finding the optical state requires solution of Maxwell's equations at each interface for each polarization. However, if we ignore the reflections and refractions and just consider the polarization state, then we can consider the light to be given by a 2×1 Jones vector:

$$\begin{pmatrix} E_x \\ E_y \end{pmatrix} = \begin{pmatrix} A_x e^{-i(\omega t+\delta_x)} \\ A_y e^{-i(\omega t+\delta_y)} \end{pmatrix}, \quad (6.20)$$

where A is the amplitude, x, y orthogonal components for light traversing in the z direction, and δ represents the phase related to the wavelength λ through the refractive index by

$$\delta_j = \frac{2\pi n_j t}{\lambda} \; ; j = x, y. \quad (6.21)$$

If the light is incident on a birefringent layer, which has refractive indices n_ε and n_o and thickness t, then the state of the polarization of the transmitted light will depend on the orientation of the incident light with respect to the optic axis and the phase difference between the x and y components. The maximum retardation due to the birefringence is Γ, given by:

$$\Gamma = \frac{2\pi \Delta n t}{\lambda}. \quad (6.22)$$

Transformation of the Jones vector requires the operators to be 2×2 matrices called a Jones matrices [62].

Consider a simple uniform planar sample with parallel rubbing directions and a uniform cell spacing d. Ignoring the reflections and absorption in this fashion is reasonable for thin ($d < 25\,\mu m$), well-aligned samples, and it means that the polarization state may be calculated using the Jones matrix for a birefringent retarder:

$$\begin{pmatrix} e^{-i\frac{\Gamma}{2}} & 0 \\ 0 & e^{i\frac{\Gamma}{2}} \end{pmatrix}. \quad (6.23)$$

Placing this retarder at an arbitrary angle ϕ between crossed polarizers gives the Jones vector for light transmitted by the device:

$$\begin{pmatrix} E'_x \\ E'_y \end{pmatrix} = \begin{pmatrix} 1 & 0 \\ 0 & 0 \end{pmatrix} \begin{pmatrix} \cos\phi & -\sin\phi \\ \sin\phi & \cos\phi \end{pmatrix} \begin{pmatrix} e^{-i\frac{\Gamma}{2}} & 0 \\ 0 & e^{-i\frac{\Gamma}{2}} \end{pmatrix} \begin{pmatrix} \cos\phi & -\sin\phi \\ \sin\phi & \cos\phi \end{pmatrix} \begin{pmatrix} 0 & 0 \\ 0 & 1 \end{pmatrix} \begin{pmatrix} 0 \\ 1 \end{pmatrix}$$

$$= \begin{pmatrix} -2 i\sin\phi\cos\phi\sin\frac{\Gamma}{2} \\ 0 \end{pmatrix}. \quad (6.24)$$

The transmission T is then given by

$$T = \frac{E'^* \cdot E'}{E^* \cdot E} = \frac{E'^2_x + E'^2_y}{E^2_x + E^2_y}$$

$$= \sin^2 2\phi \sin^2 \frac{\Gamma}{2} = \sin^2(2\phi)\sin^2\left(\frac{\pi \Delta n d}{\lambda}\right), \quad (6.25)$$

where E^* is the conjugate if E (which may be complex) and $t = d$ is used for the spacing of the liquid crystal cell.

Equation 6.25 suggests that the transmitted light is maximum if the cell gap and birefringence are at the quarter-wave plate condition $\Delta n \cdot d = \lambda/4$, with the polarizers oriented at $\pm45°$ to the alignment direction. The eye is most sensitive to green wavelengths and setting $\lambda = 550\,nm$ gives broad transmission across the wavelength range such that the device will appear white; for a typical liquid crystal, this condition occurs for cells spaced at about 2–3 μm.

6.3.3 Basic operation: electrically controlled birefringence mode

Applying an electric field to the aligned liquid crystal will tend to align the liquid crystal director parallel to the field direction, if the material has a positive $\Delta\varepsilon$ or is perpendicular to the field if negative. Considering the pixel in one dimension only and the director at the containing surfaces anchored strongly, the total free energy of the liquid crystal F is given by combining Equations 6.3 and 6.12 as

$$F = \frac{1}{2}\int_0^d \left[\begin{array}{l} k_{11}(\nabla \cdot n)^2 + k_{22}(n \cdot \nabla \times n)^2 \\ + k_{33}(n \times \nabla \times n)^2 - \varepsilon_0 \Delta\varepsilon(n \cdot E)^2 \end{array} \right] dz. \quad (6.26)$$

For the simple case shown in Figure 6.6a, the electric field is applied normal to the surfaces such that $E = (0, 0, V/d)$ and initially $n = (1, 0, 0)$. As the field is increased, there is a torque on the director acting to increase the tilt angle θ:

$$F = \frac{1}{2}\int_0^d \left[\left(k_{11}\sin^2\theta + k_{33}\cos^2\theta\right)\frac{d^2\theta}{dz^2} - \varepsilon_0\Delta\varepsilon E^2 \sin^2\theta \right]dz .$$

(6.27)

This integral may be solved using the Euler–Lagrange equation:

$$\left(k_{11}\cos^2\theta + k_{33}\sin^2\theta\right)\left(\frac{d\theta}{dz}\right)^2 + \varepsilon_0\Delta\varepsilon E^2 \sin^2\theta = C,$$

(6.28)

where the constant C is found from realizing that the maximum tilt θ_m must occur at the cell center $z = d/2$ because of symmetry, at which point $d\theta/dz = 0$ and hence:

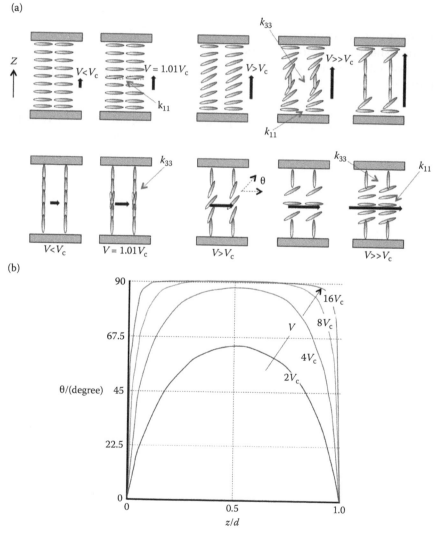

Figure 6.6 (a) Fréedericksz transitions in the planar homogeneous geometry for a positive $\Delta\varepsilon$ nematic (top) and the homeotropic geometry with a negative $\Delta\varepsilon$ nematic (bottom). The dominant elastic constants at different positions in the device are indicated. (b) Calculated tilt profile versus V/V_c for a material with $\varepsilon_\parallel = 20.25, \varepsilon_\perp = 5.36$, $k_{11} = 11$ pN, $k_{33} = 17$ pN in a cell of gap $d = 10\,\mu m$ [63].

$$C = \varepsilon_0 \Delta\varepsilon E^2 \sin^2 \theta_m. \tag{6.29}$$

Substituting back into the Euler–Lagrange Equation 6.28 gives

$$\frac{d\theta}{dz} = \frac{V}{d}\sqrt{\varepsilon_0 \Delta\varepsilon \frac{\left(\sin^2\theta_m - \sin^2\theta\right)}{\left((k_{33}-k_{11})\sin^2\theta + k_{11}\right)}}. \tag{6.30}$$

At low-field strengths, the small angle limits to the tilt angles and allow Equation 6.30 to be reexpressed as

$$V \cdot dz = d\sqrt{\frac{1}{\varepsilon_0 \Delta\varepsilon}\left[\frac{k_{11}}{\left(\theta_m^2-\theta^2\right)} + \frac{(k_{33}-k_{11})\theta^2}{\left(\theta_m^2-\theta^2\right)}\right]}d\theta, \tag{6.31}$$

which tends toward

$$V \cdot dz = d\sqrt{\frac{k_{11}}{\varepsilon_0 \Delta\varepsilon}} \cdot \left(\theta_m^2-\theta^2\right)^{-\frac{1}{2}} d\theta \tag{6.32}$$

as $\theta^2 \to 0$. Integrating Equation 6.24 to find the threshold field gives

$$V_C \int_0^{\frac{d}{2}} dz = \frac{d}{2}\sqrt{\frac{k_{11}}{\varepsilon_0 \Delta\varepsilon}} \cdot \int_0^{\theta_m}\left(\theta_m^2-\theta^2\right)^{-\frac{1}{2}} d\theta,$$

$$V_C[z]_0^{\frac{d}{2}} = d \cdot \sqrt{\frac{k_{11}}{\varepsilon_0 \Delta\varepsilon}} \cdot \left[\tan^{-1}\frac{\theta}{\sqrt{\left(\theta_m^2-\theta^2\right)}}\right]_0^{\theta_m},$$

$$V_C\frac{d}{2} = d\sqrt{\frac{k_{11}}{\varepsilon_0 \Delta\varepsilon}} \cdot \left[\tan^{-1}\frac{\theta_m}{\sqrt{0}} - \tan^{-1}\left(\frac{0}{\theta_m}\right)\right],$$

$$V_C = \pi\sqrt{\frac{k_{11}}{\varepsilon_0 \Delta\varepsilon}}. \tag{6.33}$$

Hence, there is a critical voltage that is independent of cell gap at which the electric field induced distortion begins, a threshold that depends on the root of the ratio of the relevant elastic constant and the dielectric anisotropy. By analogy, the threshold for a negative $\Delta\varepsilon$ liquid crystal in the homeotropic geometry is

$$V_C = \pi\sqrt{\frac{k_{33}}{\varepsilon_0 |\Delta\varepsilon|}}. \tag{6.34}$$

Figure 6.6b shows the situation where a positive $\Delta\varepsilon$ material is used but the field is applied in the plane of the cell; for example, using interdigitated electrodes. Similarly, a planar sample with the preferred alignment direction normal to the applied field will cause the director to twist at the electrode surface. In such instances, the electric field is not uniform, though it may be approximated at the electrode surface to be V/l, where l is the separation between the electrodes of opposing voltage. In such cases, the cell gap does not cancel in Equation 6.33 and the threshold depends on both cell gap and electrode spacing

$$V_C \approx \pi\frac{l}{d}\sqrt{\frac{k_{33}}{\varepsilon_0 \Delta\varepsilon}}; \quad V_C \approx \pi\frac{l}{d}\sqrt{\frac{k_{22}}{\varepsilon_0 \Delta\varepsilon}} \tag{6.35}$$

for the homeotropic and planar cases, respectively. This latter geometry is utilized with in-plane switching (IPS) devices, as will be described in Section 4.6. If the field is applied normal to a uniform director, the threshold is well defined. In practice, the alignment layers induce a small pretilt, and the director profile of the quiescent state is not uniform, neither is the electric field applied normal to the director. Thus, the threshold is rather second order in nature and begins to occur somewhat below the voltage predicted by Equations 6.33 through 6.35.

Above the threshold, the solution to Equation 6.30 must include the higher angle terms. Again considering the planar-aligned case shown in Figure 6.6a, dividing equation through by V_C gives

$$\frac{V}{V_C} \cdot dz = \frac{d}{\pi}\sqrt{\frac{\kappa\sin^2\theta+1}{\left(\sin^2\theta_m - \sin^2\theta\right)}}d\theta, \tag{6.36}$$

where $\kappa = (k_{33}-k_{11})/k_{11}$. Immediately above the transition, a reasonable solution is found by integrating Equation 6.28:

$$\frac{V}{V_C} = \frac{2}{\pi}\int_0^{\theta_m}\sqrt{\left[\frac{1+\kappa\sin^2\theta}{\left(\sin^2\theta_m - \sin^2\theta\right)}\right]}d\theta \tag{6.37}$$

using a Fourier analysis to solve the complete elliptical integral of the third kind,

$$\frac{V}{V_C} = \left(1 + \frac{1}{4}(\kappa+1)\sin^2\theta_m + \cdots\right), \quad (6.38)$$

which has the first-order solution:

$$\frac{z}{d} = \frac{1}{\pi}\sin^{-1}\left(\frac{\theta}{\theta_m}\right) - \theta\sqrt{\left(\theta_m^2 - \theta^2\right)}\cdot$$

$$\frac{1 + 3\kappa + \cdots}{12\pi\left[1 + \frac{1}{4}(\kappa+1)\theta_m^2 + \cdots\right]}, \quad (6.39)$$

where

$$\theta_m = \sin^{-1}\sqrt{\left(\frac{4}{(\kappa+1)}\left[\frac{V}{V_C} - 1\right]\right)}. \quad (6.40)$$

This indicates that, immediately above the threshold, the tilt of the director increases linearly, with a gradient that is inversely related to the elastic ratio k_{33}/k_{11}. This is indicated in Figure 6.6a, where the bend elastic constant becomes increasingly pertinent at the center of the device, and splay is increasingly pushed to the alignment surfaces.

Above the electric field-induced Fréedericksz transition, the situation is complicated by the effect of the nonuniform dielectric properties of the distorted director profile on the applied electric field. The electric torque is related to the electric displacement vector **D** and above the transition the director profile is no longer uniform such that **E** has a direction that is dependent on the distortion. That is, the effect of the field depends on the elastic energy, but conversely, the elastic energy depends on the field. Ignoring the effects of charge, $\nabla\cdot\mathbf{D} = 0$ and assuming **D** is a function of z only, solutions to the free energy expression require **E** to be calculated self-consistently with the director profile:

$$D_z = \frac{V}{\int_0^d \frac{1}{\varepsilon_0\left(\Delta\varepsilon\sin^2\theta + \varepsilon_\perp\right)}dz} \quad (6.41)$$

and thus

$$\frac{V}{V_C} = \frac{2}{\pi}\sqrt{\left(\frac{\Delta\varepsilon}{\varepsilon_\perp}\sin^2\theta_m + 1\right)}\cdot\int_0^{\theta_m}$$

$$\sqrt{\frac{1 + \kappa\sin^2\theta}{\left(\frac{\Delta\varepsilon}{\varepsilon_\perp}\sin^2\theta + 1\right)\left(\sin^2\theta_m - \sin^2\theta\right)}}d\theta. \quad (6.42)$$

Solutions to this integral are done numerically. Indeed, display manufacturers use commercial packages, such as "LCD Master" from Shintech, "TechWiz" from Sanayi, or "DIMOS" from Autronic Melchers [64], to calculate the director profile as a function of applied field for their devices.

Consider the case of Figure 6.6a where the cell retardation is set to the quarter-wave plate condition at 550 nm (the peak of the eye's response, $\Delta n.d = 550\,\text{nm}/4 = 137.5\,\text{nm}$) and the alignment direction is set at $\phi = 45°$. As the voltage is then applied, the overall birefringence of the cell decreases as the optic axis of the director at the center rotates to toward the field and viewing direction and the effective birefringence is reduced to $\Delta n'$. Simplistically, we can consider the director to be uniformly at some average tilt angle $\bar{\theta}$, the new refractive indices are given by

$$n_o' = n_o; \frac{1}{n_e'^2} = \frac{\sin^2\bar{\theta}}{n_o^2} + \frac{\cos^2\bar{\theta}}{n_e^2} \quad (6.43)$$

such that the effective birefringence $\Delta n'$ is

$$\Delta n' = n_e' - n_o = \frac{n_e n_o}{\sqrt{n_e^2\sin^2\bar{\theta} + n_o^2\cos^2\bar{\theta}}} - n_o. \quad (6.44)$$

From Equation 6.25, the transmission of polarized light (i.e., after the first polarizer) becomes

$$T = \sin^2\left(\frac{\pi\Delta n'd}{\lambda}\right). \quad (6.45)$$

With sufficient field, $\Delta n'd \to 0$ and Equation 6.45 predicts $T \to 0$. Thus, above V_C the cell changes from white toward black. Figure 6.7a shows this decrease in transmission for red, green, and blue wavelengths, calculated using Equations 6.40 and 6.45 for a 0.67 mm cell filled with a liquid crystal material with $n_e = 1.7$, $n_o = 1.5$, and $k_{33}/k_{11} = 2.0$.

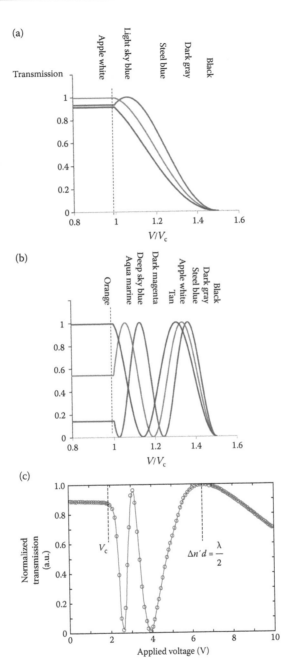

Figure 6.7 Theoretical transmission characteristic for an electrically controlled birefringence LCD. (a) The retardation at the quarter-wave plate condition for green light ($d = 0.67\,\mu m$, $\Delta n = 0.2$); (b) a thicker cell with retardation at the full wave condition for $\lambda = 500\,nm$ ($d = 2.5\,\mu m$, $\Delta n = 0.2$). Other fitting parameters used: $n_e = 1.7$, $n_o = 1.5$, and $k_{33}/k_{11} = 2$. (c) Experimentally determined transmission curve for a cell close to the full plate condition. The results are taken for a white light source imaged through an eye-response filter.

Figure 6.7b shows the behavior for a sample with a higher initial retardation; the same material with a 2.5 μm cell gap. Each wavelength has transmission peaks corresponding to odd multiples of the quarter-wave plate condition (i.e., $\Delta n'd/\lambda = \tfrac{3}{4}$ and $\tfrac{1}{4}$) because these peaks occur at different voltages for different wavelengths, the cell appears to change color as the voltage is increased.

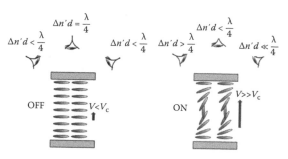

Figure 6.8 Explanation for the poor viewing angle for the electrically controlled birefringence effect.

There are a number of reasons why such a simple device is not used as a display effect, illustrating the thinking that the display engineer needs to follow.

- The optimum cell gap is very small, typically less than 1 μm, and always much lower than the cell gap required for a high production yield (typically 4 μm or higher).
- The device cannot be passive matrix multiplexed because the saturation voltage is typically several times V_C. This will be explained in the following section.
- The optical configuration is very sensitive to changes in cell gap. For example, a typical LCD tolerance is ±0.15 μm; such a change would cause a 12% reduction in the transmission and coloration for the Electrically controlled birefringence (ECB) device.
- The viewing properties of the device are very poor in both high and low voltage states, as illustrated in Figure 6.8. In the 0 V state, the effective birefringence decreases off-axis, the quarter-wave plate condition is lost and the device looks dark instead of bright (in the direction orthogonal to this, there is little change, and the device still looks white). In the high voltage state, the off axis transmission is even darker than the full ON state when viewed in the direction of the director tilt. When viewed in the other quadrant, the retardation increases rather than decreases as desired on switching causing contrast inversion, and the device begins to appear colored.

The response time for even such a simple device is complex due to the effects of field-induced flow and the need for the anisotropic viscosities to be considered. If a flow is ignored, the response time depends solely on the twist viscosity γ_1 that describes rotation of the director. The Euler–Lagrange equation for the time-dependent free energy of Equation 6.28 then becomes

$$\left(k_{11}\cos^2\theta + k_{33}\sin^2\theta\right)\frac{\partial^2\theta}{\partial z^2} + \left(k_{33} - k_{11}\right)\sin\theta\cos\theta\left(\frac{\partial\theta}{\partial z}\right)^2$$

$$+ \varepsilon_0\Delta\varepsilon E^2 \sin\theta\cos\theta = \gamma_1\frac{\partial\theta}{\partial t}.$$

$$(6.46)$$

For simplicity, the elastic anisotropy is discarded and a single elastic constant k is assumed:

$$\varepsilon_0\Delta\varepsilon E^2 \sin\theta\cos\theta + k\frac{\partial^2\theta}{\partial z^2} = \gamma_1\frac{\partial\theta}{\partial t}. \qquad (6.47)$$

Applying a small angle approximation [65], the characteristic ON and OFF times are

$$\tau_{ON} = \frac{\gamma_1}{\varepsilon_0\Delta\varepsilon E^2 - \left(\frac{\pi}{d}\right)^2 k} = \frac{\gamma_1 d^2}{\varepsilon_0\Delta\varepsilon\left(V^2 - V_C^2\right)}$$

$$= \frac{\gamma_1 d^2}{\pi^2 k\left[\left(\frac{V}{V_C}\right)^2 - 1\right]}; \tau_{OFF} = \frac{\gamma_1 d^2}{\pi^2 k}. \qquad (6.48)$$

These times are made short by using materials with low viscoelastic ratios γ_1/k but most effectively by keeping the cell gap d low. Examples of mixtures providing response times as low as 30 ms at −20°C and 10 ms at +20°C in a 3 μm cell have been produced [66]. Given the 3–4 μm lower limit on cell gap set by clean-room quality and device uniformity, it would seem advantageous for high Δn

materials to be used to enable the required optical effect with a lower cell gap. In practice, however, γ_1 tends to increase with Δn and so the efficacy of this is limited. Ensuring that the voltage is made high can quicken the ON time significantly. However, it is the sum of the ON and OFF times that is relevant for display applications. Moreover, if the display requires intermediate grey levels or some degree of passive matrix addressing (see Section 3.4), the ON voltage is limited and even the ON time can be slow. For example, switching to the gray level closest to V_C is inherently slow.

In the case of a passive matrix addressed device, even the OFF pixels have a voltage V_d applied and the response time is given by [67]

$$\tau_{OFF} = \frac{\gamma_1 d^2}{\pi^2 k \left[1 - \left(\dfrac{V_d}{V_C} \right)^2 \right]}; \qquad (6.49)$$

In practice, the response is complicated by induced flow of the liquid crystal and the aligning effect of the director in response to such flows. For example, the OFF response is often slowed considerably by backflow that tends to initially reorient the director at the center of the cell in the opposite direction to that which gives the eventual lowest energy state [68]. Avoiding this "optical bounce" effect is considered further in Section 6.6.2.

6.3.4 Passive matrix addressing and the multiplexing limit

For low information contents, it is satisfactory to form a display where each pixel is driven directly from the driver. Usually, one of the electrodes is shaped into the desired pattern, such as a pixel for a seven segment or alphanumeric image and the other electrode is a common electrode. A display of N pixels then requires $N + 1$ electrodes. An example arrangement is shown in Figure 6.9a, where a single seven segment number is displayed using just eight electrodes. Rather than have separate connections to the two plates of the LCD, access to all of the electrodes is through a single bonding ledge: the opposing common electrode is connected through electrodes that are connected through the one-dimensional conducting seal. The access electrodes to each segment is kept small, to minimize overlap

with the common electrode. It is essential for such a design that the common plate has only a single point of overlap with ITO on the electrode plate and conducting seal to prevent unwanted shorting of pixels. A typical transmission versus voltage characteristic for an LCD is shown in Figure 6.9b. With a direct drive scheme, the OFF voltage must be below the threshold T_{10} (usually 0 V) and the ON voltage should be above the saturation level T_{90}.

A direct drive approach is satisfactory for devices with a small number of pixels. More complex images require some degree of matrix addressing (or multiplexing), where appropriate signals are applied to electrodes on both top and bottom plates. The electrodes may also be shaped into alphanumeric characters and designed for a low level of multiplexing or indeed into a more complex $r(\theta)$ arrangement for the display of polar coordinates. The most common design, however, is for the electrodes to comprise N rows and M columns to form a rectangular $M \times N$ display. Time division multiplexing is used to apply appropriate signals to the rows and columns, when driven from only $M + N$ connections.

Consider the simple 7×6 matrix display illustrated in Figure 6.10, where each pixel is addressed by sequentially scanning through the six rows, although the appropriate data are synchronously applied to the columns to discriminate between the ON and OFF states. Figure 6.10 shows an example instant within the image frame, where the data voltages are being applied to seven columns and are those for the third row electrode. The data signal on the columns is $\pm D$ volts, where it is the sign of the signal that discriminates between the ON and OFF states of the pixel on the third row. At this instant, $+S$ volts is being applied to the third row while all other rows are kept at ground 0 V. The data being applied at this instance are the sequence of pulses of differing polarity given by $(++--+-+)D$. The potential difference at the pixels is defined as Row–Column, such that the third row experiences the voltage:

Row 3: $+S - D; +S - D; +S + D; +S + D;$

$+S - D; +S + D; +S - D,$

Each of the other rows experiences the following:

Row 1, 2, 4, 5, 6: $-D; -D; +D; +D; -D; +D; -D$

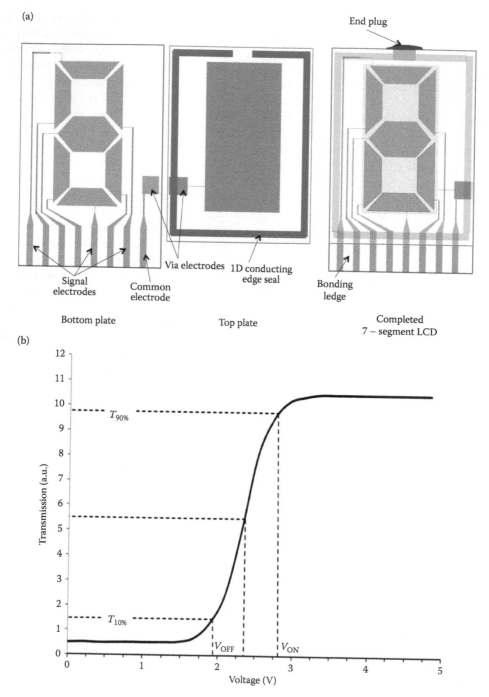

Figure 6.9 Design of (a) single element seven-segment LCD, showing front and rear plates face-up, and the completed module. (b) A typical transmission voltage characteristic for an LCD.

Once the third row is addressed, the scan signal moves to the fourth row and the data signals change sign appropriately to supply the signal for that row. The aim of the multiplexing scheme is to ensure that the high voltages $|(S + D)|$ are sufficient to discriminate from the low voltages $|(S-D)|$ despite the fact that each row is only addressed for a $1/N$ fraction of the frame time and for the remainder of the frame $(N-1)/N$ each pixel experiences the data signal $\pm D$.

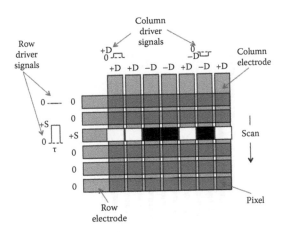

Figure 6.10 Example of passive matrix multiplexing for a 7 × 6 pixel graphic display. The row and column signals are for the 1/6 proportion of the frame that corresponds to the third row being addressed.

Nematic displays respond to the root-mean-square (RMS) voltage applied over the whole frame. As soon as the signal has been removed, each pixel starts to relax to the off state, so constant refreshing is done at a frame rate sufficiently fast to prevent this relaxation (*or frame response*) and maintain the image. For a simple scheme, the frame time will equal the slot time for each row τ, multiplied by the number of rows N. It is independent of the number of columns, and hence it is common for the rows to be chosen as the lower number in the matrix to ensure the faster frame rate.

The RMS of the voltage applied over N rows for the ON and OFF signals is

$$\overline{V_{ON}} = \sqrt{\frac{(S+D)^2 + (N-1)D^2}{N}};$$

$$\overline{V_{OFF}} = \sqrt{\frac{(S-D)^2 + (N-1)D^2}{N}},$$

(6.50)

where D and S should be chosen to ensure that V_{ON} is above the transmission saturation (i.e., $>T_{90\%}$) and V_{OFF} is below the threshold (i.e., $<T_{10\%}$). The steepness of the electro-optic response is related to $R = V_{ON}/V_{OFF}$:

$$R^2 = \left(\frac{\overline{V_{ON}}}{\overline{V_{OFF}}}\right)^2 = \frac{(S+D)^2 + (N-1)D^2}{(S-D)^2 + (N-1)D^2}$$

(6.51)

$$= \frac{b^2 + 2b + N}{b^2 - 2b + N},$$

where $b = S/D$ is called the bias ratio. The maximum number of lines that can be driven for a given S and D is found by differentiating

$$\frac{\partial(R^2)}{\partial b} = \frac{\partial\left(\frac{b^2 + 2b + N}{b^2 - 2b + N}\right)}{\partial b} = \frac{4(b^2 + N)}{(b^2 - 2b + N)^2} = 0.$$

(6.52)

This has solutions when either $b = \infty$ or $b = S/D = \sqrt{N}$. Substituting the latter into Equation 6.52 gives the Alt–Pleshko [69] relationships:

$$\left(\frac{\overline{V_{ON}}}{\overline{V_{OFF}}}\right)_{max} = \sqrt{\frac{N + 2\sqrt{N} + N}{N - 2\sqrt{N} + N}} = \sqrt{\frac{\sqrt{N} + 1}{\sqrt{N} - 1}}$$

(6.53)

or:

$$N_{max} = \left[\frac{\left(\frac{\overline{V_{ON}}}{\overline{V_{OFF}}}\right)^2 + 1}{\left(\frac{\overline{V_{ON}}}{\overline{V_{OFF}}}\right)^2 - 1}\right]^2.$$

(6.54)

This relationship suggests that, as N increases, the maximum $\overline{V_{ON}}/\overline{V_{OFF}}$ must tend toward unity; that is, the threshold characteristic must increase in steepness to allow more lines to be addressed, as shown in Figure 6.11a. The typical TN characteristic

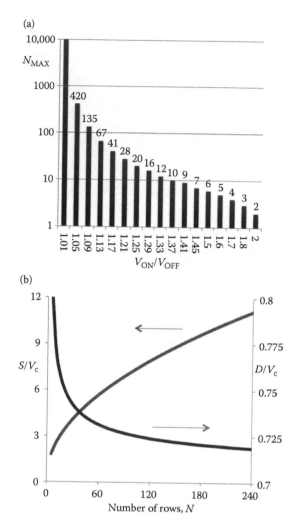

Figure 6.11 (a) The Alt–Pleshko multiplexing limit and (b) addressing voltages.

shown in Figure 6.9c has a response suited to up to six levels of time division multiplexing.

The signal voltages S and D that are required for the $N \times M$ display are found by substituting the optimum multiplexing ratio $S = D\sqrt{N}$ into the RMS voltages of Equation 6.50:

$$V_{OFF} = \sqrt{\frac{\left(D\sqrt{N} - D\right)^2 + (N-1)D^2}{N}}$$

$$= D\sqrt{\frac{2\left(\sqrt{N} - 1\right)}{\sqrt{N}}} \qquad (6.55)$$

$$V_{ON} = \sqrt{\frac{D^2 N + 2D^2\sqrt{N} + D^2 + D^2 N - D^2}{N}}$$

$$= D\sqrt{\frac{2\left(\sqrt{N} + 1\right)}{\sqrt{N}}},$$

which leads to the following data D and strobe S voltages:

$$D = V_{OFF}\sqrt{\frac{\sqrt{N}}{2\sqrt{N} - 1}} = V_{ON}\sqrt{\frac{\sqrt{N}}{2\sqrt{N} + 1}} \qquad (6.56)$$

and

$$S = V_{\text{OFF}} \sqrt{\frac{N\sqrt{N}}{2\sqrt{N}-1}}. \qquad (6.57)$$

Equations 6.56 and 6.57 are plotted in Figure 6.11b; they indicate that, at high N, $D \approx 1/\sqrt{2}\,V_C$ and $S \approx \sqrt{N/2}\,V_C$ so that the maximum voltage that needs to be delivered by the driver chip ($S + D$) increases with the level of multiplexing N, while the discriminating voltage D decreases. For this reason, STN displays with $N = 240$ typically require drivers capable of delivering $S + D = 20\,\text{V}$.

During the addressing frame, the liquid crystal director for the ON pixels in the addressed row will begin to decay as soon as the strobe voltage moves on to the subsequent rows. Thus, it is important that the slot time of the addressing scheme is arranged to minimize any flicker while at the same time the pixels are ready to display new information in the following frame; this occurs approximately where

$$\tau \approx \frac{\tau_{\text{ON}} + \tau_{\text{OFF}}}{2N}. \qquad (6.58)$$

In practice, the scheme shown above is unsuitable for long-term use because there is no DC balance to the waveform: this leads to eventual electrical breakdown of the liquid crystal and, if the image does not change, it can result in image sticking issues associated with the ionic conductivity. In practice, therefore, the waveforms will be inverted periodically to maintain a net zero DC voltage. Given the data changes from frame to frame, DC balancing can only be guaranteed if the voltages are inverted twice per frame. That is, if the slot time is halved, and the frame is divided into a positive field (+S, ±D) and a negative field with (−S, ∓D). The slot time should not be made too short because the power dissipation of charging and discharging the capacitances each field P_f is approximately related to

$$P_f \propto \frac{\varepsilon_{\text{av}} A}{d} \cdot \langle V \rangle^2 \cdot f \approx \frac{\overline{\varepsilon} A}{d} \cdot \frac{D^2}{t}. \qquad (6.59)$$

Given that power should be kept low, inversion is usually done only once per frame, often randomly within the frame time to help ensure DC balance regardless of image pattern.

6.3.5 Thin film transistors and active matrix addressing

Contrary to the approach taken by *Sharp* who targeted the new market of portable electronic calculators, *RCA* concentrated on replacing the cathode-ray tube for televisual display with an LCD from the outset [2]. The potential application of thin-film transistors as active elements in displays had been recognized since their invention at *RCA* in the mid-1960s [57]. By 1971, the *RCA* team believed [70] that active components were required to overcome the slow speed and poor electro-optic properties of the dynamic scattering mode device, proposing solutions using dual diodes, field effect transistors (FETs), and storage capacitors to ensure that the charge across the pixel remained constant across the frame. It was the team at *Westinghouse* [71] that were the first to implement this in practice, producing a TN 6 × 6 matrix driven by CdSe TFT. In the UK, the team led by Hilsum at *Royal Signals and Radar Establishment (RSRE)* understood the handling, reliability, and lifetime issues presented by CdSe and Te, and in 1976 approached Spear and LeComber at the University of Dundee, who were working on the use of hydrogenated amorphous silicon (α-Si:H) for use in photovoltaic cells. The Dundee team produced α-Si:H with electron mobility μ_e of $0.4\,\text{cm}^2/\text{V·s}$ and proved that the material was suited to address LCDs [72]. The Dundee and *RSRE* teams fabricated insulated-gate field effect TFTs onto a glass substrate using photolithography and fabricated the world's first α-Si active matrix LCD in 1980 [73]. This was followed in 1982 by a 240 × 240 α-Si TFT TN from *Canon* in Japan [74] and the immense ensuing effort and investment in Asia to bring TFT LCDs to the dominance of the display market that it enjoys today.

Very high levels of multiplexibility are possible using thin-film transistors fabricated onto the rear plate of the LCD behind each pixel (or sub pixel in a color display). A 4k UHDTV, for example, has $2160 \times 3840 \times 3 = 24.9\text{M}$ TFT per panel and is 2160 ways multiplexed. Each TFT supplies charge to a (sub-) pixel to switch the liquid crystal to any of 256 gray levels. The TFT is an FET formed from thin metal–insulator–semiconductor layers as shown in Figure 6.12a. Most LCDs use a bottom-gate configuration, where the amorphous silicon is deposited onto the prepatterned gate electrodes, *n*+ doped with hydrogen, and source and drain electrodes patterned on top of this. The electrodes are usually made from aluminium,

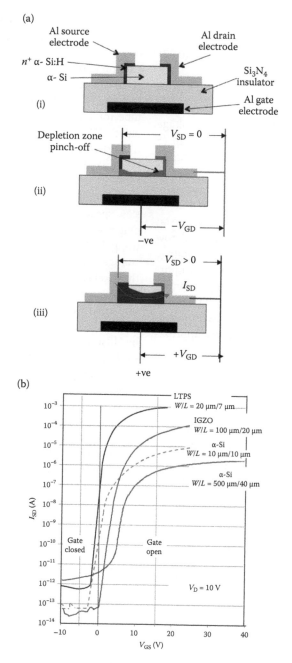

Figure 6.12 (a) Basic construction and operation of a TFT. (b) Current–voltage transfer characteristics for the original Dundee/RSRE α-Si:H TFT [72], together with contemporary results for a-Si:H [74], LTPS [75], and IGZO TFT [76].

chromium, or molybdenum. With a negative voltage applied between the gate electrode and the drain- V_{GD}, electrons are depleted from the semiconductor at the interface with the insulator, preventing current between source and drain electrodes (Figure 6.12a ii). When a positive-bias V_{GD} is applied to the gate, however, electrons accumulate at this interface allowing current flow from source to drain I_{SD} when the drain is positive and from drain to source when the drain is negative (Figure 6.12a iii). Figure 6.12b shows the transfer characteristic originally produced by the Dundee/RSRE team in 1981.

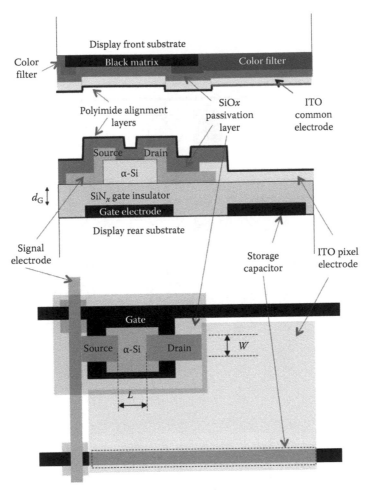

Figure 6.13 Cross section and plan views of a typical TFT-driven subpixel on a backlit color LCD.

A schematic of how a TFT is used in a typical LCD pixel is shown in Figure 6.13. The LCD has an active matrix back-plane and a front plane with a single common electrode. For backlit color displays, the front plate will also comprise the color filters, with a black matrix to ensure sufficiently high contrast, prevent color leakage, and shield the TFT from incident light and unwanted photo-induced charge generation. The TFT plate has row and column metal bus lines that allow the TFT to be addressed, with the row electrode connected to the transistor gate and the columns the transistor source. The ITO electrodes that form the pixel are connected to the drain electrode. The TFT is also protected using a SiOx barrier layer onto which the liquid crystal alignment layer is deposited. The presence of bus lines, shielded transistors, and storage capacitor reduces the active area of each pixel; aperture ratios as low as 30%–40% are typical. This

reduces the transmissivity of the panel and necessitates brighter backlighting to provide an attractive appearance. It also negates the use of the color TFT LCDs in reflective mode because of the light losses associated with the polarizers, color filters, and the low aperture ratio.

Addressing a TFT matrix is done line-by-line in a similar fashion to that described previously for passive matrix displays, but now the row and column signals are applied to the same substrate, as shown in Figure 6.14a. From the α-Si:H TFT characteristic of Figure 6.12b, it is clear that applying $+15V_{\mathrm{GD}}$ to the gate electrode allows current I_{SD} to flow between the source and drain due to the signal voltage V_{SD} applied to the column, whereas 5 V on the gate turns the TFT off and prevents current from flowing. With the scheme of Figure 6.14a, the transistors in each of the unaddressed rows (rows $n-1$ and $n+1$) are turned OFF by the negative

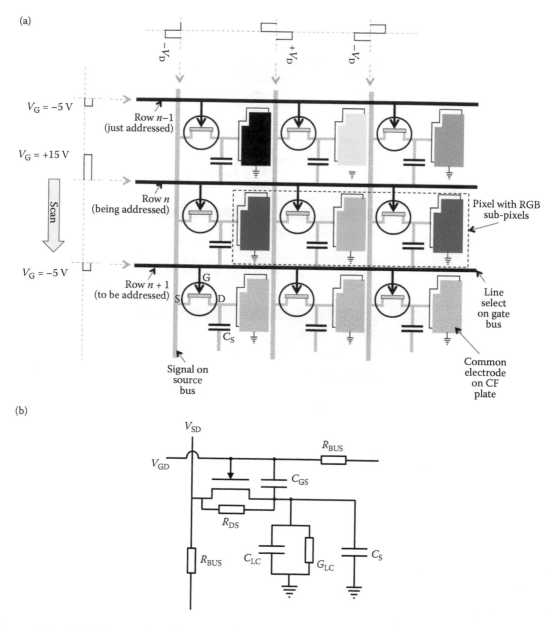

Figure 6.14 (a) Active matrix addressing of an LCD panel and (b) equivalent circuit for the pixel including the TFT parasitic losses, line losses, and the conductance of the liquid crystal.

signal applied to the gate bus lines, whereas the transistors in the addressed row (row n) are opened using $+15V_{GD}$ to the gate electrodes and switched ON. Synchronously, data voltages are applied to the source bus; there is no current for any of the OFF rows, but the pixels on the ON row are charged to the appropriate voltage by the current I_{SD}. After sufficient time for the pixel to charge to the new level τ, the gate is reclosed (with-$5V_{GD}$) and the gate pulse ($+15V_{GD}$) scans to the next row and new data on the source columns. The frame is completed when all N rows have been scanned, after the frame time τN. It is important to ensure that the liquid crystal material is exposed to the minimum net DC over several frames. To help ensure this, the polarity of the data signal is reversed, using frame inversion, scan-line inversion, column inversion, or pixel inversion (shown in Figure 6.14a).

To understand some of the important design rules for a TFT LCD, consider the example of a

16.3 in. QSXGA monitor. It has 2048×2560 pixels at 202 dpi, with the RGB subpixels on the columns (to maximize the time required to address each row). Such a panel typically has an aperture ratio of 27% and operates at 60 Hz ($\tau = 8.1\,\mu s$). That is, the subpixel dimensions are about $22\,\mu m \times 65\,\mu m$ and a pixel capacitance C_{LC} of 0.16 pF (assuming a cell gap of $d = 4$ um and permittivity $\varepsilon_{LC} = 5$).

First, the pixel should not fully discharge within the 16.7 ms frame time due to the off current of the transistor. Assuming an exponential decay and a 1% voltage tolerance gives the condition [78]:

$$\tau_{OFF} = C_{LC} \cdot R_{OFF} \approx C\frac{V_{SD}}{I_{SD}} > 200\tau N. \quad (6.60)$$

The α-Si characteristic from Figure 6.12b shows that I_{SD} is 20 pA when the TFT is OFF and 10 μA when ON. Hence, Equation 6.60 suggests that the TFT had a sufficiently low OFF current to drive a pixel of capacitance 6 pF, easily sufficient for the 40 pF pixels used in the original work of Reference [74], but is far too leaky to drive the 0.16 pF of a modern QSXGA monitor. Simply employing a storage capacitance C_S in parallel with the pixel, as shown in Figures 6.13 and 6.14, increases the capacitance to the desired level and ensures that the pixel remains charged throughout the frame. Similarly, I_{SD} in the ON condition for the highest signal level V_{SD} should be sufficiently high to charge the pixel to at least 99% of that voltage level within the addressing time τ, which occurs when τ_{ON} is 10% of the addressing time τ:

$$\tau_{ON} = C \cdot R_{ON} \approx (C_S + C_{LC})\frac{V_{SD}}{I_{SD}} < 0.1\tau. \quad (6.61)$$

Equations 6.60 and 6.61 can be combined to find the maximum number of lines that can be addressed by a TFT, N_{max}, with a given ration of ON to OFF currents:

$$N_{max} = \frac{1}{2000} \cdot \frac{I_{SD}(ON)}{I_{SD}(OFF)}. \quad (6.62)$$

The characteristic of Reference [73] shown in Figure 6.12b suggests that the original TFT could address up to 250 lines (as was claimed) but required improvement to address the 2048 rows of

the monitor display. Some of the improvements to TFT performance made to achieve such high levels of multiplexing, 256 gray levels and high contrast ratios are outlined in the following discussion.

The equivalent circuit for a more realistic pixel is shown in Figure 6.14b. The TFT includes an inherent resistance for the semiconductor R_{DS} and the capacitance per unit area of the gate C_{GS}. Also important is the conductivity of the liquid crystal G_{LC} and the resistance of the bus lines R_{BUS}. The simplest model for FETs predicts that for sufficiently high gate voltages, well above the transistor threshold V_{th} and drain voltage V_{SD}, the current I_{SD} is given by Reference [78]:

$$I_{SD} = \mu_e C_{GS} \frac{W}{L} \cdot \left(V_{GD} - V_{th} - \frac{1}{2}V_{SD}\right)V_{SD}, \quad (6.63)$$

where μ_e is the electron mobility, W is the width of the TFT channel, and L is the length, as defined in Figure 6.13. The TFT threshold is typically about +3 V and is directly related to the charge density of free electrons n_0:

$$V_{th} = -en_0\frac{d_G}{C_{GS}}, \quad (6.64)$$

where the gate capacitance per unit area C_{GS}:

$$C_{GS} = \frac{\varepsilon_0\varepsilon_{\alpha Si}}{d_G} \quad (6.65)$$

and d_G is the thickness of the SiN$_x$ insulator layer (Figure 6.13). The α-Si leakage resistance dominates the OFF current:

$$I_{SD} = \frac{W}{L} \cdot \frac{V_{SD}}{R_{DS}}. \quad (6.66)$$

Substituting Equations 6.63 and 6.66 into Equation 6.62 gives the following relationship:

$$N_{max} \sim \mu_e C_{GS}R_{DS}. \quad (6.67)$$

That is, achieving the low OFF current is done by reducing the transistor width to length ratio W/L, but the high ON to OFF current ratio, and hence maximum number of lines that can be addressed, is achieved by reducing the thickness of the insulating and semiconducting layers, leading to a typical C_{GS} of about 0.1 $\mu F/cm^2$. A low insulator thickness has the additional benefit of reducing the area of the storage capacitor and correspondingly increases the aperture ratio. Figure 6.12b also includes the characteristic of a TFT used for modern LCD panels, such as the QSXGA monitor described earlier (where $W/L = 10\,\mu m/6\,\mu m$).

Decay of the voltage across the pixel is not just related to the leakage current of the transistor and Equation 6.60: loss of charge across pixel occurs if the conductivity of the liquid crystal G_{LC} is too high. This is quantified by the voltage holding ratio (VHR), which represents the time it takes for the pixel voltage to decay to 50% [79]:

$$\text{VHR} = \sqrt{\frac{R_{ON}C}{2\tau N}\left(1 - e^{-\frac{2\tau N}{R_{ON}C}}\right)}. \qquad (6.68)$$

Ionic impurities in the liquid crystal must be minimized to maximize VHR and hence maintain a high display contrast. This cannot be done with nitrogenated compounds, preventing the use of highly polar materials such as the cyanobiphenyls. Instead, perfluorinated compounds are essential. Although mixtures produced from such compounds have a much lower $\Delta\varepsilon$, this is compensated by the low viscosity and hence fast switching speeds that can be achieved with these materials [79].

The pixel aperture ratio and hence the transmission efficiency of the backlight is dictated by the target display resolution, the area of the panel (due to the losses caused by R_{BUS}) and the fabrication tolerances. A typical TFT is fabricated as follows [78]:

1. The gate metal, usually Cr or Mo, is sputtered onto clean glass to a thickness of about 200 nm.
2. The first set of electrodes are wet etched: photoresist is printed over the glass area and exposed through a large area chrome mask placed with very high accuracy using a mask aligner. This is then developed to form the gate electrodes and bus line, and the bottom electrode of the storage capacitor.
3. Plasma-enhanced chemical vapor deposition (PECVD) is used to deposit 400 nm of Si_3N_4, followed by 130 nm of intrinsic α-Si and 50 nm of n^+ α-Si:H.

4. The source, drain, and storage capacitor electrodes are then formed by sputtering Cr over the surface and wet etching the appropriate patterns, again using a wet etching process and mask aligner.
5. The chrome electrode then acts as a self-aligned mask for plasma etching of the n^+ α-Si:H to complete the source and drain electrodes (often using an etchant stopper).
6. The remaining intrinsic α-Si:H is plasma etched away using a third mask and mask aligner step, thereby forming the TFTs.
7. The ITO is sputtered and plasma etched to form the pixels using a fourth mask and alignment step.
8. The last step uses PECVD to deposit the 350 nm of SiN_x or SiO_x barrier layer and uses a fifth mask to provide access to the bonding pads. A mask aligner is not usually required for this step that is far less critical than the previous photolithographic steps.

Each of the four critical mask alignment steps needs very high resolution to prevent panel variability. These steps are expensive, and various attempts to reduce the number of mask steps have been attempted, including the use of back-to-back diodes. High tooling costs lower the design flexibility and so TFT panels tend to be available only in a range of standard sizes and resolutions. If the requirement is for nonstandard size in a niche market, the end-user may need to choose a passive matrix approach, which is why the market for passive matrix displays remains strong despite the poorer performance. For a given display diagonal, the aperture ratio decreases linearly with increasing resolution. Various other pixel designs to that shown in Figure 6.13 have been suggested, often involving alternative placement of the storage capacitor. However, very high display resolutions require a different approach, such as the use of low-temperature polysilicon (LTPS).

Polycrystalline silicon (p-Si) was among the first semiconductors to be used for LCDs [80] and found in the first applications for TFT by *Canon* as the watch used in the 1983 film Octopussy and *Sharp's* 1991 hang-on-the-wall TV [7]. The material has a high mobility of 200–400 $cm^2/V\cdot s$, which is intermediate between the 1.5 $cm^2/V\cdot s$ of amorphous silicon and 1400 $cm^2/V\cdot s$ for crystalline. Such high mobilities allow far smaller transistors, higher ON currents [particularly important for organic light emitting diode (OLED) displays], and potentially integrating the display drivers onto the glass itself. This latter advantage potentially leads to significant overall cost savings because the drivers would be produced in the same process steps as the pixel TFT. The problem with producing p-Si TFTs was the very high processing temperatures, requiring those early demonstrators to be produced on quartz substrates. In the mid 1980s [81], LTPS TFTs were fabricated using excimer laser annealing of the α-Si to form the polycrystalline structure while keeping the processing temperature to 260°C, equivalent to that used for α-Si. Today, many smart phones benefit from the excellent properties of LTPS, which allows resolutions above 400 dpi and better battery life due to the reduction in backlight power that the high aperture ratio allows. However, the cost of LTPS is high because the fabrication of the top-gate transistors required uses 9–11 critical mask steps: this typically adds about 20% cost to the panels over equivalent α-Si LCDs.

Together with other disadvantages such as high leakage current, the high production cost of LTPS has a driven research into other semiconductors, including various metal oxides. ZnO is particularly interesting because it retains a high mobility and combines a very high ON to OFF state current ratio with optical transparency over the visible region. Recently, the 1:1:1 combination of indium gallium and zinc oxide (IGZO) [77,82] to form active element has been put into production by *Sharp*. A 50 nm amorphous IGZO layer is nitrogenated to form the n+ doped semiconductor with a mobility of $\mu_e = 10\,cm^2/V{\cdot}s$. Figure 6.12b includes a comparison of the TFT transfer characteristics for α-IGZO against both LTPS and α-Si. Although IGZO has a lower mobility than LTPS, it combines a very high ON/OFF current ratio (>10⁹) with the same low-cost fabrication designs of α-Si. It is likely that IGZO will play a leading role in future display devices, as displays continue to move to ever higher pixel contents, resolutions, and power efficiencies.

6.4 STANDARD LCD MODES

6.4.1 Overview

There is a wealth of different liquid crystals modes possible, some of the important ones being shown in Figure 6.15. Some modes have niche applications, such as the cholesteric temperature sensor or

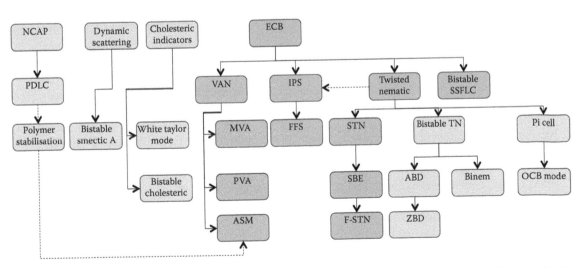

Figure 6.15 Dendrogram of the important LC modes. The LCD mainstream described in Section 6.4 is indicated by the darker grey coloration. Abbreviations: PDLC: polymer dispersed liquid crystal; ECB: electrically controlled birefringence; VAN: vertically aligned nematic; IPS: in-plane switching; MVA: multidomain vertically aligned; PVA photoaligned vertical alignment; ASM: axially symmetric mode; FFS: Fringe field switching; STN: supertwist nematic; SSFLC: surface stabilized ferroelectric liquid crystal; SBE: supertwisted birefringence Effect; F-STN: film/foil compensated STN; ABD: azimuthal bistable display; ZBD: zenithal bistable display; OCB: optically compensated bend-mode.

the ZBD and these will be described in Section 6.5. Modes that have contributed to the mainstream LCD markets from the early watches and calculators to today's full color video rate displays are covered in this section.

The evolution of the modern LCDs used in mobile phones, computer monitors, and HDTV has been undertaken in several stages, exemplified by listing the major advances, as in Table 6.4. After the basic principles had been evaluated, as described in the previous sections, the most important step was to increase the complexity of the display to allow hundreds of lines of information to operate at video frame rate, with gray scale and color. By the mid-1990s, the technological steps to achieve this had largely been satisfied through the adoption of the TFT active backplane. At that point, the race moved from complexity to appearance and in particular improving poor viewing angles. Several new modes were developed, each finding market success for different manufacturers, first allowing LCDs to replace CRT in computer monitors and in 2007, surpass CRT sales for TV. However, the war shifted to new battle grounds and yet further improvement of appearance, including ultrahigh resolution and wide-color gamut drive the competition in the mobile telecommunications marketplace. Before describing the individual modes, it is apposite to discuss the technological and market forces that dictated each shift in stage.

Following the discovery of the Alt–Pleshko multiplexing limit for passive matrix addressing [69] maximizing display content was the key driver for the LCD industry throughout the 1980s and early 1990s. Several approaches were taken:

1. Increasing the gradient of the LC transmission-voltage response, either through material improvement (controlling the elastic constant ratio k_{33}/k_{11}) or new LCD modes (STN).
2. Designing bistable LCD modes, where pixels no longer respond to the RMS signal over the frame, but are rapidly latched between the ON and OFF states and remains in the desired state after the signals applied.
3. Introducing a nonlinear element, such as TFT, that isolates and delivers the appropriate signal to each pixel.

The first of these approaches was successful for low-cost, black and white STN displays but was limited by the slow frame time, low number of gray levels, and a maximum of 480 multiplexed lines. Many novel methods for making an LCD inherently bistable were invented at the end of the 1970s and beginning of the 1980s [101]. The most promising bistable technology was the surface stabilized ferroelectric LCD (SSFLCD), invented in 1980 [88], launching an immense worldwide effort for the next 15 years in a two-way battle between the FLC and TFT approaches. The complexity of TFT fabrication seemed to promise higher costs over the passive matrix bistable approach. However, despite the launch of a color Surface Stabilised Ferroelectric Liquid Crystal (SSFLC) monitor by *Canon* in 1992 (Figure 6.16d), the TFT-driven TN display was already beginning to enable the image quality required for a new market: the laptop. Early laptop screens were black and white only (such as *Apples's MacIntosh Powerbook 100*, released in October 1991) but the sensitivity of SSFLC to shock made the technology unsuited to portable products. Although there was no superior alternative to the TFT TN LCD for laptop displays, the competition for the SSFLC monitor came from the superior performance and much lower cost of the CRT, and well before the end of the 1990s, the SSFLC development effort targeting mainstream displays was minimal.

Buoyed by early commercial success from laptops, the TFT TN would attract sufficient investment to explore new modes, targeting shifting the incumbent CRT from the monitor, and eventually TV markets. The use of TFTs brought with it a new and very important advantage: the flexibility of LC mode design. Modes that had previously been impossible to multiplex became potentially useable. In the late 1990s, modes such as IPS [92,93] and multidomain vertically aligned (MVA) nematic [94,95] offered sufficiently wide viewing angles to displace the CRT. By the beginning of the new millennium, the LCD was dominant from watches and calculators to large area monitors [102], as shown in Figure 6.16.

The battle for the television market was complicated by another new technology: the plasma display panel (PDP). As an emissive display, the viewing angle and contrast of PDP could not be surpassed by LCDs. This gave PDP an early lead in the flat-panel television market. However, ongoing LCD improvements, such as the invention of the fringe field switching (FFS) mode [96,97], and the inability to fabricate high-resolution PDP below

Table 6.4 Significant advances on route to LCD market dominance

Year	Invention	Protagonists	References	Stage
1967	Dynamic scattering mode	Heilmeier et al., *RCA*; US	[3,4]	Basics
1971	Active matrix addressing	Lechner et al., *RCA*; US	[70]	
1971	Twisted nematic mode (TN)	Schadt and Helfrich, *BBC*; CH. Fergason, *U. Kent*; US	[14] [83]	
1971	Vertically aligned nematic mode (VAN)	Kahn, *Bell Labs*; US Schiekel and Fahrenschon, *AEG-Telejunken*; W. Germany	[84] [85]	
1973	Cyano-biphenyl nematics	Gray et al., *Hull, RSRE, BDH*; UK	[15]	
1973	Formulation of wide temperature range eutectic LCD for TN	Raynes, *RSRE*; UK	[16]	
1973	In-plane switching (IPS) proposed	Kobayashi, *U. Tokyo*; JP Soref, *Sperry*; US	[86] [87]	
1973	First commercial LCD products in calculator and wristwatch	Wada et al., *Sharp*; JP *Seiko*; JP	[7]	
1974	Defect-free TN	Raynes and Waters, *RSRE*; UK	[58]	
1979	First amorphous silicon TFT used to address LCD	Spear and Le Coomber, *U. Dundee*; UK Hilsum, Hughes, *RSRE*; UK	[72] [73]	Increased complexity
1980	Bistable surface stabilized ferroelectric liquid crystal (SSFLC)	Clark and Lagerwall, *U. Göteborg*; SE	[88]	
1982	Supertwist mode (STN)	Raynes, *RSRE*; UK	[89]	
1984	Supertwist birefringence effect (SBE)	Scheffer and Nehring, *BBC*; CH	[90]	
1987	Foil compensation for STN	Katoh, *Asahi Glass*; JP	[91]	
1992	IPS mode	Baur et al., *Merck*; DE Kondo et al., *Hitachi*; JP	[92] [93]	Wide viewing angle
1997	Multidomain vertically aligned mode (MVA)	Koike and Okamoto, *Fujitsu*; JP	[94,95]	
1998	Fringe field switching mode (FFS)	Lee et al., *Hyundai*; KR	[96,97]	
2000	Patterned vertical alignment mode	Kim et al., *Samsung*; KR	[98]	
2001	Axially symmetric mode (ASM)	Yamada, Ishii et al., *Sharp*; JP	[99]	
2004	IGZO TFT invented	Nomura et al., *Tokyo Institute of Technology*; JP	[82]	Improved resolution and color
2010	QD enhanced color backlights	Jang et al., *Samsung*; KR	[100]	

Figure 6.16 State-of-the-art LCD at the turn of the millennium. (a) Watches and calculators using the reflective TN, (b) a dual display phone and PDA using foil compensated STN, (c) a laptop computer with TFT-driven TN, (d) a 14 in. monitor display based on bistable SSFLCD, and (e) one of the first IPS mode TFT monitors [100].

40 in., led to the eventual triumph of LCDs in this most important of markets too.

Coincident with LCD domination of the TV display market, the first i-Phone was launched in June 2007, heralding the age of the smart phone. The goal of providing resolution at the limit of visual perception helped drive LTPS into the marketplace. However, LTPS also had the advantage of providing higher ON currents than was possible, thereby meeting the requirement for another emissive display technology: OLED. Not only do OLED surpass LCD for contrast and viewing angle but also for color saturation. Again, the adaptability of LCD technology provides a solution: replacing the cold cathode backlights with a blue LED and adding a film containing red and green quantum dots (QD), provides sharper colors than is possible using color filters alone [100]. Once more, LCD performance increases to meet the market challenge. Midway through the 2010s, OLED is making grounds for portable displays, where features such as form factor and weight also play an important role but LCDs continue to dominate in all other markets (Figure 6.17). Will OLED eventually replace the LCD altogether? At each stage of LCD evolution, new markets have been enabled by

particular LCD modes. Many of these modes have retained their market share despite more advanced options become available. Whether it is the simple TN in watches and calculators, the STN for instrumentation, or TFT-driven TN for low-cost monitors, the successful technologies become difficult to supplant once established.

6.4.2 Dynamic scattering mode

The first operating liquid crystal device was an electro-optic shutter, devised by Heilmeier's predecessor at *RCA*, Richard Williams, in 1962 [11,103]. Williams showed that strong turbulence could be induced in a roughly planar sample of a negative $\Delta\varepsilon$ nematic with a DC or low-frequency AC field. This turbulence was induced by ionic flow in the liquid crystal disrupting the liquid crystal, thereby causing strong optical scattering in the birefringent medium. Heilmeier used this switching between scattering and nonscattering states in *RCA*'s first display demonstrators in 1968. Unlike most of the other modes described, dynamic scattering mode devices did not need polarizers, operating in a 20–100 μm spaced cell at typically 30–50 V. The onset voltage for scattering is approximately [104]

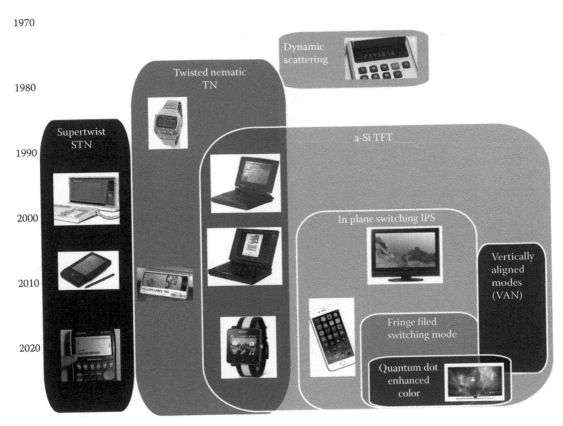

Figure 6.17 History of the mainstream of LCD from the calculator to QD 4k FHD TV.

$$V_C \sim T \frac{\eta\mu}{\varepsilon}, \tag{6.69}$$

where η is the bulk flow viscosity, μ is the ion mobility, and T is a constant typically about 100. For $\eta = 0.5$ Pa·s and $\mu = 10^{-4}$ cm²/V·s then V_C is about 5–10 V. The resistivity of the liquid crystal could be made lower than 10 GΩ·m through deliberate addition of mobile ionic impurities. The original displays of Heilmeier used DC fields to induce the ion flow, presumably because of the desire to address the devices using transistors. However, the lifetime was too short due to electrostatic breakdown effects and so the first commercial devices used low-frequency AC instead.

6.4.3 The twisted nematic LCDs

6.4.3.1 INTRODUCTION

By 1980, practically all commercial LCDs were based on the TN. This device set the benchmark for future developments, operating using alignment layers to obtain a uniformly aligned director profile, a cell gap of a few microns sandwiched between polarizers either side. The electro-optic effect is the basic Fréedericksz transition described in Section 6.3.3, wherein the director responds to the RMS voltage coupling to the dielectric anisotropy affecting a change to the birefringence profile through the device, and hence altering the polarization state of the transmitted light. Each of the remaining devices described in this section uses these principles (although the FLC is a field effect device, with a polar coupling of the field to the ferroelectric spontaneous polarization). The TN uses a positive $\Delta\varepsilon$ material, and usually a 90° twist between top and bottom surfaces, with either crossed or parallel polarizers, to operate in normally white (NW) or normally black (NB) modes, respectively. Applying a voltage three or four times greater than the threshold (i.e., applying typically 5 V) causes sufficient reorientation of the director in the bulk of the cell to cause the NW cell to appear dark, and the NB cell bright. TN may be used in transmission, usually driven by TFT and including color filters, or may be used as a two-polarizer reflective mode. In this latter device, the rear polarizer incorporates a diffusive reflector so that light incident from the front of the panel is reflected back through the device for a second pass. This leads to a much higher contrast ratio, since any light leakage in the dark state from the first pass is dramatically reduced on the second. However, it also means that the coloration of the white state is more important. Also possible are single-polarizer reflective modes, as described at the end of this section.

A key part of any display design is to ensure uniformity of texture for the ON and OFF states. For a 90° TN this means ensuring that degeneracy of both twist and tilt are removed [58]. For twist alone this is simply done using slightly uncrossed alignment directions or, preferably, by inducing a natural sign of twist with the addition of a small amount of cholesteric to give a pitch of a few hundred micron. Removing tilt degeneracy requires that there is pretilt on both alignment surfaces. However, it remains essential that the pretilt and sign of twist are matched, so that there is minimal splay from one surface to the other when undergoing the correct twist. Otherwise, reverse tilt and twist domains may still form as the director relaxes to the quiescent state after switching.

6.4.3.2 THE OFF STATE AND TN DESIGN

The polarization optics from a uniformly aligned birefringent liquid crystal cell was calculated in Section 3.2. It was shown that the polarization state after transmission through multiple elements is calculated simply by multiplying the appropriate Jones matrices together. The TN has a director twist angle that varies linearly from one surface to the other. Thus, it can be described as a series of very thin birefringent retarders, each with a slightly different orientation angle. Slicing the device into N layers, the retardation and orientation of each is given by

$$\frac{\Gamma}{N} = \frac{2\pi \Delta n d}{\lambda N}; \phi_{j+1} = \phi_j + \frac{\Phi}{N}, j = 0, 1, 2, \ldots, N-2,$$

(6.70)

where Φ is the total twist angle. The overall Jones matrix M is the multiplicative sum of each of these elements M_j:

$$M = \prod_{j=1}^{N} \left(R(\phi_j) \cdot M_j \cdot \overline{R(\phi_j)} \right) \equiv \prod_{j=1}^{N} \left(R(-\phi_j) \cdot M_j \cdot R(\phi_j) \right)$$

$$= R(-\Phi) \left[\begin{pmatrix} e^{-i\frac{\Gamma}{2N}} & 0 \\ 0 & e^{i\frac{\Gamma}{2N}} \end{pmatrix} R\left(\frac{\Phi}{N}\right) \right]^N$$

$$= R(-\Phi) \left[\begin{pmatrix} \cos\frac{\Phi}{N} e^{-i\frac{\Gamma}{2N}} & \sin\frac{\Phi}{N} e^{-i\frac{\Gamma}{2N}} \\ -\sin\frac{\Phi}{N} e^{i\frac{\Gamma}{2N}} & \cos\frac{\Phi}{N} e^{i\frac{\Gamma}{2N}} \end{pmatrix} \right]^N.$$

(6.71)

As N tends to infinity, then each retarder becomes infinitesimally thin, and [105]

$$M = \begin{pmatrix} \cos\Phi & -\sin\Phi \\ \sin\Phi & \cos\Phi \end{pmatrix} \begin{pmatrix} \cos X - i\frac{\Gamma}{2}\frac{\sin X}{X} & \Phi\frac{\sin X}{X} \\ -\Phi\frac{\sin X}{X} & \cos X + i\frac{\Gamma}{2}\frac{\sin X}{X} \end{pmatrix},$$

(6.72)

where

$$X = \sqrt{\Phi^2 + \left(\frac{\Gamma}{2}\right)^2}.$$

Solving for the transmission

$$T = \left\{ \cos X \cos\left(\Phi + \phi_1 - \phi_2\right) + \frac{\Phi}{X}\sin X \sin\left(\Phi + \phi_1 - \phi_2\right) \right\}^2 + \left\{ \left[1 - \left(\frac{\Phi}{X}\right)^2\right] \sin^2 X \cos^2\left(\Phi - \phi_2 - \phi_2\right) \right\},$$

(6.73)

where the input and output polarizer directions with respect to the input director are ϕ_1 and ϕ_2, respectively. As expected, Equation 6.73 reduces to the transmission of a retardation plate, Equation 6.25 if the overall twist $\Phi = 0$, and crossed polarizers are used ($\phi = \phi_1$ and $\phi_2 = \phi_1 + 90°$).

For the NW mode TN with $\Phi = 90°$, the polarizers are crossed ($\phi_2 = \phi_1 + 90°$) and oriented with either $\phi_1 = 0°$ or 90°. The transmittance from

Equation 6.73 then simplifies to the Gooch–Tarry expression [106]:

$$T = 1 - \frac{\sin^2\left(\frac{\pi}{2}\sqrt{1+\left(\frac{2\Delta n \cdot d}{\lambda}\right)^2}\right)}{1+\left(\frac{2\Delta n \cdot d}{\lambda}\right)^2} \qquad (6.74)$$

and the value of $\Delta n \cdot d$ is chosen to ensure that the right-hand term is zero for $\lambda \approx 550\,\mathrm{nm}$, so that the TN appears white. If the polarizers are kept parallel instead, then the transmittance becomes

$$T = \frac{\sin^2\left(\frac{\pi}{2}\sqrt{1+\left(\frac{2\Delta n \cdot d}{\lambda}\right)^2}\right)}{1+\left(\frac{2\Delta n \cdot d}{\lambda}\right)^2} \qquad (6.75)$$

and the TN is NB at the same retardation conditions. Equations 6.74 and 6.75 are plotted versus retardation in Figure 6.18. There is minimum with NB, maximum with NW, where the twisted structure of the birefringent material transmits linearly polarized light that is orthogonal to the input polarization and perpendicular (for NB, parallel if NW) to the output polarizer. This occurs at a series of conditions, given by

$$\frac{\Delta n \cdot d}{\lambda} = \sqrt{m^2 - \frac{1}{4}}, \quad m = 1, 2, 3\ldots \qquad (6.76)$$

including the first minimum at $\Delta n \cdot d = 1/2\sqrt{3}\,\lambda$ and second at $\Delta n \cdot d = 1/2\sqrt{15}\,\lambda$. At any one of these minima (or maxima for the NW, although the term minima is used conventionally regardless of the polarizer orientations), the transmission is least sensitive to changes in retardation (including viewing angle and cell non-uniformity) and polarizer orientations. As the retardation is increased (i.e., for high cell gaps) successive minima are increasingly less sensitive to retardation changes until the Mauguin condition is approached at the high retardation limit [45], where polarized light transmitted by the cell is rotated through 90° regardless of cell gap, wavelength, temperature, or polarization angle.

The choice of retardation is dictated by fabrication limitations on cell gap, the operating speed that is required, the coloration of the white state and the required contrast. Response times depend on the square of the cell gap, through Equation 6.48. This usually limits practical operation to either the first or second minimum condition. The wavelength dependences for first minute and second minute devices are shown in Figure 6.19. Again, choosing the wavelength to occur at the peak eye response of $\lambda = 550\,\mathrm{nm}$, the first and second Gooch–Tarry

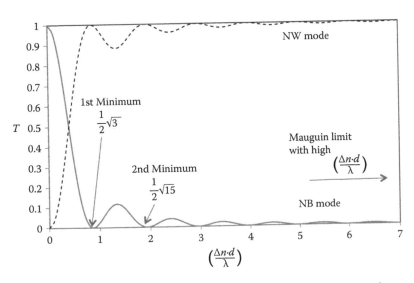

Figure 6.18 The Gooch–Tarry curve for the quiescent transmission of a 90° twisted nematic display versus retardation, for NB mode (parallel polarizers) and NW mode (crossed polarizers).

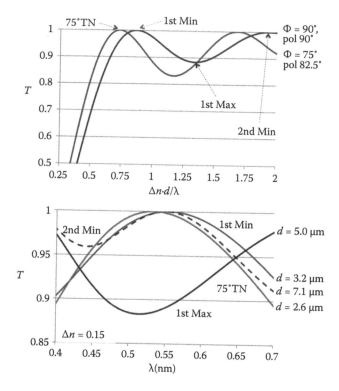

Figure 6.19 Choices of white state and cell gap for the TN. (a) The Gooch–Tarry curve for the 90° and 75° twist angle TN. (b) Theoretical wavelength dependences of the transmission for a material with a typical birefringence ($\Delta n = 0.15$) for 90° TN cells operating at the first minimum, first maximum, and second minimum and for a 75° TN with uncrossed polarizers at 85°. The cell gaps are also indicated.

minima for a TN filled with 5CB are 2.6 μm and 5.9 μm, respectively. Most TFT LCDs operate at the first minimum, using lower birefringence materials, $\Delta n \approx 0.1$ and $d \approx 4.5$ μm because this cell spacing is conducive to a high manufacturing yield and the switching speed is satisfactory. If switching speed is the priority, rather than use higher birefringence materials with the concomitant increases of viscosity, an alternative approach is to use a 75° twist [67]. The first minimum peaks at a lower transmissivity than for a 90° cell (98.4%) but this can be corrected by reorienting the polarizers through 7.5°, as shown for the results in Figure 6.19. Typically, the cell gap for the 75° TN mode is 80% that of the standard 90° TN, potentially leading to a 50% speed increase, provided the reduced cell gap remains suitable for manufacture.

For each of the modes, it is important to consider the coloration of the white state. For a backlit color TN, this is less important than for a black and white reflective device because any loss of white color balance can be compensated in the addressing signals, back light spectrum, and color filters. For a dual-polarizer reflective display, the second pass of the light through the cell exaggerates any unwanted coloration.

Figure 6.19b shows the theoretical wavelength dependences in transmission for four TN modes. Converting these spectra to Commission internationale de l'éclairage (CIE) 1931 x and y color coordinates, as shown in Figure 6.20, gives a locus of white hues depending on the retardation and the coloration of the polarizers [107]. Two loci are plotted, corresponding to a standard polarizer centered on green and a slightly bluer variety. These polarizers represent the range of colors that can be targeted while retaining a black OFF state; narrower spectrum polarizers can also be used to modify the ON state coloration, but with a noticeable coloration of the dark state too. Also shown on the chart are the coordinates for the D65 standard, representing a target for the ideal white state. Operating at the first minimum tends to give a greenish hue to the white state, whereas the second minimum

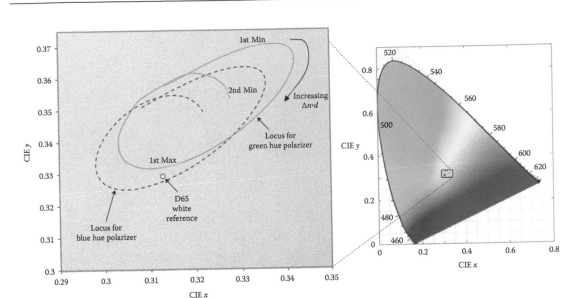

Figure 6.20 Central region of the CIE color chart (shown in full on the right) indicating the region of white transmission for a dual-polarizer reflective mode TN. Two loci are shown, corresponding to polarizers with a slightly green tint (continuous line) and a slightly blue tint (dashed). The D65 white state reference is shown as a target for pure white. The location on the color chart of the first min, second min, and first maximum are indicated.

gives an improved white due to the increased blue transmittance shown in Figure 6.19b. Thus, a second minimum TN is preferred for applications where the response time is unimportant (being five times slower than the equivalent first minimum mode) but the attractive bluish white is preferred. Alternatively, an intermediate retardation close to the point of the first maximum (i.e., at $\Delta n \cdot d / \lambda = \frac{2}{7}\sqrt{22}$) has been used [107] to produce a neutral white as close to the D65 standard as possible, particularly when combined with the slightly blue tinged polarizer. The optimum cell gap for this is

$$d \approx \frac{1}{4\Delta n}\left(0.45\sqrt{15} + 0.68\sqrt{3}\right), \quad (6.77)$$

where the blue light is close to the second minimum and red wavelengths close to the first minimum and the green transmittance is decreased somewhat (Figure 6.19b).

The other important consideration for the OFF state is the viewing angle. Figure 6.21a shows a polar plot of transmissivity for the azimuthal and zenithal directions for a NW first minimum TN operating in transmission. The OFF-state viewing angle is good: light at any azimuthal angle experiences the same retardation profile due to

the twisted structure and deviations of zenithal angle away from the display normal (shown as the central point) represent reduction of the effective retardation to which devices operating at the Gooch–Tarry minima are insensitive.

6.4.3.3 THE ON STATE AND OPTICAL COMPENSATION

The ON state transmission is very viewing angle dependent. This is evident from the contrast ratio of OFF/ON transmission shown in Figure 6.21b because contrast is dominated by the dark ON state transmission. The blackest ON state occurs when the light is transmitted parallel to the liquid crystal optic axis, the direction where the liquid crystal behaves as an isotropic medium and any light transmitted is due to leakage of the polarizers. Contrast is highest for on-axis light because this is the direction of the applied field toward, which the director reorients, although it remains affected by the distorted regions close to the surface. Off-axis light experiences a retardation with a direction dependence that is related to the director profile through the cell, which in turn depends on the applied voltage.

The Fréedericksz threshold voltage for a TN is given by

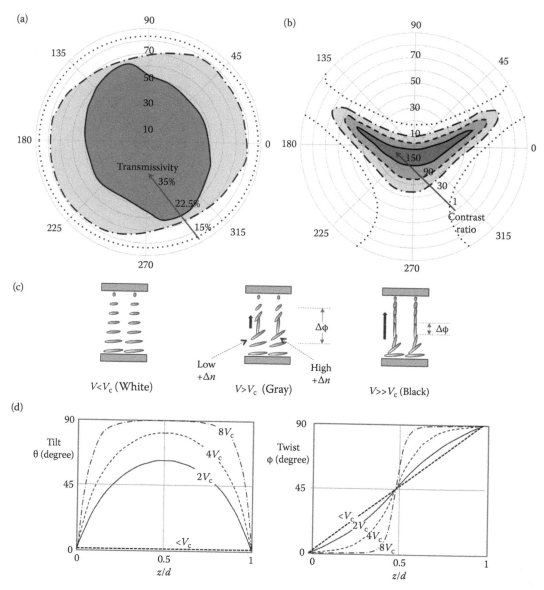

Figure 6.21 Measured viewing angle characteristics of a first minimum NW TN operating in transmission, showing polar plots of (a) OFF state transmissivity and (b) OFF/ON contrast ratio; (c) schematic of the TN with low, intermediate, and high voltages; (d) twist and tilt director profiles for the TN versus voltage.

$$V_C = \sqrt{\frac{\pi^2 k_{11} + \Phi^2 (k_{33} - 2k_{22})}{\varepsilon_0 \Delta \varepsilon}}, \quad (6.78)$$

which simplifies to

$$V_C = \pi \sqrt{\frac{k_{11} + (k_{33} - 2k_{22})/4}{\varepsilon_0 \Delta \varepsilon}} \quad (6.79)$$

for a standard $\Phi = 90°$ TN. Often, $k_{22} \approx k_{33}/2$ in practice, and so the Fréedericksz threshold is similar to that of the ECB mode, and typically around 1 V. Figure 6.21c shows schematically what happens as the voltage is increased above V_C, and numerical calculations of the response for typical elastic constants are given in Figure 6.21d. As the voltage increases, the initially linear twist from one surface to the other becomes increasingly concentrated into the center of the cell where the director approaches the vertical condition, whereas regions of splay and bend are increasingly pushed toward the surfaces. Thus, the director profile approaches a vertically aligned central region, with twist-free splay–bend regions oriented at 90° to each other close to the two surfaces.

The optical transmission through the cell is complex and requires numerical modeling. The voltage dependence of the normal incidence transmittance can be estimated by considering the midplane tilt of the director θ_m [108,109]. Immediately, above V_C this is

$$\theta_m^2 = 4\left(\frac{V}{V_C}-1\right)\Bigg/\left(\frac{k_{33}-\left(\frac{\Phi}{\pi}\right)^2\left[\frac{k_{33}^2}{k_{22}}+k_{22}-k_{33}\right]}{k_{11}+\left(\frac{\Phi}{\pi}\right)^2\left[k_{33}-2k_{22}\right]}+\frac{\Delta\varepsilon}{\varepsilon_\perp}\right)$$

(6.80)

and for the 90° TN cell it is

$$\theta_m^2 = 4\left(\frac{V}{V_C}-1\right)\Bigg/\left(\frac{5k_{33}-\left[\frac{k_{33}^2}{k_{22}}+k_{22}\right]}{4k_{11}+\left[k_{33}-2k_{22}\right]}+\frac{\Delta\varepsilon}{\varepsilon_\perp}\right)$$

$$\approx 4\left(\frac{V}{V_C}-1\right)\Bigg/\left(\frac{5k_{33}}{8k_{11}}+\frac{\Delta\varepsilon}{\varepsilon_\perp}\right)$$

(6.81)

for the reasonable approximation $k_{33} \approx 2k_{22}$. Given that the surface pretilt is small, this expression represents the linear part of the transmission-voltage characteristic above V_C shown in Figure 6.9b. The multiplexibility of the TN is related to the steepness of this characteristic through the Alt–Pleshko expression (Equation 6.54). Thus, to ensure the highest number of lines that a passive-matrix TN can be multiplexed, the denominator of Equation 6.81 should be kept low. This is done in practice using hybrid mixtures of polar and nonpolar compounds [110], in which k_{33}/k_{11} is lowered due to short-range order effects, offered a route to multiplexing of up to 20 lines.

Off-axis calculations of the transmissivity for the ON state certainly require numerical modeling. However, the form of the contrast ratio curve shown in Figure 6.21b is anticipated by considering the director profile of the high voltage state as a vertically aligned nematic (VAN) in the bulk, with two orthogonal hybrid aligned states in the surface regions. The directors in these two regions are oriented parallel to the polarizers, at azimuthal directions 45° and 135° in Figure 6.21b, and hence appears dark at all angles. This understanding led to pronounced improvements of transmissive mode TN viewing angles through optical compensation. Discotic liquid crystals are similar to the standard rod-like calamitic mesogens used in displays but exhibit negative birefringence. Combining discotic and calamitic layers with equal but opposite retardations leads to a net optically isotropic medium, black at all angles when between crossed polarizers. Figure 6.22 shows how this works in practice [111,112]. Polymer film fabricated from a hybrid aligned discotic are laminated either side of the LCD with the optic axes parallel to the fully ON state director at the adjacent surface, and the splay antisymmetric. This leads to a greatly improved off-axis contrast ratio, as shown in Figure 6.22b, and helped lead to sufficiently good performance for TN displays to be used in laptops and monitors.

Two-polarizer reflective mode TN devices are usually used in low-cost applications such as watches and calculators where the cost of extra optical compensation layers is prohibitive. However, the viewing angle characteristic of such devices is far more symmetrical than that of the transmissive device shown in Figure 6.21b, due to a self-compensating effect. Off-axis light that experiences a lower Δn from the director tilted toward it on the first pass through the device, experiences a correspondingly higher Δn when traversing in the other direction on the second pass after reflection.

With all reflective displays, it is important to consider the illumination conditions; in particular the color balance of the incident light and its degree of diffusivity. Indeed, the viewing angle characteristic can be considered both as an optical output for diffuse illumination and as a means for ensuring the maximum light input. In practice, the LCD reflectivity is controlled to some extent by the diffusivity of the reflector and/or front polarizer. Indeed, adaptions of the reflective layers can be used to deliberate trade-off viewing angle and reflectivity. The appearance of the display can be remarkably different when viewed by the diffuse light of a cloudy day or by the highly directional light on a sunny day or in a dark room with a single

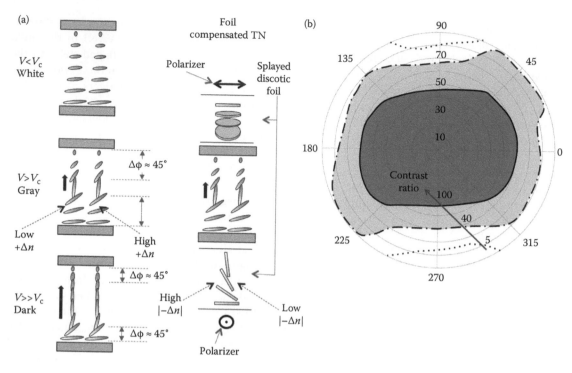

Figure 6.22 Foil compensated TN. (a) Use of discotic foils to compensate the viewing angle of the OFF state for a transmissive TN LCD; (b) experimental measurements for the resulting viewing angle.

light source. In the former case, the appearance of a scattering mode display will easily surpass that of the polarized LCD but the situation is reversed for directional lighting and viewing closer to the specular angle.

6.4.3.4 SINGLE POLARIZER REFLECTIVE MODE TN

Conventional two-polarizer reflective TN and STN LCDs have two further optical limitations. Parallax caused by the separation of the rear reflecting polarizer from the image plane by the thickness of the rear plate, leads to shadowing of the image when viewed off-axis. This can be distracting for black and white devices but is severely detrimental to reflective color devices due to color leakage between subpixels. Second, the optical efficiency of the white state is relatively low because the light passes four times through the polarizers. The transmission of the highest quality LCD polarizer is 43%. However, a further 5% is absorbed on each pass so that the maximum reflectance possible with a two-polarizer reflective display is less than 28%, typically 23%. This is also a severe limitation for pixelated reflective color LCDs, where

the color filters and decreased aperture ratio then lead to prohibitively low 7% reflectivity.

Single polarizer reflective TN reduces these optical losses by using a front polarizer only and using an internal reflector on the inside surface of the rear substrate of the display to remove the parallax. The reflector is made slightly diffusive to scatter the reflected light in different directions, so that the OFF state has an attractive appearance and good viewing angle even when lit by a point source. The reflector can form the rear electrode as shown in Figure 6.23a.

Single polarizer operation requires that the liquid crystal profile is designed to rotate the input polarized light through 90° after both passes through the liquid crystal, thereby being absorbed by the single front polarizer and appearing black. That is, the polarization is elliptically polarized after the first pass, the indicatrix is rotated on reflection and the light becomes linearly polarized after the second pass orthogonal to the input polarizer: the display appears dark. For a positive $\Delta\varepsilon$ material, this will always correspond to the OFF state because in the ON state the director approaches the vertical condition and appears optically isotropic:

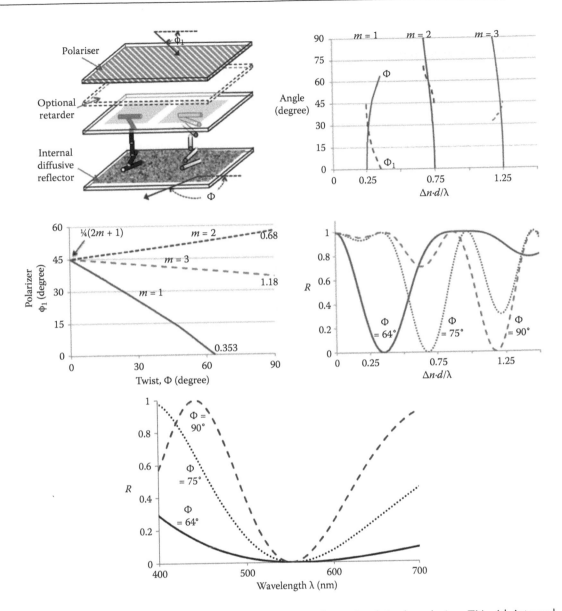

Figure 6.23 Options for single polarizer TN modes. (a) Schematic of single polarizer TN with internal diffusive reflector. (b) Solutions for the NB mode single polarizer TN. Continuous lines show the twist angle Φ for the first three solution sets to Equation 6.85, and the dashed lines show the corresponding polarizer angles ϕ_1. The numbers indicated represent the retardation with 90° twist, whereas the untwisted cells give $\Delta n \cdot d / \lambda = 0.25, 0.75$, and 1.25, for each mode, respectively, each with $\phi_1 = 45°$. (d) Calculated reflectivity for $\Phi = 64°, 75°$, and 90° twists, with polarizer angles set to give $R = 0$ for $m = 1, 2$, and 3, respectively. (e) Dark state transmission for $\Phi = 64°, 75°$, and 90° for a low Δn materials (0.09).

no change to the polarization occurs on either pass and the light is transmitted. The theoretical curves shown in Figure 6.23 ignore losses: in practice the reflectivity may reach up to 35% in the single polarizer modes. The image may be inverted and NW

operation achieved by the addition of a quarter-wave plate between the polarizer and the liquid crystal, as shown in Figure 6.23a.

Consider a single polarizer display using the simplest LCD geometry, the ECB mode of Section

6.3.3. Here, the liquid crystal acts as a switchable wave plate. Multiplying the Jones matrices for a polarizer, wave plate with optic axis at angle ϕ_1 to the polarizer, and the same wave plate and polarizer after reflection gives

$$R = 1 - 4\sin^4 \Gamma \sin^2 2\phi_1, \qquad (6.82)$$

where the retardation is approximately

$$\Gamma = \frac{\pi \Delta n \cdot d}{\lambda} \cos^2 \overline{\theta} \qquad (6.83)$$

and $\overline{\theta}$ is the average director tilt and ϕ_1 is the polarizer orientation measured from the input director. The OFF state appears dark $(R = 0)$ if $\Gamma = \phi_1 = m\pi/4$ $(m = 1, 3, 5...)$. That is, at the quarter-wave plate condition light is circularly polarized with one handedness, which is swapped to the other handedness after reflection and no longer transmitted by the retarder. This mode suffers from poor dark state leakage and coloration, due to wavelength and angular dispersion away from the quarter-wave condition. As for transmissive state devices, improved performance is achieved using the TN geometry.

There are several options for single polarizer reflective TN operation [67], including twist angles ranging from 60° to 90°. Replacing the wave plate with the Jones matrix for a TN in the derivation above gives [113]:

$$R = \left(1 - \frac{2\alpha^2}{1+\alpha^2} \sin^2\gamma\right)^2 + \left(\frac{2\alpha \cdot \sin\gamma}{1+\alpha^2}\right)^2$$

$$\left[\left(\sqrt{1+\alpha^2}\right)\cos\gamma \cdot \cos 2\phi_1 - \sin\gamma \cdot \sin 2\phi_1\right]^2, \quad (6.84)$$

where $\alpha = \Gamma/\Phi$ and $\gamma = \Phi\sqrt{1+\alpha^2}$. This gives dark state solutions $(R = 0)$ when the following two conditions are met simultaneously:

$$\Phi = \frac{1}{\sqrt{1+\alpha^2}} \sin^{-1} \sqrt{\frac{1+\alpha^2}{2\alpha^2}}$$

$$\phi_1 = \frac{1}{2} \tan^{-1}\left(\sqrt{\alpha^2 - 1}\right). \qquad (6.85)$$

These solutions represent the cases where the liquid crystal is at the quarter-wave condition, rotating the polarization through 90° over the two passes and tends to those of Equation 6.82 as Φ tends to zero $(\alpha \rightarrow \infty)$, with $\phi_1 = \pm\pi/4$ and $\Delta n \cdot d/\lambda = (2m + 1)/4$ for $m = 0, 1, 2...$. Figure 6.23b and c shows the solutions to Equation 6.85 for the first three orders.

Consider the solution for $\phi_1 = 0$, shown in Figure 6.23d and e: Equation 6.85 predicts that $R = 0$ occurs when $\alpha = \pm 1$, for which $\Phi = \pi/(2\sqrt{2}) = 63.64°$ and $\Delta n \cdot d/\lambda = 0.353$. This is a rather low retardation; even if a commercial liquid crystal with the lowest birefringence available $(\Delta n \approx 0.09)$ is used. This LCD mode requires a 2.2 μm cell gap, potentially lowering production yield and increasing cost. Second- and third-order examples are also listed in Table 6.5. Although the cell gap becomes more typical, the coloration of the dark state is far too high for use in a conventional display. Such modes, however, are useful for liquid crystal on silicon (LCOS) projection systems, where three panels are tuned to operate at the RGB wavelengths.

The 63.6° single-polarizer TN [113] was successfully deployed in Nintendo's *Color Game Boy* games console from 1998 to 2003. The display was manufactured by *Sharp*, who included achromatic retarders to invert the display to NW operation. This minimized dark state leakage, which is essential for any color display, whereas the resulting wavelength dependence of the white state was readily compensated through the color filters. Although the design gives the highest reflectivity possible with a polarized light mode LCDs, the introduction of color filters still leads

Table 6.5 Examples of normally black solutions for single polarizer TN

Order of minimum	Director twist Φ (degree)	Polarizer angle ϕ (degree)	$\Delta n \cdot d/\lambda$	Cell gap d (μm) ($\Delta n = 0.09$)
1st	63.6	0 (or 90)	0.353	2.2
2nd	45	56.4 (or −33.6)	0.683	4.2
	60	63.6 (or −26.4)	0.606	3.7
3rd	90	32.7 (or −57.3)	1.175	7.2

to a maximum reflectance of less than 10%. The approach taken by *Sharp* was to introduce prismatic elements into the back reflector, thereby directing off-axis light into the viewing direction; the reduced viewing angle display was considered suitable for this single viewer application. The adoption of better performing backlights and high aperture ratio LTPS back planes eventually led to transmissive displays only becoming acceptable for color portable applications. Interestingly, full color reflective and transflective displays remain an area where no technology, liquid crystal, or otherwise, has yet met the performance required for market success.

6.4.4 Supertwist nematic LCDs

The TN device usually includes a trace amount of cholesteric dopant, simply to impart a natural handedness and prevent domains impairing the appearance. Increasing the cholesteric content so that the natural pitch of the chiral nematic P is much lower allows twists of greater than 90° to be reached [114]. The range of conditions for such "supertwist" states is [89]

$$\frac{\Phi}{2\pi} - \frac{1}{4} \leq \frac{d}{P} \leq \frac{\Phi}{2\pi} + \frac{1}{4} \qquad (6.86)$$

or

$$\left(1 - \frac{\pi}{2\Phi}\right) \leq \beta \leq \left(1 + \frac{\pi}{2\Phi}\right), \qquad (6.87)$$

where $\beta = 2\pi \cdot d / P\Phi$. The STN range of twist angles is $90° < \Phi \leq 270°$, which corresponds to differences in the rubbing direction $\phi_1 - \phi_2$ of

$$\phi_1 - \phi_2 = \Phi - \pi, \qquad (6.88)$$

where it is important to ensure that the sign of the liquid crystal helix matches the pretilt of the two surfaces, in the same fashion as the TN. In practice, higher pretilts are needed for STN, typically $2° \leq \theta_S \leq 8°$.

Figure 6.24 shows the transmission characteristics for a set of devices with twist angles operating across the supertwist range [109]. As the twist increases, both the Fréedericksz threshold voltage

and the steepness of the curve increase, the latter allowing an increase of the number of lines that can be passive matrix addressed [89].

The inherent twist of the chiral nematic effectively reduces the twist elastic constant by the term $(1 - 2\pi/P)$ and the Fréedericksz threshold becomes

$$V_C = \pi \sqrt{\frac{1}{\varepsilon_0 \Delta \varepsilon} \left[k_{11} + \left(\frac{\Phi}{\pi}\right)^2 \left\{ k_{33} - 2k_{22}(1 - \beta) \right\} \right]}. \qquad (6.89)$$

Hence, V_C increases with twist angle Φ. Similarly, the voltage dependence of the midplane tilt angle is also affected, with Equation 6.80 now given by [109]

$$\frac{\theta_m^2}{4\left(\frac{V}{V_C} - 1\right)} = 1 / \left(\frac{k_{33} - \left(\frac{\Phi}{\pi}\right)^2 \left[\frac{k_{33}^2}{k_{22}} + k_{22}(1 - 4\beta + \beta^2) \right. \\ \left. - k_{33}(2\beta - 1) \right]}{k_{11} + \left(\frac{\Phi}{\pi}\right)^2 \left[k_{33} - 2k_{22}(1 - \beta) \right]} + \frac{\Delta \varepsilon}{\varepsilon_\perp} \right) \qquad (6.90)$$

As for the TN case, it may be assumed that $k_{22} \approx \frac{1}{2} k_{33}$ such that the denominator becomes

$$\frac{k_{33} - \frac{1}{2}\left(\frac{\Phi}{\pi}\right)^2 k_{33}(7 - 8\beta + \beta^2)}{k_{11} + \left(\frac{\Phi}{\pi}\right)^2 k_{33}\beta} + \frac{\Delta \varepsilon}{\varepsilon_\perp}. \qquad (6.91)$$

Assuming that the chiral doping is chosen to be in the central range for the given twist ($\beta = 1$), then the denominator of Equation 6.82 becomes zero and the $\theta_m(V)$ gradient infinite when that twist is set to Φ_∞:

$$\Phi_\infty \approx \pi \sqrt{\frac{\varepsilon_\perp}{\Delta \varepsilon} + \frac{k_{11}}{k_{33}}} . \qquad (6.92)$$

Low V_C requires highly positive $\Delta \varepsilon$, for which typically $0.3 \leq \varepsilon_\perp / \Delta \varepsilon \leq 0.5$. Ensuring that Φ_∞ is maintained at an attainable twist then necessitates k_{33}/k_{11} is as low as possible, contrary to the requirement for the standard TN. For example, a typical mixture suitable for STN may have $\varepsilon_\perp / \Delta \varepsilon \approx 0.4$ and

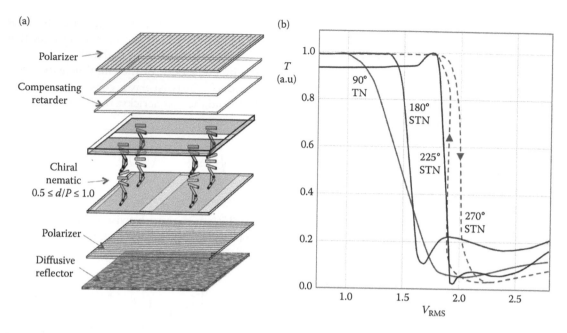

(a)

Polarizer

Compensating retarder

Chiral nematic
$0.5 \leq d/P \leq 1.0$

Polarizer

Diffusive reflector

(b)

T (a.u)

90° TN

180° STN

225° STN

270° STN

V_{RMS}

Figure 6.24 (a) Schematic of a typical two-polarizer reflective STN and (b) STN transmission characteristics for increasing twist angle.

$k_{33}/k_{11} \approx 0.85$ [79] such that the gradient is highest for $\Phi_\infty \approx 200°$. Equation 6.92 only acts as a guide for material design but it shows that keeping both k_{11}/k_{33} and $\Delta\varepsilon/\varepsilon_\perp$ low is likely to give the highest degree of STN multiplexing. Therefore, the material scientist targets a high k_{33}, $\Delta\varepsilon$ and $\varepsilon_\perp$, whereas keeping k_{33} low to combine a low threshold voltage with a high multiplexibility. Where the gradient exceeds Φ_∞ the voltage response becomes hysteretic, as shown for the 270° STN in Figure 6.24. This is unusable in RMS addressed displays and so typically twists of $\Phi = {}^4/_3\, \pi = 240°$ are used in practice. Even then, switching can be disrupted through the formation of stripe domains and electrohydrodynamic instabilities that cause the helical axis to rotate into the plane of the cell [115]. These require a lower d/P to be used than the $\beta = 1$ condition, together with ensuring the pretilt is high.

For a 240° STN, $5/8 \leq \beta \leq 11/8$ and the range of Fréedericksz thresholds is

$$\pi\sqrt{\frac{1}{\varepsilon_0\Delta\varepsilon}\left[k_{11}+\frac{10}{9}k_{33}\right]} < V_C < \pi\sqrt{\frac{1}{\varepsilon_0\Delta\varepsilon}\left[k_{11}+\frac{22}{9}k_{33}\right]}.$$

(6.93)

For the typical STN mixture with $k_{33}/k_{11} \approx 1.2$, V_C is 20–70% higher than that for the equivalent TN

operation, with the higher threshold more practical if β is kept low and the stripe voltage is to be avoided. Also, the elastic term that appears in the square brackets of Equation 6.89 is applicable to the response times: inserting into Equations 6.48 show that the STN response compared is inherently quicker than the TN. However, this is rarely found in practice because the STN is usually highly multiplexed and the response much slower due to the much smaller changes of RMS voltage.

The first STN demonstrators used a Guest–Host mode, in which anisotropic pleochroic dyes give contrast between the ON and OFF states [89]. However, optical contrast could also be achieved using optical retardation effects, in what is called the supertwist birefringence effect (SBE) [90]. STN also has a Gooch–Tarry-type dependence of the polarization optics, with the first minimum mode occurring at increasing $\Delta n \cdot d/\lambda$ with twist, as shown in Figure 6.25, and the crossed polarizer angles set to $\varphi_1 = 1/2(\Phi - \pi/2)$. The retardation for the first minimum is double that of the TN, potentially decreasing the speed of the device by a factor of 4. Even when high birefringence materials are chosen, the effect is too slow for animation, with >100 ms response times typical. However, the most significant drawback from the original technology was the optical properties: the OFF state is a

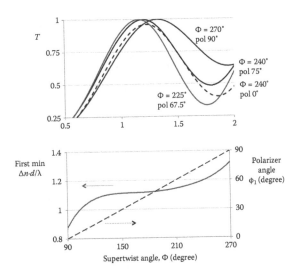

Figure 6.25 Optics of the supertwist birefringence effect mode.

prominent shade of yellowy green or blue if the NB polarizer orientations are chosen, the viewing angle is poor and the contrast is much lower than that of the TN because the twist remains more evenly distributed through the cell in the STN case as the field is applied [116]. Various attempts were made to improve the appearance, such as the use of blue polarizers to give a white on blue display with high transmissivity, but the most successful were in the late-1980s, when optical compensators were added. The first method was to mount an inactive dummy STN panel onto the front of the LCD, equivalent to the active panel but with the opposite handedness [91]. This corrected both the coloration and the viewing angle and would do so across the whole temperature range. However, the additional cost and weight prohibited this approach commercially and polymer film retardation plates are now used instead [117]. The best color compensation is achieved through the use of two polymer films [7] as indicated in Figure 6.24.

Even when the transmission characteristic is made infinitely steep, the effect of nonuniformities, including electrode resistive losses and temperature variations, prevents multiplexing much beyond 240 lines. Driving alternate rows from opposite sides of the panel doubles the maximum number of rows. Such high degrees of multiplexibility represent a considerable improvement to the passive matrix TN and for that reason the STN continues to find a market where there is demand for a high black and white image content combined with low cost. However, the lack of gray scale and slow response speed prohibits video applications and it was these factors that fed ongoing research efforts for highly multiplexed displays.

6.4.5 Ferroelectric liquid crystal displays

Although making only a small commercial impact, FLCs commanded major R&D efforts through the 1980s and 1990s, immediately after the publication by Clark and Lagerwall of the Surface Stabilized Bistable FLC mode [88] (SSFLC). Companies across Europe and the Far East each produced demonstrators to rival the nascent TFT technology [118]. However, the only panel to receive notable sales in the mainstream displays market at that time was the 15 in. 1280 × 1024, 16-color monitor produced by *Canon* (Figure 6.16).

Unlike the other mainstream LCDs described in this section, FLC does not use nematic liquid crystals but rather a particular type of smectic, the tilted smectic C phase, denoted as SmC. Unlike the nematic case, reorientation of the director for smectics is constrained by the presence of the smectic layers. The SmC phase is constrained in this fashion too but the director is relatively free to reorient about the layer normal in an imaginary cone of possible orientations. The component of the director in the layer plane is described by the unit director **c**, which acts as a two-dimensional nematic. It is the reorientation of the c-director in response to applied electric fields that yields the potential for electro-optic effects suited for display operation. However, what makes the SmC so interesting is the inherent ferroelectricity allowed by the reduced symmetry of the SmC* phase. Although the nematic phase has cylindrical symmetry, the SmC is monoclinic: there is a single C_2 symmetry axis corresponding to the direction orthogonal to the layer normal **a** and director **n**. Such symmetries are inherently biaxial, meaning that they have two optic axes and three principal permittivities. The difference between the two directions orthogonal to the **n** director is called the biaxiality. The refractive index biaxiality is negligible and FLCs are best treated as optically uniaxial materials, with a single optical axis along the director. Thus, the polarization optics can be treated in the same way as a conventional nematic. However, the dielectric biaxiality $\partial \varepsilon = \varepsilon_2 - \varepsilon_2$ is significant [119,120] and, in essence, acts in a similar fashion to the nematic $\Delta \varepsilon$, dictating the RMS response of the **c**-director. Reflections of the SmC molecules in the tilt plane do not disrupt the phase symmetry: molecular dipoles aligned along the C_2 axis have equal probability of being parallel or antiparallel. If the phase is chiral, this mirror symmetry is lost and there is a net dipole in the direction of $\pm C_2$: the ferroelectric spontaneous polarization P_S. This unique feature provides a polar switching torque many times greater than that possible in nematic liquid crystals, resulting in switching times that can be faster than $10\,\mu s$.

For simplicity, a single elastic constant B and the flow-free viscosity γ_1 are used, although the effect of the elastic anisotropy and the uniaxial dielectric anisotropy $\Delta \varepsilon$ is important too [121,122]. The switching torque of the FLC is then given by

$$\gamma_1 \sin^2 \theta_C \frac{\partial \phi_C}{\partial t} = B \cos^2 \delta_C \frac{\partial^2 \phi_C}{\partial z^2} + P_S E_z \cos \delta_C \sin \phi_C$$

$$- \varepsilon_0 \partial \varepsilon \cdot E_z^2 \sin \phi_C \cos \phi_C \cos^2 \delta_C. \tag{6.94}$$

The ferroelectric torque differs from the dielectric having a linear dependence on the applied field E, rather than the RMS E^2. This means that the torque depends on the field polarity as well as its strength. A typical FLC may have a $P_S = 50$ nC cm^{-2} and dielectric biaxiality $\partial \varepsilon = +0.5$. For a typical $\pm 10\,V$ signal applied across a $2\,\mu m$ spaced

cell, the ferroelectric torque is 50 times greater than the dielectric, dominating the electro-optic behavior. Assuming the material has a positive P_S, then $+E$ will tend to reorient the **c**-director on one side of the cone (toward $\phi = 0$) and $-E$ to the other ($\phi = \pi$). Simplistically, if these two conditions equate to the **n**-director being oriented $\pi/4$ apart and the device has a spacing set to give the quarter-wave plate condition and crossed polarizers parallel and perpendicular to the director in one of the states, Equation 6.25 predicts that the device will switch between minimum and maximum transmissivity.

A second important aspect of the SSFLC is bistability. As for any LCD, the starting point is to achieve the desired alignment, uniformly over the whole sample [123]. There are several steps to consider for ferroelectric LCDs (FLCD). The usual SSFLC device geometry relies on the N*-SmA-SmC* sequence, as shown in Figure 6.26a. The device requires parallel alignment with the required pretilt θ_S. The cell gap must be sufficiently low to unwind the cholesteric helix and provide an almost uniform nematic texture, as shown: devoid of twist and with only a slight splay and bend associated with the surface pretilt. Furthermore, the helicity of the SmC **n**-director must also be suppressed in a similar fashion. This uniform nematic texture should be retained on cooling into the SmA phase, where the layers will align uniformly perpendicular to the cell walls in what is termed "bookshelf" geometry. On cooling into the SmC* phase, the director tilts from the layer normal by the cone angle θ_C, which grows continuously from $0°$ at the second-order SmA to SmC* phase transition to typically $22°$–$25°$ at ambient temperatures (Figure 6.26b). The smectic layer spacing contracts as the director tilts from the layer normal, causing the layers to tilt by an angle δ_C and form a symmetric chevron-like structure. The layer tilt remains a constant fraction of the cone angle, typically $\delta_C \approx 0.85\theta_C$, so that the director remains continuous across the sharp chevron interface at the cell center. It is this interface that gives the SSFLC its bistability. The **c** director can be at either of two orientations, ϕ_i and $\pi - \phi_i$ in one arm, and $-\phi_i$ and $\pi + \phi_i$ in the other. These correspond to just two orientations of the **n** director, as shown in Figure 6.26b. Application of a DC field couples to the ferroelectric polarization, eventually causing such a high torque at the chevron interface that

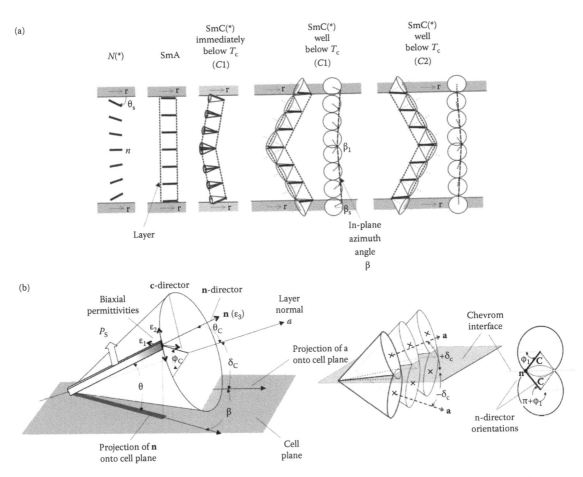

Figure 6.26 (a) Ferroelectric liquid crystal alignment on cooling through the sequence N*–S$_A$–S$_C$* for parallel aligned surfaces. C1 and C2 chevron layer textures with low and high tilt and triangular director profiles are shown. (b) Definition of angles for S$_C$* devices and the source of bistability from the chevron interface.

latching from one state to the other occurs. After the pulse, the director remains in that state, with the director relaxing back to one of the two quiescent states.

The surface pretilt and anchoring energies play fundamental roles in SSFLC devices. Conventional polyimides for nematic alignment are used. These have azimuthal and zenithal anchoring energies in the range of 0.3–1 mJ·cm^{-2}; high enough to be considered infinite for nematic LCDs, but insufficient to cause changes to the layer orientation or SmC* cone angle in FLC. Immediately below the SmA–SmC* transition, the layers always tend to tilt in the direction favored by the surface pretilts. This is the C1 state, shown in Figure 6.26a. As the temperature is cooled further, θ_C and δ_C increase, forcing the surface director to lie further from the rubbing

direction and pretilt angle. Close to the temperature where the difference between the cone and layer tilt angles (θ_C–δ_C) approaches the surface pretilt θ_S, the tilt of the layers swaps direction to form the C2 layer structure. For low surface tilt angles ($0.1° \leq \theta_S \leq 2°$) samples usually form a mixture of both C1 and C2 states. The lack of uniformity and zigzag defects that separate the regions of opposite layer tilt severely damages the device appearance. However, with intermediate pretilts ($2° \leq \theta_S \leq 8°$), the layers form the C2 state completely and uniformly, whereas for higher pretilts still ($12° \leq \theta_S \leq 35°$) the layers do not undergo the alignment transition at all and the sample is uniformly C1. Both layer geometries have been used in SSFLC devices [101,118], though the *Canon* monitor used the high pretilt C1 approach.

The out-of-plane tilt of the director θ and in-plane tilt angle of the projection of the **n**-director into the cell plane are shown in Figure 6.26b. They are given by the expressions:

$$\beta = \tan^{-1}\left(\frac{\cos\phi_C \sin\theta_C}{\sin\phi_C \sin\delta_C \sin\theta_C + \cos\delta_C \cos\theta_C}\right);$$

$$\theta = \sin^{-1}\left(\sin\phi_C \cos\delta_C \sin\theta_C + \sin\delta_C \cos\theta_C\right). \tag{6.95}$$

At the chevron interface, there is no out-of-plane tilt and the director has an in-plane twist angle β_m given by

$$\beta_m = \pm\cos^{-1}\left(\frac{\cos\theta_C}{\cos\delta_C}\right), \tag{6.96}$$

which is typically about 8°. For the C1 layer geometry, the surface tilt is chosen to be high, typically 25°. For typical values of θ_C 52 25° and $\delta_C = 22°$, Equation 6.95 gives an in-plane tilt of $\beta_s \approx 27°$. Thus, the director twists from 27° at one surface to 8° in the cell center and back out to 27° at the other surface. Assuming that the twist is approximately linear then the director profile is triangular [124], as shown in Figure 6.26a. The transmission of polarized light through such a cell is similar to that of the reflective TN mode given by Equation 6.84, except the opposite polarizer is crossed to the input polarizer in the FLC case, rather than being parallel. This has the simple solution for the angle to align the polarizers with respect to the rubbing direction, β_{ext}:

$$\tan 2(\beta_{ext} - \beta_S) = \frac{\tan\left[(\beta_m - \beta_S)\sqrt{1 + \frac{1}{4}\alpha^2}\right]}{\sqrt{1 + \frac{1}{4}\alpha^2}}, \tag{6.97}$$

where

$$\alpha = \pi\frac{\Delta n \cdot d}{\lambda(\beta_s - \beta_m)} \tag{6.98}$$

as before. Equation 6.97 suggests that the device will appear highly colored in its quiescent state for retardations close to the full wave plate condition.

However, if the cell spacing is reduced to the half wave plate condition, the device appears black with the polarizers aligned at the angel:

$$\beta_{ext} = \pm\frac{\beta_m + \beta_S}{2}. \tag{6.99}$$

The example above gives $\beta_{ext} \approx \pm18°$. Thus, if the polarizers are placed at +18° and +108° to the rubbing direction, the cell will appear black for one state and transmit most of the light when in the other domain, where the optic axis is about 36° from the polarizers. Approximating the structure to a uniform retarder, and setting the retardation to be at the half-wave plate condition for green light, then the FLC switches between states with the optic axis either parallel or at $2\beta_{ext}$ to the polarizer. Equation 6.25 suggests that the transmittance should be 91%. This could be maximized so that $\beta_{ext} = 22.5°$ (and the director reorients through 45°), for example, by using a material with a higher SmC* cone angel θ_C. However, this also causes a decrease in switching speed and so *Canon* used the lower optical efficiency to help achieve a fast, flicker-free frame for their monitor.

FLCD panels are addressed in a similar fashion to most passive matrix displays and in the same line-scanning method described in Section 6.2.4. However, the response is no longer to the RMS over the frame but rather the signal applied to each row must be sufficient to latch the pixels into a new state within the line time. In this fashion, the information is built up line-by-line. The row waveform has one particular sign of operation, allowing only one set of states to be selected appropriately. Both states are addressed, either using two subframes of apposite polarities (with a cost of increasing overall frame time) or by preceding each addressing pulse by a blanking pulse that selects the black state regardless of the data being applied to the previous lines and then latching selectively or not in the addressing line (with the cost of reduced brightness for pixels that should remain white in consecutive frames). Various addressing schemes are possible [125], influencing speed, operating window, and appearance.

A second approach to operating the SSFLC was also attempted jointly by *RSRE* (then *DERA*) and *Sharp* Corporation. This maximized device speed by using the C2 geometry and multiplexibility using a lower P_S ($\approx$10 nCcm^{-2}) and higher dielectric biaxiality $\partial\varepsilon$ ($\geq$+1) [119,121]. Equation 6.94 predicts that the dielectric and ferroelectric terms become equal at about 32.5 V for such high biaxiality, low P_S materials. If the field has the correct polarity to reorient the director from one side of the cone to the other, the dielectric biaxiality suppresses switching and above the voltage where the torques balance ($\approx$32.5 V for these values), the director will remain unswitched indefinitely. In fact, electrical pulses at about 70% of this voltage [120,121] start to slow the response rapidly, creating a minimum in the switching characteristic (τV_{MIN}). Operating close to this voltage (i.e., about 23 V, for the high biaxiality SmC* material in this example) gives a highly nonlinear response, thereby enabling thousands of lines to be addressed with a high degree of insensitivity to temperature variations and line losses. Using this τV_{MIN} mode, *Sharp* created a prototype color ¼ HDTV, operating with a 12 μs line address time, to give 256 gray levels and a 60 Hz frame rate [101].

By the mid-1990s, the key advantages of the SSFLC over TFT TN were its perceived lower cost and excellent viewing angle. The good viewing angle was inherent to the SSFLC mode due the fact that the director remains in the same plane for both switched states, as shown in Figure 6.27. Such IPS gives excellent viewing properties, surpassing that of the foil compensated TN of Figure 6.27. However, in the mid-1990s, SSFLC lost the war with TFT-driven nematics because of two reasons. First, the number of critical mask steps grew with FLC complexity. To achieve microsecond pulses across a passive matrix required metal bus lines to be prepared. Moreover, smectics are fundamentally sensitive to shock because any flow in the panel disrupts the carefully aligned layers irreparably. Second, to prevent mechanical damage to the LC alignment, polymer walls were defined photolithographically, again introducing a critical mask step. Together with the poorer yield associated with achieving perfect alignment, any cost advantage was severely eroded. The final battle came with the introduction of in-plane switched TFT nematics that matched, and eventually surpassed, the viewing angle of even the SSFLCD (Figure 6.16).

6.4.6 In-plane switching LCDs

6.4.6.1 INTRODUCTION

Operating with TFT removes steepness of the electro-optic transition as a critical design issue.

This enables LCD modes to be chosen that meet the more stringent optical requirements for large area monitor and televisual displays. The crucial weakness of TN LCDs was the viewing angle, even with optical compensation. Viewing angle is particularly important for large area displays, where images must appear uniform from the center of viewing to the corners. The viewing angle target is harsher still if the image is to satisfy multiple viewers. The television market also requires fast response times not just for black to white transitions but also between adjacent gray levels. Extremely high contrast ratios are needed to compete with emissive technologies such as CRT, PDP, and, most recently (OLED).

Achieving pixel contrasts in excess of 10,000:1 and >160° horizontal viewing angles requires cylindrical symmetry of the director profile in both of the ON and the OFF states. Two approaches were developed during the late 1990s and early 2000s: IPS and vertical aligned (VA) modes as shown in Figure 6.28. Different manufacturers championed each mode and numerous modifications were tried. This section will concentrate on the IPS mode, and its derivative FFS, whereas VA modes are dealt with in the following section.

6.4.6.2 BASICS OF IPS OPERATION

In-plane fields are impractical for passive matrix displays but are readily achieved with active matrix LCDs by moving the common electrode to the active plate (in addition to the Source and Gate lines), in-plane electrodes are etched onto the active back-plane in either a comb-like structure, as shown in Figure 6.29a, or more commonly a zig-zag configuration. The field produced by comb-like electrodes is complex, varying both in the plane of the cell in the direction normal to the electrodes. Ignoring the effect of the liquid crystal permittivity, and taking the first Fourier component only, the field has components [126]:

$$E_y \sim \frac{V}{l}\sin\left(\frac{\pi y}{l+w}\right)\exp\left(-\frac{\pi z}{l+w}\right);$$

$$E_z \sim \frac{V}{l}\cos\left(\frac{\pi y}{l+w}\right)\exp\left(-\frac{\pi z}{l+w}\right) \quad (6.100)$$

That is, the electric field components have a periodic form in the plane of the cell but decay

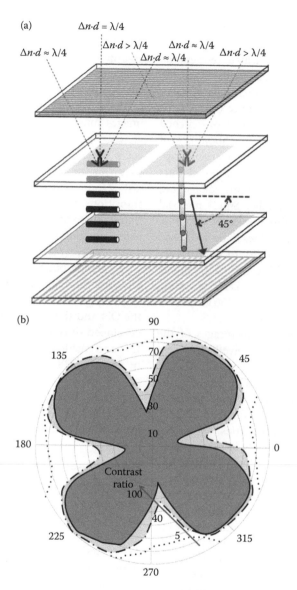

Figure 6.27 Schematic representation of a switchable quarter-wave plate with in-plane switching. One direction of viewing self-compensates the change reduction of birefringence with the increased optical path. The increase in retardation in the other direction (usually set to be the horizontal direction for the white state) is relatively small and easily compensated. A polar plot for the contrast is sketched on the right.

exponentially across the bulk of the cell. Rather unsatisfactorily, most treatments ignore the complex field behavior and assume that, if the cell gap and electrode widths are sufficiently low, the field is uniform and restricted to the cell plane. If the quiescent state of the nematic is aligned with the rubbing direction orthogonal to that of the applied field, the IPS mode has a Fréedericksz threshold field that is approximately given by Equation 6.34,

although the errors for this can be severe in practice [126]. Assuming a uniform in-plane field, the Euler–Lagrange equation for the elastic distortion above the threshold is [127]

$$k_{22}\frac{\partial^2\phi}{\partial z^2}+\varepsilon_0\Delta\varepsilon\cdot E_z^2\sin\phi\cos\phi=0, \qquad (6.101)$$

which has the simple solution

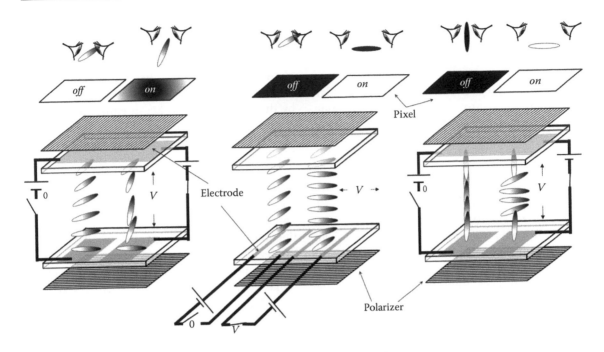

Figure 6.28 Improvement of viewing angle using IPS or VAN modes.

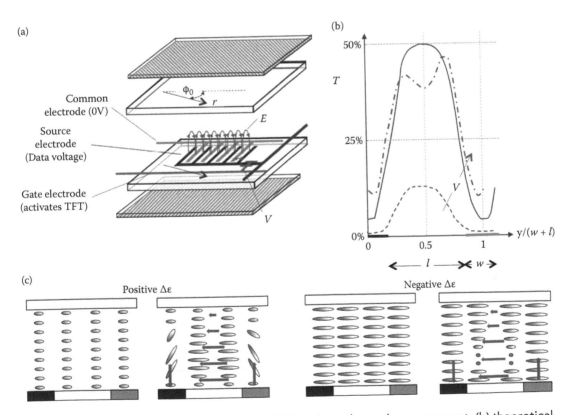

Figure 6.29 IPS principles. (a) Schematic of the TFT in-plane electrode arrangement, (b) theoretical transmission calculated for low and high switching voltages for one electrode period [124]. (c) Mode choices associated with positive Δε and negative Δε materials.

$$\phi = \phi_m \sin\left(\frac{\pi \cdot z}{d}\right) \qquad (6.102)$$

for fixed boundary conditions. Aligning the polarizers parallel and crossed to the rubbing direction, the first-order solution for the optical transmission is given by that of a wave plate (Equation 6.25). Approximating the sinusoidal dependence in Equation 6.102 to the triangular form of Equation 6.97 [124] and setting the device thickness at the half-wave retardation gives an effective in-plane tilt of the director of $\langle\phi\rangle \approx \phi_m/2$ and the transmission approaches 50% with increasing voltage. At fields higher than that which gives $\phi_m = 90°$, the twist deformation becomes increasingly concentrated toward the surfaces. Thus, the effective twist angle $\langle\phi\rangle$ exceeds the optimum 45°, thereby causing the transmission to decrease, as shown in Figure 6.29b.

The IPS mode can operate with either positive or negative $\Delta\varepsilon$ materials, Figure 6.29c, depending on the orientation of the rubbing direction with respect to the electrodes ϕ_0. In this diagram, it is assumed that the anchoring on the surfaces is strong and director reorientation occurs in the bulk of the cell, although somewhat closer to the electrode plate due to the transverse decay of the field. Clearly, the director in the positive mode will tend to orient in the field direction. For the negative mode, the director reorients to be orthogonal to the applied field, either remaining in the cell plane or tilting out of that plane. IPS is always favored in this case because the elastic energy associated with the twist elastic constant is lower than the splay-bend that would be induced by the tilt (i.e., $k_{22} < k_{11} < k_{33}$). Although the early demonstrators used negative $\Delta\varepsilon$, positive mode is now more common because the materials combine higher $\Delta\varepsilon$ with lower viscosity and hence give lower operating voltages and faster response times.

With the TN, STN, and SSFLC modes, analytical calculations for the on-axis optical and electro-optic behavior are reasonably accurate. Optimization of viewing angle required numerical solutions due to the importance of off-axis refractive and reflection effects. However, with the adoption of the IPS mode and the variety of multidomain VAN modes covered in the following section, both the electric field and the director profile vary in two or even three dimensions. This necessitates numerical simulation to optimize the LCD [31,64]. For example, Figure 6.29b shows the calculated transmission for a 2D simulation of the IPS mode [128]. Above the electrodes, the field is almost vertical and the director either reorients vertically for positive $\Delta\varepsilon$ or remains unaffected by the field if negative $\Delta\varepsilon$; in either instance the electrodes appear as unwanted dark bands in the pixel. This necessitates that the gap should be much larger than the width of the electrodes $l \gg w$. If too high, the field is reduced and the switching voltages become higher. Typically, the electrodes are $w = 3\,\mu m$ wide and have a gap $l = 6$–$8\,\mu m$. Therefore, the banding represents a major reduction in the optical efficiency of IPS mode.

Ideally, the transmission–voltage characteristic should approach linearity across the voltage range, readily giving gray levels. The response of the IPS mode is strongly dependent on the angle of the rubbing direction with respect to the electrodes, ϕ_0 [128]. For positive mode IPS, $\phi_0 \approx 30°$ is used typically [129], not only leading to suitable transmission characteristic and near linearity of the gray scale response times, but also significantly improving the response time, which decreases linearly with increasing ϕ_0.

The viewing angle characteristics of the IPS mode exceed even that of the SSFLCD, shown in Figure 6.27. The first improvement was to orient the electrodes in a small-angled zigzag, to help widen the viewing cone. However, the stringent requirement to maintain very high contrast well off-axis still necessitates the use of optical compensators. For example, the crossed polarizers themselves leak at the high angles in the four quadrants centered at $\pm45°$ and $\pm135°$. A typical IPS mode LCD used in television achieves pixel contrast ratios in excess of 2000:1 for direct viewing, and contrast in excess of 1000:1 over 175° horizontal and vertical viewing. The lowest pixel contrasts still surpass 200:1 at 140° viewing in the 45° quadrants. Such impressive viewing angle figures are achieved using a front uniaxial wave-plate with its axis oriented crossed to the rubbing direction (and input polarizer) combined with a negative uniaxial wave plate formed from a homeotropic discotic liquid crystal polymer [130].

6.4.6.3 FFS MODE

Despite the impressive performance of IPS, the mode suffers from poor optical efficiency due to the banding structure caused by lack of director

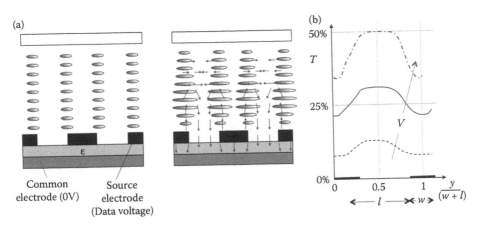

Figure 6.30 Fringe field switching mode. (a) Basic structure and operation and (b) typical transmission distribution with varying voltage within a period of the data electrodes, shown in blue.

switching above the electrodes. For high-resolution portable display applications, this is very important because increasing the output from the backlight to compensate for optical inefficiency is too costly for battery life. Many mobile phones and tablets deploy FFS mode LCD [96,97]. This is a modification to IPS mode, where a uniform counterelectrode is placed beneath a dielectric layer with the interdigitated electrodes on top. The pitch of the electrodes is much smaller than the cell gap (Figure 6.30). There is little electric field directly above the central line of the data electrodes, but immediately either side of this there is an in-plane field component that causes the director to reorient. This arrangement greatly reduces the transmission loss caused by the banding and can be operated with electrodes that are closer together, therefore, allowing lower operating voltages to be used. The fringing fields break the symmetry sufficiently to allow operation without alignment pretilt, again helping to improve viewing angle toward the limit. This is difficult with conventional rubbed polymers but can be achieved using photoalignment.

Typically, the dielectric layer is 500 nm thick, providing much higher fields close to the electrode surface than typical for the IPS mode [131]. Together with the reduced electrode spacing, this lowers operating voltages, and hence power. The dielectric layer acts as the storage capacitor for the pixel, thereby increasing the aperture ratio and concomitantly increasing optical efficiency still further. Typically, FFS mode has double the optical efficiency of the IPS mode, with the only disadvantage being that two transparent electrodes need to

be deposited and etched onto the rear plate during fabrication process. The success of FFS mode can be measured by its adoption for many portable products, notably the *Apple* iPad from 2011.

6.4.7 Vertically aligned nematic modes

As for IPS, the invention of VAN occurred early in the history of LCDs [85] but was reinvigorated with the widespread adoption of TFT and the need for wide-viewing angle technologies in the mid-1990s. Homeotropic alignment gives a near perfect black state at normal incidence and a viewing angle easily compensated using a negative uniaxial retardation plate. Being dominated by the dark state, the contrast of VAN mode devices is generally exceptional at all angles and is independent of temperature or cell gap variations. The director is switched into the plane of the cell when a negative $\Delta\varepsilon$ liquid crystal is used. The direction of tilt will be degenerate and form scattering domains unless some preferred orientation is imparted to the cell. For example, if one of the homeotropic surfaces is rubbed, it gives a pretilt of typically 89.8°, and the director will tilt uniformly in this direction with increasing field. However, this will give a poorer viewing angle for the white state, as indicated in Figure 6.8. To overcome this limitation, *Fujitsu* invented the multidomain vertically aligned mode in 1997 [94,95], termed MVA mode. The aim of the invention was to maintain the vertical cylindrical symmetry as the director reorients with applied voltage, using domains of opposing tilt. Typically, each pixel is subdivided into two or four areas with orthogonal tilt directions for each. Different area ratios may be used for the horizontal and vertical directions, provided that the area for opposing pairs is equivalent.

Initial attempts to produce multidomain alignment used the rather impractical approach of dual rubbing, where a rubbed surface was protected during a second antiparallel rubbing by a photolithographically defined mask that was subsequently removed. The approach that *Fujitsu* took was to arrange dielectric protrusions onto the electrodes and underneath the homeotropic alignment layer, as shown in Figure 6.31a. Each protrusion has a convex shape with side walls angled to the surface but is sufficiently small and rarefied to have negligible effect on the overall pixel alignment. When the voltage is applied, the slight field fringing around the protrusion causes tilting in opposite directions on either side of the protrusion, thereby breaking the symmetry and automatically causing domains of the opposite tilt sense. Early modules included protrusions on both inner surfaces, as shown in Figure 6.31, but including

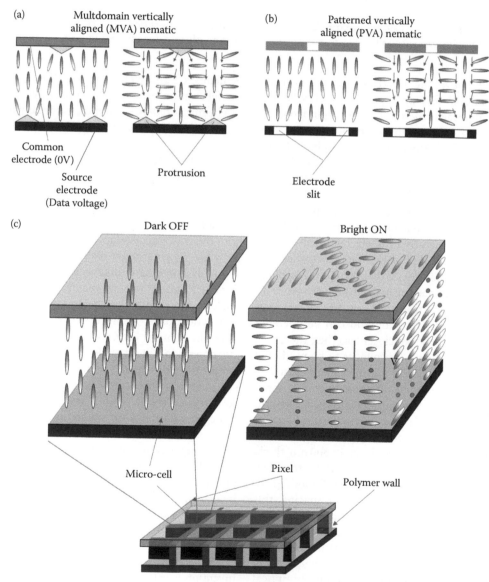

Figure 6.31 Vertical aligned nematic modes. (a) Multidomain vertical aligned (MVA) mode, (b) patterned vertical aligned (PSA) mode, and (c) axially symmetric multicell (ASM) mode.

the protrusion on a single surface only saved additional photolithographic steps and was found to give satisfactory performance.

Samsung made a further advance for the VAN mode in 2000. Similar to the MVA mode, the patterned vertically aligned mode (PSA) replaced the protrusions with slits in the electrodes to create the fringing fields (Figure 6.31b) [98]. *Sharp* [99] designed the axially symmetric microcell (ASM) mode LCD, where polymer walls surround the nematic, forming a microcell container for the liquid crystal (Figure 6.31c). The walls are formed by photoinduced phase separation of a monomer during the polymerization that occurs on UV exposure of a grid like pattern. A chiral dopant is added to the nematic with the correct pitch to ensure a twisted orientation of 90° for the given cell gap. Unlike the MVA and PVA modes, disclinations are avoided altogether because the director forms a monodomain while retaining the axial symmetry that results in the exceptional viewing angle.

These VAN modes share the properties of wide viewing angle, very high contrast, and fast operation. High switching speeds are common to each of the vertically aligned modes because the field-induced distortion is dominated by the bend elastic constant k_{33}, which is usually significantly higher than both k_{11} and k_{22} (Equation 6.48). The mode also has the advantage of not requiring the rubbing step during fabrication, which can give improved yield and costs. However, the mode is slower than modern IPS mode panels because full switching is needed to obtain the bright state, whereas IPS switches the director through a lower angle if $\phi_0 \gg 0$.

6.5 LCDS FOR NONMAINSTREAM AND NICHE MARKETS

6.5.1 Introduction

The wealth and diversity of different device modes that exploit liquid crystals is a measure of both the variety of phases and behaviors exhibited by these organic materials and by the ingenuity of the scientists, engineers, inventors, and innovators working in this field over the past half century. The path to providing flat screen monitors and full high definition television displays is littered with unsuccessful attempts. However, many of those technologies found, or indeed created, niche markets, offering unique selling points in other applications. A selection of some of the more important LCDs used outside the conventional direct view displays are summarized in this section.

6.5.2 Cholesteric LCDs

6.5.2.1 TEMPERATURE, STRAIN, AND GAS SENSORS

From the first discovery of liquid crystals by Reinitzer in 1888 [18], the bright colors of cholesteric liquid crystal textures have fascinated observers. Indeed, the first liquid crystal application was not an electronically addressed display but rather as sensors that deployed this coloration. In the early 1960s, prolific inventor James Fergason observed that the peak wavelength of cholesteric selective reflection depended strongly on both temperature and trace amounts of chemical vapor [132], envisaging device applications. His work inspired many applications from detecting minute temperature changes on human skin due to the influence of nicotine or underlying tumor to visualization of faulty electronic circuitry. Photographer Henry Groskinsky, inspired by the vivid colors of the cholesteric, recorded these applications for *Life Magazine* in 1968 [133]. The article also mentioned that the liquid crystal could be encapsulated into a polymer, another of Fergason's inventions that will be described in Section 6.5.5, that was used by *National Cash Registers* to detect the mechanical strain within a loaded spring, an application that was later extended to aeronautical testing of planes, rockets, and turbine blades [134]. Indeed, it was encapsulation that eventually led to the mass production of liquid crystal thermometers that remain popular today.

The cholesteric has a natural helical structure shown in Figure 6.1. Light traveling along the helical axis undergoes Bragg reflection due to the repeating nature of the optic axis. This occurs for circular polarized light with the same handedness as the cholesteric pitch P and at a band of wavelengths centered on λ_0 given by

$$\lambda_0 = \bar{n} \cdot P \cdot \cos\theta, \tag{6.103}$$

where θ is the angle of incidence with respect to the helical axis and $\bar{n}$ is the average refractive index ($= \frac{1}{2} n_e + \frac{1}{2} n_o$). Strong coloration occurs when λ_0 matches a visible wavelength, from 400 to 700 nm. The width of the reflection band $\Delta\lambda$ is approximately

$$\Delta\lambda \approx \Delta n \cdot P. \qquad (6.104)$$

The color is, therefore, made more vivid by using weakly birefringent cholesteric materials. Light of wavelengths outside the band of selective reflection is transmitted through the sample, as is all light of the opposite handedness. Optical contrast then requires the liquid crystal layer to be mounted onto a dark backing material, to absorb the transmitted light. Tuning of the temperature range over which the pitch varies in the visible regime usually requires two or more components with different pitches and temperature dependences. The pitch diverges as the second-order cholesteric to smectic phase transition is approached, leading to rapidly changing color from red to blue with increasing temperature. Varying the concentration of components with strong smectic local ordering in the cholesteric mixture provides control over the operating temperature range for the thermometer.

6.5.2.2 BISTABLE CHOLESTERICS

Selective reflection of colored light from cholesteric liquid crystals can be used for electronic displays too. Moreover, the cholesteric electro-optic characteristic can be arranged to be bistable, allowing unlimited multiplexibilty using low-cost passive matrix addressing. Among the first optical switching modes to be studied at *RCA* was the bistable cholesteric [135], the switching mechanism for which was elucidated during the 1970s [136,137]. However, the success of the bistable cholesteric is largely due to the team at Kent State University headed by Doane [138] and the many innovations made by the engineers at the spin-out company *Kent Displays Inc. (KDI)* [139,140].

The basic operation of the display is to switch the cholesteric liquid crystal between the Grandjean texture, where the axis of the helix is largely normal to the display that, therefore, appears reflective, and the focal conic texture, where the helix lies in the plane of the cell and light is forward scattered to be absorbed by the rear (black) substrate. Latching between the states is done via an intermediate homeotropic state, where the field is unwound by a high electric field coupling to the positive $\Delta\varepsilon$

of the liquid crystal, as shown in Figure 6.32a. If the field is switched off immediately, the helix forms with its axis vertical to the pixel, thereby appearing colored due to selective reflection. If the field is reduced more gradually through an intermediate level, the helix forms in the plane of the cell, in the focal conic state, and the pixel appears dark. The degree of hysteresis and the sensitivity of the device to mechanically induced damage is controlled using polymer stabilization.

Most displays are designed to be monochrome, using materials with high birefringence to give the broadest reflectivity. The highest practical levels of Δn give yellowish-green coloration against black, although some customers prefer the inverted optics of white and blue when a blue background is used. Together with the slow response speed, high voltage and sensitivity to shock, the poor appearance means that the devices are not commercially successful and have been largely superseded by other choices. However, bistable cholesterics remained of interest for full color bistable reflective displays. Lower birefringent materials are used to give individual layers with sharp red, green, and blue reflection bands. Stacking three layers in series then allows full reflective color. This type of technology was successfully applied to large area signage by the company *Magink*, providing bright billboards for advertising purposes [141]. As a reflective display, the panels were ideal for bright sunlit conditions, where they could outperform LED electronic signage. *KDI* also used a triple stack to produce full color reflective displays for portable products. Parallax between the layers was minimized through polymer stabilization of the individual active layers mounted directly onto a backing foil [140], as shown in Figure 6.32b. Not only did this enable good optical performance, but also the resulting display was very flexible indeed, as is apparent from the demonstrator shown in Figure 6.32c. Although not successful commercially, this also remains true of all other reflective color display modes and remains an important gap for future developments.

The technical advances made by *KDI* for flexible color plastic displays promised new applications outside the display field, such as electronic skins, Figure 6.32d, and electronic writing tablets (Figure 6.32e). The *Boogie board* is an electronic writing pad that uses mechanical pressure to induce the reflective Grandjean texture, on a black background.

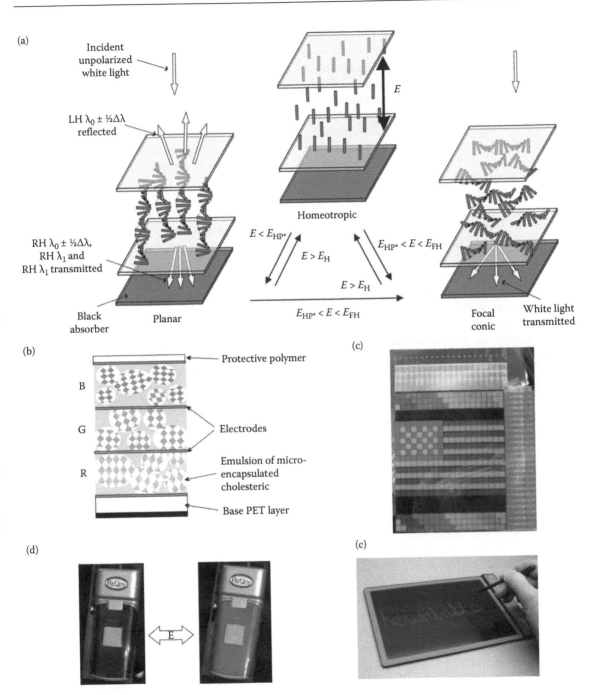

Figure 6.32 Bistable cholesterics. (a) Operating principle, (b) trilayer stack for full color, (c) prototype of the film backed triple N* stack, (d) switchable electronic skin using a single layer cholesteric, (e) the successful Boogie Board product from KDI, using bistable cholesteric to form a writing tablet.

When the image needs to be refreshed, an electrical blanking pulse is applied to erase the page. This product continues to be successful in a niche market that the technology has created.

6.5.3 Bistable nematic LCDs and ZBD

Bistable operation of an LCD allows many lines of information to be passive matrix addressed, where each frame is written line by line and each line retains its information until it receives the next addressing signal. Before the large-scale adoption of TFT back planes and active matrix addressing, various bistable nematic modes were invented, complementing the contemporary efforts to develop bistable cholesteric and FLCD. Few of these modes made it even to demonstrator stage. However, in the late 1990s, there was a resurgence of interest in bistable nematics [126], notably the work at *Seiko-Epson* on the bistable twisted nematic (BTN) mode, the 180° BTN or *Binem* developed by *Nemoptic* and the ZBD by *ZBD Displays Ltd.* (now *Displaydata*). Several factors drove this renaissance:

- Simple passive matrix displays based on TN and STN did not lose market share with the introduction of TFT; rather, the markets that they had created, such as watches, calculators, instrumentation, etc., continued to need low cost, usually reflective displays. Whereas new, high investment production plants in Japan, Korea, and Taiwan developed TFT LCDs for high end applications, a plethora of manufacturers in China continued to serve what remained a \$1B passive matrix market well into the 2000s.

- Bistability promised to create new niche LCD markets, particularly where ultralow power was required. Markets included electronic shelf-edge labels with high information content, electronic book readers, smart card displays, secondary displays for mobile phones, and an enormous range of indicator displays, from car key fobs to razor blades. Indeed, the image storage offered by bistability continues to be a unique selling point for many applications associated with the Internet-of-things, where displays can be updated automatically using radio frequency (RF) signals, rather than rely on costly batteries and associated circuitry.

- Bistable displays often have good potential for plastic displays. TFT on plastic remained a challenge to fabricate through the 2000s, due to the difficulties of registering high-resolution patterns on a flexible backplane, and the low fabrication temperatures required for plastic substrates. Bistable nematics offered the possibility of high image content based on simple, low temperature, low cost fabrication on plastic.

However, these drivers were also attractive to other non-LCD modes, such as Janus colloids from *Gyricon*, interference mode micro-electro-mechanical-systems (MEMS) from *Iridigm/Qualcomm*, electrowetting from *Liqua-vista/Amazon* and *Gammadynamics*, and electrophoretics from *Bridgestone* and *E-ink*. In 2005, *Amazon* selected *E-ink* displays for its *Kindle* e-Book readers, based largely on the appearance of its white state and despite the significantly higher cost due to the active matrix. This created demand for scattering mode devices and many new markets chose electrophoretics over LCDs. One bistable LCD technology that survived was the ZBD, marketed by *ZBD Displays Ltd.*, now *Displaydata*. This spinout from *R.S.R.E* (then *DERA*) in the UK targeted the retail signage market, and in particular, electronic shelf-edge labeling. For a retailer to replace tens of thousands of electronic labels per store, cost was paramount, and the bistable LCD has a significant advantage over electrophoretics, whereas offering superior performance and higher image content than the incumbent directly addressed segmented TN LCD. The company's success was not only due to the bistable LCD but also by development of a novel RF communications protocol [125] that allowed small to midsector retailers to take advantage of the labeling, whereas only hyperstores had been able to afford the infrastructure required for previous labeling systems. The company has sold several million labels worldwide and now offers a combination of the ZBD LCDs alongside E-ink for higher-end application.

Bistability results where a device has two stable states with similar free energies that are separated by an energy barrier, wherein transitions from one state to the other are discontinuous, or first order. An early approach was the BTN mode [142], which followed similar principles to that of the STN, but set the d/P ratio of the chiral nematic to lie halfway between states of low and high twist angles Φ. For example, setting $d/P = 0.5$ with parallel surface alignment should give a π twist state. However, if the pretilt on both surfaces is sufficiently high, the cost of the induced splay energy becomes greater than that for twist. Thus, the chiral nematic may either unwind to a uniform 0π state to match the surface condition or may wind further to form the $\Phi = 2\pi$ twist state (Figure 6.33a). Switching from one state to the other then relies on whether or not flow is induced immediately after a high electrical

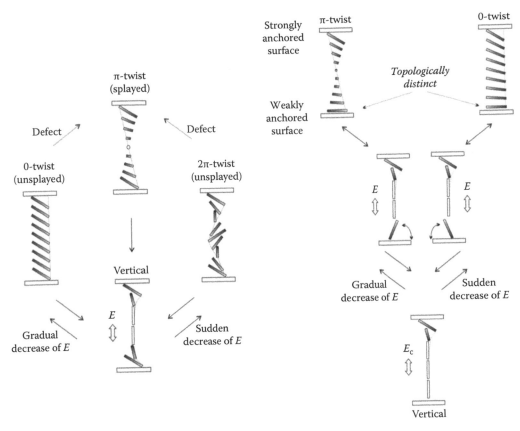

Figure 6.33 Operation of bistable twisted nematic. (a) 0–2π mode and (b) 0–π mode or Binem.

pulse coupling to a positive Δε. If the pulse returns to 0 V via an intermediate voltage, there is little induced flow and the 0π state is formed, whereas a direct transition to 0 V induces flow that encourages director twist at the cell center and the Φ = 2π state is formed. The two states are metastable, so the texture relaxes back to the intermediate π-state after a second or two on removal of power. This means that the device was not suited to zero power applications. Rather, *Seiko Epson* used it as the display for Hi-Fi Graphic Equalizer displays due to its very fast optical response [143].

A similar approach was taken by *Nemoptic*, who also used a BTN configuration, but with d/P = 0.25 to give either 0 or π twist states (Figure 6.33b). In this instance, the director cannot change its twist from one state to the other without breaking the anchoring at one of the surfaces, requiring that one of the surfaces be deliberately weakly anchored [144]. Switching of the device again utilized back flow depending on the trailing shape of the addressing pulses. Marketing the device under

the trade name *Binem, Nemoptic* produced various demonstrators [145], including full color reflective displays, TFT-driven panels, and, intriguingly, a pixelated switchable quarter-wave plate mounted onto the front of an OLED display to switch between high-power emissive video frame rate display and ultralow power black and white E-reader mode [146]. Despite the excellent optical appearance, 0-π BTN required cell gaps below 2 μm and suffered from manufacturing tolerances that were difficult to achieve. Perhaps the biggest cause of the company's eventual demise in 2010 was that it failed to find the correct niche for its product.

The ZBD is rather unusual in several aspects [147,148]. It uses a grating as a surface alignment layer, designed to impart bistable pretilts of the contacting nematic regardless of the overall geometry chosen (Figure 6.34). This allows many different LCD designs to be used, including VAN–HAN mode [147], HAN–TN mode [147,149], multistable VAN–HAN–TN modes [150], gray-scale displays [151], single polarizer mixed TN modes [152],

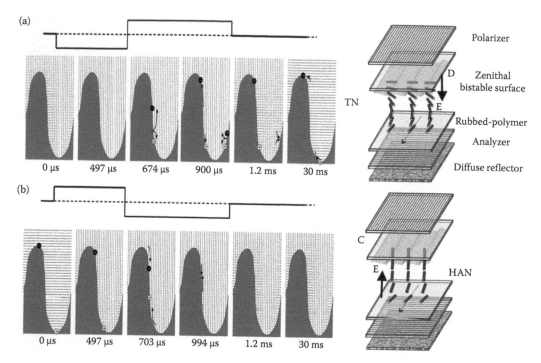

Figure 6.34 The zenithal bistable display mode, ZBD. (a) Latching from the C to D state occurs with a bipolar pulse with the trailing part positive with respect to the grating. The applied field nucleates defects at the bifurcation point on the grating sidewall, and the positive polarity favors the D state with $-1/2$ disclinations at the convex surface and $+1/2$ defects at the concave surface. The resulting low surface tilt creates a 90° TN state when placed opposite a conventional monostable rubbed polymer surface as shown. Operating in the NW mode, with the grating aligned parallel to the front polarizer and with n matched to the liquid crystal ordinary refractive index gives an excellent white state. (b) Applying pulses of the opposite polarity causes the defects to retrace their path along the grating surface until they annihilate. The resulting C state is continuous and homeotropic, thereby creating HAN alignment. This appears black when between the crossed polarizers as shown, and the viewing angle self-compensates when in reflective mode.

scattering modes [153], and ultrafast Pi cell modes [150]. The device is an early example, and the only LCD currently on sale, that utilizes the flexoelectric effect for latching between the two states. Moreover, the device deliberately uses disclinations at the deep, homeotropic-grating surface to stabilize the low-tilt or D state. The potential barrier between this state and the high-tilt defect free or continuous state (C state) is mediated by the creation and annihilation of these surface defects. The voltage for latching from one state to the other V_{CD} is related to the pulse width τ by Reference [154]:

$$\frac{V_{CD}}{d} \approx \frac{\gamma_1 l_s}{(e_1+e_3)\tau} + \frac{2W_\theta}{(e_1+e_3)+\sqrt{\varepsilon_0 \Delta \varepsilon K_{33}}}$$

$$= \frac{\gamma_1 l_s}{(e_1+e_3)\tau} + V_{th}, \quad (6.105)$$

where W_θ is the zenithal anchoring energy of the homeotropic grating surface, l_s is the coefficient of slip for defects moving across that surface, $e_1 + e_3$ is the sum of the splay and bend flexoelectric coefficients for the liquid crystal material, and the dielectric effect of the grating has been ignored. Controlling the anchoring energy in the range 0.2 mJ·cm^{-2} to 1 mJ·cm^{-2} allows the threshold voltage V_{th} to be adjusted to typically about 1 V μm^{-1}, typically 7 V.

The device is usually configured with the grating opposite a standard parallel aligned rubbed polymer surface to create a 90° TN when in the low-tilt D state. This state is always formed first on cooling from the isotropic to nematic phase because the defects are stabilized at the surface when the S order parameter is low. Thus, the interpixel gaps remain in the TN state and good display reflectivity results when operating in the NW TN mode. Typical cell gaps are 7 μm, operating at 15–20 V using standard STN driver electronics, LC mixtures, and polarizers. Operating temperatures from −25°C to 40°C and−5°C to +60°C were achieved using low and high anchoring energies, respectively.

The main technical challenge for *ZBD Displays Ltd.* was to introduce a low-cost and reproducible manufacturing method for a 0.8 μm pitch, 1 μm high blazed sinusoidal grating into a standard passive matrix LCD production line using Gen 2 glass. This was done by copying a photolithographically defined master grating into a lacquer on polyethylene terephthalate (PET) film via a nickel sputtering and electroforming replication technique [148,155]. The film was shipped from the UK to manufacturers in China, where it is used to emboss the grating into a homeotropic photopolymer deposited on the glass surface, thereby replacing the conventional rubbing step for that plate. This method allows the technology to meet the same price point as conventional STN, where the costs of the compensation foils required for the STN offset that of the ZBD grating film. Having achieved such low price points for its chosen niche market, it remains to be seen if ZBD can replace conventional TN and STN displays in other markets.

6.5.4 Polymer dispersed liquid crystals

A weakness of LCDs that is particularly evident in reflective mode devices is the constraint for polarizers, which absorbs more than half of the available light. This inefficiency was avoided in the first dynamic scattering displays but the contrast and lifetime of those devices were far inferior to retardation-based LCDs such as the TN that their period of success was very short lived. Hilsum [156] produced a scattering device by mixing glass microspheres into a nematic, creating a scattering texture that could be switched to a nonscattering state by an electric field. A more practical device was the polymer-dispersed liquid crystal (PDLC). This is another example of an LCD that initially aimed to produce bright displays but which found success when the technology was applied to a novel product with a niche market; for PDLC this was privacy glass.

Following his successes with cholesteric sensors and the invention of the TN LCD soon after Schadt and Helfrich (claiming precedence in the United States), Fergason invented a method of encapsulating liquid crystal droplets into a polymer matrix in the early 1980s [157]. Originally called NCAP by its inventor, PDLC usually takes the form of a plastic layer that can switch between scattering and nonscattering states with an applied electric field [158,159]. An ITO-coated glass or transparent plastic cell is filled with a nematic liquid crystal mixed with a monomer. Curing of the monomer into a solid polymer form is done by evaporating the monomer solvent, by applying a thermal treatment, or, most commonly, through the initiated polymerization of constituent photoreactive groups. Phase separation of the liquid crystal occurs as the polymer forms from the monomer, creating droplets within the polymer matrix. Often, a surfactant is included both to help control the dispersion of droplet size and align the nematic into a radially symmetric or bipolar state, as shown in Figure 6.35. Droplets in the 0.5–5 μm range scatter incident light intensely due to the refractive index mismatch between the polymer and the liquid crystal. Application of an electric field across the sample then causes reorientation of the liquid crystal director, leaving the cross-linked polymer undisturbed. Usually, the liquid crystal has a positive $\Delta\varepsilon$ and an ordinary refractive index n_o matched to that of the polymer. Thus, the scattering is reduced as the liquid crystal aligns parallel to the field and the light incident close to the normal direction is transmitted unchanged.

Droplets with radial alignment have a single defect at the droplet centroid (called a *"Boojum"*) leading to spherical symmetry, whereas bipolar droplets have point defects on opposing surfaces and cylindrical symmetry of the director field. Typically, droplets are between 1 and 10 μm and form the bipolar structure with randomly oriented symmetry axes. The electrical field behavior is complex. The field required to align the director parallel to the field, and hence approach the minimum deflection of incident light, is inversely proportional to the droplet radius, a [160]. For a radial droplet, the critical field is approximately related to [161]

$$E_c = \frac{4}{a}\sqrt{\frac{\bar{k}}{\varepsilon_0\Delta\varepsilon}}, \qquad (6.106)$$

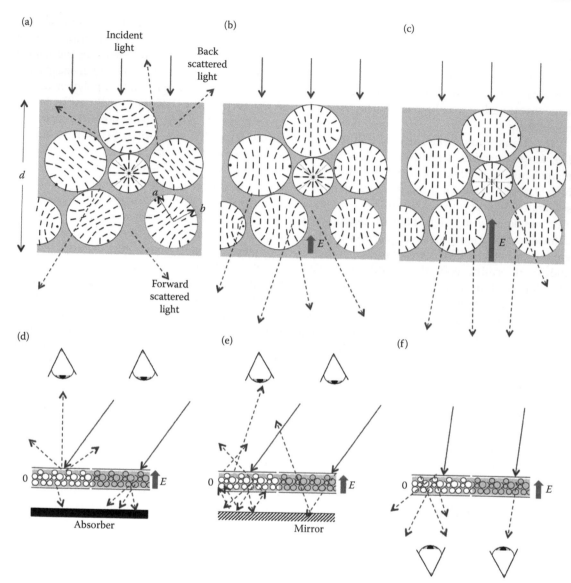

Figure 6.35 Polymer dispersed liquid crystal for a positive Δε operating in the scattering to nonscattering mode. (a) Droplets of different sizes with random orientations of the director profile, forming either a radial or bipolar configuration. Light is strongly scattered, including back scattering. (b) Application of the field initially causes little distortion of the director profile within each droplet, but the profile adjusts to align the disclinations within the plane of the film and the director with a net direction parallel to the field. The scattering reduces somewhat. (c) Increasing the field causes distortion of the director field within each droplet toward the vertical orientation, where the refractive index of the droplet and polymer matrix approaches the matched condition, and little scattering occurs. (d) Conventional back-scattering mode of PDLC; (e) novel, low-voltage mode using specular reflection as the black state; and (f) common mode for privacy windows, where the plastic film switches between translucent and transparent.

where $\bar{k}$ is a mean elastic constant. The symmetry of a perfectly spherical bipolar droplet prevents analytical solutions being made and needs the symmetry to be broken. A practical approach is to assume that each droplet is slightly elongated with different semimajor axis a and semiminor axis b. For liquid crystals with positive $\Delta\varepsilon$, weak applied fields reorient the symmetry axis parallel to the direction of the field. The critical field is then given by [161]

$$E_C = \frac{1}{3ab}\left(\frac{\sigma_{LC}}{\sigma_P} + 2\right)\sqrt{\frac{k_{33}\left(a^2 - b^2\right)}{\varepsilon_0 \Delta \varepsilon}}, \quad (6.107)$$

where σ_{LC} and σ_p are the low frequency conductivities of the liquid crystal and polymer, respectively. The response times are given by

$$\tau_{ON} = \frac{\gamma_1}{\varepsilon_0 \Delta \varepsilon E^2 + \frac{k_{33}}{a^2 b^2}\left(a^2 - b^2\right)},$$

$$\tau_{OFF} = \frac{\gamma_1 a^2 b^2}{k_{33}\left(a^2 - b^2\right)}. \quad (6.108)$$

Typically, droplets of $a = 1\,\mu m$, $b = 0.9\,\mu m$, $k_{33} = 10$ pN, $\gamma_1 = 0.05\,kg\,m^{-1}s^{-2}$ and $\Delta \varepsilon = 15$, and $\sigma_{LC}/\sigma_p \approx 28$ give $E_C \approx 1.5\,V\,\mu m^{-1}$ and $\tau_{OFF} \approx 20\,ms$. Substantial scattering needs 20 or so droplets in the direction of incidence, so film thicknesses of $20\,\mu m$ are typically used. Thus, the voltage of such PDLC films can be high, typically much higher than $30\,V$ to achieve saturation of the transparent, nonscattering state.

If the droplets are sufficiently small, and the device spacing high, then the OFF state can lead to significant back scattering. This can be used in a reflective mode display by mounting the PDLC film onto a black background to provide optical contrast (Figure 6.35d). However, the switching voltage inherently increases both with smaller droplets and higher device spacing. Moreover, the shallow switching response necessitates the use of TFT to matrix address the pixels, limiting the switching voltage to $15\,V$ and preventing any useful degree of optical effect. Recently, *Sharp* has produced low-voltage TFT PDLC displays where the layer is mounted instead onto a specular mirror [162]. Relatively large droplets keep the voltages sufficiently low for TFT addressing ($\approx 6\,V$) and the forward scattered light is scattered more strongly after being reflected back toward the observer a second time by the mirror. In the ON state, the forward scattering is removed and the viewer sees the specular reflected light. In applications, where the display is illuminated by a point source, this will appear black in all directions except where the display is oriented directly between the viewer and light source (Figure 6.35e).

Although never successful for mainstream displays, one market where the PDLC has found a niche is for privacy windows. The PDLC can be made as a laminated plastic roll, which can be applied adhesively to existing infrastructure, and trimmed to the appropriate size and shape. The window can then switch between a translucent, "milky white" appearance in the OFF state, to being clear and fully transparent when ON. Although this adds costs to the window, it is increasingly popular in high-end architectural projects for both interior and exterior applications. Uses include privacy control of conference rooms, intensive care areas, or bathroom doors, for example, to providing architectural design features. Moreover, the material can also be adjusted to act as an environmental smart window, switching between heat retention and loss as required [163]. Other applications such as temporary projection screens, active camouflage and switchable diffusers are all possible [164].

Another potentially profitable application for PDLC is switchable holographic media [165]. If a coherent image is used to cure the polymer during the fabrication process, switchable diffractive optical elements can be fabricated with droplets ranging from 100 nm to 1 μm forming holographic gratings from the interference pattern of the illumination. Rather than cause scattering, the droplets locally modulate the refractive index to generate a diffraction grating. Holographic optical elements (HOE) are used for a variety of optoelectronic applications, including focusing, beam-steering, filtering, and optical multiplexing. The use of holographic PDLC provides the means for switchable HOE, allowing dynamic beam steering, tunable filtering, and optical signal processing. For holographic PDLC, the droplet size a is typically submicron and the OFF time is very fast: typically 50–500 μs. However, E_C also increases with decreasing droplet size and so typically 100–200 V is required. This high field dominates the ON time, which again is fast, typically 50 μs.

6.5.5 Liquid crystal on silicon

Rather than apply a semiconductor onto glass to provide transistors, LCOS places the liquid crystal element directly onto a CMOS integrated circuit [166,167]. Of course, as the silicon is opaque, the devices must operate in reflection; usually, silver electrodes patterned directly onto a CMOS integrated circuit to provide both the pixelated electric field pattern and the highly reflective rear substrate. The chip not only provides the drive signals for the contacting liquid crystal but can also deliver extra functionality, such as gray-scale gamma correction, temperature compensation, edge detection. The devices are typically only 1–2 cm in diagonal and are used for projectors and camera viewfinders. Keeping such low dimensions enables the ultrahigh resolution of the silicon circuitry industry to be applied, with typical pixel pitches for an extended graphics array, 1024×768 projection display being below $3\,\mu m$.

LCOS spatial light modulators are the active element in light projectors for conferences and home cinema systems. This is one of the largest niche markets for LCDs, despite strong competition from MEMS projection displays (digital light projectors). Recently, there has been great interest in near-eye displays and pico-projectors, such as those used in virtual and augmented reality headsets, and for Google glass. Such displays tend to use a small polysilicon transmissive TN LCD, rather than reflective LCOS.

The projector systems either use a single LCOS panel or use three panels tuned for red, green, and blue wavelengths and the final image combined using a prism system. A 45° single polarize TN can be used but VAN mode provides the high contrast needed for home cinema, as utilized by the top of the range projectors from *Sony* and *JVC* [167]. Considerations such as viewing angle and color balance become inconsequential in such projection systems, where speed and contrast are the dominant factors and performance can be optimized for individual color bands.

A different approach to color uses a ferroelectric LCOS. Rather than using three separate modulators, the fast response time of theFLC is used to provide operation at 360 Hz, thereby allowing frame sequential color when illuminated by 120 Hz alternating color band illumination [168,169]. This is used for helmet mounted virtual reality (VR), binocular displays, and viewfinders by *Forth Dimension Displays* and *Cinoptics*.

LCOS has also proven successful for nondisplay applications, particularly for spatial light modulators in optoelectronic systems such as wavelength selective switching, structured illumination, and optical pulse shaping [170].

A related technology to LCOS is the optically addressed spatial light modulator [171]; rather than electrically addressing each pixel, the device is addressed using incident light. The LCD includes unpixelated ITO electrodes sandwiching a photoconducting semiconductor and liquid crystal layers. In the dark state, the dielectric permittivity of the photoconductor is low, and most of the electric field applied across the ITO electrodes is dropped across this layer, leaving the liquid crystal unswitched. However, charge is transferred to the liquid crystal interface where light is incident on the photoconductor, which switches the liquid crystal accordingly. Early devices used GaAs photoconductor and a nematic. Recently [172,173], an α-Si Optically addressed spatial light modulator with an FLC modulation layer has been used to produce dynamic computer generated hologram with over 10^8 pixels. The speed and the bistability of the FLC allowed the image to be built successfully in a projection system using a series of electrically addressed nematic SLM. The resulting system remains arguably the most complex display to be produced, and the nearest to providing full 3D holographic dynamic images, which remains the ultimate display goal.

6.6 LCD MODES FOR POTENTIAL FUTURE APPLICATIONS

6.6.1 Introduction

The variety of different LCDs is huge and only a handful has made it to commercial success, whether mainstream or niche. Some of these modes are important to describe because they include principles yet to be utilized (e.g., the Pi-cell, V-shaped switching FLC, and antiferroelectric liquid crystals (AFLCs); others are recent modes that are yet to find a market (such as the blue phase or flexoelectric cholesteric modes). It was suggested earlier that the principal motivation for LCD developments has evolved from increasing image complexity to achieving the widest viewing angle, high resolution, and recently color depth. Alongside these developments has been a constant need for increasingly fast LCD switching. Further speed improvement continues to be important for gaming, VR, augmented reality (AR), and ultimately for glasses-free 3D displays with eye tracking for multiple viewers and frame sequential color.

Moreover, nondisplay applications of liquid crystals continue to grow in importance and variety. A good example of this is the advent of liquid crystal lasers that have more far-reaching possibilities for optoelectronics. Brief appraisals of each of these modes are included in this section.

6.6.2 Pi-cell

Nematic liquid crystals are simple to align, usually maintain alignment quality after receiving mechanical or thermal shock, and are well understood and characterized. The down side is often a slow response speed, particularly at low operating temperatures where the Arrhenius form of the viscosity dominates (Equation 6.14). An early example of a fast nematic mode is the pi-cell [174], where ON and OFF times approaching 1 ms are achieved.

The pi-cell is the most well-known example of a surface mode liquid crystal [175]. It relies on the parallel alignment of high pretilt surfaces, between which is a positive $\Delta\varepsilon$ nematic. For low surface pretilts, the lowest energy quiescent state is predominantly splayed from one surface to the other. For high pretilts, a bend state of the director becomes favored energetically. This occurs above the pretilt given by [176]

$$\frac{4\theta_s - \pi}{\sin 2\theta_s} = \frac{k_{33} - k_{11}}{k_{33} + k_{11}}, \quad (6.109)$$

which predicts that the splay and bend states are energetically equivalent states for $40° < \theta_s \leq 50°$ for typical calamitic nematics ($0.5 \leq k_{33}/k_{11} \leq 2$). Pretilts of this magnitude have been hard to achieve historically, with $\theta_s \approx 25°$ typically being the maximum. With pretilts of this magnitude, the splay state is energetically favored but the bend state is metastable and can exist alongside the splayed state separated by a π disclination. With an applied voltage coupling to the positive $\Delta\varepsilon$, the central director of the splayed state can tilt in either a clockwise or anticlockwise direction, concentrating the elastic distortion close to the top or bottom surface (Figure 6.36b). If a sufficiently high field is applied, the elastic distortion may become sufficiently high to break the anchoring at one of the surfaces and allow a first-order alignment transition to the bend or π state [177]. On removal of the field, the director relaxes back to the metastable

state where the director remains vertical in the cell center. Whereas in this state, the director field close to the surfaces can be affected by the application of a lower field in a continuous fashion, thereby modulating the retardation and causing optical contrast between ON and OFF states, with a fast response.

To understand why the pi-cell gives this fast response, the effect of viscous back-flow needs to be considered. Figure 6.36a shows the situation for a uniform director produced by antiparallel surface alignment. When the field is removed, the director is subject to both elastic and viscous restoring forces. Lateral flow of the material is set up in opposite directions in either half of the cell, acting to kick the director in one-half in the opposite direction to that of the other, thereby slowing the relaxation to the final state. This backflow, or "optical bounce" effect slows the response for ECB and TN devices significantly [68]. However, for the π-state, the final tilt remains vertical in the cell center and the flow acts in both halves in the same direction as the relaxing director, thereby hastening the relaxation process. Thus, viscous flow supports switching so that the device is very fast—typically switching in about 2–5 ms for both ON and OFF switching.

The device is oriented with the alignment directions at 45° to crossed polarizers, and the cell gap set to give the half-wave plate condition. This occurs for a higher cell spacing for the bend state due to the higher overall tilt of the director. The viewing angle of the bend state is better than the splayed state because of the symmetry of the director arrangement. That is, the change in retardation for off-axis light is increased in one-half of the cell and decreased in the other half, so that there is little retardation change. This self-compensation effect leads to the alternative term for the Pi-cell as optically compensated bend (OCB) mode [178]. Further improvements to the viewing properties can be achieved using film compensators [179].

The device is yet to be used commercially. It cannot achieve the high contrast ratio enjoyed by the IPS and VAN modes, due to the remnant surface retardation; even with optical compensation, dark state leakage is too high to meet the high contrast ratio requirement of modern TV displays. Moreover, the bend state is metastable, and usually decays back to the splayed state after removal of the field. Unless a constant bias voltage is maintained,

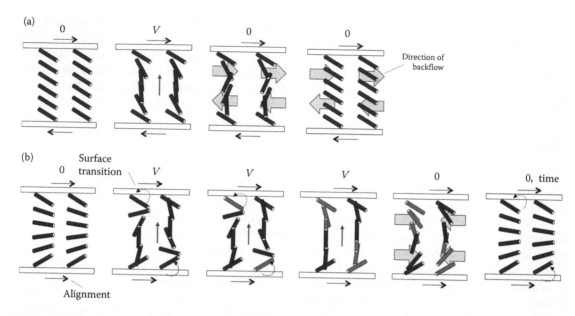

Figure 6.36 The effect of backflow and the Pi cell. (a) With antiparallel alignment, there is flow alignment in opposing directions on removal of the field, hindering the relaxation back to the 0V uniform state. (b) Parallel alignment initially gives a splayed state, but application of sufficiently high field to break the surface anchoring allows a bend, or p-state to be formed. When the field is removed, relaxation to the unswitched bend state is supported by the flow throughout the cell. However, the bend state is metastable, so the splay state reforms after sufficient time, usually nucleated from the pixel edges or LCD spacers.

domains of the unwanted splay state spread from unswitched areas such as the interpixel gaps and close to spacers. However, the principles deployed in this mode remain of interest. For example, a transflective display has recently been suggested that uses subpixellation for the bend state operating in reflection and the splay state operating with in-plane electrodes in transmissive mode [180].

6.6.3 Analog ferroelectric liquid crystal modes

The SSFLC mode described in Section 6.4.5 is bistable: either the black or white states is retained after the addressing pulse. Grey scale is provided using spatial or temporal dither [125]. Bistability was integral to the original interest in FLCDs because it provided an alternative to TFT. However, once TFT had become sufficiently low cost, interest was retained in utilizing the fast optical speed of FLCs but using an analog response combined with the active matrix.

Various analog FLC modes were studied as outlined below. Each of the modes described suffers from the need to achieve and maintain uniform smectic layer structures, which has prevented display applications to date. However, the high speeds that are possible mean that these modes may yet find utility in nondisplay and optoelectronic applications.

6.6.3.1 N*-SMC*

Conventional SSFLC uses an N*-SmA-SmC* to give a chevron layer arrangement and the implicit bistability that brings. If the FLC has a phase sequence where the SmC* cools directly from the unwound chiral nematic, the director remains in the rubbing direction at the two surfaces and the layers form in a uniform layer bookshelf geometry but with the layers angled at the cone angle θ_C to the rubbing [181]. Cooling with an applied DC field ensures that a single sign of orientation is achieved. Thus, the device is monostable, with an analog response as the director switches about the cone in the bulk of the cell with one polarity of field, and no-response with the opposite polarity. This "half V-shaped" switching can be addressed by TFT to give a fast, analog response.

A modification is the twisted FLCD [182], which combines a material with a first-order N*-SmC* transition with rubbing directions crossed to each other. The unwound N* forms a conventional TN but, as the twist elastic constant diverges on approaching the SmC phase, the smectic layers form uniformly at 45° to the rubbing directions. The director still forms a 90° twist, but moves about the SmC* cone from one surface to the other with the applied DC field, leading to a "V-shaped" switching response. This mode works best with a strongly first-order transition, wherein the SmC* cone angle is independent of temperature and is typically close to the optimum 45°; this allows the director to lie parallel to the crossed rubbing directions.

6.6.3.2 ANTIFERROELECTRIC LIQUID CRYSTALS

Certain compounds that form chiral tilted smectic phases exhibit higher ordered phases, where there is correlation of the director orientation between adjacent layers. With AFLC, the SmC* c-director and polarization directions alternate by π from one layer to the next. Application of a DC field switches each alternate layer parallel to the field, thereby forming a V-shaped switched response [183].

6.6.3.3 DEFORMED HELIX MODE FERROELECTRIC LIQUID CRYSTAL

Although formed from chiral liquid crystals, most FLC modes use sufficiently low spaced devices to unwind the pitch. This is not the case for the deformed helix mode, where the pitch is made sufficiently low that even devices at the quarter-wave plate condition wind continuously in a helical manner. Application of the field then causes distortion of the sinusoidal director variation and a shifting of the optic axis from that of the helix toward $\pm\theta_C$ [184].

6.6.3.4 THE ELECTROCLINIC EFFECT

SmA phases formed from chiral molecules undergo a field-induced director tilt with an applied DC field, forming a structure similar to the SmC* phase [185]. The strength of this electroclinic effect is greatest immediately above an SmA to SmC* transition, where the induced tilt is linear with the applied field. As with the AFLC and DH FLC modes, the switching is within the cell plane, and hence it is not only fast but has good viewing angle. However, the limited temperature range and temperature dependence have so far discounted electroclinics from application.

6.6.4 Blue phase TFT displays

Attendees at the 2008 exhibition that occurs each year alongside the Society of Information Display annual conference were surprised to see an unheralded novel LCD mode based on the blue phase at the *Samsung* stand. The company demonstrated a 15 in. TFT monitor operating at a ground breaking 240-Hz frame rate. Liquid crystals were known for their slow response and most attendees were unaware of the blue phase. Those that were most likely believed that the blue phase would never receive such serious interest from a manufacturer due to the notoriously narrow temperature ranges exhibited by these phases.

Blue phases are a subset of cholesteric liquid crystals, which occur when the cholesteric natural helicity is strong and the pitch is very short, close to the transition to the isotropic phase [186]. Rather than spontaneously twist along a single axis perpendicular to the local director, the director twists along two mutually orthogonal axes to form a double helix cylindrical structure, as shown in Figure 6.37a. The director at the center of each cylinder lies parallel to the cylinder axis and the diameter of the cylinder is $P/4$ so that the director twists through 45° from one side to the other. This means that the director remains continuous across adjacent cylinders oriented with their axes orthogonal to each other, as shown in Figure 6.37b. These double twist cylinders pack into a cubic array as shown in Figure 6.37c, mediated by disclination lines throughout the structure that occur at the interstices of the cylinders. Of the three possible blue phases, the simple cubic blue phase I, Figure 6.37f, and body-centered cubic blue phase II, Figure 6.37d have the widest temperature ranges. The occurrence of the disclinations means that the cubic structure is only stable where the pitch and order parameter are very low, typically for about 1°C–2°C below the isotropic to cholesteric phase transition. Optically, the structure appears an iridescent color due to Bragg reflection from the defects arranged on the regular cubic lattice, Figure 6.37e; hence, the phase became known as the "blue" phase, although other colors are possible depending on the lattice. Otherwise, the phase

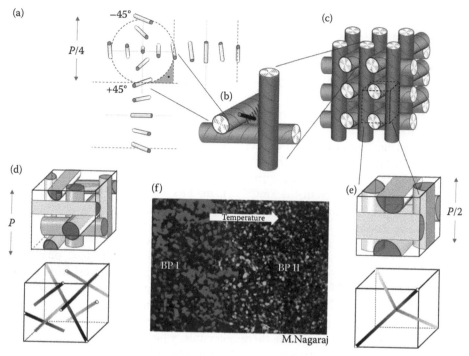

Figure 6.37 Blue phase liquid crystal. (a) The double helix structure; (b) the intersection of three double twist cylinders and the direction of the −1/2 disclination that occurs close to the cylinder intersections; (c) cubic packing of the double twist cylinders; (d) packing of the cubic blue phase II, with associated defect lattice; (e) photomicrograph of the transition between blue phase I and II; (f) body centered packing of the cubic blue phase I and the associated defect lattice.

is optically isotropic due to the cubic symmetry of the phase, and it appears dark between crossed polarizers. Blue phases exhibit a Kerr effect with an applied field due to deformation of the local cholesteric structure coupling to the dielectric anisotropy of the liquid crystal [187,188]. This induces a net birefringence Δn_E with the optic axis in the direction of the applied field [189]:

$$\Delta n_E = \lambda K E^2 \sim -n^3 E^2, \qquad (6.110)$$

where n is the isotropic refractive index of the blue phase and the Kerr constant K is also wavelength dependent.

Although the Kerr effect provided means for electro-optic modulation, the extremely limited temperature range of the blue phase seemed severely prohibitive to application. This was particularly true for displays, where operation from −20°C to +70°C is usually the minimum specification. In 2002, Kikuchi et al. succeeded in greatly extending the temperature range by photopolymerizing a monomer added to the liquid crystal

at temperatures where the material formed the blue phase [188]. The resulting polymer formed at the defect sites stabilized the lattice structure over a wide temperature range, without affecting the electro-optic properties of the liquid crystal significantly.

The *Samsung* display used in-plane electrodes to switch the polymer stabilized blue phase from an optically isotropic structure to a birefringent medium with the optic axis approximately in the plane of the cell (Figure 6.38). The device does not require alignment layers and gives excellent high contrast and wide viewing angle due to the isotropic nature of the dark state and IPS effect. The most important issue to resolve was achieving suitably low operating voltages. This was done using shaped protrusions as the electrodes [190], with 2 μm wide electrodes and 2–4 μm electrode gaps. Modeling showed that using angled electrode walls produced higher transmissivity by reducing the angle of the optical axis at the electrode edges. Using these structures, operating voltages below the 15 V target for TFT addressing were achieved. Indeed, since

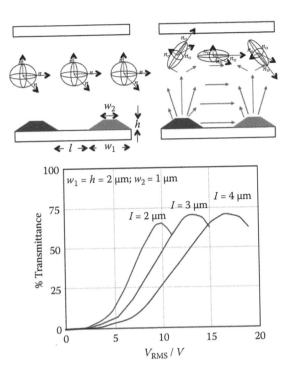

Figure 6.38 Principal of operation for in-plane switching mode blue phase LCD. The applied field induces optical birefringence (shown by the ellipsoid shape of the refractive index indicatrix). Good performance is achieved using shaped protrusions allowing 10 V operation to be achieved [186].

that original work, a host of different device layouts and liquid crystal materials have been studied [191], indicative of the ongoing interest in the blue phase mode for applications.

6.6.5 Cholesteric flexoelectricity and the uniform lying helix mode

Recently, there has been much renewed interest in the flexoelectro-optic effect in chiral nematic liquid crystals, originally proposed by Patel and Meyer [192] in 1987. This interest stems from the very fast IPS effect, typically 100 µs. If an electric field is applied perpendicular to the helical axis of a cholesteric liquid crystal, coupling between the field and the flexoelectric polarization causes splay and bend of the local director field, which in turn causes the local director to tilt away from the helical axis in the direction normal to the field (Figure 6.39a). Without the field, the uniaxial optic axis of the cholesteric lies parallel to the helical axis. The field-induced local tilt of the director causes the optic axis to tilt through angle β in the plane orthogonal to the applied field.

If the cholesteric can be uniformly aligned with the helical axis parallel to the plane of a standard LCD with transverse electrodes (uniform lying helix, ULH), then applying the electric field causes rotation of the optic axis in the cell plane, giving contrast when observed through polarizers parallel and crossed to the helical axis. This in-plane change of retardation axis provides a switchable half-wave plate.

Assuming negligible dielectric anisotropy, the twist angle is approximately linear with the applied field:

$$\tan\beta = \frac{(e_1 - e_3)P}{2\pi(k_{11} + k_{33})}E \qquad (6.111)$$

and the response time τ is

$$\tau = \frac{\gamma_1 P^2}{2\pi^2(k_{11} + k_{33})}E. \qquad (6.112)$$

Flexoelectric switching is polar, so the ULH can be driven in either direction by swapping the field

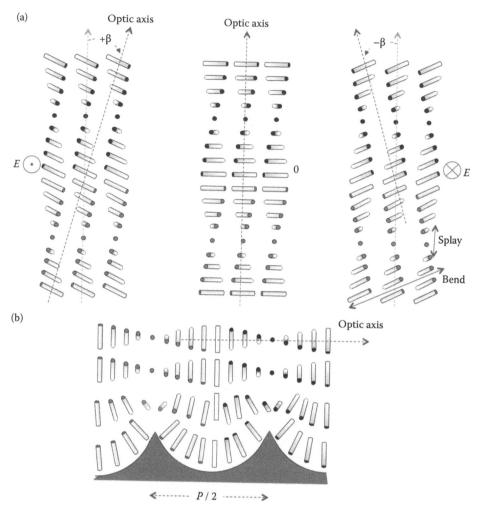

Figure 6.39 The cholesteric flexoelectric effect. (a) Rotation of the optic axis and (b) alignment of N*
into the ULH texture using cusped grating surfaces [192].

direction. Thus, Equation 6.112 is appropriate for
both ON and OFF switching because the optic axis
can be driven back to the OFF state using signals of
the opposite polarity to those for switching. Both
expressions show that the strength of the effect
is optimized for small pitch lengths P, promising
speeds below 100 μs and greater than 45° switching angles for strongly flexoelectric materials
[193], and IPS for good viewing angle. The pitch
is unwound by the RMS effect of the field cou-
pling to $\Delta\varepsilon$, so liquid crystal materials with strong
flexoelectric effect but low $\Delta\varepsilon$ are required. This
has been achieved using nematic dimers with odd
alkyl spacer groups [194].

Fabrication of devices based on this effect
requires a method for obtaining the required ULH

alignment uniformly and stably over the panel.
Strongly planar anchoring at the cell walls leads
to the Grandjean texture being formed, where
the helical axis lies parallel to the cell normal
(sometimes called the uniform standing helix).
Homeotropic alignment gives degenerate focal
conic domains, and a scattering texture. Various
methods have been investigated, including using
periodic planar and homeotropic surfaces with the
periodicity matched to $P/2$ [195], surface relief gratings [195,196], and polymer walls [197]. The director profile for a cholesteric liquid crystal aligned
on a homeotropic grating with a pitch of $P/2$ and
a cusped shape to give the lowest elastic distortion
of the ULH state [196], as shown in Figure 6.39b.
Although yet to be proven over large areas, these

methods show considerable promise for future applications of the ULH mode LCDs.

6.6.6 Liquid crystal lasers

With the successful commercialization of high-resolution, high-frame rate LCDs for TV, interest in nondisplay applications for liquid crystals has continued to grow at an amplified rate. One such application particularly relevant to optoelectronics is the invention [198] and demonstration [199,200] of the liquid crystal laser. This allows straightforward tunability of laser light across the optical wavelength range combined with simple low-cost fabrication and offers the potential for electric field tuning of the lasing wavelength.

Selective reflection from the cholesteric liquid crystal is an example of a one-dimensional photonic band-gap, wherein circularly polarized light of the correct wavelength range and handedness cannot propagate through the material in the direction parallel (or antiparallel) to the helical axis. This means that spontaneous emission from a fluorophore within the cholesteric is inhibited within the photonic band-gap, leading to photons being emitted at the band edge [201]. That is, optical pumping of a cholesteric with either

a fluorescent dye dissolved into it or as constituent part of the mesogenic molecule causes emission of lower energy photons that overlap with the forbidden band of wavelengths corresponding to the range of selective reflection. Where emission occurs across the band-gap lasing will occur at both edges. In the example system shown in Figure 6.40, a tetraaryl-pyrene derivative dye [202] pumped at 410 nm gives emission peaking in the range 450–500 nm. Once dissolved into a cholesteric with a photonic band from 500 to 520 nm lasing occurs on the lower band edge. Either lowering the cholesteric pitch or changing to a dye with a longer wavelength emission swaps the lasing wavelength to the upper photonic band edge.

The stimulated emission occurs in a single direction, as shown in Figure 6.40. Such devices offer potential for forming laser arrays, competing against III–V semiconductor vertical cavity surface emitting lasers but without the need for the multiple fabrication steps. Indeed, the simplicity of fabrication allows ink-jet printing of the liquid crystal to form an array of laser dots, each with a signature emission [203]. The low cost and ability to print onto a plastic backing layer has already earned the technology commercial application for anticounterfeiting.

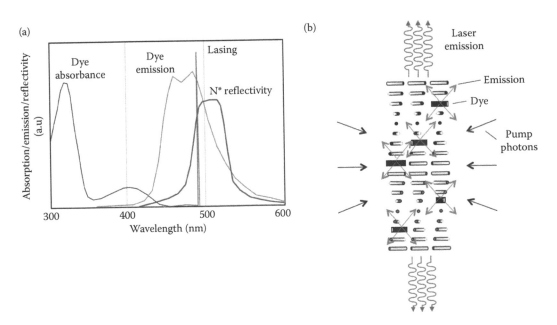

Figure 6.40 Lasing in dyed cholesteric. (a) Example of the optical properties for the fluorescent dye and cholesteric liquid crystal, leading to lasing on the lower band edge of the selective reflection and (b) schematic representation of cholesteric lasing.

Lasing is possible in other liquid crystal phases that include Bragg type structures with optical length scales, including the ferroelectric chiral SmC* and blue phases [204]. The former offers the potential for fast electric field modulation, whereas the latter produces lasing in multiple directions due to the cubic structure of the Bragg lattice.

6.7 CONCLUSION

The market success of LCDs is predicated by scientific and technological achievements from across the globe, advances that have needed collaborative efforts of mathematicians, chemists, physicists, engineers, and entrepreneurs. The breadth of the subject is unusually broad and involving. A major part of the ongoing success of LCDs is the continual evolution and adaptation that these technological advances allow. Throughout the history of the LCDs, the naysayers have maintained a mantra that LCDs cannot achieve the complexity, cannot achieve the viewing angle, cannot achieve the resolution, cannot achieve the speed, and cannot achieve the color balance of the latest competing technology. In each case, they have proven incorrect. Often, this has been through the invention of new modes, sometimes associated with different arrangements, sometime with different materials and phases. Often, the improvements have come from the use of partnering technologies; the active backplanes, polarizer, and optical films, back-light units, light guides, or manufacturing advances.

As it has always been, the competition is strong. OLED displays are already making inroads into small, high-resolution displays for mobile phones. If performance, cost and lifetimes can all exceed those of the LCD, then they will gain an ever-increasing share of the market, from low to high diagonal displays. Low-cost reflective passive matrix LCDs retain a large market presence too. However, as the cost of the TFT backplane continues to reduce, the competition from electrophoretic displays is strong and growing. Winning these battles will take yet more ingenuity from the LCD scientists and engineers.

There remain several display markets where no display technology has yet delivered, whether LCD or alternative. Although reflective color and transflective displays have been marketed, solutions to date have not been adequate to gain market acceptance. LCDs have a natural advantage for transflective mode operation, being based on transparent media that modulates ambient light. Electrophoretics absorb light preventing transmissive operation and OLEDs emit and so cannot modulate reflected light. Solutions to achieving the required performance at a suitably low cost have yet to be commercialized; perhaps the developments of new nematic modes [e.g., 180] or the application of a new LC phase, such as the blue phase III [205], will prove successful.

An area of enormous growth is the use of liquid crystals for nondisplay applications [206]. In addition to the conventional applications described in the previous sections, liquid crystal main-chain polymers such as *Kevlar* continue to prove one of the biggest LC applications, liquid crystal elastomers are used as the active element in nanomachines and molecular motors [207], liquid crystal semiconductors are attracting interest for photovoltaic and OLED [208], and the great interest in liquid crystals in chemical and biological sensors continues with unabated enthusiasm [209]. One of the largest areas of growth, however, is in the field of optoelectronics. Whether this is for smart windows, beam steering and light guiding, printable lasers, switchable lenses, optical computing, adaptable photonic structures, or Terahertz modulators, successful solutions to such applications in the future will require both the ingenuity and the multidisciplinary approach that made our predecessors so successful.

ACKNOWLEDGMENTS

The author wishes to thank Professors Peter Raynes, Cyril Hilsum, Phil Bos, Tim Wilkinson, and Dr. Mamatha Nagaraj for invaluable discussions.

REFERENCES

1. S. Kubrick (Dir.) (1968). *2001: A Space Odyssey*, Metro-Goldwyn Mayer, Released 2nd April 1968. Premiered in New York from 3rd April 1968 at the Loews Capitol Theatre, Broadway, NY.
2. B. J. Lechner (2008). History crystallized: A first person account of the development of matrix-addressed LCDs for Television at RCA in the 1960s. *Inf. Disp.*, 08(1), pp. 26–30.

3. G. H. Heilmeier and L. A. Zanoni (1967). Electro-optical device. US Patent 3499112 A.

4. G. H. Heilmeier, L. A. Zanoni and L. A. Barton (1968). Dynamic scattering: A new electro-optic effect in certain classes of nematic liquid crystals. *Proc. IEEE*, 56, pp. 1162–1171.

5. B. Levin and N. Levin (1936). Improvements in or relating to light valves. UK Patent 441, 274.

6. J. A. Castellano (2005). *Liquid Gold: The Story of Liquid Crystal Displays and the Creation of an Industry*. World Scientific, Singapore.

7. H. Kawamoto (2001). The history of liquid crystal displays. *Proc. IEEE*, 90(4), pp. 460–500.

8. G. W. Gray (1962). *Molecular Structure and Properties of Liquid Crystals*, Academic Press, London.

9. G. H. Brown and W. G. Shaw (1957). The mesomorphic state: Liquid crystals. *Chem. Rev.*, 57, pp. 1029–1157.

10. R. Williams (1963). Domains in liquid crystals. *J. Chem. Phys.*, 39(2), pp. 384–388

11. R. Williams (1962). U.S. Patent 3, 332, 485 A. Filed Nov. 9, 1962 and issued May 30, 1967.

12. C. Hilsum (2010). Flat-panel electronic displays: a triumph of physics, chemistry and engineering. *Philos. Trans. Ser. A, Math., Phys. Eng. Sci.*, 368(1914), pp. 1027–1082. doi.org/10.1098/rsta.2009.0247.

13. N. Koide (2014). *The Liquid Crystal Display Story: 50 Years of Liquid Crystal R&D That Lead the Way to the Future*. Springer, Japan.

14. M. Schadt and W. Helfrich (1971). Voltage-dependent optical activity of a twisted nematic liquid crystal. *Appl. Phys. Lett.*, 18(4), pp. 127–128.

15. G. W. Gray, K. J. Harrison, and J. A. Nash (1973). New family of nematic liquid crystals for displays. *Electron. Lett.*, 9(6), pp. 130–131.

16. A. Ashford, J. Constant, J. Kirton, and E. P. Raynes (1973). Electro-optic performance of a new room-temperature nematic liquid crystal. *Electron. Lett.*, 9(5), pp. 118–120.

17. G. H. Heilmeier (1976). Liquid crystal displays: An experiment in interdisciplinary research that worked. *IEEE Trans. Electron Devices*, 23(7), pp. 780–785.

18. F. Reinitzer (1888). Contributions to the understanding of cholesterol. *(transl.)*, *Monatshefte für Chemie (Wien)*, 9, pp. 421–441.

19. O. Lehmann (1889). Über fließende Krystalle (On flowing crystals) *Zeitschrift für Physikalische Chemie*, 4, p. 462.

20. D. Dunmur and T. Sluckin (2011). *Soap, Science, and Flat-Screen TVs: A History of Liquid Crystals*. Oxford University Press, New York.

21. I. Haller, (1975). Thermodynamic and static properties of liquid crystals. *Prog. Sol. State Chem.*, 10(2), pp. 103–118.

22. J. W. Goodby, P. J. Collings, T. Kato, C. Tschierske, H. Gleeson, and P. Raynes (eds.) (2014). *Handbook of Liquid Crystals*, Wiley VCH, Weinheim.

23. R. B. Meyer, L. Liébert, L. Strzelecki, and P. Keller (1975). Ferroelectric liquid crystals. *J. Phys.*, 26, pp. L69–L71.

24. M. Klasen-Memmer and H. Electro-optical 3,322,485 elements utilizing an organic nematic compound Hirschmann (2012). Liquid crystal materials for devices. In J. Chen, W. Cranton, and M. Fihn (eds.), *Handbook of Virtual Display Technology*, Springer, Berlin Heidelberg.

25. D. Dunmur and G. Luckhurst (2014). Tensor properties and order parameters of anisotropic materials. In J. W. Goodby et al., *Handbook of Liquid Crystals: Volume 2*, Wiley VCH, Weinheim, pp. 189–281.

26. J. Constant and E. P. Raynes (1985). RSRE Internal Memorandum, DSTL, Farnborough, UK

27. F. C. Frank (1958). I. Liquid crystals. On the theory of liquid crystals. *Discuss. Faraday Soc.*, 25(1), pp. 19–28.

28. J. C. Jones (2001). Orientational elasticity. In K. Buschow, R. W. Cahn, M. C. Flemings, B. Iischner, E. J. Kramer, & S. Mahajan (eds.), *Encyclopaedia of Materials: Science and Technology*. ISBN 0-08- 0431526, Elsevier Science, Oxford, UK, pp 6076–6086.

29. F. M. Leslie (1966). Some constitutive equations for anisotropic fluids. *Q. J. Mech. Appl. Math.*, 19, pp. 357–370; F.M. Leslie (1968).

Some constitutive equations for anisotropic fluids. *Arch. Ration. Mech. Anal.*, 28, pp. 265–283.

30. M. G. Clark and F. M. Leslie (1978). A calculation of orientational relaxation in nematic liquid crystals. *Proc. R. Soc. London A*, 361, pp. 463–485.

31. D. W. Berreman (1983). Numerical modelling of twisted nematic devices. *Philos. Trans. R. Soc. Ser. A.*, 309(1507), pp. 203–216.

32. S. T. Wu and C. S. Wu (1990). Rotational viscosity of nematic liquid crystals a critical examination of existing models. *Liq. Cryst.*, 8(2), pp. 171–182. doi:10.1080/02678299008047339.

33. R. B. Meyer (1969). Piezoelectric effects in liquid crystals. *Phys. Rev. Lett.*, 22, pp. 918–921.

34. E. P. Raynes (2014). Mixed systems: Phase diagrams and eutectic mixtures. In J. W. Goodby, P. J. Collings, T. Kato, C. Tschierske, H. Gleeson, and P. Raynes (eds.), *Handbook of Liquid Crystals: Volume 1*, Wiley VCH, Weinheim, pp. 351–363.

35. W. Martienssen and H. Warlimont (eds.) (2005). *Springer Handbook of Condensed Matter and Materials Data*, Springer, Heidelberg.

36. K. Tarumi, B. Schuler, E. Bartmann, and D. Pauluth(1996). Liquid crystalline medium. US Patent 5, 744, 060.

37. A. Buka and N. Éber (eds.) (2013). *Flexoelectricity in Liquid Crystals: Theory, Experiments and Applications*. Imperial College Press, London.

38. G. W. Gray and D. G. McDonnell (1976). The relationship between helical twist sense, absolute configuration and molecular structure for non-sterol cholesteric liquid crystals. *Mol. Cryst. Liq. Cryst.*, 34(9), pp. 211–217.

39. M. Hird (2011). Ferroelectricity in liquid crystals—Materials, properties and applications. *Liq. Cryst.*, 38(11–12), pp. 1467–1493.

40. F. C. Saunders, K. J. Harrison, E. P. Raynes, and D. J. Thompson (1983). New photostable anthraquinone dyes with high order parameters. *IEEE Trans. Electron Device*, 30, pp. 499–502.

41. I. Jánossy (1999). Optical reorientation in dye doped liquid crystals. *J. Nonlinear Opt. Phys. Mater,* 8(3), pp. 361–377.

42. M. Schadt, K. Schmitt, V. Kozinkov, and V. Chigrinov (1992). Surface-induced parallel alignment of liquid crystals by linearly polymerized photopolymers. *Jpn. J. Appl. Phys.*, 31(7), pp. 2155–2164.

43. D. Broer, G. Crawford, and S. Žumer (2011). *Cross-Linked Liquid Crystalline Systems: From Rigid Polymer Networks to Elastomers*, CRC Press, Boca Raton, FL.

44. R. J. Bushby and N. Boden (2014). Responses of discotic liquid crystals to mechanical, magnetic and electrical fields. In J. W. Goodby et al. (eds.), *Handbook of Liquid Crystals: Volume 3; Part III: Discotic, Biaxial and Chiral Nematic Liquid Crystals*, Wiley VCH, Weinheim, pp. 1–33.

45. C. V. Mauguin (1911). Sur les Cristaux Liquides de Lehmann. *Bull. Soc. Fr. Mineral*, 34, pp. 71–117.

46. V. Fréedericksz and A. Repiewa (1927). Theoretisches und Experimentelles zur Frage nach der Natur der anisotropen Flüssigkeiten. *Zeitschrift für Physik*, 42(7), pp. 532–546.

47. V. Fréedericksz and V. Zolina (1933). Forces causing the orientation of an anisotropic liquid. *Trans. Faraday Soc.* 29, pp. 919–930.

48. P. Châtelain (1941). Sur l'orientation des cristaux liquides par les surfaces frottées. *Comptes rendus de l'Académie des Sci.*, 213, pp. 875–76

49. B. Levin and N. Levin (1936). Improvements in or relating to light valves. British Patent 441, 274.

50. H. J. J. van Boort and R. Groth (1968). Low-pressure sodium lamps with indium oxide filter. *Philips Tech. Rev.*, 29, pp. 17–18.

51. M. Mizuhashi (1980). Electrical properties of vacuum-deposited indium oxide and indium-tin oxide films. *Thin Solid Films*, 70(1980), pp. 91–100.

52. Y. Katsube and S. Katsube (1966). The Tin-oxide films deposited by vacuum evaporation (I). *Shinku: J. Vac. Soc. Jpn.*, 9(11), pp. 443–450.

53. F. M. Wanlass (1963). Low stand-by power complementary field effect circuitry. US Patent 3, 356, 858.

54. E. H. Land and J. S. Friedman (1933). Polarizing refracting bodies. US patent 1, 918, 848 A.

55. J. F. Dreyer (1946). Dichroic light-polarizing sheet materials and the like and the formation and use thereof. US Patent 2544659.

56. T. Nagatsuka (2014). Chapter 4.1— Development of polarizers to support the development of the liquid crystal industry. In N. Koide (ed.), *The Liquid Crystal Display Story: 50 Years of Liquid Crystal R&D That Led the Way to the Future*, Springer, Japan, pp. 81–95.

57. P. K. Weimer (1966). Thin film transistors. In J. T. Wallmark and H. Johnson, (eds.), *Field Effect Transistors, Devices*, Prentice-Hall, Englewood Cliffs, NJ.

58. E. P. Raynes (1974). Improved contrast uniformity in twisted nematic liquid-crystal electro-optic display devices. *Electron. Lett.*, 10, pp. 141–142.

59. A. A. Sonin (1995). *The Surface Physics of Liquid Crystals*. Gordon and Breach Publishers, Luxembourg.

60. A. Hirai, I. Abe, M. Mitsumoto, and S. Ishida (2008). One drop filling for liquid crystal display panel produced from larger-sized mother glass. *Hitachi Rev.*, 57(3), pp. 144–147.

61. M. O'Neill and S. Kelly (2012). Chapter 6: Optical properties of light-emitting liquid crystals. In R. J. Bushby, S. Kelly, and M. O'Neill, (eds.), *Liquid Crystalline Semiconductors: Materials, Properties and Applications*, Springer Science and Business Media, London, pp. 163–195.

62. R. C. Jones (1941). New calculus for the treatment of optical systems. I. *J. Opt. Soc. Am.*, 31, pp. 488–493.

63. K. R. Welford, and J. R. Sambles (1987). Analysis of electric field induced deformations in a nematic liquid crystal for any applied field. *Mol. Cryst. Liq. Cryst*, 147(1), pp. 25–42.

64. "LCD Master," Shintech; http://shintech.jp/wordpress/en/software-2/lcdm3d.html; "TechWiz" Sanayi http://www.sanayisystem.com/eng/index.asp; *DIMOS*, Autronic Melchers.

65. E. Jakeman and E. P. Raynes (1972). Electro-optical response times in liquid crystals. *Phys. Lett.*, 39A(1), pp. 69–70.

66. H. Chen, M. Hu, F. Peng, J. Li, Z. An, and S. T. Wu (2015). Ultra-low viscosity liquid crystal materials. *Opt. Mater. Express*, 5(3), pp. 655–661.

67. S. T. Wu and D. K. Yang (2001). *Reflective Liquid Crystal Displays*, Wiley, New York.

68. D. W. Berreman (1975). Liquid-crystal twist cell dynamics with backflow. *J. Appl. Phys.*, 46, pp. 3746–3751.

69. P. M. Alt and P. Pleshko (1974). Scanning limitations of liquid-crystal displays. *IEEE Trans. Electron Devices*, 21, pp. 146–155.

70. B. J. Lechner, F. J. Marlowe, E. O. Nester, and J. Tults (1971). Liquid crystal matrix displays. *Proc. IEEE*, 59(11), pp. 1566–1579.

71. T. P. Brody, J. A. Asars, and G. D. Dixon (1973). A 6×6 inch, 20 lines-per-inch liquid crystal display panel. *IEEE Trans. Electron Devices*, 20, pp. 995–1001.

72. P. G. LeComber, W. E. Spear, and A. Ghaith (1979). Amorphous silicon field effect device and possible application. *Electron. Lett.* 15, pp. 179–180.

73. A. J. Snell, K. D. Mackenzie, W. E. Spear, P. G. LeComber, and A. J. Hughes (1981). Application of amorphous silicon field effect transistors in addressable liquid crystal display panels. *Appl. Phys.*, 24, pp. 357–362.

74. Y. Okubo, T. Nakagiri, Y. Osada, M. Sugata, N. Kitahara, and K. Hatanaka (1982). Large-scale LCDs addressed by α-Si TFT array. SID Dig., 82, pp. 40–41.

75. F. R. Libsch (2009). Thin-film Transistors in Active Matrix Displays. In C. R. Kagan and P. Andry (eds.), *Thin-Film Transistors*, Dekker, New York, pp. 187–269.

76. M.-K., Kang, S. J., Kim, and H. J. Kim (2014). Fabrication of high performance thin-film transistors via pressure-induced nucleation. *Sci. Rep.*, 4, p. 6858.

77. X. Huang, C. Wu, H. Lu, F. Ren, D. Chen, R. Zhang, and Y. Zheng (2013). Enhanced bias stress stability of a-InGaZnO thin film transistors by inserting an ultra-thin interfacial InGaZnO:N layer. *Appl. Phys. Lett.*, 102, p. 193505.

78. E. Lueder (2001). *Liquid Crystal Displays: Addressing Schemes and Electro-Optic Effects*, John Wiley and Sons, Chichester, UK.

79. S. M. Kelly and M. O'Neill (2000). Chapter 1: Liquid crystals for electro-optic applications. In H. S. Nalwa, (ed.), *Handbook of Advanced Electronic and Photonic*

Materials, Volume 7: Liquid Crystals, Displays and Laser Materials, Academic Press, London.

80. S. W. Depp, A. Juliana, and B. G. Huth (1980). Polysilicon FET devices for large area input/output applications. Technical Digest International Electron Devices Meeting, Washington DC, December 1980, IEEE, Paper 27.2, pp. 703–706.

81. T. Sameshima, S. Usui, and M. Sekiya (1986). XeCl Excimer laser annealing used in the fabrication of poly-Si TFT's. *IEEE Electron Device Lett.*, 7(5), pp. 276–278.

82. K. Nomura, H. Ohta, A. Takagi, T. Kamiya, M. Hirano, and H. Hosono (2004). Room-temperature fabrication of transparent flexible thin-film transistors using amorphous oxide semiconductors. *Nature*, 432(7016), pp. 488–492.

83. J. Fergason (1971). Liquid crystal non-linear light modulators using electric and magnetic fields. US Patent 3, 918, 796.

84. F. J. Kahn (1971). Nematic liquid crystal device. US Patent 3694053 A

85. M. F. Schiekel and K. Fahrenschon (1971). Deformation of nematic liquid crystals with vertical orientation in electrical fields. *Appl. Phys. Lett.*, 19, pp. 391–393.

86. S. Kobayashi, T. Shimojo, K. Kasano, and I. Tsunda (1972). Preparation of alphanumeric indicators with liquid crystals. SID Digest of Technical Papers, pp. 68–72.

87. R. A. Soref (1973). Transverse field effect in nematic liquid crystals. *Phys. Lett.*, 22, p. 165.

88. N. A. Clark and S. T. Lagerwall (1980). Submicrosecond bistable electro-optic switching in liquid crystals. *Appl. Phys. Lett.*, 36, pp. 899–901.

89. C. M. Waters and E. P. Raynes (1982). Liquid crystal devices with particular cholesteric pitch/cell thickness ratio. US Patent 4, 596, 446.

90. T. J. Scheffer and J. Nehring (1984). A new, highly multiplexable liquid crystal display. *Appl. Phys. Lett.*, 45, pp. 1021–1022.

91. K. Katoh, Y. Endo, M. Akatsuka, M. Ohgawara, and K. Sawada (1987). Application of retardation compensation; A new highly multiplexable black-white liquid crystal display with two supertwisted nematic layers. *Jpn J. Appl. Phys.*, 26(11), pp. L1784–L1785.

92. R. Kiefer, B. Webber, F. Windscheid, and G. Baur (1992). In-plane switching of nematic liquid crystals. *Proceedings Japan Displays'92*, pp. 547–550.

93. M. Oh-e and K. Kondo (1995). Electro-optical characteristics and switching behavior of the in-plane switching mode. *Appl. Phys. Lett.*, 67, pp. 3895–3897.

94. A. Takeda, S. Kataoka, T. Sasaki, H. Chida, H. Tsuda, K. Ohmuro, T. Sasabayashi, Y. Koike, and K. Okamoto (1998). A super-high image quality multi-domain vertical alignment LCD by new rubbing-less technology. *SID Dig.*, 29(1), pp. 1077–1100.

95. Y. Koike and K. Okamoto (1999). Super high quality MVA-TFT liquid crystal displays. *Fujitsu Sci. Tech. J.*, 35, pp. 221–228.

96. S. L. Lee and S. H. Lee (1996). Liquid crystal display device with improved transmittance and method for manufacturing same. KR Patent No. 19960059509, US Patent No. 5, 914, 762.

97. S. H. Lee, S. L. Lee, and H. Y. Kim, (1998). High-transmittance, Wide-Viewing-Angle Nematic Liquid Crystal Display Controlled by Fringe-Field Switching, *Proceedings 18th International Display Research Conference (Asia Display)*, pp. 371–374.

98. J. O. Kwag, K. C. Shin, J. S. Kim, S. G. Kim, and S. S. Kim (2000). Implementation of new wide viewing angle mode for TFT-LCDs. *SID Tech. Dig.*, 31, pp. 256–259.

99. Y. Ishii, S. Mizushima, and M. Hijikigawa (2001). High performance TFT-LCDs for AVC applications. *SID Dig.*, 32(01), pp. 1090–1093.

100. E. Jang, S. Jun, H. Jang, J. Lim, B. Kim, and Y. Kim (2010). White-light-emitting diodes with quantum dot color converters for display backlights. *Adv. Mater.*, 22(28), pp. 3076–3080.

101. J. C. Jones (2012). Bistable LCDs. In J. C. Janglin, W. Cranton and M. Fihn, (eds.), *Handbook of Visual Display Technology*, Springer, Berlin Heidelberg, pp. 1507–1543.

102. J. C. Jones (1999). The evolution of liquid crystal displays. *Opt. Laser Eur.*, 69(12), pp. 41–42.

103. R. Williams (1963). Domains in liquid crystals. *J. Chem. Phys.*, 39(2), pp. 384–388.

104. L. M. Blinov (1998). Behaviour of liquid crystals in electric and magnetic fields. In D. Demus, J. W. Goodby, G. W. Gray, H. W. Spiess, and V. Vill, (eds.), *Handbook of Liquid Crystals, Volume 1: Fundamentals*, Wiley VCH, Weinham, Germany. pp. 437–544.

105. S. Chandrasekhar (1992). *Liquid crystals*, Cambridge University Press, Cambridge, UK.

106. C. H. Gooch and H. A. Tarry (1975). The optical properties of twisted nematic structures with twist angles ≤90°. *J. Phys. D: Appl. Phys.*, 8, pp. 1575–1583.

107. J. C. Jones and G. P. Bryan-Brown (2010). Low cost Zenithal bistable device with improved white state. *Proc. SID Int. Symp. Dig., Tech. Pap.*, 41, pp. 207–211.

108. E. P. Raynes (1986). The theory of supertwist transitions. *Mol. Cryst. Liq. Cryst. Lett.*, 4(1), pp. 1–8.

109. E. P. Raynes (2014). TN, STN and guest host liquid crystal display devices. In J. W. Goodby et al., (eds.), *Handbook of Liquid Crystals, Volume 8: Applications of Liquid Crystals*, Wiley VCH, Weinheim, Germany, pp. 3–20.

110. M. J. Bradshaw and E. P. Raynes (1983). The elastic constants and electric permittivities of mixtures containing terminally cyano substituted nematogens. *Mol. Cryst. Liq. Cryst*, 91, pp. 145–148.

111. T. Sergan, S. Jamal, and J. Kelly (1999). Polymer negative birefringence films for compensation of twisted nematic devices. *Displays*, 20(5), pp. 259–267.

112. H. K. Bisoyi and S. Kumar (2010). Discotic nematic liquid crystals: Science and technology. *Chem. Soc. Rev.*, 39, pp. 264–285.

113. M. D. Tillin, M. J. Towler, K. A. Saynor, and E. J. Beynon (1998). Reflective single polariser low and high twist LCDs. *SID Tech. Dig.*, 29, pp. 311–314.

114. S. Aftergut and H. S. Jr., Cole (1978). Method for improving the response time of a display device using the twisted nematic liquid crystal composition. US Patent 4, 143, 947.

115. V. G. Chigrinov, V. V. Belyaev, S. V. Belyaev, and M. F. Grebenkin (1979). Instabilities of cholesteric liquid crystals in an electric field. *Sov. Phys. JETP*, 50(5), pp. 994–999.

116. I. C. Sage (1998). Displays. In D. Demus, J. W. Goodby, G. W.Gray, H. W. Spiess and V. Vill, (eds.), *Handbook of Liquid Crystals, Volume 1: Fundamentals*, Wiley VCH, Weinham, Germany, pp. 727–762.

117. O. Okamura, M. Nagata, and K. Wada (1987). Neutralized STN (NTN)—LCD full colour picture image. *Inst. Television Eng. Jpn, Tech. Rep.*, 11(27), pp. 79–84.

118. S. T. Lagerwall (2014). Ferroelectric liquid crystal displays and devices. In J. W. Goodby et al., (eds.), *Handbook of Liquid Crystals, Volume 8: Applications of Liquid Crystals*, Wiley VCH, Weinheim, Germany, pp. 213–236.

119. J. C. Jones and E. P. Raynes (1992). Measurement of the biaxial permittivities for several smectic-C host materials used in ferroelectric liquid crystals devices. *Liq. Cryst.*, 11(10), pp. 199–217.

120. J. C. Jones (2015). On the biaxiality of smectic C and ferroelectric liquid crystals. *Liq. Cryst.*, 42, pp. 732–759.

121. J. C. Jones, C. V. Brown, and P. E. Dunn (2000). The physics of τVmin ferroelectric liquid crystal displays. *Ferroelectrics*, 246, pp. 191–201.

122. J. C. Jones, M. J. Towler, and J. R. Hughes (1993). Fast, high contrast ferroelectric liquid crystal displays and the role of dielectric biaxiality. *Displays*, 14(2), pp. 86–93.

123. M. J. Bradshaw, V. Brimmell, and E. P. Raynes (1987). A novel alignment technique for ferroelectric smectics. *Liq. Cryst.*, 2(1), pp. 107–110.

124. M. H. Anderson, J. C. Jones, E. P. Raynes, and M. J. Towler (1991). Optical studies of thin layers of smectic C materials. *J. Phys. D: Appl. Phys.*, 24, pp. 338–342.

125. J. C. Jones (2014). Bistable nematic liquid crystal displays. In J. W. Goodby, P. J. Collins, T. Kato, C. Tschierske, H. F. Gleeson, and P. Raynes, (eds.), *The Handbook of Liquid Crystals. Volume 8: Liquid Crystal Devices*, John Wiley & Sons, Weinheim, Germany Chapter 4.

126. V. Y. Reshetnyak, O.V. Shevchuk, and M. Osyptchu (2004). Director profile in the in-plane switching of nematic liquid crystal cell with strong director anchoring, *Proceedings of SPIE*, (July 2004), pp. 283–292.

127. M. Oh-E and K. Kondo (1997). The in-plane switching of homogeneously aligned nematic liquid crystals. *Liq. Cryst.*, 22(4), pp. 379–390.

128. H. Ki Hong and H. Ho Shin (2008). Effects of rubbing angle on maximum transmittance of in-plane switching liquid crystal display. *Liq. Cryst.*, 35(2), pp. 173–177.

129. S. Jung, S. Jang, H. Park, and W. Park (2003). Optimization of the electro-optical characteristics of in-plane-switching liquid crystal displays by three-dimensional simulation. *J. Korean Phys. Soc.*, 43(5), pp. 935–940.

130. D. K. Yang and S. T. Wu (2006). *Fundamentals of liquid crystal devices*, John Wiley and Sons, Chichester, UK.

131. D. H. Kim, Y. J. Lim D. E. Kim, H. Ren, S. H. Ahn, and S. H. Lee (2014). Past, present, and future of fringe-field switching-liquid crystal display. *J. Inf. Disp.*, 15(2), pp. 99–106.

132. J. L. Fergason (1964). Liquid crystals. *Sci. Am.*, 211, pp. 76–85.

133. H. Groskinsky (1968). The chameleon chemical. *Life Mag.*, 64(2), pp. 40–45.

134. P. Bonnett, T. V. Jones, and D. G. McDonnell (1989). Shear stress measurements in aerodynamic testing using cholesteric liquid crystals. *Liq. Cryst.*, 6(3), pp. 271–280.

135. G. H. Heilmeier and J. E. Goldmacher (1968). A new electric-field-controlled reflective optical storage effect in mixed liquid crystal systems. *Appl. Phys. Lett.*, 13(4), pp. 132–133.

136. W. Greubel, U. Wolf, and H. Krüger (1973). Electric field induced texture change in certain nematic/cholesteric liquid crystal mixtures. *Mol. Cryst. Liq. Cryst.*, 24, pp. 103–106.

137. W. Greubel (1974). Bistability behavior of texture in cholesteric liquid crystals in an electric field. *Appl. Phys. Lett.*, 25(1), pp. 5–7.

138. D.-K. Yang, J. L. West, L.-C. Chien, and J. W. Doane (1994). Control of reflectivity and bistability in displays using cholesteric liquid crystals. *J. Appl. Phys.*, 76, pp. 1331–1333.

139. A. Khan, I. Shiyanovskaya, T. Schneider, N. Miller, T. Ernst, D. Marhefka, F. Nicholson, S. Green, G. Magyar, O. Pishnyak, and J. W. Doane (2005). Reflective cholesteric displays: From rigid to flexible. *J. SID*, 13(6), pp. 469–474.

140. T. Schneider, G. Magyar, S. Barua, T. Ernst, N. Miller, S. Franklin, E. Montbach, D. J. Davis, A. Khan, and J. W. Doane (2008). "A flexible touch-sensitive writing tablet," *Proc. SID Int. Symp. Dig., Tech. Pap.*, 39, pp. 1840–1842.

141. D. Coates (2009). Low-power large-area cholesteric displays. *Inf. Disp.*, 25(3), pp. 16–19.

142. D. W. Berreman and W. R. Heffner (1980). New bistable liquid crystal twist cell. *Appl. Phys. Lett.*, 25, pp. 109–111.

143. H. Nomura, T. Obikawa, Y. Ozawa, and T. Tanaka (2000). Recent studies on multiplex driving of BTN-LCDs. *J. SID*, 8(4), pp. 289–294.

144. I. Dozov, M. Nobili, and G. Durand (1997). Fast bistable nematic display using mono-stable surface switching. *Appl. Phys. Lett.*, 70(9), pp. 1179–1181.

145. S. Joly, P. Thomas, J. Osterman, A. Simon, S. Lallemant, L. Faget, J.-D Laffitte, M. Irzyk, L. Madsen, J. Angelé, F. Leblanc, and P. Martinot-Lagarde (2010). Demonstration of a technological prototype of an active-matrix *Binem* liquid-crystal display. *J. SID.*, 18(12), pp. 1033–1039.

146. https://www.youtube.com/watch?v=j3Ob93XGM6A

147. G. P. Bryan-Brown, C. V. Brown, and J. C. Jones (1995). Bistable nematic liquid crystal device. Patent US 06249332.

148. J. C. Jones (2008). The Zenithal bistable display: From concept to consumer. *J. SID.*, 16(1), pp. 143–154.

149. E. L. Wood, G. P. Bryan-Brown, P. Brett, A. Graham, and J. C. Jones (2000). Zenithal bistable device (ZBD) suitable for portable applications. *Proc. SID, Int. Symp. Dig., Tech. Papers*, 31, pp. 124–127.

150. J. C. Jones (2006). Novel geometries of the Zenithal bistable device. *Proc. SID Int. Symp. Dig., Tech. Pap.*, 37, pp. 1626–1629.

151. J. C. Jones, S. M. Beldon, and E. L. Wood (2003). Gray scale in zenithal bistable displays: The route to ultra-low power color displays. *J. SID*, 11(2), pp. 269–275.

152. J. C. Jones, P. Worthing, G. Bryan-Brown, and E. Wood (2003). Transflective and single polariser reflective Zenithal bistable displays. *SID Int. Symp. Dig., Tech. Pap.*, 34(1), pp. 190–193.

153. J. C. Jones (1999). Bistable nematic liquid crystal device. US Patent 7, 371, 362.

154. T. J. Spencer, C. Care, R. M. Amos, and J. C. Jones (2010). A Zenithal bistable device: Comparison of modelling and experiment. *Phys. Rev. E.*, 82, 021702, pp. 1–13.

155. G. P. Bryan-Brown, D. R. E. Walker, and J. C. Jones (2009). Controlled grating replication for the ZBD technology. *SID Int. Symp. Dig., Tech. Pap.*, 40, pp. 1334–1337.

156. C. Hilsum (1976). Electro-optic device. UK Patent 1, 442, 360.

157. J. L. Fergason (1984). Encapsulated liquid crystal and method. US patent 4, 435, 047.

158. D. Coates (1995). Polymer-dispersed liquid crystals. *J Mater. Chem.*, 5, pp. 2063–2072.

159. P. Dzraic (1995). *Liquid crystal dispersions*, World Scientific, Singapore.

160. P. Dzraic (1986). Polymer dispersed nematic liquid crystal for large area displays and light valves. *J. Appl. Phys.*, 60(6), pp. 2142–2148.

161. B.-G. Wu, J. H. Erdmann, and J. W. Doane (1989). Response times and voltages for PDLC light shutters. *Liq. Cryst.*, 5(5), pp. 1453–1465.

162. Y. Asaoka, E. Satoh, K. Deguchi, T. Satoh, K. Minoura, I. Ihara, S. Fujiwara, A. Miyata, Y. Itoh, S. Gyoten, N. Matsuda, and Y. Kubota (2009). 29.1: Polarizer-free reflective LCD combined with ultra low-power driving technology. *SID Digest*, pp. 395–398.

163. M. Kim, K. J. Park, S. Seok, J. M. Ok, H. T. Jung, J. Choe, and D. H. Kim (2015). Fabrication of microcapsules for dye-doped polymer-dispersed liquid crystal-based smart windows. *ACS Appl. Mater. Interfaces*, 7(32), pp. 17904–17909.

164. For example, http://intelligentglass.net/categories/switchable-glass/

165. T. J. Bunning, L. V. Natarajan, V. P. Tondiglia, and R. L. Sutherland (2000). Holographic polymer dispersed liquid crystals (H-PDLC). *Annu. Rev. Mater. Sci.*, 30, pp. 83–115.

166. D. Armitage, I. Underwood, and S.-T. Wu (2006). *Introduction to Microdisplays*, John Wiley & Sons, Chichester, UK.

167. M. S. Brennesholtz and E. H. Stupp (2008). *Projection Displays*, John Wiley & Sons, Chichester, UK.

168. M. J. O'Callaghan and M. A. Handschy (2001). Ferroelectric liquid crystal SLMs: From prototypes to products. *Proceedings SPIE 4457, Spatial Light Modulators: Technology and Applications*, p. 31.

169. E. E. Kriezis, L. A. Parry-Jones, and S. J. Elston (2003). Optical properties and applications of ferroelectric and antiferroelectric liquid crystals. In L. Vicari (ed.), *Optical Applications of Liquid Crystals*, CRC Press, Cambridge, UK.

170. N. Collings, T. Davey, J. Christmas, D. Chu, and W. Crossland (2011). The applications and technology of phase-only liquid crystal on silicon devices. *IEEE J. Disp. Technol.*, 7(3), pp. 112–119.

171. D. Armitage, J. I. Thackara, and W. D. Eades (1989). Photoaddressed liquid crystal spatial light modulators. *Appl. Opt.*, 28(22), pp. 4763–4771.

172. M. Stanley, P. B. Conway, S. Coomber, J. C. Jones, D. C. Scattergood, C. W. Slinger, R. W. Bannister, C. V. Brown, W. A. Crossland, and A. J. Travis (2000). A novel electro-optic modulator system for the production of dynamic images from giga-pixel computer generated holograms. *Proc. SPIE*, 3956, pp. 13–22.

173. C. Slinger, C. Cameron, and M. Stanley (2005). Computer-generated holography as a generic display technology. *Computer*, 38, pp. 46–53.

174. P. J. Bos and K. R. Koehler-Beran (1984). The pi-cell: A fast liquid crystal optical switching device. *Mol. Cryst. Liq. Cryst.*, 113, pp. 329–339.

175. J. L. Fergason and A. L. Berman (1989). A push/pull surface-mode liquid-crystal shutter: Technology and applications. *Liq. Cryst.*, 5(5), pp. 1397–1404.

176. X. Y. Yu and H. S. Kwok (2004). Bistable bend-splay liquid crystal device. *Appl. Phys. Lett.*, 85(17), pp. 3711–3713.

177. E. P. Raynes (1975). Optically active additives in twisted nematic devices. *Rev. de Phys. Appl.*, 10, pp. 117–120.

178. Y. Yamaguchi, T. Miyashita, and T. Uchida (1993). Wide-viewing-angle display mode for the active-matrix LCD Using bend-alignment liquid-crystal cell. *SID Digest*, pp. 273–276.

179. P. J. Bos and J. A. Rahman (1993). An optically 'self-compensating' electro-optical effect with wide angle of view. *SID Digest*, pp. 277–280.

180. C. G. Jhun, M. S. Park, P. K. Son, J. H. Kwon, J. Yi, and J. S. Gwag (2011). Transflective liquid crystal display with Pi-cell. *Mol. Cryst. Liq. Cryst.*, 544, pp. 88–94.

181. M. J. Bradshaw and E. P. Raynes (1986). Smectic liquid crystal devices. US Patent 4, 969, 719.

182. J. S. Patel, (1992). Ferroelectric liquid crystal modulator using twisted smectic structure. *Appl. Phys. Lett.*, 60, pp. 280–282.

183. A. D. L. Chandani, E. Gorecka, Y. Ouchi, and H. Takezoe (1989). Antiferroelectric chiral smectic phases responsible for the tristable switching in MHPOBC. *Jpn. J. Appl. Phys. Lett.*, 28, pp. L1265–L1268.

184. B. I. Ostrovsky, A. Z. Rabinovich, and V. G. Chigrinov (1980). Behavior of ferroelectric smectic liquid crystals in electric field. In L. Bata, (ed.), *Advances in Liquid Crystal Research and Applications*, Pergamon Press, Oxford, UK, pp. 469–482.

185. S. Garoff and R. B. Meyer (1977). Electroclinic effect at the A-C phase change in a chiral smectic liquid crystal. *Phys. Rev. Lett.*, 38, pp. 848–851.

186. D. Coates and G. W. Gray (1973). Optical studies of the amorphous liquid-cholesteric liquid crystal transition: The blue phase. *Phys. Lett. A*, 45, pp. 115–116.

187. Y. Haseba and H. Kikuchi (2006). Electro-optic effects of the optically isotropic state induced by the incorporative effects of a polymer network and the chirality of liquid crystal. *J. SID*, 14(6), pp. 551–556.

188. H. Kikuchi, M. Yokota, Y. Hisakado, H. Yang, and T. Kajiyama (2002). Polymer-stabilized liquid crystal blue phases. *Nat. Mater.*, 1(1), pp. 64–68.

189. Z. Ge, S. Gauza, M. Jiao, H. Xianyu, and S. T. Wu, (2009). Electro-optics of polymer-stabilized blue phase liquid crystal displays. *Appl. Phys. Lett.*, 94(10), pp. 9–12.

190. L. Rao, Z. Ge, S.-T. Wu, and S. H. Lee (2009). Low voltage blue-phase liquid crystal displays. *Appl. Phys. Lett.*, 95(23), p. 231101.

191. M. D. A. Rahman, S. Mohd Said, and S. Balamurugan (2015). Blue phase liquid crystal: Strategies for phase stabilization and device development. *Sci. Tech. Adv. Mat.*, 16(3), p. 033501.

192. J. Patel and R. B. Meyer (1987). Flexoelectric electro-optics of a cholesteric liquid crystal. *Phys. Rev. Lett.*, 58, p. 1538.

193. P. Rudquist, M. Buivydas, L. Komitov, and S. T. Lagerwall (1994). Linear electro-optic effect based on flexoelectricity in a cholesteric with sign change of dielectric anisotropy. *J. Appl. Phys.*, 76(12), pp. 7778–7783.

194. K. L. Atkinson, S. M. Morris, M. M. Qasim, F. Castles, D. J. Gardiner, P. J. W. Hands, S. S. Choi, W. –S. Kim, and H. J. Coles (2012). Increasing the flexoelastic ratio of liquid crystals using highly fluorinated ester-linked bimesogens. *Phys. Chem. Chem. Phys.*, 14(47), pp. 16377–16385.

195. L. Komitov, G. P. Bryan-Brown, E. L. Wood, and A. B. J. Smout (1999). Alignment of cholesteric liquid crystals using periodic anchoring. *J. Appl. Phys.*, 86(7), pp. 3508–3511.

196. G. Carbone, D. Corbett, S. J. Elston, P. Raynes, A. Jesacher, A. R. Simmonds, and M. Booth (2011). Uniform lying helix alignment on periodic surface relief structure generated via laser scanning lithography. *Mol. Cryst. Liq. Cryst.*, 544(1), pp. 37–49.

197. C. C. Tartan, P. S. Salter, M. J. Booth, S. J. Morris, and S. J. Elston, (2016). Localised polymer networks in chiral nematic liquid crystals for high speed photonic switching. *J. Appl. Phys.*, 119(18), p. 183106.

198. N. V. Kukhtarev (1978). Cholesteric liquid crystal laser with distributed feedback. *Sov. J. Quantum Electron.*, 8(6), pp. 774–776.

199. V. I. Kopp, B. Fan, H. K. M. Vithana, and A. Z. Genack (1998). Low-threshold lasing at the edge of a photonic stop band in cholesteric liquid crystals. *Opt. Lett.*, 23, pp. 1707–1709.

200. B. Taheri, A. Munoz, P. Palffy-Muhoray, and R. Twieg (2001). Threshold lasing in cholesteric liquid crystals. *Mol. Cryst. Liq. Cryst.*, 358, pp. 73–82.

201. E. Yablonovitch (1987). Inhibited spontaneous emission in solid-state physics and electronics. *Phys. Rev. Lett.*, 58(20), pp. 2059–2062.

202. M. Uchimura, Y. Watanabe, F. Araoka, J. Watanabe, H. Takezoe, and G. I. Konishi (2010). Development of laser dyes to realize low threshold in dye-doped cholesteric liquid crystal lasers. *Adv. Mater.*, 22(40), pp. 4473–4478.

203. D. J. Gardiner, W.-K. Hsiao, S. M. Morris, P. J. W. Hands, T. D. Wilkinson, I. M. Hutchings, and H. J. Coles (2012). Printed photonic arrays from self-organized chiral nematic liquid crystals. *Soft Matter*, 8(39), pp. 9977–9980.

204. H. Coles and S. Morris (2010). Liquid-crystal lasers. *Nat. Photonics*, 4(10), pp. 676–685.

205. H.-Y. Chen, S. -F. Lu, P.-H. Wu, and C. S. Wang (2016). Transflective BPIII mode with no internal reflector. *Liq. Cryst.*, 8292(August 2016), pp. 1–6.

206. Q. Li (2012). *Liquid Crystals Beyond Displays: Chemistry, Physics, and Applications*, John Wiley & Sons, Edison, NJ.

207. W. H. De Jeu (ed.) (2012). *Liquid crystal elastomers: Materials and applications*, Springer, New York.

208. R. J. Bushby, S. M. Kelly, and M. O'Neill (eds.) (2012). *Liquid Crystalline Semiconductors: Materials, Properties and Application*, Springer, Berlin Heidelberg.

209. R. J. Carlton, J. T. Hunter, D. S. Miller, R. Abbasi, P. C. Mushenheim, L. N. Tan, and N. L. Abbott (2013). Chemical and biological sensing using liquid crystals. *Liq. Cryst. Rev.*, 1(1), pp. 29–51.

RECOMMENDED FURTHER READING

R. H. Chen. (2011). *Liquid Crystal Displays*, John Wiley & Sons, Edison, NJ.

N. Koide. (2014). *The Liquid Crystal Display Story: 50 Years of Liquid Crystal R&D That Lead the Way to the Future*, Springer, Japan.

D. Dunmur and T. Sluckin. (2011). *Soap, Science, and Flat-Screen TVs: A History of Liquid Crystals*, Oxford University Press, Oxford.

Technology and applications of spatial light modulators

UZI EFRON
Holon Institute of Technology

7.1 INTRODUCTION

Spatial light modulator (SLM) technology has made tremendous progress in the past few years. Thus, device technologies such as active matrix-driven liquid crystal devices (LCDs) and ferro-electric LC (FLC) devices as well as micro-mirror SLMs, regarded as exotic novelties less than two decades ago, are now commercially available and constitute a significant share of the display market.

Regarded as "dream-devices" only a decade ago, the multiple quantum well (MQW) SLMs are now a reality with array sizes of 256×256 and staggering frame rates of over 100 kHz.

It is interesting to note that the trend in applications continues to be dominated by the display applications. Optical interconnects (OIs), an almost unknown application at the time, have moved strongly forward in the last few years with the explosive development of communications, while the exotic optical data processing application is still awaiting the opportunity for a break-through.

This chapter, aimed at describing the main SLM technologies and applications is organized as follows.

The scope and definition of the main SLM types is detailed in Section 7.2. The bulk part of this review contained in Section 7.3 describes the main types of SLM currently available or those in an advanced development stage. Particular emphasis is placed on the recent emerging technologies including LC, micro-mirror or micro-electro–mechanical-systems-(MEMS-) based arrays as well as MQW devices. Short summaries of the "older" SLM technologies, including solid state electro-optic, magneto-optic as well as acousto-optic and photorefractive devices are also presented in Section 7.3. Novel SLM concepts including electro-holograms and photonic band-gap (PBG) materials are presented in Section 7.4. Section 7.5 presents the main SLM applications including displays, optical communication, optical data processing, programmable diffractive optical elements, adaptive optics and wavelength image converters. Finally, Section 7.6 presents some of the fundamental limits as well as the future trends of this technology.

There have been numerous reviews in the form of books, book-chapters and conference proceedings covering SLM technology, since the mid-1970s. Some of the more recent reviews (since 1990) are given in [1–11].

7.2 SLM DEFINITION AND GENERAL DESCRIPTION

SLMs are devices that spatially modulate the amplitude, phase or the polarization state of an optical beam. By "spatial modulation", we mean either a one-dimensional (1D; linear array) or a two-dimensional (2D) spatial modulation. Although not yet practically implemented, a three-dimensional (3D) spatial modulation of an optical beam (e.g., a dynamic 3D hologram) is also possible. A combination of the modulation parameters such as phase–amplitude or polarization–amplitude modulators has also been demonstrated. Since the most important modulation parameter is the amplitude or intensity of the optical beam, the modulation of either the phase or the polarization state of the beam is often used to affect amplitude (intensity) modulation through the conversion of the spatial phase or polarization modulation into intensity modulation using interferometric arrangement or crossed polarizer–analyser configurations, respectively. The control or addressing of the SLM is another important parameter. The main choices here are electrical addressing (e.g., electro-optical or electro-absorption SLM) and optical addressing (e.g., optically-addressed SLM or OASLM); since electronic addressing is the most practical choice, this form is sometimes used to excite an intermediary field that in turn affects the optical property of the beam. Examples for this

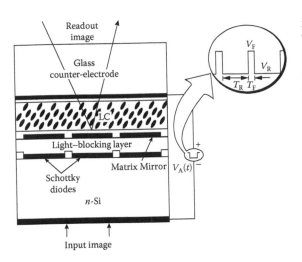

Figure 7.1 A reflective-mode, Si-Schottky-diode, LC SLM. (After Sayyah, K. and Efron, U., *Opt. Lett.*, 21, 1384–1386, 1996.)

nematic LC layer as the electro-optical modulator. The device accepts visible–IR input imagery (400–1100 nm) and can modulate the readout beam to form dynamic imagery from UV to long IR, with frame rates of up to 1 kHz. This particular device has been operated in various LC configurations allowing both phase and amplitude modulations of the readout beam.

Table 7.1 summarizes the main characterization parameters of SLM devices. The main physical mechanisms utilized in SLMs are listed in Table 7.2.

Table 7.3 summarizes the main specification and performance parameters of SLMs.

7.3 MAIN SLM TYPES AND THEIR PERFORMANCES

7.3.1 Liquid crystal devices

7.3.1.1 LC-SLMS: SCOPE

As pointed out in the introduction, the main application of LCDs and, for that matter, of most SLM devices in general, is still that of displays (see Section 7.5 later). LCs are particularly suitable for display applications featuring a combination of extremely high electro-optic coefficients, broad-band visible operation and response times, which will match human vision. Although developed mostly for displays, LC-SLMs were and are still being studied for other applications as well. These include optical data processing, OIs, adaptive optics and programmable Kinoform (binary-optical) elements, such

intermediary addressing are (electro)-magneto-optic SLM (MO-SLM) and (electro)-acousto-optic SLM. The next important parameter is the readout beam configuration. We normally distinguish between transmission-mode and reflective-mode devices. Next, we specify the readout operational wavelength of the device. Thus, in addition to the commonly used visible regime, SLM devices exist and operate in the IR and UV regimes of the electro–magnetic spectrum.

Figure 7.1 shows a cross-section of a liquid crystal SLM. According to the categories defined above, the device is a reflective-mode, photo-activated (PA) SLM, based on a crystalline Si photo-substrate (in a Schottky-diode configuration) and a

Table 7.1 Main characterization parameters of SLMs

SLM parameter	Parameter range	Example
Spatial dimension	1D, 2D, 3D	
Modulation type	Amplitude, phase, polarization conversion	
Addressing mode	Electrical, optical (PA); (electro)-acousto-optic	(Electro-) magneto-optic
SLM configuration	Transmissive, reflective	
Operational wavelength	Visible, IR, UV, (mm-wave)	
Optical modulator type	Liquid crystal, MEMS, MQW, photo-refractive, acousto-optic, magneto-optic	LCD, Bragg cell
Driver type	Active matrix (Si), passive matrix (Si), CCD (Si, GaAs), photo-conductor (Si, GaAs, BSO, …)	Si-backplane

Table 7.2 Physical mechanisms employed in SLMs

Physical mechanism	Examples
Electro-optic (linear)	PLZT [13], (KD*P) [14,15], MSLM (LiNbO3) [16,17], ferro-electric liquid crystal (FLC) [18]
Electro-optic (quadratic)	Nematic-LCD [19], cholesteric liquid crystals [20]
Electro–mechanical (MEMS)	Deformable mirror (DMD-TI) [21,22], grating LV [23,24], deformable membrane devices [25–27]
Electro-absorption	GaAs (CCD-SLM) [28,29], (MQW-SLM) [30]
Electrostatic phase distort	Thin oil film LV [31], gel-SLM [32]
Electro-capillarity	Bistable mercury-mirror devices [33,34]
Photo-refractive	PROM/PRIZ/PICOC [35–40] photo-MQW-SLM [41,42]
(Electro)-acousto-optic	Bragg cells (1D) [43], acousto-MQW [44]
(Electro)-magneto-optic	Magneto-optic (MO-SLM) [45]
Thermal phase changes	VO_2 modulator (IR) [46], LC-smectic/cholesteric transition [47], bubble cross-connect switch [48]
Plasmon-electro-optic enhancement	[49]
PBG modulation	[50]
Active light emission	OLED-array [51,52], VCSEL array [53,54]

Table 7.3 Main SLM specifications and performance parameters

Performance parameter	Description	Units
Array size	No. of horizontal/vertical pixels	N/A
Resolution		LP/mm @ MTF
Temporal response	Rise/decay time, frame rate	ms, Hz
Spectral bandwidth	Input/output beam	nm
Input sensitivity (PA Device)	Min. photo-activating power	$cm^2\ mW^{-1}$, $cm^2\ J^{-1}$
Contrast ratio/phase dynamic range		dB; bits; radians
Diffraction/output modulation efficiency		%
Output uniformity		%
Physical array size		cm × cm
Power consumption		W
Operating temperature range		°C
Relative humidity range		%
Shock resistance		G

as variable focal length lenses and beam-steering devices (see later in Section 7.5, SLM-applications).

Historically, the nematic materials were the first to be used for SLMs and other display applications (such as watch displays). As described earlier, a nematic LC layer operating as the electro-optic modulator is used in either a twisted-nematic (TN) [55] or in a controlled-birefringence configuration [56]. The main advantage of the TN configuration is in its optically broad-band operation and the relatively lower sensitivity to the LC thickness variations.

Nematic LCDs are mainly used as either small (1–3 in.) panels (or "light valves") for projection displays or as larger (10–15 in.) direct view panels for personal digital assistants (PDAs), lap-top and hand-held PCs (HPCs), or mobile phone displays. The LCDs for both applications are manufactured in both transmissive and reflective configurations. Earlier versions of LC projection displays were based on PA configurations, consisting of a photo-conductive layer used to convert the incoming, low-light-level input image (e.g., from a small-size CRT) into a spatial voltage distribution across the nematic LC layer. A separate beam is then used to read out the LC modulation and project the image onto a large screen. Due to the rapid progress in large CMOS array technology, most projection LCDs of today are based on electrically driven or electrically addressed (EA) LCDs. The older generation, PA LCDs, also called "liquid crystal light valves" (LCLVs) [57], are still in use for some of the high-end projection display applications requiring a combination of high resolution and high brightness such as very large screen displays and electronic cinema.

Electrically addressed LCDs were first based on the use of a passive X–Y addressing scheme [58]. See Volume II, Chapter 6 by J. Cliff Jones. Although this concept relies on the threshold voltage property of LC materials, the MULTIPLEXING level (e.g., the number of pixels in a line) is quite limited. With the advance in VLSI technology allowing large transistor driver arrays to be manufactured at high yield, these passively addressed LCDs are now replaced by actively addressed LCDs (based on active transistors in each pixel) in most applications.

Two other important developments in LCD technology took part in the 1980s and 1990s. These are the discovery and development of FLC [59] and the development of a scattering mode, polymer-dispersed LC (PDLC) [60,61]. The FLC, as the name suggests, is based on a LC material class (smectic C*) having a permanent electrical dipole moment and thus the molecules can physically be flipped with the reversal of the electric field. This is in contrast to the nematic LC, which responds only to the magnitude of the field (E) and not to its polarity, with subsequent longer response time. The PDLC is based on LC "bubbles" or "droplets" immersed in a polymer matrix such that upon the application of an electric field, their average director orientation changes from a random distribution to a fully-field-aligned one. By properly matching the refractive index of the LC to that of the polymer, the result is the formation of a scattering mode in the off-state and a transparent mode, with application of the field. The most important aspects of this new LC form are (1) the polarization-independent

optical modulation capability and (2) the simple packaging allowing pre-processed PDLC sheets to be pre-manufactured and assembled into the LCD without the need for the cumbersome vacuum-filling of cells with liquid crystal material.

The last decade (1990–2000) has witnessed additional important developments in LCD technology among which are the development of an efficient poly-silicon-based driver circuitry [62], as well as the in-plane switching mode (IPS) [63] and more recently a novel photo-polymerization alignment technique [64]. The poly-silicon driver technology allows an efficient electronic driver to be fabricated on the top of a glass-based substrate, thus enabling the fabrication of high performance, large panel LCDs for lap-top computers. The IPS mode significantly increases the viewing angle for an LCD display. Finally, the novel contact-less, photo-alignment (PAL) method allows the formation of multi-LC domains to reduce polarization dependence.

7.3.1.2 THE PRINCIPAL LC TYPES AND MODES OF OPERATION

As the name implies, LCs represent an intermediate phase combining the physical properties of both solids and liquids. Specifically, this class of materials behave as anisotropic liquids, as they have no translational or long range lattice order and in this regard they are fluids or liquids. However, both their electrical and optical properties are directionally dependent or anisotropic. It is important to keep in mind that the LC phase is an intermediate phase, which exists only within a narrow temperature range (a few tens of degrees). The LC material solidifies below this range of temperature, while undergoing a liquid–liquid transition above this temperature range, where it becomes a normal isotropic liquid. The LC anisotropy is tied to the directionality in the (average) physical orientation of the molecules.

The degree of anisotropy can be described by introducing the LC director and its orientation L, which indicates the average, macroscopic orientation of the LC molecules.

There exist three main types of LC class, namely, nematic, cholesteric and smectic mixtures. A brief description of each class and its mode of operation is given in the following. For an extensive review of LC classes and their associated physical and electro-optic properties, see references [65–70].

7.3.1.2.1 Nematic liquid crystals

In this class (or phase) the aggregate of the LC molecules show directional anisotropy in both their dielectric and optical properties. We thus have longitudinal and transverse polarizabilities $\alpha_{\parallel}$ and $\alpha_{\perp}$, dielectric constants $\varepsilon_{\parallel}$ and $\varepsilon_{\perp}$, as well as refractive indices $n_{\parallel}$ and $n_{\perp}$. All these parameters are defined with respect to the long molecular axis, which practically coincides with the average orientation of the LC director. Positive-dielectric anisotropy NLCs are characterized by having: $\Delta\varepsilon = \varepsilon_{\parallel} - \varepsilon_{\perp} > 0$, and similarly for $\Delta\alpha$ and Δn.

It should be noted that both positive and negative dielectric and refractive index anisotropies can be found in the LCs. Of particular interest are NLCs having a negative $\Delta\varepsilon$ but positive Δn.

These are sometimes referred to as "negative-anisotropy NLCs".

The director orientation coincides with the optical axis of the refractive index ellipsoid. Thus, one usually substitutes the ordinary (n_0) and the extraordinary (n_e) refractive indices for $n_{\perp}$ and $n_{\parallel}$, respectively.

NLCs do not have a permanent dipole moment. Thus, the application of an electric field E beyond a threshold level E_c induces a dipole moment μ, proportional to E. The field then exerts a torque proportional to $|\mu \times E| \sim |E|^2$, which re-orients the LC molecules or, macroscopically, the LC director L. This re-orientation results in an effective tilting of the refractive index ellipsoid, thus varying the effective birefringence, as experienced by the incident polarized optical beam.

The threshold field, E_c, is given by $E_c = (\pi/d)\sqrt{[k/\varepsilon_0\Delta\varepsilon]}$. Here, d is the thickness of the LC cell and k is the appropriate elastic constant. One can define a thickness-independent threshold (or critical) voltage, $V_{th} = E_c d$ (rather than a threshold field), beyond which the applied voltage starts re-orienting the LC director. The turn-on and turn-off response times, τ_{ON} and τ_{OFF}, are given by ton $\tau_{ON} \propto \tau^* d^2/[(V/V_{th})^2 - 1]$, where $\tau^* = \gamma/\pi^2 k\lambda\tau_{OFF}$.

The quadratic dependence of the torque exerted on the LC director by the applied voltage results in two important characteristics of NLCs:

1. The NLC's orientation responds to the magnitude or RMS value of the applied voltage and is therefore insensitive to the polarity of the applied voltage or field.

2. The turn-on response time is inversely proportional to V^2, whereas the off-state or the turn-off time is independent of the applied voltage. This sets the lower limit for the total response time of NLCs in practical devices to milliseconds. Such response is ideal for display devices, but is too slow for optical processing, as well as most telecommunication switch (cross-connects) applications.

There exist three basic configurations of NLC cells:

1. Planar configuration in which the NLC molecules are homogeneously aligned parallel to both electrode planes in a certain alignment direction, which can be physically imposed by a particular treatment of the LC cell electrodes.
2. Homeotropic configuration in which the NLC molecules are uniformly aligned perpendicular to the cell electrodes.
3. A twist configuration in which the parallel alignment orientations of the NLC molecules at the entrance and exist cell electrodes or windows differ by an angle ϑ_T.

This configuration is formed by fabricating alignment layers with different orientations on the two cell electrodes. The NLC molecules then align in planes parallel to the cell electrodes, where in each plane the director orientation is rotated (or twisted) relative to the preceding plane. This results in the director re-orientating or twisting continuously from the alignment orientation at the entrance electrode, to the alignment direction at the exit electrode. Typical twist angles are 45° (hybrid-field effect); 90°, 180° (π-cell) and 270° or above (super-twisted NLC = STN configurations) [71].

7.3.1.2.2 LC alignment

As is obvious from the previous discussion, one must define the LC alignment within the cell in order to make use of its electro-optical properties. This is accomplished by generating alignment "marks" at the boundary cell surfaces, which then "anchor" the adjacent LC molecules in those predefined directions, thereby affecting the alignment of the bulk LC material within the cell. Obviously, such alignment via anchoring to the surface cannot be sustained over large distances. Thus, practical LCDs are physically constrained to a thickness of a few tens of micrometres (seldom exceeding 20 μm).

There exist several methods of LC alignment. The most important ones are

Table 7.4 Summary of LC alignment methods

Alignment method	Advantages	Disadvantages
SAD	Excellent tilt control	Complex process, difficult to mass-produce
BUF	Simple, demonstrated production	Requires mechanical contact, difficult tilt control
PAL	Non-contact, amenable to mass-production, remotely programmed orientation, multi-domain	Low anchoring energy, tilt control, long-term stability
LBL	Natural method for homeotropic alignment	Difficult to mass-produce, complex process, tilt control
IBE	Good tilt control, demonstrated production	
GRS	Lithographic process	Complex process, difficult to mass-produce, optical losses (scattering)

1. Shallow angle (SiO_x) deposition (SAD) [72]
2. Mechanical rubbing or buffing (BUF) [73]
3. Photo-alignment (PAL) [74,75]
4. Langmuir–Blodgett (LBL) process [76]
5. Ion-beam etching (IBE) of polymer films [77]
6. Formation of a grating relief structure (GRS) [78]

The relative advantages/disadvantages of each are given in Table 7.4.

Parallel or homogeneous alignment. This is the simplest form of alignment in which the surfaces of both electrodes are aligned in the same direction. This alignment usually employs positive-dielectric anisotropy LC material, whose director (normally parallel to the surface and in the direction of the alignment marks) would tend to tilt towards the electrodes, upon the application of an electric field or voltage.

In order to avoid the formation of opposite tilt-angle domains, a small (1°–2°) pre-tilt angle with respect to the electrode surface is usually defined to be at a 180° difference between the two electrodes. Since in this alignment the LC molecules are anchored parallel to the electrode surfaces, the alignment mechanism makes use of the largest footprint of the elongated LC molecules and is therefore the most powerful (highest anchoring energy).

Homeotropic LC alignment. The homeotropic or tilted-perpendicular alignment (TPA) of LCs is a particularly difficult alignment, both due to the small foot-print of the molecules on the aligning

surface, and the difficulty of a precise control of the pre-tilt angle (around 88°–89°), which is crucial for a proper operation of LCDs[*].

This form of alignment, as well as some of its derivatives (such as the hybrid alignment), are attractive for LCD display technology, as they can lead to a very high contrast and reduced angular sensitivity. Several methods of implementing the homeotropic alignment are given in [79–88].

7.3.1.2.3 The electro-optic effect in NLC cells

The nematic LC molecular orientation becomes distorted under the application of an applied voltage (the so-called Fredericksz transition), with the LC director tilting at an angle $\vartheta(V)$. Although the tilt-angle varies across the LC cell thickness, one can approximate the variation of the effective birefringence $\Delta n_{eff}(V)$ of the cell by

$$\Delta n_{eff}[\theta(V)] = n_e n_o / \{n_o^2 \cos^2[\theta(V)]$$
$$+ n_e^2 \sin^2[\theta(V)]\}^{1/2} - n_o$$

[*] For homogeneously aligned (or planar-aligned) LC configuration, the pre-tilt angle guards against the formation of multiple domains in the LC layers, which is cosmetically unacceptable for displays. However, for the perpendicular TPA alignment, the pre-tilt control is much more critical, in that it determines the final in-plane orientation of the LC director. This orientation must be predetermined and controlled in order for a display device to function properly.

assuming a spatially uniform tilt-angle, $\vartheta(V)$, throughout the cell.

In the case of a parallel-aligned configuration, assuming a positive dielectric anisotropy LC, the director will be re-oriented from a parallel alignment ($\theta(V) = 0°$, for: $V < V_{th}$) to an almost perpendicular alignment with respect to the cell electrodes ($\theta(V) = 180°$), at sufficiently high field or voltages (typically: $V > 10\ V_{RMS}$). This change which results in $\Delta n_{eff}[\theta(V)]$, changing from: $\Delta n_{eff} = \Delta n_{max} = n_e - n_o$ to $\Delta n_{eff} = 0$, can be used to effectively rotate the polarization plane of the incoming optical beam. Thus, the intensity of an incoming beam polarized at 45° to the alignment direction of the (parallel-aligned) LC will be proportional to $I = I_0 \sin^2(\Delta\varphi/2)$, following a crossed-analyser behind the LC cell. The retardation angle, $\Delta\varphi$, can be approximated by $\Delta\varphi > = 2\pi d\{\Delta n_{eff}[\theta(V)] - n_o\}/\lambda$, using the uniformly aligned LC cell approximation, with a tilt-angle given by $\theta(V)$.

By aligning the input polarization along with the LC alignment direction, this configuration can be used for a pure phase modulation, where the voltage-modulated phase, $\Delta\varphi$, is given by the same expression as above.

The same basic applications for intensity (polarization-rotation) modulation or a pure phase modulation can be accomplished using the homeotropically aligned NLC arrangement discussed earlier. In this case, one needs to use a negative-anisotropy LC in order to accomplish a director re-orientation by the application of an electric field.

The effect of continuously controlling the effective birefringence in a homogeneously aligned cell (whether parallel or, perpendicularly aligned) is termed the "controlled birefringence" or CB-effect.

The TN configuration is formed by adding a chiral (helically structured) chemical agent into a nematic mixture. In this configuration, the orientation of the polarization plane of an optical beam, polarized either parallel or perpendicular to the alignment at the entrance window, will follow the director orientation rotation caused by the twisted structure across the LC cell, exiting with the plane of polarization aligned with the LC alignment direction at that exit window.

This optical rotation effect, which occurs in the absence of an electric field, is subject to the condition: $U = \pi d\Delta n/\lambda\theta_T \gg 1$ (limit), where θ_T is the total twist angle in the cell.

Defining the rotatory power, P_R, as the fraction of the optical beam intensity whose polarization gets rotated along with the twist angle as previously described, it can be shown that this rotatory power can be approximated by $P_R \approx U^2/(1 + U^2)$, for both regimes of: $U \gg 1$ and $U \ll 1$ [68, p. 372,89,90]. In particular, for $U \gg 1$, the rotatory power level is almost unity, showing that the polarization plane of the incoming beam is almost perfectly rotated with the twist. The effect of applying an electric field is to tilt the director perpendicular to the (almost plane-parallel) layers, towards the direction normal to the cell electrodes. This results in the twist arrangement becoming increasingly distorted and eventually breaking up at sufficiently high fields. The destruction of the twist structure at high field destroys the polarization property and thus allows the use of the locally applied voltage to modulate the polarization rotation or the intensity with the use of a proper polarizer–analyser arrangement.

It should be pointed out, however, that in this arrangement one cannot simply use the effect for pure phase modulation except for very low applied fields [91].

Another important difference between the controlled birefringence and the TN effect is that the latter can be used (subject to the above condition) to rotate the plane of polarization in a wide spectral window (e.g., throughout the entire visible spectrum of 450–650 nm). The CB-effect, on the other hand, is by its very nature a wavelength-dependent effect as can be seen by the λ-dependence of the retardation $\Delta\varphi$ on the previous page.

7.3.1.2.4 Cholesteric LC

A cholesteric LC (CLC) mixture is formed by the introduction of a relatively large concentration of a chiral agent, resulting in a typical enlargement (numerous 2π cycles) of the twist angle. In the CLC configurations, the degree of the rotational twist is characterized by the period of the twist rotation, namely, the pitch P_0, measured in micrometres, and related to the twist angle θ_T by $\theta_T = 2\pi d/P_0$. It has been shown [92,93] that

1. The pitch, P_0, of the helix formed is inversely proportional to the concentration of the chiral component in the nematic mixture. Typical values of the pitch are between 0.5 and 50 μm and can be attained by varying the chiral component concentration between 20% and 0.5%.

2. The transmission of the planar cholesteric configuration (where the helical axis is perpendicular to the cell surfaces) is high, except in the spectral regions where $\lambda \approx P_0$, and where a high reflectivity of a circularly polarized light is observed.

3. The twist or helix structure, of the cholesteric phase can be completely suppressed by a sufficiently large electric field. In this case, the LC director (for a positive-anisotropy nematic, $\Delta\varepsilon > 0$) lines up with the field (perpendicular to the cell electrodes), with an effective refractive index of n_o, relative to the incoming optical beam.

Upon application of a low electric field, the planar structure turns into a group of randomly oriented helical clusters that strongly scatter light. This scattering phase is termed "focal conic" texture. The capability of forming both a planar texture (at zero field), which with sufficiently long pitch can be used as a transparent state, and the use of the field-excited focal conic is the basis for the recent interest in using cholesteric materials as a bistable optical modulator. In order to obtain stable, zero-field states for both focal-conic and planar textures, polymer stabilization of the LC structure is used [94,95].

7.3.1.2.5 Smectic materials

Smectic materials possess an ordered layered structure. While there exist several subclasses of smectic phases, the two important ones are smectic A and C (SmA, SmC). These differ by the angle in which the molecules tilt with respect to the normal layers. A crystalline order, namely, a well-defined layer period, exists in both the smectic phases. It should be pointed out that the smectic and nematic phases may be based on the same material, of which SmA and SmC may constitute different phases, at different temperature ranges. The degree of disorder in the phases, is increased with increasing temperature, SmC being at the lowest temperature end and, hence, the most ordered phase; SmA follows as a more disordered smectic phase, ending at the highest temperature range with the nematic as the lowest ordered, anisotropic LC phase.

As for the optical properties, SmA is a uniaxial material with its extraordinary refractive index perpendicular to the layers' planes, while the refractive indices in both in-plane axes are equal

to the ordinary index. This symmetry, however, is lost in the SmC phase due to the molecular tilt in the smectic planes. Of particular importance is the chiral SmC or SmC* which is formed by adding a chiral (helically structured) chemical agent to the SmC mixture. In this case, the mirror symmetry of the SmC is lost and the formation of a spontaneous dipole moment is possible without the presence of an external electric field. The existence of a permanent dipole moment μ (which is reversible by the application of an electric field) results in a directional (polarity) dependence of the LC director on the externally applied electric field, through the generated torque $\mathbf{T}$, where $\mathbf{T} = \mu \times \mathbf{E}$. This then leads to on- and off-response times both being inversely proportional to the field: $\tau(E) = \gamma/\mu|E|$. This case, in contrast to the nematic LC behavior, where the off-time is independent of the applied field, having an effectively lower (field-induced) dipole moment, leads to significantly faster response times of ~10–100 μs.

The first concept and demonstration of a usable FLC configuration was published in 1980 by Clark and Lagerwall [96]. In this configuration, the researchers constructed a thin FLC cell in which the helical structure was suppressed, resulting in a uniform distribution of the director across the LC cell. This arrangement, termed "surface stabilized FLC" (SSFLC), allows two field-dependent, LC director orientations separated by $2\theta_L$ where θ_L is the tilt-or cone-angle of the FLC. The LC director can switch between both states by the application or reversal of the externally applied electric field. The response times for switching are of the order of 100 μs for typical applied fields (5–10 V μm^{-1}). Despite being limited to binary operation, the SSFLC configuration has been the principal mode of operation for FLC-SLMs for a variety of applications, in particular for LCDs. In this last application [97], pulse width modulation (PWM) methods making use of the fast response of the FLC relative to the required frame time of ~20 ms are used to enable the required grey scale operation.

Another mode of operation is based on using the SmA very near to its SmC transition point. In this configuration (electro-clinic effect) [98], the LC director orientation, which is still uniform throughout the cell, tilts proportionately to the applied electric thereby allowing continuous grey scale operation. At the same time, one can attain response times comparable to those of the FLC devices since the LC molecules already possess a permanent dipole moment.

A mode of operation employing SmC* or FLC materials was developed by Funfschilling and Schadt [99] in Switzerland, in 1989. This mode, termed "deformable helix FLC" (DHF), does not constrain the natural helix formation in a SmC cell, but makes use of the fact that the helix will tend to get distorted upon the application of an electric field. Thus, in the short-pitched helix cell formed (such that $p = \lambda$), this field-induced distortion results in a change in the orientation of the LC director, averaged over the wavelength size LC section observed. This orientation change can be used, similar to the controlled-birefringence effect in a nematic LC cell, for polarization rotation operation.

Very fast speed of response (a few microseconds), as well as grey scale operation, can be attained in this DHF mode, at the expense of a relatively low dynamic range. The latter is a consequence of the averaging effect over the LC director.

7.3.1.2.6 Polymer-based LC configurations

Recent advances in both polymerization and lithographic techniques resulted in the introduction of polymers into LC structures. In general, the advantage of introducing polymers into LC mixtures is the formation of mechanically self-supported structures, as opposed to regular LC fluids requiring external mechanical support in the form of cell window glasses or plastics. The first concept, introduced over a decade ago, was that of a PDLC [100–104]. This concept is based on the phase separation which occurs between the LC and polymer materials and which results in the formation of micrometre or, wavelength-size LC droplets within the surrounding polymer matrix. The original concept was based on the use of nematic LC materials. However, in later PDLC versions other LC phases such as FLCs and CLCs are also utilized. In the NLC-PDLC version the electro-optic modulation effect is based on matching the ordinary NLC droplet index of refraction to that of the surrounding polymer matrix. Since, in the absence of an external electric field, the LC director orientation in the droplets is randomly distributed, the droplet index will differ (on average) from that of the polymer matrix. Thus, the PDLC will strongly scatter in this state. Upon application of an external electric field, the LC directors in the NLC droplets become oriented with the field (e.g., parallel to the field direction, for a positive-anisotropy LC). In this state the droplets now appear to have an ordinary index matching to that of the surrounding polymer matrix, for optical beam incident perpendicular to the cell window. This arrangement therefore allows a two-state, polarization-independent, transparent/scattering optical modulation.

Other types of polymer-stabilized LC structure, in particular, a reflective, polymer-stabilized cholesteric texture (PSCT) for a scattering/narrow-band reflective, dual-mode bistable LC modulator, have recently been intensively studied for low-power-consumption displays [105]. One potential application is that of electronic paper.

7.3.1.3 LIQUID CRYSTAL SLMS

In the following, we will present a brief description of the main LC-SLM structures. For an extensive review of this technology, the reader is referred to the latest symposia of the Society for Information Displays over the last few years [106] where numerous papers on these structures were published.

7.3.1.3.1 Electrically addressed LCDs

A schematic of a reflective-mode, electrically addressed, LCD is shown in Figure 7.2.

A polarized readout beam enters the device from the top and is reflected from one of the electrode/mirrors placed at the bottom of the device. These electrodes addressed with different voltages affect the LC whose director tilts as shown in proportion to the increasing voltage (from left to right). The alignment layers control the orientation of the LC director at the electrode surfaces.

The LC driving is accomplished using mostly "active matrix" schemes which are based on the use of thin-film circuitry with one or more MOS-based transistors in each pixel node. The circuitry typically employs a row/column drive timing scheme in which each row is sequentially addressed such that within each of the addressed rows the column pixel transistors are sequentially addressed. The MOS transistors in each pixel allow the sequential gating of the signals into each of the pixel's storage capacitors. The addressing signal, which must be alternating in its polarity (AC) to avoid LC deterioration, is typically 1–10 V RMS.

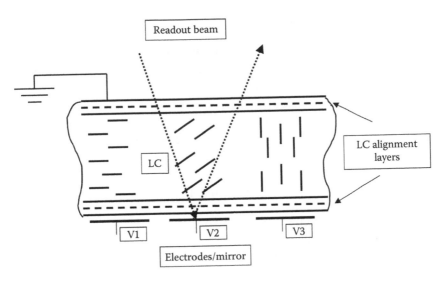

Figure 7.2 Schematic of a electrically addressed, reflective-mode LCD.

The silicon substrate in the early direct view LCD panels was amorphous silicon which, due to its limited performance (carrier mobility) has recently been replaced by poly-crystalline Si, grown on a high-temperature glass substrate.

An important consideration for a Si-based driving circuitry is that it must be well shielded from the readout beam. In a transmissive-mode device this implies a mandatory loss in the area- or aperture-factor of the pixels, and therefore in the optical efficiency, as the circuitry region of the pixel cannot be used for light transmission.

This results in an increasingly high loss of optical efficiency for smaller pixel-size arrays, as the (optically shielded) circuitry occupies an increasingly larger portion of the total pixel area. Reflective-mode LCDs, on the other hand, can be made optically efficient regardless of their pixel size.

The LC employed in the direct view displays is typically in a TN configuration, which allows a cell-thickness-independent, broad-band (450–650 nm) operation. Pixel-level-RGB filters deposited on the front glass allow color operation to be accomplished in conjunction with spatially aligned activation of the RGB sub-pixel drivers. Finally, a crossed analyser at the exit plane, in conjunction with the input polarizer, converts the polarization-rotation at the pixel-level into spatially modulated intensity (with full color).

As pointed out above, there is a significant structural difference between direct view LCDs (e.g., a lap-top or note-book screen) and projection panel LCDs. While lap-top displays are large-aperture (25–40 cm), transmissive type, projection display panels are typically of smaller size (2–8 cm in linear dimension), usually of a reflective type. These relatively smaller size panels are currently based on a single crystal silicon substrate (the so-called LC on silicon technology (LCOS)). The driver electrodes are usually utilized as the reflection layer with an appropriate metallic coating. The LC mode can either be a hybrid field effect mode [107] first used in the early LCLVs, vertically aligned nematic exhibiting a relatively wide field of view (FOV), or more recently, in-plane-switching configuration [63]. FLC as well as PDLC devices were also demonstrated in small LCD panels for projection displays. FLCs are also currently produced for video camera direct view applications [108]. An ultra-thin-crystalline Si-substrate technology recently developed [109] allows very small transmissive panels to be used for projection and head-mounted display applications.

The above class of electrically addressed LCDs encompasses the use of the different LC configurations discussed earlier: in particular, TN cells, controlled-birefringence operation, as well as FLC [110] and PDLC [111] configurations. Finally, one should mention that in addition to the traditional use of MOS- or CMOS-based active matrix driving schemes, the use of a CCD-based circuitry matrix to address an LCD has also been demonstrated with an array size of 256×256 [112].

Table 7.5 Typical performance of electrically addressed LCD projection panel

Performance parameter	Description	Units
Input/output	Electrically addressed, output intensity modulation	N/A
Device configuration	Reflective mode	N/A
Array size	Up to 2000 × 2000	Pixels
Optical modulator	Nematic liquid crystal (TN configuration)	
Driver array	Single-crystal CMOS array	
Resolution	25	LP mm^{-1} @ 20% MTF
Temporal response	Rise/decay time: 0.5–10 ms, frame-rate: 30–300 Hz	ms, Hz
Spectral region	Visible, 450–650	nm
Input sensitivity (PA device)	N/A	N/A
Contrast ratio	100:1 (40 dB)	dB
Output modulation efficiency	10–30 (unpolarized, collimated beam)	%
Output uniformity	1–3	%
Aperture size	2 × 2 (for a 1024 × 1024 array @ 20 μm	cm × cm (pixel-size)
Power consumption	~500	mW
Operating temperature range	10–40	°C
Shock resistance	1–5	G

Typical performance of an electrically addressed LCD panel for projection displays is shown in Table 7.5.

7.3.1.3.2 Photo-activated LCD SLMs

As pointed out earlier, the first type of LC-SLM developed was actually the PA SLM configuration developed at Hughes Aircraft Research Laboratories during the late 1970s [57]. These devices were the first LC-SLM products for both commercial and military uses. A typical reflective-mode, PA SLM structure is shown in Figure 7.3. The structure of such an SLM commonly known as an LCLV consists of four layers: a photoconductor (PC); a light blocking layer (LBL); a mirror (typically dielectric) (DM) and the LC layer. This structure is sandwiched between two glass electrodes providing both the mechanical support and the necessary sheet-conductivity or surface-electrode function. The operation of the device is as follows: under low illumination level at the input (PC port) most of the bias voltage drops on the (high-resistivity) PC layer, and as a result the LC is biased below threshold level and is consequently not activated. Upon illumination of a region or spot on the photoconductor, this section of the PC

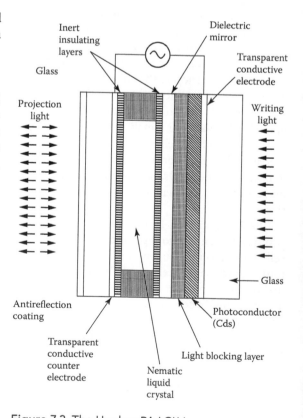

Figure 7.3 The Hughes PA LCLV.

Table 7.6 Performance parameters of the PA, a-Si LCLV

Performance parameter	Description/value	Units
Device configuration	Reflective-mode	N/A
Optical modulator	Nematic liquid crystal	N/A
Driver array	Amorphous Si-photo-substrate	N/A
Array size	2000 × 2000—equivalent	Pixels
Resolution	25	LP mm^{-1} @ 20% MTF
Temporal response	Rise/decay time: 0.5–10 ms, frame-rate: 30–300 Hz	ms, Hz
Spectral bandwidth	Input beam: 550–700, output beam: 450–650	200 nm
Input sensitivity (PA device)	0.1–0.3	m W cm^{-2}
Contrast ratio	200:1 (40 dB)	dB
Output modulation efficiency	10–30 (unpolarized, collimated beam)	%
Output uniformity	1–3	%
Physical array size	Diameter ~50	mm
Power consumption	~100	mW
Operating temperature range	10–40	°C
Shock resistance	1–5	G

becomes conductive and the voltage now shifts to the LC and the LBL/DM layers. The latter, being thin, takes only a relatively small fraction of the AC voltage, which now drops mostly on the LC layer activating those areas aligned with the illuminated sections of the photoconductor. The activation of the LC layer results in a change in the polarization rotation of the LC leading, with the use of a polarizing beam splitter, to an intensity modulation of the out-going optical beam.

In the particular case of the Hughes LCLV device shown in Figure 7.3, the two LC configurations commonly used are the hybrid-field-effect mode using a 450 TN configuration [57] or, alternatively, the tilted-perpendicular mode based on the controlled-birefringence effect in a vertically aligned, negative-anisotropy LC, which has been in use more recently [113]. These devices initially employed CdS as the photoconductive layer which has been replaced by an amorphous-silicon layer. A fast-response version of the device for color-sequential mode operation based on single-crystalline silicon photo-substrate was also demonstrated [114]. Due to the simplicity of an electrically addressed system on the one hand, and the rapid improvement in the yield of large CMOS arrays on the other, the trend today is towards electrically addressed LCD systems for projection displays. However, high-end applications requiring exceedingly large array size (>2000 × 2000) and high brightness (>5000 lm) are still a hard reach for electrically addressed LCDs. The "old-time" PA LCLVs are therefore still potential candidates for such demanding projection display applications such as electronic cinema [113]. In addition to reflective-mode devices, transmissive-mode PA SLMs have also been demonstrated [115–117], using a variety of semi-transparent photoconductors including amorphous silicon and BSO. The latter was developed at the Thales Research Laboratories, under Professor J-P Huignard. These devices were constructed using nematic, FLC and PDLC LC-modulators.

The typical performance of a PA a-Si LCLV is given in Table 7.6.

7.3.2 MEMS-based SLMs

7.3.2.1 GENERAL

The origin of these devices occurred after various attempts in the late 1970s to use continuous membrane mirrors, locally deformed under the application of local electric fields, as a means to modulate the reflectivity or the phase of the reflected optical beam [118]. This concept was expanded in the early 1980s to include the first MEMS-type structures in silicon [119,120]. This MEMS structure is essentially an array of silicon-based pixel-level metallized cantilevers which tilt in response to an electric potential applied to the pixel electrode. This MEMS structure which started with analogue tilt response design was later refined by the Texas instruments developers as a binary (digital) device with only two states. This recent design, combined with a complex driving circuitry to allow grey scale operation based on a PWM scheme, is the basis for today's high-brightness, high-resolution projectors aimed at such ambitious goals as electronic cinema [121]. The other principal MEMS-type SLM is based on the pixels constructed in a shape of inter-digitated fingers. In this configuration, the application of an electric field across this digitated finger structure results in the formation of a diffraction grating with a variable field-dependent depth. Thus diffraction of the incoming optical readout beam, rather than its deflection, constitutes the novelty of this MEMS structure. This "grating light valve" (GLV) device [122,123], conceived and developed by Bloom, is now under development.

Finally, another recent related development has been reported, which is developing a similar structure of a MEMS-silicon-based micro-mirror SLM. However, rather than using an electrostatic field to drive the micro-mirrors, the device uses a piezo-electric micro-transducer fabricated in each pixel [124]. Finally, for completeness, we should also mention a recent effort by researchers to develop mechanical continuous membrane implementations of SLM similar in concepts to the earlier attempts mentioned [125] above.

7.3.2.2 THE MICRO-MIRROR SLM (DIGITAL MICRO-MIRROR DEVICE) [120,121]

The digital micro-mirror device (DMD) SLM (Figure 7.4) is based on a MEMS structure that is fabricated using a CMOS-compatible processes over a driver array which is based on a CMOS memory.

Each pixel modulator consists of a 16×16 μm^2 aluminium mirror, which can reflect light in one of the two directions depending on the state of the underlying driver circuit. In the on-state, the mirror is rotated to $\vartheta_L = 0°$ (into the system's FOV). In the off-state, the mirror swings to $\vartheta_L = -10°$ (i.e., out of the system's FOV cone). Thus, by using the DMD in conjunction with a light source and a

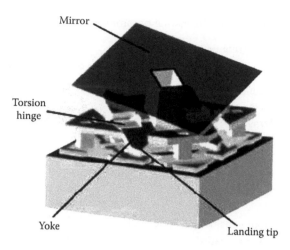

Figure 7.4 The pixel structure of the digital micro-mirror device. (After Van Kessel, P.F. and Hornbeck, L.J., *Proc. IEEE*, 86, 1686–1704, 1998.)

suitable projection optics (Figure 7.5), the mirror reflects incident light either into or out of the pupil of the projection lens.

Thus, with the above arrangement, the on-state ($\theta_L = 0°$) of the mirror appears bright, while the off-state of the mirror ($\theta_L = -10°$) appears dark.

The DMD pixel mirror is constructed over a CMOS driver array similar to an SRAM cell. An air-gap between the driver circuitry and the mirror is formed using an organic sacrificial layer.

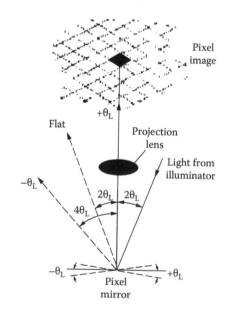

Figure 7.5 The optical system arrangement for the DMD modulator. (After Van Kessel, P.F. and Hornbeck, L.J., *Proc. IEEE*, 86, 1686–1704, 1998.)

Table 7.7 Summary of the DMD-SLM performance

Performance parameter	Description	Units
Input/output	Electrically addressed, readout beam, intensity modulated	N/A
Device configuration	Reflective mode	N/A
Optical modulator	Micro-electro-mechanical mirror	N/A
Driver array	CMOS driver array	N/A
Array size	Up to 1920 × 1080	Pixels
Resolution	25 LP mm^{-1} @ 50% MTF	LP mm^{-1} @ MTF
Temporal response	Opt. switching time: 20 μs; frame-rate: 30–300 Hz (color, 8-bit); 10 kHz binary (with custom addressing)	ms, Hz
Spectral bandwidth	Output beam: 450–650	200 nm
Input sensitivity	N/A	N/A
Contrast ratio	100:1 (20 dB)	dB
Output diffraction efficiency	60	%
Output uniformity	1	%
Physical array size	~40 × 20	mm
Power consumption	~1	W

The air-gap allows the mirrors to rotate about two flexible torsional hinges. The mirror is rigidly attached to an underlying yoke, which, in turn, is connected by two flexible torsional hinges to support posts that are formed on the underlying substrate.

An electrostatic field generated by the underlying pixel results in a mechanical torque applied on the micro-mirror. This torque, applied against the restoring torque of the hinges, produces a mirror rotation in the positive or negative direction. The mirror and yoke rotate up to the point that the yoke comes to rest against the mechanical stops.

The DMD mirrors are 16 μm^2 made of aluminium for high reflectivity. They are arrayed to form a matrix having a high fill factor (approximately 90%) to attain high optical throughputs.

Grey scale is achieved by using a PWM technique. This is an addressing method in which the effective frame rate is M times (e.g., $M = 256$) faster than the actual, visual frame rate (e.g., 30 Hz). The binary-modulating pixels are activated in duty cycle ratios between 1/M and 1.0 during the modulation of each visible frame. This, combined with the integrating response of the eye, creates an effective grey scale perception. Color operation is achieved by using color filters.

The system can also operate in a color-sequential mode using sequentially flipped color filters, in conjunction with a single device. Both the PWM grey scale method and the color-sequential operation are possible due to the relatively fast electro-mechanical response of the micro-mirror which features typical switching times of ~20 μs. Thus, assuming a 3 × 8-bit operation for a 256-level RGB, color-sequential video projection, the pixel switching requirements are $T_{switch} \sim$ (16 ms)/ [(256 levels/color) × (3 colors)] ~ 21 μs, which can still be met by the ~20 μs switching time of the micro-mirror.

As to the reliability issues related to the mechanical failure of the micro-mirror, due to the continuous flipping, TI reports that testing of hinge fatigue resulted in over 1×10^{12} (1 trillion) cycles without mechanical failure. This 20 year equivalent operation, performed in an accelerated cycling test, seems to indicate that hinge fatigue is not a reliability concern for the life of an ordinary DMD product.

In the brighter, color-parallel mode, two or three DMD arrays are used with stationary color filters to produce a full color image table (see Table 7.7).

7.3.2.3 THE GRATING LIGHT VALVE [122,123]

The GLV technology, originally developed by Bloom, similar to the DMD, is based on the MEMS

Figure 7.6 The pixel structure of a GLV device.

techniques, to form pixels in a silicon chip. Each of these pixels (typically, 25 μm size) is made up of multiple ribbon-like structures (Figure 7.6), which can actually be moved up or down over a very small distance (only a fraction of the wavelength of light) by controlling electrostatic forces. The main advantage of the GLV-MEMS technology over the DMD-MEMS technology is in the reduced displacement required of the strips (1/4λ or around 0.15 μm), relative to the DMD pixel displacement (around ±1.5 μm for the pixel tip). This leads to a substantially faster response speed of the order of 1 MHz or better.

The ribbons are arranged in such a way that each pixel is capable of either reflecting or diffracting light. An output image is formed by collecting the reflected or diffracted light with an appropriate lens system, either projected onto a front- or rear-screen system, or to be viewed directly by the eye. A Schlieren optical system is used to discriminate between the two optical states. By blocking reflected light and collecting diffracted light, contrast ratios of a few hundreds to one can be achieved. In an ideal square-well diffraction grating, 81% of the diffracted light energy is directed into the ±1st orders. By adding multiple Schlieren stops and collecting more orders, practical systems can achieve greater than 90% diffraction efficiency (DE). The gaps between GLV ribbons (defined by the minimum lithographic feature) control the optical diffraction efficiency. Thus, the theoretical DE varies between 82 and 98% for ribbon gaps between 1.2 and 0.35 μm, respectively.

The GLV technology can be employed in either digital or analogue modes. In the digital addressing mode, the switching is based on the PWM described earlier for the DMD device operation.

In the analogue mode, the depth to which ribbons are deflected is controlled by the driving circuitry. When the ribbons are not activated (deflected), the pixel is in its off-state. When the ribbons are deflected to (1/4)λ, the pixel is fully on. Any particular grey level can be generated by deflecting the ribbons to positions between these two limits.

For display applications, the GLV is operated in the scanned mode, where a linear array of GLV pixels is used to project a single vertical column of the image data. This column is optically scanned at a high rate to produce a complete 2D image. As the scan moves horizontally, GLV pixels change states to represent successive columns of video data, forming one complete 2D image per scan. The relatively fast switching speed of GLV devices (of the order of 1 MHz) allows full HDTV scanning (1920 × 1080 image) at video rates of up to 96 Hz.

The main performance parameters of the device are summarized in Table 7.8.

7.3.2.4 THE THIN-FILM MICRO-MIRROR ARRAY [124]

A thin-film micro-mirror array (TMA) display is currently under development by researchers at Daewoo Electronics in South Korea.

The modulator technology of the TMA device is similar to that of the TI DMD discussed earlier. However, rather than using electrically activated micro-cantilevers, it uses micromachined thin-film piezoelectric actuators to control the micro-mirror tilt mechanism in an analogue fashion, thus enabling grey scale operation with over 256 levels. Each pixel consists of a mirror and an actuator. In previous designs, the two had been co-planar, but the improved design has the actuator situated below the mirror, increasing the fill factor up to 94% and the contrast ratio to 200:1.

The TMA uses thin-film piezoelectric actuators in the form of micro-cantilevers, which consist of a supporting layer, bottom electrode, piezoelectric layer and a top electrode.

7.3.2.5 THE MICRO-MECHANICAL SLIT-POSITIONING SLM [126]

This is a recent concept based on the technology of MEMS described earlier, where an actuator can programmably shift an optical slit of width between 8 and 100 μm, thus enabling the position of the transmitted light regions to be changed as well as modulate the incoming light beam.

Table 7.8 Performance parameters of the grating light valve

Performance parameter	Description/value	Units
Input/output	Electrically addressed, readout beam, intensity modulated	N/A
Device configuration	Reflective/diffraction mode (1D)	N/A
Optical modulator	Micro-electro-mechanical grating	N/A
Driver array	CMOS driver array	N/A
Array size	1D: 1920 × 1 (2D also possible)	Pixels
Resolution	25	LP mm^{-1} @ 50% MTF
Temporal response	Opt. switching time: down to 20 ns; line-rate: 30–300 Hz (color, 8-bit); 10 kHz binary (with custom addressing)	ns, Hz
Spectral bandwidth	Output beam: 450–650	200 nm
Input sensitivity	N/A	N/A
Contrast ratio	>200:1 (40 dB)	dB
Output diffraction efficiency	Up to 80	%
Output uniformity	1	%
Physical array size	~50 × 0.1 (for 25 × 100 μm pixels)	mm
Power consumption	<100	mW

The proposed SLM architecture can be 1D or 2D (for imaging). The applications include micro-spectrometer systems as well as the next generation space telescope with multi-object spectrometer under development by the European and US space agencies.

When an electric field is applied to the two electrodes, the mechanical strain in the piezoelectric layer causes a vertical deflection of the mirror. The response time of each pixel is 25 μs, making it fast enough for field-sequential color display applications. The display, which is used to make a high-brightness XGA-format projector, has an optical efficiency of 20% at a panel size of 2.54 in.

In order to modulate the light intensity of the individual mirror pixels projected on the screen, a field-stop is used as a light valve. When a mirror does not tilt, all the light reflected by the mirror is blocked by the field stop and the pixel is at its off-state (dark).

When the mirror is fully tilted, all the light goes out through the projection stop and the pixel is at its brightest state (white).

7.3.3 MQW modulators

7.3.3.1 INTRODUCTION

While this is the most recently developed SLM technology, it is also the most promising one for demanding, ultra-fast-frame-rate (>MHz) applications such as optical data processing. It was pioneered in the early 1980s by Miller, who demonstrated, for the first time, the potential of artificially made stacks of quantum-size layers of alternating GaAs (quantum well material) and GaAlAs (quantum barrier materials) for a highly efficient, ultra-fast electro-absorption (EA) effect [127,128] with potential use for very fast SLMs. The effect was named the "quantum confined stark effect" (QCSE) by Miller. The interest at that time was that of fast switch or switch arrays for communications. While the EA effect can be observed in direct-gap semiconductors such as GaAs and InP (the so-called Franz–Keldysh effect), it is relatively inefficient in requiring high-voltage switching to accomplish a relatively poor contrast modulation. In fact, an attempt to use this effect in constructing a CCD-driver-based, GaAs-SLM was attempted in the early 1980s [129]. The low contrast accomplished (~1.2:1) convinced the technical community that this is not the right path to a fast, efficient SLM technology. The subsequent development of MQW-based SLM technology [130] and the, additional use of Fabry–Perot structure subsequent, combining the EA effect with the related electro-refraction (ER) effect, resulted in an SLM technology capable of gigahertz response with contrast ratio of 100:1

and higher, at moderate addressing voltage levels of ~5 V. This SLM technology constitutes the only potential solution known today for the demanding optical data processing applications.

7.3.3.2 MQW MODULATORS: PHYSICAL BACKGROUND

The MQW structure, as its name implies, consists of a stack of quantum wells (QWs) i.e., a few molecular monolayers of a low-energy-gap "well" structure (e.g., GaAs), sandwiched between thin wide-bandgap "barrier" layers (e.g., GaAlAs). This structure "compresses" the excitonic wave function (bound electron–hole pair) due to the ultra-thin dimension of the well layer in which the excitons reside, forcing the electron–hole pair to be much closer to each other. This in turn, results in a much higher coulombic energy and, therefore, higher ionization energy for the excitons compared to their ionization energies in a bulk semiconductor material. This allows excitons to exist in these QWs at room temperatures. Furthermore, the application of an electric field of the order of $100\,\mathrm{k\,V\,cm^{-1}}$ causes a partial separation of the electron–hole wave functions, thereby altering the energy levels of the excitons. This, in turn, alters the effective optical resonance of the exciton thereby shifting the resonant absorption wavelength. Since the wavelength dependence of the absorption curve is quite steep near the resonant wavelength, this shift can effectively be used to modulate the optical absorption of the optical beam (Figure 7.7). The importance of this effect for SLM application is the fact that such modulation can be effectively

performed with only a few volts of bias resulting in an EA effect with picosecond response times.

The theoretical treatment of the EA effect in a QW predicts a quadratic effect of the resonant wavelength dependence upon the applied electric field [131]. The resonant excitonic EA effect in the MQW structure is expected to result in a corresponding resonant ER effect through the Kramers–Kronig relationship which links the spectral dependence of both the absorption coefficient and the refractive index [132].

Another important type of MQW structure, the compositionally doped "nipi" structure, was initiated and developed by Ploog and Doehler [133,134]. This structure is based on alternating compositionally doped quantum-size layers, e.g., n-GaAs/p-GaAs. Often an insulating layer ("I"-layer) is inserted between the n- and p-layers and hence the name "n–i–p–I". The resulting stack of quantum-size p–n junctions, when selectively contacted with all the n- and the p-layers in parallel, can result in a very large EA or ER effect [135].

Finally, in reviewing related structures and physical effects of MQW, lower-dimensionality quantum well structures should also be mentioned. The formation of quantum well structures results in the confinement of carriers to 2D "wells", as pointed out earlier. From the standpoint of the dynamic behavior, such confinement results in the quantization of the density of energy states which now behave as a series of step functions. This effect, which actually contributes to the steep spectral absorption curve and hence to the enhancement of the EA, can be enhanced even more by the confinement of the carriers to a single dimension ("quantum lines") or ultimately to zero-dimensional "quantum dots". In particular, quantum dot structure has been extensively researched in the last few years due to the theoretical potential of yielding extremely sharp delta-shaped absorption curves. These could result, in turn, in an extremely effective switching and therefore in very efficient SLM structures. The main issue in the manufacturing of these structures has been the finite spread in the dot size distribution, which sharply reduces the steepness of the absorption curve [136].

7.3.3.3 MQW MODULATOR: DEVICE STRUCTURE

The basic device structure of an electro-absorbing MQW modulator in a transmissive mode is shown

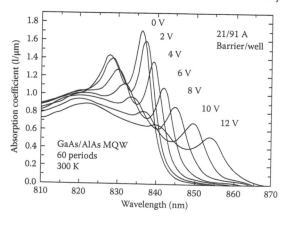

Figure 7.7 Field-induced absorption changes in an MQW sample. (After Goossen et al., *Appl. Phys. Lett.*, 64, 1071, 1994.)

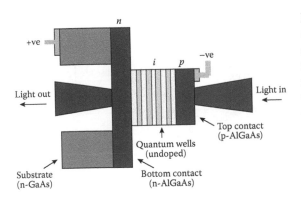

Figure 7.8 MQW modulator. (After Miller, D.A.B., *Int. J. High Speed Electron.*, 1, 19–46, 1990.)

in Figure 7.8. The MQW structure is fabricated (using either MBE or MOCVD epitaxial growth techniques) within a p–i–n structure. This configuration allows the generation of a relatively high electric field created by back-biasing the p–n junction and extending into the low-doped, high-resistivity intrinsic (I) region where the MQW structure is situated.

The medium–high-doped n- and p-AlGaAs regions constitute, in effect, transparent conductive electrodes, as these AlGaAs-based layers have a wider bandgap than the QW excitonic line and, at the same time, feature relatively high sheet-conductivity due to their high doping levels.

The transmission contrast ratio of an MQW sample can be roughly estimated as CR ~ exp($\Delta\alpha L$), where $\Delta\alpha$ is the field-induced (negative) change in the excitonic absorption coefficient, and L is the optical path length or the MQW sample's thickness.

As can be seen from Figure 7.7, an absorption coefficient change of: $\Delta\alpha \sim (-)1\ \mu m^{-1}$ is induced for a voltage change of approximately 10 V over a sample thickness of approximately 0.7 μm, which translates to a field variation of approximately $\sim 1.4 \times 10^{5\,V}\ cm^{-1}$. Based on the estimated CR above, this absorption change translates to a contrast ratio of around 2:1. This is a typical value for the performance of a transmissive-type MQW device, having around 50 QW periods, with around 10 V of applied voltage bias. For a reflective-type device, one can theoretically attain up to a quadratic enhancement of the CR since the optical path L in the CR equation above is doubled.

This translates to around 5:1 in reflectivity contrast for modulators operating in a reflection mode. To go beyond these figures necessitates either a large QW stack, (e.g., 200 periods allows over 10:1 contrast in transmission [130]) or other means of amplifying the quantum-confined Stark effect. In the following, we will shortly describe two such configurations. Historically, the first attempt was to use the photo-current generated inside the MQW modulator to affect the device's bias voltage in a positive feed-back mechanism, so as to enhance nonlinearly the QCSE by having the voltage bias of the sample be varied in proportion to the photocurrent. These early self-electro-optic effect devices (SEEDs) [137] were designed to get a low-field nonlinear, bi-stable behavior in PIN-type MQW devices.

A switching energy as low as 180 pJ for a 60 μm × 60 μm SEED has been demonstrated [138]. A second, common modification of the QW device is the integration of the device within an optical resonator cavity. This essentially "amplifies" the EA and/or ER effect by effectively extending the optical path via the multi-pass effect of the cavity.

The asymmetric Fabry–Perot QW modulator [138–140] is based on the insertion of the MQW stack within an optical resonant cavity composed of a semi-transparent top mirror (Figure 7.9) and a bottom, high-reflectivity, 1/4A-stack dielectric mirror. In this case, the resonator's reflectivity, RR, is given by $R_R = [R_T (1 - R_\alpha/R_T)^2/(1 - R_\alpha)^2]$, where $R_\alpha = \sqrt{R_T R_B}\exp(-\alpha d)$ and R_T, R_B, are the reflectivities of the top and bottom mirrors, respectively.

As can be seen, if the effective bottom-mirror reflectivity ($R_B \exp[-2_a(E)d]$) is sufficiently reduced by the field-induced absorption $\alpha(E)$, to the point where it matches the top-mirror reflectivity: $R_B \exp(-2\alpha(E)d) = R_T$, we have a zero reflectivity for the resonator. This capability of bringing down the reflectivity close to a zero level implies that we can expect high reflectivity contrast ratios for the relatively modest bias levels. The above estimate did not take into account the ER effect that must also be considered [138].

Contrast ratios in excess of 100:1 were obtained for moderate sized QW stacks and biases lower than 10 V [140]. It should be pointed out, however, that such use of a thickness-sensitive, resonant cavity makes it difficult to accomplish adequate spatial uniformity of the modulator array.

7.3.3.4 PHASE MODULATION USING MQW

An earlier attempt to demonstrate phase modulation in MQW was reported in 1988 [141]. The demonstration of 0.21 @ 852.5 nm was accompanied as

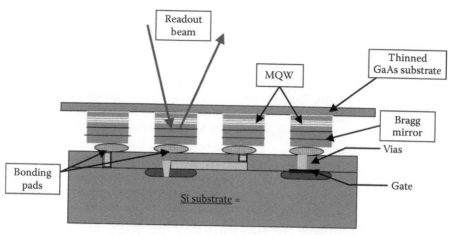

Figure 7.9 Schematics of a Fabry–Perot MQW-SLM (Lenslet Laboratories-Ramat-Gan Israel: Private Communication 2002. www.lenslet.com.)

expected by significant attenuation. As pointed out earlier, the attenuation is expected as a direct result of the Kramers–Kronig relationship. Thus, since the optical modulation effect in an MQW modulator is based on the exitonic resonance effect, one expects the maximum ER effect to always occur in the vicinity of the absorption peak. To help to maximize the ER effect while minimizing the absorption, one can define a figure of merit by FM = $\Delta n(E)/\alpha(E)\lambda$, which can be used to search for material systems (and/or spectral regions) with a high ratio of refractive to absorptive optical modulation [142,143]. This formulation implies that operation at a shorter-wavelength region may hold advantage for this type of modulation. However, earlier attempts to use this expression in search of material systems exhibiting significantly high FM were not too encouraging [142]. A more recent attempt indicates a possibility of performing phase modulation for beam-steering applications [144]. However, the authors do not indicate the absorption-associated insertion losses in this device.

7.3.3.5 MQW SLMS

Earlier attempts to construct small-size arrays of MQW-SLM devices were made [145] (6 × 6 array) and [146] (128 × 1 array).

More recent attempts of constructing and demonstrating larger-size arrays of MQW-SLMs were made [147] (128 × 128 array) and the successive effort [148,149] (256 × 256 array). The latter was reportedly operated at up to 300,000 frames per second, allowing up to 600,000 correlations per second to be performed on a 128 × 128 array. The device was based on GaAs/GaAlAs MQW stacks in a PIN geometry, inserted within an asymmetric FP cavity discussed earlier, with pixel sizes of around 40 μm × 40 μm. The group claimed up to 6 bits of grey levels achievable on their 2048 × 1 1D devices. Summary of the MQW-SLM performance is given in Table 7.9.

The demonstration of a 16 × 16 cross-bar switch "Amoeba", for optical communications was recently demonstrated [150]. The switch is based on flip-chip bonding of a 0.8 μm technology CMOS chip to an MQW array of detectors/modulators.

Optically addressed MQW-SLMs were also conceived [138] and demonstrated [151].

The recent demonstration [151] is based on a reflective-mode, GaAs/GaAlAs MQW modulator constructed inside a Fabry–Perot cavity, and combined with a free-carrier trapping layer. The latter is sensitive to the incident, near-IR to visible optical beam. A near-excitonic resonance probe beam reads out the spatial optical signal-modulated reflectivity of the Fabry–Perot cavity to form an output image. Other efforts in the development of MQW-SLM arrays that should be mentioned are [152] in developing GaInAs-based MQW modulator arrays for cross-bar switch applications.

A recent effort is based on MQW modulator-array prototypes fabricated using the GaAs/AlGaAs material system. The first prototypes fabricated are designed with an array size of 128 × 128 pixels, made with a pixel pitch of 38 μm [153].

Table 7.9 The main performance parameters of a state-of-the-art, electrically addressed MQW-SLM

Performance parameter	Description	Units
Input/output	Electrically addressed, readout beam, intensity modulated	N/A
Device configuration	Reflective-mode	N/A
Optical modulator	MQW-electro-absorption mode	N/A
Driver array	CMOS driver array	N/A
Array size	256×256	Pixels
Resolution	(est.) 25 LP mm^{-1} @ 50% MTF	LP mm^{-1} @ MTF
Temporal response	Opt. switching time: ~10 ps; frame-rate (CMOS limited): 300 kHz	ps, kHz
Spectral bandwidth	Output beam center-A~850, bandwidth~5	nm
Input sensitivity	N/A	N/A
Contrast ratio/grey levels	100:1 (20 dB)/6 bits	dB
Output modulation efficiency	~30	%
Output Uniformity	1	%
Physical array size	~10 × 10	mm

Source: Trezza, J.A. et al., Proc. SPIE, 3490, 78–81, 1998; Kang, K. et al., Proc. SPIE, 3715, 97–107, 1999.

Finally, significant improvements in the uniformity of FP-based MQW arrays (Figure 7.9), reducing optical nonuniformity to 3.3 nm across the 4 in. wafer, have recently been demonstrated [154]. These devices operate at low voltage (2.1–3.9 V), producing a reasonably high contrast ratio (16:1–98:1) and are designed for a frame rate of up 300 kHz. The array was hybridized to an 8-bit 0.25 CMOS technology, Si driver.

7.3.4 Solid-state electro-optic SLMs

In general, there has not been much progress in solid-state electro-optic SLMs. These devices, which constituted the "first-generation" SLMs in the late 1960s to early 1980s, were quite complex and bulky due to the very high bias voltage required, and the related large pixel size/SLM apertures required (see discussion on the fundamental limitations of SLMs in Section 7.6 later). Thus, devices based on KD*P and LiNbO$_3$, turned out to be commercially unviable for the main display applications, and gave way to LCDs as the SLM technology of choice for displays and to some extent for optical processing. Some references to these technologies can be found in the earlier conferences on SLM technologies [155–157].

One electro-optic SLM that should be mentioned is the PLZT-based device. There have been recent efforts in trying to form 1D and 2D products using this technology [158]. The interest in this modulator technology is in the relatively high electro-optic coefficient with half-wave voltages of the order of 100 V [159,160].

7.3.5 Acousto-optic SLMs

The acousto-optic (AO) modulator technology has been largely implemented in the form of either bulk devices such as the Bragg cell or as surface acousto-optic (SAW) devices. This is a fairly mature technology of 1D modulator arrays which has been in existence since the early 1930s [161].

The general structure of an AO modulator is shown in Figure 7.10. A piezo-electric transducer modulated by RF (100–3000 MHz) generates an acoustic wave that propagates within the AO crystal (e.g., quartz). The propagating density grating formed in the crystal gives rise to a corresponding, propagating refractive index grating. The latter diffracts an optical beam launched at the AO crystal perpendicular to the grating. The grating constant, which is essentially the acoustic wavelength A, varies inversely with the modulating RF frequency f, i.e., $\Lambda = C_s/f$, where C_s is the acoustic (sound-wave) velocity in the material. The Bragg diffraction angle ϑ, which is inversely proportional to the acoustical grating period, can be varied by modulating the RF frequency of the piezoelectric transducer. This device can therefore deflect an

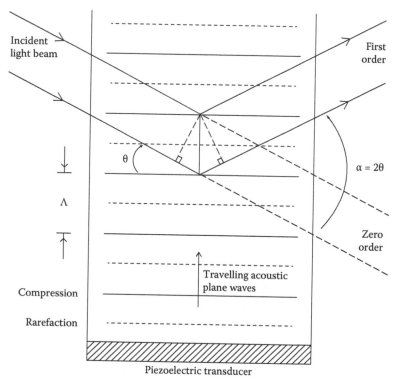

Figure 7.10 Schematic of the acousto-optic effect.

incident optical beam into a range of diffraction spots thereby acting as a 1D SLM.

A typical performance parameter of an AO Bragg cell is given in Table 7.10. From the standpoint of the SLM technology, a key parameter specifying the number of available spots or "pixels" is the time-bandwidth product. This parameter is roughly the product of the frequency bandwidth (200 MHz in this case), and the time aperture (5 pus), which is determined by the "time of flight"

Table 7.10 Typical performance parameters of an AO Bragg cell

Operating wavelength	Any within the range 442–850 nm
Time–bandwidth product	1000
Center frequency	300 MHz
3 dB bandwidth	200 MHz
Active aperture	1.5 mm H × 19 mm L
Time aperture	5 μs
Interaction medium	$PbMoO_4$
Acoustic velocity	3.63 mm μs^{-1}
Diffraction efficiency	10% at 1 W RF power (633 nm)
Electrode	Apodized to minimize acoustic walkoff
Optical surface flatness	wavelength/10 or better
Input impedance	50 Ω
Input VSWR	2:1 across RF bandwidth
Optical reflectivity	5%/surface

Source: Courtesy of Isomet Co.

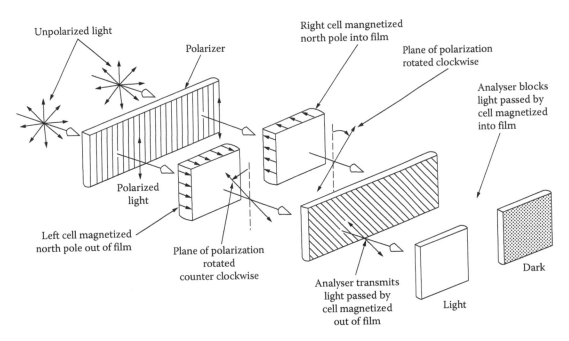

Figure 7.11 The magneto-optic effect for spatial light modulation. (After Pulliam, G.R. et al., *J. Appl. Phys.*, 53, 2754, 1982.)

of an acoustic wave (C_s = 3.63 mm μs^{-1}) across the crystal length (19 mm), giving around 1000 resolved spots.

For further information on the AO technology, we refer the reader to several extensive technology surveys which have been recently published [162–164].

7.3.6 Magneto-optic SLMs

The employment of the magneto-optic effect for spatial modulation is shown in Figure 7.11 [165–167]. The device, which is based on the magneto-optic effect associated with the magnetic domain reversal in a ferromagnetic material [gadolinium–gallium-garnet (GGG)], is electrically driven using conducting X–Y mesh-lines which form the boundaries of the magneto-optical pixels. As a particular mesh-node is cross-activated (i.e., both the X-line and the Y-line for that node) the resulting current flowing around that pixel corner generates a magnetic field, which causes a magnetic domain reversal in that magneto-optical pixel.

Although initially developed as a binary (phase or amplitude) SLM, the presence of a "neutral" state of the magnetic domains later allowed the realization of a ternary-state device [166]. Due to its relatively high insertion losses in most of the

visible spectrum, the device was targeted mainly for optical processing applications. The relatively high current required to switch to the on-state is one of the main drawbacks of this device, effectively limiting its frame rate to the low kilohertz range (see Table 7.11). Recently, a renewed effort in the development of the MO-SLM was reported [168,169]. An overall improvement by a factor of three in power consumption using a modified shape driveline with detailed device simulation was reported.

7.3.7 Photorefractive SLMs

This modulation technology is based on the refractive index modulation in response to an incident optical beam. The effect has been extensively studied in the past 30 years [170–173] and, hence, only a brief description will be presented. Figure 7.12 describes the process of photo-refraction upon the illumination by an interference pattern generated by the illumination of two coherent beams (Figure 7.12a). The spatial periodic illumination results in the generation of periodic, local photo-carriers (e.g., electrons—Figure 7.12b). These drift apart due to diffusion resulting in the formation of a space-charge field (Figure 7.12c, d). Finally, the generated space-charge field results in a periodic modulation

Table 7.11 Summary of the magneto-optic SLM performance parameters

Performance parameter	Description	Units
Input/output	Electrically addressed, readout beam, intensity/ phase modulated (ternary state: 0, ±1)	N/A
Device configuration	Transmissive/reflective modes	N/A
Optical modulator	Magneto-optic GGG	N/A
Driver array	CMOS driver array	N/A
Array size	128 × 128	Pixels
Resolution	Approximately 20 LP mm^{-1} @ 50% MTF	LP mm^{-1} @ MTF
Temporal response	Frame rate: 0.5–1.0	kHZ
Spectral bandwidth	630 (15%)–788 (45%)	nm
Input sensitivity	N/A	N/A
Contrast ratio/grey levels	> 10,000:1 @ 788 nm	
Output modulation efficiency	15–45	%
Physical array size	~30 × 30	mm
Power consumption	~2–4	W

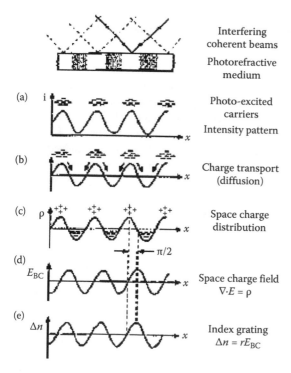

Figure 7.12 Schematics of formation of the photorefractive effect.

of the refractive index due to the electro-optic effect in this material. The resulting diffraction efficiency is therefore proportional to the appropriate linear electro-optic coefficient r, with a figure of merit for the material which can be written as [174]: $Q = n \times r/\varepsilon$, where s is the dielectric constant and n is the refractive index of the material. As it turns out, the response time of the effect is inversely proportional to the same figure of merit, Q [174].

While the effect has largely been studied for holographic storage applications, as well as adaptive optical corrections by phase conjugation (see Section 7.5), the photo-refractive (PR) crystal can actually serve by itself, as a photo-addressed or PA SLM. Thus, with the formation of a photo-induced grating by the optical input signal, the device acts to modulate a probe (readout) beam. Since both the input and the output beams can contain spatially encoded information, the device actually acts as a PA SLM, although in this arrangement another input SLM is often used. This property of PR crystals has been used for numerous applications including phase conjugation [175], optical storage [176], photorefractive correlators [177], laser beam cleaning [178,179], information processing [179,180] and novelty filter processors [181].

Early attempts to demonstrate spatial modulation included the photorefractive incoherent-to-coherent optical conversion (PICOC) device [182]. The process is based on modulating the preformed index grating generated in the photorefractive crystal by the interference of the two coherent beams, by the illumination of the incoherent (signal) image. Another demonstration of an incoherent-to-coherent spatial modulator was based on a grating-encoded phase modulation in a Ce-doped SBN crystal [183]. A more recent demonstration of

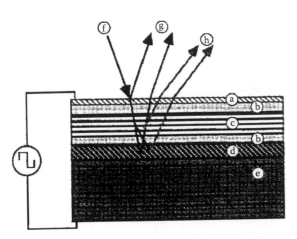

Figure 7.13 Photo-refractive-based MQW-SLM. (a) p-doped conductive layer; (b) free carrier trapping layer; (c) intrinsic MQW layer; (d) n-doped Bragg mirror; (e) n-doped GaAs wafer; (f) incident probe beam; (g) modulated reflected beams and (h) diffracted beams. (After Bowman, S.R. et al., *J. Opt. Soc. Am. B*, 15, 640–647, 1998.)

this effect was based on the self-phase conjugation in an SBN crystal, in which a 28 lp mm^{-1} resolution of spatial modulation was demonstrated with a 6 mm × 12 mm crystal [184].

A more recent employment of the photorefractive effect was in the use of an optically addressed MQW-SLM by a US Naval Research Laboratory group [185]. The researchers developed an GaAs–AlGaAs, reflective-mode MQW device, shown in Figure 7.13. This device is capable of operating both as a "regular" PA SLM and as a photorefractive or holographic mode device. In this mode, two coherent pump beams generate a periodic photocharge pattern in the carrier absorption layer (Figure 7.13b). The resulting space-charge field induces refractive index modulation within the MQW layers which in turn modulates the probe beam (Figure 7.13f). Diffraction efficiencies of 1.5% were obtained with down to a 7 μm spot-size resolution @ λ = 856 nm. A principal advantage of this SLM technology is its very high spatial resolution, while the main drawback is its inherently slow response due to the energy (time × power) to form the photo-refractive grating pattern, including the use of fast MQW configurations.

7.3.8 Smart-pixel SLMs

A smart-pixel SLM (SPS) can be described as an SLM array consisting of opto-electronic pixel circuits where each of this circuits is capable of the following:

1. Performing optical detection of the input beam.
2. Performing some level of signal processing on the incoming signal.
3. Optically modulating the output beam.

Such "smart pixel" arrays clearly integrate some of the signal processing functions into the SLM itself [186,187]. The main variable parameters of this SLM technology are

1. Detector array technology.
2. Modulator array/modulated active emitter array technology.
3. Signal processing functionality in each pixel.

Earlier concepts were based on silicon-PLZT modulators [188] where the silicon circuitry was used to provide both photo-detection and drivers for the PLZT electro-optic modulators. With the advent of liquid crystals and more recently QW modulators, smart SLMs were demonstrated with Si-FLC [189] as well as Si-VLSI MQW [187,190] combinations. More recent configurations for SPSs involve the use of a combined Si-driver and VCSEL source/modulator [191], shown in Figure 7.14.

In a recent development of a CMOS-based imaging smart-array system, a 2.5 MB s^{-1} bandwidth, over 30 dB in dynamic range with 150 μm smart pixels in an array size of 64 × 64 has been demonstrated [192].

Finally, the combination of both functions of imaging and display in a single SLM device has been demonstrated using CCD circuits operating both as an imager and a display [193]. A recent development effort based on CMOS technology has been proposed with the combined imaging and display functions implemented in each pixel of the image transceiver array [194]. The application of this image transceiver device is targeted at the "smart goggle".

The main application for the SPS technology is the OI or cross-connects [187,195–197]. Another important application is that of optical processing, where operations such as motion detection, pattern recognition, SIMD-parallel processing, neural net processing and analogue to digital conversion have been demonstrated [198–202].

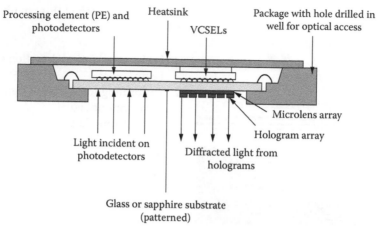

Processing element (PE) and photodetectors

Heatsink

VCSELs

Package with hole drilled in well for optical access

Microlens array

Hologram array

Light incident on photodetectors

Diffracted light from holograms

Glass or sapphire substrate (patterned)

The smart pixel array module. The original packaging for the VCSEL/Si scheme is dislayed above.

Figure 7.14 VCSEL-Si-based smart pixel. (After Neff, J. et al. http://wwwocs.colorado.edu/research/fsoi/research6.html.)

7.4 NOVEL DEVICES AND EMERGING TECHNOLOGIES

The very successful application of SLMs in the booming area of displays, and the more recent interest in their potential use in the exploding communication field, has motivated the research and development of novel spatial optical modulator technologies and devices.

7.4.1 Electro-holograms

This method is based on controlling the reconstruction process of volume holograms by means of an externally applied electric field [203–205]. The electro-holography (EH) effect exploits the voltage dependence of the photo-refractive effect in the para-electric phase of a material such as KLTN, which results in controlling the process of hologram reconstruction. Thus, the use of an electric field can result in the activation of pre-stored holograms in the photo-refractive material, which would otherwise appear optically homogeneous.

Such an effect when used in an array of such devices can be used to modulate, two-dimensionally, the optical intensity of a wave front or of an array of optical beams. Alternatively, it can be used to steer individually a 1D or 2D arrays of optical beams.

Furthermore, since this is essentially a controlled photo-refractive effect, the direction of

the diffracted beam clearly depends on the beam's wavelength. This is of particular importance in the optical cross-connect (OXC) switch application for communications (e.g., for a WDM system) where one needs to have a wavelength-dependent, routing control of the optical beams. This topic is addressed in Section 7.5 later.

Switching speeds of ~10 ns, and diffraction efficiencies of over 30% in 32 angularly multiplexed, volume holograms in a 3 mm × 3 mm × 3 mm KLTN crystal, which were switched on and off electrically, have been demonstrated.

7.4.2 PBG devices

This is an extremely interesting optical-physics phenomenon introduced in the late 1980s. The main feature of photonic band-gap (PBG) structures [206,207] is the periodic modulation of the refractive index n (or the dielectric constant), along one, two or three directions of space. In a composite formed by two dielectrics, the periodic modulation of one of the dielectrics effectively creates scattering centers, which are regularly arranged in the second medium, resulting in the coherent scattering of light. In this case, interference will eventually inhibit some frequencies that will not be allowed to propagate, thus giving rise to "forbidden bands". Under certain conditions, regions of frequency may appear to be forbidden regardless of the propagation direction in the PBG. In such a case, this region is said to present a full PBG. On the other hand, if the forbidden photonic band varies with the propagation direction in the region, a photonic pseudo-gap is spoken of. Furthermore, we can also introduce defects into the structure, resulting in the introduction of allowed energy levels into the gap, analogously to a doped semiconductor. All these facts permit us to establish a parallelism between the formalism used for electrons in ordinary crystals and that for photons in a photonic crystal (PX).

Continuing with this similarity, it is also possible to control the "optical conductivity", i.e., the optical transmission of a photonic crystal, by modulating the relative indices of diffraction of the two materials of which the crystal is composed. In particular, if an electro-optical material, such as liquid crystal, is infiltrated into the optically periodic crystal (e.g., periodically etched, porous silicon), then controlling the effective refractive index of the LC via voltage bias can result in a significant shift of the forbidden band frequency. This in turn can drastically change the optical transmission. Hence, here we have another potential efficient candidate for SLM—since minute changes in the effective refractive index could lead to very significant variations in the optical transmission. This was actually demonstrated with a matrix of porous silicon, having an air-pore pitch of 1.58 μm, infiltrated with the nematic LC E7. The photonic band-gap transmission of this structure then becomes quite sensitive to temperature variations, through their effect on the effective refractive index of the LC as was demonstrated by the Japanese as well as the Canadian–German groups which studied this effect [208,209].

7.4.3 Bubble cross-connect arrays

A modified technology, originally developed for inkjet printers, to be used in an all-optical switch [210], is based on the formation of bubbles of gas by electrical heating, which then propel droplets of ink toward the printer paper. This printer technology is now quite reliable following two decades of development and perfection. The device generates bubbles in the same way, but uses them instead as a quickly appearing and disappearing gas–fluid optical interface.

At each switching point, two silica-based single-mode wave guides intersect at a fluid-filled trench such that the angle between each wave guide and the normal to the trench is greater than the angle at which total internal reflection begins for a gas-to-fluid interface. When a bubble is created at the intersection, light reflects off it; when the bubble disappears, light passes straight through. Switching time is of the order of 1 ms; the cross-talk was measured to be −70 dB. The researchers have fabricated an array of such elements into a 32 × 32 optical switch. The device is claimed to achieve up to a 20 year lifetime.

From the standpoint of SLM technology, we have a 2D array of binary-modulator switches capable of routing an incoming (1D) array of beams into an output array with a switching speed of 1 ms. This is a particular example of a "cross-bar switch" implemented in a planar configuration. Similar functionality has been offered by using LC and micro-mirror (MEMS) technologies for optical communication (see Section 7.5).

7.4.4 Bio-chemical SLMs

This novel class of SLM technology is based on the spatial optical modulation induced as a result of chemical or bio-chemical spatial variations, which are usually converted to spatial modulation of the electric field/potential.

An earlier attempt to use bio-chemical agents to induce spatial optical modulation was made by using bacterio-rhodopsin molecules [211,212]. However, while having scientific curiosity, this type of modulator has not yet been practically implemented.

Other more recent attempts have focused on the use of bio-chemical material used as bio- or chemical sensors to affect the optical property of the underlying optical modulator either directly or by spatially modulating an electric field or potential. A fiber-optic DNA sensor for nucleic acid determination was demonstrated using a fluorescent DNA stain [213]. Such a bio-chemo-optical sensor can be replicated in an array allowing the parallel detection of multiple forms of nucleic acids, and hence can be considered as a candidate technology for the "bio-chemical SLM". Another such candidate based on an evanescent wave in a 1024 fiber optic array for the detection of oligo-nucleotides was demonstrated in 1996 [214]. Demonstrations of actual operation of arrays of bio-chemical optical modulators started to appear in the late 1990s. A fiber-optic-based micro-optode (i.e., optical micro-electrode) for the measurement of oxygen distribution was demonstrated in 1997 [215]. The researchers used a phase-modulation technique to determine the phase shift produced by the fluorescent stain, in response to a sinusoidal optical excitation. A biological sensing technique for the detection of specific bacteria was demonstrated using a micro-mechanical array of silicon nitride cantilever beams to which a specific immobilizing antibody of a certain bacterium (*E. coli*) was attached. The detection mechanism is based on the shift of the resonant frequency of the cantilevers which is measured by the optical beam. A limiting detection of 16 *E. coli* cells was demonstrated [216]. The simultaneous detection of six bio-hazardous agents using a cocktail of fluorescent antibodies coupled with the analysis of fluorescence intensity was demonstrated [217].

A surface-plasmon-based resonance was used to construct a bio-sensor array based on multiple analytes. By performing parallel, multiple spectroscopic analysis of the plasmon resonance for the various analytes, researchers demonstrated the bio-sensing ability to detect the affinity of the peptide sequence YGGFL to human β-endorphin, as well as the affinities between other types of bio-agent. The 1D array technology allows the simultaneous evaluation of up to 160 samples [218].

An inexpensive technology for the production of a colorimetric resonant optical bio-sensor, based on the use of multiple micro-titer wells and an array of fiber optic transceivers, was demonstrated [219]. The demonstration using 96 micro-titer plates showed the detection of a protein-protein affinity with an antibody detection sensitivity of 8.3 nM.

Finally, the use of a LC-based optical shutter array coupled to an aligned array of bio-sensing elements constructed on a planar wave guide was demonstrated [220]. The LC shutter array selectively contrOIs the transmission of the fluorescence generated by the biosensor array, activated via the optical wave guide.

7.5 SLM APPLICATIONS

7.5.1 Optical communication, signal processing and interconnects

7.5.1.1 OVERVIEW

The most dramatic recent technology development relevant to the SLM field, since the last comprehensive report on SLM technology was demonstrated in 1995 [4], has been the explosive interest and use of communications. The most relevant applications of the SLM technology for this field are twofold:

1. OXC switch arrays.
2. SLM-based adaptive optical aberration correction systems for point-to-point optical communication (a part of the so-called "last mile communication" problem).

A third application that is currently under development and which may be quite important for optical communication (OC) is that of adaptive ultrashort (femtosecond) pulse shaping. The latter is also an example for the use of optical signal processing (OSP) in optical communication. Indeed, since the communication field is by far the most dominant application for OSP and OI technologies, we will include a brief description of other OSP and OI applications as well within this section on optical communication.

7.5.1.2 OPTICAL INTERCONNECT AND CROSS-CONNECT

In a typical operation, the optical cross-connect (OXC) unit enables a programmable interconnection between an array of N input optical channels (possibly implemented by N input fiber-optic channels) to an N-output channel array. It is desirable that the cross-connect switch allows spatial and wavelength-dependent routing operation. The role of an OXC unit for communication is to route each of the optical signals spatially to its designated end location, based on the signal (header) information and the particular operating wavelength of the channel. An obvious candidate of SLM technology to carry out this dynamic cross-connect function is LC technology. An earlier work used a PA LC-SLM for a 256-channel system [221]. In order to improve the performance of the slow-response NLC devices, an eight-channel device was demonstrated using FLC in 1995 [222].

A significant pioneering work in the area of LCDs for OI was done using both nematic [223,224] and FLCDs [225], as well as a demonstration of polarization-insensitive switches [226], thereby overcoming one of the major weaknesses of the LC technology.

A study on the use of sub-wavelength LC-based diffractive optical elements for OXC applications was recently published [227].

The MEMS technology with its DMD representative detailed in Section 7.3 earlier is another potential candidate for the OXC technology [228,229], with numerous companies currently attempting to develop MEMS-based cross-connect switches. The relatively slow response coupled with the reflective-mode constraints is the main drawback of this technology.

The bubble-SLM technology mentioned in Section 7.4 earlier is being developed for crossconnect application. A 32 × 32 switch has been demonstrated with a 10 ms response time [209,230,231].

The electro-hologram technology mentioned in Section 7.4 earlier is under development for optical cross-connect applications. The main advantages of this technology over MEMS, LC, or the bubble technique is its relatively fast response (of the order of nanoseconds) as well as its unique wavelength selectivity.

An 8 × 8 optical crossbar is under development [232]. The OXC will consist of 64 inputs that can be routed to any one of the 64 outputs, using a hybrid III–V semi-conductor and Si CMOS flip-chip bonded technology. The optical switch will have 4096 detectors configured as a 64 × 64 array, each operating at >200 Mbit s^{-1}. Another effort has developed a 16 × 16 CMOS/GaAs OI operating at 0.83 μm [233].

Finally, there have been numerous demonstrations of VCSEL-based smart-pixel OI systems [234–237]. A free space, 64 × 64 optical cross-connect switch based on FLC-SLM was recently demonstrated [238]. The cross-connect interface can serve as a programmable parallel array processor for a variety of data processing applications.

7.5.1.3 FEMTOSECOND PULSE SHAPING

An important as well as an interesting application of the SLM technology for optical communication is the shaping of ultra-short, femtosecond optical pulses first suggested by Weiner [239] using a fixed mask, and later refined to include an LC-programmable mask or an SLM [240]. The concept is to spatially disperse the optical pulse using an appropriate grating and then spatially phase-modulate the dispersed signal pattern. The re-collection of the spatially phase-modulated pulse using a second grating, leads to a temporal re-shaping of the optical pulse (in particular, pulse compression), which can then be used for optical communication. In more recent developments,

temporal as well as spatial pulse shaping was accomplished using SLMs [241,242] with a recent demonstration of a 5 fs pulse compression [243].

7.5.2 Display applications

This is obviously a very wide application field, with a huge billions of dollars market, encompassing numerous technologies. As we are interested only in reviewing the field from the SLM technology standpoint, we will confine ourselves to nonemissive or passive display technologies, omitting technologies such as LEDs/OLEDs, electroluminescent and plasma displays. The discussion will therefore be limited to innovations in the area of LC- and MEMS-based displays.

One of the very recent applications of the PA LCLV discussed in Section 7.3 earlier is that of electronic cinema [244]. Here, only high-brightness, high-resolution projection systems are required. A projection system based on combined LCLV technologies can provide up to 3000 ANSI lumens at a resolution of 2048 × 1536 pixels. A DMD-based projection system is in close competition.

Another important display technology for large-screen, video projection displays and for the electronic cinema application is the DMD, MEMS device discussed earlier [245]. The use of SLM panels (whether LCDs or digital mirrors) is implemented in a three-panel configuration, where either red, green and blue panels or some form of color-subtraction configuration is used. This allows high brightness to be accomplished as the full bandwidth of the (white) light source is utilized simultaneously for all three channels. The disadvantage is in the complexity of the color-separation system and the cost of three panels and their drivers. The other more compact and cost-effective method is to use a "color-sequential" scheme, where one broad-band SLM panel (either LCD or DMD) is used in conjunction with a color-sequencing scheme where the red, green and blue portions of the filtered broad-band source are used sequentially to read out the panel. This requires a minimum of 180 Hz operation, which, although quite demanding, has been successfully demonstrated on LCD panels [246].

Finally, we should also mention the fast-growing field of head-mounted displays (HMDs) in which miniature LCD panels play an increasingly growing role [247–249]. LC-based, 3D stereoscopic displays are a derivative of both LCD and HMD technologies [250,251] where the stereoscopic or, 3D depth visualization can be accomplished, e.g., by ascribing orthogonal polarizations to each of the ocular views using appropriate LC shutters [250]. Auto-stereoscopic systems where the stereoscopic imagery is embedded in the displayed imagery, obviating the need for the shutter glasses, has been a major thrust recently with DMD-type SLMs serving as the image source [251]. All digital cinema projection is now done via micro-mirror SLMs from TI.

7.5.3 Optical data processing

As pointed out in the previous section, novel methods and technologies related to the application field of OSP are covered within the former section on optical communication. This section covers innovations and updates related to the optical processing of spatial data information (the latter may either be optical or electronic) with an emphasis on image processing applications. One area, albeit not new, has been continuously attracting attention as one of the most efficient optical methods of image computation, namely the optical correlation technique. With the introduction of ultra-fast SLM technologies such as the MQW-SLM, a natural progression has been made to utilize this fast SLM in demonstrating high-throughput image correlation.

The group [148,149] claims a correlation throughput of up to 600000 correlations s^{-1} with 128×128 images. This translates into an equivalent computational throughput of: $P_{comp} = 2N^2 \log Nf \sim 115$ GOPS, where $N = 128; f = 600,000$, and GOPS = giga binary operations s^{-1}.

While this obviously is a very respectable computational throughput, it is uncomfortably close (to within a factor of approximately ×10) to throughputs attainable by current dedicated ULSI, system-on-chip (SOC) processors. This state of affairs represents the on-going dilemma of optical processing—the very fast moving electronic competition.

7.5.4 Adaptive and programmable optics applications

Here, we focus on the use of dynamic, adaptive optics and programmable optic techniques for noncommunication applications including astronomical and vision systems.

7.5.4.1 ADAPTIVE OPTICAL CORRECTION

The use of adaptive optics for atmospheric aberration corrections has been intensively researched in the last four decades or so. The particular use of SLM technology as a key tool in adaptive, real-time correction systems was given in detail by Pepper et al. [4, Chapter 14].

Phase conjugation, whereby a phase-aberrated wave-front is "cleaned up", is an attractive way of correcting phase aberrations of optical beams. There have been several attempts to use the PA LCLV technology, through its capability of real-time phase conjugation, to demonstrate such an action first by using nematic LCs [252,253] and later by using FLC modulators [254]. More recent use of SLM technology for real-time aberration correction appears in references [255,256].

A recent application of this technology has emerged in the area of visual system examination and correction. A retinal imaging system, which uses a deformable mirror-type SLM for the correction of the eye-lens aberration was recently developed [257]. Here, a Hartmann-Shack-type wave-front sensor senses the ocular lens aberrations. These wave-front distortion data are subsequently fed to a deformable mirror SLM, to correct adaptively the retinal image. This system allows a clear, nonaberrated imaging of the retina for performing retinal examination and subsequent medical treatment.

The next stage in the use of this technology is an adaptive visual correction system, in which this real-time SLM-based, adaptive optics technique will be used to attain a "super-normal vision", thus allowing a near-diffracted-limited vision to be achieved.

7.5.4.2 PROGRAMMABLE OPTICAL ELEMENTS

This application, which is related to the previous "adaptive optics" one, is based on the use of phase-modulating SLMs to perform programmable optical element functions e.g., variable focal-length lenses or variable slope prisms used for beam-steering. Such devices often make use of the fact that a given optical phase profile required for a desired shaping of a monochromatic optical beam wave-front can be substituted by its "modulo-2π" representation. Thus for example, a linear phase profile, $\phi(x)$, of a glass prism is described by $\phi(x) = \tan(\alpha)xn_0$, where α is the apex angle of the prism, n_0 is the glass refractive index, and x represents a point along the length of the prism, will have the same effect on a monochromatic beam at λ_0 as a phase profile which "resets" to $\phi(x) = \lambda_0$, at x-locations in which $\phi(x)$ satisfies: $\phi(x)/\lambda_0 = 2\pi n$, where $n = 1, 2, \ldots$. Note that this allows relatively thin optical elements with thickness of the order of λ, namely a few micrometres, to be substituted for the traditional thick (mm–cm) glass elements. Also note, however, that such substitution is only valid for the particular wavelength λ_0 for which $\phi(x)/\lambda_0 = 2\pi n$. These optical elements are commonly referred to as "binary optics" or "kino-form optical elements" [258,259].

The ability to structure wavelength-thin, optical elements opened the way for using phase modulating SLMs thereby allowing programmable optical elements to be developed.

Applications of such dynamic kino-form optical elements include LC beam-steering devices [260,261], as well as variable focal-length lenses [262–264]. The main drawback of these dynamic programmable elements, as in the static kino-form elements, is their limitation in monochromatic beam shaping applications.

7.5.5 Wavelength image converters

An additional important application of SLMs is the conversion of an imagery from one wavelength to another. This can be done using either a single, photo-activated SLM whose photo-substrate and optical modulators operate at different wavelengths. Alternatively, one can use an imager device to record the imagery at a certain wavelength and relay the electronic signal to an electrically addressed SLM operating at a desired output wavelength.

A commonly used example of the latter is the regular TV monitor which relays the image picked up by a video camera and displays the color version video on the screen. Although in this case both the input and the output imagery are formally based on red–green–blue (RGB) visible channels, we can easily manipulate the TV color range, and thus can effectively perform an image wavelength conversion operation with this system.

PA LCLVs of the type discussed in Section 7.3 earlier have been successfully used in converting visible imagery to IR video scenes for IR scene simulation applications [265]. Such devices were also used for visible- to near-IR image conversion [266].

The high spatial resolution and array sizes attainable in some SLM technologies, combined with the capability of UV modulation, naturally calls for the use of this technology to perform programmable, real-time photo-lithography [267]. The use of an NLC-SLM with 600 × 800 pixel resolution, 27 mm × 20 mm aperture, over 100:1 contrast and 7/25 ms rise/decay response times at the argon ion wavelength of 351.1 nm, were used to form a 3D hologram in a photopolymer material. The hologram was subsequently read using the 633 nm of a He–Ne laser [268].

7.6 FUNDAMENTAL LIMITS AND FUTURE TRENDS

7.6.1 Performance trade-offs and fundamental limits

Some important interplays or trade-offs exist among the main SLM performance parameters. It is important to understand those limits in order to realistically design an SLM or make a sound projection of its expected performance. The main parameter trade-offs of SLMs are briefly described below.

7.6.1.1 DYNAMIC RANGE/SPEED

This is perhaps the most significant trade-off in the SLM technology. It is somewhat intuitive in that a large electro-optic coefficient, which is related to the dynamic range attainable by the optical modulator system, will be associated with a reduced speed of response, and vice versa. A good example for two modulator systems showing this trade-off would be, on the one hand, the class of nematic liquid crystal material featuring an enormously high (second-order) electro-optic effect with an effective half-wave voltage of around 2 V, but with a typically low frequency response of around 100 Hz. (The half-wave voltage can roughly be taken as inversely proportional to the electro-optic coefficient.) This is contrasted with solid-state electro-optic crystals such as the KD*P or $LiNbO_3$ capable of responding to bias frequencies well above 1 MHz, but with the penalty of a very large, 1000 V half-wave voltage. To gain some insight into this trade-off—reminiscent of the well-known gain–bandwidth product (GBWP) in electronic devices—it can be argued that under certain hypotheses this trade-off may in fact, obey a "universal" form of the GBWP behavior, for a variety of optical modulator material systems, where the gain is actually represented by the generalized susceptibility relevant to the particular modulator material system (e.g., the electro-optic coefficient, for solid state EO modulators), and where the bandwidth is the total frequency bandwidth of the modulator system, comprised of both the optical (spectral) bandwidth of the modulator and the electric field, RF-modulating frequency (see Appendix).

7.6.1.2 RESOLUTION-SPEED TRADE-OFF

This constraint is originated in the dynamic range–speed trade-off, as discussed earlier. Thus, for very fast-responding materials (e.g., solid-state electro-optic crystals) the high biasing voltage required, due to their low EO coefficients, as a result of the above, dynamic range–speed trade-off, will necessitate the use of large pixel sizes to avoid field break-downs.

7.6.1.3 DYNAMIC RANGE–RESOLUTION

A common manifestation of this trade-off is the familiar "blooming" effect, which occurs in the "over-activated" regions of the SLM pixels. This is usually due to the finite charge-holding capacity of the driver array (both in PA and electrically addressed devices). The excessive pixel driver charge, resulting from over-driving the pixel in attempting to attain high output signal (or equivalently, high dynamic range), spills over to adjacent pixels, resulting in the "blooming effect". However, this trade-off can also be "utilized" in a constructive manner by using a cluster of pixels in a high-resolution, binary-mode device to produce effectively a grey-level modulation at the lower resolution defined by the pixel cluster, in a similar way to half-tone image techniques.

7.6.1.4 SENSITIVITY–SPEED

This trade-off (for PA devices) is usually the consequence of the constancy of the energy flux required for the photoactivations (J cm^{-2}), rather than that of the power flux (W cm^{-2}). Examples are PA SLMs, in which we must compensate for low input power flux levels, by reducing their frame rate or integration time. Another well-known example is the reciprocal relationship between

the response time for the formation of a grating in a photo-refractive material, and the power level of the input signal. In general, the consequence is the necessity to use longer integration periods to accumulate sufficient signal charge and thus to lower the speed or the frame rate of the device. This is similar to the limitation of imagers under low-level illumination.

7.6.2 Future trends in SLM technology

The last decade has witnessed tremendous changes in three main optics-related areas, namely, communications, information processing and display technologies. The explosive growth and the prospects of even more dramatic future developments are certainly expected to impact the trend in the related SLM technology. In addition, the trend of developing artificial man-made optical materials, which started around two decades ago, has led to a very successful development of, in fact, the fastest optical modulator in existence today, namely, the MQW modulator. Therefore, a continuation of all these trends in the next decade is predicted. The development of OXC-SLM technology, which has already seen the development and adaptation of MEMS-based devices, will continue to expand with a thrust towards the development of novel SLM technologies such as the Bubble Array or the electro-hologram devices. We have already witnessed the redirection of the historic optical data processing technology into the use of OIs. It is predicted that this will continue as a thrust for the development of SLMs in information processing systems. The development of relatively slow (~1 ms) phase-modulating devices for adaptive and programmable optical element applications will continue to grow. At the same time, it is expected that the effort in developing faster modulators will be expanded. In the booming and attractive field of displays, active arrays such as organic LEDs (OLEDs) may well challenge the traditional passive, SLM-type devices such as LCDs. Finally, with the strong interest in bio-chemical data processing and biological sensors—enhanced by the explosion in genetic research as well as the recent threats of weapons of mass destruction, one can see a significant thrust in the area of bio-chemical SLMs (see

Section 7.4.4) to allow a fast, parallel processing and analysis of bio-chemical data.

APPENDIX: ON THE DYNAMIC RANGE–SPEED TRADE-OFF

The following derivation represents an intuitive argument rather than a rigorous proof and should be treated as such. We chose to include it despite its approximate nature, as the consequences of this hypothesis may be far reaching in terms of the fundamental limits of SLMs.

First let us generalize the term dynamic range of the system to be the range of values which the appropriate optical response function of the particular modulator system can take. It is the range of the electro-optic coefficient for an electro-optic modulator system or the range of the electro-absorption coefficient for the MQW modulator system. The speed of response is obviously directly related to the frequency bandwidth of the system. With this in mind, taking the electro-optic effect as an example, the change in the refractive index of the material Δn, upon the application of an electric field E, is given in general by [269]

$$\Delta n \sim n^3 r E \qquad (7.1)$$

where r is the appropriate electro-optic coefficient. This behavior can be presented more generally, as a response of a system parameter $P(t)$, to an external driving field $F(t)$.

It can be shown that the temporal response function $\chi(t)$ (or the generalized susceptibility) of the system acts essentially as an impulse response function to the external driving field $F(t)$, such that

$$P(t) = \int \chi(t-\tau)F(\tau)dt \qquad (7.2)$$

where in the frequency domain the convolution operation of Equation 7.2 turns into a simple product, namely [270,271],

$$P(\omega) = \chi(\omega)F(\omega) \qquad (7.3)$$

This is a generalization of Equation 7.1 above with P, χ, F, replacing Δn, r, E, respectively.

Now, the DC value of χ, namely, $\chi_{DC} = \int \chi(t)\,dt = \chi(\omega = 0)$, must be finite, based on the material stability considerations. This is so since an infinite value would imply that a vanishingly small fluctuation in the external field δF would result in a finite change of the system parameter P (e.g., the refractive index) [272]. However, since the latter is determined by the structure of the particular material system (e.g., an electro-optic crystal), a finite constant (DC) change in its value implies a finite structural change. Such a finite structural change, with a vanishingly small perturbation, indicates an inherent instability. Now, by the Parseval identity:

$$\left| \int \chi(t) \right|^2 dt = \int |\chi(\omega)|^2 d\omega \qquad (7.4)$$

and hence the finiteness of $\chi_{DC} = \int \chi(t)dt$ implies a finite value for $\int |\chi(t)|^2 dt$, and hence for $\int |\chi(\omega)|^2 d\omega$ Assuming a finite bandwidth, $\Delta\omega$, for $\chi(\omega)$, and making the approximation:

$$I\chi = \int \chi(\omega)d\omega \sim \chi_{AV}\Delta\omega \qquad (7.5)$$

where χ_{AV} is an average value for $\chi(\omega)$ within the finite bandwidth $\Delta\omega$, we conclude that the finite value of $I\chi$ indicated by Equation 7.4 implies that the susceptibility χ_{AV} must, in general, be traded off against the system bandwidth, $\Delta\omega$, or speed of response. Thus, we conclude that there must exist, indeed, a trade-off between the dynamic range of the generalized electro-optic response function and the speed of response of that system.

Next, we need to consider the frequency range of the system, where there are basically two physically different frequency regions. The first, $\omega_1 = \omega_{RF}$, is the RF field (e.g., the low-frequency externally applied electric field that controls the electro-optic effect). The second frequency region, $\omega_2 = \omega_{opt}$, is the optical frequency range encompassing the spectral (wavelength) region in which this particular susceptibility (say, the EO effect) is nonzero. Typically, we have $\omega_{opt} \gg \omega_{RF}$. Let us consider in detail the particular EO example with its associated second-order susceptibility, $\chi(2)$, related to the EO (Pockel) effect. Using results from nonlinear optics, we have for the second-order polarization $P^{(2)}(t)$ [273].

$$P^{(2)}(t) = \varepsilon_0 \int_{-\infty}^{\infty} d\omega_1 \int_{-\infty}^{\infty} d\omega_2 \chi^{(2)}(-\omega; 0, \omega) E_1(0) E_2(\omega) e^{j\omega t}$$

$$(7.6)$$

where we take: $\omega_2 = \omega_{opt} = \omega$; $\omega_1 = \omega_{RF} = 0$ (low frequency); $E_1(0) = $ const (=amplitude of the external bias field); $E_2(\omega) = E_2^{(0)} = $ const (=amplitude of the optical field). We now assume, as in the previous argument, that $\int \chi^{(2)}(\omega) d\omega \sim \chi_{AV}^{(2)} \Delta\omega$ is finite over the finite (optical) bandwidth $\Delta\omega$ and thus we have an upper limit to the polarization given by

$$\left| P^{(2)}(t) \right| \leq \Delta\omega_2 \Delta\omega_1 \chi_{AV}^{(2)} E_1(0) E_2^0. \quad (7.7)$$

now the polarization, or its variation with the applied field, is directly related to the variation in the electronic or ionic displacement $\langle x \rangle$ via [274]

$$P = Ne\langle x \rangle \quad (7.8)$$

and so again due to material stability considerations P, or $P^{(2)}(t)$, must remain finite, given a finite $\langle x \rangle$, for any finite field levels below the breakdown limit of the material. Therefore, the conclusion is that the quantity:

$$SBP = \Delta\omega_2 \Delta\omega_1 \chi_{AV}^{(2)} = \Delta\omega_{opt} \Delta\omega_{RF} \chi_{AV}^{(2)} \quad (7.9)$$

must be finite for this case of an EO material system.

So, we see that the GBWP constant takes on a new form for the "susceptibility–bandwidth product" (SBP), which involves two frequency ranges rather than one, due to the two operating fields: the RF and the optical field.

We can conveniently express this new SBP, in units of frequency bandwidth (Hz) × optical bandwidth (cm^{-1}) × optical susceptibility, which we take as proportional to the linear electro-optic coefficient or $\Delta n/\Delta E$ (cm V^{-1}). So, for the EO case we have the SBP quantity in units of Hz V^{-1}.

It can be argued that the above considerations can be extended to a generalized susceptibility (e.g., AO, magneto-optic). This is so since the polarization modulation resulting from the application of the appropriate external field and with it, the associated ionic or electronic displacement,

must remain below the structural damage level, regardless of the optical modulation mechanism (e.g., acoustic or magnetic). Now, since this range of acceptable ionic or electronic displacement in solids, although not constant, varies within one to two orders of magnitude, we expect the SBP for all solid state material systems to be also roughly within one to two orders of magnitude.

Let us examine this hypothesis for three material systems:

Liquid crystals: $\Delta\omega_{RF} \approx 104$ Hz (FLC); $\Delta\omega_{opt} \approx 2.5 \times 10^4$ cm^{-1} (400–10,000 nm); $\Delta n/\Delta E \approx 0.2/10^4$ (cm V^{-1}) Here, we get for the SBP$_{LC}$:

$$SBP_{LC} = \Delta\omega_{opt} \Delta\omega_{RF} \chi_{AV}^{(2)} \approx \Delta\omega_{opt} \Delta\omega_{RF} [\Delta n / \Delta E]$$

$$\approx 5 \times 10^3 (Hz\ V^{-1})$$

MQW modulator (electro-refraction): $\Delta\omega_{RF} \approx 10^{10}$ Hz $\Delta\omega_{opt} \approx 30$ cm^{-1} (850–852 nm); $\Delta n/\Delta E \approx 0.01/10^5$ (cm V^{-1}). Thus we get for the SBP$_{MQW}$:

$$SBP_{MQW} \approx \Delta\omega_{opt} \Delta\omega_{RF} [\Delta n / \Delta E] \approx 3 \times 10^4 (Hz\ V^{-1})$$

Solid-state electro-optic modulator ($r = 30$ pm V^{-1} = 3×10–9 cm V^{-1}): $\Delta\omega_{RF} \approx 107$ Hz (FLC); $\Delta\omega_{opt} \approx 1.5 \times 10^4$ cm^{-1} (400–1000 nm); $\Delta n/\Delta E \approx n^3 r \approx 30 \times 3 \times 109$ cm V^{-1}

$$SBP_{EO} \approx \Delta\omega_{opt} \Delta\omega_{RF} [\Delta n/\Delta E] \approx 1.4 \times 10^4 (Hz\ V^{-1})$$

We thus see that the SBPs of all these three markedly different material systems, with orders of magnitude variations in their RF frequency responses, spectral bandwidths and optical susceptibilities, come to roughly the same level, within an order of magnitude. This finding supports the "universality" hypothesis for this quantity.

We also note that for systems with comparable spectral (optical) bandwidths such as liquid crystals and solid-state EO materials, the SBP constancy simplifies approximately to a trade-off between the EO susceptibility or $\Delta n/\Delta E$ and the RF bandwidth $\Delta\omega_{RF}$ in analogy to the well-known GBWP.

Finally, an interesting situation arises when we can invoke a trade-off between the optical susceptibility and the frequency response within the same material system, e.g., by incorporating an optical feedback in a PA nematic LCLV [275,276]. In this case, it can be shown that the product of the open-loop optical gain defined for the phase modulating LC-SLM by

$$\Delta\phi_{out} = G_0\Delta\phi_{in} \qquad (7.10)$$

and the LC frequency bandwidth,

$$\Delta\omega = 1/\tau_0 \qquad (7.11)$$

the GBWP G_0/τ_0, remains unchanged as the optical feedback mechanism is turned on, namely when we have

$$\Delta\Phi_{out} = G_{CL}\Delta\Phi_{in} \qquad (7.12)$$

where for the closed loop system with a negative feedback fraction, β, we have

$$G_{CL} \approx \frac{G_0}{1+\beta G_0} \qquad (7.13)$$

and the associated LC frequency response is

$$\Delta\omega = 1/\tau_{CL}. \qquad (7.14)$$

It can be shown that [275]:

$$G_{CL}/\tau_{CL} \approx G_0/\tau_0.$$

This GBWP constancy allows a 50 Hz (open-loop) PA nematic LCLV to attain kilohertz frequency response using a negative feedback, closed loop system [276]. This is in close analogy to the constant GBWP behavior in electronic systems (e.g., operational amplifiers).

REFERENCES

1. Efron, U., ed. 1990. *Spatial Light Modulators and Applications-III, Proc. SPIE*, vol 1150 (Bellingham, WA: SPIE).
2. Optical Society of America (OSA). 1990. *Topical Meeting on Spatial Light Modulators*, Sept. 1990, Incline Village, NV.
3. Optical Society of America (OSA). 1993. *Topical Meeting on Spatial Light Modulators*, March 1993, Palm Springs, CA.
4. Efron, U., ed. 1995. *Spatial Light Modulator Technology* (New York: Dekker).
5. Roy, A. and Singh, K. 1995. Spatial light modulators and their applications: A bibliographical review for the years 1990–1991. Atti dela Fondazione *Giorgio Ronchi*, 51, 529–601.
6. Burdge, G. (Chairman). 1997. *Topical Meeting on Spatial Light Modulators* (Washington, DC: Optical Society of America (OSA)).
7. *Applied Optics: Special Issue on Spatial Light Modulators*, vol 37, November 1998.
8. Sutherland, R.L., ed. 1998. Spatial light modulators, *Proc. SPIE* 3292.
9. Diffractive and holographic technologies, systems, and spatial light modulators, IV 1999, *Proc. SPIE*, vol 3633.
10. Diffractive/holographic technologies and spatial light modulators, VII 2000, *Proc. SPIE*, vol 3951.
11. Efron, U., ed. 2001. Spatial light modulators: Technology and applications, *Proc. SPIE*.
12. Sayyah, K. and Efron, U. 1996. Optically addressed spatial light modulator with high photosensitivity and intensity adaptation range. *Opt. Lett.*, 21, 1384–1386.
13. Esener, S.C., Wang, J.H., Drabik, T.J., Title, M.A., and Lee, S.H. 1987. One-dimensional silicon PLZT spatial light modulator. *Opt. Eng.*, 26, 406–413.
14. Donjon, J., Dumont, F., Grenot, M., Hazan, J.P., Marie, G., and Pergrale, J. 1973. *IEEE Trans. Electron Dev.* 20, 1037.
15. Casasent, D. 1978. *Opt. Eng.*, 17, 344.
16. Warde, C., Weiss, A.M., Fisher, A.D., and Thackara, J.I. 1981. *Appl. Opt.*, 20, 2066–2074.
17. Schwartz, A., Yang, X.Y., and Warde, C. 1984. Electron beam addressed, microchannel spatial light modulator. *Proc. SPIE*, 465, 23–28.
18. Moddel, G. 1995. Ferroelectric liquid crystal spatial light modulators. In *Spatial Light Modulator Technology*, ed. U. Efron (New York: Dekker), pp. 287–359.
19. Chigrinov, V.G. 1999. *Liquid Crystal Devices* (Boston: Artech House).
20. Huang, X.Y., Miller, N., Khan, A., Davis, D., Doane, J.W., and Yang, D.K. 1998. Greyscale of bistable reflective cholesteric displays. *SID '98 Digest*, pp. 810–813.

21. Hornbeck, L.J. 1990. Deformable mirror spatial light modulator. In *Spatial Light Modulators and Applications III*, ed. U. Efron (Proc. SPIE, vol 1150), pp. 86–103.

22. Van Kessel, P.F. and Hornbeck, L.J. 1998. A MEMS-based projection system. *Proc. IEEE*, 86, 1686–1704.

23. Bloom, D.M. 1997. Grating light valve: Revolutionizing display technology. *Proc. SPIE*, 3013, 165–171.

24. Kubota, S.R. 2002. The grating light valve projector. *Opt. Photonics News*, 13, 50–60.

25. Fisher, A.D., Ling, L.C., Lee, J.N., and Fukuda, R.C. 1986. Photoemitter membrane light modulator. *Opt. Eng.*, 25, 261–268.

26. Hornbeck, L.J. 1983. *IEEE Trans. Electron Dev.* 30, 539.

27. Pape, D.R. 1984. An optically addressed membrane spatial light modulator. *Proc. SPIE*, 465, 17–22.

28. Kingston, R.H., Burke, B.E., Nichols, K.B., and Leonberger, F.J. 1982. Spatial light odulation using electro-absorption in GaAs CCD. *Appl. Phys. Lett.*, 41, 413–415.

29. Kingston, R.H., Burke, B.E., Nichols, K.B., and Leonberger, F.J. 1984. An electro-absorptive CCD spatial light modulator. *Proc. SPIE*, 465, 9–11.

30. Miller, D.A.B. 1987. Quantum wells for optical information processing. *Opt. Eng.*, 26, 368.

31. Mast, F. and Waser, R. 1980. Optischer bild-verstärker. European Patent Specification 0029-006 (Gretag AG).

32. Hess, K., Dändliker, R., and Thalman, R. 1987. Deformable surface spatial light modulator. *Opt. Eng.*, 26, 418–422.

33. Lea, M.C. 1983. *Appl. Phys. Lett.*, 43, 738.

34. Lea, M.C. 1984. Optical modulators based on electro-capillarity. *Proc. SPIE*, 465, 12–16.

35. Horwitz, B.A. and Corbett, F.J. 1978. *Opt. Eng.*, 17, 353.

36. Petrov, M.P. 1980. Diffractive and dynamic properties of photosensitive electro-optic media. *Am. Inst. Phys. Conf. Proc.*, 65, 493–507.

37. Casasent, D. 1981. Soviet PRIZ spatial light modulator. *Appl. Opt.*, 20, 3090–3092.

38. Petrov, M.P. 1984. Physical basis of operation of the PRIZ spatial light modulator. *Optik*, 67, 247–256.

39. Marrakchi, A., Tanguay, A.R., Yu, J., and Psaltis, D. 1985. *Opt. Eng.*, 24, 124.

40. Psaltis, D., Yu, J., Marrakchi, A., and Tanguay, A.R. 1984. Photorefractive inco-herent to coherent optical conversion. *Proc. SPIE*, 465, 2–8.

41. Rabinovich, W.S., Bowman, S.R., Katzer, D.S., and Kyono, C.S. 1995. Intrinsic MQW spatial light modulators. *Appl. Phys. Lett.*, 66, 1044–1046.

42. Lahiri, I., Kwolek, K.M., Nolte, D.D., and Melloch, M.R. 1995. Photorefractive p–i–n diode MQW spatial light modulator. *Appl. Phys. Lett.*, 67, 1408–1410.

43. Pape, D.R. 1995. Acousto optic Bragg cell devices. In *Spatial Light Modulator Technology*, ed. U. Efron (New York: Dekker), pp. 415–442.

44. Jain, F.C., and Bhattacharjee, K.K. 1989. MQW optical modulator structures using surface acoustic wave-induced Stark effect. *IEEE Photon. Technol. Lett.*, 1, 307–309.

45. Davis, J.A. and Waas, J.M. 1990. Current status of the magneto-optic spatial light modulator. *Proc. SPIE*, 1150, 27–45.

46. Strome, D.H. 1984. Cinematic infrared scene simulator based on vanadium dioxide spatial modulator. *Proc. SPIE*, 465, 192–196.

47. Sasaki, A. et al. 1980. *Proc. Soc. Inf. Display*, 21, 341.

48. Wallace, J. 2000. Laser Focus World, 36 (5).

49. Xu, H., Davey, A.B., Wilkinson, T.D., and Crossland, W.A. 1999. Plasmon effect: Optically enhancing the small electro-optical effect of a fast-switching liquid crystal mixture. *Opt. Eng.*, 39, 1568–1572.

50. Leonard, S.W., Mondia, J.P., VanDriel, H.M., Toader, O., John, S., Busch, K., Birner, A., Gosele, U., and Lehman, V. 2000. Tunable two-dimensional photonic crystals using liquid crystal infiltration. *Phys. Rev. B*, 61, R2389–R2392.

51. Friend, R.H. 1998. Organic electrolumi-nescent displays. *Soc. Inf. Display Lecture Notes*, 2, 1–27.

52. Burroughes, J.H. et al. 1990. Light emit-ting diodes based on conjugated polymers. *Nature*, 347, 539–541.

53. Iga, K., Koyama, F., and Kinoshita, S. 1988. VCSEL array surface emitting semiconductor lasers. *IEEE J. Quant. Electron.*, 24, 1845–1855.

54. Chang-Hasnain, C.J. et al. 1991. Multiple wavelength tunable, surface emitting laser arrays. *IEEE J. Quant. Electron.*, 27, 1368–1376.

55. Efron, U. 1991. Liquid crystals materials devices and applications. In *Handbook of Microwave and Optical Components*, vol 4, ed. K. Chang (New York: John Wiley & Sons), pp. 372–374.

56. Efron, U. 1991. Liquid crystals materials devices and applications. In *Handbook of Microwave and Optical Components*, vol 4, ed. K. Chang (New York: John Wiley & Sons), pp. 370–371.

57. Grinberg, J. et al. 1975. *Opt. Eng.*, 14, 217.

58. Chigrinov, V.G. 1999. *Liquid Crystal Devices* (Boston: Artech House), pp. 234–240.

59. Clark, N.A. and Lagerwall, T. 1980. *Appl. Phys. Lett.*, 36, 899.

60. Fergason, J.L. 1985. *Soc. Inf. Display Digest*, 16, 68.

61. Vaz, N.A. 1989. *Proc. SPIE*, 1080, 2.

62. Ichikawa, H., Kataoka, H., Oka, T., Iida, M., Fujioka, T., and Ino, M. 2000. Low power poly-Si reflective colour AMLCD with 1024 × 480 pixels. *Soc. Inf. Display Digest*, 31, 1203–1207.

63. Mishima, Y., Nakayama, T., Suzuki, N., Ohta, M., Endoh, S., Iwakabe, Y., and Kagawa, H. 2000. Development of a 19-in.-diagonal UXGA super TFT-LCM applied with super-IPS technology. *Soc. Inf. Display Digest*, 19, 260–265.

64. Schadt, M., Scmitt, K., Kozinkov, V., and Chigrinov, V.G. 1992. Photo-polymer alignment methods. *Jpn. J. Appl. Phys.*, 31, 2155.

65. de Gennes, P.G. and Prost, J. 1993. *The Physics of Liquid Crystals* (Oxford: Oxford University Press).

66. Blinov, L.M. 1983. *Electro-Optical and Magneto-Optical Properties of Liquid Crystals* (New York: John Wiley & Sons).

67. Chandrasekhar, S. 1992. *Liquid Crystals*, 2nd edn. (Cambridge: Cambridge University Press).

68. Efron, U. 1991. Liquid crystals materials devices and applications. In *Handbook of Microwave and Optical Components*, vol 4, ed. K. Chang (New York: John Wiley & Sons).

69. Chigrinov, V.G. 1999. *Liquid Crystal Devices* (Boston: Artech House).

70. Khoo, I.C. and Wu, S.T. 1993. *Optics and Non-Linear Optics of Liquid Crystals* (Singapore: World Scientific).

71. Chigrinov, V.G. 1999. *Liquid Crystal Devices* (Boston: Artech House), pp. 100–134.

72. Janning, J.L. 1972. Thin film surface orientation for liquid crystals. *Appl. Phys. Lett.*, 21, 173.

73. Becker, M.E. et al. 1986. Alignment properties of rubbed polymer surfaces. *Mol. Cryst. Liq. Cryst.*, 132, 167–180.

74. Schadt, M. et al. 1992. Surface induced parallel-alignment of liquid crystals by linearly polarized photopolymers. *Jpn. J. Appl. Phys.*, 31, 2155–2164.

75. Kataoka, S., Taguchi, Y., Iimura, Y., Kobayashi, S., Hasebe, H., and Takatsu, H. 1997. *Mol. Cryst. Liq. Cryst.*, 292, 333.

76. Ikeno, Y. et al. 1988. Electrooptic bistability of a ferroelectric liquid crystal device prepared using polyimide Langmuir-Blodgett orientation films. *Jpn. J. Appl. Phys.*, 27, L475.

77. Lien, S.-C.A. et al. 1998. Active-matrix display using ion-beam-processed polyimide film for liquid crystal alignment. *IBM J. Res. Dev.*, 42 (3), 537–542.

78. Nakamura, M. and Ura, M. 1981. Alignment of nematic liquid crystals on ruled grating surfaces. *J. Appl. Phys.*, 52, 210.

79. Uchida, T. et al. 1980. *Jpn. J. Appl. Phys.*, 19, 2127.

80. Seki, H. et al. 1990. Tilted-homeotropic alignment of liquid crystal molecules using the rubbing method. *Jpn. J. Appl. Phys.*, 29, L2236–L2238.

81. Lackner, A.M. et al. 1990. *Dig. Soc. Inf. Display SID 90*, XXI, 98.

82. Bleha, W.P. 2000. D-ILA technology for electronic cinema. *Dig. Soc. Inf. Display SID 2000*, XXXI, 310–313.

83. Miller, L. et al. 1991. Method for tilted alignment of liquid crystals with improved photostability, US Patent specification 5011267.

84. Lu, M. et al. 2000. Homeotropic alignment by single oblique evaporation of SiO2 and its application to high resolution microdisplays. *Dig. Soc. Inf. Display SID 2000*, XXXI, 446–449.

85. Yoshida, H. and Koike, K. 1997. *Jpn. J. Appl. Phys.*, 36, L428–L431.

86. Furumi, S. et al. 1999. *Appl. Phys. Lett.*, 74, 2438–2440.

87. Park, B. et al. 1999. *J. Appl. Phys.*, 86, 1854–1859.

88. Kimura, M. 2000. New photo alignment technology based on (4-chalconyloxy) alkyl groups. *Dig. Soc. Inf. Display SID 2000*, XXXI, 438–441.

89. Gooch, C.H. and Tarry, H.A. 1975. *J. Phys. D*, 8, 1575.

90. Efron, U. and Wu, S.T. 1985. *Opt. Eng.*, 24, 111.

91. Konforti, N., Marom, E., and Wu, S.T. 1988. Phase only modulation with twisted nematic liquid crystal–spatial light modulators. *Opt. Lett.*, 13, 251–253.

92. Blinov, L.M. 1983. *Electro-Optical and Magneto-Optical Properties of Liquid Crystals* (New York: John Wiley & Sons), Chapter 6.

93. de Vries, H. 1951. *Acta Crystallogr.*, 4, 219.

94. Dierking, I. et al. 1996. *J. Appl. Phys.*, 81, 3007–3014.

95. Yang, D.-K. et al. 1994. *J. Appl. Phys.*, 76, 1331–1333.

96. Clark, N.A. and Lagerwall, S.T. 1980. *Appl. Phys. Lett.*, 36, 899.

97. Displaytech FLC Devices.

98. Garoff, S. and Meyer, B. 1979. Electroclinic effect at the A–C phase change in a chiral smectic liquid crystal. *Phys. Rev. A*, 19, 338–347.

99. Funfschilling, J. and Schadt, M. 1989. *J. Appl. Phys.*, 66, 3877–3882.

100. Fergason, J.L. 1985. *SID Int. Symp. Dig. Tech. Papers*, 16, 68.

101. Doane, J.W., West, J.L., Golemme, A., Whitehead, J.B. Jr., and Wu, B.-G. 1988. *Mol. Cryst. Liq. Cryst.*, 165, 511.

102. Crawford, G.P. 1996. Liquid crystal polymer dispersions for reflective flat-panel displays. *Soc. Inf. Display Tech. Digest*, F/4-1–F/4-47.

103. Kitzerov, H.S. 1994. Polymer dispersed liquid crystals, from the nematic curvilinear phase to ferroelectric films. *Liq. Cryst.*, 16, 1–31.

104. Coates, D. 1995. Polymer dispersed liquid crystals. *J. Mater. Chem.*, 5, 2063–2072.

105. Xu, M. and Yang, D. 1999. *PSCT. Proc. SID*, 950–953.

106. *Soc. Inf. Display, Int. Symp. Digest* 1996; *Soc. Inf. Display, Int. Symp. Digest* 1997; *Soc. Inf. Display, Int. Symp. Digest* 1998; *Soc. Inf. Display, Int. Symp. Digest* 1999; *Soc. Inf. Display, Int. Symp. Digest* 2000.

107. Grinberg, J. et al. 1975. *Opt. Eng.*, 14, 217.

108. Displaytech Co. (Longmont, CO, USA), LightView—QVGA. Display Module, Model QDM-0076-MV5. www.Displaytech.com.

109. Kopin Co. (Taunton, MA). CyberDisplay-1280-Mono. www.kopin.com.

110. Akimoto, O. and Hashimoto, S. 2000. A 0.9-in UXGA/HDTV FLC micro-display. *Soc. Inf. Display, Tech. Digest*, 31, 194–197.

111. Date, M., Hisaki, T., Naito, N., Nakadaira, A., Suyama, S., Tanaka, H., Uehira, K., and Koshiishi, Y. 2000. Direct-viewing display using alignment-controlled PDLC and holographic PDLC. *Soc. Inf. Display Tech. Digest*, 31, 1184–1188.

112. Efron, U., Sayyah, K., Byles, W.R., Goodwin, N.W., Forber, R.A., Wu, C.S., and Welkowsky, M.S. 1991. The CCD-addressed liquid crystal light valve—An update. *Proc. SPIE*, 1455, 237–247.

113. Sterling, R.D. and Bleha, W.P. 2000. D-ILA™ technology for electronic cinema. *Soc. Inf. Display Digest*, 31, 310–313.

114. Sayyah, K., Efron, U., and Forber, R.A. 1995. Color-sequential crystalline Si-LCLV-based projector for consumer HDTV. *Proc. Soc. Inf. Displays (SID) Digest*, 520.

115. Kanichi, J., ed. 1991. Amorphous Si for optically-addressed spatial light modulators. In *Amorphous and Microcrystalline Semiconductor Devices: Optoelectronic Devices* (Norwood, MA: Artech House), pp. 369–412.

116. Aubourg, P., Huignard, J.P., Hareng, M., and Mullen, R.A. 1982. Liquid crystal light valve using bulk, monocrystalline $Bi_{12}SiO_{20}$. *Appl. Opt.*, 21, 3706–3712.

117. Moddel, G. 1995. Ferroelectric LC spatial light modulators. In *Spatial Light Modulator Technology*, ed. U. Efron (New York: Dekker), pp. 310–311.

118. Hornbeck, L.J. 1983. *IEEE Trans. Electron. Devices*, 30, 539.

119. Brooks, R.E. 1984. Micro-mechanical light modulators for data transfer and processing. *Proc. SPIE*, 465, 46–54.

120. Hornbeck, L.J. 1990. Deformable mirror spatial light modulator. *Proc. SPIE*, 1150, 86–102.

121. Van Kessel, P.F., and Hornbeck, L.J. 1998. A MEMS-based projection system. *Proc. IEEE*, 86, 1686–1704.

122. Bloom, D.M. 1997. Grating light valve: Revolutionizing display technology. *Proc. SPIE*, 3013, 165–171.

123. Kubota, S.R. 2002. The grating light valve projector. *Opt. Photonics News*, 13, 50–60.

124. Hwang, K.-H., Song, Y.-J., and Kim, S.-G. 1998. Thin film micromirror array for high-brightness projection displays. *Jpn. J. Appl. Phys.*, 37, 7074–7077.

125. Sakarya, S., Vdovin, G., and Sarro, P.M. 2002. Technology of reflective membranes for spatial light modulators. *Sens. Actuators A*, 97–98, 468–472.

126. Riesenberg, R. 2001. Micro-mechanical slit positioning system as transmissive SLM. *Proc. SPIE*, 4457, 197–204.

127. Chemla, S. et al. 1983. *Appl. Phys. Lett.*, 42, 864–865.

128. Miller, A.B. et al. 1983. *Appl. Phys. Lett.*, 42, 925.

129. Kingston, R.H. et al. 1982. Spatial light modulator using electro-absorption in a GaAs CCD. *Appl. Phys. Lett.*, 41, 413–415.

130. Hsu, T.Y. et al. 1988. *Opt. Eng.*, 27, 372–384.

131. Miller, D.A.B. et al. 1985. *Phys. Rev. B*, 32, 1043–1060.

132. Chang, Y.C., Schulman, J.N., and Efron, U. 1987. *J. Appl. Phys.*, 62, 4533.

133. Ploog, K. and Döhler, G.H. 1983. Compositional and doping superlattices in III-V semiconductors. *Adv. Phys.*, 32, 285–359.

134. Döhler, G.H. 1986. Light generation, modulation and amplification by nipi doping superlattices. *Opt. Eng.*, 25, 211–218.

135. Kiesel, P. et al. 1993. High speed and high contrast electro-optic modulators based on nipi doping superlattices. *Superlattices Microstruct.*, 13, 21–24.

136. Wu, W.-Y. et al. 1987. *Appl. Phys. Lett.*, 51, 710–712.

137. Miller, D.A.B. et al. 1986. *Appl. Phys. Lett.*, 49, 821.

138. Efron, U. and Livescu, G. 1995. Multiple quantum well spatial light modulators. In *Spatial Light Modulator Technology*, ed. U. Efron (New York: Dekker).

139. Yan, R.H., Simes, R.J., and Coldren, L.A. 1989. Electroabsorptive Fabry–Perot reflection modulators with asymmetric mirrors. *IEEE Photon. Technol. Lett.*, 1, 273.

140. Lin, C.H. et al. 1994. *Appl. Phys. Lett.*, 65, 1242.

141. Hsu, T.Y., Efron, U., and Wu, W.Y. 1988. Amplitude and phase modulation in a 4-|j, m-thick GaAs/AlGaAs multiple quantum well modulator. *Electron. Lett.*, 24, 603–604.

142. Efron, U. and Livescu, G. 1995. Multiple quantum well spatial light modulators: Materials devices and applications. In *Spatial Light Modulators and Applications*, ed. U. Efron (New York: Dekker), pp. 243–247.

143. Hsu, T.Y. and Efron, U. 1989. Review of multiple quantum well spatial light modulators. *Proc. SPIE*, 1150.

144. Ahearn, J.S. et al. 2001. Multiple quantum well spatial light modulators for optical data processing and beam steering applications. *Proc. SPIE*, 4457, 43–53.

145. Livescu, G. et al. 1988. *Opt. Lett.*, 13, 297.

146. Hsu, T.Y. and Efron, U. 1989. *Proc. SPIE*, 1150, 80.

147. Worchesky, T.L. et al. 1996. *Appl. Opt.*, 35, 1180–1186.

148. Trezza, J.A. et al. 1998. *Proc. SPIE*, 3490, 78–81.

149. Kang, K. et al. 1999. *Proc. SPIE*, 3715, 97–107.

150. Krishnamoorthy, A.V. et al. 1999. *IEEE J. Sel. Top. Quant. Electron.*, 5, 261–275.

151. Bowman, S.R. et al. 1998. *J. Opt. Soc. Am. B*, 15, 640–647.

152. Walker, A.C. et al. 1999. *IEEE J. Sel. Top. Quant. Electron.*, 5, 236–249.

153. Junique, S. et al. 2001. MQW-SLM for optical information processing. *Proc. SPIE*, 4457, paper No. 10.
154. Lenslet Laboratories-Ramat-Gan Israel: Private Communication 2002. www.lenslet.com.
155. Efron, U., ed. 1984. *Spatial Light Modulators and Applications*, vol 465 (Bellingham, WA: SPIE), pp. 2–9, 23–29, 82–97.
156. Efron, U. and Warde, C., eds. 1986. Materials and devices for optical information processing: Special issue. *Opt. Eng.*, 25, 250–261.
157. Efron, U., ed. 1987. *Spatial Light Modulators and Applications*, vol 825 (Bellingham, WA: SPIE), pp. 106–113, 88–94, 198–206.
158. Esener, S. 1995. Smart pixels: Technology and applications to parallel computing. In *Spatial Light Modulator Technology*, ed. U. Efron (New York: Dekker), pp. 449–453.
159. Yariv, A. and Yeh, P. 1984. *Optical Waves in Crystals* (New York: Wiley-InterScience), p. 231.
160. Esener, S.C., Wang, J.H., Drabik, T.J., Title, M.A., and Lee, S.H. 1987. One dimensional silicon-PLZT spatial light modulator. *Opt. Eng.*, 26, 406–413.
161. Debye, P. and Sears, F.W. 1932. *Proc. Natl Acad. Sci. USA*, 18, 409–414.
162. Saleh, B.E.A. 1991. *Fundamentals of Photonics* (New York: Wiley-InterScience), Chapter 20.
163. Yariv, A. and Yeh, P. 1984. *Optical Waves in Crystals* (New York: Wiley-InterScience), p. 231. Chapter 8.
164. Pape, D. 1995. Acousto-optics Bragg-cell devices. In *Spatial Light Modulator Technology*, ed. U. Efron (New York: Dekker).
165. Pulliam, G.R., Ross, W.E., MacNeal, B.E., and Bailey, R.F. 1982. *J. Appl. Phys.*, 53, 2754.
166. Davis, J.A. and Waas, J.M. 1990. Current status of the magneto-optic spatial light modulator. *Proc. SPIE*, 1150, 27–45.
167. Ross, W.E., Psaltis, D., and Anderson, R.H. 1983. Two-dimensional magneto-optic spatial light modulator for signal processing. *Opt. Eng.*, 22, 485–490.
168. Park, J., Cho, J., Nishimura, K., and Inoue, M. 2002. Magneto-optic spatial light modulator for volumetric digital recording system. *Jpn. J. Appl. Phys.*, 41, 1813–1816.
169. Park, J., Cho, J., Nishimura, K., and Inoue, M. 2002. New drive line shape for reflective magneto-optic spatial light modulator. *Jpn. J. Appl. Phys.*, 41, 2548–2551.
170. Ashkin, A. et al. 1966. *Appl. Phys. Lett.*, 9, 72.
171. Glass, A.M. et al. 1972. *Nat. Bur. Stand. Spec. Publ.*, 372, 15.
172. Guenter, P. 1982. *Phys. Rev.*, 93, 199.
173. Wood, G.L. et al. 1995. Photorefractive materials. In *Spatial Light Modulator Technology*, ed. U. Efron (New York: Dekker).
174. Yeh, P. 1989. *Proc. SPIE*, 825, ed. U. Efron, pp. 96–100.
175. For a more comprehensive review of the use of PR materials for phase conjugation see e.g., Feinberg, J. 1985. Optical phase conjugation in photo-refractive materials. In *Optical Phase Conjugation*, ed. R.A. Fisher (New York: Academic).
176. This technology area has been extensively published. For a recent article see e.g., Burr, G.W. et al. 2001. Volume holographic storage at an area density of 250 Gpixels/in². *Opt. Lett.*, 26, 444–446.
177. Iemmi, C. and La Mela, C. 2002. Phase only photo-refractive joint transform correlator. *Opt. Commun.*, 209, 255–263.
178. Choiou, A.E. and Yeh, P. 1986. *Opt. Lett.*, 11, 461.
179. O'Meara, T.R., Pepper, D.M., and White, J.O., 1985. Applications of nonlinear, optical phase conjugation. In *Optical Phase Conjugation*, ed. R.A. Fisher (New York: Academic).
180. Yau, H.F., Lee, H.Y., and Cheng, N.J. 1999. *Appl. Phys. B*, 68, 1055.
181. Delaye, P. and Roosen, G. 1999. Evaluation of a photo-refractive two-beam coupling novelty filter. *Opt. Commun.*, 165, 133–151.
182. Marrakchi, A., Tanguay, A.R. Jr., Wu, J., and Psaltis, D. 1984. *Proc. SPIE*, 465, ed. U. Efron, pp. 82–96.
183. Ma, J., Liu, L., Wu, S., Wang, Z., and Xu, L. 1989. *Opt. Lett.*, 14, 572.
184. Sharp, E.J., Wood, G.L., Clark, W.W. III, Salamo, G.J., and Neurgaonkar, R.R. 1992. *Opt. Lett.*, 17, 207.
185. Bowman, S.R., Rabinovich, W.S., Beadie, G., Kirkpatrick, S.M., Katzer, D.S., Ikossi-Anastasiou, K., and Adler, C.L. 1998. *J. Opt. Soc. Am. B*, 15, 640–647.

186. Esener, S. 1995. Smart pixels: Technology and applications to parallel computing. In *Spatial Light Modulator Technology*, ed. U. Efron (New York: Dekker).

187. Krishnammorthy, A. et al. 1997. Progress in optoelectronic VLSI smart pixel technology based on GaAs/GaAlAs MQW modulators. *Int. J. Optoelectron.*, 11, 181–198.

188. Lin, T.H. et al. 1990. *Appl. Opt.*, 29, 1595.

189. Cotter, L.K., Drabnik, T.J., Dillon, R.J., and Handschy, M.A. 1990. *Opt. Lett.*, 15, 291.

190. D'Asaro, L.A., Chirovsky, L.M.F., Laskowski, E.J., Pei, S.S., Woodward, T.K., Lentine, A.L., Leibenguth, R.E., Focht, M.W., Freund, J.M., Guth, G.D., and Smith, L.E. 1993. Batch fabrication and operation of GaAs/AlGaAs field effect transistor self-electro-optic effect device (FET-SEED) smart pixel arrays. *IEEE J. Quant. Electron.*, 29, 670–675.

191. Neff, J. et al. http://wwwocs.colorado.edu/research/fsoi/research6.html.

192. Leibowitz, B., Boser, B.E., and Pister, K.S.J. CMOS 'smart pixel' for free-space optical communication. In *Proceedings of SPIE—The International Society for Optical Engineering*, vol 4306A (Electronic Imaging '01), San Jose, CA, January 2001.

193. Efron, U., Sayyah, K., Byles, W.R., Goodwin, N.W., Forber, R.A., Wu, C.S., and Welkowsky, M.S. 1991. The CCD-addressed liquid crystal light valve—An update. *Proc. SPIE*, 1455, 237–247.

194. Efron, U., Davidov, I., Sinelnikov, V., and Friesem, A. 2001. CMOS/LCOS-based image transceiver device. *Proc. SPIE*, 4457, 188–196.

195. Walker, A.C. et al. 1998. Opto-electronic systems based on InGaAs-complementary metal oxide semiconductor smart pixel arrays and free space optical interconnects. *Appl. Opt.*, 37, 2822–2830.

196. Krishnamoorthy, A.V. et al. 1997. Dual function detector–modulator smart pixel. *Appl. Opt.*, 37, 4866–4870.

197. Zhang, L., Hong, S., Min, C., Alpasan, Z.Y., and Sawchuk, A.A. 2001. Optical multi-token-ring networking using smart pixels with field programmable gate arrays. *Proc. SPIE*, 4470.

198. Kane, J.S., Kincaid, T.G., and Hemmer, P. 1998. Optical processing with feedback using smart pixel spatial light modulator. *Opt. Eng.*, 37, 942–947.

199. Cassinelli, A., Chavel, P., and Desmulliez, M.P.Y. 2001. Dedicated optoelectronic stochastic parallel processor for real time image processing. *Appl. Opt.*, 40, 6479–6491.

200. Wu, J.-M., Kunzia, C.B., Hoanca, B., Chen, C.-H., and Sawchuk. A.A. 1999. Demonstration and architectural analysis of CMOS/MQW smart pixel cellular logic processor for SIMD parallel pipeline processing. *Appl. Opt.*, 39, 2270–2281.

201. Shoop, B.L. and Das, P. 2002. Mismatch-tolerant distributed photonic analog-to-digital conversion using spatial oversampling and spectral noise shaping. *Opt. Eng.*, 41, 1674–1687.

202. Kane, J.S. 1998. Smart pixel feedforward neural network. *IEEE Trans. Neural Netw.*, 9, 159–164.

203. Agranat, A.J. et al. 1989. Opt. Lett., 14, 1017.

204. Agranat, A.J. et al. 1992. Opt. Lett., 17, 713.

205. Agranat, A.J. 1999. IEEE-LEOS Summer Topical on WDM Components.

206. Yablonovitch, E. 1987. *Phys. Rev. Lett.*, 58, 2059.

207. John, S. 1987. *Phys. Rev. Lett.*, 58, 2486.

208. Yoshino, K. et al. 1999. Tunable optical stop-band and reflection peak in synthetic opal infiltrated with liquid crystal and conducting polymer as photonic crystal. *Jpn. J. Appl. Phys.*, 38, L961–L963.

209. Leonard, S.W. et al. 2000. Tunable, two-dimensional photonic crystals using liquid crystal infiltration. *Phys. Rev. B*, 61, R2389–R2392.

210. Wallace, J. 2000. Laser Focus World, 36.

211. Lindvold, R.L. and Lausen, H. 1997. Projection display based on optically-addressed SLM using bacteriorhodopsin thin film. *Proc. SPIE*, 3013, 202–213.

212. Reddy, K.P.J. 1997. Analysis of bacteriorhodopsin and its applications in photonics. *Proc. SPIE*, 3211, 2–13.

213. Piuno, P.A.E., Krull, U.J., Hudson, R.H.E., Damh, M.J., and Cohen, H. 1995. Fibre-optic DNA sensor for fluorometric nucleic acid determination. *Anal. Chem.*, 67, 2635–2643.

214. Abel, A.P., Weller, M.G., Duveneck, G.L., Ehrat, M., and Widmer, H.M. 1996. Fibre-optic evanescent wave bio-sensor for the detection of oligonucleotides. *Anal. Chem.*, 68, 2905–2912.

215. Holst, G., Glud, R.N., Kuehl, M., and Klimant, I. 1997. A micro-optode array for fine scale measurements of oxygen distribution. *Sens. Actuators B*, 38/39, 122–129.

216. Illic, B., Czaplewski, D., and Craighead, H.G. 2000. Mechanical resonant immunospecific biological detector. *Appl. Phys. Lett.*, 77, 450–452.

217. Rowe-Taitt, C.A., Hazzard, J.W., Hoffman, K.E., Cras, J.J., Golden, J.P., and Ligler, F.S. 2000. Simultaneous detection of six bio-hazardous agents using a planar waveguide array biosensor. *Biosens. Bioelectron.*, 15, 579–589.

218. O'Brien, M.J. II, Perez-Luna, V.H., Brueck, S.R.J., and Lopez, G.P. 2001. A surface plasmon resonance array biosensor based on spectroscopic imaging. *Biosens. Bioelectron.*, 16, 97–108.

219. Cunningham, B., Lin, B., Qiu, J., Li, P., Pepper, J., and Hugh, B. 2002. A plastic colorimetric resonant optical biosensor for multiparallel detection of label-free biochemical interactions. *Sens. Actuators B*, 85, 219–226.

220. Lundgren, J.S., Watkins, A.N., Racz, D., and Ligler, F.S. 2000. A liquid crystal pixel array for signal discrimination in array biosensors. *Biosens. Bioelectron.*, 15, 417–421.

221. Collings, N., Latham, S.G., Chittick, R.C., and Crossland, W.A. 1990. Reconfigurable optical interconnect using an optically addressed light valve. *Int. J. Opt. Comput.*, 1, 31–40.

222. Patel, J.S. and Silberberg, Y. 1995. Liquid crystal and grating-based multiple wavelength cross-connect switch. *IEEE Photon. Technol. Lett.*, 7, 514–516.

223. Manolis, I.G., Wilkinson, T.D., Redmond, M.M., and Crossland, W.A. 2002. Reconfigurable multi-level phase holograms for optical switches. *IEEE Photon. Technol. Lett.*, 14, 801–803.

224. Wilkinson, T.D. and Crossland, W.A. 2001. Optical routing with LC arrays. *Proc. SPIE*, 4534, 64–69.

225. Crossland, W.A. et al. 2000. Holographic optical switching: The ROSES demonstrator. *J. Lightwave Technol.*, 18, 1845–1854.

226. Crossland, W.A., Holmes, M.J., Robertson, B., and Wilkinson, T.D. 2000. Liquid crystal polarization independent beam steering switches for operation at 1.5 microns. In *LEOS 2000*, vol 1 (Piscataway: IEEE) pp. 46–47.

227. Apter, B., Acco, S., and Efron, U. 2001. A study of LC-based sub-wavelength diffractive optical elements for optical cross-connect applications. *Proc. SPIE*, 4457, 20–30.

228. Kannie, T. et al. 2002. A highly dense MEMS optical switch array integrated with planar lightwave circuit. In *International Conference on Micro-Electro-Mechanical Systems (MEMS)* (Piscataway: IEEE), pp. 560–563.

229. Bakke, T., Tigges, C.P., and Sullivan, C.T. 2002. 1 × 2 MOEMS switch based on silicon on insulator and polymeric waveguides. *Electron. Lett.*, 38, 177–178.

230. Haven, V.S. et al. 2001. Recent advances in bubble-actuated cross-connect switches. In *CLEO/Pacific Rim—Technical Digest*, vol 1 (Piscataway: IEEE), pp I-414–I-415.

231. Chen, D. et al. An optical cross-connect based on micro-bubbles. In Micro-Electro-Mechanical Systems (MEMS), *ASME International Mechanical Engineering Congress*, ASME 2000, pp. 35–37.

232. Walker, A.C. et al. 1998. *Appl. Opt.*, 37, 2822–2830.

233. McCarthy, A., Tooley, F.A., Laprise, E., Plant, D.V., Kirk, A.G., Oren, M., and Lu, Y. 2000. Free space optical interconnect system using polarization-rotating modulator arrays. *Proc. SPIE*, 4089, 272–277.

234. Liu, Y. 2002. Heterogeneous integration of OE arrays with Si electronics andmicrooptics. *IEEE Trans. Adv. Packag.*, 25, 43–49.

235. Kasahara, K.M., Kim, T.J., Neilson, D.T., Ogura, I., Redmond, I., and Schefeld, E. 1998. Wavelength division multiplexing free-space optical interconnect networks for massively parallel processing systems. *Appl. Opt.*, 37, 3746–3755.

236. Tuantranont, A., Bright, V.M., Zhang, J., Zhang, W., Neff, J.A., and Lee, Y.C. 2001. Optical beam steering using MEMS-controllable microlens array. *Sens. Actuators A*, 91, 363–372.

237. Huang, D. et al. 2001. Free space optical interconnection of 3-D opto-electronic VLSI chip stacks. *Proc. SPIE*, 4292, 95–104.

238. White, H.J. et al. 1999. Optically connected parallel machine: Design, performance and application. *IEE Proc. Opto-Electron.*, 146, 125–136.

239. Weiner, A.M., Heritage, J.P., and Kirschner, E.M. 1988. *J. Opt. Soc. Am. B*, 5, 1563.

240. Weiner, A.M. 2000. Femto-second pulse shaping using spatial light modulators. *Rev. Sci. Instrum.*, 71, 1929–1960.

241. Wefers, M.M., Nelson, K.A., and Weiner, A.M. 1996. *Opt. Lett.*, 21, 746.

242. Nuss, M.C. and Morrison, R.L. 1995. *Opt. Lett.*, 20, 740.

243. Karasawa, N., Li, L., Suguro, A., Shigekawa, H., Morita, R., and Yamashita, M. 2001. Optical pulse compression to 5 fs by use of only a spatial light modulator for phase conjugation. *J. Opt. Soc. Am. B*, 18, 1742–1746.

244. Bleha, W.P. 2000. D-ILA technology for electronic cinema. *Dig. Soc. Inf. Display (SID)*, 2000, 310–313.

245. Van Kessel, P.F. and Hornbeck, L.J. 1998. A MEMS-based projection system. *Proc. IEEE*, 86, 1686–1704.

246. Sayyah, K., Efron, U., and Forber, R.A. 1995. Color-sequential crystalline Si-LCLV-based projector for consumer HDTV. *Proc. Soc. Inf. Displays (SID) Digest*, 520.

247. Morrissy, J.H., Pfeiffer, M., Schott, D., and Vithana, H. 1999. Reflective microdisplays for projection or virtual-view applications. *Soc. Inf. Display (SID) Digest*, 808–811.

248. Gleckman, P., and Schuck, M. 2001. Optical characteristics of a high performance LCoS virtual display. *Proc. Soc. Inf. Display (SID)*, XXXII, 62–65.

249. Spitzer, M.B., Zavracky, P.M., Crawford, J., Aquilino, P., and Hunter, G. 2001. Eyewear platforms for miniature displays. *Proc. Soc. Inf. Display (SID)*, XXXII, 258–261.

250. Peli, E., Reed Hedges, T., Tang, J., and Landmann, D. 2001. A binocular stereo-scopic display system with coupled convergence and accommodation demands. *Proc. Soc. Inf. Display (SID)*, XXXII, 1296–1299.

251. Haseltine, E.C. 2000. Displays for location based entertainment. *Proc. Soc. Inf. Display (SID)*, XXXI, 962–965.

252. Garibyan, O.V. et al. 1981. Optical phase conjugation by microwatt power of reference wave via liquid crystal light valve. *Opt. Commun.*, 38, 67–70.

253. Marom, E. and Efron, U. 1987. Phase conjugation of low power optical beams using liquid crystal light valves. *Opt. Lett.*, 12, 504–506.

254. Johnson, K.M. et al. 1990. High speed, low-power optical phase conjugation using a hybrid amorphous silicon/ferroelectric liquid crystal device. *Opt. Lett.*, 15, 1114–1116.

255. Shirai, T., Barnes, H., and Haskell, T.G. 2001. Real time restoration of blurred image with liquid crystal adaptive optics system, based on all-optical feedback interferometry. *Opt. Commun.*, 188, 275–282.

256. Neil, M.A.A. et al. 2002. Active aberration correction for the writing of three-dimensional optical memory devices. *Appl. Opt.*, 41, 1374–1379.

257. Miller, D.T. 2000. Retinal imaging and vision at the frontiers of adaptive optics. *Phys. Today*, January, 31–36.

258. McHugh, T.J., and Zweig, D.A. 1989. Recent advances in binary optics. *Proc. SPIE*, 1052, 85–90.

259. Sheppard, C.J.R. 1999. Binary optics and confocal imaging. *Opt. Lett.*, 24, 305–306.

260. Matic, R. 1994. Blazed phase liquid crystal beam steering. *Proc. SPIE*, 2120, 194–205.

261. McManamon, P.F. et al. 1996. Optical phased array technology. *Proc. IEEE*, 84, 268–297.

262. Suguira, N. and Morita, S. 1993. Variable focus liquid-filled optical lens. *Appl. Opt.*, 32, 4181–4186.

263. Commander, L.G., Da, S.E., and Selviah, D.R. 2000. Variable focal length micro-lenses. *Opt. Commun.*, 177, 157–170.

264. Paige, E.G.S. and Sucharov, L.O.D. 2001. Enhancement of the imaging performance of a variable-focus Fresnel zone plate based on a single binary phase-only SLM. *Opt. Commun.*, 193, 27–38.

265. Efron, U., Wu, S.T., Grinberg, J., and Hess, L.D. 1985. Liquid crystal-based, visible to infrared dynamic image converter. *Opt. Eng.*, 24, 111–118.

266. Wu, S.T., Efron, U., and Hsu, T.Y. 1988. Near infrared to visible image conversion using a Si-liquid crystal light valve. *Opt. Lett.*, 13, 13–15.

267. Bertsch, A. et al. 1997. *Photobiology A*, 107, 275.

268. Chatwin, C. et al. 1998. UV microstereo-lithography, *Appl. Opt.*, 37, 7514–7522.

269. Yariv, A. and Yeh, P. 1984. *Optical Waves in Crystals* (New York: John Wiley & Sons), Chapter 7.

270. Saleh, B.E. and Teich, M.C. 1991. *Fundamentals of Photonics* (New York: John Wiley & Sons), pp. 928–929.

271. Lipson, S.G., Lipson, H., and Tannhauser, D.S. *Optical Physics*, 3rd edn. (Cambridge: Cambridge University Press), pp. 397–399.

272. Using the system impulse response function h(t) theory, a commonly used stability criterion is that ∫|h(t)| dt must be finite, Oppenheim, A.V., and Willsky, A.S. 1997. *Signals and System* (New York: Prentice-Hall), p. 114.

273. Butcher, P.N. and Cotter, D. 1990. *The Elements of Nonlinear Optics* (Cambridge: Cambridge University Press), Chapter 2.

274. Moss, T.S., Burrell, G.J., and Ellis, B. 1973. *Semiconductor Opto-Electronics* (New York: John Wiley & Sons).

275. Efron, U. 1991. Liquid crystals; materials, devices and applications. In *Handbook of Microwave and Optical Components*, vol 4, ed. K. Chang (New York: John Wiley & Sons), pp. 419–421.

276. Pepper, D.M., Gaeta, C.J., and Mitchell, P.V. 1994. Real-time holography, innovative adaptive optics and compensated optical processors using spatial light modulators. In *Spatial Light Modulator Technology*, ed. U. Efron (New York: Dekker), pp. 627–631.

8

Organic electroluminescent displays

EUAN SMITH
Light Blue Optics Ltd.

8.1 INTRODUCTION

Over the past few years organic light emitting diode (OLED) technology has emerged as a major player in the established field of displays. Since the demonstration of efficient light emission from small organic molecules [1] and conjugated polymers [2], built on earlier studies of organic electroluminescence in polymers [3] and small molecules [4], their potential application to displays has been technically compelling.

OLEDs have several advantages over other display technologies. Compared to inorganic semiconductors, OLEDs offer large-area manufacturing on cheap substrates with easy integration of different color devices. Compared to liquid crystal displays (LCDs), OLEDs avoid the need for separate back lighting and for color filters (although they are often still used, for example, to make RGB displays using white OLEDs), enabling thinner and lighter construction with higher power efficiency. OLEDs only emit light where needed, which improves both efficiency and contrast. OLEDs also naturally offer a good angular distribution of light, which is possible but complex for LCDs. However, an OLED is not a panacea. OLEDs arc current controlled, which requires more complex backplane circuitry (than an LCD) and tracking able to deliver higher currents to the center of the display. An OLED will never offer lower power consumption than an electrophoretic display and may not attain comparable lifetime to the discrete inorganic LEDs used in very high-brightness stadium-sized displays. Most critically, OLED technology is still relatively young and cannot yet compete on cost with the mature LCD industry because of the latter's high yields and huge volumes. However, despite recent substantial improvements, LCDs still struggle to match OLEDs for performance. Certainly you can get LCDs with high contrast, or amazingly accurate color and gray scale, or high speed, or wide viewing angles, or high efficiency, but rarely all of these at the same time.

Organic light emitting materials divide into two main structural categories (small molecules and polymers) and two light generation mechanisms (fluorescence and phosphorescence). These divisions are important for several reasons. The material structure determines how it can be applied to the substrate: solution processing for polymers, vacuum deposition for small molecules. This has significant implications for displays. The light generation mechanism determines the ultimate quantum efficiency, fluorescence having a quarter to a half the efficiency of phosphorescence [5].

This chapter aims to show how OLED devices can be applied as a display technology. There are many similarities between small molecule and polymer-based technologies, so the two are covered together, with their differences only brought out as necessary. The basic device physics, the chemical and material properties of specific organic materials, are not covered in detail here.

The first half of this chapter will discuss what distinguishes OLEDs from other display technologies, working up from the key properties of an OLED pixel to the performance of different types of display. The second half then concentrates on the construction of an OLED display panel, explaining the structure, fabrication methods, and important properties of the constituent components.

8.2 OLED DISPLAY TECHNOLOGY

An OLED is a thin-film self-emissive device that converts an applied electrical current into light.

By arraying a sufficient number of these devices and by providing a suitable means of controlling the current through such devices, a display device can be constructed. The performance of an OLED display can be entirely understood from the characteristics of the OLED devices and the method of driving.

8.2.1 OLED devices

OLEDs are thin, large-area LEDs. This statement points toward the most important properties that govern how an OLED display should be driven, i.e., an LED requires current to emit light and is, most appropriately, current controlled. That it is a thin-film large-area structure means that the transport time through the device is fast. However, it also means that it has a significant capacitance and is prone to leakage. The total electro-optic response of the OLED device is split into three sections—electrical response, optical response, and aging effects.

8.2.1.1 ELECTRICAL RESPONSE

An OLED is a stack of various organic and inorganic layers. While the precise layers used vary a great deal from device to device, as far as the electrical properties are concerned the layers can be split into three groups, shown in Figure 8.1a: hole conduction, light emission, and electron conduction. Figure 8.1b shows the equivalent circuit for this structure. The emission layer(s) typically has much lower conductivity than the others and the device should only conduct strongly once sufficient forward bias is applied, so that both electrons and holes can reach the emission layer(s). Charge accumulation in this thin structure is represented in the equivalent circuit as a parallel capacitance. The fabrication of real, slightly less than ideal, devices also gives rise to other important properties, in particular, reverse bias current leakage that adds a parallel resistance to the equivalent circuit and conduction through all transport layers adding series resistance.

8.2.1.1.1 Capacitance

Unless the device is forward biased above the threshold voltage, no carriers are injected into the emission layer and so the OLED structure reduces to a thin insulator (<100 nm) sandwiched

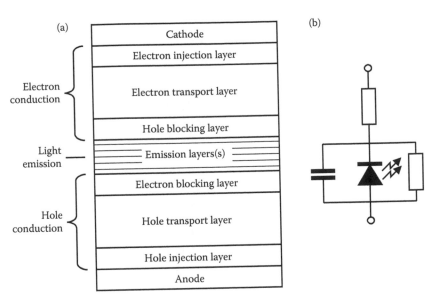

Figure 8.1 (a) OLED device layers in functional groupings and (b) a simplified OLED equivalent circuit.

between two large-area conductors, i.e., the structure for a capacitor, typically with a typical capacitance of the order of 300 pF mm^{-2}. Under steady-state operation (active matrix driven), this property matters little; however, when pulse drive is required (see the later section on passive matrices), the capacitance becomes very important. The capacitance could be reduced by increasing the thickness of the emission layer; however, this would also increase the drive voltage of the diode. As the energy required to charge the capacitance varies as the voltage squared, a thicker device will tend to have an increased power consumption.

8.2.1.1.2 Diode properties

An OLED device has a well-defined threshold voltage, above which the current flowing through the device turns on rapidly. The threshold voltage depends on the charge injection at the contacts (i.e., the voltage required between the contacts and the organic layers for charge injection to occur) and on the bandgap of the emission layer. In general, the bandgap depends on the wavelength emitted and on the type of emission. Shorter wavelengths (blue rather than red) require a larger bandgap and hence a higher threshold voltage, and at any wavelength a phosphorescent emitter requires a larger bandgap than a fluorescent emitter. Typical

turn-on voltages might range between 2 and 4 V. It should also be noted that, in general, the threshold voltage will reduce with temperature. Unlike typical inorganic LEDs, the (steady-state) current–voltage response of the OLED is not exponential—at least, not once there is a significant current flowing.

Figure 8.2 shows a typical I–V curve for an OLED device. Under low fields, the device is largely injection limited and this does indeed give rise to an exponential response. However, once the voltage is taken a little above threshold, the I–V response follows an approximate power law, similar to an old-fashioned thermionic diode, due to the current flow becoming space charge limited [i.e., the injection of charges into the emission layer is inhibited by the concentration of like charges in the low conduction emission layer(s)]. Unlike a thermionic diode, the power law is not necessarily quadratic but can be anything from a power of 2–4 due to the carrier mobility being field dependent. The full I–V properties of an OLED device can be described by using the Murgatroyd equation [6]. However, over a specific region of interest, the response can usually be adequately approximated as $J \propto (V - V_t)^n$, where J is the current density, V the applied voltage, V_t the threshold voltage, and n an exponent depending largely on the field-dependent mobility.

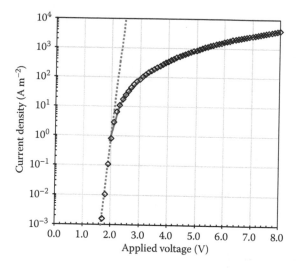

Figure 8.2 A typical I–V curve for an OLED device (diamonds) with an exponential fit of current density (dashed line) for low fields and a power law fit (solid line, see text) for high fields.

8.2.1.1.3 Reverse leakage

An OLED is a very thin large-area structure. The thin structure and large area results in an inevitable leakage current even if the device is perfectly

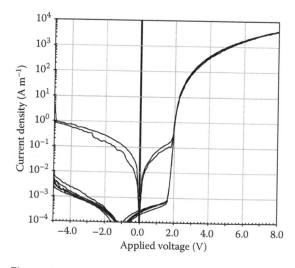

Figure 8.3 I–V characteristics of a number of similar devices, six with good and two with poor reverse leakage characteristics. The voltage offset, at low currents, of the good devices is a measurement artifact.

fabricated with uniform layers and no defects. Any nonuniformity will increase the local electric field and therefore the leakage. This problem is termed "reverse leakage" because it is only under reverse bias that leakage effects become noticeable. Figure 8.3 shows the I–V characteristics of a number of similar devices, two of which show poor reverse leakage properties. For an active matrix-driven device, this is not a problem as devices are never strongly reverse biased. For a passive matrix display, however, leakage has a large impact on display performance (see the section on passive matrix driving for more details), and so care must be taken with the selection of materials and fabrication process to minimize this effect.

8.2.1.2 OPTICAL RESPONSE

8.2.1.2.1 Light output: Device driving

The current efficiency of a device is usually expressed in units of cd A^{-1} and is typically reasonably stable against drive level, device-to-device variation, temperature, and only slowly degrades as the device ages. This results in a light output that is, over a normal operating range, linear with current. In contrast, the light output against voltage is highly nonlinear and varies a great deal from device to device, with temperature, and as the device ages. Figure 8.4 shows the light output against voltage and current for three nominally identical devices, demonstrating the reproducibility (and linearity) of current drive.

It is clear that in order to produce displays that are uniform, have a long life, low burn-in, and accurate gray-scale control, current drive is required.

8.2.1.2.2 Response time

OLED devices can switch extremely quickly. The specific response time varies with materials, device geometry, and applied field, but is typically of the order of 10 ns–1 μs, i.e., essentially instantaneous as far as a display application is concerned. Any observed motion artifacts will, therefore, be a function of the drive scheme rather than a fundamental property of the OLED. This will be covered in more detail in the section on display performance.

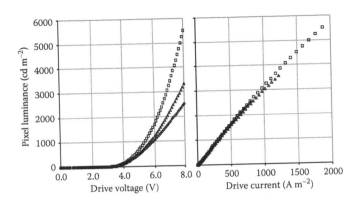

Figure 8.4 Light output against voltage and current for three nominally identical devices, demonstrating the reproducibility of current drive and its superiority over voltage drive.

8.2.1.2.3 Light output coupling

OLED devices can be designed to emit light through either the anode (known as bottom emission as the anode is typically formed on the substrate), the cathode (top emission), or even both for transparent displays. This choice of device type has a strong impact on the optical output, both color and efficiency, but in either case the out-coupling problem for an OLED is the difficulty of extracting light from a high-index emissive layer out into air, whether this is through the substrate or out through the cathode.

Organic semiconductors, whether small-molecule or polymer, are media of high refractive index. When light propagating through such a medium meets its interface with a low-index medium such as air, if its angle of incidence is too shallow then it will undergo total internal reflection and not escape. The fraction of light thus trapped will depend on the angular emission profile from the emitting layer(s). Unless steps are taken to reduce this effect, the fraction of emitted light that leaves the device is on the order of 17% for small-molecule OLEDs and 30% for polymer OLEDs [7].

In principle, there are two approaches to increasing the optical out-coupling: coupling light out of the trapped optical modes or increasing the fraction of light emitted into the out-coupled modes. To extract light from trapped modes typically some kind of scattering structure is required. For display applications, the efficiency benefits are often outweighed by the side effects of increasing the spatial cross talk (guided light from one pixel out-coupled at the next) and ambient reflection, so that both the dynamic and static contrast are reduced (see Section 8.2.3 on display performance), and so this approach is only really considered for lighting applications (see, e.g., Reference [8]) and not used in displays.

Increasing the proportion of light emitted into out-coupled modes is now part of the standard device design optimization and involves the careful tuning of the device thickness and emission zone with further enhancements possible through the addition of partially reflective layers [9]. In this way, efficiency and color can be tuned to better match the display requirements (e.g., producing a deeper blue). Side effects can be some color and luminance variation with viewing angle, although usually much less than often observed with LCD displays.

8.2.1.3 OLED AGING AND BURN-IN

As a device is driven, there is a gradual loss of performance due to degradation of the functional elements of the device. This primarily manifests as an increase in the voltage required and a reduction in the light output at given current density.

Usually the only quoted lifetime figure is a luminous intensity half-life for a given current drive or initial brightness; however, voltage aging can be as important for some applications. In particular, even if a device has a huge half-life, the voltage increase can reduce the system lifetime if there is insufficient headroom on the supply voltage.

In all cases, it is the system life, determined by the driving conditions and the minimum acceptable display performance, which should be calculated to

Table 8.1 Required subpixel luminous intensities to achieve a Rec.709 standard white

	Color set A		Color set B		Color set C	
	CIE x, y	Y (cd/m²)	CIE x, y	Y (cd/m²)	CIE x, y	Y (cd/m²)
Red	0.64, 0.33	96.1	0.64, 0.33	85.1	0.64, 0.33	121.2
Green	0.30, 0.60	220.3	0.30, 0.60	286.1	0.20, 0.70	249.2
Blue	0.15, 0.15	83.6	0.15, 0.06	28.9	0.15, 0.06	29.5

Assumptions include a 100 cd/m² white average luminance, a CIE white point of 0.3127, 0.3290, and a 25% aperture ratio per subpixel.

compare potential material sets. This is particularly the case for RGB display systems (which form an image using separate red, green and blue emitting sub-pixels) where the luminance half-life is often not the most relevant figure, as it is differential aging (usually referred to as burn-in), both between the colors and between different areas of the screen, which is usually most apparent to the user, and can become apparent at very modest decreases in device efficiency, for example, a degradation of only 3%–5% [10]. Furthermore, in comparing different options of available materials, it is important to calculate the system lifetime of each complete RGB material set rather than taking one component in isolation. A common example of this is the choice of blue material. It is typically the case that deep blue emitters will exhibit a shorter lifetime, at a given luminous intensity, than lighter blue alternatives. However, the proportion of red, green, and blue subpixel light output required to produce a target white point depends on the color points of each of the subpixels and on the desired white point.

Table 8.1 shows three examples of RGB color sets and the luminous intensities these would require. Color set A includes a light blue and set B includes a deeper standard blue. Although the deeper blue may have a shorter nominal lifetime than the lighter blue, the deeper blue will not need to be driven as hard and, therefore, will have a longer lifetime than the headline figure might suggest. The color coordinates of the red and green emitters will also affect the result (see, for example, color set C, which has a deeper green). The impact of a given OLED lifetime on the performance of a display can only be properly assessed as part of a complete RGB color set.

A discussion of the potential causes of aging effects is beyond the scope of this review, and a statement of "current" specifications, considering the pace of development, is liable to be seriously out of date

by the time of reading. However, the reader needs to be aware that aging of a device depends strongly on drive level, duty cycle, and environmental conditions (particularly temperature). Furthermore, the lifetime under one set of conditions cannot always be extrapolated reliably to another. It is, therefore, important to obtain the aging characteristics for the conditions specific to an application.

8.2.2 Display driving

8.2.2.1 PASSIVE MATRIX DISPLAYS

A passive matrix is the simplest possible matrix display and consists of arrays of rows and columns forming pixels where they intersect. One row is selected and all the pixels on that row are driven from the columns. The other rows are selected in turn until the entire frame has been scanned. From their inception OLEDs were, due to their diode nature, touted as ideal for the fabrication of passive matrix displays, with visions of huge screens for very low cost. Of course, life is never that simple. As we shall see, the size of a passive matrix display is limited by the peak current required for pulsed light emission, by current leakage and, in particular, by the effects of device capacitance.

8.2.2.1.1 Passive matrix operation

Figure 8.5 shows (1) a functional diagram of an OLED passive matrix and (2) the current, voltage, and light output of one pixel during a line scan. The line scan sequence (assuming current drive) is as follows, with the step numbers corresponding to the labels in Figure 8.5b:

1. The row is selected (usually switched to ground while other rows are held high) and the current, I, is driven into the column.
2. The voltage across the device will increase linearly until the threshold voltage is reached. The

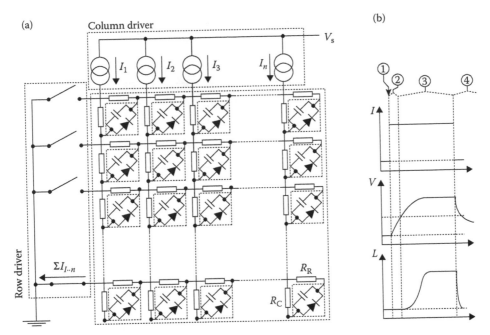

Figure 8.5 (a) Functional diagram of an OLED passive matrix and (b) current, voltage, and light output versus time of one pixel during a line scan.

time this will take is $t_C = CV_t/I$, where C is the capacitance of the column and V_t the threshold voltage.

3. The pixel is now emitting light and the voltage gradually tends toward the nominal drive level. The time constant for this is harder to derive as it depends on the specific I–V relationship of the device.

4. The current supply is turned off and the voltage across the device rapidly drops to the threshold voltage, and then very gradually drops back to zero, assuming no other pixels are driven.

Electrically, to complete the picture, the resistive losses from the pixel's row and column, plus the driver compliance, need to be added. Taking all of these into account, a simple model can be constructed, which, despite its simplicity, can be a powerful predictor of the expected display performance.

Figure 8.6 shows results from an OLED passive matrix model, predicting power consumption per pixel versus number of rows. This is an iterative model based on the electrical response of an OLED pixel as described in Section 8.2.1.1. These results show that the potential dominance of capacitive power consumption in larger displays. With more rows comes more column capacitance and more

charging cycles leading to an approximately quadratic increase in capacitive losses with row count. Furthermore, there is the knock-on effect of an increased current demand causing more resistive losses and driver compliance losses. It is these capacitive losses that often limit the practical size, resolution, and/or brightness of a passive OLED display.

8.2.2.1.2 Leakage and defects

All OLED devices in a passive matrix are connected to one another via the common anode and cathode lines. Therefore, a single pixel's defect or current leakage can have an impact on the performance of the whole display. The effects on display operation of current leakage (localized or over the display) and other display defects depend on the details of the specific drive scheme.

The simplest possible driver with no gray scale, an off-state anode clamped to 0 V and off-rows open circuit, is relatively unaffected by display defects. Pixel shorts on their own will, largely, only prevent the operation of that specific pixel and reverse leakage will, if not excessive, cause a slight loss of brightness of the pixel being driven (assuming current drive is used). With more significant leakage, spatial cross talk, apparent as streaking along the row and column from the active pixel,

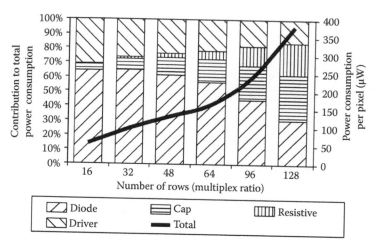

Figure 8.6 Power consumption per pixel versus number of rows in a passive matrix display. Results are calculated from an analytical model.

can become increasingly visible, although the acceptable threshold for this depends upon the application and displayed images. The opposite extreme might be a driver that uses methods to actively drive the deselected rows and columns off to minimize the effects of capacitance and improve contrast and gray scale but where a shorted pixel would cause all other pixels on a common row or column to cease functioning.

The selection of a driver for a display needs to be informed not only by the target application but also by the level and type of defects and leakage likely to be present in the display and the sensitivity of the specific drive scheme to such defects. A high-fidelity gray scale and contrast capable driver can only work well with a zero defect and low leakage display—if the display in question tends to suffer from defects of any form then, in general, the simpler the driver the better.

8.2.2.1.3 Intensity-scale methods

There are two ways to modulate the quantity of light emitted by a pixel during a line scan—by changing the height of the current pulse (analog drive) or its duration (pulse width modulation [PWM]).

Driving an OLED at a lower current for a longer time (analog drive) might seem, on first inspection, the better option. A lower average drive voltage lowers both the direct OLED power consumption and the (often considerable) capacitive losses. A lower drive level also reduces aging. Unfortunately, analog gray scale can also result in poor image quality as described below.

PWM always has a full charge–discharge cycle per frame, whereas, with analog drive, the charging and discharging are always only the difference in drive level between pixels, resulting in another power advantage for analog. However, this also results in the initial state of the voltage on a pixel being dependent (for analog drive) on the state of the previous pixel causing potential cross talk in the form of vertical edge blurring.

The most serious difficulty for analog drive, though, is dynamic range. Human perception of brightness is highly nonlinear and this is accommodated for in displays by making the pixel light output proportional to the gray level raised to some power (referred to as a gamma), typically 2. A display with a modest 32 levels of brightness and a gamma of 2 requires a 1024:1 ratio in actual light output between its brightest and darkest levels. On top of this, there is also often a global brightness setting that might require a further 100:1 dynamic range. For an analog-driven display, this becomes a ~100,000:1 ratio in drive currents, which is extremely challenging. For a PWM-driven display, however, it is relatively simple to obtain the 1024:1 dynamic range through the PWM pulse timing with a cruder setting of reference current for the global brightness setting. Almost all passive matrix OLED drivers use PWM control.

8.2.2.1.4 Capacitance charging

Pixel capacitance can cause three effects: a power loss, an offset in the light output to a given drive level and cross talk through residual capacitance.

The power loss has been dealt with previously, the offset in light output is due to the time taken to charge the pixel up to the operating voltage, and the cross talk is due to the residual charge left at the end of a line scan (and present at the start of the next).

The simplest way to minimize the power loss is simply to do nothing, as corrective actions for the other two effects tend to increase power consumption. However, for some displays the cross talk, in particular, can cause problems, and in this case some form of precharge and discharge is often used.

Pixel precharging is typically implemented by applying a voltage to the pixel column, prior to the current drive signal, which charges the pixel to a set voltage. This injects current more rapidly than the current drive would be able to and thus keeps the charge-up "dead" time to a minimum. Pixel discharging ensures that the voltage is rapidly reduced at the end of the PWM pulse.

Using precharge and discharge can substantially improve the visual performance of a display. Unfortunately, it can also substantially increase power consumption. It might be possible to use charge recycling methods, such as are used with LCD displays, to reduce the power consumption of an OLED passive matrix. The limiting factor in this case is the anode [typically indium tin oxide (ITO)] resistance. If Q is the charge, R is the anode resistance, and t is the time that is available to recover the charge, then resistive energy loss during recovery is Q^2R/t. This means that, although such recycling could help reduce power consumption, in large high-resolution displays, where capacitive losses are greatest, there is also the highest anode series resistance and least time in which to recover charge. The benefit of charge recycling would be limited unless thicker ITO or anode bus bars were used to reduce the series resistance.

8.2.2.1.5 Multiplex control

There are a number of controller enhancements that can be achieved through multiplex control and all take advantage of the graph in Figure 8.6—the lower the multiplex rate (the number of rows over which the column drive is shared), the lower the power consumption.

The simplest method of making a display more efficient involves splitting the columns across the center of the display and driving them from top and bottom, in essence splitting the display into two passive matrices. This comes at the cost of an increased component count (double the column drivers, although they need handle less than half the power) and possible lower yield (double the column connections to make). Nevertheless, this is a valuable method in achieving larger displays, doubling the row count being practically achievable.

Selective scanning modes, reducing the multiplex rate by driving only a subset of the rows, can also prove to have substantial benefits. This could be a fixed display mode, for example, only driving the central quarter of a display, or dynamically skipping blank rows.

Multiline addressing schemes have in the past provided large improvements to the performance of passive matrix LCD screens and some attempts [11,12] have been made to achieve the same in OLED. Unfortunately, both the algorithms and drive schemes required to implement multiline addressing are complex, and they were largely overtaken by the adoption of active matrix backplanes for OLED displays.

8.2.2.2 ACTIVE MATRIX DISPLAYS

From the above, it is clear that OLED passive matrices, while capable of providing high-quality image reproduction for small display formats, are unsuitable for expansion to larger display formats. The solution to this is to form an active matrix of simple per-pixel thin-film transistor (TFT) drive circuits patterned onto the display glass, building on the active matrix technology developed for LCDs. However, as with passive matrices, there are particular factors that need to be considered when applying active matrix technologies to OLEDs, the primary ones being TFT carrier mobility, TFT threshold voltage nonuniformities, and OLED nonuniformities. The first of these restricts our choice of TFT technology, and the other two lead us to prefer current-programmed drive circuits over voltage-programmed despite their greater complexity.

Table 8.2 shows typical carrier mobility for the various TFT technologies available, along with crystalline silicon for reference. Figure 8.7 shows the driver channel width requirements of a transistor to control a current through an OLED pixel to achieve a 300 cd m^{-2} display luminance assuming typical device dimensions and OLED efficiencies. The >100 µm driver channel width requirement of amorphous silicon (a-Si) is significant when compared to the 300 µm pixel pitch. This does

Table 8.2 Mobilities of available transistor technologies

TFT technology	Majority carrier mobility (cm² V⁻¹ s⁻¹)
a-Si (amorphous Silicon)	≤1
Oxide TFT (e.g., IGZO)	10–40
LTPS (low-temperature polysilicon)	30–150
CGS (continuous grain silicon)	300–500
Bulk silicon	~1000

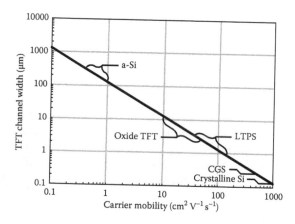

Figure 8.7 Channel widths of transistors capable of driving an OLED pixel at 300 cd m⁻², assuming an efficiency of 10 cd A⁻ⁱ, pixel pitch of 300 μm, and typical values for TFT electrical characteristics.

Possibly the most promising TFT technologies are the relatively new, transparent oxides, for example, indium–gallium–zinc oxide (IGZO) [13]. These seem to offer the simplicity of a-Si with a performance approaching that of LTPS.

TFT threshold voltage variation is an issue with most TFT types. Polycrystalline devices (e.g., LTPS) often have grain boundary variations from device to device, whereas amorphous devices (e.g., a-Si) typically show threshold voltage shifts with age. Such variations can, if not compensated for, cause significant brightness nonuniformities in a display. Of course, variations in the OLED pixel response can also cause nonuniformities that, although not dependent upon the TFT technology directly, is dependent upon the design and operation of the pixel circuit, which in turn is limited by the choice of TFT technology.

There are a number of circuit types (and a great many variations thereof) that can be used to drive an OLED pixel, and these are covered in the following sections. However, there is one other design factor important to the operation of an active matrix display and this is the choice of top or bottom emitting structure.

not rule a-Si out as a potential TFT technology; however, it certainly makes the case borderline. Low-temperature polysilicon (LTPS) and its variants [e.g., continuous grain silicon (CGS)] are the favored current production technology but the processing required is complex and expensive.

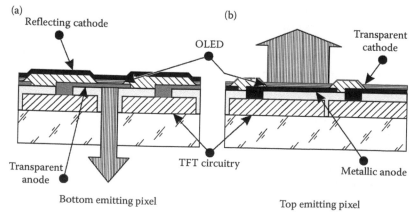

Figure 8.8 Cross sections of bottom emitting and top emitting devices.

The standard OLED structure uses a transparent anode (e.g., ITO) and reflective metallic cathode (see Figure 8.8a). This means that the device must emit light through the substrate and therefore, as with LCDs, the pixel area must be shared between the pixel circuitry and emitting area. However, there is another option that uses a metallic anode and thin transparent cathode, typically referred to as a top emitter (see Figure 8.8b). The resulting increase in aperture ratio (the ratio of the emitting area to the total pixel area) gives a boost to display lifetime as a lower current density in the OLED device suffices for a given brightness and can also increase efficiency primarily by allowing larger TFT areas (less TFT voltage drop). These advantages are offset by significant disadvantages: the encapsulation must both be transparent and possess better barrier properties than in the standard case as the thin, highly reactive, electron injector is no longer protected by a thick metal layer.

8.2.2.2.1 Voltage programming

Figure 8.9a shows a schematic of a typical active matrix OLED (AMOLED) display panel with Figure 8.9b showing the simplest OLED pixel

drive circuit. With an LCD all that is required is a method to fix a voltage over a capacitor, so the only transistor required is a select device acting as a switch to connect the cell to the data line when the row on which the pixel resides is addressed. In the case of an OLED, a drive transistor is needed through which current can be controlled, thus the simplest circuit uses a select TFT to set a gate voltage on a drive transistor, and the gate voltage is set to produce the required luminous output [14]. However, the current controlled through the drive transistor is dependent not only on the gate voltage but also on all the TFT characteristics including, in particular, the threshold voltage.

The alternative way of driving the two-TFT cell is to turn on the transistor hard and use display subframes to achieve gray scale. This method does work well; however, it is essentially a voltage-driven method and so suffers from sensitivity to OLED variations and accelerated aging (compared to current drive), as discussed in Section 8.2.1.2.

Attempts have been made to compensate for variation in TFT threshold voltage; one such [15] can be seen in Figure 8.9c. During the line scan, the select transistor is first activated and 0 V applied to

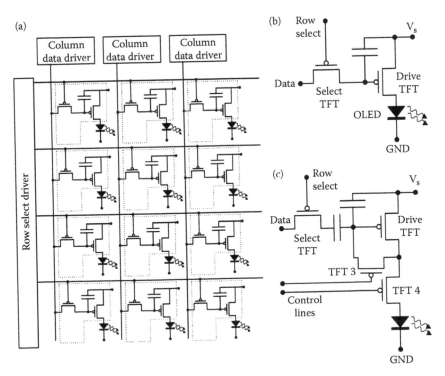

Figure 8.9 Voltage-programmed active matrix pixel circuits: (a) schematic of an active matrix, (b) simplest OLED pixel circuit, and (c) threshold voltage-compensated circuit.

the data line. TFT 3 is closed, which discharges the charge on the gate of the drive TFT, bringing its voltage down to threshold so that when TFT 3 is subsequently opened the threshold voltage is held at the drive TFT gate. The signal applied on the data line will now produce a voltage at the drive TFT gate offset by the threshold voltage. The select transistor is deactivated holding the drive level and TFT 4 is closed allowing current to flow through the OLED. Although it compensates for threshold voltage changes, this circuit still depends on the uniformity of the other parameters of the drive TFT, as well as requiring two TFTs that can handle the full drive current and two extra control lines. Both of the above circuits also suffer, when analog driven, from the highly nonlinear relationship between gate voltage and drive current.

8.2.2.2.2 Current programming

All current drive schemes have the advantages of linearity and relative insensitivity to transistor and OLED variations. Two general methods are possible in principle—the setting of OLED current directly or indirectly through a current mirror.

Figure 8.10a shows a typical example of the first of these circuits of which there are many variations [16,17]. At the start of the addressing period, the deselect TFT is opened and the select TFTs are closed, diverting the output from the drive TFT to the data line. The display controller sinks a current through the data line and any mismatch between this current and that supplied by the drive TFT will modify the charge on the

capacitor until the currents are balanced. At the end of the addressing period, the select TFTs are opened and the deselect TFT is closed, holding the charge on the capacitor and redirecting the drive current to the OLED. This pixel circuit has excellent uniformity and extremely good linearity resulting in a high fidelity of gray scale. Its primary disadvantage is that it requires three TFTs that must be capable of handling the full drive current as well as needing two complementary control lines.

Figure 8.10b shows the current mirror circuit that was used in an early 13-in. OLED demonstrator [18]. The operation of this circuit relies upon an amplifying current mirror. The relationship between the geometries of the drive and mirror TFTs results in a scaling factor between their drain-source currents when a common voltage is applied to the gate. This relationship can only work when other TFT parameters are sufficiently similar between the two devices. However, this is the case when the TFTs have similar layouts and are in close proximity. During the addressing period, the erase TFT is initially closed to clear the charge on the capacitor. The scan TFT is then closed, a current is sunk through the data line and the voltage on the common gate capacitor adjusts so that the mirror TFT supplies this current. The erase and scan TFTs are then opened and the voltage is held on the gate, producing a scaled current in the drive TFT. This circuit possesses the advantages of the previous circuit without the problems of two control lines or three large TFTs. Use of either of

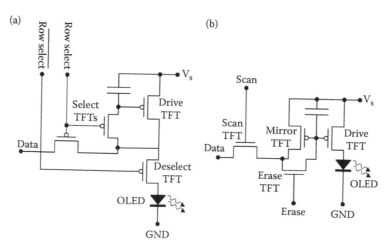

Figure 8.10 Current-programmed pixel circuits: (a) current sampling circuit and (b) amplifying current mirror.

these pixel cells can produce displays with excellent uniformity and a high fidelity of gray-scale reproduction.

The circuits shown and described above are examples of the most common approaches taken when designing OLED drive circuits. Further subtleties in the design of the circuitry and drive scheme, not discussed here, can relate to the particular choice of TFT technology, OLED devices or the specific requirements of the display design, for example, to reduce the effects of burn-in and pixel aging [19].

8.2.3 Display performance

This section is inevitably much briefer than it could be as this is a very big subject. The key display performance metrics will be discussed in the context of AMOLED displays, as passive matrix (PMOLED) displays are typically now only used in applications where high performance is not an issue. For more information, see the further reading section at the end, in particular, the Displaymate website, which has some excellent analyzes of the performance of different display technologies.

8.2.3.1 STATIC IMAGE PERFORMANCE

There are six properties that determine how well a static image is reproduced on a display:

Luminance is the maximum brightness, which a display can produce. With an emissive display (including OLED) there will always be a trade-off with lifetime, and with any display a trade-off with power consumption. As a headline figure, it can also often be misleading—a display with twice the luminance does not appear twice as bright. A useful approximation is that perceived brightness varies as the cube root of luminance, so twice the luminance means only 25% more perceived brightness. LCDs can often have a much higher luminance than an equivalent OLED panel, although these days this is often just a design decision (see black level) rather than a limitation.

Black level determines the contrast of the display. What makes OLEDs stand out is that, under dark conditions, a black pixel has a near zero luminance, even if nearby pixels are on full white. Contrast is the white-to-black luminance ratio and has a very strong impact on perceived image quality. A high contrast makes the image appear brighter and more colorful. A really high contrast,

that is, a truly black background on a display, removes the gray rectangle we are used to watching images within and makes some scenes look much more realistic. The contrast often quoted with displays is the "dynamic contrast," which is the ratio of an all-white screen luminance to an all-black screen, which can be >100,000:1 for an LED backlit LCD display. This has little relevance to actual performance. Static contrast, that is, the white-to-black ratio of patches shown simultaneously, is what really matters. For a good LCD, this is often around 1000:1 and for OLED displays, this is usually quoted as around 1,000,000:1 [20].

The color gamut, that is the range of color, which can be reproduced, is determined by the display's RGB color primaries. There is little need to reproduce colors beyond what can be described in the image or video format; however, there is also little harm so long as the color reproduction is managed properly. OLEDs have gained somewhat of a reputation for artificially saturated color reproduction, but in reality this is a function of the image processing rather than the OLED pixels. Both LCD and OLED displays can be made with virtually perfect color primaries and most good screens have accurate color modes, although it is also true that some users, and some markets, seem to prefer displays with unnatural supersaturated colors (much to the despair of displays experts). It also may be the case that, historically, products with OLED displays were set up to stand out from competing LCD-based devices. Both superaccurate and supersaturated colors are possible with both OLED and LCD.

Gray-scale reproduction determines how well all of the colors between black, white, and the RGB primaries are reproduced. Displays can have a perfect color gamut but still fail to achieve good image reproduction because they have poor gray-scale control. LCD devices are often highly nonlinear, requiring careful image processing and signal control to achieve good gray-scale control. OLEDs, in contrast, are fundamentally linear devices, and so very accurate gray-scale control should always be achievable with a suitable drive scheme.

Resolution is the one area where LCD has an advantage. An active matrix LCD requires just one transistor per pixel, whereas an AMOLED needs at least two, and often four, per pixel. A 1080p (1920×1080 pixel) display would need 6.2 million transistors if it were an LCD, but possibly 25 million if it were an OLED. This is not a fundamental issue, but does make the fabrication of OLED backplanes much more challenging, and means that some high-resolution OLED screens must have a top emitting structure so that the transistors sit under the OLED allowing both to use the full pixel area.

Viewing angle has always been quoted as an OLED advantage, and indeed OLED displays can be viewed from any angle with minimal color or luminance changes. Modern LCD screens can, however, have wide viewing angles if the application requires it (although this may be at the expense of other attributes) and indeed only emitting light in a restricted viewing range can offer a power consumption saving.

8.2.3.2 DYNAMIC PERFORMANCE AND 3D

OLED pixels can switch extremely fast. Phosphorescent pixels do switch more slowly than fluorescent pixels, but still in microseconds. Therefore, any motion artifacts are most likely to be due to the driving scheme, and in particular the screen refresh rate, than anything fundamental to an OLED.

Active matrix displays are hold-type displays—they hold an image static on the screen for a frame before changing it on the next frame (typically on a rolling basis taking up to a frame time to update the image). If an eye tracks an object moving across a screen, the image of the object will be held, for the frame time, at a sequence of positions; this will cause a blurring or a judder of the moving image, as is shown in Figure 8.11a. It would be possible to hold the image for a shorter time to reduce this effect (Figure 8.11b), but that would either reduce the display luminance or, if the luminance were increased to compensate, reduce the display life. The better option is to use a higher refresh rate and indeed many modern TVs have refresh rates of 120 or 240 Hz (Figure 8.11c). For this to work, the data source needs to have high frame rate data or, more commonly at present, intermediate frames need to be interpolated. LCD screens are often sold with 120 or 240 Hz refresh rates with the same goal of reducing motion artifacts, but the LCD response time will still blur the tracked image more than in the OLED case (Figure 8.11d).

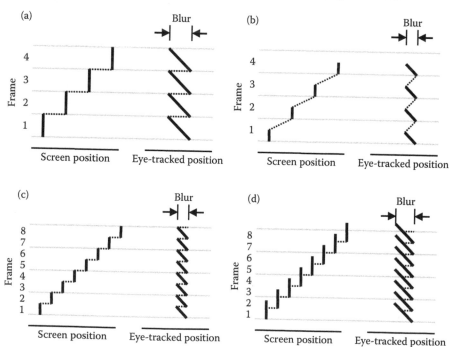

Figure 8.11 Motion artifacts on hold-type active matrix displays. Each figure shows the displayed position of an object (thick black line) for a number of frames relative to the display and to the path an observer follows as they track an object across the screen, and the blur length this produces. (a) Shows a standard frame rate display with a full-frame position hold, (b) shows a standard rate with a half-frame hold, (c) shows a double standard rate display, and (d) shows the case when the display pixels have a long response time.

The switching time of a display is especially important for time multiplexed 3D. Here a user wears a special pair of glasses, which have electronic shutters in front of each eye synchronized to a screen showing alternating left-eye and right-eye images (with blanking frames inserted to deal with the switching time of the glasses and the rolling update of the screen). If the screen is at all slow to react in switching between images, there will be bleed (cross-talk) between the left-eye and right-eye views, which can cause significant eye strain in the viewer. This is not an issue with an OLED display, and indeed current Samsung OLED TVs, for example, the Samsung KN55S9, are able to display two completely separate video streams and synchronize these to two pairs of shutter glasses so two people can watch separate TV content simultaneously on the same set with no discernible cross talk.

8.2.3.3 SYSTEM PERFORMANCE

There are a number of other significant performance factors not covered by the above two categories:

Power consumption is determined by the display specification and by the performances of the optical display structure, the OLED, and the backplane. The typical current consumption will be the peak luminous intensity times the average relative luminance content of a typical image and the display area, divided by the efficiency of the optical stack and the cd/A efficiency of the OLED pixels themselves. The voltage over which this current needs to be supplied is the sum of the (worst case) voltages across the OLED, the drive TFT, the panel supply tracks, and any overheads needed, e.g., temperature variations and aging. Broadly speaking, the OLED is responsible for the current and some of the voltage drop and the backplane (TFT and tracking) for the rest of the voltage, so to get a low-power consumption it is essential to manage both. The power also depends strongly on image content: power consumption will be lower for user interfaces with a dark background.

At the time of writing, it would seem that for typical images OLED screens on mobile devices consume somewhat more power than an alternative LED backlit LCD, although the figures are comparable and any detailed comparison is masked by a large number of design decisions and trade-offs, which differ between the two technologies. Of course, this is a rapidly changing field for both LCD and OLED so the situation will certainly change in future. For TVs, there are too few OLED devices so far to make much of a comparison, although it is clear that both OLEDs and LCDs consume much less power per unit area than plasma or cathode ray tube (CRT) screens did.

Lifetime and burn-in are of course two areas of performance, which have been the focus of much attention. Although lifetime has been, historically, the most discussed, it is really burn-in, which is the issue. Due to eye adaptation, a uniform drop in luminance, unless very large, is hardly noticed under normal viewing conditions. What hurts is differential aging between two areas of the screen leaving the negative of the average viewed content superimposed over the content being observed, typically referred to as burn-in or image sticking. Compared to the typical 50% half-life figures quoted, as little as a 2%–3% burn-in [10,21] is perceptible to a user. Therefore, OLED displays not only need long lives, they also need to ensure that this first few percentage drop does not happen quickly. This will typically be achieved through both improved intrinsic performance and drive schemes, which correct for OLED aging (see, e.g., Reference [19]).

OLEDs seem to perform (in terms of lifetime and burn-in) well enough for at least mobile screens. These devices typically have a short product life (are replaced frequently) but are also used intensively with repeated fixed graphics (e.g., the "home" screen of the device). While burn-in has been observed with some OLED smartphones, it does not seem to be a significant problem, whether through OLED performance or good screen management or both.

Transient effects are related to burn-in. Aging and burn-in are long-term permanent effects, but some OLED material systems also exhibit short-term effects, which can have a similar appearance, but from which an OLED will recover. For example, it could be observed on the first commercial AMOLED screen, on the Kodak LS633 camera, that the white icons could appear inverted over a bright scene, but this apparent burn-in would fade over a few seconds. These effects were most likely due to charge trapping in a device and seem to be much less apparent in more recent material systems.

8.3 DISPLAY FABRICATION

This section describes the structure of an OLED display and the key aspects of the components and fabrication process. The discussion will be kept as generic as possible, with the different fabrication options, e.g., OLED vapor vs. solution deposition, brought out in the relevant subsection. The only construction option that needs to be brought out at the start is the choice between active and passive matrix addressing as this has an impact on the overall structure. The layers that constitute an OLED display will be discussed in the order they would be encountered in a fabrication process, i.e., from the substrate up, and grouped as layers prior to OLED deposition (the substrate stack: substrate, backplane, and anode), the organic layers themselves, and finally the cathode and encapsulation.

8.3.1 Substrate stack

8.3.1.1 SUBSTRATE

Glass is the obvious choice of substrate, being used in all established display technologies from liquid crystal to CRT. Its stiffness and dimensional stability make it straightforward to handle through the manufacturing process, and it can easily handle the required process temperatures. (For the highest conductivity, ITO requires processing above 400°C, which is too hot for many plastic substrates.)

Glass is essentially SiO_2 with other materials added to reduce its melting point while ensuring that it remains amorphous (does not crystallize). There are very many different kinds of glass, but, for the relatively straightforward requirements of display substrates, the main contenders are standard soda lime and borosilicate glass. Soda lime is usually used because it is the cheapest but in displays it typically has a layer of SiO_2 deposited on its surface to prevent out-diffusion of sodium. Borosilicate has a higher melting point, does not have an out-diffusion problem and is tougher, particularly against thermal shock. The choice of which to use is usually determined by the processing temperature requirements; for example, an LTPS backplane would require the use of borosilicate. The main disadvantages of glass in general are brittleness (displays have a tendency to break when dropped), weight, and lack of flexibility.

Plastic becomes a possibility due to the inherent flexibility of OLEDs (if not necessarily the other display components). Plastics present a difficult challenge for liquid crystal-based devices because of the need to accurately maintain the thickness of the liquid crystal film. The sensitivity to changes in this liquid-layer thickness can easily be seen by gently pressing on an LCD. Organic light emitters, being entirely solid, do not suffer from this problem.

The two main requirements on a flexible substrate are to withstand the processing temperature (backplane and electrodes) and offer a sufficiently good diffusion barrier to oxygen and water. No plastic material can meet the barrier requirements on its own. This is well known in the food packaging industry, where the shelf-life of plastic-wrapped products (e.g., potato crisps/chips) can be significantly increased by using a metallized plastic bag. Organic displays are incredibly moisture sensitive and also require the barrier coating to be transparent. Suitable barrier-layer technology is being developed, but devices with lifetime close to those on glass have not yet been demonstrated. Single-layer barrier coatings of transparent materials, such as silicon dioxide, that are thin enough to be flexible, tend to suffer from pinhole defects, so the general principle is to apply a multilayer stack of alternating barrier layers and plastic layers [22]. Care has to be taken to ensure low stress in the finished structure.

It should also be noted that, even without the cost of barrier layers, the high-quality plastic materials required for displays (typically polyethylene terephthalate [PET] or polyethylene naphthalate [PEN]) are more expensive than glass, although that higher material cost may be compensated by lower handling costs.

8.3.1.2 BACKPLANE

The practice of making LCDs with a simple electronic drive circuit at each pixel is well established. The pixel drive circuits are usually made from TFTs of amorphous or polysilicon. a-Si has good enough performance for most LCD displays (charge carrier mobility up to 1 cm² V⁻¹ s⁻¹) and can be deposited at low temperature (300°C). Higher performance TFTs can be made by crystallizing the a-Si, but this requires a temperature of ~600°C, beyond the capability of typical display substrates. Instead localized processes have been developed, for example, using lasers to heat only the top layer of silicon to the required temperature to produce so-called LTPS, which can be formed on borosilicate display glass giving a huge boost to performance (40–120 cm^2 V^{-1} s^{-1}). OLED displays have more stringent carrier mobility requirements (see Section 8.2.2.2) and at present LTPS is the most common active matrix backplane material for OLED displays, but it is expensive, can suffer from nonuniformities, and is difficult to scale. In 2004, a new backplane option based on transparent oxides was reported [23]. The performance of oxides, particularly IGZO, looks increasingly promising for OLED display fabrication [24] and has the advantage of a very similar process flow to a-Si.

8.3.1.3 ANODE

For displays emitting through the substrate, the most obvious requirements of the anode layer are sufficient transparency and conductivity. Inevitably, for a layer of a given material, there is a trade-off between these two properties controlled by the layer thickness, thicker layers being more conductive but less transparent. For a material with resistivity ρ (Ω cm) and optical absorption coefficient α (cm^{-1}), a layer of thickness t has a sheet resistance ρ/t and optical transmission e$^{-\alpha t}$. An appropriate figure of merit for the material, indicating its suitability as a transparent conductor, is, therefore, the product ρα, the sheet resistance of a layer, which would transmit 1/e of the light. There are other important criteria for anode materials. It must be possible to pattern the material, usually with a wet chemical etch. The anode material must also be sufficiently smooth. A localized material spike of height comparable to the thickness of the organic layer will make it thinner, producing a large local electric field across the layer, and hence increased current and accelerated degradation. This may lead to a black spot in the display or even a short circuit. There has also been concern that the anode material can act as a diffusion source of species that might degrade the device performance with time. The conductive metal oxides can be thought of as highly doped semiconductors—self-doped due to crystal defects partially caused by nonstoichiometry. The species in excess (e.g., indium in ITO) can diffuse out and interact with the organic layers.

Although the work function of the anode material might be thought to be an important property, this is not critical when a highly doped conductive polymer such as polyethylenedioxythiophene (PEDOT)/polystyrenesulphonic acid (PSS) (see Section 8.3.2.2) is in contact with the anode. As both anode and highly doped polymer exhibit metallic conduction, there is a negligible barrier to charge flow regardless of work function.

ITO is almost universally used as the anode. It can achieve resistivity of 1.6 × 10^{-4} Ω cm and α ∼ 10^4 cm^{-1}, although it does need annealing at 400°C to reach this. Where ITO is not the primary conductor, for example, in an active matrix backplane with a metal track process, the full temperature anneal may not be required. ITO can be etched in concentrated hydrochloric acid.

Other candidates for the anode material include zinc oxide and tin oxide. For a thorough review of transparent conductors, see Hartnagel et al. (1995) in the further reading list.

8.3.1.4 SUBSTRATE STACK PREPARATION

For a passive matrix display, the display manufacturer generally buys glass sheets, already uniformly coated with ITO. The desired pattern is imparted to the ITO layer using standard lithographic techniques involving a wet chemical etch. For an active matrix display, the process and patterning steps will depend on the technology used and are beyond the scope of this chapter, but the result—an array of anodes to be coated with the OLED layers and a cathode—is essentially the same.

To avoid short circuits between anode and cathode, it is vital that the organic layers are continuous. However, with organic layers typically less than 100 nm thick, it does not take a large piece of contaminant material to break the continuity of the layer. This means that to make a display with reasonable yield very thorough cleanliness must be maintained throughout the process. The cleaning of the substrate is therefore a critical process. This is typically done by a combination of mechanical agitation (brushing, ultrasonic, and megasonic) in a series of liquids (water with detergent and then organic solvents). The final step in substrate preparation is an oxidizing surface treatment. This can be carried out either by oxygen plasma or by UV–ozone treatment. It has the combined effect of removing any residual organic contamination (e.g., photoresist) and changing the surface energy of the ITO so that the first organic layer easily wets it.

8.3.2 Organic layers

8.3.2.1 VAPOR PHASE DEPOSITION

Thermal evaporation in a high vacuum is the standard method of depositing all small-molecule organic layers. It is a well-established and reasonably straightforward process and has the benefit of also being a purification process (most contamination will evaporate at a different temperature). Unfortunately, as OLED materials are not readily compatible with patterning after deposition the only way to pattern these organic layers is to do the deposition through a metal shadow mask held very close to the substrate. Shadow masking is adequate for simple low-resolution displays but becomes increasingly difficult as the substrate size increases and the pixel size diminishes. Examples of the types of difficulties include shadow mask thermal expansion during processing or OLED deposition on the mask causing aperture clogging and mask distortion. For example, a 4.5″ diagonal HD mobile display (1920 pixels wide) has subpixels 16 μm across and might require a position tolerance of 4 μm. If the shadow mask is made of steel (thermal expansion coefficient of $13 \times 10^{-6}/°C$), then the mask temperature would need to stay within 3°C to keep that position tolerance.

Despite this, shadow masking is used in volume production. To avoid the problems of shadow masking, an alternative, originally pioneered by Kodak and then LG, is to deposit a white emitting stack and then use color filters to obtain red, green, and blue subpixels in addition to an unfiltered white subpixel.

8.3.2.2 SOLUTION PROCESSING

Solution processing has been used for fabricating some commercial displays, but so far this has been used only for simple spin-coated monochrome devices and has largely, if not entirely, been supplanted by vacuum deposition. There is, however, substantial activity in developing printing techniques for producing large area RGB OLED screens, driven by the difficulties of yielding shadow-masked vacuum-processed screens on a large scale. Traditionally, solution processing has been used for polymer OLED materials only but there is now active development of soluble small molecule materials systems [25]. The rest of this section will discuss polymer solution processing;

however, much of it should be relevant to suitable small molecule material systems. Throughout this section, the term ink will be used to describe the solution of the organic layer in a solvent.

Spin coating is possibly the simplest way of forming an OLED device. The simplest polymer devices can use only two layers. The first, known as the hole conduction layer, is usually PEDOT and PSS, with a large excess of PSS [26], in an aqueous ink. It is spin coated to give a layer typically 70 nm thick and then baked to speed up the drying process.

The second layer is the emissive polymer, which is spun from an organic solvent, usually toluene or xylene. To achieve uniform appearance over the area of a display, the light emitting polymerlayer must be of uniform thickness, typically 70±5 nm. To achieve this level of uniformity on large substrates, a multistage spin coating process (multiple speeds and times) is used. As well as simplicity, spin coating also has the advantage that the action of spinning the ink over the substrate and throwing off the excess also tends to clear off some remaining contaminants. Its disadvantages are the high material wastage (spun-off ink is not recoverable) and no patterning ability.

Inkjet printing is the most promising method for manufacturing multicolor polymer displays [27]. As explained later, there are tighter constraints on properties of the polymer solution than for spin coating and typically two or more solvents are used to make a polymer ink. An inkjet print head is essentially an ink-filled reservoir connected to a small nozzle. In equilibrium, the ink does not flow out of the nozzle due to a combination of surface tension forces and the reduced pressure in the head resulting from having the main ink supply below the head. To print, a pulsed pressure wave is generated insider the reservoir, typically using a piezoelectric actuator, and this causes a drop of ink to be ejected from the nozzle and propelled toward the substrate. For this process to work well in a given design of inkjet head, the surface tension and particularly the viscosity of the ink must lie within narrow limits. As the total mass of polymer ejected in the drop will determine the final film thickness, the concentration of the ink is fixed by the drop volume, the pixel area and the number of drops printed per well. These combined constraints make developing good inks a complex process. There is a further complication relating to the requirement that, within each pixel, the polymer

(a)
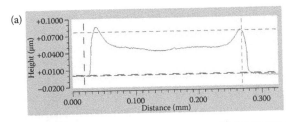

(b)
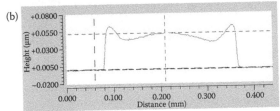

(c)
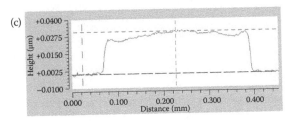

Figure 8.12 Cross sections of inkjet printed polymer films from a range of different solvent blends. Height measurements were taken using a white light interference scan. Solvent blends were (a) 90:10, (b) 70:30, and (c) 50:50 (low boiling point:high boiling point). In this case, the 50:50 blend gives the flattest film.

film must be flat over a good proportion of its total area. Generally, when drops of solution are allowed to dry on a surface, the resulting film is not flat. This is commonly seen when spilt coffee is allowed to dry leaving a dark perimeter, and the result is often referred to as the "coffee ring effect." The actual profile produced depends on a number of effects, namely the solvent evaporation rate across the drop surface, the speed of liquid flow within the drop, and the speed of solute diffusion in the solvent. As a rule of thumb, low-boiling point solvents tend to give a peak at the edge and high-boiling point solvents give a peak in the middle. This means that a sufficiently flat film usually requires a blend of two appropriate solvents. Cross sections of polymer pixels are shown in Figure 8.12.

8.3.2.2.1 Other printing techniques

Various other printing techniques, including (among many others) Laser Induced Thermal

Imaging (LITI) [28], Slot Coating [29], Gravure printing [30], and Nozzle coating [31], have been proposed and demonstrated for patterning the organic layers in a display. Each has shown some promise but each has also its own, often unique, demands on OLED ink properties. At the time of writing, inkjet printing remains the main concentration of both materials and process development for solution-processed patterning of OLED displays.

8.3.3 Cathode and encapsulation

8.3.3.1 CATHODE

8.3.3.1.1 Cathode materials

There are, broadly, two approaches to injecting electrons into an OLED device. A cathode metal can be chosen with a work function suitable for direct injection of electrons into the organic layer, or a highly doped organic layer can be used between the cathode and the organic layer. In either case it is desirable to ensure that all OLED devices in a display can use a common cathode structure to avoid the need for cathode patterning.

The energy level of electrons injected into an organic semiconductor is known as the LUMO (lowest unoccupied molecular orbital) and is equivalent to the conduction band of a regular semiconductor. The work function (the energy required for an electron to escape from this level into a vacuum) is typically relatively low (2–3 eV). Metals that have a similarly low work function are typically the group I and II metals, such as calcium (2.9 eV), barium (2.7 eV), and lithium (2.9 eV). These would usually be deposited as a very thin layer, just thick enough to ensure a continuous film (~5 nm), and then coated over with, e.g., aluminum, both for protection and to provide more conduction. The protection is required because a low work function also means that these metals are very reactive precisely because it is easy for a reactant (e.g., oxygen or water) to liberate an electron causing the metal to oxidize.

Improved device performance has been reported for both small molecules and polymers, using an interface layer of lithium fluoride (or other group-one fluorides) between the organic layer and the cathode metal. These materials are insulators in the bulk but act to enhance electron injection when less than a critical thickness (around 10 nm).

The mechanism of this improvement is not fully understood but probably relates to interaction with both the organic layer and the cathode metal [32].

A more recent alternative to highly reactive metal cathodes is to use a highly doped organic layer [33]. This can be done just on the cathode side or can be used to construct a complete PIN diode structure (P-type doped at the anode, Intrinsic (undoped) in the middle, and N-type doped at the cathode). For this to work well, other layers are often required, such as selective carrier blocking layers, and indeed some designs of vacuum deposited small molecule devices can have 15 or more layers where the doped layers are used to stack multiple OLED devices in series often to use multiple emitters to obtain white-light emission (see, e.g., Reference [34]).

8.3.3.1.2 Additional metallization

In many display designs, additional metal tracking is needed to reduce the lateral resistance of the transparent anode. Without this, pixels a long way from the edge would experience a large voltage drop. This metal can be deposited and patterned using conventional semiconductor processing methods. It is usual to use aluminum with an adhesion layer of chrome.

8.3.3.1.3 Processing

The standard means of providing a metallic cathode contact is thermal evaporation in high vacuum, using a separate metal shadow mask in close contact with the display plate to define separate connections to each row of pixels in the display. OLED materials are sensitive and it has been found that the simplest method of depositing metal without damaging the polymer is thermal evaporation with a low evaporation rate. If other techniques are used, such as electron beam evaporation or sputtering, great care must be taken to prevent exposure of the organic layers to high-energy particles (electrons or ions). The injection of electrons into the OLED structure depends on contact with a low-work function electrode, for example, calcium or barium. However, the low work function means that these materials are inevitably highly reactive. It is usual, therefore, to evaporate the minimum thickness of the primary electrode material required to guarantee a continuous film and then provide a backing layer of a less reactive metal, such as aluminum or silver. The backing metal acts as a diffusion barrier and provides a low resistance to lateral current flow.

8.3.3.2 ENCAPSULATION AND FINISHING

The active layers in a display require encapsulation to provide both mechanical and chemical protection. Providing mechanical protection is straightforward, but to chemically protect the OLED device from reacting with atmospheric oxygen and moisture is extremely challenging. The typically quoted requirement is for a water vapor transmission rate below 10^{-6} g/m²/day [22].

The standard solution is to attach a metal or glass "can" over the active display, using a UV-cure resin seal around the perimeter. The edge seal is the weak point, which allows some ingress. To reduce the impact on the display to an acceptable level, it is necessary to include a desiccant or "getter" inside the package. This is a reactive material with a high surface area that captures most of the offending water molecules before they can attack the display.

A potentially lower cost solution to encapsulation is well established in the semiconductor industry to provide the required diffusion barrier by vacuum deposition of a thin conformal hermetic coating. Silicon nitride, aluminum oxide, and aluminum nitride are examples of suitable material. The organic materials in an OLED device require that the encapsulation is deposited at relatively low temperatures, which makes it much harder to produce continuous films with no defects such as pinholes. The typical technique for OLED displays is to deposit multiple alternating layers of an inorganic barrier and organic buffer layer [35] to prevent propagation of pinholes in the same location. Such an encapsulation stack could be considered on its own if good enough or in combination with other encapsulation methods.

After encapsulation, there are a variety of operations, the details of which are outside the scope of this chapter. A typical sequence would be test the displays, scribe and break the substrate into individual displays, attach electrical connections to the drive electronics, and test again.

8.4 SUMMARY

In the 30 years since invention, OLEDs have developed from a barely visible laboratory curiosity to full commercial production for a range of display sizes from mobile devices to large TVs.

By going step by step through OLED display operation, performance and fabrication process, we hope to have shown the reader where the strengths and weaknesses of the technology lie, both for organics in general and the differences between polymers and small molecule-based devices.

Organic devices make bright, attractive, and colorful flat panel emissive displays. They have the fundamental advantages over other display technologies of a simple structure and low-voltage operation coupled with an intrinsically wide viewing angle, near-ideal contrast ratio, and an extremely fast response time.

Vacuum-processed small molecule devices currently represent the vast majority of commercial OLED production. Vacuum processing, however, has a significant hurdle to cross to achieve a high yield on large RGB-patterned OLED displays, particularly with the push to 4K and even 8K resolutions—at present all large format displays in production use an unpatterned white OLED layer plus color filters to produce RGB emitting pixels. Solution-processed polymer and small molecule devices patterned with a printing process such as inkjet printing has, perhaps, the greater potential for low-cost large area manufacture, but still needs to be proven as a viable display manufacturing technology.

REFERENCES

1. Tang, C. W. and Van Slyke, S. A. 1987. Organic electroluminescent diodes. *Appl. Phys. Lett.* 51: 913–915.
2. Burroughes, J. H., Bradley, D. D. C., Brown, A. R., Marks, R. N., MacKay, K., Friend, R. H., Burn, P. L., and Holmes, A. B. 1990. Light-emitting diodes base on conjugated polymers. *Nature* 347: 539–541.
3. Partridge, R. H. 1983. Electroluminescence from polyvinylcarbazole films. *Polymer* 24: 733–762.
4. Helfrich, W. and Schneidere, W. G. 1965. Recombination radiation in anthracene crystals. *Phys. Rev. Lett.* 14: 229.
5. Wilson, J. S., Dhoot, A. S., Seeley, A. J. A. B., Khan, M. S., Kohler, A., and Friend, R. H. 2001. Spin-dependent exciton formation in π-conjugated compounds. *Nature* 413: 828–831.
6. Murgatroyd, P. N. 1970. Theory of space-charge-limited current enhanced by Frenkel effect. *J. Phys. D: Appl. Phys.* 3: 151.
7. Kim, J.-S., Ho, P. K. H., Greenham, N. C., and Friend, R. H. 2000. Electroluminescence emission pattern of organic light-emitting diodes: implications for device efficiency calculations. *J. Appl. Phys.* 88: 131.
8. Yamae, K., Tsuji, H., Kittichungchit, V., Ide, N., and Komoda, T. 2014. Highly efficient white organic light-emitting diodes with over 100 lm/W for next-generation solid-state lighting. *J. Soc. Inf. Disp.* 21: 529–540.
9. Brütting, W., Frischeisen, J., Schmidt, T. D., Scholz, B. J., and Mayr, C. 2013. Device efficiency of organic light-emitting diodes: progress by improved light outcoupling. *Phys. Status Solidi A* 210: 44–65.
10. Laaperi, A. 2008. OLED lifetime issues from a mobile-phone-industry point of view. *J. Soc. Inf. Disp.* 16: 1125–1130.
11. Smith, E. C. 2008. Total matrix addressing. *J. Soc. Inf. Disp.* 16: 201–209.
12. Xu, C., Karrenbauer, A., Soh, K. M., and Codrea, C. 2008. Consecutive multiline addressing: a scheme for addressing PMOLEDs. *J. Soc. Inf. Disp.* 16: 211–219.
13. Mo, Y. G., Kim, M., Kang, C. K., Jeong, J. H., Park, Y. S., Choi, C. G., Kim, H. D., and Kim, S. S. Amorphous-oxide TFT backplane for large-sized AMOLED TVs. *J. Soc. Inf. Disp.* 19: 16–20.
14. Brody, T. P., Luo, F. C., Szepsi, Z. P., and Davies, D. H. 1975. A 6 × 6-in. 20-lpi electroluminescent display panel. *IEEE Trans Electron. Dev.* ED-22: 739–748.
15. Dawson, R. M. A., Shen, Z., Furst, D. A., Connor, S., Hsu, J., Kane, M. G., Stewart, R. G., Ipri, A., King, C. N., Green, P. J., and Flegal, R. T. 1998. The impact of the transient response of organic light-emitting diodes on design of active matrix OLED displays. IEEE International Electron Device Meeting, IEDM'98 Technical Digest., pp. 875-878.
16. Hunter, I. M., Young, N. D., Johnson, M. T., and Young, E. W. A. 1999. Design of an active matrix polymer-LED display with reduced horizontal cross-talk. *SID Proceedings of the Sixth International Display Workshop*, Vol. 99, pp. 1095–1096.

17. He, Y., Hattori, R., and Kanicki, J. 2000. Current-source a–Si:H thin-film transistor circuit for active-matrix organic light-emitting displays. *IEEE Electron. Dev. Lett.* 21: 590.

18. Sasaoka, T., Sekiya, M., Yumoto, A., Yamada, J., Hirano, T., Iwase, Y., Yamada, T., Ishibashi, T., Mori, T., Asano, M., and Tamura, S. 2001. A 13.0-inch AM-OLED display with top emitting structure and adaptive current mode programmed pixel circuit (TAC). *SID Int. Symp. Dig. Tech. Pap.* 32: 384–386.

19. Shin, D. Y., Woo, J. K., Hong, Y., Kim, K. N., Kim, D. I., Yoo, M. H., Kim, H. D., and Kim, S. 2009. Reducing image sticking in AMOLED displays with time-ratio gray scale by analog calibration. *J. Soc. Inf. Disp.* 17: 705–713.

20. Langendijk, E. H. A. and Hammer, M. 2010. Contrast requirements for OLEDs and LCDs based on human eye glare. *SID Int. Symp. Dig. Tech. Pap.* 41: 192–194.

21. Lim, K., Park, S., Lee, D. G., Kim, J., Shin, H., Park, J., and Lim, M. 2013. Estimation and evaluation of image sticking on OLED device. *SID Int. Symp. Dig. Tech. Pap.* 44: 1152–1154.

22. Burrows, P. E., Graff, G. L., Gross, M. E., Martin, P. M., Hall, M., Mast, E., Bonham, C. C., Bennett, W. D., Michalski, L. A., Weaver, M. S., and Brown, J. J. 2001. Gas permeation and lifetime tests on polymer OLEDs. *Proc. SPIE* 4105: 75–83.

23. Nomura, K., Ohta, H., Takagi, A., Kamiya, T., Hirano, M., and Hosono, H. 2004. Room-temperature fabrication of transparent flexible thin-film transistors using amorphous oxide semiconductors. *Nature* 432: 488.

24. Hayashi, R., Sato, A., Ofuji, M., Abe, K., Yabuta, H., Sano, M., Kumomi, H., Nomura, K., Kamiya, T., Hirano, M., and Hosono, H. 2008. Improved amorphous In–Ga–Zn–O TFTs. *SID Int. Symp. Dig. Tech. Pap.* 39: 621–624.

25. Seki, S., Uchida, M., Sonoyama, T., Ito, M., Watanabe, S., Sakai, S., and Miyashita, S. 2009. Current status of printing OLEDs. *SID Int. Symp. Dig. Tech. Pap.* 40: 593–596.

26. Groenendaal, L., Jonas, F., Freitag, D., Pielartzik, H., and Reynolds, J. R. 2000. Poly(3, 4-ethylenedioxythiophene) and its derivatives: past present and future. *Adv. Mater.* 12: 481–494.

27. Duinevald, P. C., de Kok, M. M., Buechel, M., Sempel, A. H., Mutsaers, K. A. H., van de Weijer, P., Camps, IG J, van den Biggelaar, T. J. M., Rubingh, J. J. M., and Haskel, E. I. 2002. Ink-jet printing of polymer light-emitting devices. *Proc. SPIE* 4464: 59–67.

28. Chung, H. K., Lee, K. Y., Lee, S. T. 2006. Alternative approach to large-sized AMOLED HDTV. *J. Soc. Inf. Disp.* 14: 49–55.

29. Faircloth, T. J., Innocenzo, J. G., and Lang, C. D. 2008. Slot die coating for OLED displays. *SID Int. Symp. Dig. Tech. Pap.* 39: 645–647.

30. Nakajima, H., Morito, S., Nakajima, H., Takeda, T., Kadowaki, M., Kuba, K., Handa, S., and Aoki, D. 2005. Flexible OLEDs poster with gravure printing method. *SID Int. Symp. Dig. Tech. Pap.* 36: 1196–1199.

31. Chesterfield, R. J., Frischknecht, K. D., Stainer, M., Truong, N., Murai, M., Oze, H., Shiota, A., and Suzuki, S. 2009. Multinozzle printing: a cost-effective process for OLED display fabrication. *SID Int. Symp. Dig. Tech. Pap.* 40: 951–954.

32. Brown, T. M., Friend, R. H., Millard, I. S., Lacey, D. J., Burroughes, J. H., and Cacialli, F. 2001. Efficient electron injection in blue-emitting polymer light-emitting diodes with LiF/Ca/Al cathodes. *Appl. Phys. Lett.* 79: 174–176.

33. Huang, J., Pfeiffer, M., Werner, A., Blochwitz, J., Leo, K., and Liu, S. 2002. Low-voltage organic electroluminescent devices using pin structures. *Appl. Phys. Lett.* 80: 139.

34. He, G., Rothe, C., Murano, S., Werner, A., Zeika, O., and Birnstock, J. 2009. White stacked OLED with 38 lm/W and 100,000-hour lifetime at 1000 cd/m2 for display and lighting applications. *J. Soc. Inf. Disp.* 17: 159–165.

35. Lewis, J. S., and Weaver, M. S. 2004. Thin-film permeation-barrier technology for flexible organic light-emitting devices. *IEEE J. Sel. Top. Quantum Electron* 10: 45–57.

FURTHER READING

Bulovic, V., Burrows, P. E., and Forrest, S. R. 1999. Molecular organic light-emitting devices. *Semiconductors and Semimetals*, vol. 64, ed. G. Mueller. New York: Academic, pp. 255–306.

Bulovic, V., and Forrest, S. R. 2000. Polymeric and molecular organic light-emitting devices: a comparison. *Semiconductors and Semimetals*, vol. 65, eds G. Mueller, R. K. Willardson and E. R. Weber. New York: Academic, pp. 1–26.

Cambridge Display Technology: www.cdtltd.co.uk.

Displaymate: www.displaymate.com especially www.displaymate.com/shootout.html.

Hartnagel, H. L., Dawar, A. L., Jain, A. K., and Jagadish, C. 1995. *Semiconducting Transparent Thin Films*. Bristol: Institute of Physics Publishing.

Novaled: www.novaled.com.

Sato, Y. 1999. Organic LED system considerations. *Semiconductors and Semimetals*, vol. 64, ed. G. Mueller. New York: Academic, pp. 209–254.

Sumitomo Chemical: www.sumitomo-chem.co.jp/english/pled/.

Tsujimura, T. 2012. *OLED Display Fundamentals and Application*. Wiley. ISBN: 978-1-118-14051-2.

Universal Display Corporation: www.udcoled.com.

Three-dimensional display systems

NICK HOLLIMAN
University of Durham

9.1 INTRODUCTION

Today's three-dimensional display systems provide new advantages to end-users; they are able to support an auto-stereoscopic, no-glasses, three-dimensional experience with significantly enhanced image quality over previous generation technology. There have been particularly rapid advances in personal auto-stereoscopic three-dimensional display for desktop users brought about because of the opportunity to combine micro-optics and LCD displays coinciding with the availability of low-cost desktop image processing and three-dimensional computer graphics systems.

In this chapter, we concentrate our detailed technical discussion on personal three-dimensional displays designed for desktop use as these are particularly benefiting from new micro-optic elements. We emphasize the systems aspect of three-dimensional display design believing it is important to combine good optical design and engineering with the correct digital imaging technologies to obtain a highquality three-dimensional effect for end-users. The general principles discussed will be applicable to the design of all types of stereoscopic three-dimensional display.

9.2 HUMAN DEPTH PERCEPTION

Defining the requirements for three-dimensional display hardware and the images shown on them is an important first step towards building a high-quality three-dimensional display system. We need a clear understanding of how a digital stereo-scopic image is perceived by an end-user in order to undertake valid optimization during the design process.

Binocular vision provides humans with the advantage of depth perception derived from the small differences in the location of homologous, or corresponding, points in the two images incident on the retina of the eyes. This is known as stereop-sis (literally solid seeing) and can provide precise information on the depth relationships of objects in a scene.

The human visual system also makes use of other depth cues to help interpret the two images incident on the retina and from these build a men-tal model of the three-dimensional world. These include monocular depth cues (also known as pictorial [18] or empirical [39] cues), whose signifi-cance is learnt over time, and oculomotor cues in addition to the stereoscopic cue [39]. We consider these in turn and introduce in detail binocular vision both in the natural world and when looking at an electronic three-dimensional display.

9.2.1 Monocular and oculomotor depth cues

Redundancy is built into the visual system and even people with monocular vision are able to per-form well when judging depth in the real world. Therefore, in the design of three-dimensional displays, it is important to be aware of the major contribution of monocular two-dimensional depth cues in depth perception and aim to provide dis-plays with at least as good basic imaging perfor-mance as two-dimensional displays. Ezra [12] suggests this should include levels of brightness, contrast, resolution and viewing range that match a standard two-dimensional display with the addi-tion of the stereoscopic cue provided by generating a separate image for each eye.

The monocular depth cues are experiential and over time observers learn the physical significance of different retinal images and their relation to objects in the real world. These include

- *Interposition*: objects occluding each other sug-gest their depth ordering.
- *Linear perspective*: the same size object at dif-ferent distances projects a different size image onto the retina.

- *Light and shade*: the way light reflects from objects provides cues to their depth relation-ships; shadows are particularly important in this respect.
- *Relative size*: an object with smaller retinal image is judged further away than the same object with a larger retinal image.
- *Texture gradient*: a texture of constant size objects, such as pebbles or grass, will vary in size on the retina with distance.
- *Aerial perspective*: the atmosphere affects light traveling through it, for example due to fog, dust or rain. As light travels long distances, it is scattered, colors lose saturation, sharp edges are diffused and color hue is shifted towards blue.

Many of these cues are illustrated in Figure 9.1 and can be considered to be two-dimensional depth cues because they are found in purely monoscopic images. Two other non-binocular depth cues are available: motion parallax and oculomotor cues.

Motion parallax provides the brain with a pow-erful cue to three-dimensional spatial relation-ships without the use of stereopsis [18,39] and this is the case when either an object in the scene or the observer's head moves. Motion parallax does not, however, make stereopsis redundant, as compre-hending images of complex scenes can be difficult without binocular vision. Yeh [66] and others have shown that both stereopsis and motion parallax combined result in better depth perception than either cue alone.

Figure 9.1 Picture illustrating the depth cues available in a two-dimensional image. (Photographer David Burder.)

Oculomotor depth cues are due to feedback from the muscles used to control the vergence and accommodation of the eye. They are generally regarded as having limited potential to help depth judgement [16,39,41] and we will move on to consider how human binocular vision works when used to view the natural world.

9.2.2 Binocular depth perception in the natural world

Extracting three-dimensional information about the world from the images received by the two eyes is a fundamental problem for the visual system. In many animals, perhaps, the best way of doing this comes from the binocular disparity that results from two forward facing eyes having a slightly different viewpoint of the world [5]. The binocular disparity is processed by the brain giving the sensation of depth known as stereopsis.

Stereo depth perception in the natural world is illustrated in Figure 9.2. The two eyes verge the visual axes so as to fixate the point F and adjust their accommodation state so that points in space at and around F come into focus.

The vergence point, F, projects to the same position on each retina and therefore has zero retinal disparity, i.e., there is no difference between its location in the left and right retinal images. Points in front or behind the fixation point project to different positions on the left and right retina and the resulting binocular disparity between the point in the left and right retinal images provides the observer's brain with the stereoscopic depth cue. Depth judgement is therefore relative to the current vergence point, F, and is most useful to make judgements on the relative rather than absolute depth of objects in a scene.

Points in space, other than F, which project zero retinal disparity are perceived to lie at the same depth as the vergence point; all points that project zero retinal disparity are described as being on a surface in space known as the horopter. The shape of the horopter shown in Figure 9.2 is illustrative only; it is known in practice to be a complex shape and to have non-linear characteristics [3,18].

Geometrically, we can define angular disparity, a, as the difference between the vergence angle at the point of fixation, F, and the point of interest. Considering Figure 9.3: Points behind the fixation point, such as A, have positive disparity.

$$\alpha_a = f - a. \tag{9.1}$$

Points in front of the fixation point, such as B, have negative disparity.

$$\alpha_b = f - b. \tag{9.2}$$

The smallest perceptible change in angular disparity between two small objects is referred to as stereo acuity, δ [65]. The advantage of defining stereo acuity as an angle is that it can be assumed to be constant regardless of the actual distance to and between the points A and B. However, it is also helpful to know how this translates in terms of the smallest perceived distance between objects at the typical viewing range of a desktop three-dimensional display. This will allow us to compare the

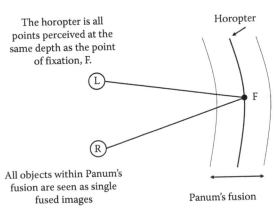

Figure 9.2 The geometry of the binocular vision when viewing the natural world.

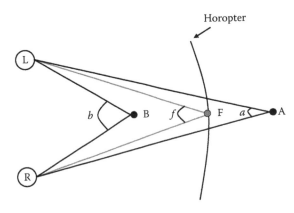

Figure 9.3 Angular disparity is defined relative to the current fixation point.

ability of the eye to perceive depth with the ability of different displays designs to reproduce it.

Considering Figure 9.4 when points A and C can just be perceived to be at a different depth, then stereo acuity will be

$$\delta = a - c. \tag{9.3}$$

Various studies [28,31,65] show the eye is able to distinguish very small values of δ, as little as 1.8 in. (seconds of arc). As the exact limits vary between people, Diner and Fender [8] suggest that a practical working limit is to use a value of stereo acuity $\delta = 20$ in. Using this value, we can calculate the size of the smallest distinguishable depth difference at a given distance from the observer. We choose $m = 750$ mm as the distance from the observer as a common viewing distance for desktop stereoscopic displays and use an average eye separation, $e = 65$ mm.

Calculating along the center line between the visual axes, we can find the minimum distinguishable depth, n, at distance m by considering points A and C. The angle a can be calculated as

$$a = 2 \times \arctan\left(\frac{e/2}{m}\right) = 2 \times \arctan\left(\frac{32.5}{750}\right) \tag{9.4}$$

by the definition of stereo acuity we know that

$$\tan(c/2) = \tan\left(\frac{a-\delta}{2}\right) = \tan\left(\frac{a - 20\,\text{in.}}{2}\right) \tag{9.5}$$

and if n is the distance between A and C we also know that

$$\tan(c/2) = \left(\frac{e/2}{m+n}\right) \tag{9.6}$$

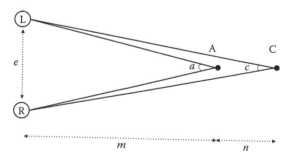

Figure 9.4 Stereo acuity defines smallest depth difference an observer can perceive.

rearranging Equation 9.6 we have

$$n = \left(\frac{e/2}{\tan(c/2)}\right) - m \tag{9.7}$$

Substituting Equation 9.4 in Equation 9.5 and using the result to solve Equation 9.7 gives $n = 0.84$ mm.

We can conclude that a person with a stereo acuity of 20 in. and an eye separation of 65 mm will be able to perceive depth differences between small objects of just 0.84 mm at a distance of 750 mm from the eyes.

It is also possible to calculate a geometric value for the furthest possible range of stereo vision which occurs when the vergence angle between the two visual axes is equal to or less than the stereo acuity.

The distance m from the observer to the point A when the angle $a = \delta$ is given by

$$m = \left(\frac{e/2}{\tan(a/2)}\right) \tag{9.8}$$

Again taking $\delta = 20$ in. and $e = 65$ mm we get $m = 670$ m.

This means that points such as C at a distance of 670 m or more from the observer will not be able to be distinguished in distance from A using binocular vision alone. Just before this limit is reached, the smallest distinguishable depth difference between points will have increased to over 300 m and it is clear only gross differences in depth will be perceived at the furthest limits of stereoscopic perception.

To summarize the above, binocular vision uses the stereoscopic depth cue of retinal disparity to perceive an object's depth relative to the fixation point of the two eyes. At close and near range this provides a high degree of depth discrimination and even at tens of metres from the observer enables relative depth perception for larger objects.

9.2.3 Depth perception in electronic stereoscopic images

Wheatstone [60] demonstrated that the stereoscopic depth sensation could be recreated by showing each eye a separate two-dimensional image. The left and right eye views should be two-dimensional planar images of the same scene from slightly different viewpoints; the difference in the viewpoints generates disparity in the images. When the images are subsequently viewed, the observer perceives depth in the scene because the image disparity creates a retinal disparity similar, but not identical, to that seen when looking directly at a natural scene.

Wheatstone was able to demonstrate this effect by building the first stereoscope and many devices have since been invented for stereoscopic image presentation each with their own optical configurations. Reviews of these devices and the history of stereoscopic imaging are available in several sources [23,30,32,40,53].

To help characterize and compare the performance of different electronic three-dimensional display designs, we will consider the perception of depth in planar stereo image pairs and how this differs from the stereoscopic perception of depth in the natural world.

A key physiological difference is that although the eyes need to verge off the stereoscopic image plane to fixate points in depth, their accommodation state must always keep the image plane itself in focus. This requires the observer to be able to alter the normal link between vergence and accommodation and is one reason why images with large perceived depth are hard to view. This suggests that the perceived depth range in stereoscopic image pairs needs to be limited to ensure the observer will find a stereo image pair comfortable to view.

While there are several studies of the comfortable perceived depth range on electronic three-dimensional displays [17,64,65], it can be difficult to factor out variables relating to display performance from the results. Display variables include absolute values, and inter-channel variations, of brightness and contrast in addition to stereoscopic image alignment and crosstalk. All of these can affect the comfortable range of perceived depth on a particular display. For example, high-crosstalk displays generally do not support deep images as the ghosting effect becomes more intrusive to the observer as screen disparity is increased.

An analysis of the geometry of perceived depth assuming a display with ideal properties helps identify the geometric variables affecting perceived depth independently of the display used. Geometric models of perceived depth have been studied by Helmholtz [23] and Valyus [53] and

more recently in [8,24,27,64]. We present a simplified model in Figure 9.5 for discussion purposes which helps emphasize the key geometric variables affecting the perception of stereoscopic images.

Figure 9.5 shows the geometry of perceived depth for a planar stereoscopic display; for simplicity, we consider the geometry along the center line of the display only; more general expressions are available [23,64]. The viewer's eyes, L and R, are separated by the interoccular distance, e, and are at a viewing distance, z, from the display plane. The screen disparity between corresponding points in the left and right images, d, is a physical distance resulting from the image disparity which is a logical value measured in pixels. Image disparity is constant for a given stereo pair; however, screen disparity will vary depending on the characteristics of the physical display. Screen disparity in a pair of aligned stereo images is simply the difference of the physical x coordinates of corresponding points in the right x_r and left x_l images:

$$d = x_r - x_l \tag{9.9}$$

Two key expressions relating screen disparity to perceived depth can be derived from the similar triangles in Figure 9.5. Perceived depth behind the screen plane, i.e., positive values of d, is given by

$$p = \frac{z}{\left(\frac{e}{|d|}\right) - 1} \tag{9.10}$$

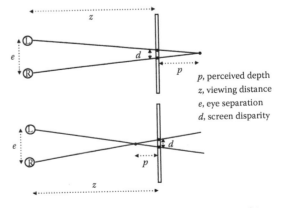

p, perceived depth
z, viewing distance
e, eye separation
d, screen disparity

Figure 9.5 Perceived depth behind (top) and in front (bottom) of the display plane.

Perceived depth in front of the screen plane, i.e., negative values of d, is given by

$$p = \frac{z}{\left(\dfrac{e}{|d|}\right)+1} \qquad (9.11)$$

Equations 9.10 and 9.11 provide several insights into the geometric factors affecting perceived depth:

- z, the viewing distance to the display. Perceived depth is directly proportional to the viewing distance, z. Therefore a viewer looking at the same stereoscopic image from different distances perceives different depth. How important this is, is application dependent, but applications such as CAD, medical imaging and scientific imaging may critically depend on accurate and consistent depth judgements.
- d, the screen disparity. Perceived depth is also directly proportional to screen disparity, d. The screen disparity for any given stereoscopic image varies if the image is displayed at different sizes, either in different size windows on the same screen or on different size screens. Again this is important to note in applications where depth judgement is a critical factor. It means stereoscopic images are display dependent and an image displayed on a larger display than originally intended could exceed comfortable perceived depth limits or give a false impression of depth.
- e, individual eye separation. Perceived depth is inversely proportional to individual eye separation which varies over a range of approximately 55–75 mm with an average value often taken as 65 mm. Children can have smaller values of eye separation and therefore see significantly more perceived depth in a stereoscopic image than the average adult. It may be particularly important to control perceived depth in systems intended for use by children, as they will reach the limits of their vergence/accommodation capabilities sooner than most adults.

For display design, controlling these variables so that the viewer sees a consistent representation of depth ideally requires tracking head position, identifying eye separation and controlling screen disparity. These are challenging goals in addition to designing a display with imaging performance as good as a two-dimensional display.

9.2.4 Benefits of binocular vision

An important question is what advantages does binocular vision provide in the real world? As a visual effect, it clearly fascinates the majority of people when they see a three-dimensional picture. Beyond the attractive nature of stereoscopic three-dimensional images, they provide the following benefits over monocular vision:

- *Relative depth judgement.* The spatial relationship of objects in depth from the viewer can be judged directly using binocular vision.
- *Spatial localization.* The brain is able to concentrate on objects placed at a certain depth and ignore those at other depths using binocular vision.
- *Breaking camouflage.* The ability to pick out camouflaged objects in a scene is probably one of the key evolutionary reasons for having binocular vision [47].
- *Surface material perception.* For example, lustre [23], sparkling gems and glittering metals are in part seen as such because of the different specular reflections detected by the left and right eyes.
- *Judgement of surface curvature.* Evidence suggests that curved surfaces can be interpreted more effectively with binocular vision.

These benefits make stereo image display of considerable benefit in certain professional applications where depth judgement is important to achieving successful results. In addition, the effect of stereopsis is compelling enough that stereoscopic images have formed the basis of many entertainment systems.

9.3 THREE-DIMENSIONAL DISPLAY DESIGNS USING MICRO-OPTICS

The possible combinations of LCD and micro-optics provide many degrees of freedom for display design; the ideal three-dimensional display design will depend on specific application requirements. However, there are characteristics that all display designs should give consideration to and we briefly review these here.

There is a need to compare the basic image quality of a three-dimensional design to that achieved by current two-dimensional displays; i.e., the two-dimensional characteristics of a three-dimensional display should match the performance of two-dimensional displays as closely as possible. Key characteristics are

- Brightness typical of a current LCD display is 150 cdm^{-2}
- Contrast typical of a current LCD display is 300:1
- Color reproduction, measured white points and measured CIE coordinates of primaries

These values are typical of current two-dimensional displays but are clearly a moving target as two-dimensional displays improve.

In addition, there are a number of important characteristics unique to three-dimensional displays. The first is that two-dimensional characteristics need to be matched between all the viewing windows of the three-dimensional display. Each viewing window should also be matched spatially and temporally so that there is no noticeable position or time difference between corresponding images.

Inter-channel crosstalk appears to an observer as a ghost image, which will be particularly visible at high-contrast edges in images. It is an unwanted feature in most display designs because high values of crosstalk are known to be detrimental to three-dimensional effect, particularly on high-contrast displays showing large values of perceived depth [42]. Ideally crosstalk measurements need to be no more than 0.3% if the ghosting effect is to be imperceptible to an observer. Crosstalk, although often due to optical effects in the display, can also result from poor separation of the two image channels in the display driving electronics, image compression formats or the camera system generating the images.

An observer of a two-dimensional display will usually expect to be able to see a good quality image at a wide range of positions in front of the display. Because of the need to direct images separately to the two eyes, many three-dimensional displays have a more limited viewing freedom. Consideration needs to be given to the targets for lateral, vertical and perpendicular freedom in a display design. Three-dimensional display systems capable of supporting multiple observers will often do so at the expense of viewing freedom. Improved viewing freedom can be found in designs with multiple viewing windows or using head tracking to steer viewing windows to follow the observers' head movements. When head tracking is used, a design needs to consider targets for the maximum supported head speed as this directly determines key tolerances.

Some displays have the capability to operate in either three-dimensional or two-dimensional modes switching electronically or mechanically between the two. In this case, the image quality in each mode needs to be considered against the performance of a standard two-dimensional display, as a display in three-dimensional mode will often have different optical performance to the same display in two-dimensional mode.

The capability of a three-dimensional display to represent perceived depth is probably the single most important design target; however, we will return to how to quantify and compare this between displays after presenting details of representative three-dimensional display designs.

We would like three-dimensional displays to provide the ability for the observer to accommodate naturally at the fixation point. However, this is not a feature supported in stereoscopic images and has been attempted in very few display designs.

9.3.1 Stereoscopic systems

Stereoscopic displays require users to wear a device, such as analysing glasses, that ensures left and right views are seen by the correct eye. Many stereoscopic display designs have been proposed and there are reviews of these in several sources [30,32,34,40,53]. Most of these are mature systems and have become established in several professional markets but suffer from the drawback that the viewer has to wear, or be very close to, some device to separate the left and right eye views. This has limited the widespread appeal of stereoscopic systems as personal displays for home and office use even when the three-dimensional effect is appealing. However, stereoscopic displays are particularly suited to multiple observer applications such as cinema and group presentation where directing individual images to each observer becomes difficult compared to providing each observer with a pair of analysing glasses.

As stereoscopic display systems are well described elsewhere, we limit ourselves here to a summary of the major types using electronic displays:

A spatially multiplexed image (SMI) with left (L) and right (R) image pixels is placed behind a patterned micro-polariser (uPol) element.

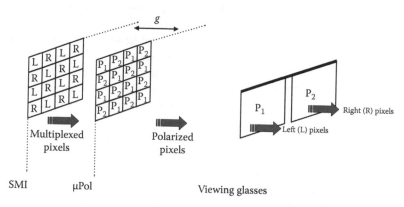

When viewed with polarized glasses the P1 polarized pixels are seen only in the left eye and P2 polarized in the right

Figure 9.6 The micro-polarizer stereoscopic

- Wheatstone mirror stereoscopes using CRT displays or LCD displays.
- Polarized glasses in combination with a method of polarizing the two views.
- Shutter glasses working in synchronization with a view switching display.
- Analglyph glasses analysing different color channels to obtain the images.
- Brewster stereoscopes, of which head mounted displays are up to date examples.

A series of stereoscopic display designs that use polarizing micro-optics have been produced [14], as shown in Figure 9.6. The micro-optics split a single display into two differently polarized views, which are viewed correctly by left and right eyes when the observer wears a pair of analysing polarized glasses. This requires two half resolution views and may be achieved using a chequerboard pattern of image multiplexing and polarization as shown in Figure 9.6 as the spatially multiplexed image (SMI) and patterned micro-polarizer (μPol).

A drawback of the design, particularly for direct view LCD-based displays, is the parallax between the display pixels and the micro-polarizer when the micro-polarizer is mounted over the LCD due to the layer of substrate between the two elements forming the gap g in Figure 9.6. If the head moves from the nominal viewing position, part of the adjacent view's pixel becomes visible resulting in crosstalk. One way to reduce this is to use interlace the images in alternate rows so at least lateral head movement is not affected by parallax. As noted by Harrold [20] this problem can only be fully solved in the long term by manufacturing the micro-polarizer element within the LCD pixel cells reducing the parallax between polarizer and pixel.

9.3.2 Auto-stereoscopic systems

Auto-stereoscopic displays are those that do not require the observer to wear any device to separate the left and right views and instead send them directly to the correct eye [6]. This removes a key barrier to acceptance of three-dimensional displays for everyday use but requires a significant change in approach to three-dimensional display design. Auto-stereoscopic displays using micro-optics in combination with an LCD element have become attractive to display designers and several new three-dimensional display types are now available commercially. The key optical reasons [62] for combining micro-optics with LCD elements are

- LCDs offer pixel position tolerances better than 0.1 μm.
- LCD pixels, unlike CRT pixels, have high positional stability.

- LCD elements have carefully controlled glass thickness.

Auto-stereoscopic displays have been demonstrated using a range of optical elements in combination with an LCD including

- Parallax barriers, optical apertures aligned with columns of LCD pixels.
- Lenticular optics, cylindrical lenses aligned with columns of LCD pixels.
- Micropolarizers are found in several auto-stereoscopic three-dimensional display designs.
- Holographic elements have been used to create real images of a diffuse light source.

In the following, we introduce how these elements are used in auto-stereoscopic three-dimensional display designs including two-view and multi-view designs. We begin by looking at auto-stereoscopic two-view designs using twin-LCD elements.

9.3.3 Two-view twin-LCD systems

A successful approach to building high-quality auto-stereoscopic displays has been to use two LCD elements and direct the image from one to the left eye and from the other to the right eye; the principle is illustrated in Figure 9.7. Several designs have adopted this approach including [12,13,22].

Ezra [12,13] describes one of the designs, which produces bright, high-quality, full color moving three-dimensional images over a wide horizontal viewing range. As shown in Figure 9.8, the display produces two viewing windows using a single

illuminator. The arrangement of optical elements generates horizontally offset images of the illuminator at a nominal viewing distance to form the viewing windows. An observer's eye placed in one of the viewing windows will see an image from just one of the LCD elements.

If a stereo pair of images is placed on the left and right LCD elements, respectively, then an observer will see a stereoscopic three-dimensional image. The image appears in the plane of the left LCD as the observer looks at the display and depth is perceived in front and behind this plane. As the two LCDs are seen separately, each eye has a full resolution image and the interface is simply two synchronized channels of digital or analogue video which can be generated at low cost on a desktop PC system.

This basic configuration can be enhanced in several ways: if the light source is moved then the viewing windows can be steered to follow the observer's head position. In order to implement window steering, new technologies for tracking head position have also been developed [25]. The effect of implementing head tracking linked to window steering is to increase the viewing freedom of the display and if the images are updated the design has been demonstrated to provide a full look-around effect. This allows the observer to look around the display and see different views of the scene as they would in the natural world. Image generation for look-around can be implemented by using a three-dimensional computer graphics system to generate the new views when given head tracking position information.

Another possible development [13] is to have multiple light sources providing multiple stereo views to multiple viewers. This could be implemented either by sending the same image pair to each viewer, or by time slicing the light source and the displays to send a different image to each viewer in rapid succession.

The system uses bulk optics and therefore has a large footprint, particularly as the LCD display diagonal size increases. This led to the micro-optic twin-LCD display [61] which provides the same effect in a smaller footprint and is more practical for scaling to larger display sizes.

The micro-optic twin-LCD display is illustrated in Figure 9.9. The two LCD elements remain in the design with a half mirror acting as a beam combiner between them. The arrangement of optical elements

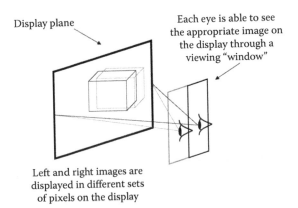

Display plane

Each eye is able to see the appropriate image on the display through a viewing "window"

Left and right images are displayed in different sets of pixels on the display

Figure 9.7 Two-view displays create two viewing windows.

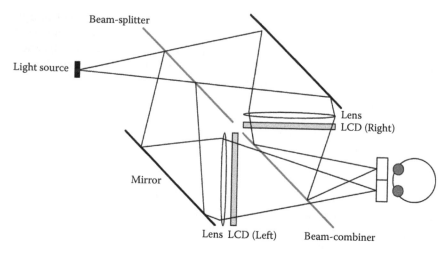

Figure 9.8 The twin-LCD display [12].

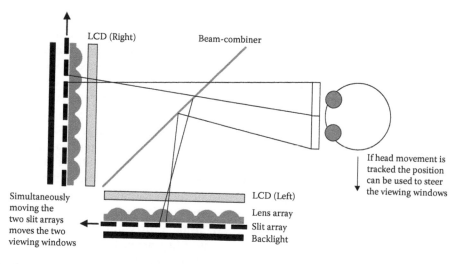

Figure 9.9 The micro-optic twin-LCD display [61].

behind one LCD panel directs light so that it forms one viewing window at a nominal viewing distance from the display, another is formed adjacent to this from the backlight of the other LCD panel. As with the bulk optic display the observer placing their eyes in the viewing windows will see the appropriate image in each eye and experience a stereoscopic three-dimensional effect.

As discussed in Reference [61] the micro-optic display produces a better viewing window profile than the bulk-optic display. This is because the micro-optics form a wider and more even illumination distribution for each viewing window so that, when steered, the windows can be moved further laterally before aberrations reduce their quality. This also results in

side lobes of better quality, which in untracked displays can be used by additional observers.

9.3.4 A note on viewing windows

One of the key influences on the perceived performance of auto-stereoscopic displays is the quality of the viewing windows that can be produced at the nominal viewing position. Degradation of the windows due to unresolved issues in the optical design can lead to flickering in the image, reduced viewing freedom and increased inter-channel cross-talk. All of these reduce the quality of viewing experience for observers in comparison to the two-dimensional displays they are used to using.

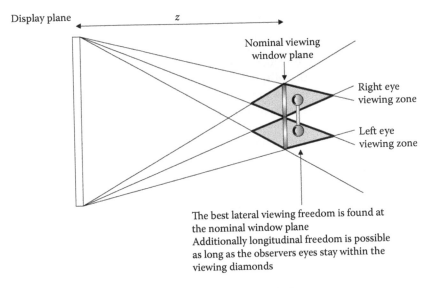

Figure 9.10 Viewing freedom in an auto-stereoscopic display [61].

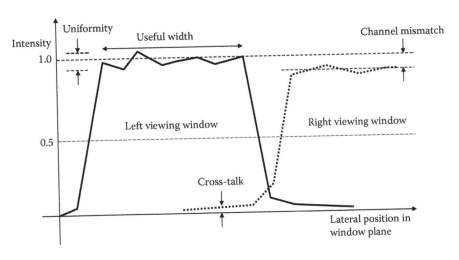

Figure 9.11 The characteristics of a viewing window [63].

In addition, in head tracked systems degraded window quality can lead to harder constraints on the accuracy and response speed of the tracking and window steering systems, increasing system costs [25].

The auto-stereoscopic displays considered so far produce two viewing windows in space typically at a nominal distance from the display in a plane parallel to the display surface, as shown in Figure 9.7. Although often illustrated in two dimensions, the viewing windows have a three-dimensional shape and from above appear as diamonds tapering away from the nominal viewing plane as shown in Figure 9.10. As long

as an observer's pupils stay within these diamonds, and the display is showing a stereo image, the observer will see a three-dimensional image across the whole of the display.

Experimentally the window intensity profile can be determined by measurements using a 1 mm pinhole, a photometric filter and a detector. To fully characterize a display performance, the profile measurements should be repeated at a range of positions vertically and longitudinally offset from the nominal window position. The variables characterizing the quality of the viewing windows are discussed in [61] and are summarized here in Figure 9.11.

The useful width of the window determines how far an observer can move before the image quality degrades. Larger useful width, up to the interoccular separation, typically 65 mm, provides more comfortable viewing in fixed position displays as there will be a small but useful lateral range of head positions at which a good three-dimensional image can be seen.

A systems benefit of wider viewing windows is that it helps relax the tolerances required for window steering and tracking mechanisms in head-tracked displays such as [12,61]. This is because a wider viewing window allows more time and/or distance before the steering and tracking mechanisms have to respond to user head movement in order to prevent the user moving out of the useful width and seeing a degraded image on the display.

9.3.5 Two-view single-LCD systems

Even with the advantages of a micro-optic design twin-LCD three-dimensional displays have a component cost that must include two LCD elements. This cost is acceptable in some applications when image quality is the key requirement; however, for the mass market, i.e., personal office and home use, it is desirable to find display designs based on a standard single LCD element.

We will group the single-LCD auto-stereoscopic designs by the type of optical element used to generate the viewing windows, beginning with the parallax barrier.

9.3.5.1 PARALLAX BARRIER DESIGNS

Typical emissive displays have pixels with diffuse radiance, that is they radiate light equally in all directions. To create a twin-view autostereoscopic display, half the pixels must only radiate light in directions seen by the left eye and half the pixels in directions seen by the right eye. The parallax barrier is perhaps the simplest way to do this and works by blocking light using strips of black mask.

The principle of the two-view parallax barrier is illustrated in Figure 9.12. The left and right images are interlaced in columns on the display and the parallax barrier positioned so that left and right image pixels are blocked from view except in the region of the left and right viewing windows, respectively. Although not illustrated the viewing windows repeat in side lobes to each side of the central viewing position and can be used by more than one observer if the optical quality remains high enough.

The pixels and barrier are arranged so the center of each pair of left and right view pixels is visible at the center of the viewing windows. The geometry defining the design of the parallax barrier pitch, b, can then be determined from considering similar triangles in Figure 9.12.

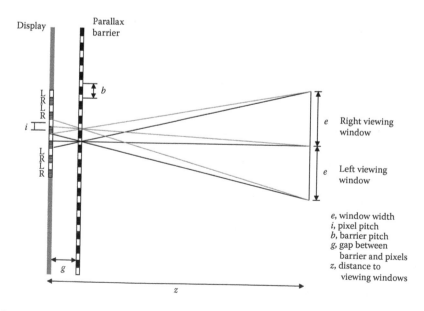

Figure 9.12 The principle of the front parallax barrier.

$$\frac{b}{z-g} = \frac{2i}{z} \qquad (9.12)$$

which can be rearranged to give

$$b = 2i\left(\frac{z-g}{z}\right) \qquad (9.13)$$

The result, Equation 9.13, is that the barrier pitch for a two-viewing window display is just less than twice the pixel pitch on the display. This small difference between the pixels and the barrier pitch accounts for the variation in viewing angle between the eyes and the pixels across the display and is often referred to as viewpoint correction.

Viewing distance, z, for the best quality viewing windows is another design factor and again from similar triangles in Figure 9.12, we can deduce a geometric relationship for this.

$$\frac{i}{g} = \frac{e}{z-g} \qquad (9.14)$$

which can be rearranged to give

$$z = g\left(\frac{e+i}{i}\right) \qquad (9.15)$$

The window width is typically set to the average eye separation, $e = 65$ mm, the pixel pitch, i, is defined by the display and the gap, g, between display and barrier is defined by the thickness of the front substrate on the LCD. For example, pixel width might be of the order $i = 0.1$ mm and the gap, including front substrate and polarizer, $g = 1.15$ mm. The result is relatively little control of the closest possible viewing distance and given current LCD substrate thickness many current parallax-barrier-based displays have optimal viewing distances of $z = 750$ mm.

More recent two-dimensional displays could use a substrate such as Corning Eagle[2000] with thickness from 0.4 to 0.63 mm and given a polarizer of thickness 0.2 mm may then be able to reduce viewing distance for a front parallax barrier to $z = 390$ mm. This compares favorably with the typical viewing distance of two-dimensional displays of 300–350 mm although care would be needed to avoid artefacts at the edges of the screen plane where the viewing angle increases with decreasing viewing distance.

Variations on the basic twin-view parallax barrier design and further practical issues are described by Kaplan [29] including a discussion of multi-view parallax barrier displays and aperture design.

Okoshi [40] notes that problems with parallax barriers include the reduced brightness due to blocking the light from pixels, reflection from the glass surface of the parallax barrier and the design of the parallax apertures to avoid diffraction problems. However, these disadvantages have been addressed and recent LCD-based designs overcome the first two problems by using bright light sources and antireflection-coated optics. The result is parallax barriers are now widely used for two-view displays such as described [62,63] and illustrated in Figure 9.13.

The diffraction problem is more serious but has also recently been addressed. An ideal display would have viewing windows described by a top hat function; however, in practice they have the characteristics shown in Figure 9.11. A number of factors determine this and an important one is the detailed design of the parallax barrier apertures, w, shown in Figure 9.13. A wider aperture results in a brighter image but reduces the geometric performance of the aperture and creates less well-defined windows. A narrow aperture results in a less bright image with better window definition; however, too narrow an aperture suffers from diffraction effects which in turn results in less well-defined windows. In both cases, the crosstalk performance, useful width and uniformity of intensity at the viewing window are affected.

A detailed study of the barrier position, aperture design and related diffractive effects is presented in [35,62]. In [62] a comparison was made between placing the parallax barrier behind and in front of the LCD element. The analysis uses a model of the parallax barrier accounting for Fresnel diffraction and compares this to a set of experimental measurements. Placing the parallax barrier behind the display results in lower crosstalk while placing it in front of the display has very much better intensity uniformity and useful width at the window plane. These factors are decisive for tracking displays and hence the front position was adopted to build a single-LCD observer tracking display [62].

In [35] several apodization modifications to the parallax barrier are analysed; these include soft aperture edges, multiple sub-apertures at aperture

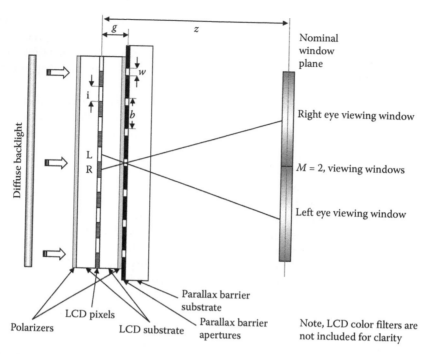

Figure 9.13 Detail of a single-LCD front parallax barrier design [62].

edges and combinations of the two techniques. The analysis concluded that choosing the correct apodization can make a substantial improvement to the window profile improving both the crosstalk performance and viewing freedom of the display. In particular, crosstalk of less than 1% is theoretically achievable using an improved parallax element; this is a significant improvement over the value of 3.5% achieved using unmodified apertures. These new studies show it is now possible to overcome the limitations of parallax barriers identified by Okoshi.

A practical problem encountered by users of two-view parallax barrier displays without head tracking is how to find the best viewing position. One reason is the parallax barrier produces not just the central two viewing windows but also repeated lobes to each side of these as illustrated in Figure 9.14a. An observer in position A will see an orthoscopic image (left image to left eye and right image to right eye) as will an observer in position D. However, an observer in the intermediate position C sees a pseudoscopic image (the left image in the right eye and the right image in the left eye). This causes problems as typically pseudoscopic images show false depth effect and it can be hard for novice observers to determine if they are seeing

a correct three-dimensional image or not. A number of devices have been proposed to help observers determine when they are in the correct viewing position; the VPI (viewing position indicator) display described in [62,63] achieves this by integrating an indicator into the parallax element.

The parallax barrier in the VPI display is divided into two regions: the image region, which is most of the display, and the indicator region, which may cover just the bottom few rows of pixels on the display. The result is shown in Figure 9.14a and b, respectively. In the image region the conventional barrier design allows the left and right views to be seen at the nominal viewing position A. In the indicator region, the display shows a pattern of red and black stripes and the barrier design is modified so that the indicator region shows black to both eyes only when the observer is in a position to see an orthoscopic three-dimensional image as at A. If the observer is approaching, as at B, or in, as at C, a pseudoscopic region he will see red in one eye in the indicator region indicating he should move laterally until returning to the orthoscopic zone. A drawback of the VPI design is that when the observer is in viewing zone D, she can see an orthoscopic image but the indicator will still show red. However, the observer is guaranteed

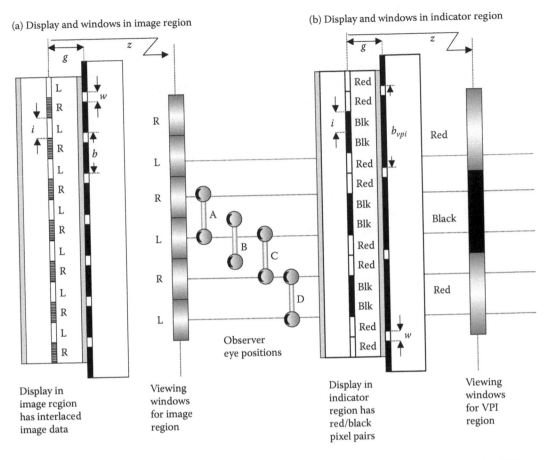

Figure 9.14 The VPI display operation (a) in the image region and (b) in the indicator region [62].

that whenever a black indicator region is seen, he will see an orthoscopic image on the display and this seems a reasonable trade-off.

The indicator region is implemented by using a barrier pitch in the indicator region double that used in the image region. As a result, the VPI display requires little additional design or manufacturing cost and uses only a few lines of pixels to display the appropriate indicator pattern. It has the benefit that once the parallax barrier is aligned for image viewing, the indicator mechanism is automatically aligned. The VPI also works to help guide observers find the best longitudinal viewing position if the aperture width, w, is kept the same in both the image and indicator regions of the parallax element.

A range of designs using parallax barrier optics in combination with LCD elements has been proposed, prototyped and commercialized.

A large range of display designs was developed using parallax barriers [19]. One example uses both

a rear and a front parallax barrier with the aim of reducing crosstalk, although no window profile measurements are given to say how successful this was. Because the combination of two parallax barriers reduces display brightness, the rear barrier was mirror coated on the side facing the illuminator to recirculate light. A further design using just a rear parallax barrier places an electronically switchable diffuser between the parallax barrier and the LCD element. This allows instantaneous switching between two-dimensional and three-dimensional modes and if the diffuser is programmable also allows three-dimensional windows to appear on a two-dimensional display. Several designs also combine a window steering mechanism and head tracker to increase lateral viewing freedom; one of these [26] uses an electronically programmable LC parallax barrier.

A design also based on an electronically programmable parallax barrier is described by Perlin [44–46]. A key goal for the design is to steer the viewing windows to track the viewer in three dimensions by varying the pitch and aperture of the parallax barrier in real time. The aim is to generate real time viewpoint correction so the viewer can vary position and still see a three-dimensional image across the whole display surface. The potential benefit of the design is in extending longitudinal movement with respect to the display and it is also capable of accounting for head rotation, which effectively varies the observer's eye separation. The design is relatively complex and before choosing this approach, it would be wise to make a comparison with the longitudinal freedom already available from a fixed aperture display with good quality viewing windows. In practice, realizing the display presents a number of challenges including the optical quality achievable from the programmable parallax element and the speed and latency targets with which the tracking and steering mechanisms need to work.

9.3.5.2 LENTICULAR ELEMENT DESIGNS

Lenticular elements used in three-dimensional displays are typically cylindrical lenses arranged vertically with respect to a two-dimensional display such as an LCD. The cylindrical lenses direct the diffuse light from a pixel so it can only be seen in a limited angle in front of the display. This then allows different pixels to be directed to either the left or right viewing windows.

The principle for a two-view lenticular element stereoscopic display is illustrated in Figure 9.15 and described in [50]. This shows the geometry for a viewpoint-corrected display where the pitch of the lenticular is slightly less than the pitch of the pixel pairs. As with parallax barrier displays the effect of viewpoint correction is to ensure pixels at the edge of the display are seen correctly in the left and right viewing windows. The lenticular pitch needs to be set so that the center of each pair of pixels is projected to the center of the viewing windows and this can be found by considering similar triangles where

$$\frac{2i}{z} = \frac{l}{z-f} \tag{9.16}$$

$$l = 2i\left(\frac{z-f}{z}\right). \tag{9.17}$$

Typically, the pixel pitch i is set by the choice of two-dimensional display and the minimum focal length, f, determined in large part by the substrate thickness on the front of the display. The viewing distance can again be derived from similar triangles:

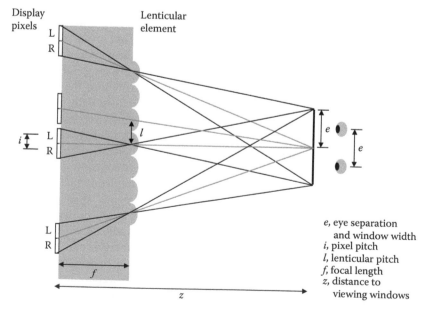

Figure 9.15 Front lenticular auto-stereoscopic display principle [50].

$$\frac{i}{f} = \frac{e}{z-f} \tag{9.18}$$

which can be easily rearranged to give

$$Z = f\left(\frac{e}{i}+1\right) \tag{9.19}$$

Typically, the window width for a two-view system is taken to be the average eye separation, $e=65$ mm, to give some freedom of movement (up to $e/2$) around the nominal viewing position. Combining this factor with the display-related values of i and f, it may be that there is again little choice over the closest possible viewing distance.

Lenticular elements have been used less often than parallax barriers in recent two-view display designs; one exception is the range of displays designed by the DTI corporation.

The DTI display design described by Eichenlaub [10,11] uses light guide and lenticular elements behind an LCD display to generate light lines that are functionally equivalent to having a rear parallax barrier. The principle of creating viewing windows using the light lines is shown in Figure 9.16. The pitch required for the light lines can be calculated using similar triangles as for the parallax barrier example discussed earlier.

$$\frac{b}{z+g} = \frac{2i}{z} \tag{9.20}$$

which can be rearranged to give

$$b = 2i\left(\frac{z+g}{z}\right). \tag{9.21}$$

In this case, the pitch of the light lines, b, is slightly larger than twice the pixel pitch to achieve viewpoint correction. Again the gap, g, will determine viewing distance and is likely to be constrained by the substrate glass thickness when using an LCD.

The backlight construction that creates the light lines is shown in Figure 9.17. A modified light guide uses a series of grooves to generate an initial set of light lines, which are then re-imaged by the lenticular element to form a larger number of evenly spaced light lines in front of the light guide.

A two-dimensional/three-dimensional switching diffuser in front of the lenticular element is made of polymer dispersed liquid crystal (PDLC) which when on is transparent allowing the display to operate in three-dimensional mode. When the PDLC is off it becomes a diffuser, scattering light and preventing the initial set of light lines reaching the lenticular lens. The result is a diffuse illumination for the display, which will operate with similar performance to a normal two-dimensional display. Various size displays have been constructed with 5.6 and 12.1 in displays having crosstalk of 3% and 6% and uniformity of 20% and 24%, respectively.

The DTI design has the advantage of being able to electronically switch between two-dimensional and three-dimensional illumination modes as well

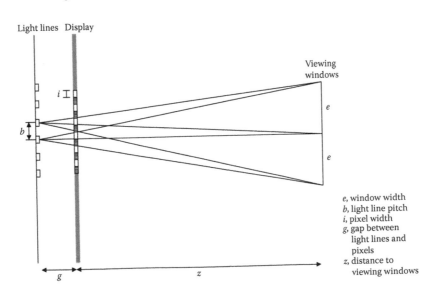

Figure 9.16 The geometry of rear parallax illumination by light lines.

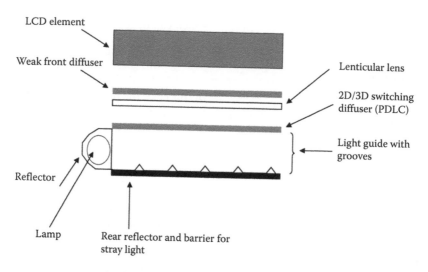

LCD element

Weak front diffuser

Lenticular lens

2D/3D switching diffuser (PDLC)

Light guide with grooves

Reflector

Lamp

Rear reflector and barrier for stray light

Figure 9.17 The DTI compact backlight allowing two-dimensional/three-dimensional illumination.

as being small enough to be used in portable display devices. In addition, there are no optical elements in front of the display surface allowing the observer to directly view the LCD display. Against this are some trade-offs and the three-dimensional mode has higher crosstalk than a well-designed parallax barrier system.

Other designs for single-LCD three-dimensional displays using lenticular optics include [37,38,43,61].

A novel design using micro-prism elements was proposed [48,49]. The D4D display uses an array of vertically oriented micro-prisms as the parallax element and the left and right images, vertically interlaced in columns, are directed to two viewing windows by the micro-prisms. A commercial display based on this principle included a head tracking device and both electronic image shifting and mechanical movement of the micro-prisms were investigated as ways to steer the viewing windows.

9.3.5.3 MICRO-POLARIZER DESIGNS

Displays using polarization to create light steering optical elements have been proposed by several groups. The stereoscopic display design described by Faris [14,15] can also be configured to have an auto-stereoscopic mode by using a series of stacked micro-polarizer elements to create a switchable parallax barrier. However, despite this potential for auto-stereoscopic operation most of the commercial products from VREX have been stereoscopic systems.

Harrold et al. describe display designs using micro-polarizers in [20,21]. The design exploits the polarized light output from an LCD element over which is created a patterned retarder array. A final polarizing layer is placed over the retarder array effectively creating a front parallax barrier and hence an auto-stereoscopic display. If the final polarizing layer is constructed so that it is removable, the display can be mechanically switched between a two-dimensional display mode and an auto-stereoscopic three-dimensional display mode (Figure 9.18).

Key to the success of this design is the construction of the patterned retarder array to an accuracy of better than 1 part in 2000 for the 13.8 in XGA display prototype. This was achieved using a process based on standard LCD manufacturing techniques to create a manufacturable patterned retarder array that is front mounted onto the LCD element.

A stereoscopic display design is also described by Harrold in [20] where the patterned retarder and polarizer are constructed inside the LCD element to avoid the parallax problems of the Faris design [14,15]. A prototype LC cell demonstrated the feasibility of this approach.

A micro-polarizer display described by Benton [1,2] uses a combination of polarization and bulk optics to create two viewing windows that can be steered electronically if a suitable head tracker is available. An LCD panel with the analysing polarizer removed acts as an electronically

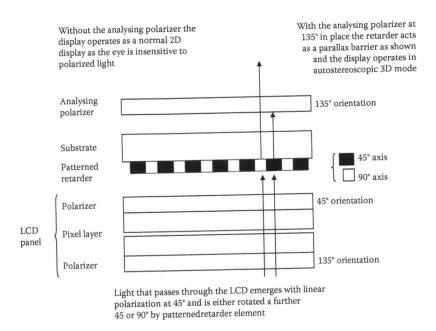

Without the analysing polarizer the display operates as a normal 2D display as the eye is insensitive to polarized light

With the analysing polarizer at 135° in place the retarder acts as a parallax barrier as shown and the display operates in autostereoscopic 3D mode

Analysing polarizer — 135° orientation

Substrate

Patterned retarder

■ 45° axis
□ 90° axis

Polarizer — 45° orientation

LCD panel

Pixel layer

Polarizer — 135° orientation

Light that passes through the LCD emerges with linear polarization at 45° and is either rotated a further 45 or 90° by patternedretarder element

Figure 9.18 The Sharp micro-polarizer display with two-dimensional/three-dimensional switching capability [20].

programmable polarizing light source: light coming from the light source LCD will be either rotated at 90° or not rotated. An illumination pattern of two blocks of light is displayed on the light source LCD, each polarized differently. A micropolarizer array arranged as rows behind an image LCD display allows alternate rows of image to be illuminated by differently polarized light and hence appear in the viewing windows for the left and right eyes. A large lens after the LCD produces an image of the viewer-tracking LCD (polarized light source) at the intended viewing distance of about 1 m creating the two viewing windows.

Benton notes there can be problems with the lens (a Fresnel lens) creating Moire patterns in association with the image display LCD. In common with many auto-stereoscopic displays the viewer has to be at or close to the nominal viewing distance, which at 1 m is significantly further than typical two-dimensional display viewing distances. No measurements of crosstalk or window brightness uniformity are given

9.3.5.4 HOLOGRAPHIC ELEMENTS

Holographic optical elements (HOEs) have been used [51,52] to create three-dimensional displays in conjunction with LCD elements. When

illuminated the HOE acts to form the viewing windows. The HOE is arranged in horizontal strips to reconstruct a real image of a diffuse illuminator; the strips are arranged so alternate strips reconstruct left and right viewing windows. When placed behind a display with two horizontally interlaced images, the observer will see an autostereoscopic image.

A number of practical problems in the optical design are discussed in [51] and in particular color fringing due to the diffractive nature of the HOE could prove difficult to overcome. Otherwise, this design has several advantages and can be modified to track users by moving the light source and also constructed so that it can be switched between two-dimensional and three-dimensional using a modified light source.

9.3.6 Multi-view systems

The viewing freedom of a three-dimensional display is a key requirement in certain applications, for example public information kiosks, where ease of viewing is needed to attract and retain the attention of passers-by. Multi-view systems, as in Figure 9.19, provide viewing freedom by generating multiple simultaneous viewing windows of which an

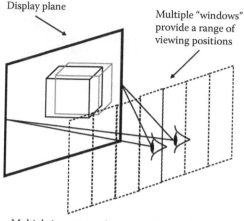

Display plane

Multiple "windows" provide a range of viewing positions

Multiple images are shown simultaneously and a single viewer sees any two of them at any time

Figure 9.19 Multi-view displays create multiple viewing windows.

observer sees just two at any time. Multi-view systems can also support more than one observer if enough horizontal viewing freedom is available.

Bulk optic multi-view displays have been developed and are reviewed in the literature [7,36]. The display was designed to use temporal multiplexing of the view images and because the basic switching speed and interface bandwidth of LCD displays were not sufficient, this led to the use of highspeed CRT technology.

Micro-optic multi-view designs using standard two-dimensional displays have been proposed where the images are spatially multiplexed. The Heinrich-Hertz-Institut has a well-established programme investigating lenticular three-dimensional displays and Borner [4] describes a number of multi-view designs.

The principle for a multi-view LCD display using a front lenticular element, similar to the two-view lenticular design described previously, is illustrated in Figure 9.20. This shows a five-view lenticular display, where each pixel in every group of five pixels is directed to a different viewing window. As with the two-view displays the system should be viewpoint corrected so that the viewing windows are aligned with pixels across the whole display.

To use the display, five images are sliced vertically into columns and interlaced appropriately. The images will then be visible separately in the five viewing windows V1–V5 in Figure 9.20. The viewing windows can be designed as shown so pairs of image separated by one image, for example V2 and V4, are seen simultaneously by the left and right eyes and if these form a stereo pair, then an observer sees an image with stereoscopic depth. In

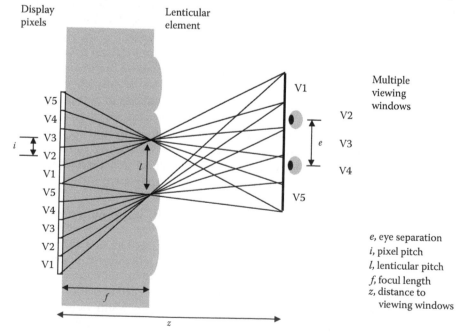

e, eye separation
i, pixel pitch
l, lenticular pitch
f, focul length
z, distance to viewing windows

Figure 9.20 The principle of amulti-view front lenticular auto-stereoscopic display.

addition, if the observer moves laterally they can see a different pair of images, for example V3 and V5, and therefore a different stereoscopic view of the scene.

Using a similar geometrical argument as for two-view lenticular displays, the pitch of the lenses can be determined by

$$l = N_v i \left(\frac{z-f}{z} \right) \qquad (9.22)$$

where N_v is the number of viewing windows required.

There are several drawbacks to the basic multi-view approach that are particularly apparent when electronic displays are used [55]. The first is there is a black mask between LCD pixels and this is imaged into dark lines between each view window which is distracting to observers when the eye crosses a window boundary. Also images with any significant depth will result in an image-flipping artefact as the observer moves the eye across one view window and into the next. Finally as more views are used the horizontal resolution of the images decreases rapidly. To overcome these problems, a new approach to multi-view LCD display was proposed [55].

Several multi-view systems based on lenticular micro-optics and single LCD displays were proposed [54–56]. A significant step forward was made by positioning the lenticular array at an angle to the LCD pixel array; this mixed adjacent views reducing image flipping problems and spreading the effect of the black mask making it less visible. The other benefit of this design is that each view has a better aspect ratio; rather than splitting the display horizontally into many views both horizontal and vertical directions are split.

The arrangement of one lenticule and the underlying pixels in the slanted lenticular design is shown in Figure 9.21. The slanted lenticular arrangement means that all pixels along "a" line such as a will be imaged in the same direction. In this case all view three pixels are seen in the same direction. The arrangement shown allows seven views to be interlaced on the display and imaged in different directions by the lenticule. As the eye moves from position "a" to "b" the eye sees a gradual transition from view 3 to view 4. At most viewing positions the eyes will see a combination of more than one view; while this inherent crosstalk

limits the depth that can be shown on the display it does hide the transition between views at boundaries and blurs the appearance of the black mask so that it is no longer an obvious visual artefact. For the seven-view display described in [56] the magnification of the lenticules is designed so that a viewer at a distance of approximately 700 mm from the display sees views 3 and 5 in left and right eyes respectively, i.e., views separated by one view form a stereo pair.

An alternative design where the pixels are slanted instead of the lenticular element is described in [57]. However, such a major change to LC display design is unlikely to happen unless there is a substantial worldwide market for three-dimensional displays or an advantage of slanted pixels for two-dimensional LCD operation is found.

The multi-view display design [33] adopts a similar solution, citing an earlier reference [59] as the source of the idea for using a lenticular slanted with respect to the vertical image axis. This display generates nine viewing regions, through which the user can see nine equal resolution images. Based on an SXGA (1280×1024) LCD display this results in each viewing window image having a two-dimensional resolution of 426 by 341 pixels.

Experience with lenticular optics [61] suggests displays based on lenticular optics have to make additional design trade-offs. An important one is

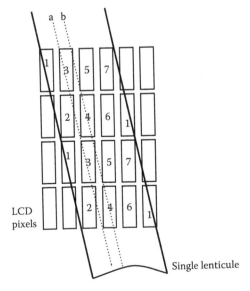

Figure 9.21 The slanted arrangement of the lenticular lens and pixels in the multi-view display [55].

the difficulty of anti-reflection coating the lenses, which can lead to distracting reflections on the display surface. Another is that scattering of light in the lenses generates a visible artefact looking to the user like a light grey mist present throughout the three-dimensional scene.

To summarize multi-view displays:

- Temporally multiplexed displays with high resolution per view suffer a number of drawbacks: they need high-speed display elements and high-bandwidth image generation and interface circuits. This seems likely to delay their widespread adoption in personal three-dimensional display applications.
- Spatially multiplexed designs have lower resolution per view than twin-view displays and recent designs build in crosstalk limiting the three-dimensional depth. Despite this they are attractive commercially because of the benefit of viewing freedom they provide and their relatively low cost and manufacturability.

A solution for the future is to build a system with an intermediate number of views, say three, not requiring mechanical view steering and use a head tracking device to keep the images up to date with the observer's head position. One such system, known as PixCon, is described in [61] and another design is presented in [48]. A similar idea for using view switching in a twin-view system was proposed in [50]. A prerequisite for this is low-cost, accurate, observer head tracking and some good progress is being made in this area [25].

9.4 THREE-DIMENSIONAL DISPLAY PERFORMANCE AND USE

9.4.1 Comparing perceived depth reproduction

Perceived depth reproduction is the single most important reason for building three-dimensional displays but system characteristics in this respect are rarely reported in the literature. In this section, we consider three generic designs, the twin-LCD and single-LCD two-view systems and a single-LCD multi-view system, and analyse their ability to reproduce perceived depth. Similar real examples of these designs are the twin-LCD display [61], the single-LCD VPI display [62] and the nine-view multi-view display, but our discussion abstracts from the details of specific display implementations for clarity. We compare the ability of the three generic designs to reproduce depth to each other and to the performance of the human eye; we also consider the demands on the graphics and imaging systems supplying the displays with content.

The generic three-dimensional display designs are assumed to be based on the same underlying LCD element, a 1280 by 1024 pixel display with a horizontal pixel width of $i=0.3$ mm approximating an 18.8 in. diagonal SXGA display. The three-dimensional displays can then be characterized by the effective pixel width in the image seen by one eye:

- The twin-LCD twin-view display has two overlaid images and the pixel width in each view is the same as the base panel at $i=0.3$ mm.
- The single-LCD twin-view display has two horizontally interlaced images and the pixel width in each view is double the base panel at $i=0.6$ mm.
- The single-LCD multi-view display has nine views, interlaced horizontally and vertically, and the pixel width in each view is triple the base panel at $i=0.9$ mm.

We assume the latter two displays overlay the left and right eye images to simplify discussion, but note in practice it will be necessary to consider the exact interlacing of RGB components.

The following set of characteristics provides a basis for comparing display designs. Our aim is to capture the characteristics that are important in the human perception of three-dimensional displays.

Total display resolution: However a stereoscopic display is designed to provide views to each eye, the total display resolution, i.e., the sum of all pixels in all views, largely determines the computational effort required to generate the images for display and the bandwidth required in interface circuits. Displays which require image interlacing will also require additional functionality in interface circuits as pixels from different views typically need to be interlaced at the RGB component level. Bandwidth requirements can be determined from total display resolution and the desired frame rate.

Resolution per view: The resolution per view is a key characteristic of a three-dimensional display. Having stereo three dimensions does not replace the need for high spatial resolution and anyone used to 1280×1024 monoscopic displays will notice the step down when dividing these pixels between two or nine views. However, a three-dimensional display can often look better than a monoscopic display with the same resolution as a single view on the three-dimensional display because the brain integrates the information received from the two views into a single image.

Perceived depth voxels: As shown in Figure 9.22, a pair of corresponding pixels in the left and right images represents a volume of perceived depth; we will call this a stereoscopic voxel or voxel as in [24]. Of particular interest is the depth of a voxel that a display can represent for a given screen disparity between corresponding pixels. We can use this to compare

the depth representation abilities of different displays in depth and to compare displays with the ability of the eye to perceive depth. The perceived voxels are arranged in planes from in front to behind the display; as they recede from the viewer the cells increase in depth [8,24].

The depth span of a voxel can be found by using Equations 9.10 and 9.11 as appropriate to calculate difference in depth of points 1 and 2 in Figure 9.22.

Consider zero pixels disparity as in Figure 9.22(o). For point 1, a pixel width $i=0.3$ mm implies screen disparity of $d=-0.3$ mm and assuming $z=750$ mm and $e=65$ mm then the perceived depth in front of the screen plane is

$$P = \frac{750}{\left(\dfrac{65}{|1-0.3|}\right)+1} = 3.45 \text{ mm} \tag{9.23}$$

For point 2 the screen disparity is $d=0.3$ mm and the perceived depth behind the screen plane is

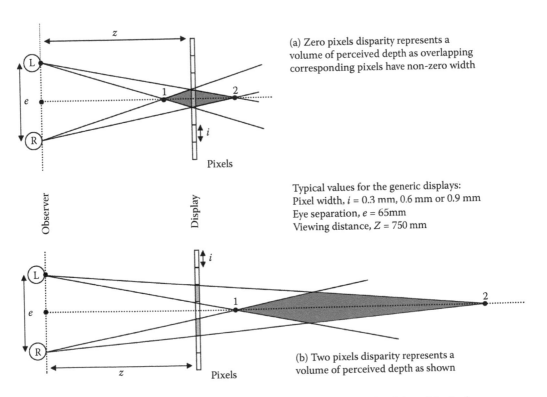

(a) Zero pixels disparity represents a volume of perceived depth as overlapping corresponding pixels have non-zero width

Typical values for the generic displays:
Pixel width, i = 0.3 mm, 0.6 mm or 0.9 mm
Eye separation, e = 65mm
Viewing distance, Z = 750 mm

(b) Two pixels disparity represents a volume of perceived depth as shown

Figure 9.22 The perceived depth represented by corresponding pixels of 0 and 2 pixels screen disparity.

$$p = \frac{750}{\left(\frac{65}{|0.3|}\right) - 1} = 3.48\,\text{mm} \qquad (9.24)$$

Therefore, the total perceived voxel depth in this case is 6.93 mm. This is the perceived depth represented by corresponding pixels with zero disparity at the screen plane. In practice, it tells us this display cannot reproduce a depth difference between objects at the screen plane of less than 6.93 mm. Results of similar calculations for all three generic displays are given in Figure 9.24.

Perceived depth range: The perceived depth range, that is the nearest and furthest points a display can reproduce, is of interest. Geometrically, this can be calculated from the maximum screen disparity available; however, for most displays of any size the geometric range is much more than can be viewed comfortably by the majority of observers. Instead, it is important to determine the comfortable perceived depth range experimentally and for our discussion we adopt results reported in [27]. This suggests a comfortable working range for the majority of people is from 100 mm in front to 100 mm behind the display surface and this range could probably be extended to 200 mm in front and 500 mm behind and still be comfortable to view for the majority of observers. We take the ±100 mm range for our calculations here without affecting the generality of the discussion.

Stereoscopic resolution: Identifying the comfortable working range of perceived depth on a display also allows us to define the resolution of perceived depth within this range. Perceived depth voxels of equal screen disparity form planes of voxels parallel to the display surface as illustrated in Figure 9.23. We will define stereoscopic resolution to be the number of planes of voxels within the range of ±100 mm.

Stereoscopic resolution can be calculated identically for each of the generic displays, which have the same viewing distance, by finding the screen disparity, d, that generates voxels at ±100 mm. The sum of these values is then divided by the width of a stereoscopic pixel, i, on the display in question.

The table in Figure 9.24 shows values of the characteristics discussed here for the three generic displays. Not surprisingly the twin-LCD display with the most pixels per view has the best results for depth reproduction with an ability to reproduce depth differences of 7 mm at the screen plane and a stereoscopic resolution of 60 planes of depth in the working depth range ±100 mm. However, the eye is much better at perceiving depth than the best display is at reproducing it with a minimum detectable depth difference of 0.84 mm and an equivalent stereoscopic resolution of 240 planes of depth in the working range ±100 mm.

This difference suggests significant improvements are still possible to the depth reproduction characteristics of stereoscopic displays. It is also important to keep in mind when using the displays if depth judgement is critical to task performance.

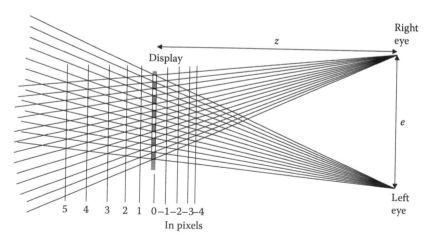

Figure 9.23 Stereoscopic resolution is defined by planes of stereoscopic voxels.

Characteristic	Twin-LCD Twin View	Single-LCD Twin View	Single-LCD Multi (9) View	Human Vision
Total resolution	2× 1280(h)×1024(v)	1280(h)×1024(v)	1280(h)×1024(v)	
View resolution	1280(h)×1024(v)	640(h)×1024(v)	426(h)×341(v)	
View pixel width	0.3 mm	0.6 mm	0.9 mm	
Viewing distance	750 mm	750 mm	750 mm	
Voxel depth: 0 pixels disparity	7 mm	14 mm	21mm	0.84 mm
Stereo resolution (in ± 100 mm)	60 voxels	31 voxels	20 voxels	~240 voxels

The calculations for the table assume an observer eye separation of 65 mm.

Figure 9.24 Table comparing characteristics of the generic displays and the eye.

In the next section, we briefly review how to create images that account for the available working depth range and resolution of three-dimensional displays.

9.4.2 Perceived depth control and image generation

As discussed three-dimensional displays have limits on the comfortable perceived depth range they can reproduce. This results in a working volume of space around the display plane that content producers can use to present a scene in. The working volume available on the display is unlikely to match the volume in the scene being captured. As a result, several approaches to mapping depth from a scene onto the available depth range of the target display have been proposed.

Traditionally, this has involved a discussion of whether to set cameras at eye separation or not and whether camera axes should be parallel or verging. However, recent methods approach the problem as a mapping of a volume of scene space onto the available working volume on the target display. These methods automatically calculate stereoscopic camera parameters given a camera position, scene volume to capture and target display specifications. Wartell describes one approach in [58] while a simpler and more general method is given by Jones [27]. These have significant benefits in ease of image generation for content creators and guarantee that depth mappings will be geometrically consistent even on head-tracked displays. The result is stereoscopic images should no longer be produced with excessive perceived depth or unwanted distortions.

Despite the long history of stereo image generation, it is only recently that new technology, in the form of three-dimensional computer graphics and digital camera technology, has been able to give enough control over the image creation process to use these methods. The methods are particularly important to apply when creating images for testing and improving a display design because poorly created images, or even just images produced correctly for a different three-dimensional display, can cause the highest-quality display to become uncomfortable to view.

9.5 SUMMARY

Advances in micro-optics, display technologies and computing systems are combining to produce an exciting range of new opportunities for three-dimensional display designers. To achieve a good three-dimensional display design requires a systems approach combining optical, electrical, mechanical and digital imaging skills along with an understanding of the mechanism of binocular vision.

The characteristics and geometry of binocular vision define limits on the maximum range of binocular vision and the minimum depth differences it is possible to perceive in the natural world. Because the perception of depth is relative to the current fixation point, binocular vision is best suited to making relative depth judgements between objects.

Stereoscopic images do not provide the same stimulus to the eyes as the natural world and the implications of this affect three-dimensional display design and use. In particular, while the eyes verge to fixate different depths in a stereo image the eye's accommodation must keep the image plane, rather than the fixation point, in focus. This places measurable limits on how much perceived depth is comfortable to view on a particular three-dimensional display.

As well as the stereoscopic depth cue the brain uses many two-dimensional depth cues to help it understand depth information in a scene. Therefore, the first aim for a three-dimensional display design needs to be to keep the same basic image quality as a two-dimensional display including values of brightness, contrast, spatial resolution and viewing freedom.

We have introduced two-view and multi-view auto-stereoscopic display designs based on micro-optic elements including parallax barriers, lenticular arrays, micro-polarizers and holographic optical elements. These provide different trade-offs in cost, system complexity and performance. Key characteristics that define the performance of different displays include

- Perceived voxel depth at zero disparity, i.e., minimum reproducible depth at the screen plane
- Stereoscopic resolution, i.e., the number of discrete voxel planes in ±100mm depth
- Viewing window characteristics particularly inter-channel crosstalk and uniformity

As three-dimensional display quality continues to improve, it becomes increasingly important to consider the quality of the stereoscopic images used to evaluate displays. This requires the adoption of new methods for image generation based on an improved understanding of the human perception of stereo images to define the mapping of depth from a scene onto the working depth range available on the three-dimensional display.

REFERENCES

1. Benton, S. A., Slowe, T. E., Kropp, A. B., and Smith, S. L. 1999. Micropolarizer-based multiple-viewer auto-stereoscopic display. *Proc. SPIE* 3639, 76–83.
2. Benton, S. A. 2002. *Auto-Stereoscopic Display System* US 6 351 280 (filed November 1998).
3. Blakemore, C. 1970. The range and scope of binocular depth discrimination in man. *Physiology* 211, 599–622.
4. Borner, R. 1993. Autostereoscopic 3D-imaging by front and rear projection on flat panel displays. *Displays* 14(1), 39–46.
5. Cumming, B. G. and DeAngelis, G. C. 2001. The physiology of stereopsis. *Annu. Rev. Neurosci.* 24, 203–238.
6. Dodgson, N. 1997. Autostereo displays: 3D without glasses. *Proc. EID* EID97.
7. Dodgson, N. A. 2002. Analysis of the viewing zome of multi-view auto-stereoscopic displays. *Proc. SPIE* 4660, 254–265.
8. Diner, D. B. and Fender, D. H. 1993. *Human Engineering in Stereoscopic Display Devices* (New York: Plenum).
9. Eichenlaub, J. 1997. A compact, lightweight, 2D/3D auto-stereoscopic backlight for games, monitor and notebook applications. *Proc. SPIE* 3012.
10. Eichenlaub, J. 1998. A compact, lightweight, 2D/3D auto-stereoscopic backlight for games, monitor and notebook applications. *Proc. SPIE* 3295.
11. Eichenlaub, J. and Gruhlke, R. 1999. Reduced thickness backlighter for auto-stereoscopic display and display using the backlighter. *US Patent* 5,897,184.
12. Ezra, D., Woodgate, G., Harrold, J., Omar, B., Holliman, N., and Shaprio, L. 1995. New auto-stereoscopic display system. *Proc. SPIE* 2409.
13. Ezra, D., Woodgate, G. J., and Omar, B. 1998. Autostereoscopic directional display apparatus. *US Patent* 5,726,800 (priority from December 17th 1992).
14. Faris, S. M. 1994. Novel 3D-stereoscopic imaging technology. *Proc. SPIE* 2177, 180–195.
15. Faris, S. M. 1993. Multi-mode stereoscopic imaging system. *US Patent* 5,264,964.
16. Glassner, A. 1995. *Principles of Digital Image Synthesis* (San Mateo, CA: Morgan Kaufmann).
17. Gooding, L., Miller, M. E., Moore, J., and Kim, S. 1991. The effect of viewing distance and disparity on perceived depth. *Proc. SPIE* 1457, 259–266.

18. Goldstein, E. B. 2002. *Sensation and Perception*. 6th edn (Belmont, CA: Wadsworth).

19. Hamagishi, G., Sakata, M., Yamashita, A., Mashitani, K., Inoue, M., and Shimizu, E. 2001. 15" high resolution non-glasses 3-D display with head-tracking system. *Trans. IEE, Jpn.* 121(5), 921–927.

20. Harrold, J., Jacobs, A. M. S., Woodgate, G. J., and Ezra, D. 1999. 3D display systems hardware research at Sharp Laboratories of Europe: An update. *Sharp Tech. J.*, 24–30.

21. Harrold, J., Jacobs, A., Woodgate, G. J., and Ezra, D. 2000. Performance of a convertible, 2D and 3D parallax barrier auto-stereoscopic display. *Proceedings SID, 20th International Display Research Conference,* September 2000, Palm Beach, FL.

22. Hattori, T., Ishigaki, T., Shimamoto, K., Sawaki, A., Ishiguchi, T., and Kobayashi, H. 1999. An advanced auto-stereoscopic display for G7 pilot project. *Proc. SPIE* 3639, 66–75.

23. Helmholtz, H. 2000. *Treatise on Physiological Optics 1867*. 1924 edn (reprinted Bristol: Thoemmes).

24. Hodges, L. F., and Davis, E. T. 1993. Geometric considerations for stereoscopic virtual environments. *Presence* 2(1), 34–43.

25. Holliman, N., Hong, Q., Woodgate, G., and Ezra, D. 2000. Image tracking system and method and observer tracking auto-stereoscopic display. *US Patent* 6,075,557.

26. Inoue, M., Hamagishi, G., Sakata, M., Yamashita, A., and Mahitani, K. 2000. Non-glasses 3-D displays by shift-image splitter technology. *Proceedings 3D Image Conference 2000*, Tokyo, July 2000.

27. Jones, G., Lee, D., Holliman, N., and Ezra, D. 2001. Controlling perceived depth in stereoscopic images. *Proc. SPIE* 4297A, 42–53.

28. Julesz, B. 1971. *Foundations of Cyclopean Perception* (Chicago, IL: University of Chicago Press).

29. Kaplan, S. 1952. Theory of parallax barriers. *J. SMPTE* 59(1), 11–21.

30. Lane, B. 1982. Stereoscopic displays. *Proc. SPIE* 0367.

31. Langlands, N. 1926. Experiments on binocular vision. *Trans. Opt. Soc.* 27, 4–82.

32. Lipton, L. 1982. *Foundations of Stereoscopic Cinema*. Now available electronically (Princeton, NJ: Van Nostrand-Reinhold).

33. Lipton, L. 2002. Synthagram: Auto-stereoscopic display technology. *Proc. SPIE* 4660.

34. McAllister, D. F. 1993. *Stereo Computer Graphics and Other True 3D Technologies* (Princeton, NJ: Princeton University Press).

35. Montgomery, D. J., Woodgate, G. J., and Ezra, D. 2001. Parallax barrier for an auto-stereoscopic display. *GB Patent* 2,352,573.

36. Moore, J., Dodgson, N., Travis, A., and Lang, S. 1996. Time-multiplexed color auto-stereoscopic display. *Proc. SPIE* 2653, 10–19.

37. Morishima, H., Nose, H., and Taniguchi, N. 1999. Stereoscopic image display apparatus. *US Patent* 6,160,527.

38. Nose, H. 1997. Rear-lenticular 3D-LCD without eyeglasses. *O plus E.* 217, 105–109.

39. Ogle, K. N., 1964. *Researches in Binocular Vision* (London: Hafner).

40. Okoshi, T., 1976. *Three-Dimensional Imaging Techniques* (New York: Academic).

41. Pastoor, S. 1991. 3D-television: A survey of recent research results on subjective requirements. *Signal Process.: Image Commun.* 4(1), 21–32.

42. Pastoor, S. 1995. *Human Factors of 3D Imaging*. Web document distributed by Heinrich-Hertz-Institut, Berlin.

43. Pastoor, S., and Wopking, M. 1997. 3-D displays: A review of current technologies. *Displays* 17(2), 100–110.

44. Perlin, K., Paxia, S., and Kollin, J. S. 2000. An auto-stereoscopic display. *Proceedings ACM Siggraph Conference*, July 2000.

45. Perlin, K. 2001. Displayer and method for displaying. *US Patent* 6,239,830 (filed May 1999).

46. Perlin, K., Poultney, C., Kollin, J. S., Kristjansson, D. T., and Paxia, S. 2001. Recent advances in the NYU auto-stereoscopic display. *Proc. SPIE* 4297, 196–203.

47. Schiffman, H. R. 2000. *Sensation and Perception: An Integrated Approach*. 5th edn (New York: Wiley).

48. Schwerdtner, A., and Heidrich, H. 1998. The Dresden 3D Display (D4D). *Proc. SPIE* 3295, 203–210.

49. Schwerdtner, A., and Heidrich, H. 1998. Optical system for the two and three dimensional representation of information. *US Patent* 5,774,262 (filed Germany 1993).

50. Susumul, N. T. and Morito 1. 1990. Technique of stereoscopic image display. *EP Patent* 0 354 851 (filed Japan August 1988).

51. Trayner, D. and Orr, E. 1997. Developments in auto-stereoscopic displays using holographic optical elements. *Proc. SPIE* 3012, 167–174.

52. Trayner, D., and Orr, E. Direct View holographic auto-stereoscopic displays. *Proceedings of the Fourth UK VR-SIG*, Brunel University.

53. Valyus, N. A. 1966. *Stereoscopy* (New York: Focal).

54. van Berkel, C., Parker, D. W., and Franklin, A. R. 1996. Multi-view LCD. *Proc. SPIE* 2653.

55. van Berkel, C., and Clarke, J. A. 1997. Characterisation and optimisation of 3D-LCD module design. *Proc. SPIE* 3012, 179–186.

56. van Berkel, C. and Clarke, J. 2000. Autostereoscopic display apparatus. *US Patent* 6,064,424.

57. van Berkel, C. and Parker, D. 2000. Autostereoscopic display apparatus. *US Patent* 6,118,584.

58. Wartell, Z., Hodges, L. F., and Ribarsky, W. 1999. Balancing fusion, image depth and distortion in stereoscopic head tracked displays. *Proceedings of ACM Siggraph99 on Computer Graphics*.

59. Winnek, D. F. 1968. Composite stereography. *US Patent* 3,409,351 (issued Nov. 1968).

60. Wheatstone, C. 1838. Contributions to the physiology of vision I: On some remarkable and hitherto unobserved phenomena of vision. *Philos. Trans. R. Soc. (Biol.)* 18, 371–395.

61. Woodgate, G., Ezra, D., Harrold, J., Holliman, N., Jones, G., and Moseley, R. 1997. Observer tracking autostereoscopic 3D display systems. *Proc. SPIE* 3012, 187–198.

62. Woodgate, G., Harrold, J., Jacobs, M., Moseley, R., and Ezra, D. 2000. Flat panel autostereoscopic displays— Characterisation and enhancement. *Proc. SPIE* 3957, 153–164.

63. Woodgate, G., Moseley, R., Ezra, D., and Holliman, N. 2000. Autostereoscopic display. *US Patent* 6,055,013 (priority Feb. 1997).

64. Woods, A., Docherty, T., and Koch, R. 1993. Image distortions in stereoscopic video systems. *Proc. SPIE* 1915, 36–48.

65. Yeh, Y., and Silverstein, L. D. 1990. Limits of fusion and depth judgement in stereoscopic colour displays. *Human Factors* 32, 45–60.

66. Yeh, Y. 1993. Visual and perceptual issues in stereoscopic colour displays. In *Stereo Computer Graphics and Other True 3D Technologies*, ed. D McAllister (Princeton, NJ: Princeton University Press).

10

Optical scanning and printing

RON GIBBS
Gibbs Associates

10.1 INTRODUCTION

Optical scanning is familiar to the general public through such widespread applications as supermarket barcode scanners, desktop color scanners, laser lightshows and desktop laser printers. However, scanning technology is also applied to commercial printing processes, thermal imaging, medical diagnostic equipment, biochemical analysis, and quality control of such diverse items as sheet steel, drawn wire, windshield glass, semiconductor wafers and rice—to name just a few of the many applications.

Optical scanning in electro-optical systems began to be developed for pre-press commercial print processes in the 1940s, and for airborne mapping and reconnaissance in the 1950s, but the field really took off after the invention and commercialization of the laser in the 1960s. Since then the field has developed rapidly, with military investment in forward-looking infrared (FLIR) scanners, and a steady stream of new commercial applications.

Scanning systems can be broadly divided into input and output systems. Input systems acquire electronic data in one, two or three dimensions from physical objects, by scanning a detector across the object. Output systems create an image from the electronic data by scanning a modulated light beam (usually a laser) across a light-sensitive medium. Input systems can be further subdivided into remote/local sensing and passive/active scanning. Although the various types of system have often developed separately, and in some cases are described by different terminologies, there is a great deal of overlap and common ground in the technology used.

Scanning technology is very varied, both in the components employed and in their configuration into optical systems, and no treatment of the technology can hope to be complete. Typically, to meet a given requirement, many possible solutions exist, and comparing potential solutions can involve analysis of a complex trade-off of optical, mechanical, electronic and software issues. In addition, non-scanning solutions may also be possible, especially those involving detector and light source arrays. In some instances, electronically scanned arrays can be considered to be part of scanning technology, although analysis of the system in this way is more meaningful when the electronic scan is combined with optomechanical scanning.

This chapter is organized as follows: An overview of the most commonly encountered scanning configurations is followed by a description of the most important scanning system performance requirements. Then follow descriptions of scanning system deflector and lens components, and finally more detailed descriptions of a few examples of practical input and output scanning systems. An attempt has been made to generalize, as far as possible, although inevitably many detailed issues are specific to a particular category of scanning system.

10.2 SCANNING SYSTEM CONFIGURATIONS

Scanning systems can take many forms, and it is helpful to group them together in related fundamental configurations. There are several commonly used ways of classifying scanning systems, which are used in different circumstances, although none of them is perfect in the sense of including all possible types of scanning system.

10.2.1 Classification by deflector position

The most commonly applied classification, due to Beiser [1], is based on the relative positions of the scanning deflector and the objective (focusing lens) subsystems that make up the basic optical configuration (Figure 10.1).

Objective scanning is defined as the coincidence of the deflector and objective, and it implies

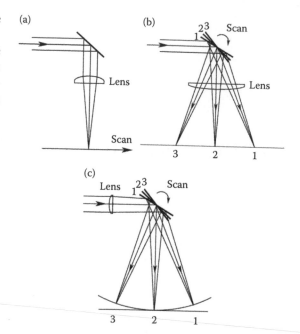

Figure 10.1 Scan configurations by deflector position. (a) Objective scan, (b) Pre-objective scan, and (c) Post-objective scan.

translation of either the complete optical system or the scanned object. Examples of scanning systems employing this classification include push broom (linear translation of a linear array), XY (orthogonal linear translation stages) and external drum scanning (see Section 10.2.2). The scanning speed is limited by the mechanical inertia of large structures, which must move at the scanning speed.

Since this particularly affects acceleration and deceleration, objective scanning systems are either slow or continuous, as in the cases of external drum scanners (see Section 10.2.2) or airborne or satellite push broom systems, where the aircraft flight over the ground provides the scan motion. To overcome this limitation, configurations based on angular deflectors, giving an optical lever advantage, must be used.

In pre-objective scanning, the deflector precedes the lens in the optical path. This usually requires a complex lens, as it must operate over the range of off-axis field angles defined by the angular scan range. However, this is the only scanning configuration that can generate a fast-scanned flat image plane via a suitably designed flat-field lens (see Section 10.5.3), which is the usual use for this configuration. Either a single-axis deflector producing a straight-line scan, or a two-axis deflector to scan a flat plane can be used.

In post-objective scanning, the deflector follows the objective lens in the optical path. This allows relatively simple on-axis optics, but leads to a curved field, which can be circular if there is no separation between the rotation axis and the deflection point. This curvature can be acceptable, provided that the system either has a large depth of focus or is imaging a curved (e.g., cylindrical) surface. These curved bed systems are described further in Section 10.2.2.

Additional optics (usually based on large aspheric mirrors) can be inserted between the deflector and the image to flatten the image field. These are often described as post-objective field flatteners, but the resulting systems have more of the characteristics of pre-objective scanners.

This classification system deals with individual scanning mechanisms for a single scan direction. In two-dimensional scanning systems, two orthogonal scanning mechanisms are combined. In the case of a raster scan pattern, fast scan and slow scan axes are defined. Often, the fast scan mechanism uses angular deflectors in a post-or pre-objective configuration, and the slow scan mechanism uses objective scanning.

10.2.2 Classification by image geometry

Particularly in applications that involve flexible object or image media, it is useful to classify scanning systems by the (input or output) imaging medium surface form, which can be either flat or curved when mounted for scanning. The medium could either be the surface of a solid object (e.g., a cylinder) or a thin, flexible sheet of material, which is constrained to take the shape of a mounting surface, or bed. This classification system deals with complete scanning system geometries.

A flatbed scanner usually has the benefits of convenient material geometric form and simple handling mechanisms, together with the possibility of a compact, lightweight system. Usually, a pre-objective scanner or linear array is used, often requiring a complex lens design (see Section 10.5). It is possible to flatten the curved field of a post-objective system by using dynamic focus of the objective, but this is limited to slow scanning speeds because of the mechanical inertia of the lens. A closely related form is the capstan scanner, which combines a straight-line scanner with a transport mechanism for the input or output sheet that moves it across the scan line using motorized rollers. Typically, this winds a photographic film from a roll to a take-up roll.

Non-flatbed scanners are based on cylindrical surfaces, either convex or concave. Convex cylinder scanners are called external drum, as the scanning optics is external to the cylinder surface. In these, the optical system is usually stationary and the cylinder rotates (Figure 10.2). In this configuration, there is great system flexibility for the optical design, which can vary from simple (single-beam, fixed resolution) to complex (multiple beams, variable resolution). In addition, the (mechanical) scanning system is separate from the optical system, so it is relatively easy to convert an existing system by changing the optics module, or to change the scanned image format by changing the size of the cylinder. As the focal length of the optical system can be small, very high resolution and good image quality can be achieved, but the scan speed is limited by the very high rotational inertia of the large cylinder drum. This causes problems for acceleration and deceleration, as well as safety and stability concerns that limit rotation speeds to about 2000 rev/min.

An alternative form of external drum scanner uses a line scanner to generate a line along the cylinder, which rotates only once for each complete scan of the drum surface. This is the configuration for typical xerographic laser scanners.

Concave cylinder surface scanning systems are based on post-objective scanners, which have simple on-axis lens optics, and which can take one of the two forms, as described above. When the deflector rotation axis is parallel to, and collinear with, the optical axis, an internal drum scanner results (Figure 10.3). This has the advantage that as the deflector is small, very high rotation rates can be attained. In addition, a large scanning angle range (up to 270°) is easily achieved, and fairly high resolution scanning is possible over a fairly large area. The image surface is stationary, which simplifies and speeds up material changeover between scans.

Two forms of internal drum scanner optical layouts may be identified. If the laser and optical systems are small, the complete optics may be

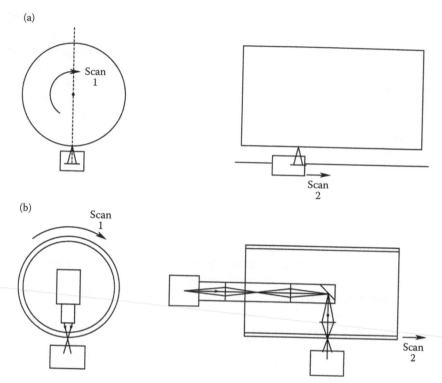

Figure 10.2 External drum configurations. (a) Solid drum output, reflective input and (b) Transparent drum, transparency input.

mounted on the linear traverse slide, in a carriage-mounted configuration. When the optics is too large or heavy, only the final focusing lens and the spinner mirror is mounted on the carriage, and the preceding optics generates a collimated beam collinear with the cylinder axis. The beam must be sufficiently large that the beam divergence is small over the carriage traverse length. This is known as a beam-riding configuration.

Multiple-beam internal drum systems are difficult to realize, because of the image rotation along the scan line inherent in the scan geometry. A half speed contra-rotating dove prism before the spinner mirror can correct this [2], although the speed control requirements are very demanding. By using a two-axis acousto-optic deflector, derotation of up to three beams is possible [3].

A post-objective scan configuration with the deflector axis perpendicular to the optical axis (such as in Figure 10.1c) is classed as curved bed. Compared with the internal drum configuration, only a small angular scan is possible, but the deflection angle is double the mirror rotation angle, resulting in a faster scan. Multiple-beam

scanning is possible, as there is no rotation on the scanned beam axis. For faster scanning, a prismatic polygon mirror may be used in place of a single-facet mirror. However, in this case, the separation between the rotation axis and mirror surface results in distortion of the image surface curvature and the scan linearity.

10.2.3 Classification by scan pattern

The main division is between raster and vector scanning systems. Raster scanning repetitively scans a beam in a regular pattern of adjoining lines over a rectangular area, like a TV display. Raster scanning systems are characterized by

- Fixed scan speeds, often set by a system electronic clock
- Orthogonal scan mechanisms—one being a fast line scan and the other being a much slower page scan

For multiple-beam or multiple-detector systems, a further sub-classification can be applied, depending

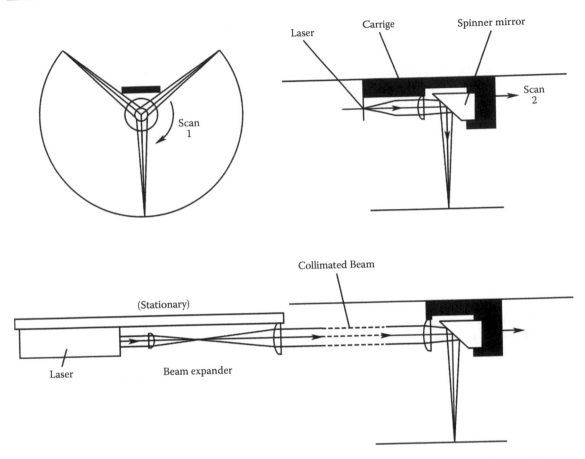

Figure 10.3 Internal drum configurations. (a) Carriage-mounted optics and (b) Beam-riding optics.

on the method of overlapping the beams. In a swathe scanning system, each set of multiple beams forms an overlapping scan pattern or swathe, as in a push broom scanner. To scan a larger width, multiple scans are made with the first beam of each swathe adjoining the last beam of the preceding swathe. Some systems allow adjacent elements of detectors (e.g., CCD) or light sources (e.g., LED), but when some physical separation of elements is necessary, the array must be angled with respect to the scan direction, to allow overlap of adjacent elements. In this case, a timing correction must be made to the electronic signals to or from the array elements.

If an array of n elements has a spacing between elements of $(n+1)p$, where p is the required separation between adjacent scan lines, an interconnecting scan pattern will eventually result if each pass of the array is advanced by a distance np in the slow scan direction. This is known as an interlaced scan.

In page width segmentation, in a single pass of the scanner the beams are equispaced across the image width. On each pass of the scanner, the beams are moved across by one beam width so that each beam adjoins the corresponding beam from the previous scan, until it reaches the starting position of its following neighbor.

Random-access and vector scanning systems require agile, low-inertia deflectors to move a scanning spot along arbitrary paths within a scanning area. Typically, orthogonal galvanometer deflectors or electro-optical deflectors are used for very high speed scanning applications. In vector scanning, the two deflectors are synchronized to move the spot at constant speed, to maintain a constant exposure. In random access, the spot is quickly switched to a new direction, requiring sophisticated control to minimize the switching time without causing oscillation.

Complex scan patterns, similar to Lissajous figures, may be generated by simple scanning systems comprising of two fixed-speed oscillating deflectors. The deflectors could be resonant galvanometer scanners, or rotating transmissive glass wedges known as Risley prisms. Different combinations of the relative speed and phase of the two deflectors produce a wide range of scan patterns [4].

10.3 SCANNING SYSTEM PERFORMANCE SPECIFICATION

In this section, the performance parameters that are relevant to scanning systems are examined. Scanning system performance requirements vary widely in terms of resolution, speed and quality, and a thorough analysis of these requirements is essential before any design is attempted.

10.3.1 Resolution

Resolution of a scanning system can be expressed in several different ways. The addressability of a digital system is determined by the sampling frequency of acquisition or output of pixel data points, and is usually measured in units of dots per inch (dpi). The image is usually sampled at finer intervals than the optical resolution of the system. Hence, addressability can give a misleadingly high value for the system resolution if not interpreted with care.

The spot size (of the scanning laser beam, or of the detector element image) is a measure of the optical resolution of the system. The cross-scan (perpendicular to the scan direction) resolution is well characterized by spot size, but the in-scan (in the direction of the scan direction) resolution of a flying spot scanner is degraded by 'smearing' of the scanned spot as a result of its motion. In the cross-scan direction, the ratio of the spot size to the scan line pitch is significant, as this determines the intensity ripple in this direction, and hence the visibility of scan lines.

In a laser scanning system, the beam usually has a Gaussian intensity profile, and the spot size is determined at either the 50% (referred to as full-width, half-maximum (FWHM)) or $1/e^2$ (13.5%) level. In some systems, the spot is elliptical in shape, with the size optimized for both in-scan and cross-scan resolutions, but where a round beam is used, the spot size is a compromise between the two requirements.

For an angular scanned system, the concept of angular scanned resolution is useful. Angular scanned resolution is defined in terms of the number of resolved spots per scan line, N.

$$N = \theta/\Delta\theta \qquad (10.1)$$

where θ is the maximum scan angle and $\Delta\theta$ the scan angle increment corresponding to the resolution limit of the system. For an optical aperture width D, the resolution is determined by the diffraction limit

$$\Delta\theta = a\lambda/D \qquad (10.2)$$

where λ is the wavelength and a a constant that depends on the aperture shape, beam uniformity and resolution criterion used. (For the common case of an untruncated Gaussian beam defined by the FWHM spot size, $a = 0.75$.) Hence, the fundamental scanning equation [5] is obtained

$$N = \theta D/a\lambda. \qquad (10.3)$$

A consequence of this equation is that by increasing the scan angle with additional optics such as an afocal beam expander does not improve the number of resolved spots per scan line, as θD is an optical invariant.

The calculated modulation transfer function (MTF) can be very useful when applied to a two-dimensional raster scanning system used for imaging [6]. Although MTF is only strictly applicable to linear systems, and some elements of scanning systems (especially electronic and software image processing) can be highly non-linear, the calculations can still provide valuable insights if treated with caution.

In-scan and cross-scan MTFs are different, because of the time-dependent factors that affect in-scan resolution, which are linearly related to spatial frequency response by the scan speed.

For an input system, the model must include

- Lens performance, or aperture, if diffraction limited; calculated or measured
- Detector element size and shape; calculated from the Fourier transform of the detector dimensions
- Scan motion; calculated from velocity as a sinc function [6]

In addition, electronic responses of scanner system components must be included

- Detector frequency response
- Detector amplifier frequency response
- Software processing of the output data (frequency response is linearly related to MTF)
- For output systems, the key optical factors in the MTF calculation for the cross-scan direction are
- Optical spot size and intensity profile—calculated from the Fourier transform
- Output medium—measured (related to the microscopic grain structure of the active material)

Additional system factors for the in-scan direction are

- Spot motion
- Modulator frequency response, due to the optical modulation device and the electronic amplifier response

Measurement of scanning system MTF is difficult for both input and output systems, due to problems in generating sinusoidal spatial frequencies and because of the complex sampled nature of the image. In practice, spatial resolution is usually assessed by subjective methods, using test images and resolution charts.

10.3.2 Scan duty cycle

Duty cycle is an important measure of the efficiency of the scanning system. It is usually defined for a line scanning subsystem as the proportion of time that the system is actually performing input or output of scanned data during the scanning operation. It arises because every scanning mechanism has a 'dead' time, which arises from the scanning geometry.

Examples of how dead time arises for specific scanning mechanisms are

- *Galvanometer*: acceleration/deceleration at start/end of linear angular scan range, and flyback time if the scan is unidirectional.
- *Polygon*: transit time of the beam across corners of the polygon mirror.
- *Internal drum*: obscuration of the deflected beam by the spinner carriage.
- *Acousto-optic deflector*: acoustic fill time.

The duty cycle has a significant effect on the system electronic bandwidth requirement, and on the radiometry. The electronic bandwidth of the system must be higher than the average data rate due to this inefficiency in the optical system. For example, the optical modulation rate f_m of an output system is related to the average data rate f_b by

$$f_m = \frac{f_d}{\eta}. \tag{10.4}$$

The effect on radiometry is that higher optical power is required to illuminate or expose the scanned object than if the scan were perfectly efficient. In some cases, this can limit the achievable scanning speed of the system.

10.3.3 Radiometry, input systems

The critical radiometric performance measure for input systems is the dynamic range of the system. This is defined as the ratio between the system's white and black levels I_w and I_b, which are the maximum and minimum possible detector signals.

The white level can depend on the following factors:

- The brightness of the optical source, including its condenser optics
- The optical properties (e.g., reflectivity or transmissivity) of the scanned object
- The efficiency (e.g., numerical aperture, transmission) of the detector optical system
- The size, responsivity and saturation level of the detector element

The dynamic range may be expressed as a simple ratio, but is more commonly expressed in density units

$$D = \log_{10}\left(\frac{I_w}{I_b}\right) \tag{10.5}$$

or as an integer number of bit levels

$$N_D \leq \frac{\ln(I_w/I_b)}{\ln 2}. \tag{10.6}$$

Optical density is often a more relevant measure of signal, because of the logarithmic response of many physical processes, including photographic emulsions and the eye. N_D, as defined here, is the maximum number of physically significant bit levels that can be achieved in a digitized output.

10.3.4 Radiometry, output systems

For exposure of an image on an area A, the optical power of the beam is given by

$$P = \frac{ks}{\eta_{\text{opt}}}\left(\frac{A}{\eta_L \tau}\right) \qquad (10.7)$$

where s is the sensitivity of the exposure material, η_L the scan duty cycle, η_{opt} the optical throughput efficiency and τ the total exposure time.

For most scanning systems, the material sensitivity must be valid for very short exposure times. This can differ significantly from manufacturers' measured values at shorter exposures because of reciprocity failure. For photographic materials, the sensitivity usually increases at shorter exposures, whereas for thermal materials it decreases.

The value of the constant k in the equation depends on the beam profile and the required overlap between adjacent scan lines. When $k=1$, a threshold exposure results, when only a thin line at the center of the scan line is exposed. A typical value for a practical system is approximately 2.

Note that, for a fixed scanner speed in terms of the number of scan lines per second, as the system resolution increases, the exposure time increases. Hence, the power requirement also increases.

10.4 SCANNING DEFLECTORS

Angular scanning systems use deflector subsystems to deflect the optical axis of a beam from a point within the scanning system in the form of a rotation about a scanning axis. A variety of deflection subsystems have been developed to meet the needs of different scanning applications. The deflector subsystem, with or without auxiliary focusing optics, is itself often described as a scanner by manufacturers.

Distinction may be made between high-inertia deflectors, which can only be used for continuous, constant speed scanning, and low-inertia deflectors, whose deflection angle can be varied quickly. The latter category often has a limited angular deflection range, which limits its application to low-resolution systems. Low-inertia deflectors include fast-steering, or tip-tilt mirrors, which are two-dimensional beam steering devices primarily used for line-of-sight stabilization or tracking applications.

10.4.1 Galvanometer mirror

This important class of deflector employs the electromagnetic effect of a permanent magnet and the magnetic field created by an electric current in stator coils to produce a rotary torque, which drives a spindle through a limited angular range [7]. A mirror is mounted on the end of the spindle to deflect the optical beam. The mirror is a significant contributor to the mechanical inertia of the rotor, which limits the maximum speed and/or angular range of the scanner. Angular range is typically no more than 40° (80° optical deflection) for low-accuracy applications, and less than 20° for high-end applications.

The rotor can be configured as moving iron, moving magnet or moving coil [8]. Originally, galvanometer scanners were based on moving iron designs [9], but modern designs are based on moving coil or moving magnet designs. Moving coil designs have relatively high inertia, but high and stable torque, and are typically used for high-accuracy applications with large mirrors (greater than 30 mm aperture). Moving magnet galvanometer scanners, aided by recent developments in high-energy-density rare-earth magnetic materials, have the highest torque-to-inertia ratio, and hence the highest speeds. The angular deflection speed and position are controlled by sophisticated closed-loop servo electronics for high-accuracy applications, angular position being accurately measured using accurate, capacitive or optical sensors with fast response times. Resolution of the order of one microradian can be achieved.

For two-dimensional scanning, two deflectors are often used, mounted close together with their axes at right angles. In this configuration, the second mirror must be larger than the first, to accommodate the beam movement. Because of the separation of the scan axes, the scan pattern is geometrically distorted, but this can be computed [10] and compensated by software correction of the drive signals to the galvanometers.

Galvanometer scanners are most often used in one of two modes: vector scanning or regular periodic form. In vector scanning, the two deflectors of a two-dimensional scanner are controlled in synchronism to move the optical beam along a programmed path in response to a required pattern (e.g., alphanumerical characters), usually at a constant spot speed. For regular (e.g., raster) scanning, the deflector is driven so that the deflection is linear with time over a large proportion of the scan period.

For a high duty cycle a sinusoidal variation is inadequate, and the deflector angle is controlled to a sawtooth form, where the beam that is scanned at linear velocity in one direction is, quickly returned to the starting point. The scan must be unidirectional because of the slight ellipticity in the scan line due to bearing hysteresis.

For very high fixed scanning rates, resonant scanners have been developed. These use flexure, torsion or taut-band mounts to enable very high scan speeds of up to 8 kHz or more. Although these deflectors have low inertia, the scan speed and sinusoidal oscillation are fixed, so electronic signal timing correction is required to obtain an undistorted scan with reasonable scan duty cycle.

10.4.2 Polygon mirror

Deflectors based on continuously rotating spindles are very commonly used in scanning systems, and may have either a single mirror facet (monogon) or many facets (polygon).

These deflectors have much higher mechanical inertia than galvanometer scanners, but can rotate at very high speeds, thanks to the development of precision air bearings. These bearings have largely superseded ball bearings for high-end applications, having much higher speeds, much lower vibration and longer lifetime. Bearings can be hydrostatic, i.e., externally pressurized, requiring a source of compressed air. However, many bearings nowadays are hydrodynamic (or self-acting), requiring no external air supply. These require much tighter manufacturing tolerances, but modern manufacturing techniques, coupled with developments in bearing design (especially materials and surface finish), make these a practical choice for most high-speed applications.

At very high rotation speeds (>30,000 rpm), air resistance (windage) and turbulent flow of air at the mirror edges limit the achievable speed, and very high-performance scanners are often enclosed in partial vacuum or in a helium atmosphere, which has much lower viscosity than air [11].

Polygon mirrors may be pyramidal or (more commonly) prismatic. Prismatic polygons have their facet surfaces parallel to the rotation axis, while the facets of pyramidal polygons are at an angle (usually 45°) the rotation axis. Prismatic mirrors deflect the optical beam by twice the mechanical rotation, whereas pyramidal mirrors deflect it by the rotation angle.

Prismatic polygons are most commonly used in a configuration that has the input and output beams in the same plane, perpendicular to the polygon rotation axis (Figure 10.4). This configuration avoids scan line bow. Careful calculation is required to design the polygon dimensions to achieve the required system performance with the minimum size [12,13].

Active facet tracking can be used to further reduce the size of the polygon facets. In this technique, a subsidiary small-angle deflector is inserted before the polygon. This high-bandwidth deflector is programmed in such a way that it always deflects the input beam to the center of the active polygon facet.

Pyramidal mirrors (Figure 10.5) are more expensive to fabricate, but can achieve very high scan speeds in a compact configuration. They are often used overfilled, i.e., with an input beam larger than the mirror facet. This results in a scanned spot of constant power and a high scan duty cycle, but the optical transmission efficiency is poor. It is possible to design a coarse two-dimensional scan system by varying the pyramidal angle between adjacent facets to produce an angular step between successive scan lines.

The single-facet (monogon) versions of prismatic and pyramidal deflectors are shown in Figure 10.6. The pyramidal type is the more commonly used, usually in a post-objective internal drum scanner, and many innovations have been applied to increase the speed. Optically transparent enclosures have been developed to overcome windage limitations [14], and special mirror shapes have been developed to reduce dynamic distortions of the cantilevered mirror surface [15].

10.4.3 Holographic

Holographic deflectors [16] are based on rotating (flat or curved) surfaces containing fine diffraction

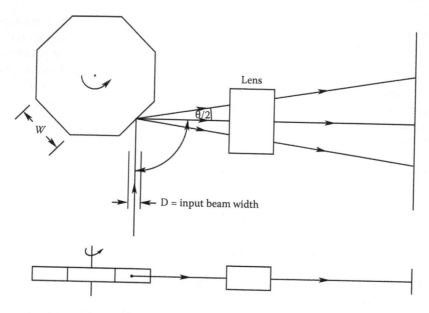

for N-sided polygon mirror

$$\theta = \frac{4}{N} \qquad \eta = 1 - \frac{D}{W \cos \alpha}$$

Figure 10.4 Prismatic polygon scanner configuration.

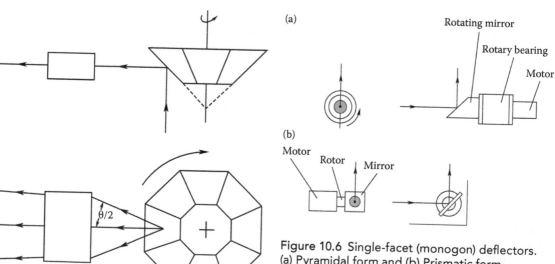

for N-sided polygon mirror

$$\theta = \frac{2\pi\eta}{N}$$

Figure 10.5 Pyramidal polygon scanner configuration.

Figure 10.6 Single-facet (monogon) deflectors. (a) Pyramidal form and (b) Prismatic form.

gratings, which can be either linear gratings to deflect collimated beams, or more complex curved structures that have optical power and act as lenses. Thick blazed transmissive gratings can have high transmission efficiency in a single diffraction order for a single wavelength. A

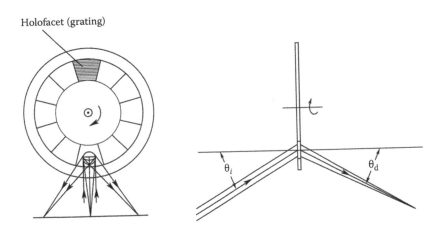

Figure 10.7 Disc hologon.

single surface that combines several holographic elements is referred to as a hologon (from holographic polygon).

The disc hologon is a form in which hologram elements are arranged radially around a flat circular disc, which rotates about its axis (Figure 10.7), and performs a similar function to a prismatic polygon [17]. The figure shows a hologon with optical power, but linear gratings and a separate lens are often employed in a pre-objective scanning configuration. In this case

$$\sin\theta_i + \sin\theta_d + \frac{\lambda}{t} \qquad (10.8)$$

where t is the grating period.

If the hologon is tilted by a small angle $d\alpha$, the change in diffracted angle is given

$$d\theta_d = \left[1 - \frac{\cos(\theta_i + d\alpha)}{\cos(\theta_d - d\alpha)}\right] d\alpha \qquad (10.9)$$

from which it can be shown that sensitivity of the output angle to wobble in the hologon tilt is minimized when $\theta_i = \theta_d$, which is the Bragg diffraction condition.

An alternative form of holofacet deflector is based on holograms arranged around a cylindrical surface that rotates about the cylinder axis (Figure 10.8), and is analogous to a pyramidal polygon.

Holographic deflectors have several potentially attractive fundamental features, compared with polygon mirrors:

- Thin lightweight elements, with improved aerodynamic shape (elimination of sharp edges in the rotating part), therefore potential for higher-speed rotation.
- Reduced centripetal deformations and/or reduced sensitivity to such deformations, therefore improved accuracy at high rotation speeds.
- Reduced wobble sensitivity (at Bragg angle), therefore potential elimination of anamorphic deflection error correction optics.

In addition, they offer increased design flexibility to enable compact configurations, and are potentially more accurate and/or economic to be manufactured in quantity.

However, they have some limitations:

- Wavelength sensitivity of deflection angle, which must be corrected by auxiliary diffractive optics when used with typical laser diodes [18].
- Specialized and technically demanding design and fabrication requirements, few suppliers having the necessary skills and facilities, and typically a lengthy, expensive development for a new system.
- The most useful configurations are radially asymmetric, leading to complex and difficult optical system design issues; in particular [19], careful consideration may need to be given to
  - Scan bow and scan linearity; arising from complex, asymmetric scan geometry.

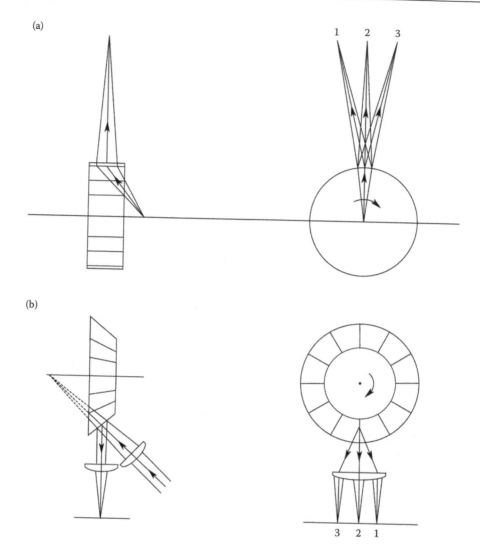

Figure 10.8 Cylinder hologons. (a) Transmissive and (b) Reflective.

- Radiometric uniformity; diffraction efficiency varies with angle of incidence.
- Polarization: diffraction efficiency depends on the relative orientations of the input beam polarization and the grating, unless the beam is circularly polarized.

A single-facet holographic deflector, comprising a single linear grating, has been successfully used as a monogon deflector [19]. This is designed to have a diffraction angle of approximately 90°. The grating is mounted at 45° to the input beam, when the wobble insensitivity is most effective, in a mount that rotates about the beam axis. The rotating mount can include a lens after the grating, for use in internal drum configuration, or the deflector can be used with a fixed flat-field lens in a pre-objective scan configuration.

10.4.4 Optoelectronic

The term optoelectronic is used here to denote any deflector that has no mechanical moving parts, and whose operation depends on the interaction of an electrical signal with the optical properties of the deflector material to produce an angular deflection. These deflectors can have effectively zero inertia, the response being limited only by the speed of the drive electronics, making them ideally suited to broadband position control and ultra-high speed scanning. However, in practice, they are only rarely viable alternatives to mechanical scanning mechanisms, due to their limited angular deflection range and optical aperture, and hence poor angular resolution, typically much less than 1000 spots per scan line. Their application is therefore generally limited to low-resolution very high-speed imaging and active error correction subsystems.

Electro-optic (EO) deflectors use the electro-optic Pockels effect of some crystal materials to deflect a linearly polarized beam by an angle proportional to an applied voltage. Electro-optic coefficients of even the best crystal materials have low values, so angular deflection is small. Because of the birefringence and thermal sensitivity of available EO materials, sophisticated designs are required, often passively compensating the undesirable effects of temperature and mechanical stress using additional crystals.

One commercial implementation produces a uniform refractive index gradient across the crystal using a quadrupole array of electrodes [20]. An alternative form of EO deflector works by inducing a refractive index change in a prism (Figure 10.9).

The deflection angle may be expressed as

$$\theta = \frac{\Delta n}{n} \frac{L}{D} \qquad (10.10)$$

where $\Delta n = (n_1 - n_2)$ and n is the refractive index at the final air interface. For an EO material, the change in refractive index with applied voltage V_z is given by

$$\Delta n = 2n^3 r_{ij} \frac{V_z}{t} \qquad (10.11)$$

where r_{ij} is the material electro-optic coefficient and t the thickness of the cell in the z direction (perpendicular to the paper).

The deflector performance is improved by fabricating the deflector in the form of several sequential prisms of alternating polarity, increasing the length of the cell while allowing practically achievable crystal sizes. This has been implemented using semiconductor-manufacturing technology, growing LiTaO$_3$ crystal prisms using photolithography [21].

Acousto-optic (AO) deflectors use diffraction in a crystal material, produced by pressure waves, to deflect the beam by an amount dependent on the applied frequency f. The deflection results from a thick diffraction grating created within a crystal material by a radio-frequency traveling pressure wave generated by a piezo-electric acoustic transducer bonded to the side of the crystal (Figure 10.10). The first order deflection angle is given by the diffraction Equation 10.8 above, where the grating period $t = f/v_a$, where v_a is the acoustic velocity of the cell material.

For high diffraction efficiency, Bragg diffraction is used, where the deflection angle of the diffracted beam at the center of the scan line is twice the angle of incidence. Varying the frequency of the acoustic wave about the mean frequency varies the deflection angle, producing a scan.

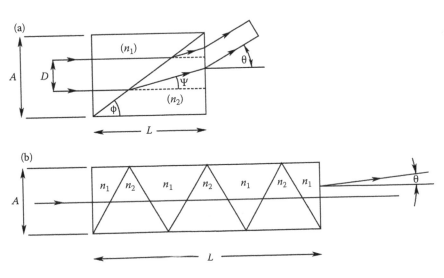

Figure 10.9 Prism-type electro-optic deflector. (a) Single prism and (b) Multiple prism.

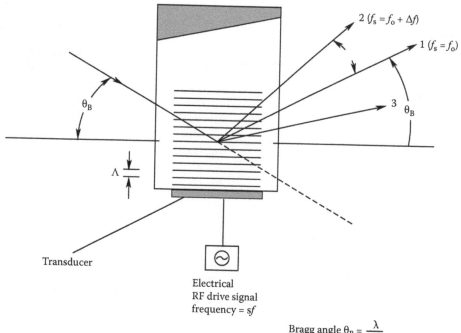

Bragg angle $\theta_B = \dfrac{\lambda}{2\Lambda}$

Relative deflection angle $= \theta = \dfrac{\lambda}{v_a} \Delta f$

Figure 10.10 Acousto-optic deflector.

The number of resolved spots per scan line for an acousto-optic deflector is given by [22]

$$N \approx \frac{\tau \Delta f}{a} \qquad (10.12)$$

where τ is the transit time of the acoustic wave across the optical aperture and Δf the change in transducer frequency from the center frequency.

To maximize τ, a material with a slow acoustic velocity (usually TeO_2) is used and the size of the laser beam in the acoustic wave direction is made as large as possible, although these factors increase the response time of the device. To keep the size of the crystal within practical limits, cylindrical lenses are used to shape the beam within the deflector. Practical considerations of crystal size, acoustic attenuation and deflector speed limit the beam width to about 50 mm.

To ensure that the deflected beam has constant power over the scan range, it may be necessary to use a phased-array transducer design [23]

to maintain Bragg diffraction over the range of deflection angles.

During scanning, the acoustic wave frequency varies across the beam aperture. The frequency gradient produces a cylindrical lens effect with a focal length of

$$f_{cyl} = \frac{v_a \tau}{\theta_T} = \frac{v_a^2}{\lambda (df / dt)}. \qquad (10.13)$$

The output beam is converging or diverging, depending on the scan direction. This can prevent bidirectional scanning, and requires the use of an external compensating cylindrical lens in the optical system.

Two-dimensional AO deflectors have been produced, using transducers on orthogonal faces of a single crystal, although practical limits to crystal size mean that the achievable angular resolution is severely limited.

Recent research has employed optoelectronic devices with high spatial resolution, such as liquid crystal or micromirror arrays, to produce deflection via phased array phase shift [24,25] or variable-pitch diffraction gratings [26]. These devices are still in the early stages of development, and show some promise of enabling fast beam steering with lower-cost devices and drivers, but show little improvement in resolution performance over the existing acousto-optic and electro-optic technology.

10.5 SCANNER OPTICS

10.5.1 Reduction-type array imaging lens

In a reduction-type CCD scanner, a high-quality multi-element imaging lens images the scanned document on the detector array. This lens must be specially designed for the purpose, because a standard photographic quality camera or enlarger lens is not suitable.

The requirements that must be addressed in the design include

- *Magnification*: consider both the ratio between document width and array length and the ratio between required spatial resolution (in terms of addressability) and detector element spacing. In some systems, the lens must be corrected for a range of magnifications, requiring multi-configuration optimization.
- *Focal length*: a compromise between mechanical layout and achievable optical performance. Short focal length increases the field angle, which makes the design more difficult.
- *Spatial resolution*: MTF must be specified for both sagittal and tangential directions, for the full range of field positions, and for each wavelength. The effect of sampled imaging must be considered; in particular, the Nyquist frequency acts as a limit to resolution.
- *Color correction*: for color scanners, excellent color correction is required, often involving secondary spectrum correction and therefore careful glass type selection. Lateral color is often a limiting aberration, which causes highly visible color fringes on off-axis image features.

- *Aperture*: often determined by the radiometric requirements of the detector. The effect on diffraction-limited resolution and lens performance must also be considered.

The complexity of lens designs ranges from relatively simple Cooke triplet types to complex double-Gauss derivatives with six or more elements.

10.5.2 Contact-type array scanner optics

Input scanning using a long array detector with small elements can be achieved without complex lenses, if the scan line length and resolution can be identical to those of the array. In this case, a configuration with the detector in contact with the scanned object is usually not feasible because

- Movement of the detector across the object will cause scratching.
- For a reflective object, there is no clear path for illumination of the object around the detector.
- Detector arrays are usually packaged in a hermetically sealed form, with a window and air space between the sensor surface and the object. This gap results in a loss of light collection efficiency and resolution.

The usual solution is to include a micro-optical system that images the object on the detector at unity magnification. This system can be based on either

- A linear array of glass microlenses and microprisms
- A linear SELFOC (gradient index lens) array lens
- An optical fiber array

The magnification in these systems must be non-inverting, so that the images from the array lenses overlap to form a linear image at the detector array. This results in a compact, low-cost configuration, and is commonly used in hand-held and desktop scanners. Compared to reduction-type systems, the numerical aperture and the resolution of these optical systems are low, and the performance of the very long arrays is inferior to wafer-scale CCD arrays.

10.5.3 Flat-field laser scanning lenses

In a pre-objective flatbed scanning system, a specially designed lens is required to focus the scanned beam to a line. A theoretically perfect thin lens produces a curved field of radius f, so a flat-field lens must be designed with the correct amount of negative field curvature to compensate the natural image curvature.

In addition, in most cases, it is highly desirable that a constant rotation rate of the scanner deflector should result in a scan spot with linear velocity at the lens focus. Thus, the linear spot deflection y at the image plane must be related to the scan angle by the relationship, known as the f–θ characteristic

$$y = f\theta \qquad (10.14)$$

where f is the lens focal length and θ the scan angle, so that

$$\frac{dy}{dt} = f\frac{d\theta}{dt}. \qquad (10.15)$$

A distortion-free lens has the characteristic

$$y = \tan\theta \qquad (10.16)$$

so the lens must additionally be designed with the correct amount of barrel distortion to produce the desired f–θ characteristic. Such a lens is therefore referred to as an f–θ lens.

Another common requirement is for telecentricity in image space, i.e., the central, or chief rays of the beam are parallel with the lens axis for all scan angles. For a thin lens it can be seen from Figure 10.11 that this is achieved when the entrance pupil of the lens is positioned at the back focus of the lens. Telecentricity ensures that the focused spot is not elongated into an ellipse by the angle of incidence at the image plane. The drawback is that at least the final lens element must be longer than the scan line. In practice, true telecentricity is often not required, and a reasonable divergence of chief rays from the scan lens can be tolerated, greatly reducing the size of lens elements. The fractional increase of spot dimension in the y-direction is

$$\frac{\Delta s}{s} = \frac{1}{\cos(\theta)} - 1 \qquad (10.17)$$

so, for example, an angle of 15° increases the spot dimension by only 3.5%. In practice, this would be acceptable in many systems.

With modern lens design software, it is quite possible to design lenses that are simultaneously corrected for field flatness, f–θ distortion and telecentricity. However, excellent optical correction is also usually required, to ensure uniform spot size

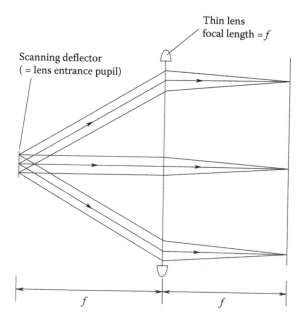

Figure 10.11 Telecentric scan lens.

over the scan line. The fewer the constraints placed on the design, the better the achievable lens resolution performance. Although it is sometimes possible to design around an existing lens design, the system-specific requirements for wavelength, scan line length and resolution most often dictate a need for a custom lens.

Conventional, spherical-surfaced glass lenses typically comprise three or more lens elements. Although these can be of a single glass type, better correction is obtained, even for a monochromatic requirement, by using more than one glass. Using a combination of crown and flint glasses, color correction can be achieved for two, or even three, wavelengths. Usually, diffraction-limited performance is required; this becomes more difficult to achieve as the spot size decreases and the scan angle increases. The number of resolved spots per scan line is a good measure of the lens quality, with anything over 25,000 spots/scan line being difficult to achieve.

For systems that are manufactured in large numbers, one- or two-element injection-molded polymer lenses are usually used. As the mound tool determines the lens shape, aspheric surfaces are economic, and the lens can therefore achieve a good optical performance with fewer surfaces than can be achieved with spherical surfaces. By using very complex, asymmetric surfaces, it is possible to correct even the effects of pupil wander—the small movement of the beam relative to the lens axis during polygon rotation [27].

Because of the limited choice and range of materials, color correction is not possible with plastic lenses, and the number of spots per scan line is usually less than 10,000. Thermal stability is often an issue, even within a normal office environment, because of the high temperature coefficients of expansion and refractive index of plastics. It is possible to passively compensate for temperature variations by carefully designing hybrid glass/plastic optics [28].

10.5.4 Deflector error correction

A number of design techniques have been developed to reduce, or eliminate, cross-scan deflection errors caused by manufacturing and alignment tolerances of prismatic polygons. These techniques allow relaxation of what would otherwise be prohibitively tight tolerances.

Active deflection error correction uses a small-angle, high-speed deflector in the optical path just before the main scanning polygon. This controls the input beam angle in the cross-scan direction, to compensate for cross-scan angle errors in the polygon. In an open-loop control system, the fixed pyramidal errors of the polygon mirror are programmed in a calibration procedure. In a closed-loop control system, a separate optical measurement system is used to measure the cross-scan deflection angle, and a servo control system dynamically adjusts the input angle to correct any deviation. In the latter case, non-repeatable errors such as bearing wobble can be corrected in addition to polygon fabrication errors.

Passive deflector correction can be achieved at a much lower cost using non-rotationally symmetric optical surfaces in the flat-field lens. The principle of the correction system is to use a cylindrical lens to focus the input beam on the polygon facet only in the cross-scan plane, and then to use cylindrical or toroidal elements within the flat-field lens to make the facet surface conjugate with the image plane in the cross-scan direction (Figure 10.12). Where the optics consists of conventionally polished glass optical elements, a toroidal lens is added to the usual flat-field lens either before the lens, or close to the image plane. However, where molded plastic optics is used, the toroidal elements, incorporating aspheric surface forms, may be integrated into the lens design, minimizing the number of optical elements and achieving better cross-scan error correction.

Other correction systems are possible, using multiple reflections to cancel the deflection errors. A pentaprism has the well-known property that the deflection angle is a constant right angle, independent of the angle of incidence, and this is generally true for double-reflection systems. A pentaprism [29], or any of its open mirror equivalents [30], might be used as the deflector element of an internal drum scanner. These internal correcting reflector systems have the disadvantage of requiring significant complexity in mechanical mounting and balancing. This disadvantage does not apply to external correction systems. An external double-reflection element may be used to correct deflection error in a polygon. In practice, this greatly compromises the system design, significantly increasing the polygon size and flat-field lens aperture and reducing scan duty cycle.

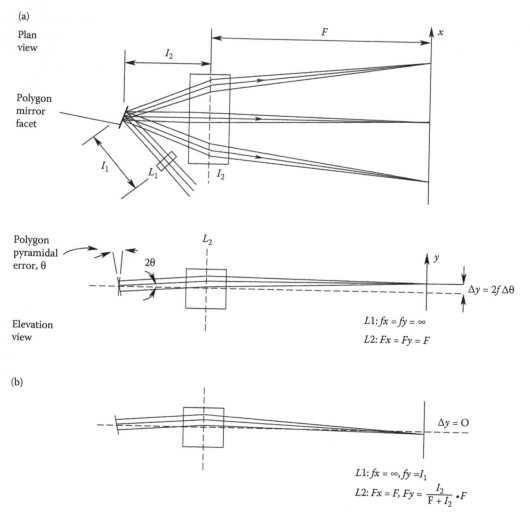

Figure 10.12 Use of anamorphic lenses for deflection error correction. (a) Uncorrected and (b) Corrected.

10.6 INPUT SCANNING SYSTEMS

In this section, we look at examples of designs of scanning systems used for input of data. These are essentially electronic imaging systems, but they can give far more information than a visual image, providing precise measurement of color, scattering properties, size or fluorescence at all points within the scanned area.

Scanning systems have been developed for one-dimensional (e.g., supermarket barcode scanners and wire diameter measurement systems), two-dimensional (e.g., desktop document scanners) and three-dimensional (e.g., reverse engineering of aerodynamic automobile models) measurements. Three-dimensional scanners are mainly one of the three types. The first type uses laser radar principles, i.e., a two-dimensional spatial scan combined with high-speed time-of-flight measurement of fast laser pulses.

The second type uses optical triangulation to measure step changes in the height of a projected laser line, measured by a two-dimensional sensor array at an angle to the line projection plane. Lastly, confocal scanning microscopy [31,32] uses a small-aperture detector at the focus of a retro-reflected beam to produce a scanned image with very small depth of focus, enabling high-resolution three-dimensional images when the object is stepped through focus on successive scans.

Input systems can be classed as either passive (i.e., using ambient or fixed illumination with a scanning detector) or active (i.e., providing illumination via a scanned beam for a fixed or scanning detector). FLIR thermal imaging systems [33] detect infrared thermal emissions in the 8–12 μm band. Active input systems that use the scanned illumination beam (usually a laser) to define the scanned image resolution are known as flying spot scanners. Examples of applications that use these two classes of scanning are discussed later.

10.6.1 CCD document scanner

CCD scanners are based on linear array CCD detectors, which image the width of the document, with a linear mechanical scan stepping along the length of the document. Scanners are classed as either contact type, where the resolution is essentially that of the detector elements, or reduction type, in which higher resolution is achieved by imaging the document on the detector array with a magnification less than unity. Reduction-type scanners have higher performance, but are more expensive and bulky.

A schematic optical layout of a typical large-format document scanner is shown in Figure 10.13. Illumination along a line of the document (which could, for example, be a poster, a map or an engineering drawing) is provided by two cylindrical fluorescent lamps, concentrated by cylindrical reflectors. In a camera assembly, a multi-element lens images the line on a tri-linear CCD device. This simultaneously produces three red, green and blue color-separated images of a line of the document, immediately after the time-delays between the lines are corrected. This provides sufficient data to accurately measure the color of the document at each point, after calibration.

The spatial resolution is determined by the number of elements in each line of the CCD array, typically 5000, and the lens magnification. The elements are typically 8 μm square, and there is no gap between elements in a line. To obtain higher resolution, two or more camera systems are optically butted together, to increase the number of resolved spots per scan line. This is achieved by slightly overlapping the images of the two CCD sensors at the document, and 'stitching' the electronic images together by software processing.

The CCD array scans the line image data electronically, and the line is scanned along the document by a motorized pair of pinch rollers that transport the document across the stationary optical system. To achieve the necessary precision in this transport system, a sophisticated control

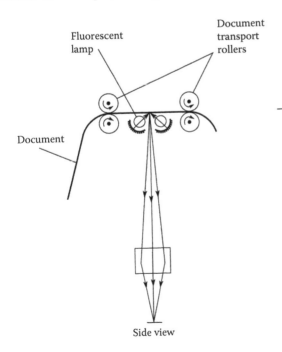

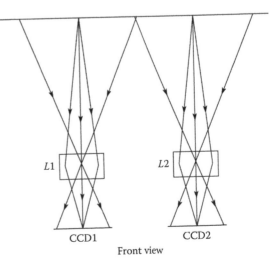

Figure 10.13 Large-format document scanner.

system is required to ensure accurate speed and constant tension in the scanned document.

10.6.2 Flying spot scanners

The term flying spot scanning was originally applied to CRT scanning, but nowadays it is more often applied to laser scanning. The resolution of the system is defined by the laser spot size at the object, and a non-imaging detector system collects the light that is transmitted, reflected or scattered by the object. Laser barcode scanners work on this principle.

A more complex system, used for on-line inspection of defects in float glass, is shown in Figure 10.14. In this system [34], a 633 nm HeNe laser beam is scanned by a rotating 12-facet prismatic polygon across the 2 m wide glass sheet during manufacture, as the glass is moved on a conveyor in a direction perpendicular to the scanning direction, producing a continuous raster scan of the sheet. Glass defects are detected and analyzed on-line by measurement of the transmission and scatter of the 0.45 mm diameter laser spot.

In this system, the detector system is contained within the scanner head, by employing a retro-reflecting screen of micro spheres to return the light transmitted through the glass to the scanning system. The reflected light forms stationary images of the scanning spot at two detector positions, returning via two facets of the scanning polygon. One of the detectors is partially masked by a circular obstruction corresponding to the laser spot size, and this measures the scattered light. The other detector simply measures the transmission through the glass.

The polygon facet size is determined by the light collection requirements of the detector system, rather than by the size of the illuminating laser beam. As the retro reflecting sheet produces a cone of about 2° angle, a large polygon with facets of 50 mm×25 mm is required, which makes the polygon manufacture challenging.

10.7 OUTPUT SCANNING SYSTEMS

In output scanning, a light source (almost always a laser) is used to write image data to a screen or

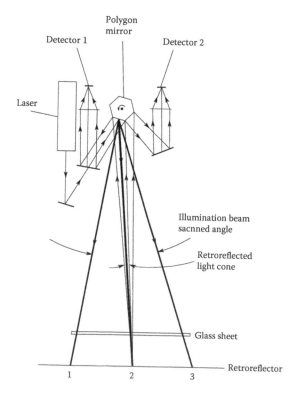

Figure 10.14 Glass sheet inspection scanner.

to a light-sensitive output medium. Systems vary greatly, depending on the size and characteristics of the laser, whose wavelength can be between 350 nm and 10.6 µm, and whose power can be between 1 mW and 1 kW.

Most output systems are two dimensional, imaging to a flat or cylindrical surface. Image size, resolution and speed are the key scanner characteristics. Laser displays take advantage of the eye's persistence of vision to create a two-dimensional image for advertising or entertainment.

The writing process can be based on thermal (e.g., laser marking) or photonic (e.g., photographic) processes. In some processes, the image density is a monotonic function of incident power density, in which case the image is described as continuous tone, or contone. Other processes are sharply thresholded, so that the image density takes one of two (or rarely, more) discrete values depending on the incident power density. In the latter case, the image is made up of a fine array of dots, and is described as halftone.

The power density variations are produced by modulating the continuous-wave laser power as it is scanned across the image surface. Modulation may be controlled by an external acousto-optic or electro-optic modulator, or by controlling the laser current in some cases, particularly in laser diodes.

10.7.1 Desktop laser printer

A typical xerographic laser printer is shown schematically in Figure 10.15. The configuration may be described as a single-pass external drum, or as a flat-field line scan, with drum rotation providing the slow axis scan. In this example, two adjacent scan lines are imaged from independent low-power 780 nm laser diodes LD1 and LD2, doubling the imaging (drum rotation) speed. The image is produced by a spatial variation of electrostatic charge on the pre-charged photoconductive drum surface, which is discharged by incident light. Toner particles are attracted to the charged areas, and transferred by contact to sheets of paper to produce the finished print.

In the example shown, an eight-facet polygon produces the scan, rotating at 30,000 rpm. A three-element f–θ lens produces an elliptical spot of 50 µm in the scan direction and 60 µm in the cross-scan direction, for a printer resolution of 600 DPI over a scan length of 312 mm. This allows a printing speed of 20 A3 copies per minute.

Key features of the design are

- The two collimated laser diode beams are efficiently combined using a polarizing beam splitter, aligned in such a way that the beams

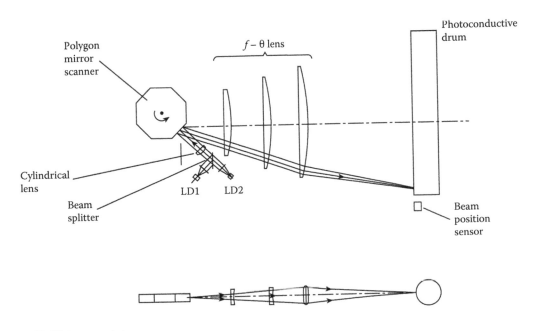

Figure 10.15 Laser printer.

overlap at the polygon—which is also the entrance pupil of the $f-\theta$ lens—and are at an angle to each other to separate the beams by exactly one scan line at the image drum.

- A photodiode just beyond the image area monitors the beam power and position, as part of the control feedback loops for the laser diode and polygon motor drivers.
- Deflector error correction is achieved using an anamorphic lens design, with a cylindrical lens before the polygon and different $f-\theta$ optical power in in-scan and cross-scan planes.
- One of the $f-\theta$ lens elements is glass, with spherical surfaces. The other two elements are injection-molded plastic with anamorphic, aspheric surfaces. The use of glass enables thermal compensation of the lens performance over a 10°C–40°C range, which would be a problem with an all-plastic lens.

REFERENCES

1. Beiser, L. 1974. Laser scanning systems. *Laser Applications*, vol. 2, ed. M. Ross. New York: Academic, pp. 53–159.
2. Levy, L. 1968. *Applied Optics*, vol. 1. New York: Wiley.
3. Aylward, R. P. 1999. Advances and technologies of galvanometer-based optical scanners. *Proc. SPIE* 3787: 158–164.
4. Marshall, G. F. 1999. Risley prism scan patterns. *Proc. SPIE* 3787: 74–87.
5. Beiser, L. 1983. Generalised equations for the resolution of laser scanners. *Appl. Opt.* 22: 3149–3151.
6. Boreman, G. 2001. *Modulation Transfer Function in Optical and Electro-Optical Systems*. Bellingham, WA: SPIE Press.
7. Montague, J. 1991. Galvanometric and resonant low-inertia scanners. *Optical Scanning*, ed. G. F. Marshall. New York: Dekker.
8. Shinada, H. 1996. US Patent 5 502 709.
9. Brosens, P. 1976. Scanning accuracy of the moving-iron galvanometer scanner. *Opt. Eng.* 15: 95–98.
10. Verboven, P. E. 1988. Distortion correction formulas for pre-objective dual galvanometer laser scanning. *Appl. Opt.* 27: 4172–4173.
11. Calquhoun, A. B., Gordon, C. S., and Shepherd, J. 1988. Polygon scanners—An integrated design package. *Proc. SPIE* 96: 184–195.
12. Beiser, L. 1991. Design equations for a polygon laser scanner. *Proc. SPIE* 1354: 60–66.
13. Sherman, R. 1991. *Optical Scanning*, ed. G. F. Marshall. New York: Dekker.
14. Ketabchi, M., Tiffanhy, B. L., and Vettese, T. J. 1997. Single-facet scanning subsystem for digital imaging. *Proc. SPIE* 3131: 290–299.
15. Ng, R. K. M., Rockett, P., and Wardle, F. P. 1999. Analysis of the distortion of a high-speed single-facet rotating mirror. *Proc. SPIE* 3787: 252–263.
16. Beiser, L. 1979. *Holographic Scanning*. New York: Wiley.
17. Cindrich, I. 1967. Image scanner by rotation of a hologram. *Appl. Opt.* 6: 1531–1534.
18. Kay, D. B. 1984. Optical scanning with wavelength correction. US Patent 4: 428–643.
19. Kramer, C. J. 1991. Holographic deflectors for graphic arts systems. *Optical Scanning*, ed. G. F. Marshall. New York: Dekker.
20. Fowler, V. J. 1964. Electro-optical light beam deflection. *Proc. IEEE* 52: 193.
21. Li, J., Chen, H. C., Kawam, M. J., Lambeth, D. N., Schlessinger, T. E., and Stancil, D. D. 1996. Electrooptic wafer beam deflector in LiTaO$_3$. *IEEE Photon. Technol. Lett.* 8: 1486–1488.
22. Gottlieb, M. 1991. Acousto-optic scanners and deflectors. *Optical Scanning*, ed. G. F. Marshall. New York: Dekker.
23. Korpel, A. 1981. Acousto-optics. *Applied Optics and Optical Engineering* VI, ed. R. Kingslake and B. J. Thompson. New York: Academic, pp. 89–141.
24. McManaman, P. F. and Watson, C. A. 1997. Optical beam steering using phased array technology. *Proc. SPIE* 3131: 90–98.
25. Stockley, J. E., Subacius, D., and Serati, S. A. 1999. Liquid crystal on VLSI silicon optical phased array. *Proc. SPIE* 3787: 105–114.
26. Levrentovich, O. D., Shiyanovskii, S. V., and Voloschenko, D. 1999. Fast steering cholesteric diffractive devices. *Proc. SPIE* 3787: 149–155.
27. Li, Y. and Katz, J. 1997. Asymmetric distribution of the scanned field of a rotating reflective polygon. *Appl. Opt.* 36: 342–352.

28. Yamaguchi, M. and Shiraishi, T. 1999. Development of four-beam laser scanning optical system. *Proc. SPIE* 3787: 2–12.

29. Starkweather, G. K. 1984. US Patent 4475787.

30. Marshall, G. F., Vettese, T. J., and Carosella, J. H. 1991. Butterfly line scanner. *Proc. SPIE* 1454: 37–45.

31. Wilson, T. (ed.) 1990. *Confocal Microscopy*. London: Academic.

32. Masters, B. R. (ed.) 1996. *Selected Papers on Confocal Microscopy*. Bellingham, WA: SPIE Press.

33. Lloyd, J. 1975. *Thermal Imaging Systems*. New York: Plenum.

34. Holmes, J. D. 1997. Inspection of float glass using a novel retro-reflective laser scanning system. *Proc. SPIE* 3131: 180–190.

Enabling technologies for sensing, data processing, energy conversion, and actuation

Energy reduction by sensing, data processing, energy conversion and actuation

11

Optical fiber sensors

JOHN P. DAKIN
University of Southampton, UK

KAZUO HOTATE
University of Tokyo, Japan

ROBERT A. LIEBERMAN
Lumoptix Inc., USA

MICHAEL A. MARCUS
Lumetrics Inc., USA

11.1 INTRODUCTION

It is said that man has five senses—sight, touch, smell, taste, and hearing—at his disposal. Clearly, the one he finds most generally valuable is his sight. This is hardly surprising, as it is excellent for remote sensing, gives an effectively instantaneous response, has truly enormous parallel information capacity, and provides far more reliable and quantitative data than any of the others—it is common to say "you believe the evidence of your own eyes." Optical sensor technology may still have some way to go to match the same compact design and all-round performance as the eye/brain combination, but

the promise is clearly there! As an aside, it is worth noting that we, as humans, are not able to remotely sense electrical or magnetic fields well (although migrational birds are believed to use magnetic sensors), so, if one believes that nature often chooses the best methods, the technology of optical sensors may perhaps evolve to overtake that of electrical ones in time. Modern camera and CD player technologies have shown that even highly complex optoelectronic systems can be manufactured cheaply in volume, so cost should not present a major barrier for large-scale application of standard devices.

This chapter reviews the many ways in which optical fibers may be used, in conjunction with optoelectronic instrumentation, to sense physical or chemical parameters. This is done using either intrinsic sensors, where the fiber itself acts as the sensing element, or extrinsic sensors, where the fiber acts in its normal communication role, to carry light to and from an external optical sensor. The fiber can enable remote and rapid sensing to be performed by sensing at specific points or over specific cable paths. This complements the capability of optical sensors such as cameras, which have valuable imaging capability, as fiber cables do not need a direct "line-of-sight" to operate and measure in inaccessible places. Because there are now very many varieties of optical fiber sensor, it has been necessary to concentrate on a few important concepts and only describe a few practical sensor types in more detail. A far more detailed account of fiber sensors is given in the textbooks mentioned in bibliography References [1,2].

Optical fibers are widely used in long distance trunk telecommunications systems (see Vol II, Chapter 1) and their use in shorter distance applications is growing rapidly. For communications systems, it is desirable to utilize a cable having transmission properties that are insensitive to environmental changes. Intrinsic fiber sensors, in contrast, rely on deliberately configuring the optical system to be sensitive to external influences on the fiber or cable. This involves the arrangement of the fiber cable, its associated light source, the optical receiver, and the chosen signal processing scheme to maximize (and detect) a change in transmission properties that is characteristic of the parameter to be sensed. This may be carried out either by using especially sensitive cables or by interrogating conventional fiber cables in a manner that highlights small changes in transmission. The resulting changes in transmission can then be used to measure features of the external environment.

An extrinsic fiber sensor usually uses a conventional optical fiber (preferably sheathed within an environmentally insensitive cable), which is only employed as a convenient light-guiding medium to transport light to and from more conventional optical sensors at the end of, or at specific points on, the cable. This sensor, in response to an external physical parameter, modifies the light coupled back from the input fiber into the return fiber and guided back to the detector system. Such a sensor may still be undesirably affected by environmentally induced changes in cable properties, unless particular care is taken with its design. Avoiding unwanted cable-induced errors is a very important practical consideration, in both intrinsic and extrinsic sensors. In a good sensor system design, care will be taken to avoid significant sensitivity to undesirable losses in the optical network, such as could occur, for example, due to fiber bending or connector variations.

The research area of fiber optic sensors is a very fruitful one for original thought, as there is not only a very large number of physical and chemical parameters that may be sensed, but also a multitude of ways in which the parameter may be arranged to modulate the optical transmission. However, in order to make a successful sensor, the normal principles of good instrument design must be adhered to [i.e., low cross talk to undesirable parameters, good signal-to-noise ratio (SNR), and repeatable and reliable operation] and the sensor must be designed to be cost effective for the intended application. Many of the successful commercial applications of such sensors are in niche market areas, which rely on the inherent advantages of optical sensors over more conventional electrically based sensors. The primary technical advantages include the following:

- Intrinsic freedom from electromagnetic interference (EMI), lightning strike, etc.
- Intrinsic safety in hazardous (explosive vapor) environments (provided optical signal power is low, which is almost invariably the case).
- High electrical isolation, enabling their use in medical applications and for data collection from points at high voltage. The fiber also gives freedom from problems of electrical short circuit and open circuit.
- Excellent resistance to chemical corrosion (can be used in highly corrosive environments, e.g., saltwater, acid, and alkali).
- Passive operation; no electronics circuitry or electrical power is required at the remote sensing point.
- May be used in high-temperature areas, where electronic systems would not survive.
- Optical fibers are smaller, lighter, and cheaper than electrical cables.
- May be used for distributed sensors of extreme length due to the low losses ($0.2 \, dB \, km^{-1}$) achievable in optical fibers.
- Where optically based data transmission is already envisaged, sensor information can be carried in the same optical fibers.

- Optical sensing can provide a very rapid response in many applications as changes in optical transmission can be detected almost instantly.

When the use of optical signal cables becomes more common for short distance routes, it is likely that optical transducers will start to find more general applications, even in areas where the above intrinsic advantages are not a major consideration. The ability to sense, communicate, and multiplex signals within an optical network is an important attribute, which will lead to their greater application.

We shall start by considering the different types of sensor, and the physical means by which each can operate.

11.1.1 Summary of intrinsic types of sensors, where the optical fiber is used directly as the sensing element

The *intrinsic* optical fiber sensor takes advantage of measurable changes in transmission characteristics of the fiber itself. The principal parameters of interest for sensing are as follows:

- Light generation in the fiber due to physical interactions (e.g., scintillation or Cerenkov radiation).
- The propagation time of light in the fiber (proportional to the length and inversely proportional to the velocity of light). This can be measured as a temporal delay or as a phase change in the light.
- The optical power transmitted by the fiber (either the total power or the spectral variations in transmission).
- The distribution of optical power between the various modes of propagation. (This can be measured from either the near-or far-field waveguide mode patterns at the end of the fiber.)
- The state of polarization (SOP) of the transmitted energy through the fiber (or backscattered energy from the fiber).
- The light scattered from within the fiber core material. This can include elastic scattering (same wavelength as incident light) or involve

Raman scattering, Brillouin scattering, or other nonlinear interaction, such as optical gain by Raman or Brillouin processes.

11.1.2 Summary of extrinsic types of sensors, where the optical fiber is used merely as light guiding medium, to address more conventional optical sensors

The main methods of operation of *extrinsic* fiber sensors are outlined as follows:

- Light due to a chemical or physical interaction is collected by the fiber (e.g., radiation due to high temperature pyrometry, optical scintillation of a semiconductor, and chemiluminescence).
- The optical power transmitted by an external optical modulator coupled to the fiber (either the total power or the spectral variations in transmission) is monitored. Examples include mechanical optical shutters, electro-optic switches or modulators, variable optical filters (such as Fabry–Perot, grating monochromator, temperature-dependent semiconductor).
- The optical power reflected by an external optical modulator coupled to the fiber (either the total power or the spectral variations in transmission) is monitored. Examples as above.
- The optical power scattered elastically (no wavelength change) by an external medium coupled to the fiber is monitored. This effectively measures turbidity of the medium, in reflective mode.
- The optical power scattered inelastically (scattered-wavelength changes) by an external medium coupled to the fiber is measured. This includes processes such as Raman scattering, fluorescence, and photon correlation spectroscopy.
- The SOP of the energy transmitted by an external optical modulator coupled to the fiber (or that reflected back into the fiber) is monitored.

11.1.3 Evanescent field sensors

Evanescent field sensors exhibit some of the characteristics of *intrinsic* sensors and *extrinsic* types but do not exactly fit into either category. As with *intrinsic* sensors, the light is guided by the fiber (or by an optical waveguide attached to the fiber) in the sensing region, but, in the case of *evanescent field* sensors, a portion of the optical energy travels outside the physical limits of the waveguide material. In the lateral direction, the field decays rapidly as distance from the waveguide increases, a behavior known as *evanescent field* decay. Because there is light energy outside the guide material, it is possible for this light to interact with (e.g., be absorbed, scattered, or excite fluorescence in) the surrounding material. Then, for example, the *evanescent field* sensor can detect the optical properties of a fluid in which it is immersed or solid material in which it is embedded, simply by measuring the light transmitted through the fiber. Unfortunately, because the evanescent field region is very thin (typically <1 pum μm thick), this type of sensor is very prone to surface damage or contamination. Also, the evanescent field penetration depends strongly on the refractive index of the surrounding material (and hence on temperature) and on whether or not the guide is bent, so it is very sensitive to the operational environment.

Instead of immersing the evanescent field sensor directly into a solution to be monitored, it can be coated with an active layer, such as a polymer or a semipermeable, glass-like, sol–gel coating. These coatings can be made sensitive to desired chemicals by incorporating (immobilizing) an indicator chemical into them. Provided the layer is semipermeable, chemicals can diffuse in from the surroundings and change the optical transmission of the indicator material. Clearly, the presence of this layer reduces the problems of contamination of the critical evanescent field region, as this now lies inside the solid material.

A further variation is to coat the light guide with a thin metal layer, which can enhance the field in the evanescent region by a mechanism called plasmon resonance, which, as the name implies, involves excitation of the "electron gas" present within all metallic conductors. This plasmon resonance mechanism can give greatly improved sensitivity to any absorption in the evanescent field region and still allow indicator layers to be coated on as before.

Because of the ease of contamination and the environmental sensitivity of evanescent sensors, they are usually better suited to qualitative testing for chemicals, rather than a means of performing quantitative chemical or spectral analysis of fluids in which they are immersed. Nonetheless, they can be a very sensitive means of detecting trace quantities of chemicals, provided they are designed and used with care.

The following types of *evanescent field sensor* (with bracketed notes to indicate where more details can be obtained from bibliography 1) are commonly used:

- Unclad fiber, with evanescent field extending directly into the gas or liquid to be sensed (bibliography [1], vol. 1, p. 607).
- Fiber with a physically sensitive (or chemically reactive) cladding layer, with the evanescent field extending into cladding. The chemical must diffuse into, and physically modify, or react with, the layer (bibliography [1], vol. 4, pp. 352–354).
- Polished half-coupler sensor. Here, the fiber is set in a slit in a glass block, with part of the cladding polished away to expose the external medium (or a reactive polymer coating layer) to the evanescent field region (bibliography [1], vol. 1, pp. 213–215).
- Integrated optics sensors, with natural propagation of evanescent field above surface of chip. Again there is exposure of the external medium (or reactive polymer layer) to the evanescent field region (bibliography [1], vol. 1, chapter 9).
- Plasmon resonance. Here, a thin metallic layer enhances the optical coupling of the light in the fiber to the evanescent field region in the external medium (or reactive polymer layer). This enhancement can be used with most of the above configurations (bibliography [1], vol. 1, pp. 203–206).

Please note that Chapter 3, by G. Stewart, in vol. 3 of bibliography 1, gives a detailed overview of evanescent field devices.

Below, we shall give more detailed descriptions and case studies of optical sensors. The remainder of the chapter is split into four sections, each written by a different coauthor. Section 11.2 (John Dakin) covers the basics of intensity-based sensors and Section 11.3 (Kazuo Hotate) treats interferometric types. Section 11.4 (John Dakin) covers how sensors may be multiplexed and Section 11.5 (John Dakin, Bob Lieberman, and Kazuo Hotate) shows how sensors that operate on a fully distributed basis can be made. The final section, section 11.6 (Mike Marcus) deals with a detailed case study of a type of interferometric sensor that has many industrial uses. Because the chapter is composed of contributions from several authors, there are small differences in style of writing and diagrams reflecting their different contributions. It should be emphasized again that not all types of sensor can be covered in such a short chapter, and there has had to be much careful selection of which types to include and which case studies to present in more detail. The reader requiring more detail is again referred to the bibliography, to the "Optical fiber sensors" and the "Europt(r) ode" international series of conferences and also to the many SPIE-organized (http://www.SPIE. org) conferences covering this area.

11.2 INTENSITY-BASED OPTICAL FIBER SENSORS

11.2.1 Physical sensors

11.2.1.1 SIMPLE OPTICAL INTENSITY SENSORS

We shall first describe the very simplest form of intensity-based sensors, where only the transmitted or reflected light level is measured. We shall start with intrinsic sensors, then discuss extrinsic types.

11.2.1.1.1 Light generation in fiber itself

A few mechanisms allow generation of light in a fiber, as a result of a direct interaction with the physical field to be monitored. There are two principal types here. One is a fiber sensor for detection of pulses of ionizing radiation, having extreme peak intensity, of the type that can occur due to thermo-nuclear events (see Figure 11.1).

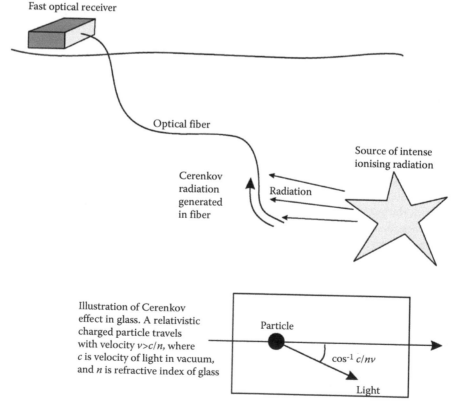

Figure 11.1 An optical sensor for detection of intense ionizing radiation (relativistic charged particles) using the Cerenkov effect.

If a relativistic charged particle (one with a velocity in free space close to the velocity of light in vacuum, c) passes through a medium such as glass, of refractive index, n, and if its entrance velocity initially exceeds the velocity of light c/n, in that new medium, then it will lose energy rapidly in the form of electromagnetic radiation. The energy loss occurs due to the Cerenkov effect [1,2] and is emitted mainly in the form of blue and ultraviolet (UV) light, but also with some energy extending to longer wavelengths. If an intense pulse of such radiation strikes a fiber, the temporal variations in the pulse intensity can be monitored, by detecting the light at the far end of the fiber.

A sensor that is more sensitive to lower levels of radiation is one that uses fiber fluorescence [3] resulting from the radiation, either fluorescence intrinsic to the silica fiber or that generated more efficiently by doping the fiber with fluorescent material. Arrays of polymer fibers, doped with fluorescent organic dyes, have been used as long linear scintillators to track the passage of ionizing radiation in particle accelerator systems, with detectors at the end of each fiber to collect the light.

11.2.1.1.2 Microbend sensors

Microbend sensors [4,5,6] are intrinsic sensors, which take advantage of the loss in optical fibers when they are bent. The word microbend is used because gradual bends, i.e., bends of large radius >50 mm, cause very little loss in a fiber, or they would not find the extensive uses that they have in communications, whereas microbends (or sharp "kinks," of small bending radius of a few mm or less) can cause high losses, even if present over very short lengths.

The losses in the fiber are due to conversion of energy in guided fiber-core modes to cladding modes, which are then usually lost by absorption or scattering in the fiber sheathing material. In the case of monomode fibers, the mechanism is relatively simple, as there is only one guided mode (strictly two, if the duality of polarization modes is taken into account) to be coupled out, and the extent of coupling, and hence the loss, can be uniquely determined by the degree of bending.

In the case of the multimode fibers, more commonly used with such sensors, the situation is much more complex. First, bending causes mode conversion, in particular coupling of lower order modes

(associated with rays in the core which travel at a small angle to the fiber axis) to higher order ones (associated with rays traveling at larger angles to the axis) and eventually to cladding modes. This "chain" of events, coupling energy from low order modes, via higher order ones, to radiation ones, depends in a complex way on not only how the fiber is bent, but also on which modes were initially present in the fiber in the section immediately before the bend. Because of the latter aspects, it is extremely difficult to derive quantitative information from multimode microbend sensors, particularly if several are cascaded along the length of a single fiber, as the bend condition of each one will affect the response of subsequent ones.

Multimode microbend sensors have a particularly strong response if they cause the fiber to be periodically bent (Figure 11.2), with a particular spatial period that corresponds to the "zig–zag" period associated with the highest order rays that the fiber is capable of guiding [7]. These highest order rays are those striking the core cladding interface at the normal critical angle associated with Snell's law of total internal reflection. Thus, a strong response can be achieved by pressing the fiber between corrugated plates, with offset corrugations (or plates covered with parallel metal pins), which will periodically deform the fiber in a wave-like manner, with the appropriate spatial period (usually around 1–2 mm, for typical multimode fibers) needed to ensure strong mode coupling. Under these conditions, very high losses can be induced in the fiber, even with a transverse displacement of only a few microns amplitude. However, the response is highly non-linear and care must be taken to avoid the possibility of exceeding the long-term mechanical bend limitations of the fiber. Clearly with soft polymer coatings, such as acrylate, slow mechanical creep will be a problem, leading to variations of sensor response with time. Attempts have been made to improve the mechanical stability and

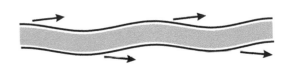

Figure 11.2 Schematic showing how light is lost at periodic bends in a multimode fiber.

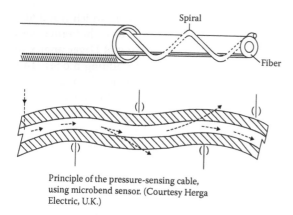

Principle of the pressure-sensing cable, using microbend sensor. (Courtesy Herga Electric, U.K.)

Figure 11.3 The pressure-sensitive cable of Herga Electric, first devised by Harmer. The spirally wound filament causes periodic bending of the inner glass fiber, when the outer cable is deformed by lateral pressure.

reproducibility, particularly at higher temperatures, by using metal-coated fibers.

Probably the most practically useful, albeit rather qualitative, microbend sensor is the distributed cable version, first devised [8,9] by Dr. Alan Harmer at Battelle labs, Geneva. This consists of an optical fiber which has a thin polymer fiber thread helically wound around it, before the combination is sheathed within an outer polymer tube (see Figure 11.3). The spatial winding period of the helically wound fiber is designed to correspond with the mode-coupling length. When the outer tube is compressed from opposite sides, the fiber is deformed to cause high loss. This assembly would clearly also suffer from slow mechanical creep, but it finds most use as a cable to sense, essentially in an on/off manner, the presence of lateral force or weight on the cable. Applications are detection of the pressure due to a human foot (e.g., for safety mats near machinery or for intruder detection on fences or perimeters) or as a safety device to detect, and hopefully prevent, trapping of a hand or head in a sliding door or window (e.g., lift door, automobile window, and sun-roof). A variety of such cable and cable-in-mat sensors have been sold commercially by Herga Electric, UK [6].

A more recent development in microbending sensor technology [10,11] has been to produce a sensor to detect the presence of water around a fiber cable, for example, for the detection of floodwater in cable ducts. This is important, as water can freeze and break fibers, or can cause hydrogen generation by corrosion of metals. The sensor cable design (Figure 11.4) has a hard inner cylindrical core, coated with a hydrogel layer (so this hydrogel layer has a hollow cylindrical

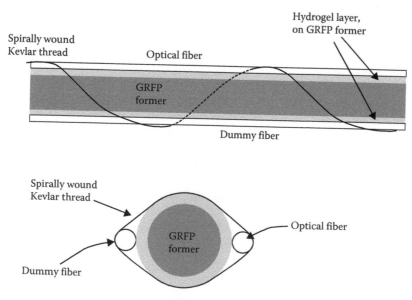

Figure 11.4 Illustration of a cable designed for sensing presence of water, which also responds to areas of high humidity [11]. The hydrogel layer absorbs water and swells, causing the glass fiber to be pushed outwards. The outer spiral filament constrains it at regular spatial intervals, causing it to suffer periodic microbending, and hence loss.

geometry). A light guiding fiber is held against the outer cylindrical surface of this hydrogel layer using a helical plastic fiber to keep it in place. If the hydrogel layer gets wet, it swells dramatically, forcing the optical fiber outwards, except where held in place by the helical fiber. This periodically deforms the fiber to make another form of microbending sensor, in this case one sensitive to water. The diagrams in Figures 11.2 through 11.6 illustrate the microbend sensor, plus a number of other loss-based fiber sensors.

11.2.1.1.3 Other intrinsic fiber-loss sensors

There are several other types of intrinsic sensor that are based on attenuation in optical fibers.

An attractive sensor [18] is the radiation dosimeter (Figure 11.5), in which the attenuation of the fiber is increased by ionizing radiation. Silica fibers are suitable for high levels of radiation, as they need large doses to suffer substantial change, whereas lead-glass fibers have a significant response at lower levels. In all cases, however, the losses exhibit a time (and hence a dose rate) dependence and the losses also reduce after thermal annealing. This annealing can be significant even at room temperature.

Another commercially useful sensor [13] for detecting leaks of cryogenic fluids (e.g., liquefied natural gas or petroleum gas) was devised (Figure 11.6) using a type of commercially available polymer-clad silica (PCS) fiber. The type of fiber used consisted of a pure silica core with a lower refractive index optical cladding, composed of silicone polymer material. When cooled to a temperature well below 0°C, the polymer shrinks more than the silica, due to its higher index of thermal expansion, and its refractive index rises to exceed that of the silica core. This causes the attenuation of the fiber cable to increase dramatically, so if light is launched at one end, there is a reduction in the detected light intensity at the output end, thereby indicating a leak of cold fluid at some point along the cable.

11.2.1.2 PROPAGATION-DELAY OR TIME-DELAY SENSORS

Propagation-delay sensors (Figure 11.7) allow the monitoring of physical or chemical effects by examining the time light takes to travel through or return from the sensing element.

This element is usually the fiber itself. Its length can change if it is mechanically strained (delay changes arise due to change in physical length or due to refractive index changes arising from the strain) or if it changes temperature. If the fiber is very long, it is possible to examine small changes in the arrival time of a short optical pulse that has traveled through it. However, because of the extremely high velocity of light, it is difficult to perform such a measurement with

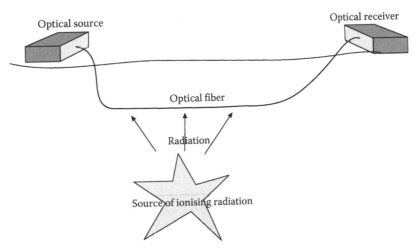

Figure 11.5 The optical fiber dosimeter. Ionizing radiation causes increased attenuation in fibers, particularly lead-glass ones, causing the detected signal to reduce.

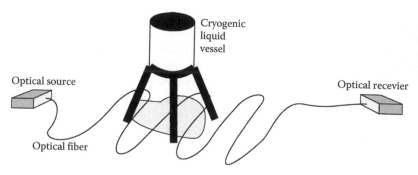

Figure 11.6 The cryogenic leak detector of Pinchbeck and Kitchen. Silicone-PCS fibers suffer attenuation when strongly cooled, because the cladding refractive index increases to equal or exceed that of the fiber core, preventing light guidance.

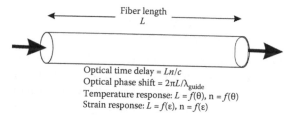

Figure 11.7 The concept of the propagation delay sensor. The delay is a function of both temperature and strain. In intensity-based sensors, the time delay has to be measured by giving the light some form of intensity modulation (Interferometric detection of delay will be discussed later.)

high precision. An alternative way is to modulate the light source at a very high frequency (typically between 500 MHz and a few GHz) and observe phase changes in the intensity modulation envelope caused by delay changes in the optical carrier signal. (Of course, by far the most precise method of monitoring is to launch a continuous lightwave through a length of fiber and examine the changes in optical phase due to the external influence, but this method comes into the area of interferometric sensors, which will be covered in detail in Section 11.3 of this chapter.)

11.2.1.3 EXTRINSIC INTENSITY-BASED SENSORS

11.2.1.3.1 Light-collecting sensors

An attractive type of extrinsic intensity-based sensor is one in which the fiber merely collects light from an external source and guides it to a detector. The simplest example of this is a fiber used to check that a light bulb (e.g., a vehicle headlight) is

operational. Fibers have also been used to monitor the temperature of hot bodies [14] such as turbine blades in jet engines, by collecting Planck hot-body infrared radiation and guiding it to a detector (remote pyrometry). Clearly, without care, there is a risk of surface contamination of the fiber. One alternative to solve this problem [15] is

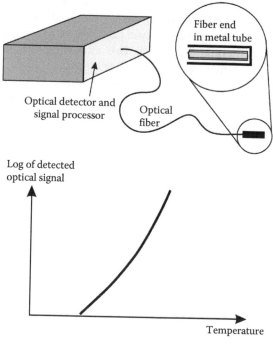

Figure 11.8 Fiber pyrometer of Dakin and Kahn, with opaque coating or closed tube over fiber end tip. Such a long coating or tube acts as an excellent black-body source, and is the way commonly used to approximate to a perfect black body. Lower curve shows how the detected signal rises dramatically with temperature.

to deliberately cover the fiber end with an opaque metal coating or sealed tube, and collect the energy emitted from the inside surface of the layer or tube, which produces an excellent approximation to a perfect black body (Figure 11.8).

Later generations of such sensors are now used in commercial instruments. With such fiber pyrometers, the simplest approach is to measure total intensity. The signal conveniently varies with temperature, but even more dramatically than would be expected from the normal fourth power law for the total radiated power, as the fiber detector combination usually only allows transmission and detection of energy from the shorter wavelength, higher energy, "tail" of the Planck radiation curve, so the measurement accuracy is reasonable even with this simple approach. However, the accuracy can be improved [16] by monitoring at two or more wavelengths, to determine the effective color temperature of the source. This then reduces the power-referencing errors that can occur when simply measuring intensity on a single detector.

11.2.1.3.2 External modulator sensors

Many forms of external modulator can cause variation of coupling between outgoing and return optical fibers. The simplest is just a mechanical shutter between the two fibers, which moves to shut off light. Such simple shutters can act as safety interlocks, e.g., for fire doors or hazardous areas, where only the safe condition allows light to be transmitted. Such sensors can be connected in series, such that light only passes if all are in a safe condition, making a logical "AND" condition to ensure all are safe. Input light can be modulated to ensure that it is not simply ambient light being collected by breaks in the system. In contrast, electrical switches can always short circuit or open circuit, particularly in corrosive chemical environments, with conductive fluids, or in seawater environments.

A liquid level sensor devised by Pitt [17] involved connecting a small right-angled microprism to the end of a fiber, or of a pair of parallel fibers, such that light was reflected back by total internal reflection when the tip was in air or gas (Figure 11.9). When wetted by the liquid, the retroreflection was greatly reduced (as only weak Fresnel refection now occurs), indicating the liquid level has reached the position where the prism lies.

An alternative, in transmissive mode, is a liquid-level sensor [18], made using fibers connected via a small gap, that is filled when the liquid reaches it, and monitoring the transmission at a wavelength where this liquid absorbs strongly (e.g., at 980 nm or 1.4 μm to monitor water). As the transmission change due to absorption can be very strong, particularly for water at 1.4 μm, and particularly because a second (unabsorbed) wavelength of light can be used as an intensity reference signal, this sensor is potentially more "fail-safe" in nature than the prism type.

A useful high-resolution proximity/distance detector (Figure 11.9) can be made simply from two or more parallel fibers, with their end faces terminated in a common reference plane [19,20]. One or more incident fiber is excited with light from a source, and the return fiber(s) collects the scattered or reflected light. Light exits the incoming fiber(s) in an illumination cone, of divergence angle determined by the fiber numerical aperture, and the receiving fiber can only receive light that is reflected or scattered back from a point lying within a similar receiving cone. If a reflective point or surface is too close, no light is received by the return fiber, as these cones do not intersect. As distance to the reflective point increases, there is first a very steep increase in signal as the cones start to overlap, then a fairly rapid decrease, this time because return signals reduce due to the much smaller signal that can be received by a tiny fiber core located a long way away from the reflective point. A simple method is to use a split (bifurcated) bundle of many optical fibers, where light is launched through one set and received by the other set. The measurement end is where all the launch fibers and receiving fibers are combined, in a randomly distributed manner, across the polished end of the bundle. Such simple intensity-based sensors can monitor surfaces to resolutions of a micron or less, depending on the core sizes of the fibers, the nature and consistency of the reflecting surface they are monitoring, and the SNR of the optoelectronic detection system. This latter aspect of SNR is very important for many intensity-based sensors and will now be quantified for simple systems.

11.2.1.4 CALCULATION OF SNR OF INTENSITY-BASED SENSORS

It is useful to be able to calculate the signal to noise ration (SNR) expected for intensity-based sensors. Most low-frequency optical receivers used in sensors are based on P-instrinsic diode (PIN) or

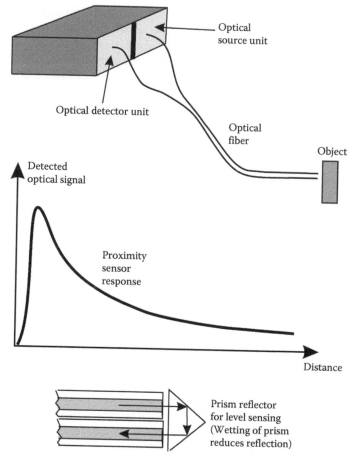

Figure 11.9 The optical fiber proximity detector and its modification to produce a liquid level sensor. If care is taken to avoid the two-way ambiguity in the detected signal from a surface, it can also be used to obtain approximate distance measurement, but errors clearly arise if the reflective properties of the surface vary. Adding a prism to the fiber ends (lower picture) produces a liquid level sensor. In this latter case, immersion prevents total internal reflection and hence greatly reduces retroreflection by the prism.

positive-intrinsic-negative silicon diodes, followed by a field effect transistor (FET) transimpedance amplifier to convert the photocurrent to an output voltage. The noise in such receivers is usually limited by the Johnson noise in the feedback resistor up to a point where the output voltage reaches about 50 mV, above which the receiver becomes shot-noise limited. Most practical intensity sensors operate in this shot-noise-limited regime, so we shall assume that this is the performance limitation here.

If we assume the mean optical signal power received at the detector is P_{det} and the fractional change in signal due to the effects of the measured parameter is Δ (Δ is an intensity modulation index):

Then the desired modulation component in the optical signal due to these is $\Delta \cdot P_{det}$ and the resulting modulation component in the detected photocurrent is $\Delta \cdot R_s\, P_{det}$, where R_s is the responsivity of the detector (in $A \cdot W^{-1}$).

The mean level of the detected photocurrent, I_{mean}, is given by $R_s \cdot P_{det}$.

The shot noise in the photocurrent is given by the standard formula: $I_n = (2 \cdot q \cdot I_{mean} \cdot B)^{-0.5}$, where q is the electronic charge and B is the postdetector-filter bandwidth of the signal.

The SNR is, therefore, given by $\Delta \cdot R_s\, P_{det}\, (2 \cdot q \cdot I_{mean} \cdot B)^{-0.5}$.

Manipulating, the $SNR = \Delta \cdot R_s \cdot P_{det} \cdot (2 \cdot q \cdot R_s \cdot P_{det} \cdot B)^{-0.5}$.

Hence, $SNR = \Delta \cdot (R_s \cdot P_{det} / 2 \cdot q \cdot B)^{0.5}$.

The SNR, therefore, increases in proportion to the modulation index, Δ, induced by the measurand, and also in proportion to the square root of detected signal power. Reducing the detection bandwidth, B, improves the SNR in proportion to the square root of the factor by which the bandwidth is reduced.

11.2.1.5 PROBLEMS WITH SIMPLE INTENSITY-BASED SENSORS AND THE NEED FOR SENSOR REFERENCING

The use of intensity-based sensors is very attractive because of the ease of measuring the intensity (e.g., from the detected photocurrent in a silicon photodiode). However, simply measuring the output intensity from a single fiber containing a sensor can lead to a number of sources of error/uncertainty in the state of the sensor. The following factors can affect the output signal and hence cause errors in the observed sensor reading:

- Light source variations
- Fiber lead and connector variations
- Variations in detector response
- Detector preamplifier noise

In practice, the noise and variations in response of the optical receiver are not usually a major problem when using silicon detectors at low frequency. However, typical changes in light source intensity can be significant and changes in the transmission of optical fibers and connectors can be a real problem, particularly where there is mechanical deformation. In order to resolve the above problems, it is usually necessary, when making a quantitative measurement, to reference the sensor. This involves providing a reference optical signal, which, ideally, has not been changed by the measurand in the same manner as the sensing signal.

The simplest method of referencing is based on the provision of more than one fiber path from the light source, leading to two separate detectors, with the sensing section included in only one of the paths. The sensing signal is then normalized to the reference signal by division. More complex systems are possible, with routing of signals via separate paths using couplers [21] or fiber switches. Alternatively, fiber paths of different length can be used to separate signals from a pulsed source. Here the separation occurs in the time domain, according to the propagation delay each signal has experienced in its separate optical path.

In some cases, the need to measure the actual intensity can be avoided if the light is intensity modulated in a periodic way by the sensor head, and, rather than measuring the intensity, the quantity monitored is the frequency (rate) or time of modulation. Measurement of modulation frequency is not only easy, but it also avoids the need for sensor referencing, and the accuracy of measurement of cycles in a given period increases linearly with the measurement period, provided no modulation-cycle counts are missed, so very high accuracy is possible over extended periods. Very simple optical tachometers (Figure 11.10) can be designed, using mechanical parts that give a periodic amplitude variation of light by virtue of variations in their light-blocking (e.g., a light-chopper wheel with radial slots) or light-reflective (e.g., a cylindrical metal shaft, with flats polished on it) properties as they rotate.

Another attractive method uses silicon resonator technology, of the type now used in many modern electronic transducers. Here, the modulator element is a tiny micromachined silicon structure, such as a flexible cantilever or bridge that is driven into mechanical oscillation (in conventional sensors, by electrostatic or piezoelectric actuation) and its period is measured. Many physical influences (e.g., temperature, mechanical tension, and pressure) can be arranged to interact and change the resonant frequency of such a structure, e.g., by changing the tension in the bridge element. The optical equivalent [22,23] is essentially the same concept, but the element is driven by light exiting from an optical fiber, which is modulated to cause periodic heating of the microstructure (Figure 11.11).

This forms a primitive heat engine, which gives rise to periodic thermal expansion and contraction of one side of the tiny silicon bridge (although photonic interactions have other means of causing mechanical changes in semiconductors), which drives this tiny element into resonant oscillation. The oscillation is of very low amplitude, unless the light modulation rate matches its resonant frequency.

The small motions of the oscillating element are also monitored optically by changes in reflection amplitude (possibly involving simple optical interferometry to enhance the effects of small

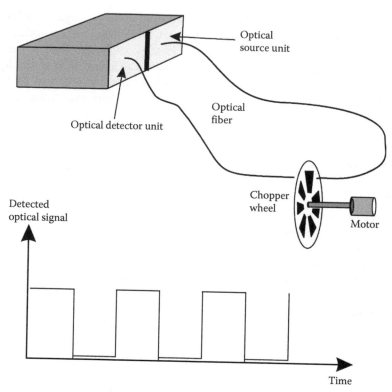

Figure 11.10 The optical fiber tachometer devised by M Johnson. The rate of beam interruption allows detection of shaft rotation speed.

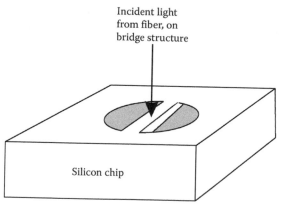

Figure 11.11 The silicon micro-resonator sensor, which is an optical equivalent of a sensor concept now common using conventional semiconductor technology. In the case of the optical sensor, excitation is by intensity-modulated optical radiation from the incident fiber. Detection of the resulting oscillations in the tiny silicon bridge can be performed by monitoring reflected intensity.

displacements) and the frequency of modulation is swept so the resonance can be detected, and even electronically locked onto. Such sensors appear at first sight to be a dream come true, seemingly solving the problems of measuring optical amplitude, and providing a "frequency-out" measurement that has very high precision due to the high resonant frequency (typically ~100 kHz) of such microstructures. Unfortunately, there is a subtle problem because the incident light heats the oscillating microstructure and so changes its resonant frequency in a manner depending on the incident optical power. This gives the sensor an unfortunate degree of undesirable dependence on incident light intensity that still needs to be taken care of in design or operation of the sensor.

11.2.1.6 SPECTRAL FILTERING SENSORS

A particularly attractive method of sensor referencing against intensity changes is to use spectral encoding in the sensor, where the sensor head provides an optical filtering function. Then, at the receiver, the relative signal strength at two or more optical wavelengths is monitored. If the sensor can filter out just one narrow spectral band, having a central wavelength dependent on the measurand, then this is even more attractive, as this wavelength can then be monitored almost independently of the optical power level or intensity.

11.2.1.6.1 Intrinsic spectral filtering sensors using doped fibers

Rare-earth-doped fibers can change their spectral attenuation significantly with temperature, as the occupancy of internal electronic energy levels changes. These fibers can be used in short lengths or coils, as point temperature sensors, or simply as sensors designed to warn of fire or overheating affecting a section of what may be a much longer fiber.

11.2.1.6.2 Fiber grating sensors

Fiber grating sensors [24–26] use optically written structures in the light-guiding core of the optical fiber, which act as very narrow band optical filters (Figure 11.12). These have a very narrow band reflection spectrum (typically between 0.1 and 1 nm bandwidth) and usually have a peak back-reflection of between 5% and 100%. In transmission, they have a similar bandwidth, but act as blocking filters over a similar narrow band.

The grating consists of a short region of fiber (typically 1–20 mm length, L), where the refractive index of the fiber core has been modified to cause it to exhibit a periodic variation with length in the axial direction. This variation is achieved by lateral illumination of a fiber with two converging beams of UV light (Figure 11.12a), either created with beam-splitting optics or created by diffraction of a single beam in a phase mask. The beams interfere to give fringes, with a predesigned spatial variation of optical intensity, depending on the wavelength and angle of convergence. This bright and dark

fringe pattern gives rise to a corresponding refractive index variation in the photosensitive fiber core as a result of a *photorefractive* effect. This effect is small, yet significant in germania-doped silica fibers. Even a very small periodic refractive index variation can build up coherently to cause a very significant (even close to 100%) reflection of light at the Bragg wavelength, λ_{Bragg}. This is the wavelength at which each low-intensity wavelet, reflected from each minor undulation in refractive index, can add coherently with all the others from other parts of the grating, and is given by

$$\lambda_{Bragg} = 2 \cdot n_{eff} \cdot \Lambda,$$

where n_{eff} is the effective refractive index of the fiber core and Λ is the grating period.

It is this need for this phase matching, or constructive addition of back-reflected wavelets, in order to give significant reflection, that results in the desired narrowband filtering effect; the longer the grating, the narrower the filter effect and the higher the grating reflection coefficient for a given index change.

The in-fiber grating is an excellent sensing element [27] as it is tiny, and even without external components, and can be configured to sense mechanical strain influences. As with traditional electrical resistance strain gauges, the wavelength of the grating depends not only on strain, but also on temperature, so there is a need to measure or compensate for temperature. The review on grating sensors by Kersey [28] and the textbooks in the bibliography cover many ways in which this may be done, but the simplest method is to have another (unstrained) grating as a reference thermometer. If it is merely required to monitor bending of a thin plate, then it is possible to bond identical gratings on opposite sides of the surface and simply monitor the differences in wavelength shifts observed [29].

From Reference [27] a typical thermal sensitivity of a grating with 1300 nm center wavelength is of the order of 0.01 nm °C^{-1}, with the fractional change per unit temperature being typically ~8.2×10^{-6} °C^{-1} at all wavelengths. The fractional response of center wavelength to strain is typically of the order of 75% of the strain.

A major application area is in monitoring [30,31] of composite materials (with the grating embedded in the composite) or of large mechanical

(a)

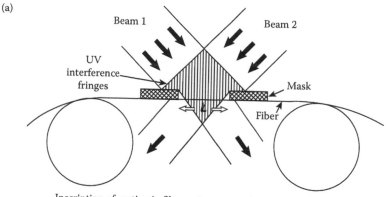

Inscription of grating in fiber, using converging uv light beams
(grating is produced over exposed length L)

(b)

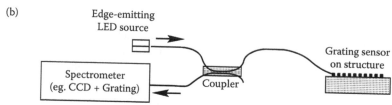

System for the interrogation of grating wavelength , using spectrometer

(c)

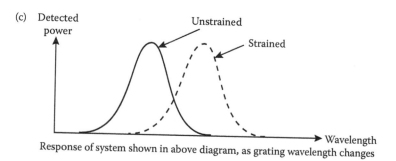

Response of system shown in above diagram, as grating wavelength changes

Figure 11.12 The in-fiber Bragg grating sensor. This shows (a) how a fiber grating can be written using interference between converging beams of light, (b) how a grating can be bonded onto the surface of a structure to be monitored (embedding into composite materials is also common) to form a strain sensor, and (c) the response of the sensor as the structure is strained.

or civil engineering [32] structures, with the gratings embedded in grooves or bonded to the surface (Figure 11.12b) or attached in surface patches. With appropriate interrogation systems, and with mechanical amplifiers, grating sensors are even being considered for optical hydrophones to detect weak acoustic signals in the sea. Clearly, there are many physical parameters that can be arranged to give a strain in the fiber (e.g., hydrostatic pressure or depth and magnetic field). Although they are only optically written, the gratings have been shown to withstand elevated temperatures of several hundred degrees Celsius, but of course, the

wavelength varies with temperature when heated or cooled.

In order to sense the grating wavelength, many interrogation methods [31,33–37] have been devised. The most common method is to launch light from a broadband source (LED or fiber superluminescent source) into the fiber lead to the sensor and then detect the peak reflective wavelength of light reflected from the in-fiber grating, using some form of spectrometer (Figure 11.12a and b). The most common types used are fixed diffraction-grating/ charge-coupled-detector (CCD) spectrometers, which usually require some form of fiber depolarizer to avoid errors [37] due to polarization effects, or scanned narrowband filters (e.g., using Fabry–Perot, as in Reference [28] or acousto-optic tunable filters, as in Reference [36]. Systems where another grating is simply stretched to form a filter that can be tuned to track the sensor grating have been reported by Jackson et al. [33]. There are also simpler passive methods where the wavelength change is converted to an intensity change using some form of optical filter having a high slope of transmission versus wavelength. Examples include interference filters and wavelength-selective fused fiber couplers [31,38], of a type often used for separation of two closely spaced wavelengths. Ideally here, the light is split into two channels, one increasing in intensity, the other decreasing, as wavelength is increased, so the two complementary outputs can be ratioed to remove the common mode intensity changes that could occur due to variations in the light source, fiber leads or connectors.

Apart from the methods where the grating merely acts as a filter, it can also be used as sensor, in a configuration where it forms part of an active fiber laser [35]. There are two basic types here, the first being where a conventional in-fiber grating (the sensor grating) acts as one mirror of an optically pumped rare-earth-doped fiber laser and the other mirror is arranged to be broadband. As the grating is stretched (or heated), changes in its Bragg wavelength cause the laser output wavelength to change. The advantages are that the linewidth is much smaller, giving improved spectral and strain resolution, and the optical output power is much larger than for the alternative simpler configuration with filtering of a broadband source. A second, more compact, form of the same idea is achieved using in-fiber distributed feedback Bragg lasers (DFB fiber lasers) as the sensing element. Here the active rare-earth-doped lasing region forms part of the actual Bragg grating sensing element. The resolution of such sensors is nothing short of dramatic, as they oscillate at a frequency around 1014 Hz, yet they can be arranged to be monitored to 1 Hz resolution, using "mixing-down" by beating or heterodyning with a reference laser, giving a potential resolution of 1 part in 1014. More usually, the beat frequency between two orthogonally polarized lasing modes has been monitored, as this is more easily used, being typically in the range of 0.5–2 GHz, where the beats can be monitored

directly using a fast optical detector [39,40]. These publications also discuss how thermal compensation is possible using monitoring of both the actual laser frequencies and the beat frequency between two mixed lasing modes.

11.2.1.7 EXTRINSIC SPECTRAL FILTER SENSORS

Many types of *extrinsic spectral filtering sensor* have been constructed [41]. One of the first methods [42] was to use a material that changes its spectral transmission characteristics with temperature (Figure 11.13).

There are many examples of such materials (e.g., crystal, glass, or polymer). As one attractive group of materials, most semiconductors have steep transmission band edges, beyond which they suddenly become reasonably transparent. At the *band edge* of all common semiconductors (e.g., Si, Ge, and GaAs), both the transmission versus wavelength slope and the wavelength of 50% transmission vary significantly with temperature. There is also a large family of commercially available long-pass optical filter glasses (e.g., Schott and Corning glass companies) with semiconductor-like optical behavior. Clearly, all of these can be used to make practical and stable sensors, and many low-cost spectrometers are now available to interrogate them. They are attractive for monitoring in remote or inaccessible locations, or in areas of high E-M field.

A class of spectral-filtering sensors is based on fiber-coupled versions of common bulk optical spectrometer components, such as Fabry–Perot filters (closely spaced mirrors), diffraction grating, zone-plate, or prism spectrometers or monochromators. The Fabry–Perot filter is particularly useful, in view of its small size and narrow linewidth, and the ability to coat fiber end surfaces to make a tiny version from a short fiber section with reflective end coatings [43].

11.2.1.8 EXTRINSIC SENSORS USING MONITORING OF FLUORESCENCE SPECTRUM

Apart from using simple transmission measurement, the fluorescence spectrum of some materials can be monitored, a method again well suited to the use of direct-bandgap semiconductor crystals [44] in the sensor probe. However, in this mode, there is now the possibility of using translucent materials, such as the types of phosphors used in

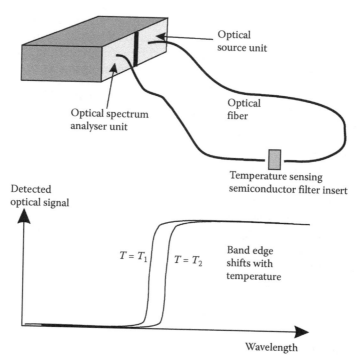

Figure 11.13 An optical fiber thermometer, where the thermal dependence of the band-edge position of a semiconductor (or special filter glass with semi-conductor-like properties) is monitored with a simple spectrometer.

fluorescent lights and in TV display tubes, which, with the aid of a directional coupler, can be conveniently monitored in back-"scatter" mode via a single fiber lead. This fiber-coupled configuration (see later Figure 11.15b) will be described again, when we discuss chemical sensors.

Decay lifetime sensors can sense various parameters by monitoring the time delay after excitation of the sensor probe material. Excitation is either by a short optical pulse or by a repetitive pulse train, or using modulated light having some other (e.g., sinusoidal) periodic intensity variation. The most common mechanism is to look, directly or indirectly, at the time decay of fluorescence in the probe (Figure 11.14).

The fluorescence decay curve, which is usually of an exponential shape, can be monitored, to determine the lifetime. Alternatively, the phase delay in the detected fluorescence intensity signal, relative to the initial intensity modulation waveform (usually sinusoidal or square-wave) of the incident light, can be monitored.

Several practical physical and chemical sensors use this mechanism, and it has proved particularly useful for temperature monitoring. Thermal changes affect the fluorescent lifetime of many substances (e.g., phosphors, semiconductors, laser crystals, or glasses), but phosphors [45] and ruby crystals [46] have been used most for thermal sensing (see also review by Gratton and Palmer [47]). Many of these materials have been used extensively for long periods under conditions of energetic electron bombardment or powerful optical illumination and have excellent long-term stability. Laser crystals, in particular, are not only optically and thermally stable, but also often (e.g., ruby) have excellent mechanical hardness and strength.

11.2.2 Intensity-based chemical sensors

There are two basic types of optical chemical sensor, one using *direct optical spectroscopy* of materials to be detected, the other making use of a *chemical indicator*, i.e., a compound that acts as an intermediary, with a strong, hopefully chemical-species-specific, change in its optical properties when exposed to a target chemical or group of

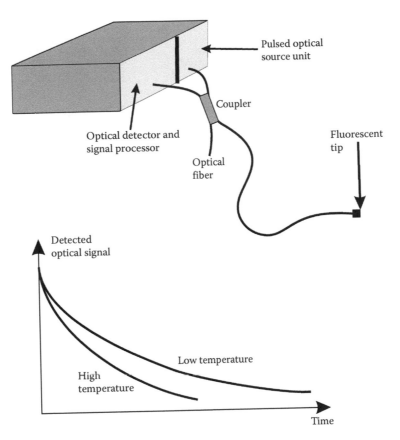

Figure 11.14 Sensor to measure temperature, where the decay of fluorescence of a phosphor (or semiconductor chip), attached on the end of an optical fiber, is monitored. The lower curves show typical responses, following pulsed excitation from the light source.

chemicals. Below, we shall cover these two types, but, in this short outline, it will not be possible to cover all the concepts concerned, so the interested reader is referred to the textbook in bibliography [1] (particularly vol. 4, Chapters 7 and 8) for further reading. For reasons of lack of space, the discussion of chemical sensors here will be shorter than that for physical sensors.

11.2.2.1 CHEMICAL SENSORS USING DIRECT OPTICAL SPECTROSCOPY

Because materials can absorb or emit light only at wavelengths corresponding to allowed internal energy level changes, characteristic of particular elements or functional groups, spectroscopy has become one of the most valuable tools of the analytical chemist. Most of the standard spectroscopic techniques used by analytical instrument designers can also be implemented in fiber optic form (see examples in Figure 11.15).

Possible methods include measurement of transmission and turbidity, attenuated total reflection, fluorescence, Raman scattering (including surface-enhanced Raman scattering, or SERS), to give just a few examples. The great advantage of fibers is the real-time, online measurement capability, allowing the instrumentation to be kept in a benign environment, away from the probe end, that may be remote or inaccessible, and could have any variety of dangerous, corrosive, toxic, or flammable materials present.

11.2.2.1.1 Transmission (absorption) and turbidity measurements

Transmission defines the fraction of light passing through a component or medium. Absorption is the total loss of light arising from conversion of light to heat, whereas turbidity is where light is merely lost from a collimated path by scattering, such that it is no longer collected.

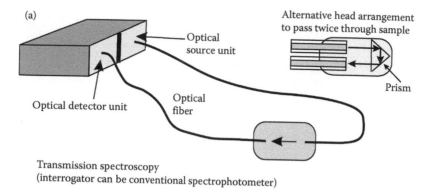

(a)

Optical
source unit

Alternative head arrangement
to pass twice through sample

Prism

Optical detector unit

Optical
fiber

Transmission spectroscopy
(interrogator can be conventional spectrophotometer)

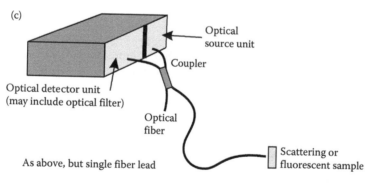

(b)

Optical
source unit

Optical detector unit
(may include optical filter)

Optical
fiber

Scattering or
fluorescent sample

Sensing of scattering (Raman or elastic) or fluorescence sample

(c)

Optical
source unit

Coupler

Optical detector unit
(may include optical filter)

Optical
fiber

Scattering or
fluorescent sample

As above, but single fiber lead

Figure 11.15 Optical fiber sensors for remote spectroscopy. This shows (a) transmission spectroscopy, where, both the source and interrogator unit can be an integral part of a commercial spectrophotometer, modified to allow fiber extension leads; (b) sensing of backscattered signals, where, a filter or spectrometer can be used to select out Raman or fluorescent signals from the incident light if desired; and (c) a modification to allow a single fiber lead to the measurement head.

The power, $P(\lambda)$, transmitted through a sample in a small wavelength interval at a center wavelength λ, is given by Lambert's law:

$$P(\lambda) = P_0(\lambda) \cdot \exp\left[-\alpha(\lambda) \cdot \ell\right],$$

where $P_0(\lambda)$ is the power entering the sample in this wavelength interval, $\alpha(\lambda)$ is the attenuation coefficient of the material at wavelength λ, and ℓ is the optical path length through the sample to the point at which $P(\lambda)$ is measured. Lambert's law does not apply if scattering is high, such that multiple scattering occurs.

The sample can be said to have a transmission $T(\lambda)$, at the wavelength λ, where

$$T(\lambda) = \exp\left[-\alpha(\lambda)\ell\right].$$

Alternatively, the sample can be said to have an absorbance $A(\lambda)$, where

$$A(\lambda) = \log_{10}[1/T(\lambda)] = \log_{10}[P_0(\lambda)/P(\lambda)] = 0.43\alpha(\lambda) \cdot \ell.$$

The factor 0.43, or log(e), has to be included to account for the use of log10 for absorbance calculations, whereas natural exponents are used for attenuation coefficients.

Transmission, absorption, or turbidity measurements can be achieved most easily by using a commercial spectrophotometer with extension leads. These have a unit that fits into the cell compartment of a standard instrument, with a first lens that takes the collimated light that would normally pass through the sample chamber, focuses it instead into a large core diameter (usually >200 μm) optical fiber down-lead and with a second lens that recollimates light coming back from the return fiber lead into a low-divergence beam suitable for passage back into the instrument. There is also a remote measurement cell, connected to the remote end of both these fiber leads, where a first lens collimates light coming from the down-lead into an interrogation beam, which passes through the remote measurement cell, after which a second lens collects the light and refocuses it into the fiber return lead going to the spectrometer instrument. Such optical transformations lead to inevitable losses of optical power, of typically 10–20 dB (equivalent to losing 1–2 units of absorbance), but as most modern instruments have a typical dynamic range of >50 dB, this is a price that many users are prepared to pay for a useful remote measurement capability.

It should be noted that the optical power losses occur mainly due to the imperfections of the focusing and recollimation optics and due to reflection losses at interfaces, rather than to fiber transmission losses. If suitably collimated beams were to be available in the instrument, if large core diameter fibers could be used to connect to and from the probe, and if all optics, including fiber ends, could be antireflection coated, there should really be very little loss penalty. Such losses, therefore, arise primarily because of the need for the fiber leads to be as flexible as possible (so hence choice of small diameter fibers) and the usual need to compromise design on grounds of cost.

There are many other probe head designs that are possible. The simplest design, for use with measurement samples showing very strong absorption, is simply to have a probe which holds the ends of the down-lead and return fiber in axial alignment, facing each other across a small measurement gap, where the sample is then allowed to enter. Losses are low for fiber end spacing of the same order as the fiber diameter or less but rapidly increase with larger gaps. The probe is far easier to miniaturize and handle if the fiber down-lead and return lead are parallel in one cable. This can be achieved using a right-angled prism or other retroreflecting device to deflect the beam in the probe tip through the desired 180° that allows it to first leave the outgoing fiber, pass through a sample and then enter the return fiber. Use of a directional fiber coupler at the instrument end allows use of a single fiber, but then any residual retroreflection from the fiber end will be present as a cross talk signal, adding light signal components that have not passed through the medium. Clearly, there are many variants of such optical probes, some involving more complex optics (e.g., multipass probes), some constructed from more exotic materials to withstand corrosive chemicals.

A very simple option that has often been used with such single fiber probes, for monitoring the transmission of chemical indicators, is to dissolve the indicator in a polymer that is permeable to the chemical to be detected, and also incorporate strongly scattering particles in the polymer. When a small piece of such a polymer is formed on the fiber end, the particles give rise to strong backscattered light, and the return fiber guides a portion of this to the detection system. This backscattered light had of course to pass through the indicator polymer in its path to and from each scattering particle, so the returning light is subject to spectral filtering by the indicator. Although this is a very lossy arrangement, it is extremely cheap and simple and has formed the basis of many chemical sensors, for example ones using pH indicators [48].

There are now many commercial types of miniature CCD spectrometers that have been specially designed to analyze the spectrum exiting from an optical fiber. These generally use the light-guiding core of the return fiber as the input "slit" of a diffraction grating monochromator, using a sensitive CCD detector array to provide a set of parallel output signals, one from each narrow-band spectral component of the received light. These have high optical efficiency, as nearly all the energy is incident on the detector. For ultimate resolution, however, the well-known Fourier transform spectrometer principle can be used, where a scanned optical interferometer is used to analyze the spectrum of the received light. Each optical spectral component gives its own electrical frequency component in the detected photocurrent signal, observed as the interferometer is scanned (a narrowband laser signal input will give a pure sinusoidal output from the scanned interferometer). The entire spectrum can, therefore, be obtained by Fourier analysis of the temporal variations of the detected signal, to extract the relative magnitude of each frequency component in this detected signal.

11.2.2.1.2 Chemical sensing by detection of fluorescence and Raman scattering

It was stated above that, for transmission and turbidity measurements, there should really be only a small additional loss penalty due to using optical fibers, provided full use is made of expensive precautions to avoid loss of light. This is unfortunately not true when fluorescence or Raman measurements are required, as light is scattered over all angles, and the return fiber can only ever collect a relatively small portion of this light. In addition, the processes where this *inelastically scattered* (i.e., different wavelength to that incident) light is generated, often have a poor *quantum efficiency* (i.e., have a low ratio of total re-emitted to total incident photon flux).

An arrangement similar to that of Figure 11.15b has been used for oxygen sensing, using Ruthenium dye complexes [49], where the fluorescence decay is quenched by the oxygen gas.

There is no space here for a full discussion of these processes, but their potential value, particularly that of Raman scattering, for chemical sensing, has meant workers will continue to persevere to get useful performance, despite the low return light levels encountered with fiber-coupled systems. Both these mechanisms involve excitation of a sample with light, usually at a wavelength shorter[*]

---

[*] Measures to reduce this will be discussed later, but one useful way to improve an OTDR system is to include optical fiber amplifiers to boost the launch power and to act as an optical preamplifier for returning signals.

than the scattered light to be observed, and then the re-emitted light is collected and narrow-band filtered. This filtering is first to separate it from the incident light, but also, in the case of Raman, to examine it for the spectral features characteristic of a target compound to be examined.

It is useful to briefly estimate the approximate magnitude of additional losses when using fibers with these processes. The loss due to launching of excitation light into a fiber is usually negligible with Raman, as powerful narrow-line laser sources are used, but ultimately the limit may be set by nonlinear processes or, in the case of large-core multimode fibers, by optical damage thresholds. Similar excitation can be used for low-level fluorescence monitoring, provided no photo-bleaching or other photodegradation of the monitored substance can occur at high illumination intensity. The main potential loss is therefore that of light collection. If we assume the medium is excited close to the end of a fiber, only a region of dimensions of the same order as the fiber core diameter will be intensely excited. Thus, if a very large $200\,\mu m$ core diameter fiber is used, a region of approximately 200 to $>500\,\mu m$ in length will be excited effectively. As a re-emitting point in the excited medium gets further from the fiber tip, the brightness of its illumination reduces and the effective angle subtended by the fiber core collecting the light (which can be approximately considered to be re-emitted in all directions) gets smaller. The collection can, therefore, be approximated as that received by the fiber numerical aperture (say NA of 0.3) from a region around $200\,\mu m$ long. In a normal Raman spectrometer, the sample can be excited by a narrow focused laser beam. There can therefore be a useful and collectable Raman emission from a thin sample region, of length equal to that of the spectrometer input slit on which it must be focused. The length of the slit may be perhaps $10\,mm$ long, and light entering this may perhaps be collected with a wide acceptance angle monochromator, perhaps having an NA as high as 0.4. Thus, the fiber-based system may perhaps have a light collection reduced by a factor of $50 \times 16/9$ (product of the useful excited length ratio and the square of the acceptance NA ratio), or approximately 90 times, when compared to a bulk optical system. A similar factor also applies to fluorescence detection.

Apart from these photometric limitations, Raman scattering has a particularly poor quantum efficiency, and the already weak scattered signals will be typically two orders of magnitude weaker when coupled into return fibers, with their poor collection efficiency for divergent light (see bibliography [1], Chapters 7 and 13). Despite this, however, several commercial fiber-coupled Raman systems are available. These combine the great advantage of Raman scattering (which, merely by careful spectral filtering, allows rejection of elastically scattered light from turbid samples) with the ability of fibers to probe into inaccessible, remote, or hazardous environments.

A few practical examples of chemical sensing using direct spectroscopy are given in vol III.

There are many more means of performing direct spectroscopy with optical fibers, as the above has only presented a few. Other methods will be discussed in the later section on distributed and multiplexed sensing. As space is limited here, it is instructive to summarize, in the form of a brief list of examples, where systems have been built for practical applications. In many cases, active research is still being pursued in many of these areas:

- Sensors examining transmission of groundwater to track pollution.
- Fiber-probe sensors to examine the transmission or reflection spectrum of blood to determine oxygenation state.
- Fiber-probe sensors to examine bilirubin in the digestive system of the body.
- Gas sensors, based on remote absorption measurements, using fiber probes.
- Sensors examining the transmission of petrochemicals to determine octane rating. These can be extrinsic sensors as in Figure 11.15a or can use evanescent field monitoring.
- Fluorescence sensors to determine oil or other aromatic hydrocarbons in water.
- Technologies using arrays of optically addressed "microdots," each dot in the array having a different composition. For example, arrays of fluorescent indicator dots, each dot having different optical properties, can detect several different substances using one fiber (or can give multiple signals to cancel cross talk from other substances, using mathematical regression models).

- Methods using evanescent wave coupling to the measured substance, for example, with a bare glass or silica fiber. Alternatively, a conventional monomode fiber with part of the cladding polished away, or a "D-type" fiber that has a very thin cladding on one side, can be used. In such sensors, the field of the guided light extends beyond the vitreous waveguide into the chemical to be measured, allowing absorption of the latter to be observed as fiber attenuation.
- Technologies as above, but using *surface plasmon resonance* to enhance the coupling to the evanescent field.
- Refractive index sensors, where a fiber with Bragg gratings is side polished, removing cladding material to allow the evanescent field to extend outside the fiber. Then, the peak reflected wavelength of the grating is affected by the effective propagation constant of the fiber, which is now also a function of the refractive index of this surrounding medium. Again, octane rating of fuels is of interest here.

The textbooks in bibliography [2] give a very comprehensive overview of many of these types of optical chemical sensors, and a few applications are presented in vol III.

11.3 INTERFEROMETRIC SENSORS

We shall now discuss interferometric sensors. Here, the sensing action involves the interaction of fiber-guided light beams, where there is coherent addition of the electric field components of their electromagnetic waves. This leads to a mixing condition that can vary from constructive to destructive interference, depending on the relative phase of the combining light beams. In order to obtain high visibility interference of interfering free-space beams, it is necessary to match their intensity profiles, their wave front shapes (i.e., beam direction and divergence) and their polarization states over the full transverse width of the interfering beams. In order to observe or detect the effects of interference, however, the light beams must also eventually fall onto a "square law" optical detector. All standard optical detectors, including the human eye, monitor optical power or intensity, which is proportional to the square of the electrical field component, hence the term "square law."

11.3.1 Interferometers using fiber optic technology

A single mode optical fiber guides not only one propagation mode, but also strictly two, if the two possible orthogonally polarized modes are considered. The fundamental guided mode is usually called the HE_{11} mode. This fiber mode has quite a simple transverse power distribution, very closely matching the well-known Gaussian TE_{00} intensity profile, typical of that of the beam from a single transverse mode gas laser, with a central (axial) peak in the intensity. The big advantage of mono-mode fibers is that (apart from the possibility of the two different principal polarization modes) only a single spatial mode is allowed, the fiber acting as a perfect mode filter to ensure that the two overlapping fields have identical spatial characteristics. The slight curvature of E-M field lines in the fiber, varying a little with transverse position, does not significantly reduce fringe visibility, as the other interfering signal (or signals) guided in the same fiber have a matching curvature.

Because of this behavior, we can fabricate interferometers using single mode fiber to define the optical paths and using compact fiber coupler components as beam splitters or combiners [50]. In Figure 11.16a and b, we show a Michelson and a Mach–Zehnder type interferometer, respectively, both implemented with single mode fibers.

Unfortunately, there is a big disadvantage when light is guided by an ordinary single-mode telecommunications-type fiber, as, unlike the situation for free-space, or in-air, beams, the polarization direction of the HE_{11} mode can easily be changed by environmental influences. For these to change the polarization of fiber-guided light, they need only to cause a significant asymmetrical physical distortion of the fiber. Such influences can occur due to, for example, fiber bends, lateral mechanical stresses, and transverse thermal gradients. In the extreme case of this so-called *polarization fading*, the visibility of the interference can fade completely, as orthogonally polarized guided beams cannot interfere. This is simply because electric fields at 90° can no longer cancel the resultant intensity now being independent of optical phase of the combining beams.

To compensate for possible signal fading due to this polarization fluctuation, it is common to use polarization controllers (PCs) when using ordinary single mode fibers. However, in real-world

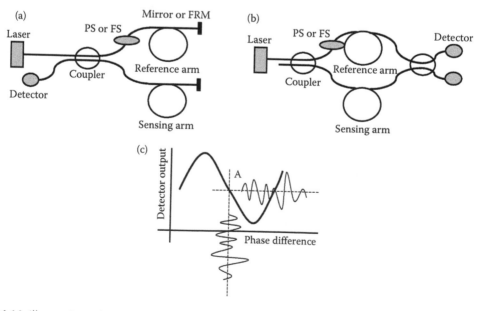

Figure 11.16 Illustration of two commonly used fiber interferometers: (a) Michelson arrangement, (b) Mach–Zehnder arrangement, and (c) the typical detector response to phase difference. Note that the zero position on the phase-difference scale, where intensity would be at a peak, is unknown in this figure.

sensors, it is not attractive having to have to continually adjust polarization in order to compensate for environmental effects, unless the PCs are themselves controlled by automated optoelectronic feedback systems. Although the latter is possible, it is still a rather complex and expensive solution. Fortunately, it is possible to greatly reduce such effects with passive solutions. One method is to use polarization-diversity optical receivers, where the combined light is first split into differently polarized components before detection on several separate detectors. A simpler and far more efficient passive solution, however, is to use polarization-maintaining fiber (P-M fiber). In this fiber, the direction of the linear polarization of the HE_{11} mode is constrained to lie along one of its principal polarization axes, as the lightwave propagates through it. Such a fiber is made with a deliberate transverse asymmetry, such that either of the two principal polarization modes, once launched, cannot readily couple to the other, unless there is a very severe localized deformation of the fiber. An even more effective method is to use single polarization fibers, which are polarization-maintaining types where one polarization mode is made to be highly lossy, such that only the other mode can effectively propagate without high attenuation. If any small degree of polarization mode conversion then takes place in this fiber, the light in the undesired mode is rapidly lost.

In the case of a dual-reflective path, optical fiber Michelson interferometer, we have another ingenious way to stabilize for any in-fiber polarization fluctuation. To achieve this, Faraday rotator mirrors (FRM) are placed at the end of the fibers. A Faraday mirror device has the property of ensuring the reflected lightwave has a returning SOP rotated by 90° with respect to the incident wave. Consequently, after a two-way propagation along the fiber, the SOP of the lightwave always remains orthogonal to the incident one, even though the actual SOP of each beam along the fiber will always fluctuate. If a wave is split, by a coupler, into two arms, each having Faraday mirrors at their far (distal) ends, *both* returning beams or guided modes return with orthogonal polarization states to that of the launched light, so *both* have identical polarization states.

In a fiber interferometer, the detected output, I_d, is given simply by

$$I_d = A + B\cos\Delta\theta,$$

where A and B are constants and $\Delta\theta$ is the phase difference between the two lightwaves returning from the reference and the sensing arms. The output intensity changes, as a function of phase difference, $\Delta\theta$, are shown schematically in Figure 11.16c. In Figure 11.16a or b, a relatively large phase drift can always be caused by temperature or mechanical fluctuation, which results in the presence of low-frequency components, or drift, in $\Delta\theta$.

For many sensors, such as acoustic wave (e.g., fiber hydrophones) or vibration monitors, it is desired to monitor or measure very tiny phase changes, usually of a cyclic or transiently varying nature, caused by the external measurand, typically having a low- to mid-AC frequency content in the range 10 Hz to 10 KHz. In order to measure these small, relatively rapidly changing signals, other sources of slower drift, for example, due to slow thermal changes or due to slowly varying mechanical strain, must be corrected for or stabilized, so as to keep the interferometer at its most sensitive operating point. This is the maximum slope of the sinusoidal response, termed the *quadrature* point.

To achieve the desired stabilization, an optical phase shifter (PS) can be placed in the reference arm, as shown in Figure 11.16a and b. In these stabilized interferometers, the detector output is fed-back to the PS, so that the output signal is held at *quadrature* point A, where the slope of the curve is steepest. In this way, tiny alternating optical phase signals, as small as 10^{-6} rad or less, can be detected. At an optical wavelength of 1 μm, this represents an optical path length change of only 10^{-12} m, corresponding to a change in the physical length of the sensing fiber by a tiny amount, corresponding to a small fraction of the diameter of a hydrogen atom! (Of course, individual atoms at the fiber ends have positional uncertainties greater than this, but the end position is the average of enormous numbers of such atoms). This means that fiber interferometric sensors can realize extremely high sensitivity and this sensitivity can be further increased by using a multiturn fiber coil as a sensing element.

For many applications, it is convenient to place an optical frequency shifter (FS) in one arm to deliberately induce a frequency difference between the two lightwaves. This is also shown in Figure 11.16. Under these conditions, the detected output, I_d, can be expressed as

$$I_d = A + B\cos(2\pi\Delta ft + \Delta\theta).$$

This new configuration is called a *heterodyne* interferometer, a term which implies that the interfering lightwaves now have different frequencies. The phase of the detected electrical signal can now be detected by conventional phase or frequency demodulation schemes, of a type commonly used in commercial or domestic radio receivers. One such electrical frequency demodulation method is to delay one signal compared to the other and use an electronic multiplier or mixer, followed by a low-pass filter, an arrangement that conveniently converts the electronic phase difference of two electronic signals to amplitude, with a linear-saw-tooth response characteristic. Another common method, again well known to radio engineers, is to use phase-lock loop demodulation technology. In order to distinguish the earlier, rather simpler, interferometer (in which the two lightwaves had the same frequency), from the *heterodyne,* or difference frequency one, we have just discussed that the earlier one has been termed a *homodyne* interferometer.

In the interferometer shown in Figure 11.16a and b, the optical angle-modulation elements, i.e., the PS or FS modulators, are placed in the reference arm, which is sometimes located adjacent to or near the sensing arm to assist with thermal compensation. If the electronically driven PS or FS devices are placed in, or near, the physical measuring environment, they might pick up electrical interference, which would reduce the advantage that an optical fiber sensor is normally not affected by, nor induces electromagnetic noise. This would remove the inherent *electromagnetic compatibility* of the sensor. To improve the electromagnetic

behavior, a configuration shown in Figure 11.17 has been proposed, where, rather than using a separate modulator element, a frequency modulation is created in the laser light source, which then, due to different optical time delays, creates a phase difference $\Delta\theta_1$, between lightwaves that have traveled over different length optical paths.

This then gives a phase difference expressed by

$$\Delta\theta_1 = \frac{2\pi}{c}\Delta L \cdot \Delta f_1,$$

where ΔL is the optical path length difference between the two arms. The lasing frequency of a semiconductor diode laser can conveniently be changed merely by modulating the drive or injection current, providing a very simple current-controlled optical oscillator. By using this property, we can compensate for the effects of slow thermally or mechanically induced drift in the interferometer, without the need for any separate PS or FS. Using this configuration, there is then no need for any electronic or electrical components in the sensing environment, giving excellent freedom from electrical interference, etc. When the laser frequency is modulated by a sinusoidal wave of frequency $\omega\rho$, we induce a set of components at frequencies $m\times\omega\rho$ in the detector output, where m is an integer. Each of these components has a phase corresponding to the optical phase difference to be measured. This configuration shows, in many ways, a similar output to that of the heterodyne interferometer, but because no separate modulator is used, it is called the *phase-induced carrier method.*

Because of the over-riding need to reduce polarization fluctuation and phase drift in highly

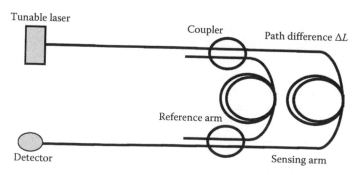

Figure 11.17 Configuration of a passive remote Mach–Zehnder interferometer. This allows remote interrogation by sweeping the laser source, avoiding the need for modulators with electrical drive signals to be included in the remote interferometer section.

sensitive fiber optic interferometer sensors, many different methods have been proposed and developed by researchers. The textbooks and review papers in the bibliography present greater details of many of these schemes, but below we will present a case study of the optical fiber gyroscope to introduce some of the methods that can be employed. The gyroscope is a very special case, where it is essential to achieve a very high degree of steady-state stability. Fortunately, many other types of interferometric fiber sensor are designed to measure only dynamic changes, such as might arise from acoustic signals (e.g., the fiber hydrophone) or from other mechanical vibrations (e.g., the fiber vibrometer or accelerometer), so these do not suffer from quite so many undesired sensitivities to slow drift from environmental aspects as the fiber gyroscope can. The discussion below illustrates that despite many fairly complex difficulties, a host of potential problems can be overcome to make a cost-effective practical sensor, although a great deal of background research was necessary to reach this point and the gyro still needs careful design.

11.3.2 High-sensitivity sensing with fiber interferometers

11.3.2.1 THE FIBER OPTIC GYROSCOPE

The fiber optic gyroscope [51–53], which we shall call "FOG" for short, was one of the earliest types of interferometric fiber sensor, and is the one which

has perhaps received the most research funding and scientific attention.

A FOG detects rotation relative to an inertial frame. The basic operating principle of this sensor is based on a concept known as the "Sagnac effect" [54], which originally used two optical beams, each directed in opposite directions around loops using mirrors, before being caused to interfere on a detector. The basic configuration of the all-fiber version is shown in Figure 11.18.

It can be considered that two lightwaves, propagating in opposite directions in the same closed optical fiber coil, exhibit a traveling-time difference, which is proportional to the rotation rate of the optical path with respect to the inertial frame. This time difference results in a phase difference, θ, between the beams at the output of the loop, given by

$$\theta = \frac{4\pi L a}{c\lambda}\Omega,$$

where L, a, c, λ, and Ω are, respectively, the length of the fiber coil, its radius, the speed of light in vacuum, the optical wavelength and the coil rotation rate. The magnitude of the Sagnac phase shift is generally extremely low for most typical rotation rates, particularly for those typical of vehicle navigation, where directional changes are usually quite slow. To overcome this limitation, a very long (and hence necessarily low loss) fiber coil needs to be used for the sensing coil. For aircraft navigation, a rotation rate resolution of only $0.01°$ h^{-1} is required. Even when

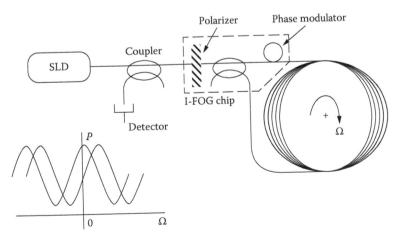

Figure 11.18 Basic configuration for a FOG. This shows the phase modulation elements used to bias the device contained on an integrated optics chip (I-FOG chip) that also contains the polarizer and a beam-splitting element.

using a sensing fiber as long as 1 km, this typically corresponds to an induced phase difference as small as 1 μrad. To measure such small and slow-changing phase changes, a successful FOG requires extremely careful control of the many subtle noise and drift factors that can otherwise occur. Fortunately, after extensive research over many years, these noise factors have been studied in detail, and many excellent countermeasures to reduce them have already been invented.

As a result of this success, FOGs are now feasible for many new applications of gyroscopes, such as car navigation, antenna/camera stabilizers, radio-controlled vehicles, unmanned track navigation, and so on. These new applications could only be realized because of unique advantages of the FOGs compared to traditional mechanical gyros, such as short warm-up time (no need to build up mechanical speed, as in normal types), less maintenance and low cost. FOGs have also now been used in traditional application fields, such as airplane navigation, rocket launching and ship navigation. A part of the navigation system of the Boeing 777 now uses FOGs.

To measure tiny phase differences, of the order of micro-radians, between the two lightwaves traveling in the fiber coil of Figure 11.18, a special processing scheme had to be established. When the system is in the rest state, the signals interfering at the detector in Figure 11.18 would normally give rise to a maximum in the sinusoidal response (dashed curve). At this peak, the gradient is of course zero, so there is no initial response to small phase changes, so hence zero sensitivity for a small rotation rate. In order to solve this problem, a phase biasing scheme is needed, in order to drive the state of the interferometer to a region of the phase response curve where the gradient is nonzero. To achieve this, a phase modulator, driven by a signal of sinusoidal or square waveform, is applied at, or near, the end of the sensing fiber coil. Due to the timing difference of the modulation between the clockwise (CW) and the counter-clockwise (CCW) waves, the two lightwaves now have a periodic phase difference when they impinge on the detector, and give a "mixed" or detected output signal having the same frequency as the applied phase-modulation waveform.

By detecting the electronic output synchronously (i.e., by multiplying it with an electronic reference signal at the same frequency as the applied phase modulation, and then low-pass

filtering), the signal shown by the solid curve can be obtained. Now it can be seen that the detected output changes, as desired, with input rotation. The polarizer shown in Figure 11.18 is required to reduce drift effects due to the polarization fluctuation in the sensing fiber coil. Essentially, it ensures that light travels in each direction in the loop in different directions, but in the same polarization state, although, of course, each individual beam still exhibits polarization changes as it propagates round the loop. The system shown in Figure 11.18 is called the "minimum reciprocal configuration" for the FOG.

Even when the minimum configuration is used, many noise, drift, and signal-fading factors can still exist. The first major problem to be solved is the need to avoid polarization fading. When the state of the polarization fluctuates in the fiber, the light power received at the detector changes and this can result in significant reduction of the SNR—in the worst case, the signal can even totally disappear! A way to avoid this problem, using only a passive component, is to insert a fiber depolarizer at some point in (most conveniently at the end of) the sensing fiber coil. The depolarizer is fabricated simply using a short birefringent polarization-maintaining fiber, in which the two orthogonally polarized propagation modes have a difference in velocity. When the usual broadband LED (or a superluminescent fiber) wide spectrum (low coherence) light source is used to excite the FOG, the variations in differential mode delay result in a different output polarization state from the polarization-maintaining fiber section for each wavelength component. This effectively reduces the degree of polarization by "scrambling" it, to give a different polarization at each wavelength, and to render it effectively unpolarized when the effect is averaged over the full bandwidth of the source (even though each individual wavelength component is still strongly polarized). Hence, using this "depolarizer," the polarization fading problem can be overcome, as some wavelengths will still interfere without fading. Unfortunately, an undesirable polarization component is induced, which is perpendicular to the polarization axis of the polarizer. Due to the finite extinction ratio of practical polarizer elements, this configuration is not suitable for realizing FOGs of very high resolution. This configuration is therefore only used for low-cost, moderate-grade gyros.

To increase the optical efficiency, and hence the sensitivity, a polarization-maintaining fiber coil is usually used. With such a coil, undesirable coupling to the orthogonal polarization component is greatly reduced. Moreover, even if a small coupling were to occur, the two polarization modes in the fiber have different propagation velocities, so the undesired component has a large optical delay compared with the desired component. Therefore, when using a low coherent source, such as an ELED (edge-emitting LED), the undesirable component cannot interfere with the desired one. This configuration is used for intermediate grade gyros. To improve the sensitivity more, a LiNbO$_3$ integrated optical circuit modulator can be introduced as shown in Figure 11.19.

This integrated optical circuit is called the FOG chip, in which one coupling branch, two-phase modulators, and a polarizer are all integrated together on a common planar substrate. This configuration is used for high-grade FOGs. When a proton-exchanged LiNbO$_3$ planar waveguide is used, it also acts as a polarizer with quite a high (>60 dB) extinction ratio. In this system of Figure 11.19, a sophisticated signal-processing scheme has also been included, which is explained below.

A very small Faraday rotation effect can occur in the fiber-sensing coil, due to the Earth's magnetic field, and this results in another error factor in the FOG. The Earth's magnetic field lines are essentially parallel over the small dimensions of the coil, so the Faraday effect induced in one half of the coil would ideally be canceled by that induced in the other half. However, when birefringence exists in the fiber coil due, for example, to bending or other mechanical stress, the FOG can suffer an error or drift. One way to reduce it is again to use polarization-maintaining fiber for the sensing coil. A polarization-maintaining fiber with high birefringence can effectively reduce the drift because it prevents the lightwave suffering polarization rotation due to the Faraday effect. Another way to reduce the drift is to place a depolarizer at each end of an ordinary single-mode fiber coil. Recently, a general formula to describe the Faraday effect induced drift has been derived, in which the mechanism of the reduction of this drift has been theoretically derived.

If the temperature distribution in the coil changes with time, in a manner asymmetric with respect to the center of the fiber length, the CW and the CCW lightwave will experience a very

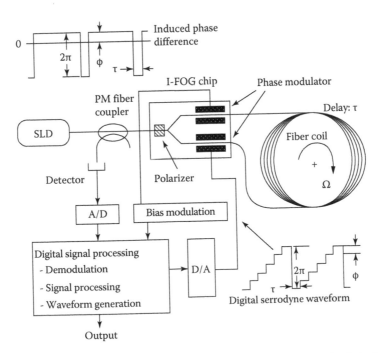

Figure 11.19 Schematic of a more sophisticated gyroscope arrangement with digital serrodyne modulation.

slightly different line-integral of temperature along the coil. Transient thermal changes of very small magnitude can give rise to a significantly different phase change in each direction in the coil, resulting in output drift. A temperature rate of change of only 0.01°C s^{-1} can induce a very substantial drift of the order of 10° h^{-1} in an FOG with a badly designed coil! Special coil winding technologies have been invented, in which every section of fiber in the coil lies adjacent, and closely thermally coupled to a similar section of coil that is located at a symmetrical position with respect to the fiber center (i.e., both sections lie at equal in-fiber distance from, but at opposite sides of, the coil center). This makes the overall temperature distribution symmetrical with respect to the center. To make a high-grade model, with 0.01° h^{-1} sensitivity, such techniques have been applied.

For high-grade applications, such as aircraft navigation, a wide dynamic range of about seven decades, and a good linearity of about 10 ppm are required. To realize these, a special signal-processing scheme, involving closed-loop operation, has been proposed and used.

The schematic configuration of this has already been shown in Figure 11.19 and we shall now describe the signal-processing scheme. The phase modulation waveform shown in the figure is used to intentionally give a more complex phase-difference modulation between the CW and the CCW lightwaves. This phase difference occurs because of the different effective position of the modulator in each path and the phase modulation waveform used is called the *digital serrodyne waveform*. The modulation waveform can be thought of as having a "saw-tooth" envelope, but with steps, as can conveniently be generated by a D/A converter. Each phase step is adjusted to correspond to one roundtrip traveling time, τ, of the lightwave in the fiber coil. The amplitude is kept at 2π. After the CW and the CCW waves have both suffered the same time delay of τ, these two waves arrive with a phase difference at the detector, this difference being equal to the phase step ϕ in the serrodyne waveform. The Sagnac phase is then compensated, using this phase difference as input to a feedback control loop. Under the condition where the amplitude of the waveform is set to 2π, the phase difference ϕ is proportional to the frequency of the serrodyne waveform. Consequently, the input rotation rate is

converted to a modulation frequency, which can then be measured with a frequency counter, to provide a wide dynamic range output from the sensor, which also has good linearity.

Applications of such gyros are expanding rapidly in various fields. The required sensitivity and dynamic range can vary greatly according to the application. FOGs have already been developed, in moderate, intermediate, and high-grade forms, with progressive increases in cost and complexity. For moderate-grade applications, the analogue output of the detector is directly measured, this is called *open-loop operation*. For high-grade applications, the *closed-loop operation* method should be used.

Japanese gyro makers have created new application fields for industrial and consumer applications, for example, car navigation systems, and control systems for cleaning-robots, forklifts, agricultural machines and unmanned dump trucks suitable for hazardous environments. A camera stabilizer, to provide stable TV pictures from a helicopter, has also been developed with FOG technology for the sensor. Radio controlled helicopters with I-FOGs have also been produced for agricultural applications, such as planting seeds and spraying chemicals. A North-finder "optical fiber compass," using an open-loop FOG has been developed. The National Aerospace Development Agency of Japan used an inertial sensor package with FOGs, for the first time, in their rockets, for micro-gravity-mission experiments. The FOG was selected for this mission as a silent gyro. The first launching took place in 1991 and was the first application of the FOG in space. The Institute of Space and Astronautical Science, Japan, has developed a rocket having an inertial navigation system (INS) with closed loop FOGs. The first successful flight of this M-V rocket was on 12 February 1997, using an additional radio-wave guidance technique. The M-V-1 launched a Satellite MUSES-B, with a mission to construct a VLBI (very long base-line interferometer) for radio astronomy, when in radio contact with other antennas on the Earth. Also, in this satellite, an open-loop FOG, having a 0.05° h^{-1} bias stability, was used for rate control.

It should be pointed out that the Boeing 777 uses a inertial navigation system that combines the use of six more conventional ring-laser gyros, of 0.01° h^{-1} grade, with four more recently developed all-polarization-maintaining-fiber open-loop FOGs, having 0.5° h⁻¹ capability.

Applications requiring a sensitivity even greater than 0.001° h⁻¹ exist, such as for space applications and ship navigation. Potential applications include deep-space and precision spacecraft navigation, and space pointing, and stabilization. However, for such higher-grade applications, the ELED that is commonly used to excite simpler FOGs has insufficient power and lacks wavelength stability. Because of this, laser-pumped superluminescent Er-doped fiber sources have been developed, and, using such a source, higher power and extremely high wavelength stability of typically a few ppm °C⁻¹ can be obtained. A rotation resolution better than 0.001° h⁻¹ has already been demonstrated by several companies, but requires carefully temperature-stabilized conditions.

11.3.2.2 FIBER OPTIC HYDROPHONES

We shall now discuss fiber sensors for detection of acoustic signals in water, as required for many military (detection of marine vehicles) and civil applications (seismic oil exploration).

When an acoustic wave impinges on a fiber in water, the sound pressure induces change in its density and length, which results in a phase change of the lightwave propagating in the fiber. The change is of course very small in quiet seas, but it can be detected using interferometric configurations with a very long length (100–400 m) of sensing fiber. Using the compensation schemes for slow drift of temperature and mechanical strain, which were described above, a highly sensitive acoustic wave sensor in water, called a hydrophone, has been developed. As we discussed above, signals, as small as 10^{-6} rad or less, can be detected, representing an optical path length change of only 10^{-12} m at 1 μm. This corresponds to a change in the physical length of the sensing fiber, which is only a small fraction of the diameter of a hydrogen atom! It is clear this will give tremendous acoustic sensitivity. Best sensitivity is achieved using mechanical amplifiers, such as mechanically compliant (e.g., made of easily deformable, perhaps even air-filled, materials) rods as coil formers. The fiber is wound around these, such that when acoustic waves interact, a compressive change in diameter of the soft rod is transferred to length changes in the fiber, giving a much greater phase change for the same acoustic influence, than would occur for bare fiber.

By using a wide variety of sensor multiplexing techniques, such as time-division multiplexing (TDM), large arrays of such hydrophones have also been realized. These will be discussed in more detail later in Sections 11.4 and 11.5 of this chapter, and the textbooks and the review paper by Kersey in the bibliography adds yet more detail.

11.3.2.3 THE FIBER OPTIC CURRENT SENSOR

High-voltage power systems often deal with enormous voltages (as high as 750 kV or more!) and very large currents, often several kA. In this area, optical fiber current measurement schemes [6] can provide excellent electrical insulation and almost total immunity to EMI. Optical fiber current sensors (OFCS) satisfying such requirements are soon expected to take the place of traditional electrical current transformers (CT). After many years of careful research, rigorous field tests of the OFCS are finally showing good performance, and the research now seems to be in the final engineering stages.

The OFCS measures current indirectly, by measuring the rotation of the SOP induced by it by the total magnetic field component along the axis of the sensing fiber. Figure 11.20a shows a typical configuration of a polarimetric OFCS. The sensing fiber coil is wound around a current-carrying conductor, and linearly polarized light is launched into the coil. The SOP rotation is due to the Faraday effect of the magnetic field, in the direction of the fiber axis, the magnetic field being induced by the current-carrying conductor inside the fiber coil. The plane of polarization of the propagating HE_{11} mode in the fiber is rotated through an angle, ϕ, given by

$$\phi = V \int H \, dl,$$

where ϕ, V, H and dl are, respectively, the Faraday rotation angle, the Verdet constant, the axial magnetic field component and the length along the fiber. Because the sensing fiber coil is formed into closed path, Ampere's law gives the relation

$$I_s = \oint H \, dl,$$

where I_s is the current in the electrical conductor passing through the fiber sensing coil. Suppose the

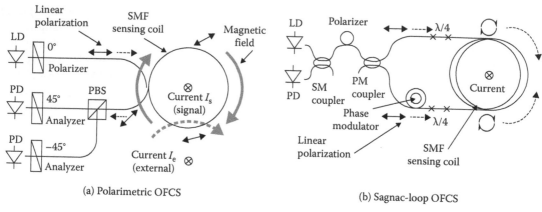

Figure 11.20 Two types of optical fiber current sensors: (a) using polarimetric interferometer and (b) using a Sagnac interferometer.

fiber has a number of complete turns, n, then, from the two equations above, the relation

$$\phi = nVI_s$$

is obtained. These form the well-known fundamental equations for the OFCS. These relationships are, however, only valid when the sensing fiber coil has no linear birefringence. If significant linear birefringence is present, the polarization state in the fiber is disturbed, and the output of the OFCS is changed, giving measurement errors. To overcome such problems, and reduce the linear birefringence of the fiber, several solutions have already been developed. Twisting a single mode fiber during the winding process is one method, and thermally annealing the fiber coil, to relieve strain induced by the winding process, is another. Additionally, a low birefringence fiber using flint glass has also been developed, which also has a stronger Verdet constant. This has been demonstrated to give a high performance, suitable for meeting practical engineering requirements [55].

The configuration shown in Figure 11.20a is only a polarimetric interferometer, i.e., mixing of polarization components is enabled in a polarization analyzer. This should ideally have its axis inclined at 45° to the axis of each interfering beam. Another configuration for the current detection has been developed, in which a Sagnac interferometer arrangement, similar to that used for the fiber optic gyro, is adopted. The configuration is schematically shown in Figure 11.20b. In this system, the SOP in the sensing fiber coil is arranged to be

circular by using quarter wave elements at both ends of the coil. In this configuration, the circular SOP is maintained throughout the propagation in the coil, but, via the Faraday effect, a phase difference is induced between the CW and the CCW traveling in the fiber coil. The phase difference is read out from the interference signal at the detector, using the same type of signal processing as the FOG. Hence, all the sophisticated and compact optical and signal processing modules, already developed for use with the gyro can also be used in this version of the current sensor.

We shall now conclude our discussions on the different types of single-point (discrete) sensors. The above discussions have made occasional reference to multiplexed or distributed sensors, but the following two sections will now concentrate on these aspects more fully.

11.4 MULTIPLEXED OPTICAL FIBER SENSOR SYSTEMS

11.4.1 Introduction

This section shows how sensors may be multiplexed [56] and the following one will deal with truly distributed [57] optical fiber sensors. It is convenient to define multiplexed sensors as those designed to collect data from a number of discrete sensing points, or sensing regions, and distributed sensor types as those that operate on a continuous length of fiber, and are capable of determining the variations of a desired parameter along the length of the fiber as a continuous function of distance. Figure 11.21 depicts the basic concepts of distributed and multiplexed sensors. The upper curve, Figure 11.21a, shows a schematic of an in-line example of both types (fan-out topologies and many more arrangements are possible for multiplexed sensors) and the lower curve, Figure 11.21b, shows the type of response to monitor the values of parameters, that each type can provide, as a function of physical location.

Extending the capability of a single measurement terminal, to address a multiplexed array of many passive sensing heads, rather than a single one, not only makes the use of a more complex monitoring station more economically viable, but it can also lead to a more accurate and reliable comparison (see Figure 11.21b) of values of the measured parameter, because the same interrogator is used to read each one. The distributed sensor (bottom curve, Figure 11.21b), however, goes a stage further, allowing the full distribution of the measured quantity to be determined with no gaps in coverage. In both multiplexed and distributed cases, one or more passive sensors, or sensing-fiber sections, with controlled environmental parameters, can be used as a calibration aid for the interrogator.

As with multiplexing in communications, the ultimate information gathering capacity is fundamentally limited by the available bandwidth and by the SNR of the detected optical signals, and similar care must be taken to avoid undesirable cross talk between signals from apparently independent information sources; represented in this case by the individual sensor elements.

In describing multiplexed sensors, we might perhaps start with a historical note. One of the earliest schemes for sensor multiplexing in optical fibers was reported by Nelson et al. [58], where a fiber optic branch-tapped network was used, in conjunction with an optical time-domain reflectometer (OTDR). This system, which will be described later, is capable of receiving and independently monitoring the separate returns from a series of reflective sensors.

The following sections will describe many more methods, outlining the most significant developments in the technology, with the sensors being grouped or classified according to the method used to address the various sensing elements.

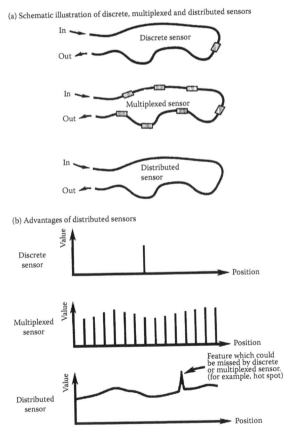

(a) Schematic illustration of discrete, multiplexed and distributed sensors

(b) Advantages of distributed sensors

Figure 11.21 Schematic illustration showing (a) an in-line configuration of multiplexed and distributed sensors and (b) the different types of spatial coverage offered by each type of sensor.

11.4.2 Spatial multiplexing (separate fiber paths)

Starting with the simplest low technology approach, the technique of using separate fibers to communicate with each separate sensing element (Figure 11.22), although trivial in scientific terms, is virtually guaranteed to avoid one of the pitfalls of multiplexing, that of cross talk between sensors.

As it is the easiest method to implement, it was one of the first to have been used for practical applications. Perhaps the most unusual and dramatic use so far has been the application of 152 separate graded index fibers, each 0.6km long, for nuclear weapon diagnostics [59]. For this particular application, light was generated at the sensor head end, so the light source and outgoing fiber paths shown in Figure 11.22 were not required, and the multiple

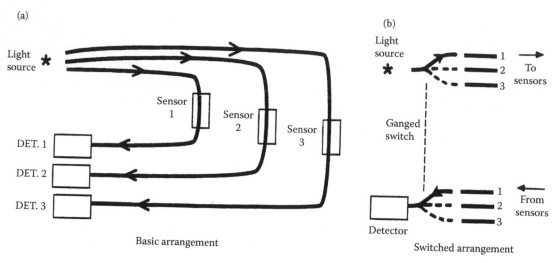

Figure 11.22 Spatial multiplexing using separate fiber paths: with (a) fixed paths and (b) with fiber switches to enable one source and detector to be used.

fiber channels were used to guide light from the event. Each individual receiving fiber guided light to produce a "pixel" of an image on a fluorescent phosphor imaging screen, presumably located in a monitoring area at a safe distance!

For cost-sensitive applications, where a slow update is acceptable, it is convenient to incorporate a fiber switch into a single optoelectronic processor unit, with a single light source and detector to permit it to "poll," i.e., sequentially connect to, each of the sensing heads in turn (Figure 11.22b).

A multiplexed system for monitoring fluorescent dye concentration in water, via separate 600 m lengths of optical fiber to each sensing head, has been reported [60]. In this system, a mechanically scanned mirror system was used to inject light from an exciting laser sequentially into each separate sensor head via separate fibers, the returning fluorescent light traveling through separate fibers to a common photomultiplier detector. This system was capable of monitoring fluorescent dye concentration down to levels of 10^{-10} by weight in water.

Although usually used to address independent sensors, multiple fibers have been used to permit precision digital sensing in systems where each "bit" of binary-coded information is carried by a separate fiber. This technique has been used in simple position-encoding schemes, using binary-coded discs of the type used in conventional optical shaft-angle encoders [61]. A more complex system has been described for the "digital" measurement of temperature, using a series of isothermal (i.e.,

all in close thermal contact) birefringent crystals, each sensing crystal having a different length according to the significance of the desired order of the "bit" of binary-coded information [62]. The temperature-dependent birefringence of each is separately monitored via a polarization analyzer, which converts these changes in the birefringence into changes in the amplitude of the transmitted light, each with a different sinusoidal response to temperature. The separate outputs were combined to produce the equivalent of a binary word, which defined the temperature of the crystals.

11.4.3 Time-division multiplexing

The concept of TDM, in combination with high-speed digital technology, has revolutionized modern-day communications. It is hardly surprising that it is also an attractive choice for multiplexing sensors! The time differences between signal returns from each sensor, necessary for a TDM system, are usually achieved by arranging differing total optical propagation delays for the signals from each sensing element, using extra coils of fiber, whenever longer delays are needed. In addition, in order to distinguish, and separate, the signals from each sensor, it is necessary to modulate the light source with a temporally varying signal.

The simplest form of encoding, for ease of both modulation and demodulation, is to use a repetitive pulse, having a short duration so that the set of returning differentially delayed pulses from each sensor does not coincide at the detector. Also, it must have a pulse repetition rate that is low enough to allow each reflected pulse "echo" to return back from the most distant element of the sensing array before the first of the subsequent set of reflected pulses returns from the nearest sensing element.

One of the first multiplexing methods suggested for use with hydrophone sensor arrays was of this type [63]. One of their proposed methods is shown in Figure 11.23a. This used two parallel fiber-optic, cross-coupled, highways, the first to distribute a transmitted optical pulse and tap a portion of it into each transmissive (loss-modulation type) sensor, and a second highway to collect signals from each individual sensor and guide the set of returning pulses to the detector.

The second of their proposed arrangements used a single tapped highway, with reflective sensors, each having a measurand-dependent reflection, connected to this highway with directional fiber couplers (Figure 11.23b). This array was interrogated using a conventional OTDR arrangement, with a semiconductor laser source and an avalanche photodiode (APD) detector. The basic method of optical time-domain reflectometry, which we mentioned briefly in the chapter introduction, was devised by Barnoski and Jensen [64] and researched further by Personick [65]. The concept is depicted in Figure 11.24, in a configuration commonly used to monitor losses in optical fibers and reflections from discontinuities. It is based on an optical radar (or LIDAR) concept, where a short pulse of light is launched into a fiber waveguide

and variations of backscattered light signal with time are monitored.

Usually, light from a pulsed semiconductor laser (or Q-switched fiber laser light source) is launched into a section of fiber via a directional coupler, which serves also to direct a portion of the back-scattered light fraction, returning from the fiber on test, to a high-speed positive-intrinsic-negative field effect transistor or APD detector. The time of flight determines the distance, and the intensity variation normally indicates properties of the fiber under test (see lower curve of Figure 11.24). It has become a standard test-gear instrument for use by optical fiber engineers and researchers, to examine the continuity and attenuation of optical fibers, and observe reflections from fiber breaks, connectors, or other discontinuities.

If used to interrogate multiplexed sensors, it is possible to measure variations in either the reflected power from, or the transmission loss in, each sensor element (distributed sensing will be discussed in more detail later). Desforges et al. [66] have also reported experimental results with reflective sensors using an OTDR, but in their case the sensors were located at regions where the optical fibers are deliberately bent, to cause noninvasive coupling of light into and out of a continuous fiber to the reflective sensors.

The OTDR may also be used to monitor discrete loss-modulation sensors, of the "microbend" type we discussed earlier, situated along the length of a continuous optical fiber. This technique has been proposed to monitor strain or deformation

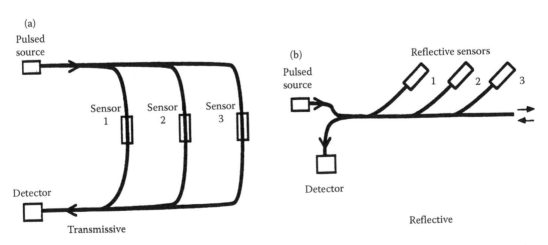

Figure 11.23 TDM of fiber sensors: (a) ladder network with transmissive sensors and (b) branched network with reflective sensors.

(a)

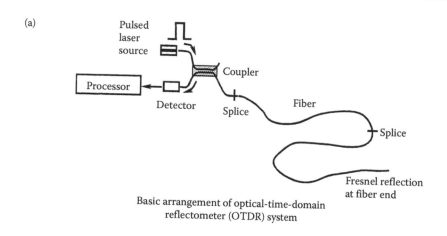

Basic arrangement of optical-time-domain
reflectometer (OTDR) system

(b)

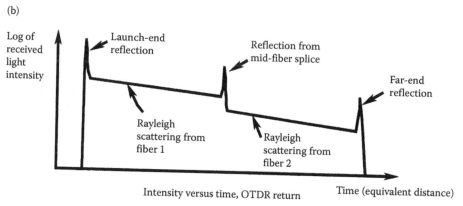

Figure 11.24 Basic concept of the OTDR, showing (a) optical arrangement and (b) typical OTDR
returns from a fiber.

in civil engineering structures [6,67,68] and for nondestructive testing of engineering composite materials [69]. However, care must be taken with such multiplexed arrangements if quantitative results are required, because such microbend sensors cause severe mode conversion in multimode fibers, and the loss of each such sensor is strongly dependent on the mode excitation of the fiber immediately before the micro-bent section. Thus, when multimode fiber is used to interconnect and form an array of closely located sensors, there will, in general, be significant interelement cross talk in the response.

The main disadvantage of a simple pulsed source for addressing multiplexed arrays is the poor optical efficiency due to the low excitation duty cycle. This problem is exacerbated, either by the peak-power limitations of semiconductor light sources, or by the onset of nonlinear optical processes in monomode fibers. A similar problem has, however, already been previously met, and dealt with, in radar systems, where it was also desired to distinguish differentially delayed signal returns without need for enormous transmitted power levels.* The radar technique most relevant to time-division optical sensor multiplexing is that where the transmitted radio signal has a pulse-code intensity modulation envelope that has a 50% duty cycle and has an autocorrelation function corresponding to that of a periodic pulse train [70]. Simple cross-correlation of the detected signal, with differently phased versions of the code originally transmitted, can then be used to separate the pulses corresponding to the individual sensors, just as if a single pulse of much higher peak power were to be used.

---

* Anti-Stokes Raman light, at wavelengths shorter than the incident light, although normally very weak, can still have a useful intensity if it is measured at a wavelength very close to the incident wavelength. It then often has less interference from fluorescence light.

An alternative means of reducing the peak power requirements for the optical source (again one commonly used in radar systems) is to produce a temporally extended "chirp" signal from an electrical pulse, using a dispersive surface acoustic wave (SAW) filter, and to use this signal to modulate the intensity (amplitude) of the light source [63]. The detected return signal is then subjected to an inverse transformation, using a second SAW filter, thereby reconstituting a replica of the originally transmitted pulse from each reflected signal; the net result from a sensor array, with delays of different lengths, being a time-division multiplexed stream of pulses, similar to those which would be obtained without the SAW filters.

As mentioned earlier, the TDM technique is also attractive when used with interferometric sensors. If an optical heterodyne method is used, where the returning signal is mixed with a reference signal of shifted frequency, the amplitude of the returning signal will, provided it remains constant, have no effect on the phase of the resulting beat signal. Such sensors, therefore, will require no amplitude referencing, as the output of the sensor is represented by the phase of the beat signal. A multiplexing scheme of this nature was first reported for hydrophone applications by Dakin et al. [71]. This particular implementation (Figure 11.25a) involved the launching of a consecutive pair of optical pulses, each having slightly different frequency, into a linear array of interferometric intrinsic fiber strain sensors, each joined by partially reflective splices. The intrinsic sensors are coils of monomode fiber, usually potted into polymer cylinders, to enhance their response to acoustic pressure waves. The initial optical pulses transmitted were consecutive, i.e., did not overlap in time, but the differential delay of returning pulses from adjacent splices gave rise to coincidence of the second pulse, from the nearer splice of a sensor element, with the first pulse, from the subsequent splice (Figure 11.25b).

Thus, with suitable pulse separation and duration, a time-division multiplexed stream of heterodyne beat signals was obtained from the receiving detector, each heterodyne pulse corresponding to an element of the array, and carrying phase modulation proportional to the changes in optical path length. These path length changes were a direct measure of the strains arising from insonification of the corresponding fiber hydrophone element. Phase demodulation of each time-demultiplexed

channel yielded the acoustic signals, free from any dependence on the amplitudes of the reflected light signals. In a patented modification, the same research group [72] first showed that it is possible to compensate the unbalanced interferometer arrangement using an optical loop of fiber as a "pre-delay," thus balancing optical paths and greatly reducing the effects of undesirable phase noise that arise from frequency fluctuations of the source laser (see later Section 11.4.7 on frequency-modulated carrier wave (FMCW) methods).

The attraction of such coherent heterodyne TDM approaches is the great sensitivity to even tiny phase changes that can be achieved using the coherent detection process and, secondly, the excellent dynamic range that is possible using electronic phase demodulation of the intermediate frequency signal. The improvement in the detection process is greatest if a returning reflected signal is mixed with a strong local oscillator signal, derived at the monitoring station from the initial source [73]. The advantage of *differential* sensing of the distance between adjacent reflective splices, achieved by the method shown in Figure 11.25, may theoretically be retained if a three-wave mixing process is performed [74]. Using this latter method, the beat signal between two weak received signals should be recoverable, after mixing with a strong local oscillator signal on the detector.

Based on many of the concepts arising from those early publications, the field of marine acoustic sensing has been one of the major practical success stories in the field of optical fiber sensors. The applications are in two main areas, firstly for naval surveillance applications (all the usual ones of towed arrays, vehicle arrays and fixed sea bed arrays) and secondly for seismic surveys, where intense sound sources, often explosive ones, are used to investigate subsea rock strata for oil-bearing features. It now appears very likely that optical fiber hydrophone sensor arrays, having a conveniently passive all-fiber "wet-end," will completely take over from older technology using piezo-electric sensors. The latter require electrical preamplifiers, complex electrical wiring to electronic multiplexers, and a sophisticated electronics communications system. Electrical systems are difficult to design and reliably maintain in a corrosive and conductive salt water environment. An excellent review of the subject has been given by Kersey in bibliography [7] (this chapter). An

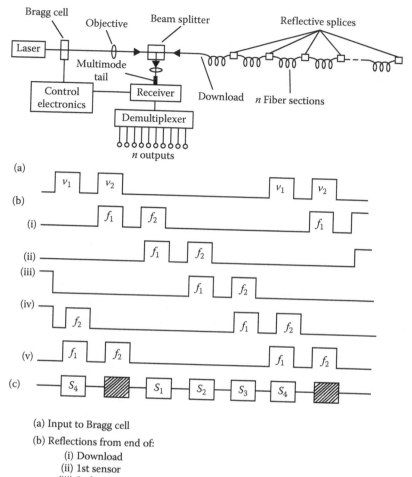

(a) Input to Bragg cell
(b) Reflections from end of:
    (i) Download
    (ii) 1st sensor
    (iii) 2nd sensor
    (v) 3rd sensor
    (iv) 4th sensor
(c) Output from photodiode

Figure 11.25 (a) Diagram of optical fiber hydrophone array. (b) Timing diagram for a four-element array with: (1) the input to the Bragg cell; (2) the reflection from the end of (i) the down-lead, (ii) the first sensor, (iii) the second sensor, (iv) the third sensor and (v) the fourth sensor and (3) the output from the photodiode.

advanced seabed sensor array system and a working sea test are described in papers by Nash et al. [75] and Cranch et al. [76].

11.4.4 Wavelength-division multiplexing

The use of wavelength-division multiplexing (WDM), unlike the alternatives of TDM and sub-carrier frequency-division multiplexing (FDM) techniques, has the advantage that there is no

theoretical loss penalty when compared with the single-fiber-per-sensor approach. The method involves guiding optical power to each sensor, and back to a corresponding sensor, via a route dependent on the wavelength designated for the interrogation of that particular channel (see Figure 11.26 for a very basic schematic). The path of light to each sensor is directed using WDM coupling components, similar to those used in communications systems. These components are in theory lossless, but will, in general, introduce

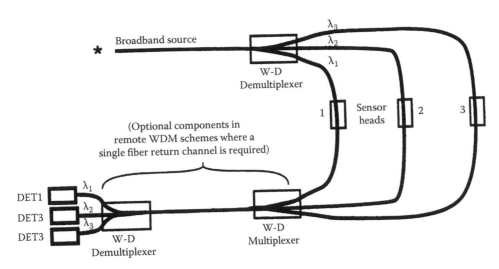

Figure 11.26 Schematic diagram of a wavelength-division multiplexed arrangement for remote detection of the state of three amplitude-modulation sensor heads.

a small loss in practice (typically 1–3 dB for each pass), imposing, therefore, a very small penalty in an otherwise lossless multiplexing method.

Although the spectral width of the fiber transmission "window" is potentially enormous, when compared with the very much lower information rate theoretically required for telemetry, the potential of multichannel WDM was originally rather limited by practical difficulties in achieving sufficient selectivity in the WDM filters. This has changed dramatically with the excellent wavelength routing components that have now been developed for multichannel WDM telecommunications systems. Using broadband LED or superluminescent fiber sources, or multichannel laser sources, the practical number of independent channels in the fiber window is now potentially very large. It is possible to use "spectral slicing" techniques with either LED [77] or superluminescent fiber sources [78]. This latter technique involves filtering out two or more separate narrow channels, from within the broad spectral linewidth of the LED (in this case for distributed sensing, but the basic concept is the same). The separation can be performed using narrow-band filters—a method that inevitably decreases the average signal in each channel by a factor at least in proportion to the number of channels required.

Another method of achieving a large number of channels is to use a number of independent narrow-linewidth laser sources, which removes this inherent loss penalty, as modern WDM combiners can, combine channels with zero loss in theory, and in practice they have close to 100% efficiency. Generally, the limit with lasers is set by practical considerations, such as the selectivity of channel filters, the availability, selection, and stabilization of laser sources, and the cost of a multilaser system of this nature. Again very dramatic developments in the light-sources (e.g., DFB fiber lasers) passive components (WDM splitters, add-drop multiplexers, etc.) and many other sub-systems have been made to facilitate use of WDM in telecommunications applications, which paves the way for such methods in sensors.

The WDM approach was first used to permit the addressing of separate "bits" in a 10-bit digital sensor, designed to measure a *single* physical parameter, in this case the angular position of a shaft in a fiber optic shaft angle encoder [79]. A broadband white light source in the transmitter/receiver terminal launched light into the outgoing fiber to the sensor head (shaft angle encoder), and two diffraction gratings were used, the first one in the encoder, first to separate the white light from the outgoing fiber lead into 10 wavelength channels, and then to recombine the reflected code-carrying channels after intensity modulation of each "bit" by the coded disc. A second similar diffraction grating was used again in the decoder, to direct each of the returning signals, according to their wavelength, onto a separate detector of an array. The digital encoder disc was a "Gray-coded" disc, as commonly used in conventional optical encoders, but in this version a different color was used to interrogate each separate coded band on the encoder disc, hence allowing transmission through a single optical-fiber cable (or two-way send/return cable) by wavelength multiplexing.

The multichannel WDM approach has also been proposed to provide additional picture elements for remote imaging [80]. In this case, a diffraction grating system was again used in the sensor head to enable the spatial position of each pixel of a one-dimensional white-light image to be uniquely coded, according to its optical wavelength, before transmission to a remote monitor station. At this point, the image is recreated, using a second diffraction grating to perform the inverse transformation. This way, one dimension of the two-dimensional image is divided according to wavelength. The other dimension can be divided easily using a one-dimensional array of separate fibers.

Now that WDM systems have become so widely accepted for fiber-optic communications systems, the practical capability of WDM systems to address large arrays sensors is becoming ever greater and yet more economically viable. It is applicable to many types of optical sensor, and there are now many key components for WDM routing, including multiline laser sources, and low-loss multichannel filters, grating filters, add-drop filters, to name but a few. An example of a multielement hydrophone array, using a combination of both TDM and WDM methods, has been presented by Vohra et al. [81].

11.4.5 Multiplexing of in-fiber Bragg gratings using TDM and WDM methods

In view of the importance of Bragg grating sensors, it is appropriate to discuss how these sensors can be multiplexed. In view of their narrow reflective spectrum, they readily lend themselves to WDM methods. Second, the ability to write them at any point in a fiber, with associated position-dependent variations in the two-way optical delay from a pulsed light source to the sensors, according to their position, allows TDM to also be used. Space here does not permit a full discussion of what is now a major research area, but the

excellent review by Kersey et al. in the bibliography [7] covers this area very extensively.

Some of the methods used to address multiple gratings, all situated in a single fiber cable, are listed below, many of which were of course mentioned in the earlier section describing how individual gratings may be addressed:

- Use of broadband source launched into fiber and a spectrometer (CCD spectrometer or Fourier-transform spectrometer) to interrogate the reflected spectrum (system as in Figure 11.12, but with multiple in-line Bragg gratings in the fiber).
- Use of a scanned narrow-band optical filter (e.g., conventional bulk-optic Fabry–Perot, all-fiber Fabry–Perot, acousto-optic tunable filter, and scanned Bragg grating) in conjunction with a broadband source launched into fiber, and a detector to receive reflected light from arrays of in-fiber Bragg gratings. The filter may either be scanned to measure the peak reflectivity of each grating in turn or with the aid of a feedback loop, may be locked on to track each grating in turn.
- Use of a mechanically-scanned (fiber is stretched with a lead-zinc-titanate (PZT) all-fiber Michelson interferometer, followed by a single optical detector, to interrogate reflected signals from arrays of in-fiber Bragg gratings. Fourier transformation of the detected output signal is performed, in the same manner as used in the well-known Fourier transform spectrophotometer using a bulk-optics interferometer [82].
- Rare-earth-doped fiber laser, operating multiwavelength, using the Bragg gratings as end mirrors.

Both the review by Kersey (bibliography [7]) and another by Dakin and Volanthen [83] cover many of the pitfalls that have to be avoided to prevent undesirable measurement errors. Examples of potential problems include the following:

11.4.6 Subcarrier FDM

In point sensors, which are connected by separate fiber paths to a common detector, the technique of FDM of the intensity modulation waveform can be used, modulating the electrical signal to drive a light source. Such a modulation may be used to facilitate separation of the outputs from individual channels in the detected signal. The simplest way of using the technique is the relatively trivial method of transmitting signals from separate light sources, each modulated by electrical signals of different frequency, via separate fibers to each of the sensor heads. The outputs may then be combined (or added) into a common output fiber and detected on a common detector, yet still be separable by frequency-selective electrical channel filtering. However, a more elegant FDM approach, using a single LED source, has been devised by workers at the University of Strathclyde in Glasgow [84]. This method, using a simple transmissive system, is shown in Figure 11.27a. The three subcarrier modulation signals add at the detector output, as shown in Figure 11.27b, with the resultant of their vector addition being dependent on their relative phase angles. These phase angles are a function firstly of the original phases of the transmitted envelope modulation and secondly of the different delays they experience in transmission through the optical network containing the point sensors.

The term FDM is generally used in optical communications systems to imply systems in which an optical carrier is amplitude modulated by a composite electronic signal, consisting of a sum of modulated *subcarrier* signals with different frequency channel allocations. In order to provide a more precise terminology, the prefix "subcarrier" has been inserted in the title of this section, but for the remainder of this review, the abbreviation FDM will, for convenience, be used to describe such systems.

These signals may be separated, first by employing multichannel phase-sensitive detectors on the detected optical signal (using the original applied modulation signals as electronic reference signals) and then by solving simple simultaneous equations (linear regression) on the resulting scalar quantities. This process allows removal of any cross talk terms from light that has passed through the sensor heads of each other channel.

The above system has the advantage over TDM methods of having 100% duty cycle and using much simpler and slower electronics. For low-frequency sensors, it should be easy to construct

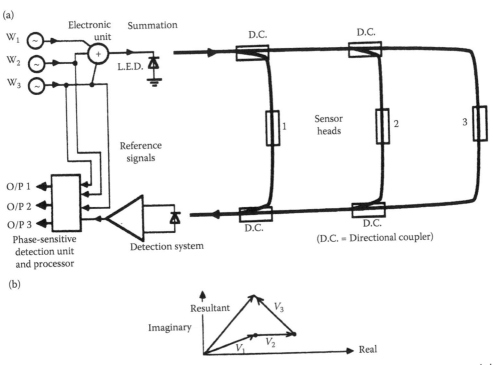

Figure 11.27 Multiplexing using frequency-division-multiplexing of subcarriers (intensity modulation signals), showing: (a) schematic diagram of arrangement for three amplitude-modulation sensor heads and (b) vector diagram for the phase-sensitive summation of subcarrier signals on an optical detector.

with a low-noise bandwidth in each of the phase-sensitive detection stages. However, although the electronics does not require a high-frequency response, care must be taken to ensure stability of electronic and optical delays, as serious cross talk could otherwise result.

11.4.7 The FMCW multiplexing scheme

This multiplexing method, like many of the preceding techniques, was adapted from the radar field and has much in common with the FDM method discussed in the previous section. However, in this case, the RF subcarriers signals only actually appear at the optical detector, as a result of "beating" or "mixing" of optical waves, where difference frequency (heterodyne) signals are generated due to the "square-law" characteristics of optical detection. The transmitted signal in the FMCW method is an optical carrier wave, the frequency of which increases (or decreases) linearly for a period, T, after which time it flies back to its initial frequency, before

repeating the process. When now the source is connected to an interferometer with a differential path delay, the return signals on the detector differ by a frequency proportional to firstly the optical path difference they have experienced and secondly the frequency slew rate of the source (see Figure 11.28).

Such a system was first proposed in single sensor form by Uttam and Culshaw [85], with a later description from Giles et al. [86]. If the source is connected via a series of interferometers, each having different delay paths [87], a series of RF carrier signals are produced at the detector by the heterodyne mixing process. The various sensor output signals may be isolated by electronically filtering out the corresponding frequencies. The sensor output signals may be represented by the phase, frequency or amplitude of the recovered RF signals, depending on the scale of the path length change (small path length changes are more readily discernable as phase changes, large path length changes more conveniently as frequency changes) or the transmission changes occurring in the sensor.

(a) Schematic of FMCW Michelson sensor

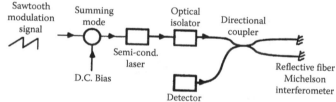

(b) Frequency versus time for optical and electronic signals

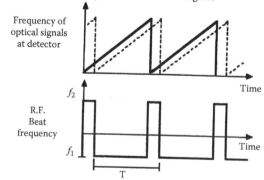

(c) Multiplexed FMCW arrangement (Al-Chalabi et al. 1985)

Figure 11.28 The FMCW method of sensing and multiplexing, showing: (a) a schematic diagram of the FMCW Michelson sensor, (b) variation of frequency with time for optical and electronic signals and (c) an arrangement for multiplexed sensors using FMCW.

The technique of FMCW for sensor multiplexing suffers from several practical difficulties, when used with a large number of sensors. The main problems arise from the large number of optical paths, which can produce a background of unwanted signals, from both the expected sensor paths and additional stray paths due to multiple reflections [88]. Also, when it is desired to extract precise phase information from the method, problems may arise from nonlinearities in the laser frequency modulation ramp. Possible "pulling" of the laser frequency by reinjection of retro-reflected light can be avoided by use of an isolator, but a semiconductor laser usually has intrinsic nonlinearities in its frequency chirp response due to the complex internal optical and thermal laser time constants.

Other problems arise because the technique necessarily requires an unbalanced interferometer for it to operate, so will eventually have reduced fringe contrast as the path differences increase, due to the finite coherence length of the source. It also necessarily suffers signal degradation from phase-noise, due to random frequency fluctuations of the laser source from the ideal saw-tooth, as, unlike certain other methods [72] the phase noise cannot be passively compensated for by using balanced optical paths. Limits on the permissible path imbalances greatly reduce the range of heterodyne frequencies that may be allocated for multiplexed sensors, a problem that may be exacerbated by nonlinearities in the frequency ramp and undesirable amplitude modulation of the source. However, despite some disadvantages, the FMCW technique is conceptually elegant and has the major advantage of being able to monitor steady-state changes of the length of an interferometric sensor, simply from the absolute magnitude of the heterodyne frequency.

More recently, a variation of FMCW has been used for addressing hydrophone arrays. Here, it is used to generate a heterodyne carrier, and has been renamed the "phase-generated carrier" method. Rather than modulating the source directly, a fiber-compatible PS, such as a mechanically strained (piezo-driven) coil, or an integrated optic FS can be used. These methods are discussed in detail in bibliography [1], vol 4, and in the excellent review by Kersey.

It should be pointed out that the FMCW method is not limited to situations in which the optical carrier is modulated. It has been shown by Mallalieu et al. [89] that FMCW may also be operated using a frequency-modulated subcarrier wave, rather than modulating the optical wave. This removes the need for both coherent sources and monomode fiber, and simplifies the optical system for sensors that do not require the use of optical interferometry, but because of the much lower frequency slew rates, will not normally achieve the phenomenal sensitivity to minute length changes that optical interferometry offers. It has excellent attractions for high-resolution interrogation of fibers, and improvements in the modulation of the source will be discussed again later in this chapter.

11.4.8 Coherence multiplexing

The technique of coherence multiplexing is an optical method, which is similar in concept to a technique used in spread-spectrum radio frequency communications. A signal, with a superimposed random (or pseudorandom) modulation, may be demodulated to recover information by correlating the received signal with a similarly encoded random (or pseudorandom), but delayed reference signal. In sensing systems, a broadband source of short coherence length may be used as a transmitted signal, as such a source will, in general, be subject to naturally occurring random phase or frequency excursions. If such a signal is guided via two equal, or near-equal, monomode fiber paths, then the signals suffer nearly equal delays, and have a strong correlation, provided the path difference is small compared with the coherence length of the source. Under these conditions, high-contrast interference fringes can be observed if the output signals are mixed on a square-law detector. If, however, the paths differ by very much more than the coherence length of the source, the fringe contrast becomes close to zero.

The arrangement in Figure 11.29, first proposed by Brooks et al. [90], shows how several remote Mach–Zehnder interferometers may be independently addressed using the coherence multiplexing technique, provided the optical path length differences, l_1-l_0, l_2-l_0, and l_1-l_2, are all much greater than the coherence length of the source. The sensing method is based on observing the fringe shifts, which occur as the path differences l_1-l_0 and l_2-l_0 in the sensors 1 and

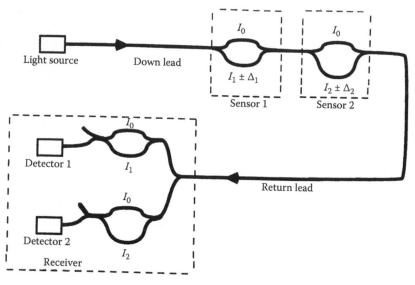

Figure 11.29 Schematic arrangement for coherence multiplexing, shown for the simplest case of separation of signals from only two remote Mach–Zehnder interferometric sensors.

2, are changed by small increments, Δ_1 and Δ_2, respectively. The changes are observed by interferometric comparison with the corresponding fixed lengths $l_1 - l_0$ and $l_2 - l_0$ in the receiver interferometers. Only close-matched paths give visible interference fringes.

As a means of separately interrogating the outputs from a small number of remote sensors, the method has attractions. However, as the number of sensors increases, the number of possible optical paths in the network, from source to detector, increases very dramatically and the use of a very short coherence length source, such as an LED, soon becomes necessary. This then presents extreme difficulty in achieving the very close path length equalization required, in order to ensure adequate fringe contrast, and results in a rather small dynamic range of measurement before fringe visibility is lost. Further practical disadvantages of the system are the need for each sensor to have a unique path length difference (and hence, without careful design, a different sensitivity) and the need for adequate stabilization of the receiver interferometers, to avoid undesirable errors due to drift or vibration in these reference interferometers.

In conclusion, therefore, it appears likely that, although elegant in conception, the method is less likely to find practical application in systems with more than perhaps 10 sensors, unless other multiplexing methods, such as WDM and TDM are also used.

A method for grating interrogation using coherence effects has been presented by Dakin et al. [91]. Here, a number of in-line Bragg-grating-pair sensors are interrogated, in turn, using a scanned Michelson interferometer. Fringes are only observed when the path length of the interrogating Michelson interferometer matches the optical spacing of the Bragg gratings, and the rate of fringe crossing gives a measure of the wavelength of the gratings.

In the final section of this chapter, M Marcus will also give a detailed case study of work at Kodak on a white-light interferometer system for decoding multiple light paths in manufactured optical products. These have been used for assessing the inter-element spacing and aspects such as focal lengths of production cameras.

11.4.9 Conventional sensors with an acoustic-sensing "fiberdyne" highway

As a final sensor multiplexing method, a hybrid technology will be described, which is really a noninvasive data collection highway. Its sensor use would be for collecting data from arrays of conventional electrical sensors, using a continuous optical fiber sensing highway. This highway is essentially a long, continuous, acoustic sensor, based on interferometric techniques, to which signals from electrical sensors could be coupled using piezoelectric transduction elements, held in close mechanical contact with the fiber or cable. In order to distinguish the separate signals applied to the common highway, it is possible to allocate electrical subcarrier signals, of different frequency for each sensor, and allow the sensor output signals to modulate (intensity or angle modulation) these electrical carriers and impart information before they are applied to the piezoelectric transducers.

The original method, referred to as a "fiberdyne" system, was first devised and built by workers at UCL, London [92]. The first version (Figure 11.30a) used a Mach–Zehnder twin-fiber interferometer arrangement, in which one fiber acted as a passive reference arm and the second was used as the strain-sensitive highway, which was subjected to an acoustic influence from the piezoelectric transducers. The method was initially proposed as a wideband fiber highway and had a high sensitivity, but, without use of a PC, would be subject to polarization fading when used with a normal monomode fiber.

A later version, from the same research team, used a single multimode fiber and relied on variation of the "speckle" interference pattern emerging from the fiber, as the PZT transducer caused mode conversion in it (Figure 11.30b). This method had the advantage of not requiring a separate reference fiber, but, as a result of the complex beating processes between many fiber modes, and the need for differential modulation of mode delays, it exhibited a somewhat lower sensitivity. It had a greatly reduced optical energy efficiency, as only a fraction of the speckle pattern was actually incident on the detector (otherwise only the *total* transmitted

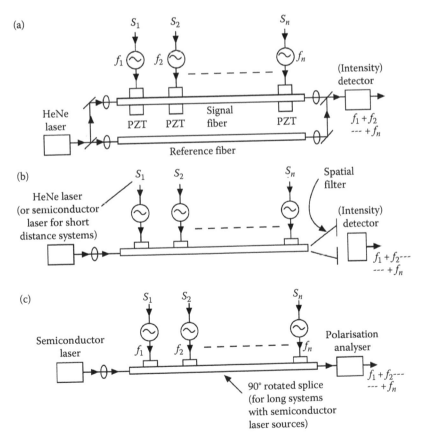

Figure 11.30 "Fiberdyne" highway systems: (a) original fiberdyne concept using monomode fiber (Mach–Zehnder configuration), (b) multimode fiberdyne highway (speckle modulation) and (c) polarimetric fiberdyne highway (polarization modulation).

power would be detected, and phase modulation would not change this) and there was a strong possibility that either multimode fading or polarization fading could occur. The fading can only be prevented by arranging for diversity of phase, polarization and spatial arrangement of detecting the speckle. This requires multiple detectors and possibly active electronics to select nonfading channels. Despite these problems associated with this first implementation [93,94], a sensor telemetry highway was constructed and tested. In this, a number of stations could be fed in, using a set of channels, with relatively low radio frequencies for the carrier signals ($\approx 1\,MHz$). The channels had an even lower baseband-channel frequency (information rate) of $\approx 10\,kHz$, which although low would be more than sufficient for many data collection requirements from sensors. This signal to highway coupling was carried out acoustically, as before, but, in these later systems, the acoustic waves were transmitted via the plastic protective sheathing of a fiber-optic highway cable, the reduction in coupling efficiency giving rise to a reduced information bandwidth.

An improved method has been devised (Figure 11.30c), using a polarization-maintaining fiber, again with lateral acoustic excitation. If suitably aligned, this mechanical excitation has the effect of modulating the phase of one polarization mode, relative to that of the orthogonally polarized mode, resulting in the modulation of the polarization of the transmitted light [95]. This method is more efficient in its use of the received optical energy, as there are only two polarization modes in the fiber, and, when these interfere on the detector, no optical energy is wasted and fading can occur provided the correct phase bias is maintained between the two received polarization modes (this phase bias can be changed by stretchers to set the quiescent operating condition at the quadrature, or maximum-slope point of the sinusoidal response of the polarimetric interferometer).

11.5 DISTRIBUTED OPTICAL FIBER SENSORS

We shall now discuss fully distributed sensing methods. Systems that permit the monitoring of not only the magnitude of a physical parameter or measurand, but also its variation along the length of a continuous uninterrupted optical fiber are particularly attractive, for both economic and logistical reasons. Distributed sensors not only allow a simple easily deployable sensing cable, where the communications link and sensor are one entity, but they also permit an easy and reliable comparison of a parameter at different points (the same interrogator takes measurements at different points) and the sensor cable measures at every point along the length with no "dead spots." This latter aspect is particularly important for safety related sensors (e.g., fire detectors or detectors for hazardous leaks of chemicals) and for sensors to detect intrusion through perimeters.

The simplest form of distributed sensor, the optical time domain reflectometer (OTDR) we discussed briefly earlier, has been in commercial use for many years as a measurement method for telecommunications, and is commercially available from many manufacturers. However, in spite of the relatively early development of the OTDR concept, it was several years before its use for the distributed measurement of parameters external to the optical fiber was envisaged [96]. Rogers suggested, for the first time, a method for the measurement of the spatial distribution of magnetic and electric fields, pressure, and temperature, using OTDR concepts in conjunction with polarized light sources and polarization-sensitive receivers. More details of such sensors are given below.

11.5.1 Backscattered sensors using the OTDR concept (general introduction)

All the sensors discussed in this section make use of radar-type (more accurately, LIDAR-type) Rayleigh backscattering (or backward-traveling light from other inelastic light generating processes, such as Raman, Brillouin or fluorescence) to make truly continuous measurements on unbroken optical fibers or fiber cables.

The basic method of optical time-domain reflectometry was, as described earlier (Section 11.4.3, Figure 11.24) the first type of distributed optical fiber sensor. In view of its importance here, we shall firstly briefly review it and then give more details of the principle of operation. A pulsed semiconductor laser is coupled into a section of fiber through a directional coupler, which serves also to couple the backscattered light fraction, captured and returned via the fiber on test, to the high-speed optical receiver.

In normal telecommunications fibers, the Rayleigh component of the scattered light represents well over 98% of the returning signal (except during the short time intervals when more intense specularly reflected pulses return from discrete discontinuities, such as connectors, air-spaced splices or distal fiber ends).

For uniform fibers, the detected temporally varying Rayleigh scattered power $I(t)$ varies as the product of at least five important physical factors, all of which can either be chosen by appropriate design or component selection. These are first the launched energy, E_0, second the scattering attenuation coefficient, α_s, in the fiber, third the fraction, S, of scattered light that is captured by the fiber in the return direction, fourth the inverse-4th power λ^{-4} of the optical wavelength, λ, and finally the two-way optical attenuation factor, $\exp (\int \alpha(x) \, dx)$, that occurs during propagation from the source to the scattering point and back to the detector:

$$I(t) = \frac{1}{2} E_0 v_g \alpha_s S \exp\left(\int \alpha(x) dx \right).$$

There are also a few other factors that are more difficult to change significantly, such as vg, the velocity of light in the guide, which has to be close to that of silica, as the doping levels are low, and the fixed numerical factor of 0.5. The factor S has a close to quadratic dependence on the numerical aperture of the fiber, and hence on the core/cladding index difference, so is higher for high numerical aperture (NA) multimode fibers. However, the significantly lower attenuation coefficient, $a(x)$, in monomode fibers can become a more important advantage when very long lengths of fiber are probed, despite their much lower S value. The SNR of an OTDR reduces very rapidly as the distance resolution improves, as firstly much shorter pulses are needed (reducing the launched energy, E_0) and secondly a high-speed receiver has a wider noise bandwidth and usually a higher noise spectral density too, both conspiring to give far worse noise performance with short pulse excitation and fast detection*. Because the scattering coefficient in high-quality fibers does not usually vary significantly along the length, the method has proved extremely useful for measuring spatial variations of fiber attenuation. However, if the geometry or numerical aperture of the fiber varies significantly, changes in the guidance properties of the fiber (primarily in the modal "V" number, that changes the number of allowable modes) will cause additional variations in the backscattered signature [97,98]. These workers showed, however, that variations may be at least partially compensated for, provided two separate OTDR signatures are taken from opposite ends of the same fiber. Clearly, this requires two instruments or a loop sensor with a two-way fiber switch.

In the following subsections, several sensing methods using OTDR, and variations based on various inelastic scattering and fluorescent methods, will be described.

11.5.2 Monitoring of variations in attenuation using OTDR

The original objective of the OTDR method was to examine attenuation variations in manufactured and installed lengths of optical fiber. One of the first suggestions that it may be possible to construct distributed sensors, using the attenuation characteristics observed was made by Theocharous [99], who proposed a method for the measurement of temperature. If an OTDR return signal is differentiated with respect to time and normalized by division by the instantaneous value of the signal, a measure of the fiber attenuation is obtained. If, therefore, a fiber with temperature-dependent attenuation is used, variations of temperature along the length may be monitored.

---

* It should be noted here that the radar field [70] is one from which numerous techniques, applicable to multiplexed and distributed optical sensors, have been drawn. This source of inspiration for optical sensor ideas will be referred to again later.

The technique appears at first sight to be highly attractive, but a simple semiqualitative analysis suggests that the number of distance-resolution elements is likely to be low. To measure the average temperature within each distance-resolution element, an accurate loss measurement is necessary, in order to determine the smaller temperature-dependent changes in loss. Because the OTDR return is usually rather noisy, a temperature-dependent loss of at least 1 dB in the resolution element (whatever its length) will probably be necessary, in order to accurately detect small changes, say of 1% (or 0.01 dB) in this loss. Thus, the round-trip attenuation in this example will be a loss of 2 dB for each resolution element, times the number of such elements, giving a likely limit to the number of measured resolution elements of perhaps ten, particularly bearing in mind the poorer SNR that will return from the later sections of fiber.

A comprehensive study of rare-earth-doped fibers, showing how they could be used to determine the line-averaged temperature over their length, was carried out by Quoi et al. [100]. Many different rare-earth fibers were considered, and the interesting concept of detecting light at two wavelengths, one where attenuation increased with temperature, and another where it decreased, was introduced as a means of compensating (by ratioing the two detected signals) for losses in bends, splices or connectors.

Because of the attenuation problem over longer lengths that was discussed above, distributed sensing methods monitoring attenuation are most likely to have practical application for sensing using fibers or cables which have a low intrinsic loss, but which then suffer significantly increased attenuation at just one point or small region due to some event it is desired to detect or locate. Examples of this are largely safety or security related, such as the detection (and location) of fire using a cable that might have high loss if it becomes hot. Several published applications for other applications will now be discussed below.

An early means of using OTDR monitored attenuation variations for distributed radiation dosimetry was presented by Gaebler and Braunig [12]. In this application, a short section of a fiber, which was exposed to ionizing radiation, suffered excess attenuation, enabling simultaneous detection and location of the radiation exposure (see Figure 11.31). If used as a sensor for the monitoring of exposure over a short section of a more extensive overall length, the quiescent attenuation before radiation is merely that of a normal low-loss communication fiber, enabling the increased attenuation of the short exposed section to be measured with reasonable accuracy. The method would be attractive for detecting leaks of highly radioactive material, but silica fibers are rather insensitive, and a large part of the radiation-induced loss recovers quickly with time. Lead-glass fibers are more sensitive, but have higher intrinsic losses.

Several other loss-modulation effects in fibers, which were discussed in the first part of this chapter, are also amenable to OTDR interrogation. The first of these is the plastic-clad silica fiber system (see earlier Figure 11.6) of Pinchbeck and Kitchen [13] where the fiber showed increased attenuation when cooled, due to increase in the refractive index of the cladding polymer. Clearly, the OTDR offers the possibility of leak location, whereas that shown in Figure 11.6 can only offer detection of leaks, as it only permits simple detection of fiber loss (Figure 11.32).

The second we shall mention is the commercially available "Herga" fiber cable (Figure 11.3), which exhibits high microbending loss when subjected to lateral pressure, as a result of its novel spiral plastic sheathing arrangement. Figure 11.33 shows a schematic of an OTDR-based distributed sensor, using a pressure sensing cable of this type.

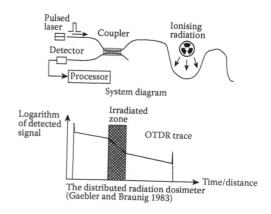

The distributed radiation dosimeter
(Gaebler and Braunig 1983)

Figure 11.31 Schematic of a distributed radiation dosimeter, using radiation-sensitive fiber, with an OTDR to detect regions of increased loss. There is potential for leak-detection system for radioactive materials.

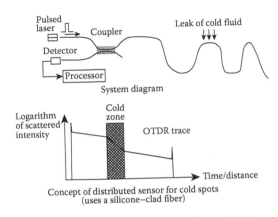

System diagram

Concept of distributed sensor for cold spots
(uses a silicone–clad fiber)

Figure 11.32 Schematic of a distributed leak-detection system for cryogenic liquids, using PCS fiber, with an OTDR to detect regions of increased loss (see also Figure 11.6).

Sensors such as these have been proposed by Alan Harmer, of Battelle research, for use as a distributed sensor to detect intruders, by monitoring losses from pressure of the intruder's foot on the cable.

The same system layout of Figure 11.33 can clearly be used with the water ingress sensor described earlier ([10,11], see also "Microbend sensors" section and Figure 11.4), which is based on the same microbending concept, and indeed this sensor type was developed by the authors using an OTDR instrument to detect the changes. One useful application is to detect leakage of water in

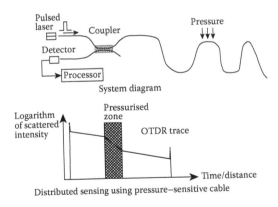

System diagram

Distributed sensing using pressure–sensitive cable

Figure 11.33 Schematic of a distributed detection system for lateral pressure, arising, for example, from an intruder treading on cable. Sensor uses pressure-sensitive cable of the type shown in Figure 11.3, with an OTDR to detect regions of increased loss.

telecommunications ducts, as this can cause damage to optical cables, particularly if it generates hydrogen, or if it freezes and breaks the fibers.

Another useful intrinsic loss mechanism that can be used is that of evanescent field absorption in the fiber cladding (Figure 11.34). Although, in principle, a bare unclad fiber could be used to detect absorbing material directly, this would easily be contaminated. It is sensible therefore to use a polymer cladding with suitable properties. Many polymer cladding materials will adsorb oil or other liquids and change their absorption [101] but if chemical indicators are available, and can be incorporated in the polymer, they can offer a more selective response to a desired target species. Blyler et al. [102] have made a sensor for measurement of ammonia using reactive cladding material.

It should be noted that it is not just absorption that may be used to form the basis of the sensor. The use of fluorescent coatings will be discussed later, after discussion of the use of other types of scattering processes in sensors.

11.5.3 Variations in Rayleigh backscatter characteristics

As already discussed, the use of OTDR to monitor fiber attenuation depends on the constancy of the Rayleigh backscattering coefficient along the length of the fiber. However, this may vary significantly in two basic ways, even in fibers of uniform geometry and composition. The first form of variability occurs in monomode fibers, using polarized illumination and polarization-sensitive detection. An arrangement as in Figure 11.24 is used, with a polarized light source, and now a polarization analyzer is used at the detection end. This diagnostic method, known as polarization optical time-domain reflectometry (POTDR), relies on the high degree of preservation of polarization exhibited by Rayleigh and Rayleigh–Gans scattered light in silica fibers, leaving the polarization changes that occur due to two-way propagation in the fiber itself to be observable.

The POTDR method was first suggested by Rogers [96], who pointed out its potential for distributed measurements of magnetic field (via Faraday rotation), electric field (via the Kerr quadratic electro-optic effect), lateral pressure (via the elasto-optic effect) and temperature (via the

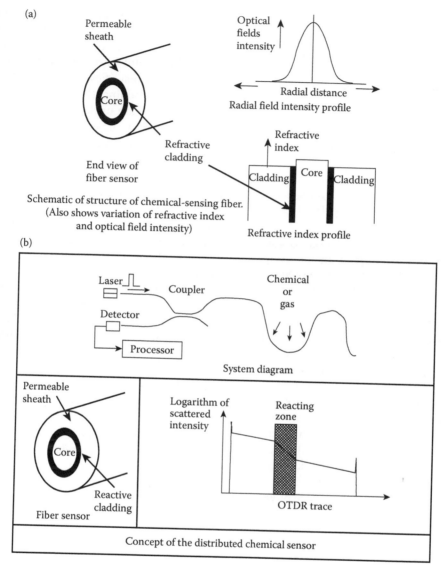

Figure 11.34 Schematic of a distributed chemical sensor system, using fiber with a reactive cladding that changes loss after exposure to the chemical. The structure of the fiber, the mode field pattern and the refractive index profile are shown in (a). Possibilities include permeation of a polymer cladding, by, for example, oil leaks, affecting either the refractive index or absorption properties, or use of a cladding with a chemical indicator designed to be selective to a desired target chemical or chemical species. The OTDR response when detecting leaks is shown in (b).

temperature dependence of the elasto-optic effect). The first experimental measurements were reported by Hartog et al. [103], who used the technique for a distributed measurement of the intrinsic birefringence of a monomode fiber, and Kim and Choi [104] who measured the birefringence induced by the bending of a wound fiber. Ross [105] carried out the first measurement of a variable external field, using the Faraday rotation of polarization as

an indication of the magnetic field environment of the fiber. A comprehensive theoretical treatment of the POTDR method has been presented by Rogers [106].

On first consideration, the POTDR technique appears to be attractive for the measurement of a large number of parameters. However, its main drawback, as with many other potentially useful sensing methods, is the variety of parameters to which it can respond, the sensitivity to strain and vibration being particularly troublesome when it is desired to measure other things. In addition, POTDR requires the use of monomode fibers, which can, when used with narrow linewidth laser sources, have particular problems due to coherent addition from multiple Rayleigh backscattering centers [107].

Moving on now from considering variations in the polarization of backscattered light, the scalar magnitude of the Rayleigh scattering coefficient may vary with temperature in some types of fiber. If so, a simple OTDR arrangement, now preferably without polarization sensitivity, can be used to perform distributed temperature measurements. Unfortunately, in normal vitreous fibers, of silica or multicomponent oxide glasses, the majority of backscattered energy arises from Rayleigh or Rayleigh–Gans scattering from frozen-in refractive index variations. These are tiny regions where high-temperature thermally induced density changes were "frozen" into the glass structure as it is cooled from the melt. As a result, once the glass has cooled, the scattered light intensity varies very little with the temperature of the glass. In many normal room-temperature liquids, however, the scattering arises from real-time thermodynamic fluctuations in the refractive index, which are dependent on the ambient temperature and, therefore, the scattering will now show a significant temperature coefficient.

This effect has been exploited for distributed sensing of temperature, in the system described by Hartog and Payne [108]. Unfortunately, the use of liquid-filled fibers presents difficulties, which, at best, complicate the method and, under some circumstances, restrict its use:

1. The use of liquid-filled fibers is inconvenient because of the need for expansion reservoirs.
2. There are temperature range restrictions imposed by the freezing and boiling points of the liquid.
3. If the central region of a long length of such a fiber is rapidly cooled, it is possible to create voids by rapid thermal contraction, before pressure differentials can refill the tube against the resistance of viscous forces.
4. Impurities or dust particles may increase the scattering cross-section or increase the localized loss of the fiber, giving the appearance of false temperature variations in the OTDR return.
5. The numerical aperture of the liquid-filled fiber shows significant temperature dependence, generally reducing as the temperature is raised, allowing hot zones in particular, to affect the calibration of results further along the fiber. (In the absence of mode-conversion, this effect may be reduced by the use of mode filters).

However, despite the above potential drawbacks, which are not uncommon with initial groundbreaking developments, the system worked well over moderate temperature ranges and was capable of monitoring the temperature distribution to ≈0.2°C resolution, over several hundred meters of fiber, with a distance resolution of the order of 2 m. This was a remarkable advance, considering that this was the first experimentally demonstrated temperature profile measuring method for use with optical fibers.

11.5.4 Distributed anti-Stokes Raman thermometry (DART)

If the spectral variation of backscattering from a germania-doped silica fiber is examined (Figure 11.35), it may be seen that there is a strong central line, primarily due to Rayleigh (or Rayleigh–Gans) scattering, but which also contains a weaker (spectrally unresolved in Figure 11.12) contribution from Brillouin scattering. At each side of the central line, however, there are side-lobes due to Raman scattering. These may be used to detect

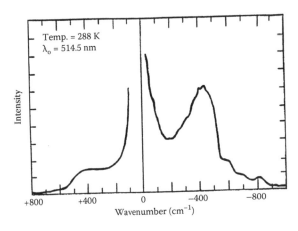

Figure 11.35 Typical spectrum of Raman backscattered light in germania-doped silica fiber (measured by N Ross, then at CERL).

temperature profiles in conventional vitreous communications fibers using modified Raman OTDR techniques [109,110].

A schematic of a basic Raman temperature sensor system is shown in Figure 11.36. The system is essentially a dual-wavelength OTDR system, where a WDM filter is used to select out the two Raman bands and direct the signals to two detectors. The elastic Rayleigh scattering of the strong incident light signal (central peak in Figure 11.35) has to be removed in the filtering process. A signal processor averages the time-varying returns, and then takes the two separate OTDR results from each detector channel, divides the two signals coming from each selected point along the fiber to determine the Raman ratio, and hence determine the temperature at each point.

From standard texts on Raman scattering, the temperature-dependent ratio, $R(\theta)$, of the anti-Stokes (higher-frequency band) and Stokes (lower frequency band) scattered intensity, at wavelengths, λ_a and λ_s, respectively, and assuming equal frequency separation from the central excitation laser line, is given by the relationship:

$$R(\theta) = \left(\lambda_a / \lambda_s\right)^4 \exp(-hc\nu/KT),$$

where h is Planck's constant, c is the velocity of light *in vacuo*, K is Boltzmann's constant, θ is the absolute temperature and ν is the frequency of the incident light.

Therefore, in addition to the distance information provided by the time delay of returning signals, a measurement of the ratio of Stokes and anti-Stokes backscattered light in a fiber can, in principle, provide an absolute indication of the temperature of the medium, irrespective of the light intensity, the launch conditions, the fiber geometry and even the composition of the fiber. In practice, however, a small correction will usually need to be made for the difference in the fiber attenuation between the Stokes and anti-Stokes wavelengths, and if convenient, it may be desired to use the dual-end measurement method discussed earlier to compensate for fiber nonuniformities.

The Raman technique appears to have only one significant practical drawback: that of a very weak return signal, the anti-Stokes Raman-scattered signal being between 20 and 30 dB weaker than the

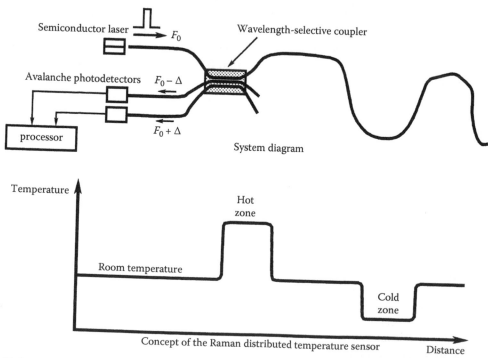

Figure 11.36 Basic arrangement of distributed temperature sensor using Raman backscatter. Range is determined by two-way time of flight of light and temperature by the calculation of the ratio of Raman Stokes and anti-Stokes signals at each range.

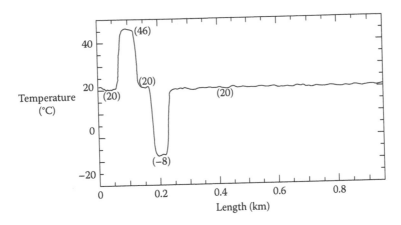

Figure 11.37 Early Raman temperature profile result, from York Ltd, United Kingdom.

Rayleigh signal, which itself is already typically 50 dB weaker than the incident light. In order to avoid an excessive signal averaging time, measurements have been taken using relatively high launched powers from pulsed lasers, and extensive signal processing is performed to average the signals. In the first experimental demonstration of the method [109], a pulsed argon–ion laser was used in conjunction with an early telecommunications-grade 50/125 µm GRIN fiber. Dakin et al. [110] described subsequent results from a considerably more compact system, based on a far more practical and convenient semiconductor laser source. A result from an early commercial prototype constructed by the group of A Hartog (previously at York Ltd) is shown in Figure 11.37. Here the result has been normalized to remove the effects of fiber attenuation, and clearly shows the hot and cold regions of fiber.

Since those early results, the Raman OTDR system has reached commercial maturity, and is another success story in the field of fiber sensors. The system has been manufactured for several years by York sensors, by the Hartog group, which is now part of the large Schlumberger company, and in Japan Hitachi cable have demonstrated engineered systems. Most systems have the capability to address multiple channels via optical fiber switches, and can operate over many km of fiber. Significant signal gains have been made using Q-switched fiber lasers to launch more light, but there are launch-power limits in monomode fibers due to the onset of stimulated Raman processes. Major applications have included fire detection, oil-well logging, chemical process plant and furnace measurements and in-line monitoring of high voltage cables and other electrical plant, although there clearly many more potential applications may become economically viable as costs become lower.

11.5.5 Time-domain fluorescence monitoring

The re-emission spectrum of most fluorescent materials generally exhibits a significant temperature variation. Thus, if an optical arrangement similar to that used in Figure 11.36 for Raman OTDR is constructed, with a laser exciting source as before, and with the detector filters now selected to examine regions of the fluorescent decay spectrum having the maximum possible differential temperature variation, a distributed temperature sensor should be possible. The potential attraction of the method, first proposed by Dakin [111], is that the fluorescent quantum efficiency may be many orders of magnitude higher than that for Raman scattering and higher doping levels may greatly enhance the signals in short distributed sensor systems. However, there remains a problem with the availability of suitable fibers.

Silica-based optical fibers, with rare-earth dopants giving high fluorescent efficiency, have been prepared [112], but sensors using these give very poor distance resolution due to the long fluorescent lifetimes. It is possible to reduce fluorescent lifetime by using materials with increased coupling to nonradiative processes, but this unfortunately also reduces the fluorescent efficiency. Polymer fibers may perhaps offer more promise in short distance systems, as these may be doped with organic dye materials with an excellent combination of high quantum efficiency (~50% or better) and fluorescent lifetimes of the order of only a few nanoseconds.

Dakin and Pratt [113] made a theoretical comparison between distributed temperature sensors based on the techniques of temperature-dependent absorption, scattering, Raman scattering and fluorescence. It was predicted that, although doping with strongly fluorescent materials will necessarily increase the loss in an optical fiber, this should, for short distance operation, be more than offset by the much higher fluorescent light levels theoretically attainable. To make such sensors in practice, it will be necessary to obtain fibers having short-lifetime fluorescent dopants showing the desired thermal variation.

Lieberman et al. [114] have reported a distributed chemical sensor for oxygen, using a fluorescent polymer cladding coating on a fiber that has its fluorescence efficiency quenched by oxygen. A low-index siloxane cladding, doped with a 9, 10-diphenylanthracene dye, was used. The evanescent field of the incident radiation coupled into the cladding dye and excited the fluorescence, and the fluorescence was coupled back, somewhat more weakly of course, by an inverse coupling process, to excite forward and backward guided modes of the fiber. Although a distributed sensor, in that continuous measurements over an extended length were taken, the method does not yet appear to have been used to determine the *variation* in oxygen concentration over the fiber length. Clearly, to do this, a dye with a short fluorescent lifetime would be necessary, and, as with Raman OTDR, significant signal averaging would probably be necessary to detect the weakly coupled signals.

11.5.6 Distributed sensors using spontaneous backscattered Brillouin light

The Raman DART system described above has evolved to reach commercial maturity relatively quickly, primarily because of its basic optical simplicity, its simple telecommunications components requirements, and because of its similarity to the already-mature conventional OTDR method. However, Raman signals are extremely weak compared to Rayleigh elastically scattered light, so relatively long electronic integration times (typically tens of seconds) are needed to achieve the necessary SNR to realize temperature resolutions of 1°C or better, particularly if it is intended to probe lengths of fiber above 5 km or so.

The total light level associated with spontaneous Brillouin scattering is typically two orders of magnitude stronger than that of Raman light, the only disadvantage being the very close frequency spacing of this light to the incident excitation line. Because Brillouin light occurs due to scattering from relatively low-energy acoustic phonons, the photon energy change is very small. An alternative simple classical physics viewpoint, which also predicts the correct optical frequency shift, is to consider that the scattering arises from thermodynamically induced moving acoustic waves in the core of the glass, so they become Doppler shifted. The resulting frequency shift in the light is therefore much smaller than that for Raman scattering, being typically only of the order of 12 GHz, so very narrow band interferometric filters are necessary to perform optical separation. Suitable narrowband filters are interferometric filters. These may be of the Fabry–Perot type, all-fiber types, based on the Mach–Zehnder or Michelson configurations, or equivalent interferometers in integrated optics form.

An alternative is to use coherent detection, heterodyning the Brillouin light with an optical local oscillator, suitably shifted from the incident laser to give a conveniently low beat frequency to process electronically. A local oscillator with a very stable frequency offset can be derived by frequency shifting a portion of light from the main excitation laser (pump laser). The signal at the detector is then a low frequency beat signal that exhibits a large fractional change for small changes in the Brillouin backscatter frequency. Once this frequency separation has been performed, a number of very useful sensor types can be constructed. The method of Kurashima et al. [115,116] allowed accurate determinations of distribution of mechanical strain in fibers from the Brillouin frequency shift. The heterodyning process makes it relatively easy to measure the Brillouin frequency shift, v_B, which is a function of the fiber core refractive index, n, the acoustic velocity, v_s, in the fiber core and the incident optical wavelength, λ:

$$v_B = n v_s \lambda^{-1}.$$

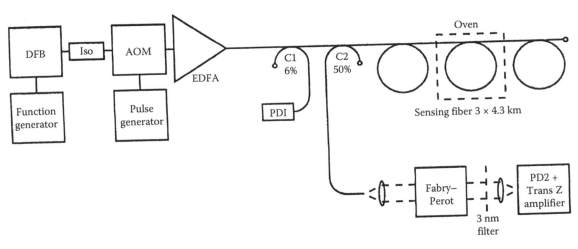

Figure 11.38 The distributed temperature sensing system of Wait and Newson, using measurement of the Landau–Placzek scattering ratio (the ratio of total Rayleigh scattered light to total Brillouin light).

ν_B is typically of the order of 12 GHz, and, according to various papers by Horiguchi et al. [117–120], varies with both temperature (typically of the order of 9.4×10^{-5} fractional change per degree kelvin) and strain (~4.6Є change, where Є is the tensile longitudinal strain).

In addition to the frequency shift, a convenient temperature-dependent intensity relationship that can be used [121] is the ratio of Rayleigh scattered intensity to total Brillouin intensity, or the Landau–Placzek ratio R_{L-P}. For a single component glass, the ratio, R_{L-P}, of Rayleigh scattered light to total Brillouin scattered light is given by

$$R_{L-P} = T_f / T(\rho_0 v^2 / \beta_T - 1),$$

where T_f is the fictive temperature of the glass, T is the absolute temperature, ρ_0 is the density, v is the acoustic velocity and β_T is the isothermal compressibility of the melt at the fictive temperature. Clearly, unlike the Raman ratio discussed before, this new ratio will depend strongly on the fiber core material properties, so renewal of a fiber by one of different type or composition is more likely to create the need for system re-calibration. However, because the SNR can be much higher in Brillouin systems, there is now scope for measurement over a much longer range. A simple arrangement for measuring temperature using the Landau–Placzek ratio method [121] is shown in Figure 11.38.

Microwave-frequency heterodyning can be used to detect the frequency shift of the Brillouin light. The frequency shift, which is generally a stronger function of strain than temperature, can be monitored, and the intensity signals can be used as a more temperature-dependent quantity, giving the possibility of distributed sensing of both strain and temperature. Maughan et al. [122] carried out simultaneous measurements of both these quantities over 30 km of fiber. It was possible to measure distributed temperature profiles over a length of 57 km. A strain resolution of 100 μЄ was achieved, and a temperature resolution of 4 K.

More methods using spontaneous Brillouin scattering with optical frequency domain processing will be discussed below. The alternative nonlinear process of stimulated Brillouin scattering will be discussed later, when systems involving interaction of counter-propagating light beams are described.

11.5.7 The optical frequency domain reflectometry technique

This distributed sensing method is essentially similar to the FMCW technique we discussed earlier as a multiplexing method. An FMCW system can be operated in OTDR-like backscattering mode in a continuous length of monomode fiber [123,124]. Now, a portion of the launched light signal, derived from the optical source with a fiber splitter, is added as a local oscillator, to coherently detect returning signals, and a beat frequency is obtained which increases in frequency as a function of the distance to the point in the fiber from which the light was backscattered. If the detected beat signal is displayed on a conventional electronic spectrum analyzer, the detected power within each small frequency interval represents the scattered light received from the section of fiber that is situated at a distance corresponding to the frequency offset observed. As the frequency slew rate of current-ramp-driven semiconductor laser diodes may be very high (100 GHz s^{-1} is easily achievable) and the frequency resolution of commercial electronic spectrum analyzers is a few Hz or less, the technique can have a far superior distance resolution capability than conventional OTDR methods. Kingsley and Davies [124] even suggested use of the technique for distributed measurements over the very small scales involved in integrated optical waveguide circuitry, where resolutions as low as a cm, or often much less, are required.

A major potential problem with optical frequency domain reflectometry (OFDR) is the coherence function of the source, which will modulate the received spectrum and therefore distort any spatial variation of scattering that it is desired to observe. Another approach to the problem uses, as with the FMCW method discussed earlier, a frequency-modulated subcarrier to amplitude modulate the source [125]. This removes the problems due to source coherence, but presents a reduced resolution due to the lower frequency slew rate possible with electronic subcarrier systems.

The FMCW technique still shows limitations when it is desired to provide high sampling rate information, or to achieve a high spatial resolution of about cm order. The method we will describe next is an improvement on this basic method.

11.5.8 Application of the "synthesis of optical coherence function" method for high resolution distributed sensing, using elastic (Rayleigh) scattering

An alternative technique to synthesize interference characteristics arbitrarily, called the "synthesis of optical coherence function," or SOCF for short, has been proposed and extensively developed by K

Hotate of RCAST, Tokyo [126–128]. In this technique, the frequency of a laser light source, connected to an unbalanced fiber interferometer, is modulated, using an appropriate electrical bias waveform, and the phase of a lightwave propagating in just one arm of the interferometer is also modulated synchronously with a similar waveform. With this method, any arbitrary shape of coherence function can be synthesized, hence setting the amplitude of the interference fringe in the interferometer to be a function of the path length difference between a lightwave returning from the remote sensing point and that returning from another defined (reference) point.

In recent years, to assist with installation and maintenance of optical fiber subscriber networks, very precise measurements of the spatial distribution of reflective discontinuities (e.g., fiber breaks or connectors) are required. For applications such as monitoring repeater stations, it is desired to measure at a distance of several km, with typically a few centimeters or less spatial resolution, a specification far too difficult to achieve with conventional time-domain methods. The standard FMCW sensing method described above, with its high spatial resolution and wide dynamic range, could be a possible candidate, but its measurement time must be made much shorter than at present possible, as otherwise the optical phase noise arising from environmental fluctuation is a problem, particularly over a long fiber.

As a more viable alternative, a solution has been developed using the synthesis of optical coherence function method [126,127,129]. Figure 11.39 shows the experimental arrangement. A switch generates a series of wide optical pulses. The optical pulse forms a window to select and determine the desired "range gate," or the boundaries of the chosen test region along the fiber under interrogation. This is because a pulse from the reference path can only overlap with a pulse from the path of the tested fiber when the two paths have nearly the same lengths. A coherence function having delta-function-like peaks can be synthesized using the FM waveform shown in Figure 11.39.

The coherence function is synthesized, in such a way that, firstly, only one coherence peak can correspond to the region under test, and, secondly, so that all other undesired periodical coherence peaks are masked by the time window. The desired peak position is then achieved by scanning, to modify the

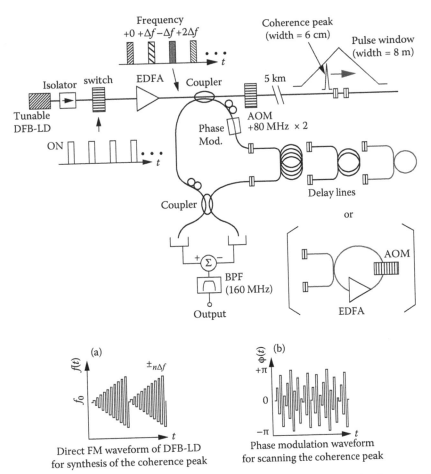

Figure 11.39 Schematic of distributed sensing using the "synthesis of optical coherence function" method conceived and developed by K. Hotate.

phase modulation waveform. Consequently, the distribution of reflectivity within the narrow selected time/range window can be measured with excellent temporal, and hence spatial resolution. In order to measure in a new range gate or window, the reference delay is changed to select, and determine, this new region. An example of the reflectivity distribution obtained by this system is shown in Figure 11.40.

Two reflections from optical connectors, of magnitude around −30 dB, can be seen clearly, whereas a strong reflection from the far end of the fiber is completely suppressed by the pulse window. Even when measuring through 5 km of optical fiber, the spatial resolution can be as good as 6 cm, using a range-gate region of 8 m. Clearly, this range gate can be moved to select and measure in any chosen region of fiber.

11.5.9 The transmissive FMCW method for disturbance location

The FMCW methods may be used to locate discrete points where mode coupling in a fiber has occurred, provided the fiber is capable of supporting two modes (e.g., different polarization modes in a high birefringence fiber) having significantly different phase velocities. External disturbances, which cause cross-coupling from the initially excited single mode, will mix or beat on a suitable detection arrangement situated at the far end of the fiber. The beat signal will have a frequency dependent on the distance from the source, at which the coupling to the second mode has taken place.

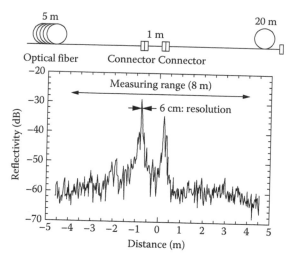

Figure 11.40 Trace showing backscatter signals from a pair of 5-km distant connectors, in a simple fiber network shown at top of figure. This was taken using the "synthesis of optical coherence function" method of K Hotate.

This approach, first suggested by Franks et al. [130], is depicted in Figure 11.41. This particular implementation used a birefringent fiber, with all the transmitted signal energy being launched into only one of the two principal polarization modes of the fiber. The disturbance to be monitored was a lateral pressure on the fiber cable, which caused coupling of light into the orthogonal polarization mode.

A convenient attribute of the technique is that the relatively close velocity matching between the polarization modes, even when using the so-called high-birefringence fiber, allows FMCW techniques to be operated over lengths very much longer than the instantaneous coherence length of the source. Two potential difficulties exist with the method, however. Firstly, mechanical strains of certain critical magnitudes may cause coupling of power from one polarization mode to the other and then completely back again, resulting in no net beat signal. Secondly, disturbances that occur exactly in the direction along a fiber polarization axis will cause no mode coupling. Otherwise, except for these somewhat unlikely conditions, the technique appears a simple and elegant method of locating the position of lateral disturbances on a continuous fiber, but is probably unsuited to measuring the magnitude of the disturbance.

11.5.10 Distributed sensing using a Sagnac loop interferometer

The use of Sagnac fiber-loop interferometers for detection of rotation has been discussed earlier. However, they can also be used to locate the longitudinal position of a time-varying disturbance acting at a noncentral location on their sensing loops [131,132].

The ability of this interferometer to locate the position of a time-varying disturbance relies on the counter-propagating nature of the light and the break in symmetry that the disturbance creates when perturbing the loop (see Figure 11.42, but please initially disregard the branching into the delay loop and to detector 2).

When a time-varying strain, $\varepsilon(t)$, acts on a Sagnac loop of optical fiber, at a distance, z, from the center of the length of fiber forming the loop,

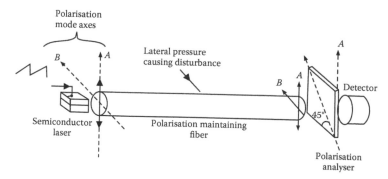

Figure 11.41 Transmissive FMCW disturbance location sensor [130]. This takes advantage of the different propagation speeds of the two principal polarization modes in a birefringent fiber, in order to provide the time delays needed for generation of the FMCW beat signals at the detector. There is no beat signal until pressure causes coupling of energy from one of these fiber modes to the other one.

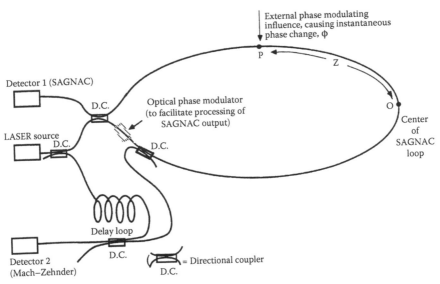

Figure 11.42 First use of a Sagnac interferometer for disturbance location. The lower part shows how a separate delay loop can be added, to form a balanced Mach–Zehnder interferometer when light taken from the Sagnac loop is added to that from the delay loop.

it perturbs the phase, $\phi(t)$ of the guided light. Due to the noncentral location of the disturbance, light travelling in one direction is phase-modulated before light traveling in the other. This results in a net phase difference, $\Delta\phi(t, z)$, between the two returning counter-propagating wave trains, when they return to the detector and interfere, at the output of the loop. It was shown by Dakin et al. [131,132] that, assuming a small slew rate, $d\phi/dt$, for the optical phase, then $\Delta\phi(t, z)$ is given by

$$\Delta\phi(t,z) \propto \frac{2z}{V_g}\frac{d\phi(t)}{dt},$$

where V_g is the group velocity of the guided light.

This, however, presents a problem, as we have a sensor response dependent on two unknowns, firstly the rate of change of the phase perturbation and, secondly, its position, relative to the center of the fiber forming the sensor loop. The initial solution, in the references above, was to separately measure the value of $d\phi/dt$. This was achieved using the delay loop in Figure 11.42, that the reader was asked to initially disregard when describing the Sagnac loop response. This additional loop enables a fraction of the light which had travelled in one direction around the Sagnac loop to be mixed with a suitably delayed portion of light derived from the same original source. This second delay loop was

chosen to be of similar length as the Sagnac loop to that point, so that a second interferometer, in this case a balanced-path fiber Mach–Zehnder could be formed, the output of which was detected to derive a measure of the actual value dependent on the phase change ϕ. Differentiation of the derived value of ϕ (or, if the phase changes are large and fast enough, simple detection of the frequency of fringe crossing by connecting the detector to a frequency counter or rate meter) yields a value for $d\phi/dt$. This new disturbance location method was applied to detect, and locate, fast thermal changes acting at different points of a 200 m length of fiber.

Following this first Sagnac disturbance location method, various new architectures using twin Sagnac configurations have been suggested, initially by Udd [133], followed by versions by Spammer et al. [134–136], Ronnekleiv et al. [137] and Fang [138]. All these avoided the need to accurately balance two independent optical paths. Such arrangements effectively used two Sagnac loops, configured to share a common fiber sensing section, but it was arranged, via several different optical routing means, that the optical path lengths of these two loops had different effective centers. These allowed two separate detector output values to be derived, now giving two response equations, and hence allowing calculation of the both of the two unknown quantities discussed earlier.

The attraction of such configurations is that now the two counterpropagating paths in each Sagnac loop have the inherent advantage of such interferometers, in that they are both intrinsically path balanced. In addition, the reference point for the center of the loop, where there is zero sensitivity, can conveniently be moved to lie outside the fiber sensing section if desired.

These new configurations were operated with either (wavelength multiplexed) twin-source configurations or with other more intrinsically lossy arrangements using directional 3 dB couplers and twin detectors. (The minimum theoretical loss of a dual-Sagnac system using 3 dB couplers is 18 dB in each Sagnac loop.) A recently reported method (Figure 11.43) by Russell and Dakin [78] has improved designs further, and for the first time allowed the use of a single source and single detector, using appropriate WDM routing components to "slice" the light from the source into two wavelength bands, and route each band of light around different Sagnac loops, again with different effective optical path center positions, and the desired shared fiber sensing section.

Describing Figure 11.43, a broadband (low coherence) Er^{3+}-doped fiber superluminescent source was spectrally sliced into two wavelength bands using wavelength-division multiplexers (WDMs). These routed the light along two, essentially independent, Sagnac interferometer loops, one for each wavelength. The first Sagnac loop was defined by the bi-directional path ABCEFHI, read clockwise around the sensor. This circuit includes a fiber delay coil, C, a sensing fiber length, E, and a piezoelectric phase modulator, F. Similarly, the second Sagnac interferometer was defined by routing light bi-directionally around a path labeled ABDEGHI. This again includes a piezoelectric modulator, D, a common sensing fiber length, E, and a fiber delay loop, G. The effective centers of each of the sensor loops are offset (in opposite directions) by half of the path length in the delay coils labeled C and G. This ensured that each Sagnac gives a different response to a common perturbation, despite sharing a common fiber section, allowing simultaneous evaluation of the position of the disturbance, its amplitude and its rate of change.

Each Sagnac was phase-biased [139] with sinusoidal strain signals of different frequencies (f_1 and f_2), each being one of a set of natural eigenfrequencies of the fiber loop. The bias frequencies (f_1 and f_2) were chosen such that the magnitude of their difference frequency $|f_1 - f_2|$ was above the frequency range of the expected disturbance signals (i.e., the base-bandwidth of the output of the sensor). This phase bias allows both of the interferometers to share a common optical detector, as it provides amplitude-modulated carriers of a different frequency for each signal generated by each Sagnac.

This system used data-acquisition hardware to sample the sensor outputs in real-time and a software system which could be instructed to either (1) locate the three largest disturbances observed in

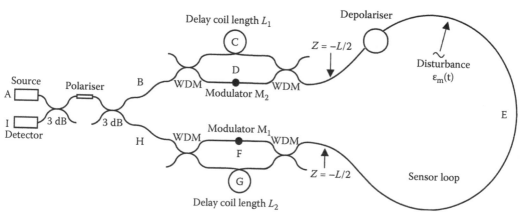

Figure 11.43 Twin-Sagnac-loop disturbance location system, using "spectral slicing" of a broadband rare-earth-doped fiber light source to obtain two wavelength channels, and WDM routing components to construct two optical loops with effectively different centers. Using modulation signals of different frequencies, at M_1 and M_2, the beat signals on one detector can be separated electronically.

the frequency domain or (2) locate a disturbance of selected character, i.e., of a specific frequency or amplitude. The positions and the frequencies of each of the disturbances were then calculated and displayed. Using the system, it was possible to monitor sinusoidally varying phase disturbances of only 0.025 radian phase change amplitude (~1.3° optical phase change, or equivalent to a few nm fiber stretching) acting on a 40 km long sensing loop and locate them with 100 m position resolution.

11.5.11 Distributed sensing using a counter-propagating optical pump pulse

If an optical signal from a steady-state, or CW source is transmitted through a fiber to a detection system, the power level received will be dependent on the total attenuation in the fiber. If, however, an intense optical pulse is now launched into the optical fiber, in the opposite direction (see Figure 11.44), the intensity of the transmitted CW lightwave will now be affected by any optical gain processes. Such effects can arise from several possible nonlinear interactions with the pump.

The first report of such a system was presented by Farries and Rogers [140]. They used a pulse, from a NdYAG-pumped dye laser at 617 nm, to provide Raman gain in a continuous length of monomode fiber. The CW beam was a 633 nm signal from a helium–neon laser source. The Raman gain is very sensitive to polarization and therefore the arrangement is capable of detection and location of lateral stresses in a fiber of low intrinsic birefringence, as these cause polarization mode conversion, hence modifying the degree of Raman gain.

The technique was a major advance, as it was the first of a class of sensors, but suffered, in its early form, from a number of practical disadvantages. As often the case for a first laboratory form, it used rather large, and hence inconvenient, laser sources, and it is likely the results would have been critically dependent on the pump power level of the dye laser source. In addition, it would also be likely to suffer badly from undesirable polarization and other variations due to environmental factors such as bends in the fiber.

More recently, sensors using stimulated Brillouin scattering have been reported, one of the most significant being the Brillouin optical time-domain analysis (BOTDA) system devised by Horiguchi [117,118], which has since formed the basis of a number of systems for sensing strain in very long fibers, using more complex versions of the simple arrangement of Figure 11.44. This sensor takes advantage of gain from the stimulated Brillouin scattering process. As with the spontaneous Brillouin scattering process mentioned earlier, the center line of the Brillouin gain curve is offset from the pump laser signal. Using a separate tunable laser (or more conveniently by deriving a

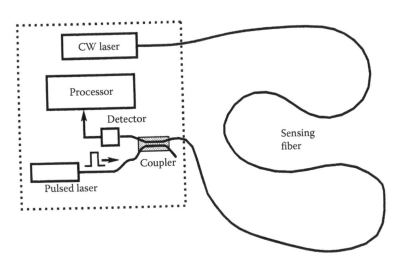

Figure 11.44 General concept of a distributed sensor using counterpropagating optical waves. The high intensity pump signal causes nonlinear optical gain processes to occur, which changes the intensity of the CW probe signal with time.

frequency-shifted signal from the one laser source) pump and probe beams can be arranged to be suitably offset in frequency, to not only ensure the desired Brillouin gain occurs, but also, by frequency sweeping, actually measure the frequency of peak gain. Taking advantage of the time-dependent nature of the signals when the pump is a short pulse, the distribution of Brillouin gain shift versus distance can be derived by a combination of sweeping the laser frequency offset and observing the temporal patterns.

11.5.12 Use of the "synthesis of the optical coherence function" method for distributed sensing by monitoring the gain process associated with stimulated Brillouin scattering (K Hotate, RCAST, University of Tokyo)

Distributed strain sensing based on Brillouin scattering, as discussed earlier, is a promising technique for "smart materials" and "smart structures" applications, and sensor systems using Brillouin OTDR are already on the market. However, the technique developed so far has a spatial resolution limit of several meters. In smart structures applications,

for example, in an aircraft's wing, this limit would be a major disadvantage. The resolution limit of the conventional technique is a consequence of its pulse-based nature. An optical pulse for generating Brillouin gain has to be longer than the damping time of the acoustic wave. With shorter pulses, the Brillouin gain spectrum (BGS) broadens out, making it more difficult to determine precisely the frequency of the spectral peak. Because the typical value of the damping time is 25 ns, the practical limit for spatial resolution turns out to be typically about 1 m.

To circumvent the resolution limit, a new technique has been developed [128]. It is based on the control of the interference between the pump and probe lightwaves that excite stimulated Brillouin scattering (SBS). SBS requires interference between two counter-propagating lightwaves. In the technique, we control their coherence, so as to localize the SBS at a specific position in an optical fiber, where their correlation is high.

Figure 11.45 shows the proposed system for measuring the distribution of the BGS along an optical fiber. The light from a 1.55 μm frequency tunable distributed-feedback laser diode (DFB-LD) is split by a coupler, to provide light sources for the pump and probe lightwaves. Light from one output of the coupler is intensity modulated (chopped) at a radio frequency, using a LiNbO₃ electro-optic modulator

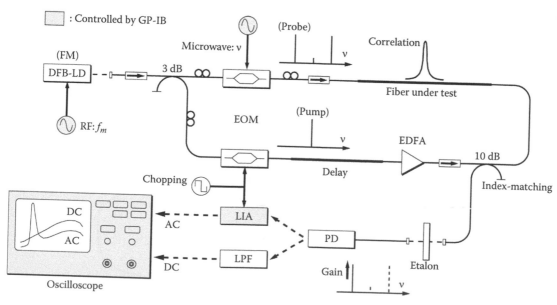

Figure 11.45 Schematic of system for high resolution sensing, using stimulated Brillouin backscatter in conjunction with the "synthesis of optical coherence function" method of K Hotate.

(EOM), with an electrical drive signal. The output is amplified by an erbium-doped-fiber amplifier, and then launched into the fiber under test to serve as the optical pump (this is shifted in frequency by only a very small amount by its intensity modulation). The other output is intensity modulated, at a much higher microwave frequency, v, by a second LiNbO$_3$ modulator, so that AM sidebands are generated around the incident lightwave of a frequency v_0. The lower sideband, at $v_0 - v$, is used to serve as the probe beam and the unwanted residual sideband and carrier are eliminated by an optical filter. This probe beam propagates in the fiber under test, in the opposite direction to the pump, until it reaches the detector.

An important point is that the pump and the probe undergo identical frequency modulation, as they both arose from the laser diode source, and the system takes advantage of its capability to be frequency modulated directly, by changing its bias current. As a result, SBS occurs exclusively at the position of peak correlation, the only point where the two lightwaves are well correlated. The correlation peak can be conveniently shifted along the fiber, simply by changing the FM frequency f_m of the laser diode source. The increase in the probe power, resulting from Brillouin gain in the fiber, is detected synchronously using a lock-in amplifier. This was referenced using the electrical drive signal used to chop the pump intensity. We obtain the BGS by varying the frequency of the microwave signal used to generate the frequency shift in the probe signal. By repeating the BGS measurement over the appropriate range of different frequency drives, f_m, to the FM, the BGS is obtained as a function of position along the fiber.

Application of the correlation-based Brillouin sensor has been investigated [128] for the measurement of strain distribution in small-scale (few cm) material samples. The sample material, chosen to demonstrate this in the laboratory, was a cylindrical acrylate-ring coil former, which could be stressed by side pressure to deform it. A single sensing coil of dispersion-shifted fiber was first wound around it, then bonded to it with epoxy cement (Figure 11.46a). The coil former could be deformed to an approximately oval cross-section to test the sensing system.

The frequency shift of the modulator was set to 3.2 GHz, corresponding to a 1 cm spatial resolution

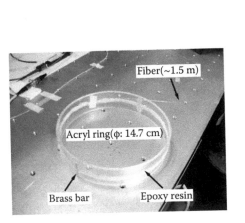

(a)

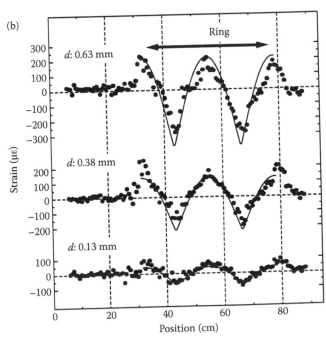

(b)

Figure 11.46 (a) Photograph of an acrylate ring sample with a single turn of fiber bonded to it. (b) The measured strain patterns using stimulated Brillouin backscatter in conjunction with the "synthesis of optical coherence function" method.

and the strain distributions for various levels, d, of ring deformations were measured. Figure 11.46b shows that, for small ring deformations, the experimental results agreed well with the theoretical value (solid lines), but deviation from the theoretically expected value became greater at high deformation values. This is because the spatial resolution became insufficient to accurately trace out the step strain gradient at high deformation values. In this scheme [128] a reasonably high measurement sampling rate, of several tens of Hz, was also been realized.

11.5.13 Distributed sensing to determine profiles of strain and temperature in long in-fiber Bragg gratings

As a final note on distributed sensing, it should be noted that there are several methods capable of resolving the effective peak reflective wavelength along a fiber Bragg grating, as a continuous function of position. Methods to produce very long fiber Bragg gratings have been discovered some years ago [141] and gratings of up to 1 m in length are now almost routinely manufactured, primarily intended for dispersion compensation in optical telecommunications systems. Methods to determine the variation of the peak reflective wavelength along the length of such gratings are available, making them highly attractive starting components for distributed sensors to measure in-fiber strain and/or temperature. Very high spatial resolution (typically 0.1 mm) over distances up to 1m is possible, allowing of the order of 10^4 sensing elements! The methods devised to address such gratings are now quite numerous and some are fairly complex, so only a brief outline will be given here.

Early methods made assumptions about the nature of the index modulation in the grating, in order to derive optical delay (two-way time of flight) information from the wavelength components reflected [142] or to get wavelength from the delay [143]. Various means, such as tunable lasers have been used to determine the wavelength information, and interferometric phase determination has been used to detect the time delay.

One attractive method of defining, more precisely, the distance to the measurement point in the grating is to use low coherence interferometry. This operates in a similar way to that of coherence multiplexing and to the distributed sensing methods described above. Volanthen et al. [144,145] used this method, where the long Bragg grating to be interrogated (initially a chirped one, 50 mm long was used) was in one arm of a Michelson interferometer, while the other arm contained a fixed-wavelength much-shorter Bragg grating. The effective distance to the grating in this second fiber arm could be changed in length, using a fiber stretcher, to determine the position of good fringe visibility in the long Bragg grating in the other arm. The attraction of this method was that no prior knowledge of the grating wavelength versus distance profile was needed, and, unlike the earlier ones, the method could work with gratings that did not have a monotonic variation of wavelength with distance.

Some recent systems have used commercial coherence domain reflectometers, which conveniently contain an in-built swept Michelson interferometer, to determine the point in the long grating [146,147]. This grating can then be monitored at any point, at any one moment in time, with an external acousto-optic tunable filter, which allows the scanning of interrogation wavelength. In these systems, an external optical amplifier was used to compensate for fiber system losses by boosting the signal strength from the ELED source in the instrument. Such sensors may have useful applications for monitoring strain distributions in complex mechanical components during factory testing, but are likely to remain rather costly and complex for many vehicular applications, such as aerospace monitoring.

In vol III chapters 2,7,20, the more like recent technologies and applications for distributed optical fiber sensors will be expanded or showing the great potential for these technologies.

In the final section of this chapter, we discuss the development of applications of optical coherence tomography (OCT) and low coherence interferometry in both medical and industrial applications.

11.6 APPLICATIONS OF OCT AND LOW-COHERENCE INTERFEROMETRY

As discussed in the OCT section of the fiber optic sensors for medical applications chapter, OCT has become an established medical imaging technique. By far its greatest commercial success has been in the field of ophthalmology. Today it is widely used, for example, to obtain high-resolution images of both the anterior and posterior segments of the eye as well as measuring the depth of eye. Today OCT is being used as a test for glaucoma and to monitor its progression [148] and to provide detailed examination of the retina, which can, for example, provide a straightforward method of assessing macular degeneration [149], diabetic macula edema [150] and axonal integrity in multiple sclerosis, [151].

Besides ophthalmology, OCT has been clinically demonstrated in a variety of other medical and surgical specialties, including gastroenterology, dermatology, cardiology, oncology and dentistry among others. Today OCT is being used throughout the world to assess coronary artery disease and is being used to image and assess arterial plaque, stents and scaffolds [152]. Quantitative screening for skin cancer is starting to occur using polarization sensitive OCT [153]. Recently OCT has been used intraoperatively during breast surgery to characterize the lymph nodes and to locate tumor margins [154]. OCT is also being investigated to assess the progress of dental caries and other structural changes in structure of teeth by observing changes in the scattering properties of the teeth over time [155]. Figure 11.47a shows a camera image of a sectioned tooth having dental caries in the region outlined by the black box and Figure 11.47b shows the corresponding OCT false color image obtained by scanning across the top of the tooth. The white region in Figure 11.1b corresponds to the decay region in Figure 11.47a. Dental caries is shown to increase the scattering of a tooth and to penetrate to a much greater depth than in a normal tooth.

OCT and low coherence interferometry are becoming well accepted as a nondestructive testing tool in a diverse set of industrial applications including tubing inner diameter, outer diameter and wall thickness measurements, single and multilayer material thickness and index of refraction measurements, medical device inspection including balloons and catheters [156] and in the field of ophthalmic metrology to characterize contact and intraocular lenses [157,158]. Materials can range from semiconductor wafers, glasses and polymers. Fiber-optic probe-based OCT and low-coherence interferometry (LCI) systems are readily adaptable to industrial environments and can be made to access and scan interiors of hard to reach surface and can operate in a number of hostile environments including radioactive, cryogenic or very hot. They are being used for both in-line and off-line process control applications. LCI with high speed data acquisition, analysis and submicron resolution with process feedback is being used to control manufacturing processes. LCI systems are also being used to determine stack layer thickness for cell phone displays and other large LCD and OLED displays [159] and have been used to evaluate the layer structure and wedge angle distribution in head-up automotive windshields [160,161]. The following section shows details of a high resolution low coherence interferometer that is used in process monitoring, inspection, and quality control applications in various fields.

11.6.1 Instrument design

The sensors described here are all variants of an interferometric sensor, based on a dual Michelson arrangement, having sub-micron distance accuracy for measuring optical distance. Several versions have been developed, for use in a variety of in-factory quality control and process monitoring applications. The basic system [162,163] includes both coherent and low-coherence light sources and employs the well-known optical configuration of the optical autocorrelator. Light from the low-coherence light source is guided to the sample under test via a single-mode fiber. Light beam components, which are partially reflected from each of the optical interfaces in the sample, mix in the interferometer section, to give a fringe pattern that is dependent on the type of source used and on the various optical path differences. The detected response, as the interrogating interferometer is scanned in length, is used to determine sample-dependent parameters, such as optical distances. When reflection occurs from input and exit surfaces, the optical thickness of sample can also be measured.

The instrument described is a dual Michelson arrangement in an autocorrelation mode as described in references [163–165] and a schematic of the instrument is shown in Figure 11.48. With reference to Figure 11.48, low coherence light from a superluminescent light emitting diode (SLED), having a center wavelength of ~1310 nm and a bandwidth of ~50 nm, passes through Port 1 and Port 2 of an optical circulator into a sample fiber. The light from the sample fiber passes through an

(a)

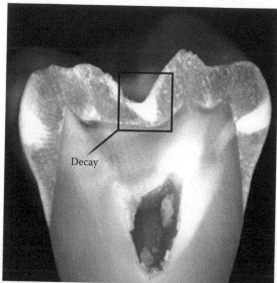

(b)

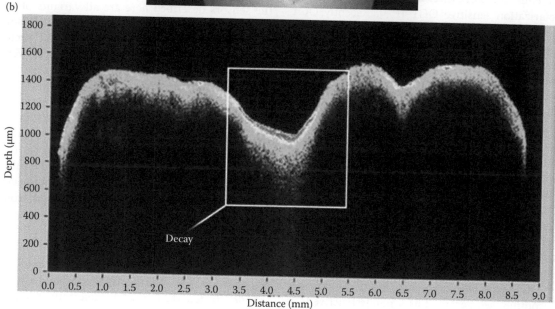

Figure 11.47 Images of a sectioned tooth having dental caries. (a) Camera image of the tooth, and (b) OCT image obtained by scanning across the top of the tooth.

optical probe and is focused onto the sample under test. Light reflecting off of each optical interfaces of the sample passes back through the optical probe, back through the sample fiber, back through Port 2, and passes through Port 3 of the circulator into an all-fiber dual Michelson interferometer. Before entering the interferometer, the light from a 1552 nm laser diode is combined with the low-coherence light that was reflected back from the sample via wavelength division multiplexer WDM

1. The combined laser and low coherence light passes through a 50/50 fiber coupler and is split into two beams. Each beam passes through a PZT fiber stretcher, and is reflected back through the fiber stretcher by a FRM.

During operation, voltage waveforms are applied to the pair of PZT fiber stretchers, 180° out of phase with each other. The voltage alternately changes the path lengths of the pair of fiber stretchers in a push-pull configuration. The 2

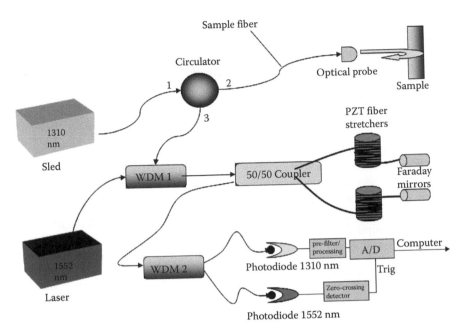

Figure 11.48 Schematic of a high resolution dual low coherence interferometer operating in the auto-correlation mode.

beams of light at each wavelength, reflected from the Faraday Mirrors, interfere with each other as they are recombined at the 50/50 coupler. The combined interfering reflected light is passed through a second wavelength division multiplexer (WDM 2) where it is separated into individual interfering beams, at 1552 and 1310 nm, respectively. The beams are then sent to respective photodiodes, which convert the interference signals to voltage levels as a function of time.

The 1310 nm interfering light passes through a 1310 nm photodiode and is preprocessed before A-D conversion. It is common practice to add balanced detectors and log amplifiers to increase the sensitivity of the measurement. The 1552 laser interferometer signal is sent to a zero-crossing detector, which is used to provide a uniform distance scale clock for acquiring the low coherence 1310 nm interference signal. The laser clock signal is used to trigger the A–D conversion of the low coherence interference signal, ensuring that data is collected at constant distance intervals. Further details of operation and performance of the interferometer can be found in Reference [165].

Application-specific components of the system include the optical probes, optical switches, sample and probe transport mechanisms and the peak processing algorithms [166,167] used to provide sample information and user interfaces. Example applications include measurement of thickness, and thickness profiles, of polymer films [168], measurement of optical retardation in films [169], measurement of liquid thickness distributions on coating hoppers [163], the length calibration of optical cells, assessment of the focus of digital camera imagers [170,171], surface profile measurements of films, wafers and imagers, head-up display (HUD) windshield assessment [160,161] and cell phone display integrity [159]. There are clearly many other potential uses of this versatile telemetry method.

For the low-coherence light source, interference with visible fringes only occurs when the path lengths of the two arms in the interferometer are equal to within a few coherence lengths. In order for any interference to occur, light must of course be reflected back into the interferometer from the sample. This will occur due to Fresnel reflection at each optical interface in the sample. The distance between adjacent interference peaks is a measure of the optical thickness (group index of refraction, n, times the true physical thickness) of the sample material. In air layers, the distance between the two adjacent surfaces is approximately the thickness of the layer, as n_{air} is very close to unity. Because the instrument uses a stabilized laser light source to provide constant distance interval measurements, the instrument measures *optical* path distance, defined, in regions of media other than vacuum, as n times the physical thickness.

As mentioned above, the measurement configuration of the interferometer is that of the optical *autocorrelation* mode, in which light reflecting from the sample is input to both arms of the Michelson interferometer. In the autocorrelation mode, light beams reflecting from the sample are made to interfere in a way that both arms of the interferometer see reflections from all of the optical interfaces in the sample (as an example the front and back surfaces of a film). As the path lengths of the coiled fibers in the interferometer are changed, a series of visible clusters of interference peaks are observed, indicating the optical path differences between adjacent optical interfaces. The zero-path-difference, or self-correlation condition occurs at the point when the two path lengths of the Michelson interferometer become equal; in which case, all optical interfaces in the sample give strong visible interference fringes. The measured distance between the largest peak, corresponding to this zero path length difference, and the first set of adjacent peaks, is a measure of the shortest optical path difference in the sample. One major advantage of using the autocorrelation configuration is that the interferometer can be located at any distance away from the sample interface, provided the fiber losses permit it. A system with the measurement head located as far as 10 km from the interferometer has been demonstrated. It is also a common practice to have a reference reflection built into the probe so that the order of peaks in the observed interferogram are in the same order as the optical interfaces in the sample.

The system processor uses a peak-location analysis technique, to find the true center of the envelope of a cluster of interference fringes. This envelope is, to a first approximation, a cosine function, having a Gaussian intensity-modulation envelope. Various peak location algorithms to find the true location of the interferogram peaks have been developed [171] including moment calculations, Gaussian-peak analysis and Fourier-phase-slope analysis [167]. When using a sampling distance interval of $\lambda/4$, they provide measurement repeatability better than 10 nm, with samples having sufficient separation between adjacent peaks. Once

the peak locations are calculated, the appropriate distances relevant to the measurements being performed must be computed. Distance calculations are noted and compared to defined acceptance ranges and thresholds.

We now review a sampling of applications of current interest to a variety of diverse fields.

11.6.2 Center thickness, sagittal depth, and refractive index measurements of contact lenses in solution

The refractive index of a lens is an important parameter in lens design. When designing a lens to have a specific refractive power, the index of refraction of the material affects the thickness and curvature of the lens. Lens thickness is an extremely important parameter in the design of contact lenses and intraocular lenses. In soft contact lenses the index of refraction also changes as a function of hydration and it is important to measure them in their hydrated form. Recently LCI and OCT have been applied to the measurement of the index of refraction of soft contact lenses in solution.

The LCI measurement geometry is shown in Figure 11.49 for a contact lens mounted in a cuvette filled with saline solution. The optical probe is located above the cuvette and the measurement location is centered on the lens. The physical distance from the inner wall of the cuvette top to the inner wall of the cuvette bottom is d_o. The saline solution has an index of refraction n_s. The LCI measurements are performed with the lens being centered over the optical probe of the LCI. The lens has physical thickness t_1 and the optical thickness $n_1 t_1$ is measured where n_1 is the refractive index of the lens. The physical sagittal depth or sag of the lens is defined as S and the optical distance $n_s S$ is measured. The top gap optical distance $n_s G$ is also measured during this step. Once the measurements are performed as shown in Figure 11.3, the lens can be removed from the cuvette without moving the cuvette so that $n_s d_o$ can be measured at the same location. The solution is then removed from the cuvette without moving the cuvette to measure $n_a d_o$ at the same location and calculating d_o using the known refractive index of air. Once d_o is measured then n_s can be calculated along with S and G. The

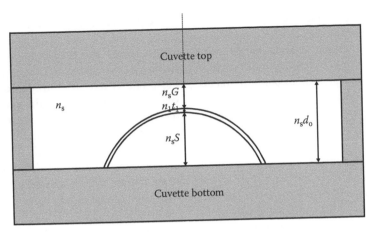

Figure 11.49 Geometry for measurement of a contact lens mounted in a cuvette containing saline solution and the LCI parameters being measured.

thickness of the lens t_1 is then calculated from the relationship $d_o - S - G$ and the index of refraction of the lens can then be calculated from the ratio of the measured optical thickness of the lens to the calculated thickness.

Figure 11.50 shows a sample LCI scan of the lens shown in Figure 11.49 measured at the center of the lens. Further details of the measurement procedure to measure index of refraction and center thickness are described in reference [157] along with the appropriate relationships. The developed procedure allows us to obtain measurement repeatability for group refractive index less than 1×10^{-3} for materials with thicknesses on the order of 100 μm, when measured in liquid. The measurement repeatability further improves for measurements in air, or for thicker materials. Combining this measurement with a digital camera to determine the diameter of the lens allows one to also measure the base curve of the lens [158].

11.6.3 Automotive HUD windshield wedge angle and thickness assessment

Automotive windshields are comprised of laminated safety glass made from two layers of glass that are bonded together with one or more layers of polyvinyl butyral (PVB). Most windshields are not simple flat structures, but include slight curvatures in both the horizontal and vertical dimensions. Many windshields today include an acoustic layer to help eliminate outside noise. HUD technology has recently become popular in automobiles as it creates inherent driver safety advantages by displaying critical information directly in the driver's line of sight, reducing eyes off road and accommodation time. This is accomplished using a system of relay optics and windshield reflection to generate a virtual image that appears to hover over the hood near the bumper. The

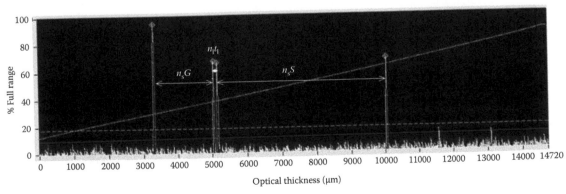

Figure 11.50 LCI scan signal of the lens shown in Figure 11.49 measured at the axial location shown in the figure.

windshield is an integral optical component of the HUD system, but in a uniform thickness windshield, the windshield-air interface causes a double image ghost effect as a result of refractive index change, reducing HUD image clarity. It is common practice to add a small constant wedge angle to totally eliminate the ghosting effect. However, this only works for the standard height driver and the appropriate wedge angle should change as a function of height of the driver due to changes in viewing angles between different drivers [161].

HUD windshields can be scanned over the surface of the windshield with a handheld optical probe [160] attached to the autocorrelator instrument described above an example of which is shown in Figure 11.51. Figure 11.51a shows how the handheld fixture is used to measure the windshield of a 2007 Chevrolet Corvette, equipped with a factory installed HUD system. The yellow line is the single

mode optical fiber cable that connects the optical probe to the low coherence interferometer. Figure 11.51b shows the thickness profile of each layer within the windshield. During the measurement, the probe is moved at an approximately constant velocity for the bottom to the top of the windshield, therefore the distance scale in the graph is approximate. It is observed that the PVB layer gets thicker as the probe moves from the bottom to the top of the windshield while the thickness of the glass layers remains constant. The change in the thickness of the PVB layer indicates presence of the wedge angle as required for HUD.

Figure 11.52 shows an interferometer single depth scan at a single measurement location of a five-layer acoustic windshield. The horizontal axis corresponds to the travel of the reference mirror of the interferometer. The peaks in the signal correspond to the interferometric signal registered by

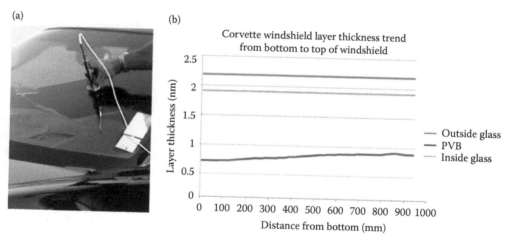

Figure 11.51 (a) Image of a 2007 Corvette with a windshield scan in progress using a hand-held probe. (b) Thickness profile of each layer as a function of distance from the bottom of the windshield.

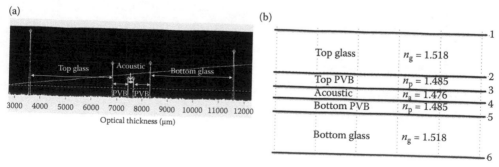

Figure 11.52 (a) Interferometer point scan for a five-layer acoustic windshield. (b) Windshield structure along with index of refraction of each of the materials.

the detector as a function of interferometer scan depth. The signal has been demodulated and compressed before it is plotted. The distance between the adjacent peaks is equal to the optical thickness of the corresponding layer. The five-layer windshield contains an acoustic layer and an additional PVB layer, as compared to the three-layer windshield shown in Figure 11.51. Figure 11.52b shows the cross-section of the windshield layers along with their corresponding refractive indices. The interfaces are numbered, with the number 1 interface corresponding to the left-most signal peak on the interferometric graph of Figure 11.52a, while peak 6 corresponds to the right-most signal peak on Figure 11.52a.

Figure 11.53 shows layer thickness distribution of a five-layer windshield having a constant wedge angle over a portion of the windshield, located between 225 and 525 mm from the bottom edge, where the thickness of the PVB and acoustic layers change linearly. This region of the windshield encompasses the viewing location of the HUD image for the nominal driver. During measurement the probe fixture was moved at a rate of 5 mm/s, resulting in 0.1 mm spacing between successive data points. The wedge angle of the combined acoustic and PVB layers was measured to be 0.588 mrad, as calculated from the slope of the best fit line for all the data points. The standard deviation of the wedge angle was calculated from the local slopes using sets of five adjacent locations to be 0.011 mrad.

11.6.4 Measurement of cell phone displays

The most common touchscreen technology used in cell phones is capacitive, an example of the layer structure is shown in Figure 11.54a. It is based on detecting the change in the capacitance between two arrays of electrodes, when the human finger approaches the surface of the screen. The touch module consists of two layers of electrodes encompassing an insulating layer (e.g., layer of glass), which acts as a spacer for the capacitor. The module is placed on top of an LCD or OLED stack, and the overall system is protected by a cover glass (e.g., Gorilla glass). In order to attach different components of the touchscreen together, manufacturers use a layer of adhesive. Unlike the thickness uniformity of the glass substrates, the uniformity of the adhesive layer is difficult to control. In the past, the nonuniformity of the layers was acceptable to touch-screen manufacturers. However, as the technology of touchscreens moves ahead at a high pace, manufacturers are experiencing more and more pressure to ensure that the layer stacks are uniform in order to improve image quality as well as the longevity and durability of their touchscreens [159].

Figure 11.54b shows the interferometric signal acquired for a smartphone touchscreen. The multitude of peaks indicate numerous layers present under the surface—one can see the top cover glass, the substrates containing indium tin oxide (ITO) electrodes, the spacer between the electrodes, and

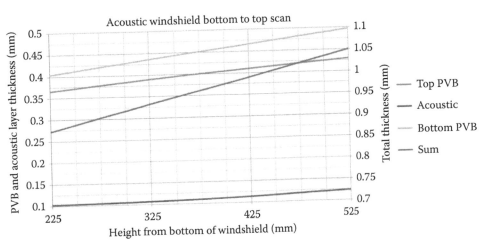

Figure 11.53 Acoustic windshield quantitative partial scan of wedged layers and their total thickness as a function of height from the bottom of the windshield.

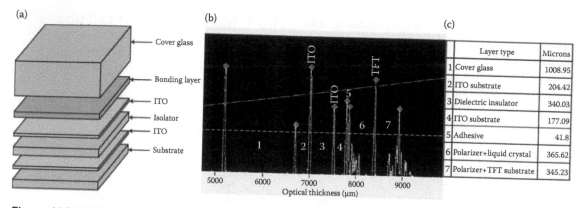

Figure 11.54 (a) Example touchscreen layer structure. (b) Interferometer point thickness scan of a touchscreen. (c) Layers observed in the structure shown in the interferometer trace of (b) and their respective measured thicknesses.

the adhesive layer between the touch module and the display. Below the adhesive layer (toward the right portion of the graph) are the polarizers and the layer containing thin film transistors (TFT). Some of the layers, such as the ITO and TFT, are much thinner than what the low-coherence interferometer is able to measure. In this case, the two interfaces of such thin layer appear as a single peak, and the measured thickness of the adjacent layers can therefore be slightly larger than the actual material. The numbered thickness regions are defined in Figure 11.54b along with their measured thicknesses. The same instrument can be adapted to generate a visual representation of cross-sectional images during a scan along a line of a screen stack as described with respect to Figure 11.55.

Figure 11.55 shows the cross-sectional OCT images obtained for different model smartphones including Samsung Galaxy S3, S4 and S5, as well as two different generations of ipad. These cross-sections were obtained while moving the optical probe in the middle of a screen, along the long dimension. The devices were lying flat, with the screens facing upward. The graphs match this orientation. The lines represent the interfaces between the inner layers, with the top line corresponding to the outer surface of the cover glass. The brightness of the lines corresponds to the reflectivity of the interfaces—the brighter the line, the more reflective is the surface. One can immediately spot two principal differences between the scans. First, the scans for ipads clearly show inconsistent adhesive layer thickness. Second, the thickness of the cover glass becomes thinner for more recent versions of the devices. It is consistent with the manufacturing trends for the display cover glass. Note,

that because the ipad cross-sections were mapped by assuming that the top-most surface is flat, the layers beyond the adhesive layer appear to be deformed due the adhesive layer inconsistency. However, most likely both the touchscreen and the image display layers experience some sort of the deformation. The cross-sectional image can be reacquired by mapping the top surface with respect to a known flat surface in order to map out the deformations exactly.

11.6.5 Imager flatness assessment and simultaneous thickness, surface and index profiling

Surface profiling is an important aspect of many industrial processes. We describe here its application to the measurement of flatness of large imagers and wafers after being glued into a package. Here, a thick optical flat is placed above an imager that has been glued into a package or a large wafer. It is coupled to an XY scanning frame as shown in Figure 11.56a. The bottom surface of the optical flat becomes the reference surface in order to determine the flatness of the imager. Figure 11.56b shows the surface profile results, at a measurement rate of 200 Hz, with the instrument set to take y-axis steps of 0.25 mm, as it is scanned at a rate of 50 mm s21 along the x-axis. The data are shown relative to the best reference plane and are inverted. It can be seen that this imager has a slightly convex surface, being bowed upwards by about 6 µm in the center. Profiles can also be obtained before, during and after curing of the imager into the package.

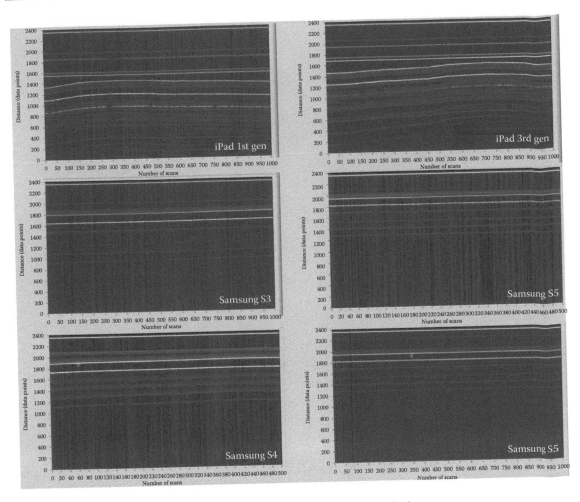

Figure 11.55 Cross-sectional images of line scans across different cell phones.

We now discuss how we have extended the concept, to perform very rapid profiling of several simultaneous properties of films, as they pass rapidly through the measurement cell shown in Figure 11.57. This optical cell design is used for simultaneous thickness, surface and (refractive) index profiling (SSTIP) of polymer films, glasses and semiconductor wafer materials [164]. In the cell shown in Figure 11.57 the sample is placed between a pair of thick optical flats and can be held flat. An XY scanning frame is located above the optical probe and the distance between the stationary optical flats as a function of position d_o can be measured and assumed to be invariant. Variations in the top gap distance d_2 between the top glass optical flat and the sample as a function of position defines the top surface profile of the film sample. Likewise, variations in the bottom

gap distance d_1 between the bottom glass optical flat and the sample as a function of position defines the bottom surface profile of the film sample. The optical thickness of the film nt is also measured as a function of position. As in the contact lens case the thickness of the film as a function of position is calculated as $t = d_o - d_1 - d_2$ as a function of position and the film index of refraction is calculated as the measured optical thickness divided by the thickness calculated at each measured location.

11.6.6 On-line coating thickness measurements

LCI and OCT can be used to assess coating hopper processes. They can be used to assess coating uniformity, and coating dynamics. They have been utilized to assess the time it takes a production

(a) (b)

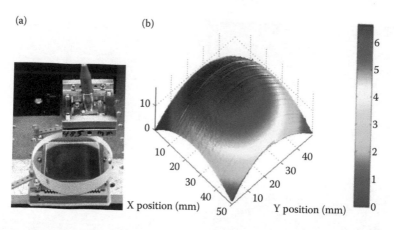

Figure 11.56 (a) Photograph of a testing arrangement for testing imager flatness. (b) 3-D map of imager surface for the imager shown in (a).

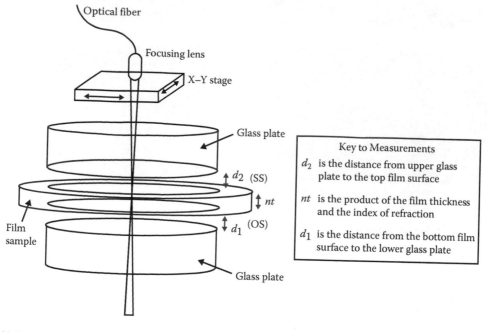

Figure 11.57 Measurement arrangement to simultaneously measure surface profiles, film thickness and refractive index of film materials.

process to settle down and reach steady state after the production fluids are first introduced into the coating stations. Figure 11.58 shows a photograph of an eight adjustable optical probe mount fixture with two optical probes mounted on the probe mounts and installed on a production multilayer coating station. During operation the line of optical probes can be scanned across the width of the hopper to assess uniformity of all of the layers simultaneously.

Figure 11.59 shows data from a hopper purging experiment in which water was replaced with a coating solution. It is important to determine when the coating solution stabilizes to its desired final concentration and flow rate. The optical thickness of the coating solution flowing down each of the slots is measured as a function of time and the stabilized thickness is noted. In no cases did it take more than 10 L of flow to reach the equilibrium thickness. Thus the 10-L point was used as the normalization

Figure 11.58 Photograph showing multiple probes for on-line assessment of liquid layer thickness during operation of a coating plant.

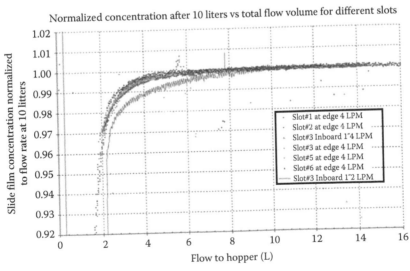

Figure 11.59 Graphic results of a hopper purging study.

point for the data. The plots show the normalized measured thickness or concentration as a function of total flow through the slots. Most of the slots and runs show very similar trends. When we lowered the flow rate from 4 to 2 L/min the fluid took longer to stabilize at its final equilibrium point.

This section has provided an overview of some industrial applications for OCT and LCI.

11.7 CONCLUSIONS

This chapter has given an overview of the field of fiber sensors, starting with simple intensity types, then more complex interferometric ones. Designs for multiplexed and fully distributed sensors have been dealt with in considerable depth. The chapter has concluded with an overview of various applications for low coherence interferometry and OCT which are being used in industrial process monitoring and control applications. Owing to the breadth of the subject, even this large chapter could not cover many concepts, so the authors apologize to any researchers if their work has been omitted. A comprehensive bibliography is given below in order to allow the reader to glean further details of this subject.

It has been shown that there are very many ways of sensing using optical fibers. Many sensors reported in scientific papers will retain only academic interest, but a few technologies are already making significant inroads into industrial, medical and military applications. The general trend, as with any relatively new technology, is first to fill niche markets, before costs can eventually be reduced to meet mass markets. Although more complex in nature, the use of techniques for achieving multiplexed and distributed sensors is increasing rapidly. It is difficult to generalize on preferred approaches, but the TDM and WDM methods of multiplexing, and OTDR-type methods (including Raman, Brillouin and fluorescent variants) for distributed sensing have many advantages. The achievement of more sensors per monitoring station, and the correspondingly greater ease of making comparisons of the measurand value at each sensor head, is a factor likely to increase the practical use of optical sensor systems. The major area requiring evolution of technology is the extension of these methods to address (and separate effectively!) several different physical parameters simultaneously on a common optical highway in an economic manner.

It will be interesting to observe which methods endure the passage of time to be developed into cost-effective and reliable system instrumentation. The factors in favor of optical sensor highways are the ever-decreasing cost of fiber cable and the dramatic improvements of cost and performance of optical components.

REFERENCES

1. Brabant, J. M., Moyer, B. J. and Wallace, R. 1957. Lead glass Cerenkov radiation photon spectrometer. Rev. Sci. Instrum. 28: 421–424.
2. England, J. B. A. 1976. Detection of ionizing radiations. J. Phys. E (Sci. Instrum.) 9: 233–251.
3. Fenyves, E. J. (conference chair). 1994. Scintillating fibre technology and applications II. Proc. SPIE 2281, San Diego, USA. (several relevant papers in these proceedings).
4. Fields, J. N. 1979. Attenuation of a parabolic index fibre with periodic bends. Appl. Phys. Lett. 36: 779–801.
5. Fields, J. N. and Cole, J. H. 1980. Fibre microbend acoustic sensor. Appl. Opt. 9: 3265–3267.
6. Bruinsma, A. J. A. and Jongeling, T. J. M. 1989. Some other applications for fibre optic sensors. Optical Fibre Sensors, vol II, eds J. P. Dakin and B. Culshaw (Boston, MA: Artech House) chapter 19, pp. 721–765.
7. Berthold, J. W. and Reed, S. E. 1979. Microbend fibre-optic strain gauge. US Patent No. 5020379.
8. Harmer, A. L. 1981. UK Patent 1584173.t
9. Harmer, A. L. 1986. Fiber-optical pressure detector. US Patent No. 4618764, October 21.
10. Michie, W. C., Culshaw, B., McKenzie, I., Moran, C., Graham, N. B., Santos, F., Gardiner, P. T., Berqqvist, E. and Carlstrom, B. 1994. A fibre optic hydrogel probe for distributed chemical measurements. Proc. SPIE 2360: 130–133.
11. Michie, W. C., Culshaw, B., McKenziei, I., Konstantakis, M., Graham, N. B., Moran, C., Santos, F., Bergqvist, B. and Carlstrom, B. 1995. Distributed sensor for water and pH measurement using fibre optic and swellable polymeric materials. Opt. Lett. 20: 103–105.
12. Gaebler, W. and Braunig, D. 1983 Application of optical fibre waveguides in radiation dosimetry. Proceedings of First International Conference on Optical Fibre Sensors, London, 185–189.

13. Pinchbeck, D. and Kitchen, C. 1985. Optical fibres for cryogenic leak detection. *Proceedings of Conference on Electronics in Oil and Gas* (London: Cahners Exhibits Ltd), pp. 469–501.

14. Harmer, A. L. 1982. Optical fibre sensors and instrumentation—a multi-client study. Interim Report by Harmer, for Battelle Geneva Research Laboratory.

15. Dakin, J. P. and Kahn, D. A. 1977. A novel fibre optic temperature probe. *Opt. Quant. Electron.* 9: 540–544.

16. Diles, R. R. 1983. High-temperature optical fibre thermometer. *J. Appl. Phys.* 54: 1198–1201.

17. Pitt, G. D. 1982. *Electron. Commun.* 57: 102–105.

18. Dakin, J. P. and Holliday, M. G. 1983. A liquid level sensor, based on O–H or C–H absorption monitoring. *Proceedings of First International Optical Fibre Sensors Conference*, OFS1, London.

19. Frank, W. E. 1966. Detection and measurement device having a small flexible fibre transmission line. US Patent 3273447.

20. Menadier, C., Kissinger, C. and Adkins, H. 1967 The fotonic sensor. *Instrum. Contr. Syst.* 40: 114–120.

21. Culshaw, B., Foley, J. and Giles. I. P. 1984. A balancing technique for optical fibre intensity modulated transducers. *Proc. SPIE* 574: 117–120.

22. Venkatesh, S. and Culshaw, B. 1985. Optically excited vibrations in a micromachined silica structure. *Electron. Lett.* 21: 315–317.

23. Culshaw, B. 1989. Silicon in optical fibre sensors, Chapter 13. *Optical Fibre Sensors*, vol 2, eds J. P. Dakin and B. Culshaw, ISBN 0-89006-376-1 (Boston, MA: Artech House), pp. 475–509.

24. Meltz, G., Morey, W. W. and Glenn, W. H. 1989. Formation of Bragg gratings in optical fibres by a transverse holographic method. *Opt. Lett.* 14: 823–825.

25. Morey, W. W., Meltz, G. and Glenn, W. H. 1989. Fibre optic Bragg grating sensors. *Proc. SPIE* 1169: 98–106.

26. St J Russell, P. and Archambault, J. L. 1996. Fibre gratings. *Optical Fibre Sensors*, vol. 3, eds J. P. Dakin and B. Culshaw, ISBN 0-89006-932-8 (Boston, MA: Artech House) Chapter 2, 475–509.

27. Morey, W. W., Dunphy, J. R. and Meltz, G. 1991. Multiplexing fibre Bragg grating sensors. *Proc. SPIE* 1586: 216–224.

28. Kersey, A. D. 1993. Interrogation and multiplexing techniques for fibre Bragg grating strain sensors. *Proc. SPIE* 2071: 30–48.

29. Xu, M. G., Archambault, J. L., Reekie, L. and Dakin, J. P. 1994. Structural bending gauge using fibre gratings, *Proc. SPIE* 2292 (Paper 48).

30. Measures, R. M., Melle, S. M. and Liu, K. 1992. Wavelength demodulated Bragg grating fibre optic sensing systems for addressing smart structure critical issues. *Smart Mater. Struct.* 1: 36–44.

31. Davies, M. A., Bellemore, D. G. and Kersey, A. D. 1994. Structural strain mapping using a wavelength/time division addressed fibre Bragg grating array. *Proc. SPIE* 2361: 342–345.

32. Nellen, P. M., Broennimann, R., Sennhauser, U. J., Askins, C. G. and Putman, M. A. 1995. Applications of distributed fibre Bragg grating sensors in civil engineering. *Proc. SPIE* 2507: 14–24.

33. Jackson, D. A., Ribeiro, A. B. L., Reekie, L., Archambault, J. L., St J Russell, P. 1993. Simultaneous interrogation of fibre optic grating sensors. *Proceedings of Ninth International Optical Fibre Sensors Conference*, OFS9, Florence, pp. 39–42.

34. Kersey, A. D., Berkoff, T. A. and Morey, W. W. 1993. Fibre Fabry–Perot demodulation for Bragg grating strain sensors. *Proceedings of Ninth International Optical Fibre Sensors Conference*, OFS9, Florence, pp. 39–42.

35. Koo, K. P. and Kersey, A. D. 1995. Fibre laser sensor with ultra-high strain resolution, using interferometric interrogation. *Electron. Lett.* 31: 1180–1182.

36. Xu, M. G., Geiger, H. and Dakin, J. P. 1996. Modelling and performance analysis of a fibre Bragg grating interrogation system using an acousto-optic tunable filter. *J. Lightwave Technol.* 14: 391–396.

37. Ecke, W., Schauer, J., Usbeck, K., Willsch, R. and Dakin, J. P. 1997. Improvement of the stability of fibre grating interrogation

systems, using active and passive polarisation scrambling devices. *Proceedings of 12th International Optical Fibre Sensors Conference, OFS12*, Williamsburg, pp. 484–487.

38. Davies, M. A. and Kersey, A. D. 1994. All-fibre Bragg grating strain sensor demodulation technique, using a wavelength-division coupler. *Electron. Lett.* 30: 75–77.

39. Hadeler, O., Richardson, D. J. and Dakin, J. P. 1999. DFB fibre laser structure for simultaneous strain and temperature measurements in concrete structures. *Proc. SPIE* 3670: 332–341.

40. Crickmore, R. I., Gunning, M. J., Stefanov, J. and Dakin, J. P. 2003. Beat frequency measurement system for multiple dual polarisation fibre DFB lasers. *IEEE Sensors J.* 3: 115–120.

41. Jones, B. E. 1981. Simple optical sensors for the process industries, using incoherent light. *Proceedings of Institute of Measurement and Control Symposium On Optical Sensors and Optical Techniques in Instrumentation*, London (London: IMC).

42. Kyuma, K., Tai, S., Sawada, T. and Nunoshita, M. 1982. Fibre optic instrument for temperature measurement. *IEEE J. Quantum Electron.* QE-18: 676–679 (see also chapter 17, by Kyuma, in vol 2 of bibliography 1).

43. Lee, C. E. and Taylor, H. F. 1991. Fibre optic Fabry-Perot temperature sensor using a low-coherence light source. *J. Lightwave Technol.* LT-9: 129–134.

44. Ovren, C., Adolfson, M. and Hok, B. 1983. Fibre optic systems for temperature and vibration measurements in industrial applications. *Proceedings of Optical Techniques in Process Control Den Hague* (Cranfield University and BHRA Fluid Engineering), pp. 67–81. (See also ASEA Innovation data sheet for semiconductor temperature measurement, 1986.)

45. McCormack, J. S. 1981. Remote optical measurement of temperature, using luminescent materials. *Electron. Lett.* 17: 630–632

46. Sholes, R. R. and Small, J. G. 1980. Fluorescent decay thermometer, with biological applications. *Rev. Sci. Instrum.* 51: 882

47. Gratton, K. T. V. and Palmer, A. W. 1986. Fluorescent monitoring for optical temperature sensing. *Fibre Opt. SPIE* 630: 256–265

48. Peterson, J. I., Goldstein, S. R., Fitzgerald, R. V. and Buckhold, D. K. 1980. Fibre optic pH probe for physiological use. *Anal. Chem.* 52: 864.

49. Lippitsch, M. E., Pusterhofer, J., Leiner, M. J. P. and Wolfbeiss, O. S. 1988. Fibre optic oxygen sensor with the decay time as the information carrier. *Anal. Chim. Acta* 205: 1.

50. Jackson, D. A. and Jones, J. D C. 1989. Interferometers. *Optical Fibre Sensors*, vol II, eds J. P. Dakin and B. Culshaw (Boston, MA: Artech House) chapter 10, pp. 329–380.

51. Hotate, K. 1997. Fibre optic gyros. *Optical Fibre Sensors*, vol IV, eds J. P. Dakin and B. Culshaw (Boston, MA: Artech House) chapter 11, pp. 167–206.

52. Hotate, K. 2000. Fibre optic gyros. *Trends in Optical Nondestructive Testing and Inspection*, eds P. K. Rastogi and D. Inaudi (Amsterdam: Elsevier Science) chapter 32, pp. 487–502.

53. Lefèvre, H. C. 1989. Fibre optic gyroscope. *Optical Fibre Sensors*, vol II, eds J. P. Dakin and B. Culshaw (Boston, MA: Artech House) chapter 11, pp. 381–429.

54. Sagnac, G. 1914. Effet tourbillonnaire optique. La circulation de l'ether lumineux dans un interferographe tournant. *J. Phys. Radium* 4: 177–195.

55. Willsch, M. and Bosselmann, T. 2002. Optical current sensor application in the harsh environment of a 120 MVA power generator. *Proceedings of 15th International Conference Optical Fibre Sensors, OFS 2002*, Portland, Paper ThD2, pp. 407–410.

56. Kersey, A. 1997. Multiplexing techniques for fibre-optic sensors. *Optical Fibre Sensors*, vol IV, eds J. P. Dakin and B. Culshaw (Boston, MA: Artech House) chapter 15, pp. 369–407.

57. Rogers, A. J. 1985. Intrinsic and extrinsic distributed optical fibre sensors. *Proc. SPIE* 586: 51–52; *Proc. SPIE* 566: 234–242.

58. Nelson, A. R., McMahon, D. H. and Gravel, R. L. 1980. Passive multiplexing system for fibre optic sensors. *Appl. Opt.* 19: 2917–2920.

59. Thayer, D. R., Lyons, P. B., Looney, L. D., Manning, J. P. and Malone, R. M. 1981. Preparation, installation and calibration of a 152 fibre imaging experiment at the Nevada test site. *Proc. SPIE* 296: 191–194.

60. Dakin, J. P. and King, A. J. 1983. Limitations of a single optical fibre fluorimeter system due to background fluorescence. *Proceedings of First International Conference on Optical Fibre Sensors, London (IEE Conference Pub. 221)*, pp. 195–199.

61. Miller, G. E. and Lindsay, T. A. 1978. Feasibility demonstration of fibre optic digital status monitoring devices. Boeing Aerospace Company (Seattle) Report No D296-10048-1.

62. James, K. A., Quick, W. H. and Strachan, V. H. 1979. Fibre optics: the way to true digital sensors. *Contr. Eng.* 30–33.

63. Nelson, A. R., McMahon, D. H. and Van de Vaart, H. 1981. Multiplexing system for fibre optic sensors using pulse compression techniques. *Electron. Lett.* 17: 263–264.

64. Barnoski, M. K. and Jensen, S. M. 1976. Fibre waveguides: a novel technique for investigating attenuation characteristics. *Appl. Opt.* 15: 2112–2115.

65. Personick, S. D. 1977. Photon probe-an optical fibre time domain reflectometer. *Bell Systems Tech. J.* 56: 355–366.

66. Desforges, F. X., Graindorge, P., Jeunhomme, L. B. and Arditty, H. J. 1986. Progress in OTDR optical fibre sensor networks. *Proc. SPIE* 718, Paper 31.

67. Asawa, C. K., Yao, S. K., Stearns, R. C., Mota, N. L. and Downs, J. W. 1982. High-sensitivity fibre-optic strain sensors for measuring structural distortion. *Electron. Lett.* 18: 362–364.

68. Bruinsma, A. J. A., Van Zuylen, P., Lamberts, C. W. and de Krijger, A. J. T. 1984. Fibre optic strain measurement for structural integrity monitoring. *Proceedings of Second International Conference Optical Fibre Sensors, OFS'84, Stuttgart* (Berlin: VDE), pp. 399–401.

69. Claus, R. O., Jackson, B. S. and Bennett, K. D. 1985. Nondestructive testing of composite materials by OTDR in imbedded optical fibres. *Proc. SPIE* 566: 243–248.

70. Skolnik, M. I. 1970. *Radar Handbook* (New York: McGraw-Hill).

71. Dakin, J. P., Wade, C. A. and Henning, M. 1984. Novel optical fibre hydrophone array using a single laser source and detector. *Electron. Lett.* 20: 51–53.

72. Wade, C. A. 1982. British Patent Application No 8207961 (Priority date 18 March 1982).

73. Dakin, J. P. and Wade, C. A. 1982. British Patent Application No 8220793.

74. Dakin, J. P., Wade, C. A. and Ellis, G. 1986. A novel 3-wave mixing approach to coherent communications. *Proceedings of 12th ECOC*, Barcelona (post deadline paper).

75. Nash, P., Latchem, J., Cranch, G., Motley, S., Bautista, A., Kirkendall, K., Dandridge, A., Henshaw, M. and Churchill, J. 2002. Design, development and construction of fibre-optic bottom-mounted array. *Proc OFS 15, 15th International Conference on Optical Fiber Sensors*, Portland, Oregon, IEEE 02Ex533.

76. Cranch, G., Kirkendall, K., Daley, K., Motley, S., Bautista, A., Salerno, J., Nash, P., Latchem, J. and Crickmore, R. 2003. Large scale remotely pumped and interrogated fiber-optic interferometric sensor array. *IEEE Photon. Technol. Lett.* 15(11): 1579–1581.

77. Pendleton-Hughes, S., Weston, N. and Carter, A. C. 1985. Forty channel wavelength multiplexing for short-haul wideband communications networks. *Proceedings of 11th European Conference on Optical Communications (ECOC 85)*, Venice, pp. 649–652.

78. Russell, S. J. and Dakin, J. P. 1999. Location of time-varying strain disturbances over a 40km fibre section, using a dual-Sagnac interferometer with a single source and detector. *Proc. SPIE* 3746: 580–583.

79. Dakin, J. P. and Liddicoat, T. J. 1982. A wavelength multiplexed optical shaft encoder. *Meas. Contr.* 15: 176–177.

80. Lear, R. D. 1981. Time dependent recordings of images transmitted over optical fibres. *Proc. SPIE* 296: 228–233.

81. Vohra, S., Dandridge, A., Danver, B. and Tveten, A. 1996. A hybrid WDM/TDM reflectometric array *Proceedings of 11th International Conference on Optical Fibre Sensors, OFS96, Sapporo, Japan*, pp. 534–537.

82. Davies, M. A. and Kersey, A. D. 1995. Application of a fibre Fourier transform spectrometer to the detection of wavelength-encoded signals from Bragg grating sensors. *IEEE J. Lightwave Technol.* 13: 1289–1295.

83. Dakin, J. P. and Volanthen, M. 1999. Review of distributed and multiplexed fibre grating sensors and discussion of problem areas (Invited). *Proc. SPIE*, vol. 3860 (paper 16).

84. Mlodzianowski, J., Uttam, D. and Culshaw, B. 1986. A multiplexed system for analogue point sensors. *Proceedings of IEE Colloquium on Distributed Optical Fibre Sensors (IEE Digest No 86/74)* (London: IEE), pp. 12/1–12/3.

85. Uttam, D. and Culshaw, B. 1982. Optical FM applied to coherent interferometric sensors. *IEE Colloquium on Optical Fibre Sensors (IEE Digest No 1982/60)* (London: IEE).

86. Giles, I. D., Uttam, D., Culshaw, B. and Davies, D. E. N. 1983. Coherent optical fibre sensors with modulated laser sources. *Electron. Lett.* 19: 14–15.

87. Al Chalabi, S. A., Culshaw, B., Davies, D. E. N., Giles, I. P. and Uttam, D. 1985. Multiplexed optical fibre interferometers: an analysis based on radar systems. *Proc. IEE* 132: 150–156.

88. Sakai, I. 1986. Frequency-division multiplexing of optical fibre sensors using a frequency-modulated source. *Opt. Quantum Electron.* 18: 279–289.

89. Mallalieu, K. I., Youngquist, R. and Davies, D. E. N. 1986. RF-band FMCW for passive multiplexing of multimode fibre optic sensors. *Proceedings of IEE Colloquium on Distributed Optical Fibre Sensors (IEE Digest No 1986/74)* (London: IEE), pp. 4/1–4/3.

90. Brooks, J. L., Wentworth, R. H., Youngquist, R. C., Tur, M., Kim, B. Y. and Shaw, H. J. 1983. Coherence multiplexing of fibre-optic interferometric sensors. *J. Lightwave Technol.* LT3: 1062–1072.

91. Dakin, J. P., Ecke, W., Rothhardt, M., Schauer, J., Usbeck, K. and Willsch, R. 1997. New multiplexing scheme for monitoring fibre optic Bragg grating sensors in the coherence domain (Invited). *Proceedings of OFS 12 International Conference*, Williamsburg, USA, pp. 31–34.

92. Kingsley, S. A., Davies, D. E. N., Culshaw, B. and Howard, D. 1978. Fibredyne systems. *Proceedings of Fibre Optic Communications Conference*, Chicago (Information Gatekeepers Inc.).

93. Culshaw, B., Ball, P. R., Pond, J. C. and Sadler, A. A. 1981. Optical fibre data collection. *Electron. Power* 27(2): 148–150.

94. Crossley, S. D., Giles, I. P., Culshaw, B. and Sadler, A. A. 1983. An optical fibre data telemetry system for use in remote or hazardous locations. *Proceedings of International Conference on Optical Techniques in Process Control*, The Hague (Cranfield: BHRA), pp. 121–132. *Appl. Opt.* 15: 2112–2115.

95. Dakin, J. P. and Pratt, D. J. 1985. Improved non-invasive fibredyne highway based on polarimetric detection of strain in polarisation-maintaining fibres. *Electron. Lett.* 21: 1224–1225.

96. Rogers, A. J. 1980. Polarisation optical time domain reflectometry. *Electron. Lett.* 16: 489–490.

97. Di Vita, P. and Rossi, U. 1980. The backscattering technique: its field of applicability in fibre diagnostics and attenuation measurements. *Opt. Quantum Electron.* 11: 17–22.

98. Conduit, A. J., Payne, D. N., Hartog, A. H. and Gold, M. P. 1981. Optical fibre diameter variations and their effect on backscatter loss measurement. *Electron. Lett.* 17: 308–310.

99. Theocharous, E. 1983. Differential absorption distributed thermometer. *Proceedings of First International Conference on Optical Fibre Sensors* (London: IEE), pp. 10–12.

100. Quoi, K. W., Lieberman, R., Cohen, L. G., Shenk, D. S. and Simpson, J. R. 1992. Rare-earth-doped optical fibres for temperature sensing. *IEEE J. Lightwave Technol.* 10: 847–851.

101. Hardy, E. E., David, D. J., Kapany, N. S. and Unterleitner, F. C. 1975. Coated optical guides for spectrophotometry of chemical reactions. *Nature* 257: 666.

102. Blyler, L. L., Ferrara, J. A. and MacChesney, J. B. 1988. A plastic clad silica fibre chemical sensor for ammonia. *Proceedings of OFS88 Technical Digest Optical Society of America*, pp. 369–373.

103. Hartog, A. H., Payne, D. N. and Conduit, A. J. 1980. POTDR: experimental results and application to loss and birefringement measurements in single mode fibres. *Proceedings of Sixth European Conference on Optical Communication, ECOC80*, York (post deadline paper) (London: IEE).

104. Kim, B. Y. and Choi, S. S. 1981. Backscattering measurements of bending-induced birefringence in single mode fibres. *Electron. Lett.* 17: 193–195.

105. Ross, J. N. 1981. Measurement of magnetic field by POTDR. *Electron. Lett.* 11: 596–597.

106. Rogers, A. J. 1981. POTDR a technique for the measurement of field distributions. *Appl. Opt.* 20: 1060–1074.

107. Healey, P. 1984. Fading in heterodyne OTDR. *Electron. Lett.* 20: 30–32.

108. Hartog, A. H. and Payne, D. N. 1982. Fibre optic temperature distribution sensor. *Proceedings of IEE Colloquium Optical Fibre Sensors* (London: IEE).

109. Dakin, J. P., Pratt, D. J., Bibby, G. W. and Ross, J. N. 1985. Distributed anti-Stokes ratio thermometry. *Proceedings of Third International Conference on Optical Fibre Sensors*, San Diego (post-deadline paper).

110. Dakin, J. P., Pratt, D. J., Bibby, G. W. and Ross, J. N. 1985. Distributed optical fibre Raman temperature sensor using a semiconductor light source and detector. *Electron. Lett.* 21: 569–570.

111. Dakin, J. P. 1984. UK Patent Application GB 2156513A (published 9 October 1985).

112. Poole, S. B., Payne, D. N. and Fermann, M. E. 1985. Fabrication of low-loss optical fibres containing rare-earth ions. *Electron. Lett.* 21: 737–738.

113. Dakin, J. P. and Pratt, D. J. 1986. Fibre-optic distributed temperature measurement—a comparative study of techniques. *Proceedings of IEE Colloquium on Distributed Optical Fibre Sensors (IEE Digest No 1986/74)* (London: IEE), pp. 10/1–10/4.

114. Lieberman, R. A., Blyler, L. L. and Cohen, L. G. 1990. A distributed fibre optic sensor based on cladding fluorescence. *IEEE J. Lightwave Technol.* 8: 212–220.

115. Kurashima, T., Horiguchi, T. and Koyamada, Y. 1992. Measurement of temperature and strain distribution by Brillouin shift in silica optical fibres. *Proc. SPIE* 1797: 2–13.

116. Kurashima, T., Horiguchi, H., Izumita, H., Furukawa, S. and Koyamada, Y. 1993. Brillouin optical-fiber time domain reflectometry. *IEICE Trans. Commun.* E76-B: 382.

117. Horiguchi, T. and Tateda, M. 1989. Optical fibre attenuation investigation using stimulated Brillouin scattering between a pulse and a continuous wave. *Opt. Lett.* 14: 408–410.

118. Horiguchi, T. and Tateda, M. 1989. BOTDA—non-destructive measurement of single mode optical fibre attenuation investigation characteristics using Brillouin interaction: theory. *J. Lightwave Technol.* 7: 1170–1176.

119. Horiguchi, T., Kurashima, T. and Tateda, M. 1990. A technique to measure distributed strain in optical fibres. *IEEE Photon. Technol. Lett.* 2: 352–354.

120. Horiguchi, T., Shimizu, K., Kurashima, T. and Koyamada, Y. 1995. Advances in distributed sensing techniques using Brillouin scattering. *Proc. SPIE* 2507: 126–135.

121. Wait, P. C. and Newson, T. P. 1996. Landau-Placzek ratio, applied to distributed sensing. *Opt. Commun.* 122: 141–146.

122. Maughan, S. L., Kee, H. H. and Newson, T. P. N. 2001. Simultaneous distributed fibre temperature and strain sensor, using microwave coherent detection of spontaneous Brillouin backscatter. *Meas. Sci. Technol.* 12: 834–842.

123. Eickhoff, W. and Ulrich, R. 1981. Optical frequency domain reflectometry in single mode fibre. *Appl. Phys. Lett.* 39: 693–695.

124. Kingsley, S. A. and Davies, D. E. N. 1985. OFDR diagnostics for fibre and integrated-optic systems. *Electron. Lett.* 21: 434–435.

125. Ghafoori-Shiraz, H. and Okoshi, T. 1986. Fault location in optical fibres using optical-frequency-domain reflectometry. *J. Lightwave Technol.* LT-4: 316–322.

126. Hotate, K. 1997. Fibre sensor technology today. *Opt. Fibre Technol.* 3(4): 356–402.

127. Hotate, K. 1999. Coherent photonic sensing. *Sensors Update* 6(1): 131–162.

128. Hotate, K. and Ong, S. S. L. 2003. Distributed dynamic strain measurement using a correlation-based Brillouin sensing system. *IEEE Photon. Technol. Lett.* 12: 272–274.

129. Hotate, K. 2002. Application of synthesized coherence function to distributed optical sensing. *IOP Meas. Sci. Technol.* 13: 1746–1755.

130. Franks, R. B., Torruellas, W. and Youngquist, R. C. 1985. Birefringent stress location sensor. *Proc. SPIE* 586: 84–88.

131. Dakin, J. P., Pearce, D. A., Strong, A. P. and Wade, C. A. 1987. A novel distributed optical fibre sensing system enabling location of disturbances in a Sagnac loop interferometer. *Proc. SPIE* 838, paper 18.

132. Dakin, J. P., Pearce, D. A., Strong, A. P. and Wade, C. A. 1988. A novel distributed optical fibre sensing system enabling location of disturbances in a Sagnac loop interferometer. *Proceedings EFOC/LAN 88* Information Gatekeepers Inc.), pp. 276–279.

133. Udd, E. 1991. Sagnac distributed sensor concepts. *Proc. SPIE* 1586: 46–52.

134. Spammer, S. J., Swart, P. L. and Boosen, A. 1996. Interferometric distributed fibre optical sensor. *Appl. Opt.* 35: 4522–4523.

135. Spammer, S. J., Chtcherbakov, A. A. and Swart, P. L. 1996. Dual wavelength Sagnac–Michelson distributed optical fibre sensor. *Proc. SPIE* 2838: 301–307. doi: 10.1117/12.259812.

136. Spammer, S. J., Chtcherbakov, A. A. and Swart, P. L. 1998. Distributed dual wavelength Sagnac impact sensor. *Microwave Opt. Technol. Lett.* 17: 170–173.

137. Ronnekleiv, E., Blotekjaer, E. K. and Kranes, K. 1993. Distributed fibre sensor for location of disturbances. *Proceedings OFS-9 International Conference*, Paper PD7.

138. Fang, X. J. 1996. Fibre-optic distributed sensing by a two-loop Sagnac interferometer. *Opt. Lett.* 21: 444–446.

139. Dandridge, A., Tveten, A. B. and Giallorenzi, T. G. 1982. Homodyne demodulation scheme for fibre optic sensors using phase generated carrier. *IEEE Trans. Microwave Theory Tech.* MTT-30: 1635–1641.

140. Farries, M. C. and Rogers, A. J. 1984. Distributed sensing using stimulated Raman interaction in a monomode optical fibre. *Proceedings of Second International Conference Optical Fibre Sensors, OFS'84, Stuttgart* (Berlin: VDE), pp. 121–132.

141. Martin, J. and Ouellette, F. 1994. Novel writing technique of long and highly-reflective in-fibre gratings. *Electron. Lett.* 30: 811–812.

142. Huang, S. H., Ohn, M. M., Leblanc, M., Lee, R. and Measures, R. M. 1994. Fibre optic intra-grating distributed strain sensor. *Proc. SPIE* 2294: 81–92.

143. Labelet, P., Fonjallaz, P. Y., Limberger, H. G., Salathe, R. P., Zimmer, C. and Gilgen, H. H. 1993. Bragg grating characterisation by optical low-coherence reflectometry. *IEEE Photon. Technol. Lett.* 5: 565–567.

144. Volanthen, M., Geiger, H., Cole, M. J., Laming, R. I. and Dakin, J. P. 1996. Low coherence technique to characterise reflectivity and time delay, as a function of wavelength, within a long Bragg grating. *Electron. Lett.* 32: 757–758.

145. Volanthen, M., Geiger, H., Cole, M. J. and Dakin, J. P. 1996. Measurement of arbitrary strain profiles within fibre gratings. *Electron. Lett.* 32: 1028–1029.

146. Volanthen, M., Geiger, H. and Dakin, J. P. 1997. Low-coherence grating characterisation scheme. *Proceedings of IEE Colloquium on Optical Fibre Gratings*, London.

147. Volanthen, M., Geiger, H. and Dakin, J. P. 1997. Distributed grating sensors using low coherence reflectometry. *IEEE J. Lightwave Technol.* 15: 2076–2082.

148. Penner, V. and Rocha, G. 2007. Use of the visante for anterior segment ocular coherence tomography. *Tech. Ophthalmol.* 5(2): 67–77.

149. Keane, P. A., Patel, P.J., Liakopoulos, S., Heussen, F. M., Sadda, S. R., Tufail, A. 2012. Evaluation of age-related macular degeneration with optical coherence tomography. *Survey Ophthalmol.* 57(5): 389–414. doi: 10.1016/j.

150. Virgili, G., Menchini, F., Casazza, G., Hogg, R., Das, R. R., Wang, X., Michelessi, M. 2015. Optical coherence tomography (OCT) for detection of macular edema

in patients with diabetic retinopathy. *Cochrane Database Syst Rev.* 10: CD008081. doi:10.1002/14651858.

151. Dörr, J., Wernecke, K. D., Bock, M., Gaede, G., Wuerfel, J. T., Pfueller, C. F., Bellmann-Strobl, J., Freing, A., Brandt, A. U., Friedemann, P. 2011. Association of retinal and macular damage with brain atrophy in multiple sclerosis, *PLoS One* 6(4): e18132. doi:10.1371/journal.pone.0018132.

152. Bezerra, H. G., Attizzani, G. F., Sirbu, V., Musumeci, G., Lortkipanidze, N., Fujino, Y., Wang, W., Nakamura, S., Erglis, A., Guagliumi, G. and Costa, M. A. 2013. Optical coherence tomography versus intravascular ultrasound to evaluate coronary artery disease and percutaneous coronary intervention. *JACC Cardiovasc Interv.* 6(3): 228–236. doi:10.1016/j.jcin.2012.09.017.

153. Marvdashti, T., Duan, L., Aasi, S.Z., Tang, J. Y., and Ellerbee Bowden, A. K. 2016. Classification of basal cell carcinoma in human skin using machine learning and quantitative features captured by polarization sensitive optical coherence tomography. *Biomed. Optics Express* 7(9): 3721–3735. doi: 10.1364/BOE.7.003721.

154. Nguyen, F. T., Zysk, A. M., Chaney, E. J., Adie, S. G., Kotynek, J. G., Oliphant, U. J., Bellafiore, F. J., Rowland, K. M., Johnson, P. A., and Boppart, S. A. 2010. Optical coherence tomography: The intraoperative assessment of lymph nodes in breast cancer. *IEEE Eng. Med. Biol. Mag.* 29(2): 63–70. doi:10.1109/MEMB.2009.935722.

155. Liang, R., Wong, V., Marcus, M., Burns, P., and McLaughlin, P. 2007. Multimodal imaging system for dental caries detection. *Proc. SPIE* 6425: 642502. doi: 10.1117/12.702131.

156. Heveron-Smith, S., Non-contact, light-based measurements for medical balloons and catheters. Medical Design Briefs, December 1, 2014.

157. Marcus, M. A., Hadcock, K. J., Gibson, D. S., Herbrand, M. E., Ignatovich, F. V. 2013. Precision interferometric measurements of refractive index of polymers in air and liquid. *Proc. SPIE.* 8884: 88841L. doi: 10.1117/12.2032533.

158. Marcus, M. A., Compertore, D., Gibson, D. S., Herbrand, M. E. and Ignatovich, F. V. 2015. Multimodal characterization of contact lenses. *Proc. SPIE* 9633: 96331W. doi: 10.1117/12.2196303.

159. Ignatovich, F., Spaeth, M., Solpietro, J., Cotton, W., Gibson, D. 2014. Measurement of film stacks in cell phones and tablets using white light interferometry. *AIMCAL Proceedings 2014 Web Coating and Handling Conference*, USA, pp. 421–427.

160. Marcus, M. A. 2016. Simultaneous head-up display windshield wedge angle and layer thickness measurements. SPIE Newsroom, July 29, 2016, doi: 10.1117/2.1201607.006610.

161. Hurlbut, J., Cashen, D., Robb, E., Spangler, L., and Eckhart, J. 2016. Next generation PVB interlayer for improved HUD image clarity. *SAE Int'l J. Passenger Cars Mech. Syst.* 9: 360–365. doi:10.4271/2016-01-1402.

162. Bush, J., Davis, P. and Marcus, M. A. 2001. All-fibre coherence domain interferometric techniques. *Proc. SPIE* 4204: 71–80.

163. Marcus, M. A., Gross, S. and Wideman, D. 1997. Associated dual interferometric measurement apparatus for determining a physical property of an object. US Patent No 5, 659, 392.

164. Marcus, M. A., Lee, J.-R., Stephenson, D. A. and Kaltenbach, T. F., Method and apparatus for combined measurement of surface non-uniformity, index of refraction variation and thickness variation. U.S. Patent No 6, 614, 534 B1, September 2, 2003.

165. Badami, V. G. and Blalock, T. 2005. Uncertainty evaluation of a fiber-based interferometer for the measurement of absolute dimensions. *Proc. SPIE*, 5879: 23–41.

166. Danielson, B. and Boisrobert, C. 1991. Absolute optical ranging using low coherence interferometry. *Appl. Opt.* 30: 2975.

167. Bracewell, R. 1978. *The Fourier Transform and Its Applications*, 2nd edn (New York: McGraw-Hill).

168. Harris, H. W., Lee, J.-R. and Marcus, M. A. 1999. Noncoherent light interferometry as a thickness gauge. *Proc. SPIE* 3538: 180–191.

169. Lee, J.-R. and Marcus, M. A. 2000. Method for determining the retardation of a material using non-coherent light interferometry. US Patent No 6, 034, 774.

170. Marcus, M. A., Trembley, T. and Uerz, D. 1998. Digital camera image sensor positioning apparatus including a non-coherent light interferometer. US Patent No 5, 757, 486.

171. Marcus, M. A., Dilella, E. A., Lee, J.-R., Lowry, D. R. and Trembley, T. M. 2003. Measurement method and apparatus of an external digital camera imager assembly. US Patent 6, 512, 587 B1.

BIBLIOGRAPHY

1. Dakin, J. P. and Culshaw, B. (eds) 1988, 1989, 1996, 1997 (in order of volume number), *Optical Fibre Sensors* vols 1–4 (Boston, MA: Artech House), ISBN 0-89006-317-6, 0-89006-376-1, 0-89006-932-8, 0-89006-940-9.

2. Wolfbeiss, O. O. *Fibre Optic Chemical Sensors and Biosensors*, vols 1 and 2 (Boston, MA: CRC Press), ISBN 08493-5508-7, 08493 5509-5.

3. Udd, E. (ed.) 1995. *Fibre Optic Smart Structures* (New York: John Wiley & Sons), ISBN 0-471-5548-0.

4. Udd, E. Organiser of SPIE series of video tutorial courses on optical fibre sensors.

5. Dakin, J. P. and Culshaw, B. 1986. Distributed fibre optic sensors. *SPIE Crit. Rev.* 44.

6. Davies DEN 1984 Signal processing for distributed optical fibre sensors. *Proceedings of International Conference on Optical Fibre Sensors*, OFS84, Stuttgart, Germany, 285–295.

7. Kersey, A. 1997. Multiplexing techniques for fibre-optic sensors, Chapter 15. *Optical Fibre Sensors*, vol. IV, eds J. P. Dakin and B. Culshaw (Boston, MA: Artech House), pp. 369–407.

8. Kersey, A. D. 1993. Interrogation and multiplexing techniques for fibre Bragg grating strain sensors. *Proc. SPIE* 2071: 30–48.

9. Kersey, A., Davies, M. A., Patrick, H. J., LeBlanc, M., Koo, K. P., Askins, C. G., Putman, M. A. and Friebele, E. J. 1997. Fibre grating sensors. *J. Lightwave Technol.* 15: 1442–1462.

10. Rogers, A. J. 1985. Intrinsic and extrinsic distributed optical fibre sensors. *Proc. SPIE* 586 51–52; *Proc. SPIE* 566: 234–242.

11. Rogers, A. J. 1988. Distributed optical fibre sensors for the measurement of pressure strain and temperature. *Phys. Rep.* 169: 99–143.

12. Culshaw, B. 1986. Distributed and multiplexed fibre optic sensor systems. *Proceedings of 11th Course of the International School of Quantum Electronics*, Erice, Sicily (Dordrecht, the Netherlands: Martinus Nijhoff).

Remote optical sensing by laser

J. MICHAEL VAUGHAN

12.1 INTRODUCTION—SENSING BY LASER RADAR (LIDAR)

12.1.1 Laser radar and laser properties

The myriad of modern applications for lasers includes the remote sensing and investigation of distant objects. When this work is conducted outdoors, at ranges of a few tens of metres to many hundreds of kilometres, the subject is usually called "laser radar". It must be admitted that this is something of a misnomer—the term radar itself derives from "radio detection and ranging", which was of course developed over 60 years ago for detection of aircraft by long-wavelength radio waves. However, we are presently considering the use of very much shorter wavelengths—that is, light waves in the visible and near-visible region. In consequence, the terms "lidar" for "light detection and ranging" and "ladar" for "laser detection and ranging" have also been introduced. While some attempts have been made to differentiate the usage of the three expressions "laser radar", "lidar" and "ladar", they are in fact generally used freely and interchangeably. However, one can do a great deal more than just "detect and range" with lasers. In consequence, two other terms are also used for more specific applications: laser Doppler velocimetry (LDV)—use of the Doppler principle for remote velocity measurements, and differential absorption by laser (DIAL)—chemical detection with lasers tuned on and off resonance absorption.

The principle of laser radar is of course very simple and straightforward. The laser beam is sent out from an optical transmitter towards the object of interest, often referred to as the "target". The target will scatter light from the beam and some of this will be reflected back towards a receiver—usually placed adjacent to the transmitter and often indeed using the same optical arrangement set up as a transmitter/receiver. This scattered and reflected light will contain information about the target so that, after the light has been detected within the receiver and converted into an electrical signal, the information may be extracted. Lasers have many very desirable properties for use in remote sensing and these are summarized in Table 12.1. Many of these laser properties and the unique characteristics of laser radiation are discussed at considerable length in other chapters of this volume, but it is worth noting a few points of particular relevance to remote sensing.

Many of the secondary characteristics derive from the primary characteristics; thus, precise focusing and pointing accuracy are due to good coherence properties and, combined with high brightness, can contribute enormous power density (it has been noted that, in certain high energy, exceptionally short-pulse lasers, the actual instantaneous power level may be greater than all the power stations of the world combined). Of particular value for lidar operation is the wide range of available lasers and wavelengths, extending from the near ultra-violet at wavelength $\lambda \sim 0.3\,\mu m$ up to the mid infra-red at $\lambda \sim 12\,\mu m$. For outdoors, any laser system must of course be operated in a way that presents no hazard whatsoever to people or equipment in the locality. At wavelengths between ~ 0.38 and $\sim 1.5\,\mu m$, this necessitates careful control of the transmitted beam to ensure laser intensity levels remain below established thresholds (taking account of scintillation in the atmosphere; see, e.g., Reference [1]). In the so-called eye-safe regime ($\lambda < 0.38$ and $> 1.5\,\mu m$), where radiation is at least

Table 12.1 Laser properties

Primary characteristics	High brightness
	Good spatial coherence
	Good temporal coherence
Secondary characteristics	Wide range of available wavelength
	Continuously tuneable in certain regions
	Continuous wave (cw) and pulsed output
	Great pointing accuracy
	Precise focusing
	Enormous power density

not transmitted through to the retina, conditions are not quite so stringent. Thus, for many applications, operation with such eye-safe lasers has been preferred. This in fact has rather profound implications for the development of lidar technology and its broad division between the two classes of laser radar discussed in the following sections.

12.1.2 Remote sensing applications of laser radar

A list of actual and potential remote sensing applications for lasers is given in Table 12.2. For such a great range of tasks, a wide variety of principles, techniques and systems have been applied and developed; some of the more important aspects of these are outlined in the following two sections. Following such basic considerations, there are a number of possible ways of discussing their application to remote-sensing problems; one could, for example, provide a categorization based on the lidar base from fixed or mobile ground platforms, from the air (by balloon or aircraft) or from a space craft. However, a categorization based on the field of application seems likely to be more informative. Accordingly, in the present report, lidar remote

sensing is discussed under six separate headings: atmospheric sensing; problems in aviation; chemical and pollution studies; military applications; geoscience; and measurements from space. Obviously, most of these individual topics overlap across several of the items listed in Table 12.2.

The present chapter deals with active remote sensing by laser. For an account of passive sensing with thermal emission in the infrared, see, for example, Chapter 19 of Volume I. For a short list of further reading including books, reviews and compendia of papers, see the Reference section at the end of this chapter.

12.2 SOME FUNDAMENTAL CONSIDERATIONS IN LASER RADAR

12.2.1 Basic principles and phenomena for remote sensing with lasers

Some of the basic principles and phenomena that may be utilized in laser radar are indicated in Table 12.3. The scattering of light from the target is of course fundamental to the technique and has

Table 12.2 Some remote-sensing applications of laser radar

Wind measurement—meteorology, airfield and aircraft shear and turbulence warning, aircraft wake vortices
Atmospheric measurement—clouds, precipitation, aerosols (dust), temperature, pressure, dispersion
Chemical species, pollutant and gas detection
Airborne operation—avionics data, true airspeed, obstacle and terrain avoidance
Scene and object imaging with various discriminants—intensity, glint, polarization, frequency shift, etc.
Range and three-dimensional sizing, depth sounding
Vibrational analysis of surfaces and structures
Spaceborne measurements—cloud, atmospheric scattering, global wind field

Table 12.3 Some basic principles and phenomena for laser sensing

Light scattering: elastic, specular, diffuse, Mie, Rayleigh, Brillouin, Raman
Differential absorption
Fluorescence
Surface heating and vaporization
Light propagation (diffraction and speckle)
Doppler effect
Interference (interferometry; light mixing/beating/heterodyning)
Timing principle

many possible forms. Of those noted, elastic (without gross change of frequency), specular (mirror-like) and diffuse (as from a rough surface) are terms relating to surface scattering. Mie refers to scattering from particles of size comparable with the wavelength λ. Rayleigh and Brillouin scattering arise from density fluctuations due to thermal effects at the surface and (more strongly) within the bulk of a material, whereas Raman scattering is re-radiation following changes of internal energy levels within individual molecules. Analysis of light scattering thus has the capability to provide physical information at many different levels.

Differential absorption, fluorescence and surface heating/vaporization are generally specific to the target and may be used with an appropriate choice of laser wavelengths to provide chemical-type information.

The available laser beam sizes and speckle effects of laser radiation are of course determined by well-established principles of optical propagation. In lidar work, the outgoing laser beam will usually be setup in the lowest order optical mode (TEMoo) with Gaussian (bell-shape) intensity distribution. At a distance z from the *minimum* beam waist diameter d_0, the beam size is given by

$$D^2(z) = d_0^2\left[1 + \left(4\lambda z / \pi d_0^2\right)^2\right] \qquad (12.1)$$

This minimum beam size, d_0, occurs where the beam is collimated (i.e., where the wave front is plane). As shown in Figure 12.1, in lidar operation, the laser may be transmitted in either one of two forms: as a focused beam on a nearby target (usually up to ~1 km) or as a collimated beam onto a distant target. In either case, from Equation 12.1, if $0 \ll d(z)$ then

$$d(z) \approx \lambda_z / \pi_0 \qquad (12.2)$$

and the angular divergence of the beam is given by

$$d(z)/z \approx 4\lambda/\pi_0 = 1.27(\lambda/0) \qquad (12.3)$$

From these expressions, it is easy to show that a 10 cm diameter beam (D_T in Figure 12.1a) at $\lambda = 10.6\,\mu m$ may be brought to a focus of $0 = 1.36\,cm$

at a range R of $z = 100\,m$. Conversely, for a collimated beam at the transmitter with $D_T = 0 = 10\,cm$ (in Figure 12.1b) the beam size D_F at 10 km range and $\lambda = 10.6\,\mu m$ would be 1.36 m. These values of beam diameter reduce very sharply with shorter wavelength λ (note the λ term in the numerator of Equation 12.2). However, the refractive effects of atmospheric turbulence are much greater at the shorter wavelength. These introduce optical aberration in the light path so that, in the lower atmosphere at least, it is difficult to focus the shorter-wavelength beams any more sharply than longer wavelengths (see Section 12.3.4 following). This, of course, no longer holds true high in the atmosphere or in space.

Figure 12.1c illustrates a laser beam of radius D_R illuminating a rough surface that is shown as an assembly of small scatterers, which reflect the beam in all directions. Typically, at a distance R, the interference of the light from all these many scatterers will produce a random speckle pattern. The characteristic size D_s of these speckle blobs is given by a similar expression to Equation 12.2 with

$$D_s \approx \lambda \cdot R / D_R \qquad (12.4)$$

Thus, as the beam illuminating the target gets bigger, the speckle size decreases. The fluctuations of speckles, and their temporal and statistical characteristics for moving targets, are a complex and fascinating topic and will be touched on in various sections of this report.

The next item in Table 12.3, the Doppler effect, is the well-known change in frequency of a wave from a moving object (as typified, for example, from a passing train whistle or a police siren). Exactly the same thing happens in the light radiation scattered from a moving target. In back scattering (i.e., that light scattered back towards the transmitter) the Doppler shift f_D is given by

$$f_D = 2V_R / \lambda \qquad (12.5)$$

This is illustrated in Figure 12.2a where V_R, equal to $V\cos\theta$, is the radial line-of-sight component of motion of the scattering target.

Laser radiation in the visible and near-visible region is, of course, of a very high frequency that cannot be followed by available detectors. Green

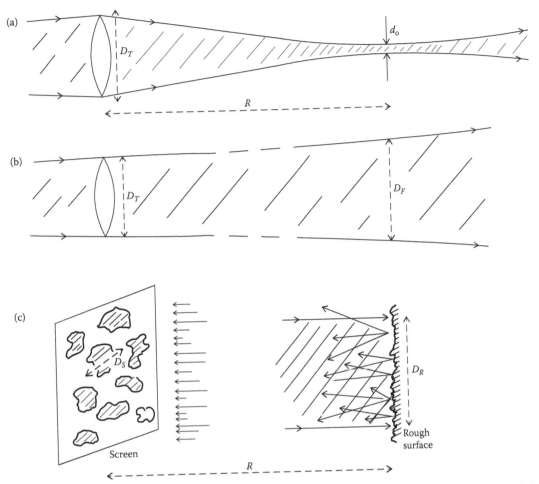

Figure 12.1 Illustration of the focusing of laser beams and laser speckle. (a) Converging beam of diameter D_T transmitted from a lidar telescope and focusing to a diameter 0 at range R. (b) Collimated beam (with plane wavefront) D_T propagating to a diameter D_F at a distant range. (c) Laser speckles (of typical size $\sim D_s$) due to a scattering from a rough surface illuminated with a laser beam of size D_R.

light of $\lambda=0.5\,\mu$m, for example, is of frequency $f=6 \times 10^{14\,Hz}$ or 600 THz and from Equation 12.5 the Doppler shift at $\lambda=0.5\,\mu$m due to a moving target with $V_R \equiv 1\,ms^{-1}$ is only 4×10^6 Hz or 4 MHz. Such a small shift may be measured with advanced classical spectroscopic techniques of high-resolution interferometry, employing, for example, Fabry–Perot etalon filters. However, the interference principle may also be employed to measure this Doppler shift as added to the much higher "carrier" frequency of the laser f_L. Thus, if the shifted and unshifted beams in Figure 12.2b are superimposed at a detector, the electromagnetic fields of the local oscillator beam at frequency f_L and the scattered beam (from a moving target) of

frequency (f_L+f_D) will beat or heterodyne together. In consequence, the detector will provide an oscillating electrical signal i_S at the difference frequency $(f_L+f_D)-(f_L)=f_D$, as described in the following section.

Finally, we consider the timing principle as indicated in Figure 12.2c. If a laser beam is transmitted as a very short, sharp pulse of length Δt_p, the measured time-of-flight t_f to and from a target gives a measure of the range R and range resolution ΔR as

$$R = ct_f/2 \tag{12.6a}$$

$$\Delta R = c\Delta t_p/2 \tag{12.6b}$$

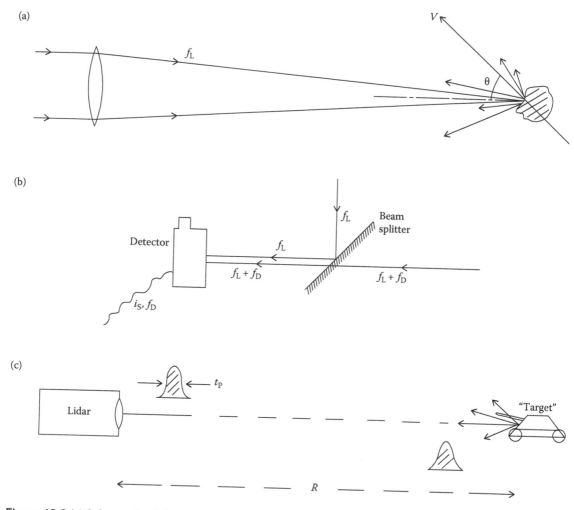

Figure 12.2 (a) Schematic of the Doppler frequency shift in light backscattered from a moving object. As shown, the line-of-sight velocity component is $V\cos\theta$ and thus the Doppler shift is $f_D = 2V\cos\theta/\lambda$. (b) Illustration of light beating or heterodyning by mixing the two beams at the surface of a detector which provides an oscillating electrical signal i_s at the difference frequency f_D. (c) Schematic of a simple time-of-flight laser rangefinder. If the measured time delay to the target and back is t_f the range R is $ct_f/2$.

where c is the velocity of light ($\sim 3 \times 10^8$ ms^{-1}). Thus, a 10 ns pulse system will readily provide a range resolution of 1.5 m, which may be adequate for many purposes; shorter pulses may be readily employed to provide centimetric precision or better.

12.2.2 Basic techniques: Incoherent direct-detection and coherent light-beating

For many applications, the scattered light from the target may be directed to a detector with no more than a simple optical filter to cut down most of the background light. However, for more advanced applications, as outlined briefly above and shown schematically in Figure 12.3, there are two basic lidar techniques for detection and spectral analysis of the scattered light field:

- Direct detection or optical/frequency domain spectroscopy in which the light field is operated on by optical elements prior to detection.
- Post-detection or time domain spectroscopy in which analysis is conducted after the light field is detected, often in conjunction with a coherent heterodyne local oscillator beam.

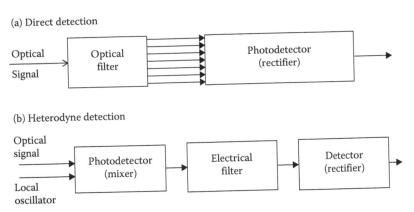

Figure 12.3 Schematic of the two basic forms of laser radar showing the flow diagram of optical to electrical processing. Direct detection lidar in which the scattered light is collected by the receiver and passed directly via an optical filter to the detector. At high resolution, interferometric stability will be required in the optical filter. Coherent heterodyne lidar in which the scattered light is mixed with an optical local oscillator beam derived from the original laser as in Figure 12.2b. In this case, very precise, interferometric control and matching optic axes and wavefronts of the two beams are required as they beat together at the surface of the detector.

The two techniques are thus different at a very fundamental level and have been extensively discussed (see, e.g., References [2,3]). In the older direct-detection techniques, the beam may be operated on by various classical interferometric devices (two-beam: Michelson, Mach–Zehnder, etc.; multiple-beam: Fizeau, Fabry–Perot, Lummer–Gehrke plate, echelle, grating, etc.) to form a visibility curve or spectrum from which spectral information may be derived. In post-detection, coherent heterodyne methods, the spectrum is formed (or equivalent information derived) by manipulation of the electrical signal as it emerges serially in time from the detector element.

For coherent operation with full mixing efficiency, very good control of the local oscillator beam and scattered signal beam is required, with very precise matching of the wavefronts and optic axes. A fundamental characteristic of such an arrangement is that, in effect, the local oscillator beam selects just a single mode of the scattered radiation field. This ensures that the full speckle characteristics of the scattered radiation (amplitude, frequency, phase), with all the dynamic and fluctuation parameters, are fully preserved in the measured electrical signal (in contrast with direct detection where these properties are necessarily smeared out by integration over a finite optical aperture). In essence, the coherent technique provides a measure of the vector sum of the amplitude

of electromagnetic field, $\breve{E}_s$, over the detector, whereas, in direct detection of the scalar sum of intensities, $\left|E_s^2\right|$ is formed. The coherent beating technique thus provides an important and perhaps little appreciated spectroscopic tool; probability distributions, higher moments, autocorrelation functions, structure functions and fractal character may be investigated for different fields (see, e.g., Reference [4]) and there is wide application to many aspects and phenomena of laser physics.

Not surprisingly, the information transfer by these two different forms of operation on an electromagnetic field is very different. The most significant factor is the rate of photons detected within a single coherence area (an étendue of λ^2) per unit band pass and, most particularly, the number of detections from one cell of phase space, equivalent to the number of counts per coherence time and usually called the photon degeneracy parameter δ. In coherent time-domain techniques operating on a single optical mode, the available etendue or light grasp is limited to λ^2. On the other hand, in direct-detection, interferometric methods, the optical filter acts directly on the light field and one measures the direct current component of the signal collected in a beam of etendue U, which can be many times larger than λ^2.

A summary comparison of the two techniques is given in Table 12.4. As will be seen in the following sections, the basic choice of technique will be

Table 12.4 Comparison of direct-detection and coherent techniques

Direct detection	Coherent heterodyne
Scattered radiation optically manipulated before being passed to detector	Scattered radiation mixed (heterodyned) with coherent local oscillator beam
Measures scattered intensity in selected frequency intervals	Full phase, frequency and amplitude information available
Light may be collected in many optical modes	Light collected in asingle optical mode
Some relaxation of laser characteristics may be available (depending on frequency resolution required)	Laser requires to be of high spatial and spectral purity
Relatively simple optical arrangement of transmitter and receiver but interferometric stability required in analyzer	Very precise, stable, optical arrangement required to maintain wavefront matching of scattered and local oscillator beams
Quasi-noise free detection at shorter wavelengths; additional noise source at longer wavelengths	Detector noise at longer wavelengths may be advantageously dominated by the local oscillator shot noise

dictated by the problem in hand and the most appropriate laser source for that problem. In broad terms, the large photon energy at shorter wavelengths means that detectors should be comparatively free of noise, so direct detection techniques may be preferred for $\lambda \lesssim 1.5\,\mu m$. At longer wavelengths with heterodyne techniques, shot noise in the local oscillator beam may be used to dominate thermal noise in the detector to provide quantum-limited detection of the signal beam. A more detailed evaluation of signal-to-noise ratio and measurement accuracy is provided in Section 12.3.3 following.

12.3 LIDAR TECHNOLOGY AND SYSTEMS

12.3.1 The building blocks of a lidar system

The basic elements of a laser radar are indicated in Figure 12.4. In the design of a system for any given task, there are many aspects to be considered and choices to be made. Immediately obvious factors include the nature of the target and the information required from it, the environment and type of platform (ground, air or space) from which the lidar must operate and not least the available financial budget. Some of the vast range of technical considerations and options are shown in Table 12.5. This list serves to establish both the intellectual challenge of laser radar and the diversity it offers of basic science, advanced technology and field measurement. In the present work, it is obviously not appropriate to consider all the various building blocks shown in Figure 12.4 and Table 12.5 in any detail. It is, however, worth outlining a number of issues. These include (1) propagation in the atmosphere, the choice of lasers and the relative merits of working at various wavelengths; (2) basic questions of signal to noise and calibration of lidars; and (3) some consideration of range resolution and optical and system design issues for operation from different platforms.

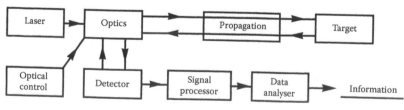

Figure 12.4 The basic components of a laser radar. See text and Table 12.5 for an outline of the many technical considerations and options for the various components.

Table 12.5 Some technical considerations and options in the design of a coherent laser radar (see also Figure 12.4)

Laser	Power level, pulsed/cw, wavelength, line stability, coherence, lifetime, mass, volume, power supply, cooling, etc.
Optics	Size, uniaxal/biaxal, optical/mechanical precision, field of view, temperature effects, longterm stability, etc.
Optical control	Scan patterns, speed, range discrimination, aiming, switchability, etc.
Propagation	Attenuation, aberrations—turbulence, absorption, etc.
Target	Scattering, signal strength, range, Doppler shift, signal fluctuations and statistics, speckle, coherence time, etc.
Detector	Bandwidth, sensitivity, noise characteristics, cooling, etc.
Signal processor	Frequency, integration times, repetition rate, thresholds, missed signals, false alarms, etc.
Data analyzer	Parameters, integration times, repetition rate, thresholds, missed signals, false alarms, etc.
Information	Range, velocity, bearing, elevation, character, signal strength, fluctuations, error rates, etc.

12.3.2 Atmospheric transmission and choice of laser

For field operation at useful ranges, one generally requires a laser with good transmission through the atmosphere (although it may be noted that, for some operations, a lidar with limited sensing out to a few tens of metres, and which thereafter is rapidly absorbed, may be valuable). Figure 19.2 in Volume I shows a low-resolution spectrum of transmission in the atmosphere near ground level in clear weather. Only certain regions in this spectrum (known as "atmospheric windows") can be used for remote sensing over extended ranges. As is obvious in Figure 19.2, Volume I, the absorption due to molecular components, primarily water vapour and carbon dioxide, in the atmosphere establishes three primary bands of good transmission: from the near-ultra violet at ~0.3 μm through the visible region and up to ~2.5 μm; 3–5 μm; and the longer-wavelength 8–13 μm region. Table 12.6 shows a brief list of lasers that may be used. At the shorter wavelengths, a vast range of lasers is available, including gas lasers (e.g., argon ion Ar⁺, helium–neon He–Ne, etc.), solid state (e.g., ruby, neodymium–yttrium aluminium garnet, Nd-YAG, etc.) and semi-conductor and fiber amplifier lasers.

The 3–5 μm range is less well served with CO and various He–Ne–Xe lasers and OPOs currently offering rather low efficiency in conversion of electrical to optical laser power (so called "wall-plug" efficiency). In the 8–12 μm band, carbon dioxide

CO_2 lasers are widely used and offer both high efficiency (often >10%) and a very wide tuning range over many possible molecular lasing transitions.

Selection of the laser type and most effective lidar for a given remote-sensing problem are obviously determined by a vast range of factors, some of which are indicated in Table 12.5 and further discussed in the following section. Any decision is often a matter of best engineering judgement and not susceptible of scientific "proof". Over the years, considerable controversy has sparked between advocates of different lasers and systems; such arguments have usually generated rather more heat than light and on occasions have delayed progress.

12.3.3 Signal strength, signal-to-noise and lidar calibration

The signal return power, collected by a laser radar from a large, solid, hard target (bigger than the laser beam), can be written as

$$P_R = P_T \cdot \varepsilon \left[A / \pi R^2 \right] T \cdot e^{-2\alpha R} \qquad (12.7)$$

where P_T is the transmitted power, e is the target reflectivity, A is the effective collection area of the lidar receiver and R is the range to the target from the lidar. Note that, if the target is small compared with the beam, the range dependence becomes $1/R^4$. The atmospheric absorption coefficient for the operating wavelength is α and T is the transmission

Table 12.6 Some of the lasers that may be employed for lidar work in the visible and near infrared

Gas lasers	He–Ne 0.63 µmm; Ar+0.45–0.55 µmm; CO, He–Ne–Xe 3–5 µmm; CO_2 9–12 µmm
Solid state	Nd-YAG—1.06 mm (and frequency doubled 0.53 mm and frequency tripled 0.35 mm); ruby 0.65 µmm; Tm, Ho: YAG: YLF 2–2.2 mm
Diode lasers	0.8–1.6 µm with power augmented in optical fiber amplifiers; distributed feedback (DFB) lasers 1.55 µmm
OPOs	Optical parametric oscillators, tunable 3–5 µmm

of the lidar optics. For a direct detection lidar (Figure 12.3a), the effective collection aperture A_{dir} is given by the area of the lidar-receiving telescope (and as dictated by the available etendue U of the analysing interferometer). However, for coherent lidar (Figure 12.3b, the situation is somewhat more complex, as discussed in Section 12.2.2. To a good approximation, with a well-adjusted system, the effective area of the collection aperture for a heterodyne system can be written as

$$A_{het} = [1/A_{SP} + 1/A_{LO} + 1/A_{AT}]^{-1} \quad (12.8)$$

where A_{SP} is the average area of a speckle element (Figure 12.1c) in the return field. The influence of the local oscillator and its spatial profile is incorporated by the A_{LO} term. Atmospheric turbulence is incorporated by the A_{AT} coherence area term. Turbulence effects will dominate, and the return signal will be greatly reduced, if A_{AT} is smaller than the combination of the unperturbed speckle and local oscillator areas A_{SP} and A_{LO}. Physically, this means that, with reference to Figure 12.1c, the speckles have been broken up by the turbulence-induced refractive index fluctuations and are much reduced in size.

In a lidar experiment, the accuracy with which a parameter may be measured will be governed by the signal-to-noise ratio (SNR). This in turn is fundamentally constrained by the quantum nature of light and the number of photons in the signal. Thus, for an experiment lasting a time t_e with a signal power P_s and effective quantum efficiency of the detector of η, the mean number N_s of signal photons detected (each of energy $h\nu$ where h is Planck's constant and the angular frequency of the light ν is equal to $2\pi c/\lambda$) will be given by

$$N_s = \eta(P_s / h\nu)t_e \quad (12.9)$$

The SNR relations for direct and coherent detection are, in fact, rather different. With reduced

spectral discrimination in direct detection, there is the possibility of collecting background light N_b (e.g., daylight) not related to the signal, and also a noise contribution N_d from the detector itself so that the average measured signal $\langle N_m \rangle$ is given by

$$\langle N_m \rangle = \langle N_s + N_b + N_d \rangle \quad (12.10)$$

From this, it might erroneously be supposed that the SNR would be given by $N_S/(N_b+N_d)$. This, however, provides no indication of the accuracy with which the signal can be measured, for which we need the fluctuations in these quantities. The magnitude of the root variance N_s is, in fact, for commonly met statistics, given by $N_S^{1/2}$, with similar expression for background and detector noise. Thus, measurements of detector output, with and without signal present, provide an SNR_{dir} given by

SNR_{dir} = (measured signal) / {sum of variances}$^{1/2}$

$$= (N_s)/[(N_s+N_b+N_d)+(N_d+N_h)]^{1/2}$$

$$= (\eta^{1/2}P_s t_e^{1/2})/(h\nu)^{1/2}[(P_s+P_h+P_d)+(P_d+P_b)]^{1/2}$$

$$(12.11)$$

In the case where $P_s \gg P_b$, P_d the SNR reduces to $N_S^{1/2}$. In this case, the "noise" is truly "noise-in-signal", as governed by the quantum nature of light detection.

In coherent heterodyne detection, due to the mixing of the local oscillator light field E_{LO} and the signal light field E_s, the detector provides an electrical signal i_s proportional to their product so that

$$i_s \propto E_{LO}E_s \quad (12.12)$$

Analysis shows that, in a well-adjusted system, the coherent SNR_{coh} in the power spectrum of the signal is given by

$$\mathrm{SNR_c} = \eta P_s / h\nu \; B = N_s / B \quad (12.13)$$

$$\delta W \geq 0.425 W_G N_S^{-1/2} \quad (12.14)$$

where η is the effective quantum efficiency of the detector and B is the operational frequency bandwidth for the lidar system. It is supposed that the local oscillator has been arranged so that its own photon shot-noise is the dominant term above any contribution from even an inherently noisy detector. The bandwidth term B requires careful consideration. In the lidar signal processing, the instrumental bandwidth B_I requires to be well matched to the frequency band B_s over which the signal is spread. If B_I is set much greater than B_s, the $\mathrm{SNR_{coh}}$ is unnecessarily diminished. If it is set much smaller, not all the signal photons are effectively utilized.

The precision with which frequency measurements (e.g., Doppler shifts) may be made can be analysed in terms of the Cramer–Rao relationship. For direct detection, interferometric spectroscopy, the Cramer–Rao relationship gives the limiting (standard deviation) accuracy δW, with which the centre of a Gaussian spectral line profile can be found, as

where N_s is the total number of photocounts distributed across such a line profile of width W_G (half width at height $e^{-1/2}$). This is illustrated in Figure 12.5 by the lower line of slope $-1/2$, showing attainable accuracy versus total photocount. These total photocounts may be accumulated by summation over many individual laser pulses and, as noted, across a collection aperture containing many optical modes (provided they lie within the allowable etendue of the interferometer).

For coherent heterodyne spectroscopy, the situation is very different (see, e.g., Reference [2]). The simple Cramer–Rao relationship of Equation 12.14 needs to be multiplied by a complex term containing the photocount degeneracy δ. The effect of this is manifest in the upper curves of Figure 12.5 for accumulation of $n=1$ and $n=100$ laser pulses. As outlined in the previous section, the light signal must be collected within a single optical mode and for a low photocount the slope (for degeneracy $\delta < 3.3$) is close to -1 and the accuracy is best improved by increasing the laser energy per pulse. Only for a

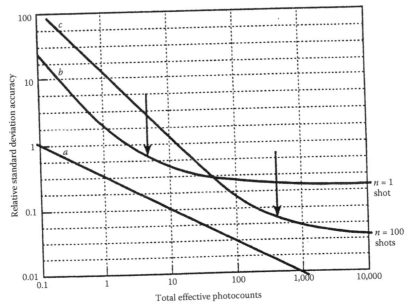

Figure 12.5 Schematic of the standard deviation of Doppler frequency estimates as a function of the total effective photocount on log-log scales. Line (a) shows the direct detection Cramer–Rao limit with slope $-1/2$. Curve (b) is the heterodyne Cramer–Rao for $n=1$ shot (i.e., pulses); note the slope of -1 at low photocount and tending to zero and saturation at high photocount. The arrow indicates the region of closest approach with photocount degeneracy $\delta \approx 3.3$. Curve (c) is the heterodyne Cramer–Rao for $n=100$ shot (i.e., pulses): this is a translation of curve (b) by a factor of 100 in effective photocount and a factor on $(n)^{1/2}=10$ in standard deviation accuracy.

higher photocount ($\delta > 3.3$, deriving from stronger scattering) is the accuracy best improved by accumulating over a number of pulses. There is thus a very clear difference in the design considerations for the two classes of lidars. For direct detection, the available laser energy may be spread without undue penalty over many pulses—the total effective photocount and accuracy depend only on the total energy transmitted (and proportionately scattered). For coherent heterodyne spectroscopy, it is essential (at lower scattering levels) to make the energy per pulse as large as possible to ensure that δ is as large as possible.

Further examination of signal statistics, processing, etc., is beyond the scope of the present chapter. The subject rapidly becomes extremely complex with many highly specialized techniques. It is, however, worth considering the topic of calibration and performance (see, e.g., References [5–8]). Equations 12.7 and following relate the lidar signal to the lidar parameters. If any of these latter are in any way defective, e.g., poor optical transmission T, inefficient detector η, reduced laser power P_T, etc., the system performance will be impaired. It is, in fact, often remarkably difficult to prove by absolute calibration test of signal strength that a lidar system (particularly a coherent system) is working to its full potential. The principle of calibration is fairly straightforward; given a test target of well-known scattering characteristics (with reliable ε in Equation 12.7) a SNR_{calc} is *calculated* for the lidar from all its individual, precisely-measured, parameters (e.g., optical transmission of lenses, beam splitters detector quantum efficiency, etc.) as inserted into the lidar equations. This value is then compared with that SNR_{obs} which is actually observed at the output of the lidar signal processing. Any major discrepancy requires explanation and a search for defective elements, alignment and signal processing components. It is of course extremely important to conduct such an exercise; a 3 dB (factor of 2) loss in SNR_{obs} for example would either require in compensation a doubling of laser power or provide a range performance reduced by at least 30%.

12.3.4 Range resolution, optical and system design issues

For pulsed lidars, the range and range resolution are given by Equation 12.6a and b and the selection of time delay t_f, length of laser pulse Δt_p and the length of the timing interval Δt_g (often called the "range gate") in the signal-processing system. Pulse length and range gate are usually set to be rather similar for most effective operation. As noted in Section 12.2.1, a 10 ns pulse gives a range resolution of 1.5 m, which may provide adequate accuracy for range measurement on a solid distant object. For an extended target, such as the atmosphere, measurements over successive volumes along the laser beam may be required, in which case, for example, a 1 µs pulse (and matched signal-processing range gate) would give ~150 m resolution. In this case, measurements up to 15 km would require 100 equal range gates, each of equivalent length 150 m in the processing.

For direct detection, the signal strength has a range dependence given by Equation 12.7 and reduces as R^{-2} (supposing that all the transmitted energy is incident on the "target"). The corresponding range dependence for the coherent heterodyne case was first evaluated in 1971 by Sonnenschein and Horrigan [9] (see also Reference [10]); it is implicit in Equation 12.8 but can be expressed more directly as

$$S(R,F) \propto [R^2\{1+(A/\lambda)^2(1/R-1/F)^2\}]^{-1} \quad (12.15)$$

where F is the range at which the lidar beam is set to focus and A is the effective area of the telescope radius (~A_{LO} in Equation 12.8). This expression has some interesting implications; for a target at a given range R, the maximum signal is attained by focusing at the target and setting F equal to R. Thus, for solid targets (so-called "hard" targets with scattering from a solid surface) expected at various long ranges, the lidar would be set with the outgoing beam almost collimated; the R^2 term dominates in the bracket on the left-hand side of Equation 12.14 and the range dependence (R^{-2}) is the same as for direct detection. Similarly, for a short range, the signal strength will peak up at ranges R close to F. This is illustrated in Figure 12.6, where a representative value of $(A/\lambda)^2$ equal to 3×10^6 m² has been used. This establishes the possibility of range selectivity with a continuous wave laser beam and is particularly useful for measurements on an extended or diffuse source of scatterers, such as small particles in the atmosphere. In this case, the measured coherent signal derives most strongly from the focal region of the beam. This comes about because the scattered radiation from this region is best matched in wave front to the local oscillator beam in the lidar. In consequence, it beats most efficiently with the local oscillator and provides the strongest signal. The curves shown in Figure 12.6 are representative of what may be achieved with a 10.6 μm coherent lidar with a 30 cm diameter transmitting/receiving telescope (with the outgoing laser beam and equivalent A_{LO} of best diameter ~16 cm). The central probe volume is commonly defined as the region within which the signal is within 3 dB of its peak position: its half-length is written as ΔR_{3dB}. Some manipulation of Equation 12.15 with the approximations $F \approx\gg \Delta R$ gives:

$$\Delta R_{3dB} = \pm F^2 (\lambda / A) \qquad (12.16)$$

Thus, at a focal setting of $F = 100$ m, a spatial resolution of about ±6 m is produced. For such diffuse, extended targets, the total observed signal S will be determined by the integral over an appropriate range interval R_1 to R_2 given by

$$S \propto \int_{R_1}^{R_2} P(R) S(R, F) \delta R \qquad (12.17)$$

where $P(R)$ is the strength of scattering at range R. In a generally clear uniform atmosphere, $P(R)$

over short ranges may be considered as constant, but obviously this is a very crude approximation in conditions of low cloud, layered fog or smoke [11].

Brief consideration of these equations indicates some of the complexity of comparing coherent lidar performance at different wavelengths. Thus, at shorter wavelength, the focal depth will be smaller and the spatial resolution will be greater; the total signal S is correspondingly reduced for the smaller ΔR. However, at shorter wavelength, the maximum useful size for the effective telescope diameter, D, is limited by refractive turbulence in the atmosphere (which destroys lateral coherence across the beam, thus reducing A_{SP}; see Section 12.2) and has an impact approximately proportional to $\lambda^{-6/5}$. Thus, if atmospheric conditions permit, an effective telescope diameter of ~50 cm at 10.6 μm wavelength would be reduced by a factor $(10.6/2.06)^{6/5}$ [12] (for an extensive discussion with many references, see also Reference [13]) at 2.06 μm to only ~7.0 cm. Such a reduced collection aperture would provide both a weaker signal and poorer spatial resolution than the admittedly larger and more expensive telescope that could be used in these circumstances at the longer wavelength. Such wavelength comparisons may be extended to many other aspects of signal collection, processing and

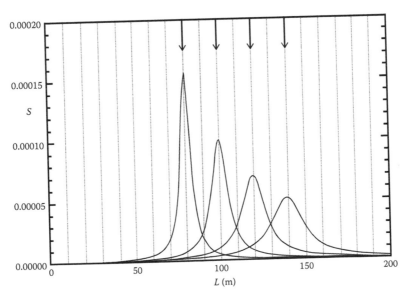

Figure 12.6 The range sensitivity for a coherent lidar as given by Equation 12.14 with the $(A/\lambda)^2$ terms equal to 3×10^6 m², and for four values of the focus $F = 80$, 100, 120 and 140 m. Note how the sensitivity peaks up very strongly around the focal range F. This permits the possibility of useful range resolution with a continuous wave laser for measurements on an extended diffuse target such as aerosols in the atmosphere.

measurement precision. Thus, in Equation 12.13, the instrumental bandwidth required to cover a given velocity interval is proportional to λ^{-1} (see Equation 12.5 for the Doppler shift) and altogether with the larger photon energy $h\nu$ gives

$$\text{SNR}_c \propto \eta P_s \cdot \lambda^2 \qquad (12.18)$$

For a given laser power, the scattering P_s from most surfaces and aerosol particles increases at shorter wavelength. To give a signal-to-noise advantage at shorter wavelength, this rate of increase obviously needs to be greater than λ^{-2} to overcome the longer wavelength advantage of Equation 12.18.

Another point of comparison is the choice of pulse length Δt_p. Fourier transform analysis of a short wave train gives a limiting frequency resolution $\Delta f_T \sim (\Delta t_p)^{-1}$. Thus, a 1 μ pulse gives a line width of ~1 MHz. Comparing this with Equation 12.5, such a line width would provide a velocity resolution ΔV given by

$$\Delta V = \Delta f(\lambda/2) = \lambda(2\Delta t_p)^{-1} \qquad (12.19)$$

Thus, at 2 μm, a pulse length of 1 μs would give a Fourier transform velocity resolution of ~1 ms⁻¹, which increases to ~5 ms⁻¹ at 10 μm. This potentially provides a significant advantage to shorter wavelengths and also offers the possibility of achieving greater spatial resolution (with shorter pulses).

Further discussion is beyond the scope of the present article; wavelength and system comparisons are generally very difficult and rarely permit a clear cut "winner". Selection of a lidar system for a given task usually requires consideration of many issues, often practical and logistical, as well as the technical options outlined in Table 12.5. As an example, the overriding constraints for an aircraft or military installation is likely to be the maximum optical aperture that can be accommodated in the structure, together with strict limits on mass, volume and available power budget.

12.4 LIDAR AND ATMOSPHERIC SENSING

12.4.1 Introduction: Atmospheric parameters and measurement

Some of the first demonstrations of lidar in the 1970s were measurements of wind and cloud. The light scattered from a laser beam in a clear atmosphere is primarily from air molecules (Rayleigh scattering) and from small airborne particles (dust, etc.), usually referred to as aerosol scattering. The strength of molecular scattering is approximately proportional to λ^{-4}, whereas the wavelength dependence of aerosol scattering typically lies between λ^{-1} and λ^{-2}. Thus, molecular scattering is much stronger at shorter wavelengths (hence the blueness of the sky) and also increases relative to aerosol scattering. Perhaps the most obvious common example of the latter is the appearance of a shaft of sunlight between clouds.

Wind and atmospheric measurements are interesting and important for many reasons: to meteorology and climatology, for example, for the information they provide on the behavior of the atmosphere, to local studies of shear and turbulence, including the planetary boundary layer and practical measurements for shipping, aviation and commerce. All these impact the four primary parameters of interest to the atmospheric physicist shown in Table 12.7.

Lidar studies make a contribution to all these areas and this is likely to increase in the future with advanced instruments operating remotely and from space.

Table 12.7 Primary atmospheric parameters

Structure	Temperature, density and pressure as a function of altitude
Dynamics	Circulation and motion at all scale sizes from global through synoptic to small local scale
Radiative properties	Distribution, characteristics and structure of clouds and aerosols and their impact on solar, ground and atmospheric radiation
Chemical concentration	Vertical structure and horizontal distribution of molecular species including minor constituents such as ozone, radicals and active molecules (Section 12.6)

12.4.2 Scattering in the atmosphere

Over the past 30 years, considerable effort has been put into the measurements of scattering in the atmosphere in many different regions, seasons and wavelengths. A brief account of molecular, aerosol and cloud scattering data is presented as an essential preliminary, and a schematic of aerosol and molecular scattering is shown in Figure 12.7. First, it is worth noting the nomenclature $\beta(\theta, \lambda, z)$ m^{-1} sr^{-1} that is commonly employed for the strength of atmospheric scattering. This expresses the fractional amount of light that would be scattered from an incident light beam along a metre length of path (m^{-1}) into the equivalent unit radian solid angle (sr^{-1}) centred at scattering angle θ, at wavelength λ and at a height z in the atmosphere. Typically, this will often be abbreviated to $\beta(\pi)$ as a general term for backscattering. It is also worth noting various lengthy publications devoted in whole or part to scattering in the atmosphere, including those by Scorer [14], Hinkley [15] (in particular the chapters by Zuev [16], Collis and Russell [17] and Inaba [18]), and also the Conference Publications of the biennial Coherent Laser Radar Meetings and International Laser Radar Conferences.

(a) *Molecular scattering.* Molecular scattering, in both theory and practice, has been extensively discussed over the years; see, for example, the early review of Fabelinski [19]. In the atmosphere, the strength of molecular scattering varies slowly, uniformly and predictably with altitude through the atmosphere. The exact exponent for wavelength scattering should be 4.09 instead of 4 to account for dispersion of the index of refraction of air. Taking the reference wavelength of $\lambda_0 = 1.06\,\mu$m and a reference height of $z_{mol} = 8000$ m, the total backscattering contribution from the air molecules in the atmosphere at height z may be modeled by an exponentially decreasing function:

$$\beta_{mol}(\pi, z, \lambda) = 10^{-7}(\lambda_0/\lambda)^{4.09} e^{-z/z_{mol}} \text{m}^{-1}\,\text{sr}^{-1}$$

(12.20)

The molecular spectrum is quite different in the IR from what it is in the visible due to the contribution of collisional effects on the molecular scattering and hence the atmospheric pressure. At low densities (or pressure), molecules scatter independently, producing a single, near-Gaussian line shape, whereas at higher densities and longer wavelengths the central component, the so-called Rayleigh line, bisects the Mandlestam–Brillouin doublet due to scattering from moving density fluctuations (sound waves). Thus, a well-defined triplet structure prevails at near infrared (2 µm) and thermal infrared (10 µm) wavelengths, and appears to offer some opportunity for Doppler wind lidar (DWL) applications with coherent heterodyne techniques, as discussed by Rye [20].

Other molecular scattering phenomena include resonant scattering and Raman scattering. Both may be used for atomic and molecular species identification and concentration measurements of, for example, water vapour and ozone (and aerosol concentration) and study of atmospheric chemistry and dynamics in the atmosphere (see, e.g., References [18,21,22] and Section 12.6) in the atmosphere.

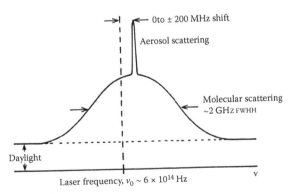

Figure 12.7 Schematic of the light collected by a lidar operating in the visible region with components due to aerosol scattering, molecular scattering and background daylight. The relative proportion of these varies quite strongly with wavelength and height in the atmosphere. In the troposphere (up to ~10 km) and in the infrared at 10 µm aerosol scattering is generally dominant.

(b) *Aerosol scattering.* The characteristics of aerosol scattering are very complex, depending on the chemical constitution of the particles, their size distribution, typical shape, etc. These are largely dictated by the previous history of the air mass. Contributory sources include material blown up by winds and convected from the Earth's surface (most dramatically, for example, the fine Saharan sand occasionally spread over Northern Europe and the loess soil from Central China which extends every spring as a plume into the Western Pacific). Other sources include volcanic eruption (e.g., El Chichon in Mexico in 1982 and Mt Pinatubo in the Philippines in 1991), forest fires, salt spray from the sea surface, man-made pollutants and the constant rain of interplanetary material (manifest, for example, in shooting stars). It has been calculated that the eruption of Mt Pinatubo alone injected about 10 km³ of material in the form of fine dust into the stratosphere, with clear impact on the global climate for a few years thereafter. Typically, the particles most important for lidar scattering lie in the size range 0.1–10 μm diameter; larger particles settle out fairly quickly, while smaller ones scatter much less strongly.

The investigation of the nature, sources, transport and chemistry of atmospheric aerosol is assuming steadily greater importance in atmospheric science. Transport and dispersion models are an important component of the numerical modelling machinery applied to investigation in these areas, and are used to simulate the concentration and dilution (typically due to turbulence) of airborne material of all kinds. These activities include the modelling of air pollution and quality involving, for example, sulphur, nitrogen and photochemistry (see, e.g., References [23,24]). Increasingly sophisticated and accurate transport modelling is being used in studies of the source strength and background concentrations of radiatively active trace gases [25,26]. Operationally,

the modelling of aerosol transports features in the simulation of the spread of volcanic ash, of critical importance to aviation [27], and pollution resulting from nuclear or major chemical accidents and conflagrations. International protocols are in place for the management of such emergencies and extensive collaborative work had been carried out on the testing and intercomparison of transport and dispersion models (e.g., Reference [28]).

Transport of airborne pathogens, and biota in general, also require suitable transport and dispersion models, which have been applied to pollution problems, and the physics of atmospheric dispersion on all motion scales from street canyons to global diffusion [29]. A second area of aerosol and gaseous transport studies of pressing importance is concerned with the evolution of climate forcing and change, in which sulphate aerosol and volcanic effluent are just two of a range of airborne substances subject to intensive investigation, typically involving general circulation models.

There have now been an enormous number of lidar measurements of aerosol scattering with ground, airborne and space platforms. The wavelength range covers the UV at ~0.35 μm to IR at 10.6 μm with direct detection and heterodyne techniques (see, for example, Tables 12.8 and 12.9).

However, there are great problems in synthesizing these into a database of global perspective, taking account of different wavelengths, regions, seasons, histories, sampling, etc., of the measurements. Two extensive airborne programmes conducted in the late 1980s and early 1990s provide a broad base of knowledge at some of these wavelengths. These were the SABLE and GABLE programmes (South Atlantic/Global Atmospheric Backscatter Lidar Experiments) of the USAF Geophysics Laboratory and the then UK Royal Signals and Radar Establishment (subsequently the Defence Evaluation and Research Agency,

Table 12.8 Recent papers on aerosol backscatter from lidar measurements over the range 0.35–1 μm

Ansmann et al. [30]	Marenco et al. [38]
Barnes and Hofmann [31]	McCormick et al. [39]
Browell et al. [32,33]	Osborn et al. [40]
Cutten et al. [34]	Parameswaren et al. [41]
Donovan et al. [35]	Post et al. [42]
Hoff et al. [36]	Shibata et al. [43]
Li et al. [37]	Spinhirne et al. [44]

Table 12.9 Principal centers for coherent lidar measurements of atmospheric backscatter $\beta(\pi, 10\,\mu m)$ and a selection of more recent references

NOAA Wave Propagation Laboratory (WPL), Boulder, CO	Groundbased, pulsed	[45–48]
NASA Marshall Space Flight Center (MSFC), Huntsville, AL	Groundbased, pulsed Airborne, CW	[34,47–54]
USAF Geophysics Laboratory (GL), Hanscom AFB, MA	Groundbased, pulsed	[55–58]
Jet Propulsion Laboratory (JPL), Pasadena, CA	Groundbased, pulsed Airborne, pulsed	[54,59–63]
DERA (Malvern), UK—formerly RSRE	Airborne CW to 16 km altitude	[47,48,55–57,64–67]
Several other smaller data bases, also SAGE extinction data	[39,68,69]	

Malvern) conducted across the North and South Atlantic, and the NASA-supported GLOBE (Global Backscatter Experiment) programme, which extended across the Pacific. These studies were conducted well after the El Chichon volcanic eruption of 1982 and prior to the Pinatubo eruption of 1991. The data were thus accumulated in an historically "clean" atmospheric period, and may thus represent the lowest levels of backscattering. This concept of a "background mode" of atmospheric backscattering had in fact been suggested by Rothermel et al. [47] from data drawn from two widely separated areas—over the UK and over the continental USA.

The measurements taken in the SABLE and GABLE trials with an airborne 10.6 μm cw lidar have been extensively analysed and discussed [56,57] and the median and upper/lower quartiles and deciles are shown in Figure 12.8. These appear to comprise the most comprehensive database currently available at 10.6 μm for the Atlantic region. The measurements amounted to nearly 200,000 individual records of backscatter made during 80 flights over the Atlantic in six different regions and/or seasons. They are put forward as a reasonable measure of global backscatter under generally clean atmospheric conditions at 10.6 μm.

At 1.54 and 2 μm, validated data sets of $\beta(\pi)$ were collected over the Pacific Ocean during the GLOBE II mission in spring of 1990 [44,70] and at 2 μm over the continental US in 1995–1996 [71], respectively. Measurements at 0.35–1 μm have been

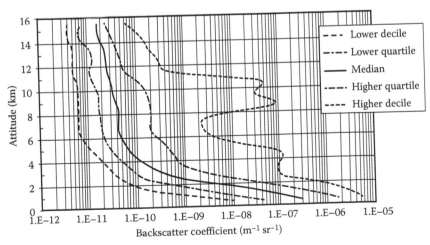

Figure 12.8 Median, quartiles and deciles of backscatter coefficient versus altitude at 10.6 μm as measured across the tropical mid-latitude and northern Atlantic and presented as a reasonable global measure under atmospherically clean conditions.

mainly directed to the estimation of the typical particle size in specific areas. However, measurements made with the NASA/LITE (Lidar In-space Technology Experiment; see also Section 12.9.2) in September 1994 [72] have been able to document large areas of aerosol in the upper troposphere and stratosphere [73], and the transport of anthropogenically produced aerosols [36], Saharan dust [74] and biomass burning aerosols [32]. The measurements were made at three different wavelengths: 0.355, 0.532 and 1.064 µm. These three wavelengths were used to derive the aerosol and cloud characteristics.

Much effort has been put into the different comparison of backscattering at different wavelengths (see, e.g., References [34,51,66,67]). On the one hand, there are very few colocated and contemporaneous measurements of backscatter at different wavelengths. On the other hand, calculations based on Mie scattering are critically dependent on the assumed aerosol size distributions, chemical constitutions and refractive indices. Generally, the atmospheric aerosol backscatter coefficient increases with decreasing wavelength. For a wavelength range excluding anomalous refractive index changes, this dependence may be expressed by a power law:

$$B(\lambda, z) = \beta(\lambda_0, z)(\lambda_0/\lambda)^\varsigma \qquad (12.21)$$

where $\beta(\lambda_0, z)$ is the backscatter coefficient at a chosen wavelength λ_0 and at a height z and where ς is the wavelength scaling exponent to derive backscatter at other wavelengths λ.

From the SABLE/GABLE data shown in Figure 12.4, and comparison with other data including modeled Mie scattering [75] and other measurements at shorter wavelengths (e.g., References [44,54,71]), the following expression for ς was derived:

$$\varsigma(0.35 < \lambda < 2.1\,\mu m, z) =$$
$$0.24\left|\log_{10}(\beta_0(10.6\,\mu m, z))\right| - 0.62 \qquad (12.22)$$

This expression is considered valid for shorter wavelengths, in particular the range 0.35–2.1 µm and may be used to translate the SABLE/GABLE data at $\lambda_0 \equiv 10.6\,\mu m$. Just such a translation from 10.6 µm is shown in Figure 12.9a and compares well with the measurements and other data at

1.06 µm shown in Figure 12.9b. Another extensive and valuable investigation has been reported by Srivestava et al. [76].

Equation 12.22 must be used with caution. At 9.1 µm, for example, the situation is rendered particularly complex by the well-known "resonance" in scattering due to the sharp changes in refractive index at this wavelength, particularly for ammonium sulphate; see, in particular, Figure 3 of Srivestava et al. [51]. In consequence, the backscatter at 9.1 µ is generally greater than at 10.6 µm and the ς component in the scaling law increases more sharply than at shorter wavelengths. The following equation encapsulates such an increase:

$$\varsigma(9.11\,\mu m, z) = 1.25\left|\log_{10}(\beta_0(10.6\,\mu m, z))\right| - 8.25$$
$$\qquad (12.23)$$

A final example of aerosol backscatter measurement and comparable atmospheric trajectory analyses is shown in Figure 12.10. Air masses with notably different aerosol backscatter, measured at three levels, 6.5, 8.25 and 10 km over the Arctic, originated over Europe, the United States and Canadian Arctic, respectively (from Reference [77]).

(c) *Scattering from the Clouds.* Clouds display large backscatter coefficient (with $\beta_c(\pi) \approx 10^{-6}$ to $10^{-2}\,m^{-1}\,sr^{-1}$) and large extinction coefficient ($\alpha \approx 10^{-5}$ to $10^{-1}\,m^{-1}\,sr^{-1}$). The strength of scattering may vary by several orders of magnitude at a given wavelength and may vary by one to several orders of magnitude for different cloud size distributions and particle number density as a function of wavelength. Dense clouds strongly attenuate at laser wavelengths so lidar measurements cannot penetrate more than tens or hundreds of metres. However, holes are present in dense clouds at the small scale (on the order of 1–100 m) and results in so-called optical porosity. Such an optical porosity allows profiling of the atmosphere beyond dense clouds depending on the relative size of the lidar footprint and optical porosity (specifically on the probability density function of optical porosity). Lidar measurements at an angle close to nadir are favorable to reach the surface, although the probability of a clear line of sight through dense clouds does not decrease too much for angles not larger than 30–40° with respect to nadir. The only cloud type with an expectation of continuous lidar penetration is cirrus. The typical thickness of cirrus

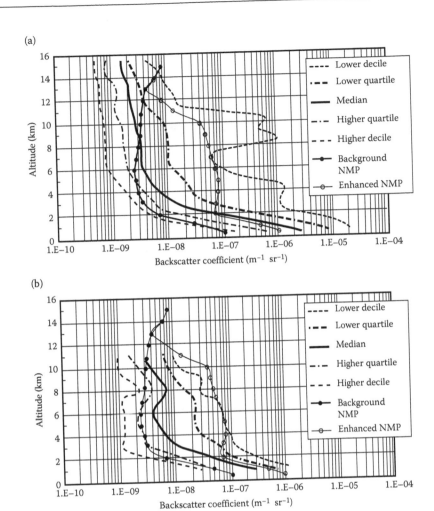

Figure 12.9 (a) Backscatter coefficient versus altitude at 1.06 μm as derived from Figure 12.8 (for 10.6 μm) with the wavelength scaling of Equation 12.21. The background and enhanced NASA New Millennium Programme (NMP) profiles are also included for comparison (see text). (b) Backscatter coefficient versus altitude at 1.06 μm for the GLOBE II measurements over the Pacific. The background and enhanced NMP profiles are also included for comparison.

clouds is <1 km but on occasion thicknesses of up to 4 km have been observed. An analysis of probability of occurrence of layered cirrus has been discussed by Vaughan et al. [66] together with calculations of the impact on spaceborne lidar operation. Backscatter and extinction coefficients have been calculated for water clouds using Mie theory or relevant approximations for spherical particles with a good accuracy (e.g., Reference [78]). This work has been conducted to derive the relevant relationships between microphysical parameters like liquid water content and optical parameters, i.e., $\beta(\pi)$, α, k (e.g., References [79,80]). As shown in Table 12.10, backscatter coefficients for a given cloud type do not change sharply in the visible and near IR up to 1–2 μm, but are typically smaller by 1–2 orders of magnitude and more variable in the 10 μm region.

12.4.3 Wind and related measurement

1. Short range up to ~1 km
   Wind measurement may be conducted with both pulsed and continuous wave (cw) lasers. The former are typically used for ranges >~1 km and, with standard techniques of range gating, the range resolution is determined by

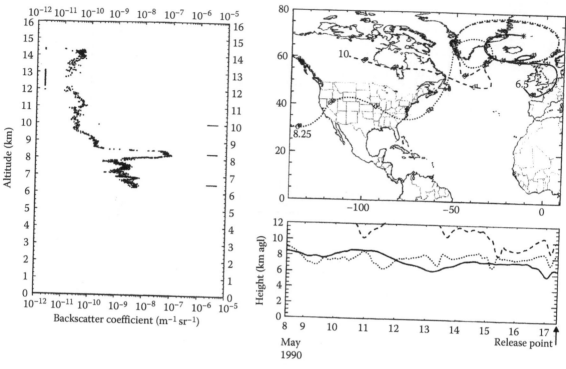

Figure 12.10 Flight 63 from Keflavik: altitude record of backscatter for the descent flying to the E along 71°N (in the region of 13°W) from 09.50 to 10.30 UTC on 17 May 1990. The strong scattering in the lower levels of the troposphere (< 9 km) shows a very strong narrow band centred at 8.25 km which was absent on other legs of the flight. There was a sharp temperature inversion at 11.5 km; in the lower strasphere the scattering peaked at ~14 km but was very weak in the band 12–13 km. Air mass back trajectories for 9 days computed over the release point 71°N, 13°W at 10.15 UTC (May 1990) for 6.5, 8.25 and 10 km. Note the anticylonic loops and the remarkably different origins of the air masses for these trajectories from, respectively, Europe, the United States and the Arctic.

Table 12.10 Approximate values of extinction and backscatter coefficient of different cloud types in the ranges ~0.4–2 and ~10 μm (PSC—polar stratospheric cloud)

Cloud type	Backscatter coefficient B_c ($m^{-1}\,sr^{-1}$)		Extinction coefficient α_c (m^{-1})		Altitude (km)
	< 2 μm	Low 10 μm High	< 2 μm	Low 10 μm High	
Cumulus	6.0×10^{-4}	1×10^{-5} to 1×10^{-4}	1.2×10^{-2}	5×10^{-3}–30×10^{-3}	2–10
Stratus	5.0×10^{-3}	3×10^{-5} to 5×10^{-4}	9×10^{-2}	1×10^{-4}–7×10^{-2}	0.2–0.7
Alto-stratus	1.0×10^{-3}	1×10^{-5} to 1×10^{-4}	1.8×10^{-3}	3×10^{-3}–2×10^{-2}	2–4.5
Cumulo-nimbus	1.0×10^{-2}	4×10^{-5} to 1×10^{-3}	1.8×10^{-1}	1.5×10^{-2}–6×10^{-2}	2–4
Cirrus	1.4×10^{-5}	1×10^{-5} to 1×10^{-5}	2×10^{-4}	5×10^{-4}–5×10^{-3}	8–16
PSC	3.0×10^{-7}		6×10^{-6}		

the laser pulse length and signal processing. For a cw beam, the atmosphere of course provides an extended target with scattering at all distances. Nevertheless, as discussed in Section 12.3.4, quite a sharp range resolution can be attained with a cw lidar by focusing the beam to give a peak sensitivity around the focal range F (Figure 12.6).

An early measurement with a cw CO_2 LDV lidar is shown in Figure 12.11. This record was

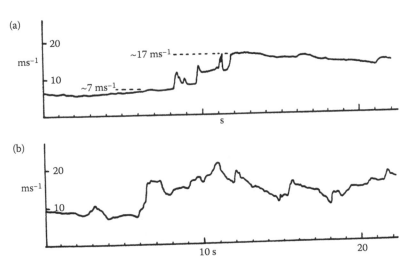

Figure 12.11 Records of strong gusts showing the line-of-sight wind component (derived from the dominant frequency in the Doppler spectrum) versus time. The CW CO_2 lidar beam was pointed at an elevation angle of 30° and the probe volume was set at a range of (a) 400 m and (b) 100 m.t

made under rather gusty conditions on an airfield with the LDV pointed at an elevation angle of 30° into the prevailing wind; the Doppler spectrum was measured every 12.8 ms. The Doppler shifted frequency of the strongest signal in the spectrum was converted to an analogue voltage and displayed as a function of time. In Figure 12.11a, the wind component changed from 7 to 17 ms^{-1} (from about 15 to 35 kt) over a period of about 5 s. Greater fluctuations are apparent at the shorter range in Figure 12.11b due to greater wind turbulence at the lower height and the reduced spatial averaging and integration of the signal over the considerably shorter probe volume.

Such a system provides a single line-of-sight wind-velocity component. However, simple extension of such equipment enables one to determine the wind field anywhere around the measuring station. The laser beam may be scanned in order to resolve various components of the wind and several different scan patterns have been developed to suit particular measurement tasks. One of the simplest is to use a conical scan about a vertical axis. In this case, if the half-angle of the cone is φ (i.e., the angle of the lidar beam from vertical) and θ_{sc} is the beam direction in the horizontal plane, the measured line-of-sight wind component V_M is given by

$$V_M = V_H \cos(\theta_M - \theta_{sc})\sin\Phi + V_V \cos\varphi$$

(12.24)

where V_H and θ_H are the horizontal wind speed and bearing and V_V is the vertical wind speed. Examination of this expression shows that, as the beam rotates around the cone, any vertical up/down draughts (usually small) contribute a constant term $V_V \cos\Phi$ and the horizontal component varies with the cosine of $(\theta_H - \theta_{sc})$. This is readily demonstrated on a polar plot of V_M (or equivalent Doppler shift), as shown by the measurement example of Figure 12.12. As the beam rotates, the magnitude of V_M traces out a figure-of-eight plot, for which the main axis gives the horizontal wind bearing (at θ_{sc} equal to θ_H) and the amplitude gives the wind speed. For the example shown in Figure 12.9, the cw CO_2 lidar beam was focused at five successive heights from 25 to 250 m with three revolutions of the beam at each height. Each revolution was completed in 1 s with recording of 64 individual Doppler spectra. With less than 2 s to alter focus, the complete cycle of wind measurement from 25 to 250 m may be completed in less than 25 s. Rapid analysis of the Doppler spectra, and fitting to Equation 12.24 to extract the wind parameters V_H, θ_H and V_V, may be carried out in real time. Figure 12.12 provides a clear example of changes of wind speed and direction with height and the data were acquired with a compact and

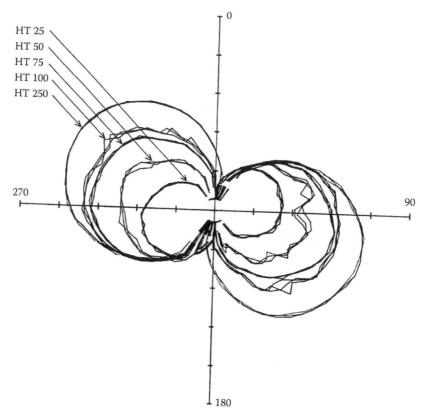

Figure 12.12 Polar plots of wind component at heights recorded at night in relatively calm conditions with a conically scanned CW CO_2 LDV. The wind strength (shown by the size of the figures-of-eight, with one scale division equal to 2 ms^{-1}) increased steadily with height from ~3.5 ms^{-1} (25 m) to ~8 ms^{-1} (250 m). A strong directional wind shear (shown by the axes of the wind plots) is also very obvious and amounts to ~36°. Data recorded with the compace, mobile DERA (Malvern LDV).

highly mobile equipment mounted in a Land Rover vehicle, which could make measurements within 5 min of arrival at a site. Another example of measurement with a cw LDV is illustrated in Figure 12.13. The conical scan was set to examine the changes of wind speed typically experienced by the revolving blades of a wind power generator. Comparison with simple models of turbulence showed reasonable agreement. Important questions of representativity and wind-velocity measurement errors for such systems have been considered experimentally and theoretically by Banakh et al. [81].

Several ground-based cw systems have now been built worldwide, mostly employing CO_2 lasers with power output in the range 4–20 W. LDV systems based on shorter wavelengths, notably with cw diodes and fiber amplifiers in the 1.5 μm region, also offer much promise

and are being developed for both monostatic and bistatic operation [82]. Such lidars should be compact, do not need detector cooling and incorporate fiber optic techniques for ease of assembly and precise wave-front matching. A recent example has been described by Karlsson et al. [83] and used to investigate signal statistics of particulate scattering in the atmosphere [82]. The non-Gaussian character with high SNR from single particles is dramatically different from the complex Gaussian scattering with multiparticles in the probe volume (see also work by Jarzembski et al. [84], who employed known particle sizes for lidar calibration).

The many applications of cw LDV include wind-flow measurement around buildings, smoke plumes from power stations and wind measurement from an oil rig to compare with satellite observations of the sea surface. Another

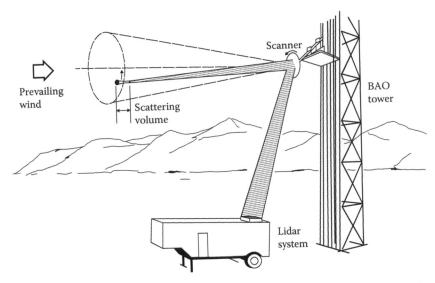

Figure 12.13 Schematic diagram for LDV measurements of rotationally sampled winds from a scanning mirror. The cone angle was fixed at 10° and the radius of the vertical measurement circle was varied by altering the beam focus. (From Hardesty, R. M. and Weber, B. F. *J. Atmos. Oceanic Technol.*, 4, 191–203, 1987.)

potentially important application is towards wind-turbine power generation; Vaughan and Forrester [85] identified three specific local tasks in addition to the strategic benefits of a global wind measuring system (see Section 12.9.3).

- *Site studies of local wind*. Wind generators require to be optimally sited. Conventional techniques with mechanical anemometers on tall towers can be very expensive. LDV, with its speed and mobility, could have a role.
- *Complex flow around wind generators*. LDV offers a unique capability for rapid, non-intrusive measurement in quite small probe volumes within the complex flows generated by the turbine blades.
- *Wind monitoring and gust warning*. Measurements made by scanning up to hundreds of meters ahead (e.g., with LDVs mounted on the nacelle of the generator) would permit advance control of blade pitch for most efficient power extraction. In addition to routine gains in efficiency, such LDVs would also provide warning of occasional large gusts or turbulence; with appropriate blade control, the risk of consequent catastrophic damage to blades or gear-train would be reduced.

Other applications of cw LDV lidars for aviation (wake vortices, true airspeed, etc.) are outlined in Section 12.5 and military tasks in Section 12.7.5.

2. Longer range measurements
   Laser Doppler wind measurements at ranges typically greater than ~1 km have been undertaken with several types of pulse lidar system:

- *Coherent heterodyne* at ~10 μm with CO_2 lasers, ~2.1 μm with Tm, Ho: YAG-YLF lasers, 1.55 μm with doped fiber amplifier lasers and 1.06 μm with Nd-YAG lasers. These all utilize aerosol scattering.
- *Interferometric direct detection* at 1.06, 0.53 and 0.35 μm with Nd-YAG lasers, utilizing both aerosol and molecular scattering.

For completeness, one should also note non-Doppler techniques, whereby time series of the strength of aerosol scattering from two beams of known separation may be compared (e.g., with correlation techniques). The time of passage of a strong fluctuation from one to the other (e.g., with peak in the correlation function) provides a measure of the cross wind but has limited temporal and spatial resolution [86–88].

Technology and application of *coherent LDV* have been well reviewed. For good frequency measurement, the transmitted laser beam requires to have minimal change of frequency through the duration of the pulse. For CO_2 lasers, the gas discharge conditions that ensure such good chirp characteristics, including gas catalysis, plasma effects and laser-induced media perturbation (LIMP), have been analysed by Willetts and Harris [89,90]. Practical use of heterodyne detectors, and attainment of good performance, has been considered by many authors; in one study, near ideal behavior with up to 13 dB local oscillators shot noise was demonstrated [91] (see also Reference [92]). Signal processing with different algorithms and procedures has also been extensively reviewed (e.g., see References [93–98]).

At the longer CO_2 wavelengths, TEA lasers have typically given larger pulses (0.2–3 J) with p.r.f. 0.1–10 Hz; for lower pulse energy (typically 1–10 mJ) and higher p.r.f, q switched and mini-MOPA lasers are employed. As an example at longer ranges and higher altitudes up to 10–20 km, the NOAA CO_2 lidar system has shown great reliability over many years. It has been used over widespread field campaigns with measurements of convective out flows, thunderstorm microbursts [99], sea breeze, flow in complex terrain [100] and downslope windstorms [101], etc. A record of measurement within the Grand Canyon taken at two levels below the rim is shown in Figure 12.14. Differential flow along the bottom of the canyon established that, on occasion, pollutants entered from the southwest USA and contributed to wintertime haze [102]. At shorter ranges, and particularly for weak winds in the boundary layer, smaller pulse systems have been widely developed and used (see, e.g., References [103,104]).

A particularly long-ranging, coherent system at 1.06 μm was reported in 1993 by Hawley et al. [105], employing a 1 J energy Nd-YAG coherent lidar. With aerosol scattering augmented by Pinatubo volcanic material, lidar measurements in a conical scan to 26 km altitude were documented, as shown in Figure 12.15, in good agreement with rawin-sonde wind speed. Coherent lidar at the eye-safe 2 μm wavelengths has employed lasers with pulse energies, typically in the range 2–50 mJ. A plan view of a 2 μm coherent system is illustrated in Figure 12.16. At the shorter wavelengths, range resolution may typically be 30–50 m (see Section 12.3); measurements on a stationary hard target showed bias errors of −3.3 cm s^{-1} with standard deviation error 11 cm s^{-1} [103]. Performance of a 2 μm coherent system was reviewed by Frehlich [106]. The compactness and potential for "turn-key" operation are very favorable considerations and such systems are now commercially available.

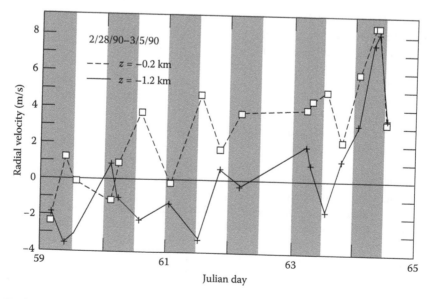

Figure 12.14 Radial flow within the Grand Canyon at two levels −0.2 and −1.2 km height relative to the canyon rim measured by CO_2 lidar. The shaded areas are night. Under the low wind (stagnant conditions) of the first few days, the flow at the top of the canyon is opposite in direction to the flow at the bottom of the canyon. In stronger winds at the end of the period, the flows were in the same direction.

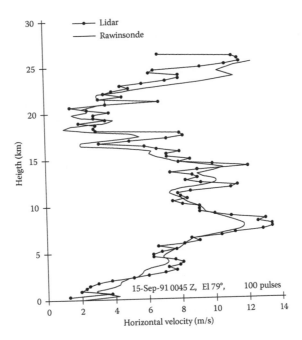

Figure 12.15 Comparison of wind speed with altitude for lidar and rawinsonde to 26 km measured above the Kennedy Space Centre Shuttle Landing Facility. The 1.06 μm lidar scanned at 20° off vertical at six fixed positions with 100 pulses averaging at each position. A complete record was completed in 3 min. (From Hawley, J. G. et al., *Appl. Opt.*, 32, 4557–4568, 1993.)

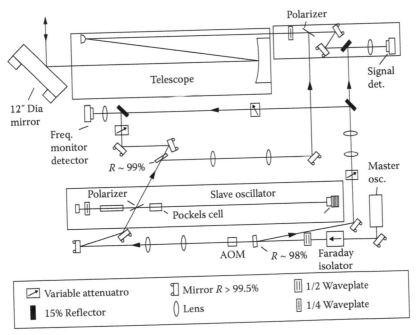

Figure 12.16 Plan view of a 2 μm injection-seeded coherent lidar reviewer. (Hardesty, R. M. and Huffaker, R. M., Remote sensing of atmospheric wind velocities using solid state and CO_2 coherent laser systems, *Proc. IEEE* @ 1996 IEEE.)

A pulsed lidar system based on fiber optic technology operating at 1–548 µ with an erbium doped fiber has recently been described by Pearson et al. [107]. Operation in both bistatic and monostatic antenna configurations was investigated.

In addition to wind measurement with such lidars, it is worth noting that the strength of aerosol scattering may also provide useful information. One proposal has, for example, suggested the early detection of forest fires from the increased scanning due to quite minor smoke plumes. A lidar beam scattering from an elevated platform could provide coverage and monitoring over quite large areas.

As discussed in Section 12.3, the design considerations for *direct detection, interferometric lidars* are very different. In particular, the signal may be accumulated from many small laser pulses without undue penalty (see Figure 12.3) and across many optical modes. In addition, at shorter wavelengths, the molecular scattering is stronger and may be utilized. A number of different interferometric techniques have been developed:

1. *Fringe imaging* (also called "multi-channel") with multi-beam interference fringes for both aerosol and molecular scattering using either Fabry–Perot or Fizeau interferometers.
2. *Double edge* (more strictly "dual channel") with two pass-band Fabry–Perot filters set on either side of the scattered spectrum.
3. *Visibility curve* with two-beam interference fringes formed in a Mach–Zehnder configuration.

Considerable controversy has been generated in the literature by the proponents of each technique. Theoretical calculations typically show calculated performance within a factor of 2–3 of the Cramer–Rao limit (discussed in Section 12.3.2; see, e.g., Reference [108]). Several fringe imaging and double-edge systems have been built and measurements made high into the atmosphere. The visibility curve (two-beam) technique is more recent but shows promising performance (see Reference [109]).

In the fringe imaging technique, the spectrum of the scattered light is imaged across a number of detector elements (typically 10–20). A high-resolution system is obviously required for analysis of the narrow aerosol return (see Figure 12.7) and the molecular spectrum is spread as a background noise across all the detector elements. The actual width of the aerosol signal is likely to be dominated by the laser line width itself. If this is of the order of 100 MHz (equivalent to ~20 ms^{-1} Doppler width at 355 nm wavelength) then the Cramer–Rao limit discussed in Section 12.3 shows that, for a measurement accuracy of 1 ms^{-1}, the minimum number of photons required is given by

$$N_p \geq (\eta T_t)^{-1}(0.425)^2(20)^2 \qquad (12.25)$$

where ηT_t is an overall instrumental efficiency made up of η-detector efficiency and T_t instrumental transmission. Supposing $\eta T t \approx 0.05$ at best gives a minimum input requirement at the receiving telescope of

$$N_p \geq 1.5 \times 10^3 \text{ photons} \qquad (12.26)$$

neglecting the impact of molecular background, daylight scattering, detector noise, etc. A lower resolution system for analysis of the much wider molecular Rayleigh line (of equivalent width ~600 ms^{-1} at 355 nm) with similar assumptions would require a minimum of ~1.3×10^6 photons.

Several fringe imaging lidars have been built and deployed in field trials (see, e.g., References [110,111]). Operation has usually been at 532 nm with frequency doubled Nd-YAG lasers. A particularly powerful system with 1.8 m diameter telescopes has been deployed at the ALOMAR Observatory (Andoya Rocket Range, Andenes, Northern Norway 69°N, 16°E). It has been used regularly for stratospheric wind measurements by day and night, through summer and winter (see Reference [112]).

In the double-edge technique, Doppler shifts are derived by calibration of the response function R given by

$$R = (I_A - I_B)/(I_A + I_B) \qquad (12.27)$$

where I_A and I_B are the measured signals passing through the two fixed filters A and B placed on either side of the scattered light spectrum, as shown in Figure 12.17. As the spectral line shifts in frequency, the relative magnitudes of I_A and I_B change. The fixed filters are formed by two Fabry–Perot cavities of slightly different spacing. With suitably chosen frequency separation of the filters, operation on the molecular spectrum can be made relatively insensitive to the narrow band, often

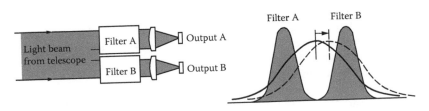

Figure 12.17 Schematic of the double edge technique with two filters A and B set on either side of the broad molecular spectrum as utilised for scattering at 0.35 and 0.53 μm. The technique can also be applied to the relatively stronger aerosol channel at 1.06 μm with appropriately spaced filters at higher resolution.

highly variable, aerosol scattering at the centre. Precise calibration of R also needs to take account of the exact form of the broad molecular spectrum, which varies with temperature through the atmosphere. Several double-edge systems have been built [113–115]; one of the first, that at Observatoire Haute Provence (OHP) in southern France, operating at 0.53 μm, has made extensive measurements into the lower stratosphere over several years [116].

A number of field trials comparing different LDV lidars have now been conducted. The VALID trial at OHP in July 1999 [120] brought together double-edge (0.53 μm), fringe imaging (0.53 μm) and heterodyne (2.1 and 10.6 μm) lidars for comparison with a microwave wind-profiler radar and balloon radiosondes. Forty-six data sets were accumulated and generally showed good agreement with cross-correlation coefficients above 60%. For some specific meteorological cases (e.g., a jet stream), measurement discrepancies were readily explained. At a trial in September 2000 at Bartlett, New Hampshire, USA, a fringe imaging lidar (0.53 μm), a double edge lidar (aerosol channel at 1.06 μm, molecular at 0.35 μm) and heterodyne lidar (1.06 μm) were deployed with microwave wind profiler and regular radiosonde launches. Generally reasonable agreement in a variety of atmospheric conditions was obtained.

12.4.4 Cloud measurement

Detailed knowledge of cloud parameters, as indicated in Table 12.11, is important for three major applications: climatological studies, weather forecasting and local information, particularly in aviation. In addition, knowledge of polar stratospheric clouds (PSCs) and their role in atmospheric chemo-dynamics, particularly for ozone destruction, is of great significance (see Section 12.6). Laser radar can be used to measure cloud height, vertical extent, structure, optical thickness, statistical distribution, classification, presence of thin

Table 12.11 Application of cloud measurement data (height, structure, etc.)

Climatological studies	Radiation balance
	Global energy balance
	Aqueous phase chemistry
	Monitoring of polar stratospheric clouds and ozone studies
Weather forecasting	Delineation of air mass discontinuities
	Assessment of frontal or convective activity
	Assimilation into numerical weather forecasting models, mesoscale, regional and global
	Improved quality of humidity and temperature retrievals
	Improvements in cloud vector winds and sea surface temperature
Local information	Objective assessment of cloud base and cloud cover, particularly for aviation and airport operation
	Precipitation and vertical visibility

and subvisual cirrus, precipitation, etc., and many studies have been conducted (see, e.g., References [59,118–121]). Spaceborne lidar systems obviously have the potential to provide such information on a global scale, as illustrated, for example, in the discussion following the first such demonstration of the NASA LITE equipment [72].

For local measurements, ground-based systems have been extensively developed and are commercially available. Relatively simple techniques with pulsed lasers are employed with direct detection of the scattered light along the beam in ranging intervals of typically 100 ns (15 m equivalent). In fair weather and well-marked clouds, the basic cloud height measurement is a straightforward ranging task. Consideration does need to be given to the precise definition of "cloud height"—should it be the initial return from the lowest fluffs of cloud, the height of the steepest increase in backscatter or the height of maximum backscattered signal? In a typical commercial lidar ceilometer (e.g., the Vailsala CT25K), the processing algorithm selects a height in the region of peak backscatter—which corresponds to the height from which pilots can usually see the ground well.

In severe weather, the problem becomes much more difficult; stray signals will be evident from rain, snow, virga (precipitation out of cloud evaporating before it reaches the ground), haze and fog. Quite complex algorithms have been developed to invert the measured backscatter profile to provide a credible vertical extinction coefficient profile. Thresholding criteria based on experience and extensive observations can then be applied to the latter to determine and classify a cloud presence. As mentioned, the Vaisala CT25K lidar ceilometer is a robust, fully automated system based on a single lens, monostatic transmitter/receiver and an InGaAs pulsed diode laser operating at 905 nm. Measurements can be made from 0 to 25,000 ft (7.5 km) with 500 range gates and range resolution of 50 ft (15 m) in a programmable measurement cycle of 15–120 s.

This equipment has been further developed with partial funding by the European Space Agency and uses an array of four standard equipments operated in synchrony. Above 1000 m, all four beams overlap so that every receiver collects scattered light from every transmitted beam. The increased SNR ensures that this more powerful equipment provides a measurement capability on

cloud up to the 75,000 ft (21 km) altitude referenced by the World Meteorological Organisation (WMO) in 1983 as the upper limit for cloud height observations.

12.5 LIDAR AND AVIATION

Following the demonstration of lidar measurements in the atmosphere, the application of such techniques to problems in aviation was rapidly developed. Immediately obvious topics were measurement of true airspeed, warning of wind shear and turbulence, calibration of pitot-static pressure-differential probes and the potential hazards of aircraft wake vortices. For operation in aircraft, equipments must obviously be compact, reliable and robust to the environment of vibration and reduced pressure.

12.5.1 True airspeed, wind shear, turbulence and pressure error calibration

In the late 1970s, a number of airborne Doppler lidars were built. One of the earliest was quite a large-pulsed CO_2 system designed to look forward of the aircraft at ranges >5 km. One of the main aims, to establish whether the equipment could detect clear air turbulence at high levels in the atmosphere, was successfully demonstrated (see, e.g., References [49,103]). For measurements at shorter ranges, the Laser True Airspeed System (LATAS) was built at the Royal Signals and Radar Establishment (RSRE) in 1980 and flown in aircraft of the Royal Aircraft Establishment (RAE). Figure 12.18 shows the installation of LATAS in the unpressurized nose of an HS125 executive jet-type aircraft. The continuous wave CO_2 laser gave a power output of ~3 W and the optics head was contained in a well-insulated temperature-controlled enclosure. All controls were operated remotely from a console in the aircraft cabin. Figure 12.19 shows a pair of typical spectra recorded at different levels in the atmosphere; one shows increased levels of turbulence in the probe volume.

From the peak of such spectra, the true airspeed of the aircraft (that is, the speed of the aircraft relative to the air it is moving in) can be determined with an absolute accuracy of typically better than 0.2 ms^{-1}. Figure 12.20 shows a record

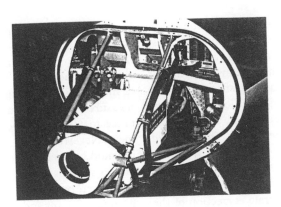

Figure 12.18 The Laser True Airspeed System (LATAS) installed in the unpressurized nose of an HS125 aircraft. For flight, this was covered with a nose cone holding a germanium window through which the lidar beam emerged.

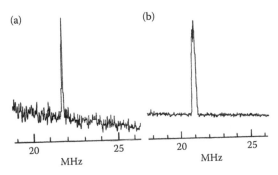

Figure 12.19 Typical Doppler spectra recorded in flight with the LATAS equipment at focal range of 100 m with 25 ms experiment time: (a) height 5000 ft, the peak frequency corresponds to 225.2 kt; (b) height 960 ft, the mean frequency corresponds to 215.0 kt; note the broadened spectral width due to the turbulence of ~5 kt.

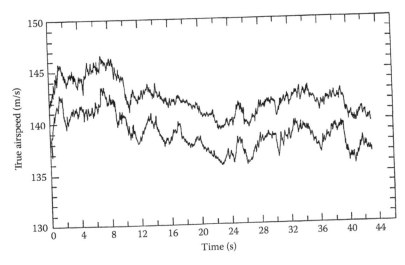

Figure 12.20 True airspeed recorded against time. The top trace is the lidar data with sampling at 25 s⁻¹. The lower trace is the speed calculated from a pressure-differential gust probe mounted on a boom. Note the close correspondence of velocity structure. (From Keeler, R. J.et al., *J. Atmos. Oceanic Technol.*, 4(1), 113–137, 1987.)

of such measurements made at 25 s^{-1} in the atmospheric boundary layer [122]. For these studies, the lidar was focused at a reduced range of 17 m (but sufficiently far enough to be away from any disturbance of the advancing aircraft) to give good comparison with a small, conventional, pressure-differential gust probe mounted in a thin boom projecting ~1 m ahead of the aircraft. The air speed derived from the pressure-differential system is also shown in Figure 12.20. The small difference between the two airspeeds (about 3 ms^{-1}) almost certainly arises from inaccurate correction of the pressure data, and provides the basis for calibrating such devices as discussed later. Closer examination of the data in Figure 12.20 shows a time offset of ~0.11 s (clearly corresponding to the 16 m forward look of the lidar), and excellent agreement of the detailed structure (and computed variances) for the two sets of data.

Although clear air turbulence at high level in the atmosphere is troubling to aircraft, with risk of injury to unbelted passengers and crew, few if any aircraft have been caused to crash. Low-level wind shear, on the other hand, often but not always associated with thunderstorm activity, has caused many serious accidents, notably with aircraft in landing or take-off phase. At this stage, aircraft have a low airspeed, typically ~120 kt (~60 ms^{-1}). In the presence of wind shear, as the aircraft passes into a region of notably different wind speed, its lift may be greatly reduced. The pilot may have insufficient time to accelerate the aircraft up to sufficient airspeed to keep it aloft. Figure 12.21 shows an early LATAS record of passage through a thunderstorm microburst during the Joint Airport Weather Studies (JAWS) trial in Colorado in 1982. In this case, the focal range was set 250–300 m ahead of the aircraft. The sensitivity extended out to 700–800 m and thus strong shear or turbulent structures entering the extended probe at longer range were evident. In Figure 12.21, the sequence of Doppler spectra in the lidar record showed a headwind that changed by over 40 kt (~20 ms^{-1}) in about 5 s. There was an additional down draft of ~6 ms^{-1}. Analysis of these measurements contributed to the development of a descending vortex-ring model for thunderstorm microburst behavior, in contrast to the more usual vertical jet and outflow model. Simulations with the LATAS parameters have shown that the ~5–10 s warning of shear from a probe range of ~300–600 m could

be useful if heeded promptly [123–125]. Wind shear and microbursts are dynamic phenomena; indeed, the simulations showed that there was significant advantage in controlling the aircraft (a medium-size passenger jet) using the airspeed measured ~300 m ahead, but increasing the distance to 600 m produced little further improvement. Nevertheless, in terms of general operation, it would be useful to look at a greater range with pulsed systems and a number of airborne equipments were built in NASA-supported programmes. Both CO_2 10 µm and solid state 1 and 2 µm lidars were successfully demonstrated at ranges to ~3 km. Comparative performance in terms of range and velocity resolution, and atmospheric factors in different conditions of rain, high humidity, etc., were assessed [126].

While wind shear remains a severe problem, the widespread deployment of lidars in civil airliners nevertheless appears unlikely in the near future. Increased awareness of the meteorological factors that precipitate severe wind shear, combined with the warning from large powerful Terminal Doppler Weather Radars (TDWR) operating at major airports, provide improved confidence and safety in airport operation. Nevertheless, it may be remarked that lidar has a very strong capability to detect so-called "dry" microbursts (i.e., without associated rain) and is thus complementary to the high sensitivity of TDWR for conditions of rain and high humidity.

In an extensive development in France in the late 1980s, a CO_2 10 µm cw lidar was similarly used for true airspeed measurement and flown in a helicopter and various transport and fighter aircraft. This was subsequently configured into a three-axis equipment called ALEV-3 by Sextant Avionique, as shown in Figure 12.22. This lidar has been used for air data calibration and aircraft certification in extensive flight tests of Airbus aircraft. With the beams focused at 70 m, the equipment gives a precise, real-time measurement of the airspeed vector and thus permits the calibration of static pressure and the angles of incidence and sideslip [127].

An advanced optical air data system (OADS) for measurements high in the atmosphere was built and flown in the early 1990s [128,129]. Good performance at ranges to a few tens of metres was obtained with a direct detection technique in which the velocity components were derived from the measured transit times of aerosols

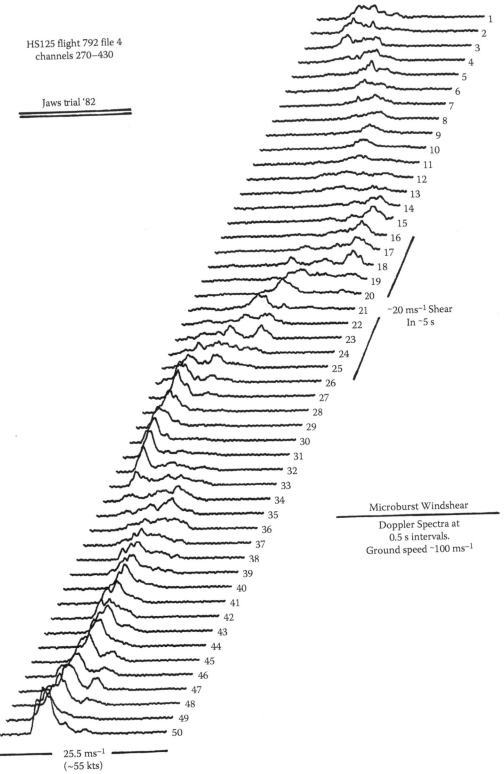

HS125 flight 792 file 4
channels 270–430

Jaws trial '82

~20 ms^{-1} Shear
In ~5 s

Microburst Windshear

Doppler Spectra at
0.5 s intervals.
Ground speed ~100 ms^{-1}

25.5 ms^{-1}
(~55 kts)

Figure 12.21 Sequence of lidar spectra recorded at 0.5 s interval during aircraft passage through a thunderstorm microburst. Note the severe wind shear (40 kt, ~20 ms^{-1} in ~5 s) in the record.

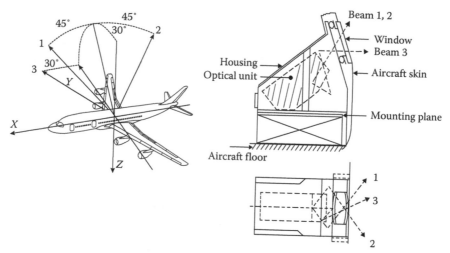

Figure 12.22 The ALEV-3 airborne lidar for air data calibration and aircraft certification. The optical axes chosen for transport aircraft. The lidar (optical unit) and installation in the aircraft. (From Morbieu, B. et al. ALEV3, a 3-axis CO_2 CW anemometer for aircraft certification. *Proceedings of 7th Conference on Coherent Laser Radar: Applications and Technology*, Paris, Paper, 1993.)

between light sheets. This programme is outlined in Chapter 18 of the present volume and used as an exemplar of optical engineering practice in a hostile environment.

More recently, an airborne wind infrared Doppler lidar (WIND) has been developed in cooperation between CNRS and CNES in France and DLR-Munich in Germany [130]. In this lidar, built more for meteorological investigations and as a precursor for study of spaceborne systems, the TE-CO_2 pulse laser operates at ~300 mJ pulse and 4 or 10 Hz prf. Successful measurements from an altitude of 10 km, with downward conical scanning at 30° from nadir, have been undertaken in international field campaigns since 1999 [131].

12.5.2 Aircraft wake vortices

In the process of generating lift (deriving from the pressure difference between the upper and lower surfaces of the wing aerofoil), all aircraft (including helicopters) create transverse rotational flow in the air that has passed over each wing. This rotational flow rapidly evolves into two powerful counterrotating vortices that extend as a pair of trailing ribbons behind the aircraft. The initial separation of the vortices is about 80% of the wing span; their rotational sense produces a strong downdraft in the region between them, and in free air they tend to sink at a rate of 1–2 ms^{-1} Their trajectory is, of course, largely determined by meteorological conditions; on approach to the ground, the sink rate is reduced and the vortices tend to separate. An extensive assessment and bibliography of the then state of knowledge of wake vortices was given by Hallock in 1991 [132].

The existence of such wake vortices, sometimes rather dramatically referred to as "horizontal tornados", represents a potential hazard to following aircraft, particularly for smaller aircraft following larger in the vicinity of airports on landing and take off. The Air Traffic Control System guards against this hazard by the application of separation minima between all pairs of aircraft operating under Instrument Flight Rules. The larger and heavier the aircraft, the more powerful the vortex it creates and hence the separation distances depends on the types of aircraft (set into various weight categories) involved. Such separation minima provide a significant constraint on airport capacity—it has been estimated, for example, that if they were not necessary the arrival capacity at Heathrow could be increased by up to five aircraft per hour. Obviously, the separations must be sufficiently large to ensure safety, but they should not be so excessively conservative as to unduly restrict the capacity of airports.

The characteristics of wake vortices—their formation, evolution, persistence, trajectory, mode of decay, etc.—have been studied for many years. In particular, powerful techniques of computational fluid dynamics and large wind tunnel measurements on reduced scale model aircraft provide information on the early stages of vortex formation and evolution. Recent water tank studies have extended this to the late stages. The positive and complementary features of lidar measurements are that they can be carried out on full-scale aircraft, in the real atmosphere and potentially at long distance (many 100s of wing span) downstream of the generating aircraft and thus on fully mature and decaying vortices.

Wake vortices were, in fact, amongst the very first subjects of study by coherent laser radar. A demonstration of coherent 10 μm LDV measurements with cw lidars was made in 1970 in a NASA-sponsored programme. Subsequently, during the

1970s and early 1980s, very extensive NASA-and Federal Aviation Agency (FAA)-funded investigations were made and helped to define vortex separation standards [132]. In the 1980s, the German DLR Institute of Optoelectronic deployed a high-performance cw lidar at Frankfurt Airport between two parallel runways with the prime aim of examining the risk due to a cross wind possibly transporting vortices from one runway to the other [133]. The runway separation of 520 m is often too small for operating both runways independently with respect to wake vortices. Figure 12.23 shows a section of the vertical measurement plane with examples of vortices moving under the prevailing cross wind from one runway to the other. A number of such vortices show a steep ascent towards the parallel runway. This bouncing effect may enhance the

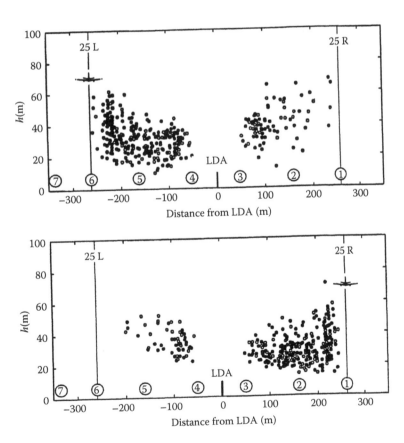

Figure 12.23 Position of wake vortices at Frankfurt airport measured by the Laser Doppler Anemometer (LDA) situated between the two parallel runways 25L (left) and 25R (right). The upper record shows the propagation of Boeing 747 vortices generated on runway 25L moving towards 25R, under the prevailing cross wind, and the lower from 25R in the opposite wind flow.

hazard since the vortex will tend to pass near the altitude of approaching aircraft. This lidar work has contributed to the development of a strategy of operation for Frankfurt Airport in various cross wind conditions.

With increasing demand in the 1990s for available runways, it is clear that wake vortices and the required separation minima limit the capacity for the world's busier airports. There has thus been increased interest recently in lidar measurements with NASA and FAA programmes in the USA, Civil Aviation Authority-funded programmes in the UK and EC-funded programmes in Europe.

This includes 10 μm cw lidars with controlled tracking of vortex position and comparison with acoustic sensors and mechanical anemometers. In addition, short wavelength 1 and 2 μm pulse lidars have been used for longer ranges.

A schematic illustration of the flow field and velocity distribution for vortices is shown in Figure 12.24. The Doppler lidar spectra expected for a beam intersecting such a flow is somewhat complex due to the changing line-of-sight velocity component along the beam and the spatial weighting function of the lidar. The form generally expected is shown Figure 12.5, as calculated

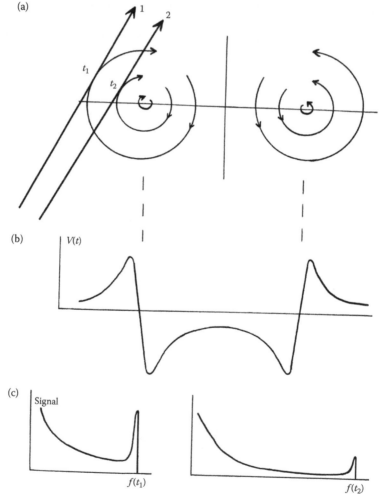

Figure 12.24 Schematic of vortex flow and lidar spectra. (a) Cross section of a pair of rotating vortices. Two positions A and B of a lidar beam intersecting the left hand vortex at tangent positions t_1 and t_2 are shown. (b) Approximate peak velocity profile $V(r)$ along the line between the vortex cores. (c) Doppler spectra for the lidar beams at A and B. Analysis shows that scattering from the immediate tangent regions t_1 and t_2 give the peaks in spectra at Doppler frequency $f(t_1)$ and $f(t_2)$ equal to $2V(t_1)/\lambda$ and to $2V(t_2)/\lambda$, respectively.

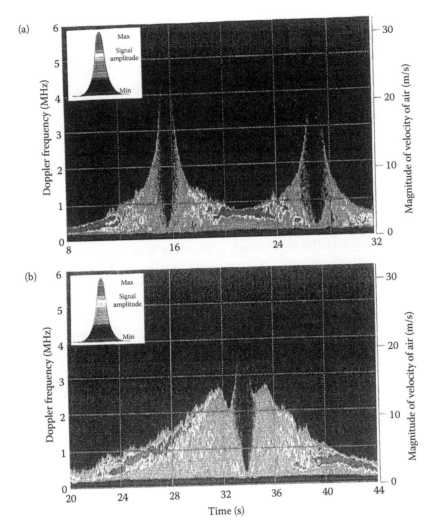

Figure 12.25 Doppler spectra (recorded at 25 s^{-1}) from wake vortices carried by the prevailing cross wind through a static, upward looking lidar. Each spectrum is plotted vertically and color coded for amplitude as shown by the insert. The time after passage of the generating aircraft is shown on the horizontal scale; the Doppler frequency (left hand) and equivalent velocity component (right hand) are on the vertical scales. (a) Pair of vortices from a B757 aircraft (CAA 320133, 3 October 1995); lidar elevation 90° and focal range 110 m. The cross wind was 4.9 ms^{-1}note the very sharp profiles of the vortices, rising close to the top of the scale, on either side of the cores. The downflow midway between the vortices is 7.4 ms^{-1}. (b) Pair of vortices from a B747–200 aircraft (CAA 320068, 30 October 1995); lidar elevation 90° and focal range 115 m. The cross wind was 6.0 ms^{-1} and downdraft ~0.7 ms^{-1} Note the characteristic shape with broad, steadily rising profile up to ~12 ms^{-1} and then rounding over into a series of inversions. The corrected downflow V_v between the vortices is 7.78 ms^{-1}.

by Constant et al. [134]. Most importantly, scattering from the immediate tangent region provides the highest frequency peak component in the spectra. The intensity of these peaks rapidly weakens as the beam gets closer to the vortex core. Nevertheless, with a lidar of adequate spatial and spectral resolution, their measurement provides a velocity profile through the vortices, as shown in the series of experimental spectra in Figure 12.25 with a fixed pointing lidar. In these pictures, the spectra are plotted vertically and coded for intensity as shown in the inset. The rapidly rising peak Doppler frequency on approach to the vortex core is very obvious.

With the lidar beam scanning to the side, many successive intersections on vortices with the characteristic cusp-like signatures may be obtained. Analysis of a record from Heathrow showed a very remarkable result in that the near-wing vortex from a B-747 aircraft pursued the unusual path shown in Figure 12.26. After initially sinking to about 40 m above ground level, it subsequently rose and, at ~70 s, was back at the glide slope with essentially undiminished strength. This is a good example where the ATC separation standards worked well—the following aircraft passed through at ~102 s, well after the vortex had moved away. The unusual character of this example must be emphasized; in the overwhelming majority of the nearly 3000 lidar records at Heathrow, the vortices were rapidly convected away from the glide slope by the prevailing wind within 30–40 s.

Much quantitative data on vortex character, strength and longevity may be obtained from such lidar records. The strength of vortex circulation or vorticity $\Gamma(r)$ in m^2 s^{-1} is given by the expression

$$\Gamma(r) = 2\pi r V(r) \qquad (12.28)$$

where $V(r)$ is the rotational speed at radius r. From Figure 12.25, values of vorticity for the B757 vortex of $\Gamma(14.3\,m) = 331\,m^2\,s^{-1}$ (at mean age 16 s) and for the B747 of $\Gamma(22.4\,m) = 545\,m^2\,s^{-1}$ (at mean age of 17 s) are obtained, in reasonable agreement with expectation for these aircraft in landing configuration. As a further example, from the scanned record for Figure 12.26, much detailed information may be extracted, including the values and changes of vorticity at different radii over the full observation period. The circulation Γ (13.5m) remained close to 510 m^2 s^{-1} out to 70 s [135].

In recent years, study of aircraft wake vortices has further intensified; EC-supported programmes have been very fruitful in combining lidar studies with wind tunnel, water tank, catapult launching of models and computer fluid dynamics (CFD) investigations (see, e.g., Reference [136]). Pulse laser systems at 2 μm wave length and spatial resolution of ~50 m may also be used for localization and measurement of vortices at ranges greater than ~1 km, as first demonstrated by Hannon and Thomson [137]. Lidar studies include measurement of military-type

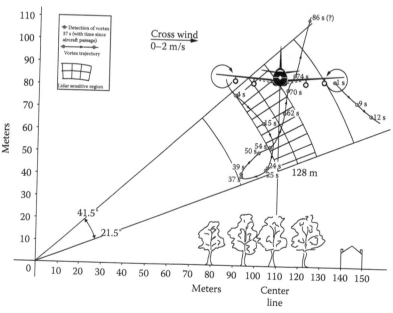

Figure 12.26 Reconstruction of vortex trajectories for record CAA 80039. The lidar was sited at a position ~110 m north of the centre line with the aircraft (B747–200) ~85 m above ground flying E to W. The lidar was set to focus at ~128 m range and the scan in the N–S plane is shown extending over 31.5±10°. The wind was WNW, about 5–6 ms^{-1} at ~295° at a height of 75 m, giving a variable cross wind of 0–2 ms^{-1} from the north. The weather was cold (3–5°) with occasional rain and sun.

aircraft in Germany [138], report of measurements at Dallas/Fort Worth airport [139] and at Heathrow [140,141]. Very recent collaborative campaigns with several lidar systems, both CW and pulsed, have been held at European airfields. Precise comparisons of wind tunnel data with lidar measurements have been made [142], and observations reported of an unusual structure with a vortex within a vortex [143]. A recent paper considered the impact on helicopter handling of vortex encounters using lidar data [144].

In summary, a broad view of the application of lidar to problems of aircraft wake vortices suggest several principal areas. First, at the most basic level, there is the important contribution to fundamental knowledge of vortices—their formation, interaction, transport, decay and dissipation and comparison with data using other techniques. This potentially should input into detailed aircraft design (e.g., wing configuration, interaction of engine, flap and tip vortices, etc.) with development of low vorticity vortices (lvv) or quickly dissipating vortices (qdv). Such progress is important particularly for the very large transport aircraft now being developed (e.g., the Airbus A380). For airport operation, two types of capacity improvement may be envisaged:

- *Strategic:* scheduled increase in arrival/departure slots. Increased knowledge of vortex decay and environmental interaction could lead to small capacity gains from refined separation standards. Even a few slots per day at a busy airport would be of great value.
- *Tactical:* considerable scope for local hour-by-hour ATC decisions to reduce separations based on meteorological and lidar (LDV) information monitoring. Potentially large fuel savings and reduction of delays, particularly for aircraft in holding patterns, should be attainable.

This last point raises the question as to whether effective forms of routine monitoring by lidar could be developed to provide useful vortex advisory systems for air traffic controllers. Any such airport advisory system must of course be cost-effective to install and operate and hence the technology must above all be robust, reliable and easily maintained.

12.6 CHEMICAL, ENVIRONMENTAL AND POLLUTION STUDY

The three primary phenomena for detection and measurement of chemical species and their environment have been noted in Section 12.2: differential absorption, Raman scattering and fluorescence. Lidar techniques using these phenomena have been widely developed for studies ranging from short range monitoring (~10–100 m) across, e.g., chemical plants to measurements high into the mesosphere even above 100 km altitude. Differential absorption lidars (DIAL) are available commercially and may be employed routinely for pollution monitoring across, e.g., industrial sites and cities. Many lidars have been incorporated in high flying, long-range aircraft and have provided vital information on atmospheric chemistry and dynamics, including analysis of polar ozone depletion. The basis of the techniques with a few recent examples are outlined. Detailed accounts may be found in, e.g., References [18,21,22,33,97]. A concise review of DIAL and Raman lidar for water vapour profile measurements has been given by Grant [145] with 133 references.

12.6.1 Differential absorption lidar (DIAL)

The simple principle of DIAL lidar is the comparison of the backscattered signal (from aerosols and molecules) for two laser beams traversing the same path; one beam is tuned to an absorption feature of the chemical species under study while the other beam is well off absorption. For a pulse lidar differential analysis of the two signals S_0 and S_{ff} with highly developed retrieval algorithms [146] will provide a range resolved concentration density map of the absorbing chemical species. For some specialist applications, cw lasers operating on fixed paths to topographic targets or retroflectors can provide a path-integrated measurement. This may be suitable for, e.g., 24-h surveillance across a chemical processing plant with the beams successively traversing various sensitive areas of processing reactors, storage tanks, etc. Such lidars can guard against any sudden catastrophic escape of materials and contribute also to the detection of low level leaks with improved operating efficiency.

A vast range of lasers have been used for DIAL equipments ranging in wavelength from near UV to mid-IR. Typically in the range 250–435 nm, species such as toluene, benzene, SO_2, NO_2 and ozone can be detected with dye lasers or frequency doubled or tripled light from, e.g., a Ti sapphire laser. Water vapour and molecular oxygen have absorption features with strong, well resolved lines in the near-IR in the region 720, 820 and 930 nm (water vapour) and 760 nm (oxygen). Dye lasers, tunable vibronic and solid-state lasers have been employed amongst others. The longer wavelengths 1–4.5 μm are well suited to detection of, e.g., methane, ethane, hydrogen chloride, hydrogen sulphide, nitrous oxide and many hydrocarbons. Varying laser dyes with sum and frequency difference mixing in non-linear crystals provides tunable coverage of much of the range, together with tunable OPO lasers. In the range 9–12 μm, covered by CO_2 laser wavelengths, there are many absorption features for hydrocarbons and various pollutant and toxic materials which can also be investigated by DIAL.

A schematic of a DIAL lidar is shown in Figure 12.27; clearly, a basic technical requirement is the accurate positioning of the laser emissions with respect to the selected molecular absorption line (typically within 0.001 cm⁻¹), with the laser line width narrower than the absorption feature. Separation of the scattered radiation at the two wavelengths (if transmitted simultaneously from two laser sources) may be achieved with narrow-band filters. Otherwise, switching between the two "on" and "off" lines must be accomplished swiftly and precisely with typical pairs of pulses at > 10 Hz rate. Attainable accuracy for concentration measurement depends on the parameters of the specific application, e.g., spatial and temporal resolution and the material under study. Typically detection limits below 10 ppb can be achieved with an accuracy of better than a few ppb. Measurements in the atmosphere for meteorological investigations have been widely developed. With molecular oxygen as the absorber and operating in the wings of the absorption lines, surface pressure and pressure profile can be measured with an accuracy of order 0.2%. Similarly, the temperature profile may be measured with an accuracy of typically better than 1 K through the tropopause using highly excited vibration lines of molecular oxygen with appropriate choice of emitted laser frequencies. Many intercomparisons and sensitivity analyses of different equipments have been made (see, e.g., References [147–149]).

For the shorter wavelengths, techniques of direct detection are employed with narrow-band optical filters for isolation of the required wavelengths and reduction of background. At longer wavelengths, heterodyne techniques have been investigated in order to improve SNR with inherently noisy detectors. The speckle-induced fluctuations

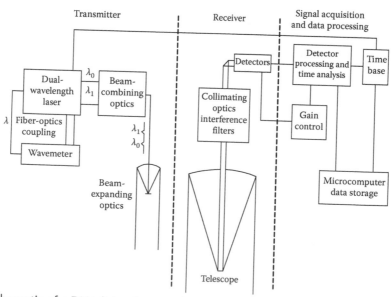

Figure 12.27 Schematic of a DIAL lidar showing the laser's transmitter and receiver.

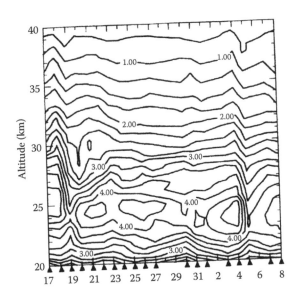
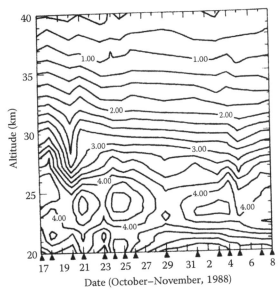

Date (October–November, 1988)

Figure 12.28 Comparison of the results for two ozone DIAL lidars spaced close together. The vertical arrows show the measurement day during the trial's period [148].

in such coherent systems require averaging, but useful improvements in range performance with CO_2 DIAL systems have been demonstrated (see, e.g., References [150–152]). Speckle correlation and averaging with two or more wavelengths for improved accuracy in DIAL have also been examined [153].

DIAL systems have now been used for many years in both fixed and mobile ground stations. An interesting comparison of results for two high-performance ozone DIAL lidars is shown in Figure 12.28 (from Reference [148]). A number of extensive campaigns have been conducted across European cities (e.g., Athens in 1994, Grenoble in 1999) bringing together a range of lidars and monitoring equipment for study of ozone smog and pollution episodes. Several large permanent ground stations (e.g., at Eureka in the Canadian High Arctic [35], at Alomar in northern Norway [112] and at OHP in southern France [154]) have been established for longer term monitoring of the atmosphere, often combining different lidar facilities with DIAL, Rayleigh, Mie and Raman lidar detection channels.

Measurements from aircraft have been particularly notable, particularly in the combination of aerosol and DIAL lidars. In the winter of 1999–2000, for example, the NASA Langley airborne equipment measured ozone at DIAL wavelengths 301.6 and 310.9 nm and aerosols at 1064, 622 and 311 nm, giving vertical profiles from 12 to 28 km

across the wintertime Arctic vortex. Further valuable information on the interaction of PSCs and their implications for chemical depletion of O_3 was obtained [22,33,155].

12.6.2 Raman lidar

Raman scattering provides an effective means for chemical species and aerosol measurement with lidar systems operating at shorter wavelengths to give the strongest scattering. Frequency doubled or tripled Nd-YAG lasers have been used; with an XeF examiner laser operating at 315 nm, Raman returns from O_2 (371.5 nm), N_2 (382.5 nm) and water vapour (402.8 nm) are created. Systems typically operate at several hundred Hz with 50–100 mJ per pulse, with collecting telescopes up to 1 m diameter and range resolution 20–100 m. The role of water vapour in atmospheric models for numerical weather prediction (NWP), and as the leading greenhouse gas, is particularly important and Raman lidar measurements offer valuable information (see, e.g., Reference [156]); a measurement example is shown in Figure 12.29 [157]. Calibration techniques for Raman lidar water vapour measurements have been widely investigated (e.g., Reference [158]) using, for example, comparison with radiosondes providing local humidity data (see, e.g., Reference [159]) and pressure and temperature data at cloud base with expected 100%

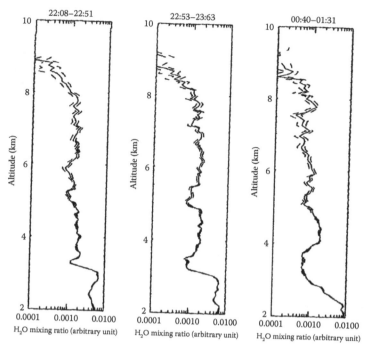

Figure 12.29 Example of Raman lidar measurement of water vapour in the atmosphere [157].

humidity (see, e.g., Reference [160]). Cloud liquid water content has also been an important study [161] for Raman lidar; liquid water has a broadened spectrum ~7 nm wide shifted about 5 nm from the narrow vapour return. Interference filters ~4 nm wide readily separate the two signals.

Another important application of Raman lidar has been in measuring aerosol extinction [162,163]. By measuring the Raman N_2 signal at the shifted wavelength and the backscatter signal (molecular and aerosol) at the laser wavelength, the height-resolved aerosol extinction σ_{aer} and the extinction to backscatter ratio $\sigma_{aer}/\beta(\pi)_{aer}$ can be determined (see, e.g., Reference [164]).

12.6.3 Fluorescence lidar

Fluorescence lidar can be applied in several areas: for detection and identification of organic and biological materials on surfaces (by land or water), for detection of fluorescent dye particles (FDP) released as tracers in the atmosphere and temperature and concentration measurement of atoms at very high altitudes.

For organic and biological materials, two diagnostics may be employed. First, the fluorescent wavelength edge above which (as the laser source is tuned to longer wavelengths) the material no longer exhibits fluorescence. Second, there is the characteristic (usually quite broad band) spectrum of the fluorescence itself. The technique has been applied to pollution studies of, for example, oil spillages (accidental or deliberate) at sea with the possibility of large area, rapid surveillance by aircraft. Similarly the technique may be applied to biological studies of water—certain algae and simple life forms have characteristic fluorescence spectra. Early detection and identification can be achieved and monitoring of development as, for example, in the phenomenon of algae blooms. The use of FDP tracers in the atmosphere has been described by Uthe [165] (see also Reference [166]). Organic resin particles of ~2 μm diameter containing a fluorescent orange dye have been used. In one example, a parcel released at 2.3 km altitude was tracked (by airborne lidar) for over 8 h and 200 km passage [165].

The first lidar measurement of sodium high in the mesosphere was made as long ago as 1969. This region is in fact difficult to study by other techniques, being inaccessible to balloons, aircraft and most satellites. It has many interesting features containing the coldest part of the atmosphere—the mesopause, metal layers formed by meteor ablation and polar mesospheric clouds (PMCs). Knowledge

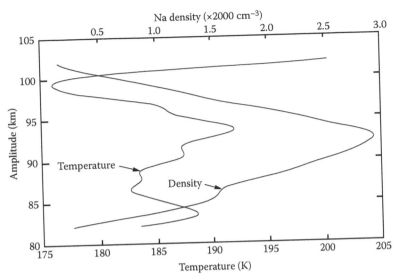

Figure 12.30 Example of fluorescence lidar measurements of temperature and Na density high in the atmosphere [170].

of the temperature, density and their variability (as a function of season, latitude and solar cycle, etc.) is important for many geophysical phenomena. Several fluorescence lidar have been built variously operating at resonance wavelengths for Na, Ca, K and Fe. Typically, a range of dye lasers pumped by Nd-YAG have been used with pulse energies in the range 5–100 mJ at typically ~20 Hz. All solid-state generation of the Na D_2 resonance line at 589 nm has also been achieved by sum-frequency mixing Nd-YAG 1064 and 1319 nm radiation with lithium niobate or triborate non-linear crystals.

For relative concentration measurements, direct detection may be employed with narrow band filters (typically ~1 nm) to reduce background light. It is also possible to simultaneously measure density and temperature and a number of techniques have been developed [167–169]. For sodium, two or more laser frequencies are transmitted in turn, one exactly at the peak of the Na D_2 line and the other shifted 800 MHz by an acousto-optic modulator. The relative scattering at the two wavelengths gives a measurement of the temperature. An example of temperature and Na density profiles in the altitude range 80–105 km is shown in Figure 12.30 [170]. For iron, the technique relies on the temperature dependence of the relative scattering for two closely spaced Fe atomic transitions at 372 and 374 nm. These two lines originate on the lowest

atomic ground state and an upper sub-ground state about 416 cm^{-1} higher; their population and hence the ratio of scattered light is determined by Maxwell–Boltzmann statistics in thermal equilibrium. Generation of the required radiation has been achieved with independent, frequency-doubled and injection-seeded pulsed alexandrite lasers. Several equipments, ground and airborne, have now been built for mesospheric studies with routine measurements over extended periods, and more concentrated trials for expected meteor showers. Figure 12.31 shows a record of such an event [171].

12.7 LIDAR AND MILITARY APPLICATIONS

Many think that in some way they (the Martians) are able to generate an intense heat in a chamber of practically absolute non-conductivity. This intense heat they project in a parallel beam against any object they choose by means of a polished parabolic mirror of unknown composition…

H G Wells: The War of the Worlds (1898)

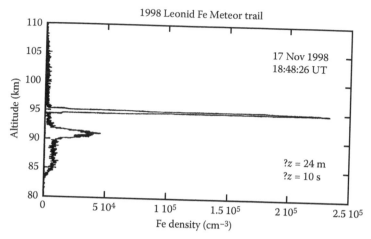

Figure 12.31 Record of a meteor ablation trail made with a Fe resonance lidar [171].

12.7.1 Introduction

In the popular mind, lasers probably represent the ultimate science fiction offensive weapon. After much media hype through the late 1980s, fueled by the United States' Strategic Defence Initiative (popularly known as "Star Wars"), the injection of very large funding and remarkable computer graphics, it is true, for example, that a leading US defence company claimed in 1996 to be able to build an airborne laser cannon that would destroy enemy missiles at ranges of several 100 km. This perhaps neatly concludes the requirement, allegedly laid down in the 1930s by the British War Office to early proponents of radio waves, that their "death ray" should be capable of killing sheep at a few miles.

Just as such early studies led to radar and its enormously important impact on the Second World War, the remote sensing capabilities of lasers are now becoming increasingly important. In fact, it may be said that later sensing is one of the few truly new (non-nuclear) military techniques introduced since the Second World War. Even modern thermal imaging and its immense capabilities had its genesis in the basic work and crude systems introduced in the late 1930s (see, e.g., R.V. Jones, *Most Secret War*, 1978).

In reviewing laser radar applied to military tasks, one can distinguish three successive phases. The first phase consists of relatively simple, direct-detection methods for very successful range finding, target marking and laser beacons employing near-IR ruby and Nd-YAG lasers. The non-eye-safe character of these laser systems gives problems of troop training in realistic conditions, and this has led in a second phase to the introduction of longer wavelength lasers such as CO_2 and Ho-YLF for basically similar tasks. In these phases, the laser is not much more than a bright light source. The third phase, which employs both advanced direct-detection and coherent techniques and exploits the full laser capabilities for wind measurement, target identification, chemical sensing, imaging, etc., is potentially of considerable military significance in the 21st century. Before outlining this work, however, it is worth reviewing the strengths and weaknesses of lidar in comparison with other military sensing technologies.

In many ways, lidar is very complementary to thermal imaging. In particular, CO_2 lasers at $10\,\mu m$ wavelengths lie in the middle of the 8–13 μm band of longer wavelength thermal imagers, and thus their atmospheric penetration and range performance is likely to be very similar. On the other hand, passive thermal imagers, with rapidly scanned multi-element detectors, provide excellent surveillance over quite large fields of view. In contrast, lidar, with few or single-element detectors and the very precise pointing capabilities of a diffraction limited laser beam, seems better suited to detailed interrogation and investigation of small regions identified as "interesting" by other means. Compared with longer wavelength centimetric radar, lidar has considerably poorer transmission, at least in the lower atmosphere, and thus most battlefield-type tasks have been confined to tactical ranges of a few tens of kilometres at most. This, of course, no longer holds true high in the atmosphere, and it is also worth appreciating the limited penetration of millimetric radar through, for example, heavy rain (due to large scattering from millimetric-size rain drops). Covertness is another important aspect; it is quite difficult for a potential target to detect that it is being illuminated with a comparatively low power, often short-pulse, laser beam, and much attention has been devoted to this. However, radar, unless used with great discretion, has been described as a beacon signalling to its adversary "here I am—come and hit me". Finally, it is worth pointing to a few unique military capabilities of lidar; these include, for example, wind sensing, detection of very small "targets" such as wires and cables and remote chemical detection.

12.7.2 Range finding, target marking, velocimetry, etc.

Simple rangefinders based on short-pulse, flash-lamp pumped, solid-state lasers (ruby, Nd-YAG) were amongst the first developments of military hardware. Using the elementary timing principle (Section 12.2.1, Figure 12.2c), very compact, robust and efficient (often battery-operated) systems were rapidly instituted at the infantry rifleman level, for tank rangefinders and also for aircraft operation. In a clear atmosphere, reliable operation over several kilometres range is easily obtained with a few metres accuracy corresponding to ~10 ns pulse length. In fog and smoke, longer-wavelength lasers generally have superior performance with the added advantage of eye safety. Over the years, very extensive studies have been carried out on relative effectiveness under varied conditions of humidity, fog, smoke, camouflage, target characteristics, etc., and laser ranging systems of many different levels of sophistication have been developed. For example, scattering from a smoke-screen laid down in front of a target is likely to give a strong signal similar to that from the obscured target itself (supposing that the laser beam can adequately penetrate the smoke). Logic that selects this later pulse signal from the target rather than the smoke signal is required (Figure 12.32a).

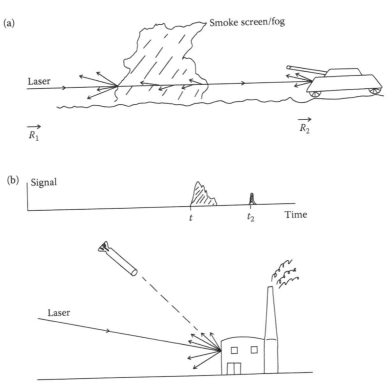

Figure 12.32 (a) Schematic of a laser rangefinder with signals from different ranges. Last-pulse logic is required to select the correct return from the target. (b) Schematic of guided missile homing on light scattered from the target marked by an incident laser beam.

In the related technique of target marking, the selected target is illuminated with a laser. Scattered radiation may then be picked up by a lidar receiver housed in a guided bomb or missile and provides information that directs in onto the target (Figure 12.32b). These methods showed their effectiveness in the 1970s in the Vietnam War; many unsuccessful attempts had been made to destroy an important bridge by conventional bombing which was eventually knocked out first time by a laser-guided device. More recently, the sophisticated development of such laser-guided missiles (LGMs) first impacted on the popular consciousness with extensive newsreel coverage of missiles apparently penetrating ventilation shafts, bunkers, etc., during the 1991 Gulf War.

For rangefinding, the use of a short pulse is but one extreme of coding the transmitted radiation. Many other forms of code may be used to acquire range or equivalent time-delay information, just as in longer-wavelength radar. These include amplitude modulation (AM) with, for example, a pseudorandom on-off code; auto-correlation

of the strength of the return signal with the AM code itself gives it a peak response at a time interval t_f corresponding to the range to the target and back. Alternatively, frequency modulation (FM) of the transmitted beam may be employed and has the potential advantage of providing Doppler velocimetry information about the target, as illustrated in Figure 12.33a with a linear chirp pulse of duration Δt_C and frequency modulation f_C. The return signals of a stationary target and moving target (with negative Doppler shift f_D) are shown. Signal-processing techniques of pulse compression ensure that the range resolution in this case is not determined by pulse length Δt_C, but rather by the rate of change of frequency (the chirp rate). Some compromise of range and Doppler frequency resolution is required but, typically, 10 m range resolution and 1m s^{-1} velocity resolution has been demonstrated in reasonably compact coherent systems at ranges up to 10 km. The sign of the Doppler shift may be established by use of the successive up and down chirps, as illustrated in Figure 12.33a.

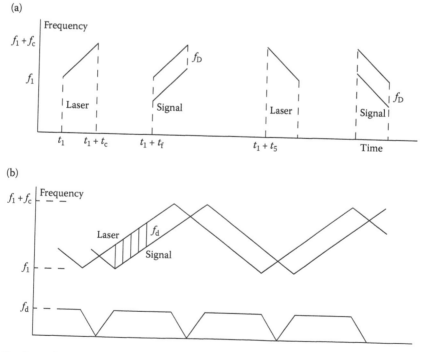

Figure 12.33 Coding of laser beams for simultaneous range/Doppler measurement. (a) Up/down linear chirp pulses of duration Δt_c, separation t_s and frequency modulation f_c. (b) Frequency modulation of a continuous wave beam (FMCW). For a distant stationary target, the range is proportional to the frequency difference f_d of the outgoing beam and return signal as shown.

Frequency modulation of a continuous wave beam (FMCW) may also be used, as illustrated in Figure 12.33b. With the delayed return signal beating against the outgoing waveform as shown, simple spectral analysis gives a frequency shift proportional to the range to the distant (stationary) target. For a moving target, this splits into two frequencies separated by $\pm f_D$. In practice, careful consideration of the various parameters indicated in Figure 12.33b of FM, code length and the target properties of range, velocity, etc., are required, including the problem of range-Doppler ambiguity. Nevertheless, several systems have been developed demonstrating high performance for simultaneous measurement of range and target velocity.

12.7.3 Micro-doppler and vibrometry

The bulk motion of a distant object may provide useful military information, as is demonstrated in many combined rangefinders and Doppler velocimeters. In addition, the information due to very small movements and vibrations within an object may potentially be of value in both civil and military fields. For a simple surface vibrating sinusoidally with amplitude a and frequency f_v, the displacement $a (\sin 2\pi f_v t)$ gives a peak speed of $\pm 2\pi f_v\, a$. From Equation 12.5, the peak micro-Doppler shift rises to $\pm 4\pi f_v a / \lambda$; with typical values of $a = 20\,\mu m$, $f_v = 200\,Hz$, the peak speed is $\pm 5.03 \times 10^{-2}\,m\ s^{-1}$ and for $\lambda = 10.6\,\mu m$ the micro-Doppler shift is only $\pm 9.51\,kHz$. Such a shift is readily detected (to provide the amplitude) but, more importantly, techniques of frequency demodulation may be employed to give the underlying vibrational frequency f_v. The thrust of much recent work has been, on the one hand, to develop the lidar systems for such measurement (coping with objects which may also be in bulk motion) but, on the other hand, to establish how characteristic and reproducible the vibrational frequencies are from particular targets and whether they can provide a key to successful identification. It is well known, for example, that their precise blade rotational frequencies are specific to particular types of helicopter and provide a means of identification. In practice, the problem is extremely complex; within a vehicle, for example, the driving frequencies originating at the primary power sources of engines, electrical motors, etc., are coupled through all manner of nonlinear mechanisms to the outside panels and frames, etc.,

from which the laser radiation is scattered. These outside surfaces are likely to have their own resonant frequencies and are further influenced for a ground vehicle by the terrain—gravel, mud, tarmac, etc., over which it is proceeding. Nevertheless, some success is being achieved and has been discussed in open literature. At the very least, the state of readiness of distant vehicles and, for example, the difference between live, active tanks and dead, knocked-out tanks or decoys is detectable. As a further example, it should be possible to match the radar techniques of aircraft identification (from signals reflected from turbine blades within the jet engine itself) with lidar methods taking signals reflected from the outer airframe skin.

12.7.4 Target imaging

The military lidar capabilities discussed to date of range finding, velocimetry and vibrometry may be considered as useful sensing discriminants in the broader task of providing a target image. These and various other discriminants available by electro-optic laser techniques are indicated in Table 12.12.

The point is that, while one is accustomed to interpreting visual grey-scale or color images, any discriminant that varies across a scene may be used to build an "image" of that scene. This may itself provide valuable information about an object (the "target") and a means of picking it out against the background.

Of the nine discriminants listed, the first five all relate to various aspects of the target region reflectivity and the signal intensity as potentially measurable in either a direct detection or coherent lidar. Under laser illumination, the scattering characteristics of different surfaces—vegetation, metal, paint, concrete, soil, grass, camouflage, etc.—vary quite widely under changing conditions of illumination (angle of incidence, polarisation, wavelength, etc.), as indicated in Figure 12.34. The term "glint" refers to the very strong, almost retro-reflection, signal that may occur on successive reflections from smooth surfaces (e.g., around a window frame or wheel arch) and is very characteristic of the presence of a man-made object as opposed to natural vegetation and soil. This is an extreme example of the difference of signal fluctuations as a laser beam is moved over different surfaces, e.g., concrete or brick compared with grass.

Table 12.12 Remote sensing discriminants for creating an image of the target region

Scattering reflectivity/intensity from strength of signal
Change of intensity at different wavelengths
Change of intensity at different polarizations
Occurrence of "glint" signals—high intensity retro-reflections
Statistical character of signal fluctuations
Range as set in different range intervals
Doppler frequency shift—bulk velocity
Micro-Doppler—frequency modulation (FM) vibrometry
Signal amplitude modulation (AM)

These and the other four discriminants listed in the table—range, bulk velocity, vibration and any amplitude modulation (indicative of moving machinery for example)—may be used to create and active "image" of the target region with suitable color coding, highlighting of significant regions, etc. The pixel size or spatial resolution in the image will be determined by the usual optical parameters—the diffraction limits and atmospheric refractive turbulence—and at 10 μm should provide a good match to co-located thermal imagers surveilling the passive scene. However, it is worth noting the prime military imperative of speed of measurement and the impact that has on optical scanning arrangements and the lidar dwell time on each pixel element. For passive imagers, whether thermal or visual, all pixel elements in the scene are radiating/reflecting energy simultaneously and the angle of view and overall speed of measurement is ultimately determined by the number of individual elements (each viewing one pixel) that can be packed into the detector. With an active lidar system, maximum sensitivity is achieved by illuminating each pixel sequentially with the laser beam, making the measurement and moving on to the next pixel. Necessarily, this is likely to be a slower process and reinforces the utility of lidar for provision of unique information from detailed interrogation of small regions as opposed to large area surveillance. The alternative of floodlighting a region, across many pixels, with a high-power expanded laser may be feasible in some circumstances but obviously mitigates against low-cost, low-power requirement and covertness.

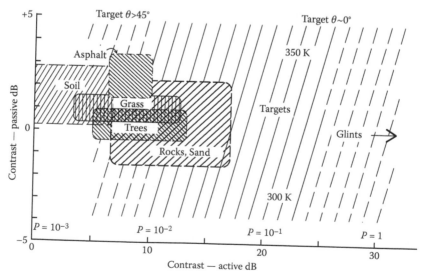

Figure 12.34 Example of the active laser radar (10 μm) and passive thermal imager (8–12 μm) contrasts to be expected between various targets and standard background (supposed at 300 K) temperature emissivity 0.999 and reflectivity $\xi = 10^{-3}$). The large variation can be attributed to different angles of incidence, degree of moisture, etc. (From Steinvall, O. et al., *SPIE* 300, 104, 1981.)

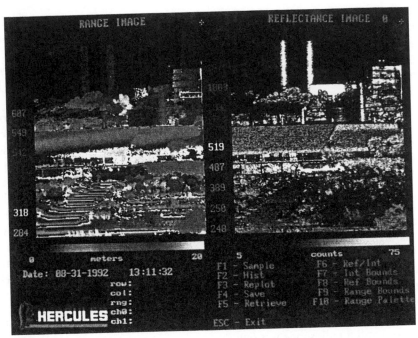

Figure 12.35 Range and reflectance imagery with a 1.047 µm direct detection lidar at 2 km, collected with the Hercules Corporation system. Much detail is lost in transcription from the original false color imagery. (Osche, G. R. and Young, D. S., Imaging laser-radar in the near and far infrared, *Proc. IEEE* @ 1996 IEEE.)

A number of lidar images are shown in Figure 12.35 and illustrate the potential. As always, extensive deployment into the armed forces requires that such systems are made cheap, compact, reliable and easy to operate and maintain.

12.7.5 Wind measurements: Rocketry, landing aids, ballistic air drops

As described in Section 12.4, rapid measurement of wind field with good spatial resolution at ranges between a few tens of metres to many kilometres is unique to laser radar. In consequence, it is not surprising that such capabilities have been extensively explored for military tasks. An immediately obvious area is the impact of wind on accuracy of munitions. Rockets, in particular during the early stages of flight, are travelling comparatively slowly and may be quite strongly influenced by the local wind conditions. For example, measurement of flow and downwash under helicopters has been measured by lidar in order to improve delivery point accuracy of rockets fired from such platforms. Very large ground-based rocket systems are also influenced

by the wind; some attempt to allow for this may be made by applying a correction in the fire control system. Conventionally, this may be a forecast wind or from a distant radiosonde measurement, possibly made several hours and tens of kilometres away. With the vagaries introduced by local terrain and natural wind fluctuations, the remaining uncertainties may contribute appreciably to the budget error for the weapon. For example, a compact coherent CO_2 lidar was developed in the late 1980s for the US Army Multiple Launch Rocket System (MLRS). The cw lidar would map the wind field at a distance of 100 m and it was said it could be produced, in volume production, at a competitive cost.

The potential of wind measurement as a landing and take-off aid has also been described. This could assist the landing of helicopters and fixed-wing aircraft on ships with monitoring of the complex wind flows around the landing deck. Such systems could also provide real, full-scale measurements around such structures and, together with wind tunnel testing and CFD modelling, assist in improved ship and platform design.

An airborne lidar system that profiles the wind field below the aircraft has recently been described.

Parachute delivery of cargo and conventional bomb drops without guidance control are obviously much affected by the prevailing winds. If these are known, the release point can be adjusted to compensate. However, such information may not be available or easily attainable in remote or hostile locations. A 2 µm lidar system based on a flashlamp-pumped Cr, Tm:YAG laser was built by Coherent Technologies and flown from the USAF Wright Laboratory. The Q-switched laser gave 50 mJ pulses at 7 Hz repetition rate and the system was easily installed in a C1411 aircraft. The scanner mirror mounted on a wedge produced a 20° × 28° (half angle) elliptically conical scan pattern with minor axis along the aircraft's longitudinal axis. Signal processing limited the first test in June 1995 to a range of 4.3 km, equivalent to a vertical distance of 3.8 km (15,000 ft) with the scan angles employed. Excellent agreement in the comparison of wind profiles measured with the airborne system, two ground-based lidars (one 2 µm and one CO_2 10 µm) and a radiosonde was found. Further studies assisted in a high-altitude bombing test above 30,000 ft (8 km) by B-52 aircraft. Using a composite wind profile, the bombers could correct for the wind and substantially improve their drop accuracy.

12.7.6 Chemical detection

The DIAL technique described in Section 12.6 suggests the possibility of detecting the poison gas and toxic nerve agent materials that might be used in chemical warfare (CW). The problem is that such materials may be chemically quite complex with rather broadband regions of absorption and transmission that makes positive discrimination and identification difficult. This contrasts with the detection of simple atmospheric constituents and pollutants with generally narrow absorption lines, as discussed in Section 12.6. In the latter case also, the lidar would usually be built with the aim of detecting certain well-defined chemical species; in practical operation, the equipment would be set to concentrate on very specific materials. In the military case, however, a range of toxic materials might be present and the lidar must be capable of seeking and detecting all of them without any prior indication of which is actually there. Nevertheless, the DIAL technique has been seriously examined and the US army at least has developed a CO_2-based chemical detection system. With isotopic

lasers, the overall tuning range may extend from 9.1 to ~11.8 µm, which offers a reasonable band of frequencies for discrimination in the longer-wavelength region. Solid-state lasers based on non-linear optical materials may also offer broad tuning ranges in wavelengths up to ~5 µm.

Finally, it is worth noting that some attention has been given to the possible remote detection by lidar of materials used for biological warfare (BW). In this case, discrimination against similar materials (pollens, spores, etc.) occurring naturally in the atmosphere would seem very difficult. However, it has been suggested that an airborne lidar searching for anomalies and discontinuities in the atmospheric aerosol scattering (Section 12.4) might provide an indication of the possible release of such BW material at several tens of kilometres range.

12.7.7 Terrain following and obstacle avoidance

Aircraft flying at moderately low level are extremely vulnerable to fast, agile missiles fired from the ground. One solution is to fly at even lower level to provide an adversary with less warning of approach. Such "nap-of-the-Earth" operation increases the obvious dangers of failing to clear the ground or flying into natural or man-made obstacles. Laser radar with its capability for very rapid measurement, fine pointing and range resolution has the potential to provide a valuable navigation and obstacle warning aid, as illustrated in Figure 12.36. Several lidar systems have been developed worldwide to explore this potential.

Figure 12.37 shows a terrain map generated by a CO_2 lidar developed in France and flown on a Puma helicopter. In this case, the lidar was an FM/CW coherent system with the beam directed into a pair of rotating prisms that generated a scanning rosette pattern. This pattern was composed of 4000 dots or pixels, each one corresponding to a pulse from the signal processing. The maximum scanning rate was ~2.5 Hz. An Anglo-French technical demonstrator programme was developed with acronym CLARA (Compact Laser Airborne Radar). This advanced 10 µm pulse heterodyne system was trialed on both a fixed-wing Tornado aircraft and a Puma helicopter and followed the earlier LOCUS (Laser Obstacle and Cable Unmasking System) developed by two groups within the GEC

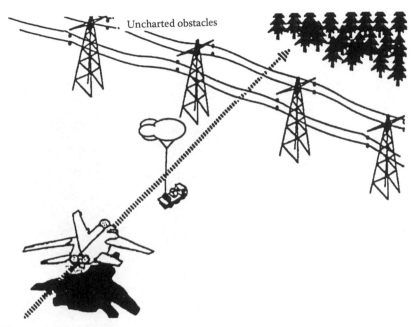

Figure 12.36 Potential obstacles in the path of a very low-flying aircraft that might be detected by a rapidly scanning laser radar.

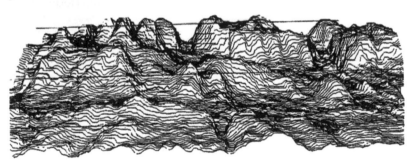

Figure 12.37 A terrain map, $10 \times 7\,km$ generated by a scanning laser radar (coherent FMCW CO_2) flown on a helicopter. (From Stephan, B. and Metivier, P., *Active Infrared Syst. Technol.*, SPIE-806, 110–118, 1987.)

Marconi Company and flown on A6-E, HS748 and Tornado aircraft. As indicated in Figure 12.38, the technology to detect power cables and display them to the aircrew was demonstrated. In addition to the two primary roles of obstacle warning and terrain following, CLARA was also designed for targeting and true air speed measurement. A pod-mounted configuration was adopted with complex optical scanner assembly to address several quite different modes of operation. Detection, classification and real-time display of obstacles, including various cables, masts, buildings and trees, must be achieved in daylight, at night and in adverse weather. In order to ensure that suitable warnings

continue to be provided when the aircraft is turning, a large field of regard, the size of which is governed by the flight envelope of the host aircraft, must be adopted for the sensor.

12.7.8 Star wars

In the original Strategic Defence Initiative (SDI), it was envisaged that lasers would have a central role, both as weapons for destroying enemy missiles and as sensors of such missiles. An important aspect of such sensing is to distinguish between decoy missiles and those actually containing warheads. A number of investigations in the United

Figure 12.38 Taken from a video record made with the LOCUS CO_{2pulse} heterodyne lidar, showing the lidar returns from cables. Much of the detail is lost in reproduction. (From Hogg, G. M. et al., *Conf. Proc.*, 563, 20, 1995.)

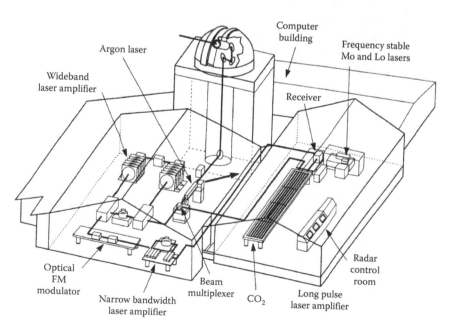

Figure 12.39 Artist's conception of the Firepond wideband imaging laser radar at the Lincoln Laboratory. Laser beams are transmitted from a 1.23 m diameter mirror Coude telescope with central abscuration of 0.21 m. (Melngailis, I. et al., Laser component technology, *Proc. IEEE* @ 1996 IEEE.)

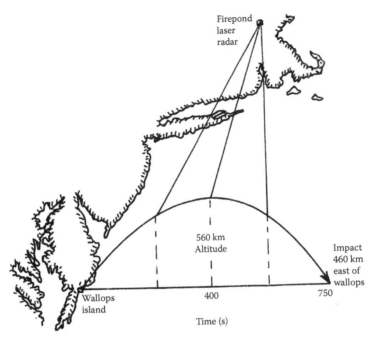

Figure 12.40 Schematic of the laser radar imaging during the SDI Firefly tests on missile detection in 1990. (Melngailis, I. et al., Laser component technology, *Proc. IEEE* @ 1996 IEEE.)

States have been reported in the open literature of the technical problems of laser pointing, tracking and examining high-flying rockets and space vehicles (see, e.g., Reference [172]). Figure 12.39 gives an artist's conception of the large installation of the Firepond Laser Radar Research Facility of the Lincoln Laboratory where much of this work has been done. In this equipment, several laser radars, but principally a large CO_2-pulsed coherent system, were used in a succession of long-range measurements. In the early 1990s, the first range-Doppler images of orbiting satellites were collected at ranges of 800–1000 km and were later extended to 1500 km. The argon laser indicated in Figure 12.39 provided the source for a visible light tracker of reflections from the satellites. Two initial SDI experiments were conducted later in 1990. These FIREFLY experiments were designed to investigate whether a laser radar could discriminate between a ballistic missile warhead and an inflatable decoy. In the tests shown schematically in Figure 12.40, a rocket was fired from the NASA Wallops Island Flight Facility several hundred kilometres to the south, rising to an altitude of 560 km in a 750 s flight. Coherent CO_2 laser radar imaging was successfully conducted over the central 50 s of the flight at a maximum range of 740 km. The prime target was a conical inflatable balloon made of carbon cloth. After approximately 360 s from launch, this was ejected in a small canister, was subsequently inflated and then set spinning in different manoeuvres to present a variety of target views to the imaging laser radar.

A second series of tests with acronym FIREBIRD (Firepond Bus Imaging Radar Demonstration) had the more ambitious objectives of investigating laser radar discrimination and countermeasure techniques. A high-performance booster rocket was used to deploy a dozen targets for study by several ground and airborne sensors along the eastern seaboard of the United States. These included passive IR sensors, UHF radars and optical sensors as well as the Firepond laser radars which also incorporated a photon-counting Nd-YAG laser radar. Generally excellent results were reported and completed the investigation of laser radar discrimination techniques. The advanced electro-optic techniques developed for this programme, including laser stabilization to the sub-Hertz level at high-power, precision coding of laser pulses, and sophisticated signal processing and analysis represent a remarkable *tour de force*. Demonstration of such precision tracking and imaging at over 1000 km ranges constituted at least a significant milestone in the history of laser radar.

12.8 LIDAR AND GEOSCIENCE

For the present purposes, lidar as applied to geoscience can be discussed in several broad areas: relatively short-range measurements to the sea surface and sea bed (few tens to hundreds of metres); sea, earth and ice surface from aircraft at up to ~15 km and finally ranging at hundreds to thousands of kilometres to satellite (and lunar) retroreflectors for studies of crustal dynamics, etc. These areas are outlined in turn.

12.8.1 Airborne mapping of terrain and ocean shelf

Ranging to the Earth's surface has been touched on in Section 12.7.7. Extensive studies with airborne systems have been conducted by several groups. Typical pulse energies of a few mJ with Nd-YAG lasers at 1.06 μm operating at up to a few hundred Hz have been employed. Figure 12.41 shows a block diagram of the lidar instrument developed at the Goddard Space Flight Centre [173] and mounted in a high-altitude aircraft; the position of the aircraft was measured to sub-metre accuracy by use of differential GPS receivers. Typical footprint size, from the beam divergence of 2.5 mrad, amounted to ~25 m at 10 km altitude. Problems of pulse waveform spreading and target signatures for such altimeters

have been considered by several authors (see, e.g., References [174,175]). Reported investigations include study of crater form structures and observations of the Mount St. Helena Volcano [176,177].

Laser depth sounding and the performance of a Swedish system has been well reviewed by Steinvall et al. [178]. Several such lidars have been built worldwide, including, for example, in the USA [179], Australia [180] and also Canada, Russia and China. In typical operation from a helicopter, two pulsed laser beams are emitted simultaneously and scanned over the surface. The longer wavelength (usually 1.06 μm, Nd-YAG) is mostly reflected from the water surface whereas the shorter wavelength (green 0.53 μm, doubled Nd-YAG) better penetrates the water and may provide a return from the bottom layer. Such lidar bathymetry provides a most promising technique for high-density, rapid sounding of relatively shallow waters typically 1–50 m deep. In addition, its use can be established for other sensing applications, as noted in Section 12.6.3—e.g., algae/chlorophyll monitoring and oil slick detection and classification. Operating from a helicopter also gives access to areas difficult for ships, e.g., around small islands, reefs and inlets. In addition, the rapid survey rate, typically 10–100 km2 h⁻¹ depending on shot density, is particularly valuable.

In evaluation of performance, the prime question is the depth penetration, as largely determined

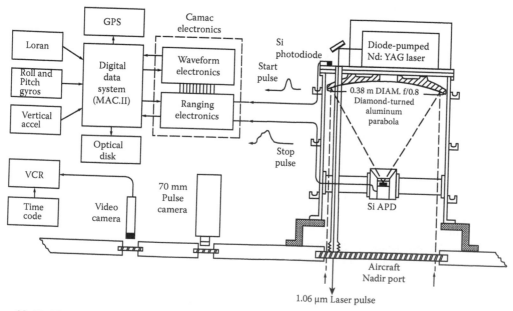

Figure 12.41 Block diagram of the airborne laser radar built at Goddard Space Flight Centre [173].

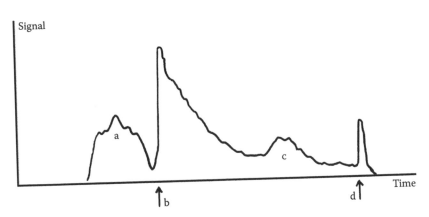

Figure 12.42 Schematic of water and seabed scattered return signals: (a) haze or "sea smoke", (b) surface return, (c) scattering layer, and (d) seabed return.

by water turbidity [181]. The maximum depth penetration D_{MAX} is given (see, e.g., Reference [178]) by

$$D_{MAX} = \ln(P^*/P_{min})/2G \qquad (12.29)$$

with $P^* = S.\varsigma/H^2$, where S is a system parameter (derived from laser power, receiver aperture and loss factors), ς is the reflectivity of the seabed and H the lidar platform altitude. P_{min} is the minimum power for detection of the bottom of the sea (usually limited in daytime by solar background and at night by detector/electronic noise). The system attenuation G is a complex function of intrinsic water parameters—beam attenuation, scattering coefficients, etc. With typical lidar parameters, values of $GD_{MAX} \approx 3$–6 can be achieved. It should be noted that a large increase in system parameter S only gives a modest increase in depth performance D_{MAX} due to the exponential attenuation expressed in Equation 12.29. Steinvall et al. [178] gave the example of a factor 10 increase in S only adding to D_{MAX} by ~4 m for $G \approx 0.3$ m^{-1} [log $10/(2 \times 0.3) \approx 4$].

Optical attenuation in sea water may be related to the Secchi disk depth D_s, which is the maximum depth at which a submerged white disk can be observed. Steinvall et al. [178] showed that values of $Ds \approx 10$ m gave $D_{MAX} \sim 20$–25 m, and $D_s \approx 30$ m gave $D_{MAX} \sim 25$–35 m for a state-of-the-art lidar bathymeter.

Other important factors limiting system performance include strong winds (affecting helicopter flight path) and associated sea state (white and broken water reduces accuracy and beam penetration).

Low-lying cloud and fog obviously inhibit performance and the thin haze (often quite dense and sometimes called "sea smoke") above the water may provide a false return for the surface itself. Within the water, narrow scattering layers and other inhomogeneities may also provide spurious returns.

Finally, the nature of the sea bed, i.e., dense vegetation, rocks and slopes, needs evaluation. Figure 12.42 shows a schematic of a lidar return, as affected by these phenomena and requiring postprocessing and evaluation for extraction of water depth to the sea bed. An important consideration is the nadir angle of the scan, which would ideally be fixed to give constant slant range and angle. However, a full conical scan at fixed nadir provides inefficient coverage with more and less dense regions. Various scan patterns have been developed; a modified semi-arc scan, part conical, ahead of the helicopter gives more uniform coverage. Typically 100–200 m swatch width is used with angle from nadir ~20°. For complete evaluation of the sea environment, lidar bathymetry provides a complementary technique to be combined with acoustic and direct mechanical methods of sampling.

12.8.2 Satellite and lunar ranging

The evolution, technology and utility of long-distance, high-precision (near millimetre) ranging to retro-reflectors mounted on satellites (and the moon) has recently been well reviewed by Wood and Appleby [182]. The present brief outline draws heavily on their report. The concept of satellite laser ranging (SLR) to improve geodetic information was first suggested by Plotkin [183]. It was considered that accuracies at the level of a few centimetres would be required to observe interesting geophysical processes of the Earth's crustal dynamics, etc. The first laser radar observations in the mid-1980s demonstrated metre-level precision and established the potential. Over the succeeding years, rapid developments of short-pulse lasers, sensitive detectors, dedicated satellites and accurate computer-guided tracking now give precision approaching a few millimetres, with more than 30 active SLR systems in a worldwide network.

For sub-centimetre precision, very short pulses are required. These are now provided by mode locking systems applied to Nd-YAG lasers giving a train of very narrow (typically ~30 ps) pulses, all equally spaced in an envelope several hundred nanoseconds long. Frequency doubling is employed to take advantage of high quantum efficiency detectors at 0.53 μm in the green. The envelope is approximately Gaussian shaped; in practice, either the most energetic pulse alone may be transmitted or this together with all the following ones. In the latter case, range ambiguity (attribution of which signal return to which pulse) is not a problem since the pulses are separated spatially by 2–3 m, whereas the range accuracy is better than 1 cm. Highly developed techniques are required for the elements of an SLR system and include

- *Transmitter and receiver telescopes:* they have been coaxial, separate co-mounted and, most recently, separate, individually mounted and controlled [184]. Very high precision, computer-directed acquisition and scanning to follow the satellite path are required.
- *Detectors:* very precise timing of the incoming signal is required. Solid-state devices offer significant improvements over the photo-multipliers used in earlier systems. Single photon detection with low-energy systems is increasingly favored.
- *Filters:* very narrow, oven-controlled, spectral filters, typically 0.1–0.2 nm wide are employed to cut down sunlight for daytime operation.
- *Retroreflectors:* due to satellite motion, the onboard corner cube retroreflectors require to be slightly "defocused". In order to ensure that the footprint of the reflected beam will cover the emitter/receiver station, one of the three mutually perpendicular surfaces is given a slight curvature. Ideally, only one retroreflector would be mounted on the satellite but this gives problems of visibility from different directions. The ultimate solution noted by Wood and Appleby [182] would be eight corner cubes joined at their apices to give reflections equivalent to a reflection from the centre of mass of the satellites.
- *Satellites:* the six dedicated geodetic satellites are typically small (~1 m diameter) and spherical, and placed in near-circular orbits at altitudes between 800 and 19,000 km. They are variously suited to different scientific tasks—low orbit: high-frequency gravity terms, ocean tides, etc.; higher orbit: crustal motion, Earth rotation, etc.

Several studies of accuracy and range corrections have been given (see, e.g., References [175,185]). The handbook article of Wood and Appleby [182] provides over 120 references to research articles in satellite laser ranging. One remarkable example was for the NASA LRS system based at Arequipa in Peru. This station was subject to a steady movement of the underlying Nazca plate causing an ENE motion measured at about 10 mm per year. After the earthquake of June 2001, analysis of the dramatic shift in the data confirmed that the station had moved ~0.5 m to the south west.

As to lunar ranging, a number of small arrays of retroreflectors were placed on the moon in the 1960s and early 1970s. Four are in regular use with ranging from two trekking stations—at Mcdonald, Texas and Grasse in France. With the R^{-4} range dependence of signal, ranging to the Moon is vastly more challenging and large telescopes (0.75 and 1.5m) and high-power pulses (~150 mJ) are required.

Several scientific tasks of astrometry can be undertaken, including accurate determination of the Earth and Moon via the tides. The Earth–Moon distance is increasing by 3.8 cm per year and the Earth is slowing on its axis by an increase in the length of day of about 2 ms per century.

12.9 LIDAR IN SPACE

12.9.1 Introduction—Why lidar in space?

During the past 30 years, many passive imagers and sounders have been placed in orbit for study of the Earth's surface and atmosphere with capabilities extending across the visible, IR and

Table 12.13 The four candidate lidar systems for space considered by ESA in 1989

A simple backscatter lidar—for measurements of cloud-top height, cloud extent and optical properties, planetary boundary layer and tropopause height, aerosol distribution—with wide applications to meteorology and climatology

A differential absorption lidar (DIAL), providing high-vertical-resolution measurements of humidity, temperature and pressure

A wind-profiling lidar with the unique capability of improved weather forecasting and global dynamics

A ranging and altimeter lidar for very accurate measurements of surface features, including ground, sea and ice cap height for solid-earth studies

microwave spectral regions. Active sensing with radar systems has also been undertaken, with very high-performance scatterometer and imaging radars carried on operational satellites such as ERS-2. Compared with ground-based or even airborne sensors of all types, the prime advantage of space lidar is the prospect of rapid global coverage, including data from otherwise totally inaccessible regions. Compared with other spaceborne sensors (both passive and active), lidar techniques provide certain unique measurement capabilities, such as cloud-top and boundary layer height and global wind fields. Other lidar capabilities (e.g., pressure, temperature and humidity and ranging/altimetry) have the potential to complement and extend existing methods and often provide improved height and spatial resolution, particularly compared with passive techniques.

Against this background, several national and international space agencies have initiated feasibility studies for lidars in space, culminating in several reports, including LASA (Lidar Atmospheric Sounder and Altimeter), Earth Observing System Vol IId, NASA 1987; LAWS (Laser Atmospheric Wind Sounder), Earth Observing System Vol IIg, NASA 1987; BEST (Bilan Energetique du Systeme Tropical), CNES, France 1988; Laser Sounding from Space—Report of the ESA Technology Working Group on Space Laser Sounding and Ranging, ESA 1989. This latter report focused on four lidar systems considered good candidates for space deployment; an extract from its executive summary is shown in Table 12.13. During the past 15 years, several evaluations of the detailed technology and logistics of space lidars for specific tasks have been funded by the space agencies. In addition, various of the critical subcomponents, such as lasers, optical arrangements, lag-angle correction, etc., have been studied and space-qualifiable prototypes constructed. Progress to actual space deployment, however, remains slow, with the exception of the NASA Lidar In-space Technology Experiment (LITE), which flew on the Space Shuttle Discovery in September 1994, and very recently the Geoscience Laser Altimetry System (GLAS) launched in 2003.

This and other steps towards spaceborne lidars are discussed in the following sections. Critical design factors for a space lidar obviously include the demands on spacecraft accommodation for mass, size, power, heat dissipation, cooling, pointing and vibrational stability, etc., all for operation in a realistic orbit. Performance factors include ultimate sensitivity limit to weak levels of scattering (largely determined by laser power and telescope aperture) and the geometric terms of range resolution (influenced by laser-pulse length), beam footprint size (beam divergence and the need to keep light levels within eye safe limits—including observation by ground telescopes) and the separation of successive footprints (laser-pulse repetition frequency, beam scanning and spacecraft velocity).

12.9.2 Backscatter lidar and LITE

The science objectives and technology requirements of a spaceborne, direct-detection, backscatter lidar were discussed in the LASA report listed above and also in a 1990 ESA report of the ATLID (Atmospheric Lidar) Consultancy Group. The latter in particular reviewed the potential for operational meteorology and environmental/climatological research. The atmospheric features that should be detectable from such an ATLID backscatter lidar are illustrated in Figure 12.43 drawn from the report. For weather forecasting (numerical weather prediction (NWP)), ATLID data should be of direct help with clouds, boundary layer height, air mass discontinuities and frontal/convective activity. For climate studies, the information on clouds and

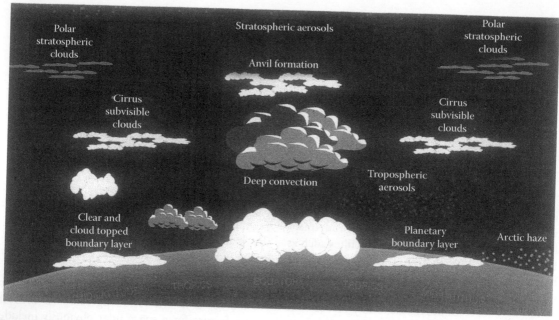

Figure 12.43 Atmospheric features of clouds and aerosols detectable by a simple backscatter lidar at various latitudes. (From ATLID Consultancy Group, *Backscatter Lidar, the Potential of a Space-Borne Lidar for Operational Meteorology, Climatology and Environmental Research*, ESA Specialist Publication, SP-1121, Noordwijk, 1990.)

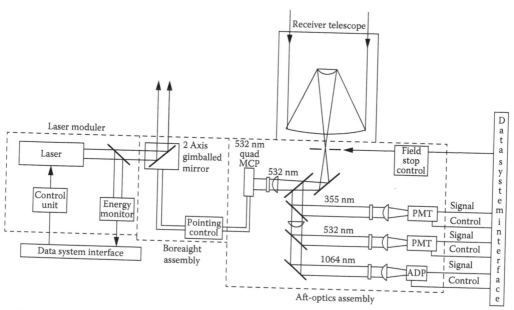

Figure 12.44 Functional diagram of the LITE system. Light at the three wavelengths was directed down to the Earth from the two-axis gimballed mirror. Scattered light was collected at the lightweight receiver telescope of 1 m aperture and passed via dichroic beam splitters to the two photomultipliers (PMT) and avalanche photodiode (ADP). Some of the 0.532 μm light was passed to the microchannel plate (MCP) quadrant detector, which provided an error signal for precise alignment of the transmitter and receiver system. (Winker, D. M. et al., An overview of LITE: NASA's lidar in space technology experiment, *Proc. IEEE* @ 1996 IEEE.)

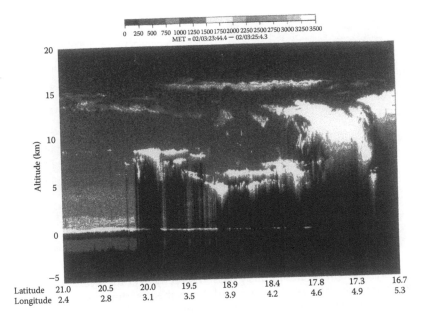

0 250 500 750 1000 1250 1500 1750 2000 2250 2500 2750 3000 3250 3500
MET = 02/03:23:44.4 — 02/03:25:4.3

Figure 12.45 The LITE return signal at 0.532 µm after offset correction showing the multilayered cloud structure associated with a tropical storm system over West Africa. Note the aerosol signal (yellow and red) on the left-hand side, particularly at the boundary layer. Ground signals are also obtained beneath the high-level thin cloud (16 km, at centre of picture), but not below the thicker cloud layers on the right hand side. (Winker, D. M. et al., An overview of LITE: NASA's lidar in space technology experiment, *Proc. IEEE* @ 1996 IEEE.)

aerosols and their effect on the Earth's radiation balance should be a valuable contribution.

As planned (see, e.g., References [186,187]), the LITE provided a valuable demonstration of the enabling technologies for such operational systems. In the 12-day shuttle mission, the LITE instrument performance was excellent and a large volume of data was generated [72]. Figure 12.44 shows a block diagram of the equipment, which was mounted on a Spacelab pallet. The flash-lamp-pumped Nd-YAG laser operated in doubled and tripled frequency mode to give simultaneous output pulses at 1.064, 0.532 and 0.355 µm. The pulse repetition rate was 10 s⁻¹ and pulse width was 27 ns. The beam footprint at the Earth's surface varied between 290 m (0.355 µm) and 470 m (1.06 µm) and successive footprints were 740 m apart. The mass of the instrument was 990 kg and the average power consumption in lasing operation was 3.1 kW.

Two of the fascinating records from this instrument are shown in Figures 12.45 and 12.46. These pictures provide some indication of the quality, depth and volume of information that could be available from an operational system, and indeed the extent of the signal processing, evaluation

and transmission systems that would need to be set up to use it effectively. For meteorological and weather forecasting purposes, such information must of course be made available in good time to the operational centres.

12.9.3 Global wind field measurement by lidar

One of the earliest feasibility studies of a satellite-borne lidar for global wind measurement was sponsored by the USAF Defence Meteorological Satellite Program. In the early 1980s, this was followed by a hardware definition study, conducted by Lockheed, of a system given the acronym WINDSAT (Windmeasuring Satellite); another title current at this time was WPLID for Wind Profiling Lidar (see, e.g., Reference [103]). These have been followed by the reports noted in Section 12.9.1, all of which considered large, multiscanning (conical or several axes) lidars. More recently, the ESA Atmospheric Dynamics Mission: Reports for Mission Selection (1999) was presented to the European Earth Observation community at a meeting attended by over 300 participants. Of

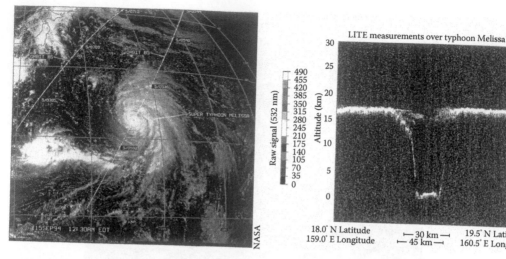

Figure 12.46 The LITE system serendipitously passed directly over the eye of typhoon Melissa. Satellite photographs were obscured by thin cloud. Nevertheless, the lidar was able to penetrate this and make measurements within the eye-wall down to the surface. (From Barnes, J. E. and Hofmann, D. J., *Geophys. Res. Lett.*, 24, 1923–1926, 1997.)

the four candidate missions, ADM was selected as the second Core Mission to be implemented. The lidar studies indicate the degree of importance attached to global scale wind measurement deriving from the unique capabilities of a Doppler Wind Lidar (DWL) compared with other spaceborne methods. For example, low-level wind strength and direction may be derived from radar scattering from the sea surface where the wave height is largely determined by the surface wind. However, such low-level winds are not of the highest value

for meteorological purposes. In the mid-levels, the winds derived from cloud-top motion are not always completely representative of the air motion in that region, and may also have some uncertainty as to relevant altitude. For the ESA ADM, the observational requirements have been chosen as shown in Table 12.14 and led to selection of a relatively simple single-axis system for the first demonstrator mission. The most suitable lidar to realize these goals has been extensively discussed with three possible candidates:

Table 12.14 Observational requirements of the ESA atmospheric dynamics mission

	Observational requirements		
	PBL	**Troposphere**	**Stratosphere**
Vertical domain (km)	0–2	2–16	16–20
Vertical resolution (km)	0.5	1.0	2.0
Horizontal domain		Global	
Number of profiles (h^{-1})		>100	
Profile separation (km)		>200	
Horizontal integration length (km)		50	
Horizontal sub-sample length (km)		0.7–50	
Accuracy (HLOS component) (ms^{-1})	1	2	3
Zero-wind bias (ms^{-1})		0.1	
Windspeed slope error (%)		0.5	
Data reliability (%)		95	
Data availability (h)		3	
Length of observation data set (yr)		3	

(a) Coherent heterodyne, with CO_2 gas lasers in the 10 um band (aerosol scattering).

(b) Coherent heterodyne, with solid-state slab lasers in the 1–2 um band (aerosol scattering).

(c) Incoherent direct detection, at ~0.3–0.5 um, and Fabry–Perot interferometer/edge technique analysis (aerosol and molecular scattering).

For the coherent systems, the laser beam must be transmitted in a collimated, single-mode beam. From a 400–500 km orbit, the beam size, or footprint, at the Earth's surface would be of order 10 m, which rules out a non-eye-safe laser of wavelength less than ~1.5 μm. For the incoherent systems at shorter wavelength, the beam may be expanded to produce an enlarged footprint with energy density within eye-safe limits. For any system, the crucial questions derive from the backscattering characteristics of the atmosphere that determine the primary Doppler signal, whether from aerosols or molecules.

As discussed in Section 12.4, molecular scattering varies uniformly and predictably whereas aerosol scattering is highly variable with occasional large increases due to volcanic activity (a suggestion for augmenting the scattering levels with a strategically placed device is shown in Figure 12.47). For both domains, a large, powerful

lidar is required and may be quantified in terms of the necessary ED^2, where E is the total laser pulse energy (Joules per individual measurement) and D is the collecting telescope diameter (metres). For successful measurement at lowest levels of atmospheric scattering (both molecular and aerosol), analysis shows that ED^2 requires to be greater than ~100 Jm^2. With $D^2 \sim 1$, for the heterodyne systems the available energy is best utilized in large pulses, ideally > 10 J, whereas for direct detection lidars the energy may be distributed over many small pulses (see Section 12.3.4).

Extensive modelling of performance for spaceborne wind lidars has been conducted. Modeled performance for different levels of aerosol scattering—median, quartiles, deciles—corresponding to the values shown in Figure 12.8 and for a coherent heterodyne lidar operating with the instrumental parameters listed is shown in Figure 12.48.

Most recently, for the ADM demonstrator, ESA has chosen a direct-detection interferometric lidar operating at 0.35 μm; the basic optical layout is shown in Figure 12.49. In a novel arrangement, the narrow band aerosol (Mie) scattering would be separately analysed by a Fizeau or possibly a two-beam Michelson interferometer; the broadband molecular (Rayleigh) scattering, which is relatively

Figure 12.47 A suggestion for augmenting aerosol scattering (β in the atmosphere to improve the performance of the spaceborne Doppler wind lidar (ALADIN). (Courtesy of Rodolphe Krawczyk, 1994.)

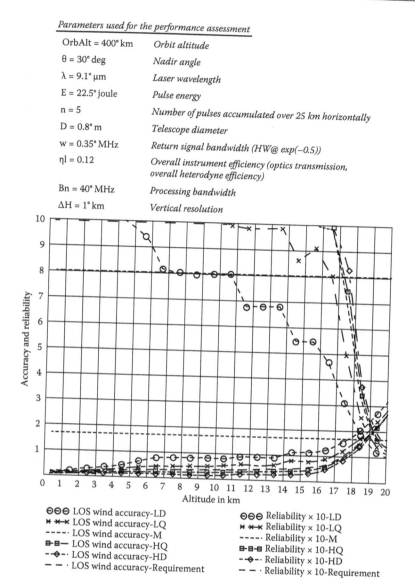

Figure 12.48 Calculated performance of a space-borne 9.1 μm coherent detection lidar with parameters shown for median and upper/lower quartiles and deciles of aerosol scattering. Note that the accuracy (ms⁻¹) curves start at the lower left-hand side and the reliability curves (0–100%) start at the top left-hand side for the different levels of scattering.

stronger at these shorter wavelengths, would be analysed by a double-edge Fabry–Perot etalon. Suggested performance is shown in Figure 12.50 with an accuracy better than 2 ms⁻¹ up to 11 km altitude. An artist's impression of this equipment in a satellite flying in a sun-synchronous polar orbit at an altitude of ~400 km is shown in Figure 12.51. Processing of the backscatter signals will provide about 3000 globally distributed wind profiles per day, above thick cloud or to the surface in clear air,

at typically 200 km separation along the satellite track. Target date for launch is 2008.

12.9.4 Chemical, ranging and solid earth studies

As noted in Table 12.13, DIAL lidar is an obvious candidate for measurements into the atmosphere from space (see, e.g., Reference [188]). The

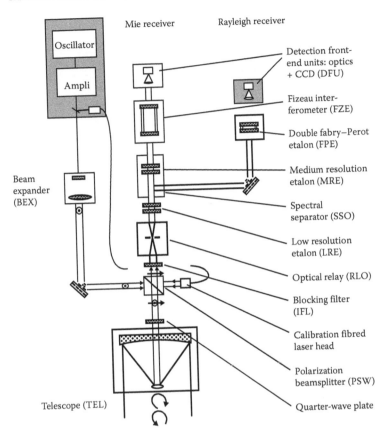

Figure 12.49 Optical arrangement for the ALADIN direct detection lidar for space-borne Doppler wind lidar. The aerosol (Mie) and molecular (Rayleigh) scattered signal are measured in the two separate interferometer systems as shown.

problems, however, are formidable: two or more laser wavelengths for each chemical species under study (ozone, water vapour, greenhouse gases, etc.), compensations for the Doppler shifts in the beams due to the large spacecraft velocity (typically $15°$ off nadir $\equiv 0.3$ cm^{-1} shift), with the beams transmitted simultaneously and co-axially to ensure return signals from the same elastic scatterers. Much study has been put into such active remote sensing from space and DIAL lidars will almost certainly be flown in the coming years. However, passive techniques over a wide range of wavelengths are providing much information. For example, in the NASA Earth Observing System (EOS), the Aura mission to study chemistry and dynamics of the troposphere and stratosphere, the instruments include sounders for measurement of thermal emission from the atmospheric limbs (microwave

and high resolution) as well as nadir viewing imaging spectrographs for ozone and trace gases.

Lidar for geoscience has been outlined in Section 12.8. Early studies for space-borne systems were given by Cohen et al. [189] and Bufton [175]. Very recently, in February 2003, the Geoscience Laser Altimetry System (GLAS) on board the NASA Ice, Cloud and Land Elevation Satellite (ICE Sat) was activated. This mission is intended to accurately measure ice sheet mass balance, cloud and aerosol heights as well as land topography and vegetation characteristics. Initial test of the space-bound lidar altimeter system was validated with data from the global Laser Reference System (LRS) noted in Section 12.8. The lidar operates on Nd-YAG at 1.06 μm; it is reported that the signal returns were as expected and the transmitted laser beam is very close to the bore sight.

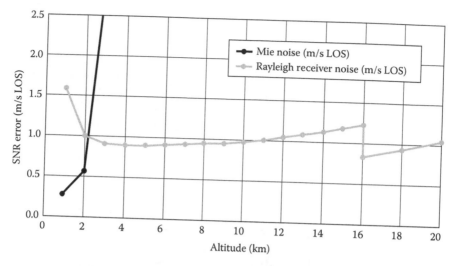

Figure 12.50 Calculated performance, by Monte Carlo simulation for the ALADIN space-borne Doppler wind lidar. For this calculation, the main parameters were: a 1.1 m diameter telescope, over 700 shots a total pulse energy per measurement of 120 J at 0.35 μm wavelength and a detector quantum efficiency of 80%. Performance may be readily scaled by the Cramer–Rao relationship. The shift at 16 km is due to a change in the required vertical resolution from 1 to 2 km at that altitude in the atmosphere [196].

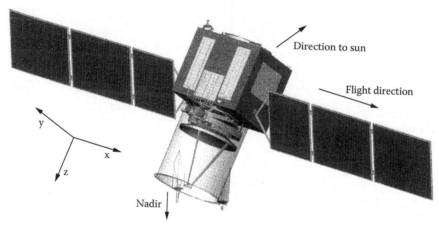

Figure 12.51 Impression of the in-orbit configuration for aspace-borne Doppler wind lidar showing the telescope pointing down at, ~35° from nadir [196].

12.10 CONCLUSIONS

From the preceding papers cited, it should be clear that the past 30 years have seen remarkable advances in laser radar technology, applications and understanding. Much fine equipment has been installed in different centres worldwide and valuable, often unique, measurements not accessible by other techniques have been made. However, there have also been significant disappointments, particularly in the wider deployment of coherent lidar equipments and techniques. This final section attempts a brief analysis of the situation and to draw some guidelines for the future.

Table 12.15 offers a crude summary comparison of the history and development of four remote-sensing technologies. It is striking that, for microwave radar, thermal imaging and sodar, the first deployment (defined as use by non-specialist scientists and technicians) took place within ~15 years, followed by widespread deployment (e.g., equipments commercially available off-shelf or bespoke) less than

Table 12.15 Comparison of remote sensing technologies

Technology	First research	First deployment (a)	Widespread deployment (b)	Comment
Microwave radar	~1935 "kill a sheep at 2 miles"	~1940—military "needs a Ph.D. in every set"	>1942—aided by compact cavity magnetrons	Enormous variety of capabilities, application wavelengths, equipment, etc.
Thermal imaging	1930–1970	~1975	~1985	Widespread applications in civil and military domains
Active acoustic sensors (SODAR)	1960–1970	~1975	~1985	Slow, limited range, low resolution etc. *BUT* relatively cheap and simple
Direct detection lidar: (a) temporal (b) spectral	1965–1975 1970–1985	~1975 ~1985	~1980 –	Military and civil ranging, target marking, cloud base, etc. Chemical detection (DIAL); limited Doppler applications
Coherent lidar	1970–1990	~2000	–	Enormous range of applications demonstrated. Why not taken up outside research?

10 years later. Examination of a radar supplier's catalogue with the dozens of different equipments, all available off-the-shelf, for many varied applications working over a large range of wavelengths provides a salutary example to the lidar scientist. Even closer is the astonishing growth in the optical technology and applications for the telecommunications industry. However, for laser radar, apart from relatively simple tasks of military and civil ranging, target marking and possibly DIAL, widespread deployment has scarcely been achieved. The question is why should this be so? Are more advanced lidar systems perceived as too difficult, too expensive, unreliable, dangerous, too uncertain, etc.?

As scientists, we tend to believe the adage 'build a better mousetrap and the world will beat a path to your door'. In the hard, commercial world, this is not true—a new technology has to be significantly superior to displace an older established one and is unlikely to do so if it is even marginally more expensive. Other relevant truisms that bear heavily on the planning, specification and design of equipments include

(1) Defined as first use by non-specialist scientists/technicians. (2) Defined as widely available off-shelf (or bespoke).

- The "best" is the enemy of the "good".
- The devil is in the detail.
- Small is beautiful—but is it achievable?
- Multifunction usually means, literally, good for nothing.

Ignoring one or more of these is usually a recipe for disaster—as is very evident in a review conducted by the author (unpublished and probably unpublishable!) of nine "failed" projects. These were all large, expensive (equivalent of several million pound sterling or more) projects funded during the past 25 years by civil and military agencies across Europe and the United States; all were intended to lead into wider development and even (in some cases) medium-scale production. The technical performances of the prototypes in these projects ranged from poor to very good indeed. For those in the latter category, lack of ongoing success was basically due in one case to failure to sufficiently interest the end user (partly due to the inadequate data/display facilities available at the time) and, for the other, an alternative microwave radar technology was preferred. In at least two projects, the basic properties of the phenomenon under investigation were poorly researched and understood (scattering strength, spectral form, etc.) so the lidars, when built to the set specification, could not provide the required information. In two others, the performance specification became overly ambitious (the "best") and multifunction leading to excessive size, complexity and cost, and poor overall performance. Other projects were bedeviled by arguments between lidar scientists over the appropriate choice of laser, not helped by inflated and premature claims from laser theorists and manufacturers as to the achievable pulse energy, prf, reliability, etc. In one bad example, basic principles of coherent operation were not properly appreciated at a high technical level, leading to poor decision making. From this brief history, a few more truisms may be drawn:

- A project will fail with too much technology "push" and not enough end user "pull".
- Bad science and over-promoted technology will breed disaster—but good science and technology will not necessarily win.
- Beware "technology jocks" bearing gifts.
- The commendation "more reliable" must mean "an in-depth engineering study of failure modes has been conducted" or it is worse than useless.

One may thus draw a few guidelines (see Section 12.5 in the present volume for a much more comprehensive analysis): the good system and project development arises from a total holistic approach incorporating:

1. A good understanding of the measurement problem with a realistic specification for the lidar based on reliable, available component technology.
2. Good science/technology in the lidar design including

- Good fundamental science of direct detection/coherent lidar.
- Translation into simple, serviceable optical design.
- Proper study of engineering failure modes.
- Good signal processing.
- Good data analysis and display which provide
  - Swift timely analysis
  - Informative graphics
  - And truly give the end user what he needs and can use

This, of course, raises various broader questions including, for example:

- Where is the basic lidar training? Few universities conduct research with coherent lidar, which seems a little puzzling, particularly with the potential for laboratory-based physics applications of coherent heterodyne techniques.
- Is there an adequate, compact and referable body of knowledge? An enormous body of research publications and reviews exists, but a more pedagogic literature is perhaps lacking.
- How to involve and convince end-users? The bulk of lidar applications research is still done by lidar scientists who initially built demonstrators but then developed an interest in, e.g., the applications to meteorology, atmospheric chemistry or wake vortices, etc., occasionally managing to interest colleagues more directly involved in these fields.
- This final point emphasizes that, without an enthusiastic body of such end-users (professional meteorologists, environmental scientists, system engineers, etc.), who can envisage worthwhile application of a lidar to "their" problem, appreciable funding is not going to be forthcoming from the research councils and agencies or from industry.

Lest this be considered too downbeat, an assessment of lidar must emphasize again some of the remarkable and unique achievements of laser radar over the past 30 years: contributions to flow measurement, wind fields, atmospheric chemistry, ozone depletion, pollution dispersal, avionics, geoscience, etc., even apart from more military-oriented applications. Looking to the future, progress is likely to be rooted in:

- Laser developments: particularly of high power and efficient, agile-frequency CO_2 lasers, precisely specified off-the-shelf solid-state 2 and 1 μm lasers, and diode and fiber amplifier lasers at ~1.5 μm with increased power and mode purity.
- Lidar construction: reduction in cost and complexity should flow from monolithic waveguide-type construction at 10 μm and, at shorter wavelengths, much use of fiber guiding, components and optical techniques borrowed from the communications industry.
- Detectors: very fast, high quantum efficiency (>80%), low-noise array detectors are becoming available at visible and UV wavelengths. Extension to the near-IR would have great impact, as would uncooled heterodyne detectors at 10 μm.
- Data analysis and display: the rapid, user-friendly access to data coupled with graphic display of results now offered by modern computing facilities provides a great advance for lidar studies and should always be budgeted for in a system programme.

Finally, it is worth noting the encouraging prospects for global wind field measurements in the Atmospheric Dynamics Mission—ESA's technology demonstrator, due to be launched in 2008, and at the other extreme the potential development of small, low-cost LDVs for flow measurement ahead of, for example, wind-turbine generators.

ACKNOWLEDGMENTS

The author expresses his deep appreciation of the constant spur of enthusiasm and interest generated by many colleagues and friends worldwide, with particular thanks to those at RSRE/DERA (Malvern)—now QINETIQ—and at ESTEC (Noordwijk). As always, any errors or omissions in the present work must be attributed to the author.

REFERENCES

1. Hermann, J. A. 1990. Evaluation of the scintillation factor for laser hazard analysis. *Appl. Opt.* 29: 1287–1292.
2. Rye, B. J. and Hardesty, R. M. 1997. Estimate optimisation parameters for incoherent backscatter heterodyne lidar. *Appl. Opt.* 36: 9425–9436.
3. Vaughan, J. M. 2002. Scattering in the atmosphere. In *Scattering*, Chapter 2.4.3, eds., E. R. Pike and P. Sabatier. London: Academic Press, 937–957.
4. Harris, M. 1995. Light-field fluctuations in space and time. *Contemp. Phys.* 3: 215–233.
5. Schwiesow, R. L. and Cupp, R. E. 1980. Calibration of a cw infrared Doppler lidar. *Appl. Opt.* 19: 3168–3172.
6. Hardesty, R. M., Keeler, R. J., Post, M. J. and Richter, R. A. 1981. Characteristics of coherent lidar returns from calibration targets and aerosols. *Appl. Opt.* 20: 3763–3769.
7. Kavaya, M. J. and Menzies, R. T. 1985. Lidar aerosol backscatter measurements: Systematic, modelling and calibration error considerations. *Appl. Opt.* 26: 796–804.
8. Vaughan, J. M., Callan, R. D., Bowdle, D. A. and Rothermel, J. 1989. Spectral analysis, digital integration, and measurement of low backscatter in coherent laser radar. *Appl. Opt.* 28(15): 3008–3014.
9. Sonnenschein, C. M. and Horrigan, F. A. 1971. Signal-to-noise relationships for coaxial systems that heterodyne backscatter from the atmosphere. *Appl. Opt.* 10: 1600–1604.
10. Kavaya, M. J. and Suni, P. J. M. 1991. Continuous wave coherent laser radar: Calculation of measurement location and volume. *Appl. Opt.* 30: 2634–2642.
11. Werner, C., Kopp, F. and Schwiesow, R. L. 1984. Influence of clouds and fog on LDA wind measurements. *Appl. Opt.* 23: 2482–2484.
12. Chan, K. P. and Killinger, D. K. 1992. Useful receiver telescope diameter of ground-based and airborne 1-, 2-, and 10-μm coherent lidars in the presence of atmospheric refractive turbulence. *Appl. Opt.* 31: 4915–4917.
13. Frehlich, R. and Kavaya, M. J. 1991. Coherent laser radar performance for general atmospheric refractive turbulence *Appl. Opt.* 30(36): 5325–5352.
14. Scorer, R. S. 1997. *Dynamics of Meteorology and Climate. Atmospheric Physics Wiley-Praxis Series*. Chichester: John Wiley & Sons.

15. Hinkley, E. D., ed. 1976. *Laser Monitoring of the Atmosphere*. Berlin: Springer.

16. Zuev, V. E. 1976. Laser light transmission through the atmosphere. In *Laser Monitoring in the Atmosphere*, ed., E. D. Hinkley. Berlin: Springer, 29–69.

17. Collis, R. T. H. and Russell, P. B. 1976. Lidar measurements of particles and gases by elastic backscattering and differential absorption. In *Laser Monitoring in the Atmosphere*, ed., E. D. Hinkley. Berlin: Springer, 71–151.

18. Inaba, H. 1976. Detection of atoms and molecules by Raman scattering and resonance fluorescence. In *Laser Monitoring in the Atmosphere*, ed., E. D. Hinkley. Berlin: Springer, 153–236.

19. Fabelinski, I. L. 1968. *Molecular Scattering of Light*. New York: Plenum Press.

20. Rye, B. J. 1998. Molecular backscatter heterodyne lidar: A computational evaluation. *Appl. Opt.* 37 27: 6321–6328.

21. Chanin, M. L., Hauchecorne, A., Garnier, A. and Nedelkovic, D. 1994. Recent lidar developments to monitor stratosphere-troposphere exchange. *J. Atmos. Terrest. Phys.* 56: 1073–1081.

22. Browell, E. V., Ismail, S. and Grant, W. B. 1998. Differential absorption lidar (DIAL) measurements from air and space. *Appl. Phys.* 67: 399–410.

23. Seinfeld, J. H. and Pandis, S. N. 1998. *Atmospheric Chemistry and Physics: From Air Pollution to Climate Change*. New York: John Wiley & Sons.

24. Malcolm, A. L., Derwent, R. G. and Maryon, R. H. 2000. Modelling the long-range transport of secondary PM10 to the UKM. *Atmos. Environ.* 34: 881–894.

25. Derwent, R. G., Simmonds, P. G., Seuring, S. and Dimmer, C. 1998. Observation and interpretation of the seasonal cycles in the surface concentrations of ozone and carbon monoxide at Mace Head Ireland, from 1990 to 1994. *Atmos. Environ.* 31: 145–157.

26. Ryall, D. B., Maryon, R. H., Derwent, R. G. and Simmonds, P. G. 1998. Modelling long-range transport of CFCs to Mace Head, Ireland. *Q. J. R. Meteorol. Soc.* 124: 417–446.

27. Heffter, J. L. and Stunder, B. J. B. 1993. Volcanic ash forecast transport and dispersion (VAFTAD) model. *Comput. Tech.* 8: 533–541.

28. Mosca, S., Bianconi, R., Bellasio, R., Graziani, G. and Klug, W. 1998. *ATMESII-Evaluation of Long-Range Dispersion Models Using Data of the First EXTEX Release*, EC EUR 17756 EN.

29. Maryon, R. H. and Buckland, A. T. 1995. Tropospheric dispersion: The first ten days after a puff release. *Q. J. R. Meteorol. Soc.* 121: 1799–1833.

30. Ansmann, A., Mattis, I., Wandinger, U., Wagner, F., Reichardt, J. and Deshler, T. 1997. Evolution of the Pinatubo aerosol: Raman lidar observations of particle optical depth, effective radius, mass, and surface area over Central Europe at 53. 4N. *J. Atmos. Sci.* 54: 2630–2641.

31. Barnes, J. E. and Hofmann, D. J. 1997. Lidar measurements of stratospheric aerosol over Mauna Loa Observatory. *Geophys. Res. Lett.* 24: 1923–1926.

32. Browell, E. V., Fenn, M. A., Butler, C. F. and Grant, W. B. 1996. Ozone and aerosol distributions and air mass characteristics over the South Atlantic basin during the burning season. *J. Geophys. Res.* 101: 24043–24068.

33. Browell, E. V., Gregory, G. L., Harris, R. C. and Kirchhoff, V. W. J. H. 1998. Tropospheric ozone and aerosol distributions across the Amazon Basin. *J. Geophys. Res.* 93: 1431–1452.

34. Cutten, D. R., Pueschel, R. E., Bowdle, D. A., Srivastava, V., Clarke, A. D., Rothermel, J., Spinhirne, J. D. and Menzies, R. T. 1996. Multiwavelength comparison of modelled and measured remote tropospheric aerosol backscatter over Pacific Ocean. *J. Geophys. Res.* 101: 9375–9389.

35. Donovan, D. P., Bird, J. C., Whiteway, J. A., Duck, T. J., Pal, S. R., Carswell, A. I., Sandilands, J. W. and Kaminski, J. W. 1996. Ozone and aerosol observed by lidar in the Canadian Artic during the winter of 1995/96. *Geophys. Res. Lett.* 23: 3317–3320.

36. Hoff, R. M., Harwood, M., Sheppard, A., Froude, F., Martin, J. B. and Strapp, W. 1997. Use of airborne lidar to determine

aerosol sources and movement in the Lower Fraser Valley (LFB), BC. *Atmos. Environ.* 31: 2123–2134.

37. Li, S., Strawbridge, K. B., Leaitch, W. R. and Macdonald, A. M. 1998. Aerosol backscattering determined from chemical and physical properties and lidar observations over the east coast of Canada. *Geophys. Res. Lett.* 25: 1653–1656.

38. Marenco, F., Santacesaria, V., Bais, A., Balis, D., di Sarra, A. and Papayannis, A. 1997. Optical properties of tropospheric aerosols determined by lidar and spectrophotometric measurements (PAUR campaign). *Appl. Opt.* 36: 6875–6886.

39. McCormick, M. P., Thomason, L. W. and Trepte, C. R. 1995. Atmospheric effects of the Mt Pinatubo eruption. *Nature* 373: 399–404.

40. Osborn, M. T., Kent, G. S. and Trepte, C. R. 1998. Stratospheric aerosol measurements by the lidar in space technology experiment. *J. Geophys. Res.* 103: 11447–11454.

41. Parameswaren, K., Rajan, R., Vijayakumar, G., Rajeev, K., Moorthy, K. K., Nair, P. R. and Satheesh, S. K. 1998. Seasonal and long term variations of aerosol content in the atmospheric mixing region at a tropical station on the Arabian sea coast. *J. Atmos. Solar-Terrest. Phys.* 60: 17–25.

42. Post, M. J., Grund, C. J., Wang, D. and Deshler, T. 1997. Evolution of Mount Pinatubo's aerosol size distributions over the continental United States; two wavelength lidar retrievals and in situ measurements. *J. Geophys. Res.* 102: 13535–13542.

43. Shibata, T., Itabe, T., Mizutani, K., Uchino, O., Nagai, T. and Fujimoto, T. 1996. Arctic tropospheric aerosols and clouds in the polar night season observed by a lidar at Eureka, Canada. *J. Geomagn. Geoelectr.* 48: 1169–1177.

44. Spinhirne, J. D., Chudamani, S., Cavanaugh, J. F. and Bufton, J. K. 1997. Aerosol and cloud backscatter at 1.06, 1.54 and 0.53 μm by airborne hard-target-calibrated Nd: YAG/methane Raman lidar. *Appl. Opt.* 36: 3475–3490.

45. Post, M. J. 1984. Aerosol backscattering profiles at CO_2 wavelengths: The NOAA data base. *Appl. Opt.* 23: 2507–2509.

46. Post, M. J. 1986. Atmospheric purging of El Chichon debris. *J. Geophys. Res.* 91: 5222–5228.

47. Rothermel, J., Bowdle, D. A., Vaughan, J. M. and Post, M. J. 1989. Evidence of a tropospheric aerosol backscatter background mode. *Appl. Opt.* 28: 1040–1042.

48. Bowdle, D. A., Rothermel, J., Vaughan, J. M. and Post, M. J. 1991. Aerosol backscatter measurements at 10.6 micrometers with airborne and ground-based CO_2 Doppler lidars over the Colorado High Plains 2. Backscatter structure. *J. Geophys. Res.* 96: 5337–5344.

49. Bilbro, J. W., DiMarzio, C., Fitzjarrald, D., Johnson, S. and Jones, W. 1986. Airborne Doppler lidar measurements. *Appl. Opt.* 25: 3952–3960.

50. Rothermel, J., Bowdle, D. A. and Srivastava, V. 1996. Midtropospheric aerosol backscatter background mode over the Pacific Ocean at 9.1 μm wavelength. *Geophys. Res. Lett.* 23: 281–284.

51. Srivastava, V., Jarzembski, M. A. and Bowdle, D. A. 1992. Comparison of calculated aerosol backscatter at 9.1 and 2.1 μm wavelengths. *Appl. Opt.* 31: 11904–11906.

52. Srivastava, V., Rothermel, J., Bowdle, D. A., Jarzembski, N. A., Chambers, D. M. and Clarke, A. D. 1995. High resolution remote sensing of sulphate and aerosols form CO_2 lidar backscatter. *Geophys. Res. Lett.* 22: 2373–2376.

53. Srivastava, V., Rothermel, J., Jarzembski, M. A., Chambers, D. M. and Clarke, A. D. 1997. Comparison of modelled backscatter using measured aerosol microphysics with focused cw lidar data over Pacific. *J. Geophys. Res.* 102: 16605–16617.

54. Gras, J. L., Platt, C. M., Jones, W. D., Huffaker, R. M., Young, S. A., Banks, S. M. and Booth, D. J. 1991. Southern Hemisphere, tropospheric aerosol backscatter measurements—Implications for a laser wind system. *J. Geophys. Res.* 96: 5357–5367.

55. Alejandro, S. B., Koenig, G. G., Vaughan, J. M. and Davies, P. H. 1990. SABLE: A South Atlantic aerosol backscatter measurement programme. *Bull. Am. Meteorol. Soc.* 71: 281–287.

56. Alejandro, S. B., Koenig, G. G., Bedo, D., Swirbalus, T., Frelin, R., Woffinden, J., Vaughan, J. M., Brown, D. W., Callan, R., Davies, P. H., Foord, R., Nash, C. and Wilson, D. J. 1995. Atlantic atmospheric aerosol studies 1. Programme overview and airborne lidar. *J. Geophys. Res.* 100: 1035–1041.

57. Vaughan, J. M., Brown, D. W., Nash, C., Alejandro, S. B. and Koenig, G. G. 1995. Atlantic atmospheric aerosol studies 2. Compendium of airborne backscatter measurements at 10.6 μm. *J. Geophys. Res.* 100: 1043–1065.

58. Gibson, F. W. 1994. Variability in atmospheric light-scattering properties with altitude. *Appl. Opt.* 23: 411–418.

59. Menzies, R. T., Tratt, D. M. and Flamant, P. H. 1994. Airborne CO_2 coherent lidar measurements of cloud backscatter and opacity over the ocean surface. *J. Atmos. Oceanic Technol.* 11: 770–778.

60. Menzies, R. T. and Tratt, D. M. 1994. Airborne CO_2 coherent lidar for measurements of atmospheric aerosol and cloud backscatter. *Appl. Opt.* 33: 5698–5711.

61. Menzies, R. T. and Tratt, D. M. 1995. Evidence of seasonally dependent stratosphere-troposphere exchange and purging of lower stratospheric aerosol from a multi-year lidar data set. *J. Geophys. Res.* 100: 3139–3148.

62. Menzies, R. T. and Tratt, D. M. 1997. Airborne lidar observations of tropospheric aerosols during the Global Backscatter experiment (GLOBE) Pacific circumnavigation missions of 1989 and 1990. *J. Geophys. Res.* 102: 3701–3714.

63. Tratt, D. M. and Menzies, R. T. 1995. Evolution of the Pinatubo volcanic aerosol column above Pasadena, California observed with a mid-infrared backscatter lidar. *Geophys. Res. Lett.* 22: 807–881.

64. Vaughan, J. M., Brown, D. W., Davies, P. H., Nash, N., Kent, G. and McCormick, M. P. 1988. Comparison of SAGE II solar extinction data with airborne measurements of atmospheric backscattering in the troposphere and lower stratosphere. *Nature* 332: 709–711.

65. Vaughan, J. M., Steinwall, K. O., Werner, C. and Flamant, P. H. 1997. Coherent laser radar in Europe. *Proc. IEEE* 84: 205–226.

66. Vaughan, J. M., Brown, D. W. and Willetts, D. V. 1998. The impact of atmospheric stratification on a space-borne Doppler wind lidar. *J. Mod. Opt.* 45: 1583–1599.

67. Vaughan, J. M., Geddes, N. J., Flamant, P. D. H. and Flesia, C. 1998. A global backscatter database for aerosols and cirrus clouds: Report to ESA 12510/97/NL/RE, p. 110.

68. Kent, G. S. and Schaffner, S. K. 1989. Comparison of 1 μm satellite aerosol extinction with CO_2 lidar backscatter. *SPIE* 1181: 252–259.

69. Rosen, J. M. 1991. A comparison of measured and calculated optical properties of atmospheric aerosols at infrared wavelengths. *J. Geophys. Res.* 96: 5229–5235.

70. Chudamani, S., Spinhirne, J. D. and Clarke, A. D. 1996. Lidar aerosol backscatter cross sections in the 2 μm near-infrared wavelength region. *Appl. Opt.* 35: 4812–4819.

71. Phillips, M. W., Hannon, S., Henderson, S. W., Gatt, P. and Huffaker, R. M. 1997. Solid-state coherent lidar technology for space-based wind measurement. *SPIE* 2965: 68–75.

72. Winker, D. M., Couch, R. H. and McCormick, M. P. 1996. An overview of LITE: NASA's lidar in space technology experiment. *Proc. IEEE* 84: 164–180.

73. Kent, G. S., Poole, L. R. and McCormick, M. P. 1986. Characteristics of arctic polar stratospheric clouds as measured by airborne *lidar. J. Atmos. Sci.* 43: 20.

74. Powell, K. A., Trepte, C. R. and Kent, G. S. 1996. *Observation of Saharan Dust by LITE Advances in Atmospheric Remote Sensing with Lidar.* Berlin: Springer.

75. Bowdle, D. A. and Menzies, R. T. 1998. Aerosols backscatter at 2μm: Modelling and validation for SPARCLE and follow-on missions. In *NOAA Working Group Meeting on Space-Based Lidar Winds*, Key West, FL, January 20–22.

76. Srivastava, V., Rothermel, J., Clarke, A. D., Spinhirne, J. D., Menzies, R. T., Cutten, D. R., Jarzembski, M., Bowdle, D. A. and McCaul, E. W. Jr. 2001. Wavelength dependence of backscatter by use of aerosol microphysics and lidar data sets: Application to 2.1-μm space-based and airborne lidars. *Appl. Opt.* 40: 4759–4769.

77. Vaughan, J. M., Maryon, R. H. and Geddes, N. J. 2002. Comparison of atmospheric aerosol backscattering and air mass back trajectories. *Meteorol. Atmos. Phys.* 79: 33–46.

78. Carrier, L. W., Cato, G. A. and von Essen, K. J. 1967. The backscattering and extinction of visible and infrared radiation by selected major cloud models. *Appl. Opt.* 6: 1209–1216.

79. Evans, B. T. N. and Fournier, G. R. 1990. Simple approximations to extinction efficiency valid over all size parameters. *Appl. Opt.* 29: 4666–4670.

80. Chylek, P. and Damiano, P. 1992. Polynomial approximation of the optical properties of water clouds in the 8–12 μm spectral region. *J. Appl. Meteorol.* 31: 1210–1218.

81. Banakh, V. A., Smalikho, I. N., Kopp, F. and Werner, C. 1995. Representativity of the wind measurements by a CW Doppler lidar in the atmospheric boundary layer. *Appl. Opt.* 34: 2055–2067.

82. Harris, M., Constant, G. and Ward, C. 2001. Continuous wave bistatic laser Doppler wind sensor. *Appl. Opt.* 40: 1501–1506.

83. Karlsson, C., Olsson, F., Letalick, D. and Harris, M. 2000. All-fibre multifunction continuous-wave 1.55 μm coherent laser radar for range, speed, vibration and wind measurements. *Appl. Opt.* 39: 3716–3726.

84. Jarzembski, M. A., Srivastava, V. and Chambers, D. M. 1996. Lidar calibration technique using laboratory-generated aerosols. *Appl. Opt.* 35: 2096–2108.

85. Vaughan, J. M. and Forrester, P. A. 1989. Laser Doppler velocimetry applied to the measurement of local and global wind. *Wind Eng.* 13: 1–15.

86. Clemesha, B. R., Kirchhoff, V. W. J. H. and Simonich, D. M. 1981. Remote measurement of tropospheric winds by ground based lidar. *Appl. Opt.* 20: 2907–2910.

87. Schols, J. L. and Eloranta, E. W. 1992. The calculation of area-averaged vertical profiles of the horizontal wind velocity from volume imaging lidar data. *J. Geophys. Res.* 97: 18395–18407.

88. Pirronen, A. K. and Eloranta, E. W. 1995. Accuracy analysis of wind profiles calculated from volume imaging lidar data. *J. Geophys. Res.* 100: 25559–25567.

89. Willetts, D. V. and Harris, M. R. 1982. An investigation into the origin of frequency sweeping in a hybrid TEA CO_2 laser radar. *J. Phys.* 15: 51–67.

90. Willets, D. V. and Harris, M. R. 1983. Scaling laws for the intrapulse frequency stability of an injection mode selected TEA CO_2 laser. *IEE J. Quantum Electron.* QE-19: 810–814.

91. Wilson, D. J., Constant, G. D. J., Foord, R. and Vaughan, J. M. 1991. Detector performance studies for CO_2 laser heterodyne systems. *Infrared Phys.* 31(1): 109–115.

92. Oh, D., Drobinski, P., Salamitou, P. and Flamant, P. H. 1996. Optimal local oscillator power for CMT photo-voltaic detector in heterodyne mode. *Infrared Phys. Technol.* 37: 325–333.

93. Lottman, B. T. and Frehlich, R. G. 1997. Evaluation of coherent Doppler lidar velocity estimators in nonstationary regimes. *Appl.Opt.* 36: 7906–7918.

94. Rye, B. J. and Hardesty, R. M. 1989. Time verification and Kalman filtering techniques for Doppler lidar velocity estimation. *Appl. Opt.* 28: 879–891.

95. Rye, B. J. and Hardesty, R. M. 1993. Discrete spectral peak estimation in incoherent backscatter heterodyne lidar. I: spectral accumulation and the Cramer-Rao lower bound II Correlogram accumulation. *IEEE Trans. Geosci. Remote Sens.* 31: 1.

96. Rye, B. J. and Hardesty, R. M. 1997. Estimate optimisation parameters for incoherent backscatter heterodyne lidar. *Appl. Opt.* 36(36): 9425–36.

97. Dabas, A. M., Drobinski, P., Flamant, P. H. and Dabas, A. M. 1999. Adaptive Levin filter for frequency estimate of heterodyne Doppler lidar returns: recursive implementation and quality control. *J. Atmos. Oceanic Technol.* 16: 361–372.

98. Rye, B. J. 2000. Estimate optimisation parameters for incoherent backscatter lidar including unknown return signal bandwidth. *Appl. Opt.* 39: 6068–6096.

99. Eberhard, W. L., Cupp, R. E. and Healy, K. R. 1989. Doppler lidar measurement of profiles of turbulence and momentum flux. *J. Atmos. Oceanic Technol.* 6: 809–819.

100. Post, M. J. and Neff, W. D. 1986. Doppler lidar measurement of winds in a narrow

mountain valley. *Bull. Am. Meteorol. Soc.* 67: 274.

101. Neiman, R. J., Hardesty, R. M., Shapiro, M. A. and Cupp, R. E. 1998. Doppler lidar observations of a downslope windstorm. *Mon. Weather Rev.* 116: 2265–2275.

102. Olivier, L. D. and Banta, R. M. 1991. Doppler lidar measurements of wind flow and aerosol concentration at the Grand Canyon. *Technical Digest Coherent Laser Radar: Technology and Applications*, July 8–12, 1991 (Snowmass, Colorado: Optical Society of America).

103. Hardesty, R. M. and Huffaker, R. M. 1996. Remote sensing of atmospheric wind velocities using solid state and CO_2 coherent laser systems. *Proc. IEEE* 84: 181–204.

104. Pearson, G. N. and Collier, C. G. 1999. A pulsed coherent CO_2 lidar for boundary layer meteorology. *Q. J. R. Meteorol. Soc.* 125: 2703–2721.

105. Hawley, J. G., Targ, R., Henderson, S. W., Hale, C. P., Kavaya, M. J. and Moerder, D. 1993. Coherent launch-site atmospheric wind sounder: Theory and experiment. *Appl. Opt.* 32: 4557–4568.

106. Frehlich, R. 1995. Comparison of 2- and 10-p.m coherent Doppler lidar performance. *J. Atmos. Oceanic Technol.* 12(2), 415–420.

107. Pearson, G. N., Roberts, P. J., Eacock, J. R. and Harris, M. 2002. Analysis of the performance of a coherent pulsed fibre lidar for aerosol backscatter applications. *Appl. Opt.* 41: 6442–6450.

108. McKay, J. A. 1998. Modelling of direct detection Doppler wind lidar. *Appl Opt.* 37(27): 6487–6493.

109. Bruneau, D. and Pelon, J. 2003. Simultaneous measurements of particle backscattering and extinction coefficients and wind velocity by lidar with a Mach-Zehnder interferometer: Principle of operation and performance assessment. *Appl. Opt.* 42: 1101–1114.

110. Abreu, V. J., Barnes, J. E. and Hays, P. B. 1992. Observations of winds with an incoherent lidar detector. *Appl. Opt.* 31: 4509–4514.

111. McGill, M. J., Skinner, W. R. and Irgang, T. D. 1997. Analysis techniques for the recovery of winds and backscatter coefficients from a multiple channel incoherent Doppler lidar. *Appl. Opt.* 36: 1253–1268.

112. Rees, D., Vyssogorets, M., Meredith, N. P., Griffin, E. and Chaxel, Y. 1996. The Doppler wind and temperature system of ALOMAR lidar facility: Overview and initial results. *J. Atmos. Terr. Phys.* 58: 1827–1842.

113. Garnier, A. and Chanin, M. L. 1992. Description of a Doppler Rayleigh lidar for measuring winds in the middle atmosphere. *Appl. Phys.* B55: 35–40.

114. Gentry, B. and Korb, C. L. 1994. Edge technique for high accuracy Doppler velocimetry. *Appl. Opt.* 33: 5770–5777.

115. Flesia, C. and Korb, C. L. 1999. Theory of the double edge molecular technique for Doppler lidar wind measurement. *Appl. Opt.* 38(3): 432–440.

116. Souprayen, C., Garnier, A., Hertzog, A., Hauchecorne, A. and Porteneuve, J. 1999. Rayleigh Mie Doppler wind lidar for atmospheric measurements. I. Instrumental setup, validation and first climatological results: II Mie scattering effect, theory and calibration. *Appl. Opt.* 38: 2410–2421.

117. Delaval, A., Flamant, P. H., Aupierre, M., Delville, P., Loth, C., Garnier, A., Souprayen, C., Bruneau, D., Le Rille, D., Wilson, R., Vialle, C., Rees, D., Vaughan, J. M. and Hardesty, R. M. 2001. Intercomparison of wind profiling instruments during the VALID field campaign. In *Advances in Laser Remote Sensing*, eds., A. Dabas, C. Loth and J. Pelon Editions. de l'Ecole Polytechnique, 101–104.

118. Pal, S. R., Ryan, J. S. and Carswell, A. I. 1978. Cloud reflectance with laser beam illumination. *Appl. Opt.* 17(15): 2257–2259.

119. Kent, G. S., Poole, L. R. and McCormick, M. P. 1986. Characteristics of arctic polar stratospheric clouds as measured by airborne lidar. *J. Atmos. Sci.* 43: 20.

120. Hall, F. F., Cupp, R. E. and Troxel, S. W. 1988. Cirrus cloud transmittance and backscatter in the infrared measured with a CO_2 lidar. *Appl. Opt.* 27: 2510–2539.

121. Krichbaumer, W., Mehneet, A., Halldorsson, T. H., Hermann, H., Haering, R., Streicher, J. and Werner, C. H. 1993. A diode-pumped Nd: YAG lidar for airborne cloud measurements. *Opt. Laser Technol.* 25: 283–287.

122. Keeler, R. J., Serafin, R. J., Schwiesow, R. L., Lenschow, D. H., Vaughan, J. M. and Woodfield, A. A. 1987. An airborne laser air motion sensing system. Part I: Concept and preliminary experiment. *J. Atmos. Oceanic Technol.* 4(1): 113–137.

123. Woodfield, A. and Vaughan, J. M. 1983. Airspeed and windshear measurement with an airborne CO_2 laser. *Int. J. Aviation Saf.* 1: 207–224.

124. Woodfield, A. and Vaughan, J. M. 1983. Airspeed and windshear measurement with an airborne CO_2 CW laser. *AGARDo-graph* 272: 7.1–7.17.

125. Woodfield, A. and Vaughan, J. M. 1984. Using an airborne CO_2 laser for free stream airspeed and windshear measurements. *AGARD Conf. Proc.* 373: 22.1–22.18.

126. Targ, R., Kavaya, M., Milton Huffaker, R. and Bowles, R. L. 1991. Coherent lidar airborne windshear sensor: Performance evaluation. *Appl. Opt.* 30: 2013–2026.

127. Morbieu, B., Combe, H. and Mandle, J. 1993. ALEV3, a 3-axis CO_2 CW anemometer for aircraft certification. *Proceedings of 7th Conference on Coherent Laser Radar: Applications and Technology*, Paris, Paper MB1.

128. Smart, A. E. 1991. Velocity sensor for an airborne optical air data system. *AIAA: J. Aircr.* 28(3): 163–164.

129. Smart, A. E. 1992. Optical velocity sensor for air data applications. *Opt. Eng.* 31(1): 166–173.

130. Werner, C., Flamant, P. H., Reitebuch, O., Kopp, F., Streicher, J., Rahm, S., Nagel, E., Klier, M., Hermann, H., Loth, C., Delville, P., Drobinski, P., Romand, B., Boitel, C., Oh, D., Lopez, M., Meissonnier, M., Bruneau, D. and Dabas, A. M. 2001. WIND infrared Doppler lidar instrument. *Opt. Eng.* 40: 115–125.

131. Reitebuch, O., Werner, C., Leike, I., Delville, P., Flamant, P. H., Cress, A. and Englebart, D. 2001. Experimental validation of wind profiling performed by the airborne 10|j.m-Heterodyne Doppler lidar WIND. *J. Atmos. Oceanic Technol.* 18: 1331–1344.

132. Hallock, J. N. 1991. Aircraft wake vortices: An annotated bibliography (1923–1990). National Technical Information Service, Springfield, Virginia. Report no. DOT-FAA-RD-90-30, DOT-VNTSC-FAA-90-7.

133. Koepp, F. 1994. Doppler lidar investigations of wake vortex transport between closely spaced parallel runways. *AIAA J.* 32: 805–812.

134. Constant, G. D. J., Foord, R., Forrester, P. A. and Vaughan, J. M. 1994. Coherent laser radar and the problem of wake vortices. *J. Mod. Opt.* 41: 2153–2174.

135. Vaughan, J. M., Steinwall, K. O., Werner, C. and Flamant, P. H. 1996. Coherent laser radar in Europe. *Proc. IEEE* 84: 205–226.

136. Harris, M., Young, R. I., Kopp, F., Dolfi, A. and Cariou, J.-P. 2002. Wake vortex detection and monitoring. *Aerosp. Sci. Technol.* 6: 325–331.

137. Hannon, S. M. and Thomson, J. A. 1994. Aircraft wake vortex detection and measurement with pulsed solid-state coherent laser radar. *J. Mod. Opt.* 41: 2175–2196.

138. Koepp, F. 1999. Wake vortex characteristics of military-type aircraft measured at Airport Oberpfaffenhofen using the DLR laser Doppler Anenometer. *Aerosp. Sci. Technol.* 4: 191–199.

139. Joseph, R., Dasey, T. and Heinrichs, R. 1999. Vortex and meteorological measurements at Dallas/Fort Worth Airport, AIAA 99-0760.

140. Greenwood, J. S. and Vaughan, J. M. 1998. Measurements of aircraft wake vortices at Heathrow by laser Doppler velocimetry. *Air Traffic Control Q.* 6: 179–203.

141. Vaughan, J. M. 1998. Wake vortex investigations at Heathrow airport, London. *Nouv. Rev. Aeronaut. Astronaut.* 2: 116–121.

142. Harris, M., Huenecke, K. and Huenecke, C. 2000. Aircraft wake vortices: A comparison of wind-tunnel data with field-trial measurements by laser radar. *Aerosp. Sci. Technol.* 4: 363–370.

143. Vaughan, J. M. and Harris, M. 2001. Lidar measurement of B747 wakes: Observation

of a vortex within a vortex. *Aerosp. Sci. Technol.* 5: 409–411.

144. Turner, G. P., Padfield, G. D. and Harris, M. 2002. Encounters with aircraft vortex wakes: The impact on helicopter handling qualities. *J. Aircr.* 39: 839–849.

145. Grant, W. B. 1991. Differential absorption and Raman lidar for water vapour profile measurements: A review. *Opt. Eng.* 30(1): 40–48.

146. Klett, J. D. 1981. Stable analytical inversion solution for processing lidar returns. *Appl. Opt.* 20(1): 211–215.

147. Ismail, S. and Browell, E. V. 1989. Airborne and spaceborne lidar measurements of water vapour profiles: A sensitivity analysis. *Appl. Opt.* 28(17): 3603–3615.

148. McDermid, I. S., Godin, S. M. and Lindqvist, L. O. 1990a. Ground-based laser DIAL system for long-term measurements of stratospheric ozone. *Appl. Opt.* 29(25): 3603–3612.

149. McDermid, I. S., Godin, S. M. and Walsh, T. D. 1990b. Lidar measurements of stratospheric ozone and intercomparisons and validation. *Appl. Opt.* 29(33): 4914–4923.

150. Hardesty, R. M. 1984. Coherent DIAL measurement of range-resolved water vapour concentration. *Appl. Opt.* 23(15): 2545–2553.

151. Grant, W. B., Margolis, J. S., Brothers, A. M. and Tratt, D. M. 1987. CO_2 DIAL measurements of water vapour. *Appl. Opt.* 26(15): 3033–3042.

152. Grant, W. B. 1989. Mobile atmospheric in pollutant mapping (MAPM) system: A coherent CO2 DIAL system, laser applications in meteorology and earth and atmospheric remote sensing. *Proc. SPIE* 1062: 172–190.

153. Ridley, K. D., Pearson, G. N. and Harris, M. 2001. Influence of speckle correlation on coherent DIAL with an in-fibre wavelength multiplexed transceiver. In *Advances in Laser Remote Sensing*, eds., A. Dabas, C. Loth and J. Pelon Editions. de l'Ecole Polytechnique, 93–96.

154. Godin, S., Marchand, M. and Hauchecorne, A. 2001. Study of the influence of the Arctic polar vortex erosion on mid-latitude from ozone lidar measurements at OHP (44°N, 6°E). In *Advances in Laser Remote Sensing*, eds., A. Dabas, C. Loth and J. Pelon Editions. de l'Ecole Polytechnique, 385–387.

155. Butler, C. F., Browell, E. V., Grant, W. B., Brackett, V. G., Toon, O. B., Burris, J., McGee, T., Schoeberl, M. and Mahoney, M. J. 2001. Polar stratospheric cloud characteristics observed with airborne lidar during the SOLVE campaign. In *Advances in Laser Remote Sensing*, eds., A. Dabos, C. Loth and J. Pelon Editions. de l'Ecole Polytechnique, 397–400.

156. Gerard, E. and Pailleux, J. 2001. Role of water vapour in Numerical Weather Prediction models. In *Advances in Laser Remote Sensing*, eds., A. Dabas, C. Loth and J. Pelon Editions. de l'Ecole Polytechnique, 285–288.

157. Riedinger, E., Keckhut, P., Hauchecorne, A., Collado, E. and Sherlock, V. 2001. Monitoring water vapour in the mid-upper troposphere using ground-based Raman lidar. In *Advances in Laser Remote Sensing*, eds., A. Dabas, C. Loth and J. Pelon Editions. de l'Ecole Polytechnique, 313–315.

158. Sherlock, V., Hauchecorne, A. and Lenoble, J. 1999. Methodology for the independent calibration of Raman backscatter water-vapour lidar systems. *Appl. Opt.* 38: 5817–5837.

159. Ferrare, R. A., Whiteman, D. N., Melfi, S. H., Evans, K. D., Schmidlin, F. J. and O'C Starr, D. 1995. A comparison of water vapour measurements made by Raman lidar and radiosondes. *J. Atmos. Oceanic Technol.* 6: 1177–1195.

160. Evans, K. D., Demoz, B. B., Cadirola, M. P. and Melfi, S. H. 2001. A new calibration technique for Raman lidar water vapour measurements. In *Advances in Laser Remote Sensing*, eds., A. Dabas, C. Loth and J. Pelon Editions. de l'Ecole Polytechnique, 289–292.

161. Whiteman, D. and Melfi, S. H. 1999. Cloud liquid water, mean droplet radius and number density measurements using a Raman lidar. *J. Geophys. Res.* 104(D24): 31411–31419.

162. Ansmann, A., Riebesell, M. and Weitkamp, C. 1990. Measurement of atmospheric aerosol extinction profiles with a Raman lidar. *Opt. Lett.* 15: 746–748.

163. Ansmann, A., Wandinger, U., Riebesell, M., Weitkamp, C. and Michaelis, W. 1992. Independent measurement of extinction and backscatter profiles in cirrus clouds by using a combined Raman elastic-backscatter lidar. *Appl. Opt.* 31: 7113–7131.

164. Schumacher, R., Neuber, R., Herber, A. and Rairoux, P. 2001. Extinction profiles measured with a Raman lidar in the Arctic troposphere. In *Advances in Laser Remote Sensing*, eds. A. Dabas, C. Loth and J. Pelon Editions. de l'Ecole Polytechnique, 229–232.

165. Uthe, E. E. 1991. Elastic scattering, fluorescent scattering and differential absorption airborne lidar observations of atmospheric tracers. *Opt. Eng.* 30(1): 66–71.

166. Eberhard, W. L. and Chen, Z. 1989. Lidar discrimination of multiple fluorescent tracers of atmospheric motions. *Appl. Opt.* 28: 2966–3007.

167. She, C. Y., Latifi, H., Yu, R. J., Alvarez, R. J., Bills, R. E. and Gardner, C. S. 1990. Two frequency lidar technique for mesospheric Na temperature measurements. *Geophys. Res. Lett.* 17: 929–932.

168. Gelbwachs, J. A. 1994. Iron Boltzmann factor lidar: Proposed new remote-sensing technique for mesospheric temperature. *Appl. Opt.* 33: 7151–7156.

169. Von Zahn, U., Gerding, M., Hoffner, J., McNeill, W. J. and Murad, E. 1999. Fe Ca and K atom densities in the trails of Leonid and other meteors: Strong evidence for differential ablation. *Meteor. Planet Sci.* 34: 1017–1027.

170. Bills, R. E., Gardner, C. S. and She, C. Y. 1991. Narrowband lidar technique for sodium temperature and Doppler windobservations of the upper atmosphere. *Opt. Eng.* 30(1): 13–20.

171. Chu, X., Gardner, C. S., Pan, W. and Papen, G. C. 2001. Recent results from the University of Illinois Iron Boltzmann temperature lidar. In *Advances in Laser Remote Sensing*, eds., A. Dabas, C. Loth and J. Pelon Editions. de l'Ecole Polytechnique, 413–416.

172. Melngailis, I., Keicher, W. E., Freed, C., Marcus, S., Edwards, B. E., Sanchez, A., Fan, T. Y. and Spears, D. L. 1996. Laser component technology. *Proc. IEEE* 84(2), 227–267.

173. Bufton, J. L., Garvin, J. B., Cavanaugh, J. F., Ramos-Izquierdo, L., Clem, T. D. and Krabill, W. B. 1991. Airborne lidar for profiling of surface topography. *Opt. Eng.* 30: 72–78.

174. Gardner, C. S. 1982. Target signatures for laser altimeters; an analysis. *Appl. Opt.* 4: 448–453.

175. Bufton, J. L. 1989. Laser altimetry measurements from aircraft and spacecraft. *Proc. IEEE* 77(3): 463–477.

176. Garvin, J. B., Bufton, J. L., Krabill, W. B., Clem, T. D. 1989. Airborne laser altimeter observations of Mount St. Helens volcano. *J. Volcanol. Geotherm. Res.*

177. Garvin, J. B., Bufton, J. L., Krabill, W. B., Clem, T. D., Schnetzler, C. C. 1990. Airborne laser altimetry of craterform structures. *IEEE Trans. Geo. Rem. Sens.*

178. Steinvall, O., Koppari, K. and Karlsson, U. 1993. Experimental evaluation of an airborne depth sounding lidar. *Opt. Eng.* 32: 1307–1321.

179. Hoge, F. E., Wright, C. W., Krabill, W. B., Buntzen, R. R., Gilbert, G. D., Swift, R. N., Yungle, J. K. and Berry, R. E. 1988. Airborne lidar detection of subsurface oceanic scattering layers. *Appl. Opt.* 27: 3969–3977.

180. Penny, M. F., Billard, B. and Abbot, R. H. 1989. LADS—The Australian Laser airborne depth sounder. *Int. J. Remote Sens.* 10: 1463–1497.

181. Lutormirski, R. F. 1978. An analytical model for optical beam propagation through the maritime boundary layer. *SPIE Ocean Opt. V* 160: 110–122.

182. Wood, R. and Appleby, G. 2003. *Earth and Environmental Sciences: Satellite Laser Ranging Article D7.1 Handbook of Laser Technology and Application*. Bristol: IOPP.

183. Plotkin, H. 1964. S66 laser satellite tracking experiment. *Proceedings of the Quantum Electronics III Conference*. New York: Columbia University Press, 1319–1332.

184. Nuebert, R., Grunwaldt, L. and Fisher, H. 2001. The new SLR station of GFZ Potsdam: A status report. *Proceedings of 12th*

International Workshop on Laser Ranging, Matera, Italy.

185. Schwartz, J. A. 1990. Pulse spreading and range correction analysis for satellite laser ranging. *Appl. Opt.* 29: 3597–3602.

186. Couch, R., Rowland, C., Ellis, K., Blythe, M., Regan, C., Koch, M. R., Antill, C. W., Kitchen, W. L., Cox, J. W., De Lorme, J. F., Crockett, S. K., Remus, R. W., Casas, J. C. and Hunt, W. H. 1991. Lidar in-space technology experiment (LITE): NASA's first in-space lidar system for atmospheric research. *Opt. Eng.* 30: 88–95.

187. McCormick, M. P., Winker, D. M., Browell, E. V., Coakley, J. A., Gardner, C. S., Hoff, R. M., Kent, G. S., Melfi, S. H., Menzies, R. T., Platt, C. M. R., Randall, D. A. and Reagan, J. A. 1993. Scientific investigations planned for the lidar In-space technology experiment (LITE). *Bull. Am. Meteorol. Soc.* 74: 205–214.

188. Megie, G. 1997. Differential absorption lidar. In *Laser Beams in Space: Application and Technology*, ed., J. Bufton. New York: Marshal Dekker.

189. Cohen, S. C., Degnan, J. J., Bufton, J. L., Garvin, J. B. and Abshire, J. B. 1987. The geoscience of laser altimetry/ranging systems. *IEEE Trans. GeoSci. Remote Sens.* 25(5): 581–592.

190. Hardesty, R. M. and Weber, B. F. 1987. Lidar measurement of turbulence encountered by horizontal-axis wind turbines. *J. Atmos. Oceanic Technol.* 4: 191–203.

191. Steinvall, O. et al. 1981. Physics and technology of coherent infrared radar. *SPIE* 300: 104.

192. Osche, G. R. and Young, D. S. 1996. Imaging laser-radar in the near and far infrared. *Proc. IEEE* 84(2): 103–125.

193. Stephan, B. and Metivier, P. 1987. Flight evaluation trials of a heterodyne CO_2 laser radar. *Active Infrared Syst. Technol.*, SPIE-806: 110–118.

194. Hogg, G. M., Harrison, K. and Minisclou, S. 1995. Nato. *Agard. Conf. Proc.* 563: 20.

195. ATLID Consultancy Group. 1990. *Backscatter Lidar, the Potential of a Space-Borne Lidar for Operational Meteorology,* Climatology and Environmental Research. Noordwijk: ESA Specialist Publication, SP-1121.

196. Atmospheric Dynamics Mission. 1999. The four candidates earth explorer core missions, Noordwijk: ESA. ESA Reports for Mission Selection, ESA-SP 1233 (4).

FURTHER READING

Ansmann, A., R. Neuber, P. Rairoux and U. Wandinger, eds., 1996. Advances in atmospheric remote sensing with lidar. *International Laser Radar Conference (ILRC)*, Berlin, Germany: Springer, July 22–26.

Conference Proceedings of the Biennial International Laser Radar Conferences (ILRC).

Dabas, A., C. Loth and J. Pelon, eds., 2001. Advances in laser remote sensing. *Selected Papers presented at the 20th International Laser Radar Conference (ILRC)*, Vichy, France: de l'Ecole Polytechnique, 2001 Editions.

Jelalian, A. 1992. *Laser Radar Systems*. Boston: Artech.

Journal of Modern Optics 41 (11), November 1994, 2063–2196 (containing a Special Section of 10 papers on Coherent Laser Radar, with short preface by J. M. Vaughan).

Killinger, D. K. and A. Moorodian, eds. 1983. *Optical and Laser Remote Sensing*. New York: Springer.

Measures, R. M., ed. 1988. *Laser Remote Chemical Analysis*. New York: John Wiley & Sons.

Proceedings of the IEEE 84 (2), February 1996, 99–320 (Special Issue of 8 paperson Laser Radar, ed., A. V. Jelalian).

Technical Digests of the Biennial Coherent Laser Radar Meetings (CLRM).

Valuable accounts of topics in light scattering and laser radar may be found in the following conference proceedings, reviews, compendia etc.

Vaughan, J. M. 1989. *The Fabry-Perot Interferometer—History, Theory, Practice and Applications Adam Hilger Series*. Bristol: IOPP.

13

Optical information storage and recovery

SUSANNA ORLIC
Technische Universität Berlin

13.1 INTRODUCTION

Optical data storage has a long history dating back to the 1960s but, with the compact disk initially, it became relevant to the consumer and industry. The success of the early laser disk indicated the possibility of data storage based on optical phenomena and materials as an alternative to magnetic storage. Optical storage offers reliable and removable storage media with excellent robustness and archival lifetime and very low cost. Today, optical disk technology covers a wide variety of applications ranging from content distribution to professional storage applications. One of the major application areas for optical storage disks is the secondary storage of computer data in personal computers (PCs) and computer networks. An optical storage system is a particularly attractive component of the data storage network because it provides fast data access times and fair storage capacities while serving as a link between different multimedia and computerized systems. Perhaps, the most enabling feature of optical storage is the removability of the storage medium that allows transportation and exchange of the stored information between desktop and laptop computers, audio, video players, and recorders. In contrast to the flying head of a hard disk drive, there are separations of a few millimeters between the recording surface and the optical "head," although active servo systems enable dynamic recording and readout from a rotating disk. Consequently, the medium can be removed and replaced with relatively loose tolerances allowing an optical disk to be handled in different drives.

The compact disk (CD) digital audio has been launched into the era of digital entertainment by Sony and Philips in 1982. CD has enjoyed unprecedented success and universal support among electronic companies and hardware manufacturers. From its origin as the music storage medium for entertainment, CD has grown to encompass computer applications (CD-ROM), imaging applications (CD Photo), and video game applications (CD Video). The CD-ROM drives became standard in PCs and CD-ROMs enhanced the efficiency of distribution and use of software, games, and video. These read-only disks containing 680 MB of information can be mass replicated by injection molding in a few seconds, and they are virtually indestructible.

Significant advances in the enabling technologies constituting the compact disk make it possible to increase capacity in digital versatile disk (DVD) format with the ability to store an entire movie in high-quality digital video on a single disk. With the DVD, optical disk storage opens a new chapter for video and multimedia applications. Laser optics, thin film, and disk replication technologies have made considerable strides in the last decade. Digital coding and compression algorithms have become more sophisticated; integrated circuits and drive mechanisms have also advanced. Furthermore, the intensive research on new optical materials and recording phenomena has resulted in the development of new technologies. The CD-Recordable (CD-R), a recordable write-once system, and the CD-Rewritable (CD-RW) drives and media have been introduced to the optical disk market. With its recordable formats, either for single recording or for expanded recording capabilities as in the case of CD-RW, optical storage became an alternative to the established hard disk and floppy disk systems, which dominated the data storage industry.

Versatility of optical storage is an additional feature of foremost importance for its applicability. Optical recording allows read-only, write-once, and rewritable (erasable) data storage. Read-only optical disks or read-only memory (ROMs) are suitable for distribution of digital contents. In read-only technology, the information is written on a master disk that is then used for printing the embossed patterns onto a plastic substrate. The printing process allows for rapid, low-cost mass reproduction making optical disks the media of choice for distribution of digital data. The write-once read-many (WORM) technology allows one to store permanently a large amount of data on a thin disk medium, to remove it, and to have fast access to it in any compatible optical drive system. Information stored on rewritable optical disks can be erased and rewritten many times. Present optical drives are designed to handle different media formats at the same time—the data stored on read-only, WORM, and rewritable media can be accessed all in one unit.

Optical information storage is based on laser–material interaction for writing and reading. More general, data storage is the conversion of raw information into physical changes in an appropriate recording medium. Data recovery is the recognition of the stored information from the storage-induced changes. The physical processes applied to different storage technologies differ widely and define their performance limits and applicability. In conventional technologies, optical storage relies on data recording on a rotating, disk-shaped medium while recovery is done by optical effects. Typically, a focused light beam is used to interact with the physical structure of the data stored on the disk. The interaction between the read beam and the medium can be due to a number of different effects but it must modify some detectable light property. The light reflected from the medium is modulated by the stored data and is then directed to a photodetector that converts the optical signal into electronic signals. All conventional optical disk systems are based on the reflection mode but often use different recording and readout mechanisms. There is a variety of optical effects and materials that have been investigated and employed in various approaches to realize an optical memory. Table 13.1 summarizes different recording mechanisms and corresponding data structure for readout in well-established optical storage concepts.

Further advances in optical memories with high storage capacity up to terabyte require research on new photoactive materials exhibiting strong laser-induced changes of their optical properties. Such materials constitute a special class of nonlinear optical materials with optical properties

Table 13.1 Optical information storage: Recording mechanism and corresponding data structure in different approaches

Recording mechanism	Data storage/readout method
Print	Surface grooves
Phase change	Reflectivity change
Magneto-optic	Polarization rotation
Photochromic	Complex refractive index change
Photorefractive	Refractive index grating

depending on the incident light intensity or energy. This chapter deals not only with the fundamentals of optical storage but also with novel approaches to higher performance storage.

13.2 OPTICAL DISK STORAGE TECHNOLOGIES

Removability of optical storage media is an attractive feature but this has made the standardizing efforts more complex compared with magnetic storage. Standardization in optical information storage was first accomplished with the appearance of CD creating the standards for CD Digital Audio format in 1982. Five years later, Sony and Philips announced the standard for CD storage of computer data—the CD-ROM. The establishment of worldwide standards has made it possible to incorporate writing technologies into CD systems. The CD-R is functionally compatible with the standard CD-ROM, even though the reflectivity changes on a CD-R are induced by physical processes different from the interference effect at the printed grooves on a conventional CD.

Following the CD-R, CD-RW, the rewritable compact disk, has been introduced based on phase-change technology. Here, the reflective layers have two states, which differ sufficiently in reflectivity to be read optically. Table 13.2 gives an overview of CD technologies starting with the CD audio that was modified to allow new applications and later advanced to exploit the new technologies with CD-R and CD-RW.

In general, an optical disk storage system consists of a drive unit and a rotating disk medium containing the information. The optical components are integrated within a small optical head allowing both optical recording and readout to be performed with the head positioned relatively far away from the storage medium, unlike magnetic hard drive heads. This allows the medium to be removable and effectively eliminates head crashes, increasing reliability. On the other hand, an optical head is heavier and leads to slower access times when compared to hard disk drives. The data recorded on an optical disk are organized in tracks that might be a single spiral or concentric rings spreading out along the radial direction.

In CD technologies, optical disks are either prerecorded or preformatted with continuous grooves or a discontinuous groove structure that is needed to position the optical head on the disk. A laser beam focused to a small spot is used for recording and readout. Information is written bit-wise by modulating the properties of the recording material under illumination. During retrieval, recorded bits are detected as changes in some optical property of the light reflected from the disk. These changes might affect the amplitude, phase, or polarization of light, and are sensed by a detector in the optical head. The rotation of the disk as well as positioning of the pickup system is provided by a

Table 13.2 Evolution of CD-based technologies from ROM to recordable and rewritable formats

Disk format	Recording technology
CD digital audio	Read-only
CD extra, CD interactive CD-ROM, video CD	Read-only
CD-R	Write-once
CD-RW	Rewritable

drive motor. In addition, the optical head contains a servo system based on some sort of optoelectronic feedback that is necessary to control the position of the focused laser spot on a rotating disk.

The optical storage systems introduced first worked as write-once, read-many systems. The information is recorded on a WORM disk by applying the laser beam to change the reflectivity. On the first ablation type WORM disks, the write laser beam burns a hole or pit into a thin absorbing layer, that is, deposited aluminum. In the early WORM systems, different lasers have been used for writing and reading. In a single laser system, the laser power is reduced for readout so that the recorded pits can be read by detecting the reflected energy. Following the ablation media, WORM disks with dye polymer media have been developed. Here, the laser beam induces a reflectivity change of dye polymer coating without material ablation. Rewritable storage has been first achieved in magneto-optical (MO) systems. The written information could be removed during an erase pass and thereby prepared for a new write cycle. To realize the MO systems, it was necessary to introduce new coating media and adapted optical systems. Consequently, there was no compatibility between WORM and MO storage media.

13.2.1 Compact disk technologies

The basic concept of an optical disk involves a reflecting layer modulated by the presence of pits that then switch the reflected signal by destructive interference. In a CD-ROM drive, the data are read out by projecting a focused laser spot on the rotating disk, then detecting the reflected light. The detected signal fluctuates with the presence or absence of pits along the track corresponding to the bit-wise stored information. Similarly, the recorded pits of CD-R and CD-RW are read out by detecting laser light reflected back from the disk. On the simple CD-ROM, the information is printed as quarter-wave surface steps leading to zero reflectivity by interference. Tracks consist of discontinuous grooves or pits that are separated by lands, whereby the pit length is determined by the bit content stored in it. The light reflected from the disk will, therefore, be modulated by the pits. Readout is done by destructive interference between the light reflected from the pit and the light reflected from the land. In this way, the data

content represented by a pit sequence is translated into the reflectivity changes that occur when this sequence is exposed to the laser beam.

On a CD-ROM, pits are preformatted into the optical substrate (typically polycarbonate) that is then coated with a high-reflective alloy (e.g., aluminum or silver). The substrate itself entails the format that defines the physical structure of the data recorded on the disk. Present optical disk technologies use different recording mechanisms and are also based on different disk formats. In the simplest case such as CD or DVD-ROM, the disk comprises a static physical structure represented by pits. Other disk formats can be categorized in two types, one of which relies on continuous grooves, and the other on wobble marks. Different disk formats also require different fabrication processes for mass production. The well-known mass production technique originates from the CD-ROM technology and is based on disk replication by stamping out copies of a disk master. This original disk is written using a short-wavelength laser that creates the pattern in a photoreactive material. The typical mastering process is based on similar photolithographic techniques as applied in the semiconductor technology. Exposing the photoresist yields the pit structure, which allows for forming the so-called stamper. The final stamper comprises a metallic layer (usually nickel) with the inverted data pattern structure, which is then used for mass disk replication. In the production of ROM disks, stamping is followed by only two additional steps—coating with a reflective layer followed by a protective layer. In contrast, recordable and rewritable disks are more complex in structure as an optically active layer is needed for data recording. Depending on the technology, different materials and layer stack configurations have been developed to provide both the functionality and compatibility.

On the write-once CD-R, the focused laser beam locally and permanently changes the complex refractive index of an organic dye polymer layer (Figure 13.1). Readout is based on signal enhancement or decrease by optical interference effects in the multilayer structure. The specification of the optical characteristics and thickness of the dye polymer layer allows the signal reflected from the reflective layer to be significantly

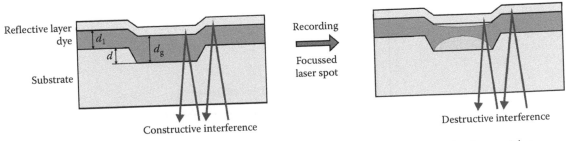

Figure 13.1 CD-R: during recording, the pit is created by heating the dye absorptive layer with a focused laser beam.

decreased by the optical change induced in the polymer layer. The reflectivity difference between recorded 1's and 0's of CD-R and simple CD-ROM media is similar, thus ensuring media interchangeability.

Rewritable optical storage is presently based on MO or phase-change (PC) media. The CD-RW drives and media introduced in 1997 by Philips represented a major breakthrough in optical disk storage. Using phase-change technology, the CD-RW systems allow disks to be written and rewritten many times over. CD-RW is, therefore, a medium of choice for both temporary and long-term data storage.

The disk structure in the case of phase-change media is more complex than the simple three-layer structure of a CD-ROM. The CD-RW disk consists of a grooved polycarbonate substrate onto which a stack of thin layers is sputtered, followed by a protective lacquer (Figure 13.2). The phase-change or recording layer is sandwiched between two dielectric layers. These are typically zinc selenide–silicon dioxide (ZnS–SiO$_2$) layers that provide thermal tuning of the recording layer. The material for erasable phase-change recording is typically a Ge$_2$Sb$_2$Te$_5$ alloy which is sputter deposited on a plastic substrate, with an undercoat and an overcoat of dielectric layers. Other phase-change materials such as PdTeO$_x$, InSnSb, AgInSbTe, etc., are also in use for both write-once and rewritable media. In addition, the stack comprises an aluminum layer from which the laser beam is reflected. As the disk comes out of the sputtering machine, the recording layer is in an amorphous state. The disk is then put into an initializer, which heats up the phase-change layer to the point where it crystallizes. Prior to recording, the phase-change layer is polycrystalline in its original state.

The phase-change recording mechanism relies on the reflectivity difference between dark amorphous zones and bright crystalline zones. For writing, a focused laser beam selectively heats small areas of the phase-change material near or above its melting temperature, which is between 500°C and 700°C. Heating the liquid state is achieved in the area under the laser beam spot. If the material is then cooled sufficiently quickly, the random liquid state is "frozen-in" resulting in the amorphous

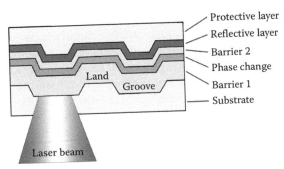

Figure 13.2 Structure of a CD-RW: The active phase-change layer is surrounded by two dielectric barriers and covered by the reflective layer and protective overcoat.

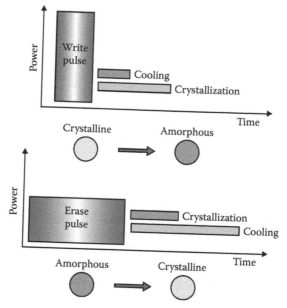

Figure 13.3 For writing, the phase-change layer is heated by a high-power laser beam to form an amorphous pit. The data are erased with a low-power laser beam by reverting the material to the crystalline phase.

state. To erase the recorded data marks, the laser heats the phase-change layer to recrystallize the material. Hereby, the temperature of the phase-change layer is kept below the melting temperature but above the crystallization temperature for longer than the crystallization time so that the atoms revert back to an ordered, crystalline state (Figure 13.3). The mark will become crystalline once again if the laser beam strikes the amorphous mark for sufficient time. Subsequent rewriting is possible when the material reverts back to an ordered crystalline state. The amorphous and crystalline states have different values of refractive index which tune the reflectivity of a multilayer stack. The recorded data can, therefore, be read out optically as reflectivity change between the low-reflective amorphous pits and high-reflective surrounding area. The phase-change process is reversible many thousands of times. Because the reflectivity difference of a CD-RW is much lower than the 70% of a CD-ROM, the sensitivity of CD drives has to be increased to allow readout of CD-RW media. In recent years, significant improvements have been achieved in both rewritable media and drive performance to increase the recording speed and cyclability.

13.2.2 Digital versatile disk

Stimulated by the success of the CD, significant developments in many related fields have been undertaken resulting in further improvements. The large potential of accumulated R&D in that area made it possible to introduce the standard for the advanced DVD format. The basic structure of DVD-ROM is similar to the conventional CD-ROM but many parameters have been refined or reinvented to increase the surface data density and thereby the storage capacity.

The surface data density on an optical disk is inversely related to the spot size of the addressing beam. By reducing the spot size of the focused laser beam, smaller pits can be resolved and the surface density increases. This can be achieved by shortening the wavelength of the laser light and/or by increasing the numerical aperture (NA) of the optical system because the spot size is related to the wavelength divided by the NA. The specification employed to expand DVD's storage capacity includes smaller pit dimensions, more closely spaced tracks and shorter wavelength lasers. Both the track pitch and the shortest pit length are nearly a half less than those of the CD (Figure 13.4). The optical system has also been refined

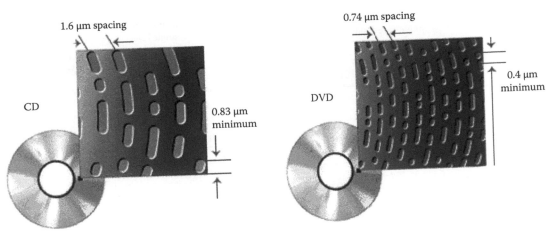

Figure 13.4 Pit–land structure of the CD compared to the refined data structure of the DVD (Sony).

with a higher NA lens, resulting in a more tightly focused laser beam. The simple single-layer DVD stores 4.7 GB per layer, which is seven times the storage capacity of the CD.

Improvements in the overall storage capacity per disk have been achieved in different DVD format extensions that use two layers or both sides of the disk. However, due to the presence of the reflective layer, the number of layers per disk side is strongly limited to two layers; one of them is semireflective and the other is high reflective. The optical signal degradation when the reading laser beam passes through the first, semireflective layer as well as crosstalk effects between the layers require two distinct wavelengths, that is, lasers, one for each layer. In general, dual-layer or double-sided configurations can provide moderate increases in storage capacity but at the same time they require more complex and more expensive systems.

Dual-layer, single-sided DVD is developed as a two-layer structure with the total storage capacity of 8.5 GB. The first semireflective layer reflects 18%–30% of the laser light, which is enough for the read-out of the stored bits. With the transmitted light, the information from a highly reflective layer can also be read out. Gold was previously used as semireflective layer material, but new silicon layers as gold replacement provide significantly reduced production costs. Double-sided dual-layer DVD promises 17 GB but the electronics for reading and decoding the multi-layer DVDs becomes much more complicated. An overview of different multilayer DVD formats is given in Figure 13.5.

A similar two-layer structure is used in Super Audio CD, the new generation of digital audio media realized as a refined version of the CD audio to provide really high-fidelity playback. This is done by adding one semireflective layer to the conventional CD structure that contains high-density information. A silicon-based high density layer is semireflective at 650 nm wavelength and has almost 100% transmittance at 780 nm, thereby allowing the standard CD laser light to be reflected from metallic reflective layer. With 650 nm wavelength, the pits smaller than those on a standard CD can be readout, leading to an enhanced storage capacity of up to 4.7 GB.

In recent years, implementation of recordable/rewriting technology in the DVD family has been the most serious challenge. Although, first recordable (DVD-R) and rewritable (DVD-RAM) products are available now, it is still not clear which format will become the standard DVD recording technology. Second generation recordable DVD format boosts capacities from 2.6 GB per side to 4.7 GB. However, rewritable DVD formats remain far away from achieving standards of compatibility, which have been the crucial issue for a successful introduction of CD-R and CD-RW. Three different DVD rewritable formats are presently competing for acceptance in the market place. The leading candidates to become the standard DVD recording technology—Sony's DVD+RW and Pioneer's DVD-RW—are incompatible so that the DVD family remains far away from CD harmony.

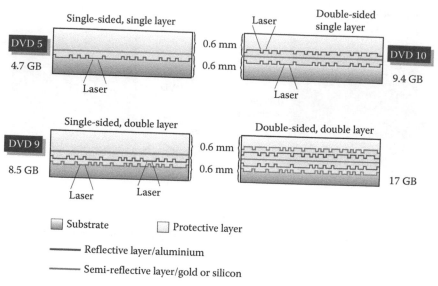

Figure 13.5 Different DVD configurations and corresponding storage capacities.

13.2.3 Blu-ray disk

The implementation of blue-violet diode lasers in DVD systems leads to a further increase of storage density. In February 2002, nine electronics companies (Philips, Sony, Pioneer, and others) have established the basic specification for a forthcoming optical disk video recording format called "Blu-ray Disk." Blue generation optical disks store six times more data than DVD. Introduced as a new video format, Blu-ray Disk enables 27 GB to be stored on a single disk. With its third generation, the optical disk technology follows the development paths established in the evolutionary step from CD to DVD. The most important issue in the development of Blu-ray was the availability of short-wavelength semiconductor lasers. The significant wavelength reduction from 650 nm in DVD to 405 nm in the new generation was enabled by the realization of gallium–nitride laser diodes emitting blue-violet light at the short-wavelength end of the visible spectrum. First introduced by Nichia Chemical Corporation, a small chemical company in Japan, violet-emitting GaN diodes have found wide applications in light-emitting diode (LED)-based systems. The invention of efficient blue LEDs was crucial in developing white light sources for illumination. The related research and development achievements have been recognized by 2014 Nobel Prize in Physics.

In addition to the wavelength reduction, the Blu-ray system comprises a high NA lens that focuses the blue-violet light of a Ga-N diode to an extremely small spot size. However, refinements of the pit-land structure only would not allow such high storage capacity. Sophisticated coding and error correction technology implemented in Blu-ray disk video recorders makes it possible to additionally enhance the storage capacity. With the blue generation, conventional optical storage is approaching physical limits beyond which the data-bearing pit-land structure may become too fine to still be detected.

13.2.4 MO disks

MO disks were the first rewritable storage media, available on the market since the beginning of the 1990s. MO disks are mainly used in professional data processing but with lower costs per megabyte, new applications are being opened. For consumer-oriented applications, the minidisk, an MO disk in the 2.5 in. format was introduced as mini portable audio medium. In the last years, the MO disk technology has been advanced continuously further to allow improved storage capacity and read/write performance.

Data storage in MO disks is based on opto-thermic magnetic effects. Information is stored as a magnetized state of the magneto-optic layer and will be read out as polarization change of the laser beam using the Kerr effect. The MO layer that stored the information as corresponding magnetized state is protected from oxidation by two dielectric barrier layers (SiN) (Figure 13.6). Together with the reflective layer, the barriers ensure an optical signal enhancement. Presently, all commercially available MO disks are based on an amorphous terbium–iron–cobalt magnetic alloy. This material belongs to a class of materials known as the rare earth–transition metal alloys.

For the recording of binary data, the MO layer is heated by the laser spot above the Curie temperature where its magnetic orientation is dissipated. When this spot cools, the new magnetic orientation is set by the magnetic head corresponding to the "0" and "1" of the digital signal (Figure 13.7). During readout, the polarization of the read beam is rotated, thereby detecting the recorded bits.

The MO media rotate the polarization vector of the incident beam upon reflection. This is known as the polar MO Kerr effect. The sense of polarization rotation is dependent on the state of magnetization of the medium. Thus, when the magnetization is pointing up, for example, the polarization rotation is clockwise, whereas downmagnetized domains rotate the polarization counterclockwise. The polar Kerr effect provides the mechanism for readout in MO disk data storage. Typical materials used today impart about 0.5° polarization rotation to the linearly polarized incident light. But, given the extremely low levels of noise in these media, the small Kerr signal nonetheless provides a sufficient signal-to-noise ratio for reliable readout. The media of MO recording are amorphous. Lack of crystallinity in these media makes their reflectivity extremely uniform, thereby reducing the fluctuations of the read signal as well as level of noise in readout.

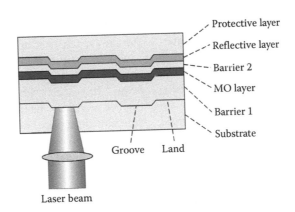

Figure 13.6 Structure of an MO disk.

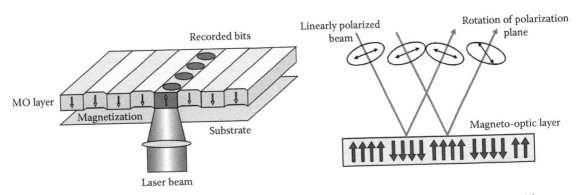

Figure 13.7 Left: Recording of an MO disk. The pit mark is formed by heating the MO layer with a focused laser light. Right: Readout of an MO disk. The data are read out by the polarization change of the read beam.

To rewrite an MO disk, all previously recorded bits must be erased before new data can be recorded. This requires either a recorder with two lasers (one to erase and one to record) or longer recording times in a single laser system because the laser must erase the data in the first rotation and then record the new data in the second one. Alternatively, one can start with an erased track, apply a reverse-magnetizing DC magnetic field to the region of the interest, and modulate the laser power to record the information along the track. This is known as the laser power modulation recording (or light intensity modulation—LIM) scheme. The laser power modulation does not allow direct overwrite of the preexisting data on the track unless a more complex media structure is employed. Exchange-coupled magnetic multilayer structure allows LIM direct overwrite in MO systems. In such systems, the top and bottom layers are exchange coupled together so that switching one layer would make the other layer to switch, too. The top and bottom layers have perpendicular magnetic anisotropy and are separated by an intermediate layer, which is in-plane magnetized and aimed to ease the transition between the top and bottom layers. In an LIM direct overwrite scheme, the recorded domains collapse under a laser beam spot of moderate power. A high-power beam is used to create the domains through the whole thickness of the multilayer stack. The recorded domains are

read out by a low-power beam, which do not disturb them so that only a laser beam of moderate power can erase prerecorded marks. Erasure is very similar to writing, in that it uses the heat generated from the laser beam and requires assistance from an externally applied magnetic field to decide the direction of magnetization after cooling down. This method is simplified in a minidisk recorder based on a magnetic field modulation overwrite system. Here, the new data are written immediately over the previous data. Writing is achieved by a continuous-wave laser beam and a modulated magnetic field.

MO disks are currently available in 3.52 in single-sided and 5.25 in double-sided formats with storage capacities up to 640 MB or 2.6 GB, respectively. Storage densities of MO media are being substantially increased using new advanced technologies like magnetic super resolution (MSR) and near-field recording (NFR). Doubling of storage capacity is expected for the first MO disks based on MSR. Further advances resulting from MSR allow storage capacities of up to 10 GB for 5.25 in. disks.

On MO disks, information is written by heating the recording material with the write laser beam. The stored data are retrieved by reflecting the read laser beam from the structured disk whereby the optical resolution is limited by diffraction. By using the MSR process, it is possible to write bits smaller than the laser spot, that is,

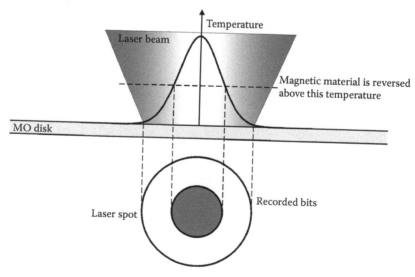

Figure 13.8 MSR. The temperature profile of a focused laser beam is used to create smaller pits than the laser spot.

below the optical diffraction limits. This method is based on specific magnetic properties of the "rare earth/transition metal" materials, which evolved as storage material for MO recording. The write process is the same as writing on conventional MO disks but the recorded bits are smaller than the laser spot. This becomes possible by using the temperature profile induced by the laser spot in a recording material (Figure 13.8). The bits are written only within the central area of the laser spot where the magnetic material is reversed by its high temperature. The advantage of MSR is that the minimum mark length readable in an MSR system is shorter than any that are readable in conventional systems.

Comparing phase-change and MO technologies, one can find several advantages and disadvantages for each. For a number of reasons, phase-change recording appears to be the ideal solution for rewritable optical storage of information. CD-RW drives are simpler than MO drives, because they do not need magnets to create external magnetic fields, and because there is no need for sensitive polarization-detecting optics in CD-RW readout. Phase-change readout makes a profit from the reflectivity difference between crystalline and amorphous states which is large enough to provide much higher signals than the relatively weak MO effect. On the other hand, repeated melting, crystallization, and amorphization of phase-change media results in material segregation, formation of submicron areas that remain crystalline, stress buildup, etc. These factors might reduce data reliability and cyclability of the phase-change media. MO disks are guaranteed to sustain over 10^6 read/write/erase cycles while the corresponding figure for phase-change media is typically one to two orders of magnitude lower. The maximum temperature reached in MO media during recording and erasure is typically around 300°C, as compared to 600°C in PC media. The lower temperatures and the fact that magnetization reversal does not produce material fatigue provide the longer life and better cyclability of the MO media.

13.3 OPTICAL DISK STORAGE SYSTEM

In conventional, well-established technologies, an optical storage system consists of a drive unit and a storage medium, which is usually a rotating disk. The disk storage medium and the laser pickup head are rotated and positioned through a drive motor. In general, optical disks are preformatted by grooves and lands that define the so-called tracks on a disk. A track represents an area along which the information is stored; an optical disk consists of a number of tracks that may be concentric rings of a certain track width separated by a land area. The separation between two neighboring tracks is the so-called track pitch. Typical track spacing on existing optical disks are on the order of micron. In the simplest recording scheme, pit marks of equal length are created along the track (either recorded or stamped) while the presence or absence of these marks corresponds to binary digits 0 and 1. Tracks might be physically existent prior to the recording or created by prerecorded data marks themselves.

On a read-only medium, for example, CD audio disk, quarter-wave deep pits are stamped into an optical disk substrate and then coated with a reflective layer to provide the readout by interference effects in the reflection configuration. In this case, pit marks define their tracks that are discontinuous grooves consisting of pits separated by lands. The length of both pits and lands within a track is defined by the encoded bit stream stored within that area. Tracks are necessary to enable the positioning of the optical head that accesses the information on the disk and to guide the laser beam during readout. In case of recordable media, the disk is typically pregrooved, that is, continuous grooves are printed, etched, or molded onto the substrate to define preexisting tracks. The grooves represent tracks and are separated by lands. The information is recorded along the tracks that can be either concentric rings or a single spiral. Alternatively, the lands may also be used for recording—in this case, adjacent tracks are separated by grooves. Moreover, land-groove recording, that is, recording in both land and groove areas, has been introduced in DVD technology to increase storage density by a more efficient usage of the available storage area.

All optical components needed for recording and retrieval of information are integrated in an optical pickup system or simply optical head. Today's optical heads are small, compact, and highly optimized systems that fly close to the disk surface. The optical head must be able to rapidly access any position on the disk for error-free recovery. Depending on the disk technology, the optical head might have different architectures but the basic configuration is the same for all present systems. Usually, the head comprises a laser diode, a collimator lens, an objective lens, a polarizing beam splitter, a quarter-wave plate, and the detector system. A typical setup for an optical disk system is shown in Figure 13.9 [1]. A linearly polarized laser beam is emitted by the laser diode. The collimator reduces the divergence and collimates the beam, which then passes through a polarizing beam splitter and a quarter-wave plate. The quarter-wave plate circularly polarizes the incident beam, which is then focused onto the disk by an objective lens that also collects the light reflected back from the disk. For recording, a focused laser beam generates a small spot within the active material to induce some kind of optically detectable changes. The recorded bit marks will then change the phase, amplitude, or polarization of the readout beam. The total reflected light is therefore modulated by the presence of pits—the light fractions reflected from the pits and the light fractions reflected from the land interfere destructively and modulate the data signal according to the stored information. The beam reflected back from the disk again passes through the objective lens and becomes recollimated afterward. The quarter-wave plate converts the circular polarization of the reflected beam to a linear one with the direction perpendicular to that of the incident beam. In this way, the incident, readout beam, and the reflected, signal beam can

be separated by the polarizing beam splitter, which directs the reflected light to the data detector. The detection system produces the data or readout signal but also optoelectronic signals needed for automatic focusing and track following. Specific servo systems are required to control the position of the optical head with respect to the tracks on the disk. Additional peripheral electronic units are used for functional drive control, data reconstruction, and encoding/decoding.

An optical disk storage system is characterized by several functional quantities that specify its performance in terms of capacity and speed. Typical parameters are storage capacity, access time, data transfer rate, and cost. The storage capacity is determined by the areal density of the stored information and the geometrical dimensions of the disk medium, that is, available storage area. The areal density characterizes the efficiency of a system in using the storage area and is a direct function of the spot size and/or the minimum dimension of a stored bit mark. It is typically given in units of gigabits per square inch or bits per square micron. The areal density is limited by the optical resolution of the laser pickup, that is, by the minimum dimensions of data marks that still can be detected by the optical system. The NA of the objective lens and the wavelength of the laser used for recording and readout determine the diffraction-limited spot size and therefore the data density but further factors also have to be taken into account such as

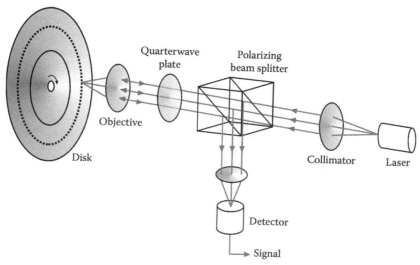

Figure 13.9 Basic configuration of an optical disk system.

track density and linear bit density. The track density (tracks/in.) expresses how close to each other neighboring tracks on the disk can be arranged. The linear bit density (bits/in.) is a metric for the spacing between optical transitions along a track.

Other parameters of interest are access time and data transfer rate, which characterize how fast the information stored on the disk can be accessed and read out. The access time depends primarily on the speed with which the optical head can move over tracks to access a given storage location on the disk. The data rate is a metric for recording and readout speed and it depends on the linear bit density and the rotational speed of the drive. The data rate becomes important when large data files, for example, images or video files, have to be stored or retrieved. Different paths to higher data rates have been proposed in CD/DVD technology including, for example, recording/readout by multiple beams, that is, usage of several laser pickup systems simultaneously. The laser beam power available for writing and reading, and the speed of the servo system controlling the optical head position are some of the parameters that limit the data transfer rate.

13.3.1 The objective

The objective lens in an optical disk system must be designed to correct for spherical aberration, which is due to the substrate thickness. Current optical disk systems use molded glass lenses to focus the laser beam to a diffraction-limited spot. There are several advantages of the molded lens which is typically a single aspheric lens over conventional objective lenses made up of multiple elements. These advantages result from the molding process itself which is much better suited to mass fabrication. Furthermore, the working distance, that is, air spacing between the objective lens and the disk surface, of a molded lens is larger, which simplifies the design and function of the head in a removable-media optical system. Also, the molded lens mass is lower than a conventional lens, which reduces actuator forces needed for automatic focusing and track following within the servo system.

In order to achieve ultimately small spots on the reflecting data layer, the objective lens must have fairly large NA, and it must be free from aberrations. Hereby, the NA of a lens is defined by $NA = n \times \sin\theta$, where θ is the half angle closed by the cone of the focused laser beam. The diameter of the diffraction-limited focused spot is then given by the ratio of the wavelength of the laser beam, and the NA of the objective lens, that is, $d_{spot} \approx \lambda_0 / NA$, where λ_0 is the vacuum wavelength of the laser beam. It becomes clear from the above relation that higher NAs are desirable if smaller spots and therefore higher storage densities are to be achieved. The smaller the spot of the readout beam, the smaller data marks can be resolved. The areal data storage density depends directly on the spot size and according to the above equation, it can be increased by reducing the wavelength and/or by increasing the NA. In practice, both approaches encounter limitations and also require adaptations and advances in a number of constituting technologies. The most important aspect in achieving greater storage densities by reducing the wavelength is the availability of short-wavelength lasers.

Reducing the spot size by increasing the NA will also reduce the focal range of the laser beam, the so-called depth of focus. The focal range can be estimated by the Rayleigh length of the laser beam and is therefore proportional to λ/NA^2, which means that the higher NA, the smaller will be the depth of focus. That clearly limits NA as an optical storage system is capable of handling the focus error only within this range. Other tolerances such as those for the disk tilt and the substrate thickness are also limited by NA and have to be considered in designing the objective lens. Presently, the NA of objectives used in CD technology is 0.45, which has been increased to 0.6 in DVD systems to achieve higher storage capacity. A further increase of NA would require far-reaching adaptations in both the disk configuration and optical system to provide that the diffraction-limited focus will be maintained with the desired accuracy on the reflective data layer.

13.3.2 The laser

Optical storage of information became first practically possible with the invention of laser. Rapid developments in the field of laser systems have supported the technological realization of existing optical storage systems. In general, lasers are used in all optical technologies for data recording and often also for data recovery. The ultimate premises for light sources in optical storage systems are small size, stability, long lifetime, and inexpensive mass production. The optical drive so as we know it became therefore possible first with the establishment of laser diodes. Conventional optical disk systems, such as CD, DVD, and MO, rely all on semiconductor laser diodes as their source of light. The diode laser technology has been the key enabling technology for optical storage while the success of CD consumer products (CD audio and CD-ROM in PCs) has pushed the diode laser to the best-selling laser products of all time. Compared to other laser types, the laser diode has many advantages with regard to the requirements defined by the design and functionality of an optical head. In general, the optical head contains one laser diode that provides light for recording, reading, and erasing (given in rewritable systems only). Each of these functions sets specific requirements on the light source.

Optical data storage systems for mainstream applications such as computer and entertainment need compact and cheap lasers that can be with ease integrated in low-weight optical heads and small, low-cost drives. In addition, specific requirements of optical data storage concern the wavelength of emitted light, optical power available for writing, reading and erasing, modulation, and life time of the laser diode which should exceed several thousand hours. The wavelength of light used to write and read bit marks is crucial for the optical resolution of the pickup system and, consequently, for the areal storage density. The shorter the wavelength the smaller bits can be resolved and the higher areal storage density can be achieved. Stimulated by the success of CD and later DVD, extensive research and industrial development efforts have been undertaken in past decades to satisfy demand for short-wavelength laser diodes with sufficient optical power for the operation in optical drives.

The power requirement for diode lasers in optical disk systems varies from several milliwatts for retrieval of stored information to several tens of milliwatts for data recording. The laser output radiation is modulated directly by modulating the input electrical current. The fast modulation of laser radiation is perhaps the most important characteristic of laser diodes for optical recording. Laser diodes can be modulated to GHz frequencies with rise and fall times of less than 1 ns. For read-only applications, low-power lasers with approximately 5–10 mW optical power can be used. In contrast, for recordable and rewritable media, more power is needed because the laser beam has to induce almost instantaneously detectable changes in the recording material. For example, amorphous pit marks on a CD-RW are thermally induced by a pulsed laser beam with an optical power of either 50 or 60 mW.

Another requirement concerns the spatial coherence and single transverse mode operation as the laser beam has to be focused to the diffraction limit. The laser cavity must be a single-mode waveguide over its operating power range to provide wave fronts that are needed to achieve ultimately required small spots. The longitudinal mode stability has not been a requirement up to now so that laser diodes incorporated in the present optical heads typically operate in several longitudinal modes. Otherwise, a single longitudinal mode operation might become important in reducing undesired wavelength fluctuations. Fluctuation of both intensity and wavelength is one of the characteristics of diode lasers. Although intensity fluctuations reduce the signal-to-noise ratio of the readout process and generate noise in the servo signals, wavelength fluctuations set additional requirements on achromatic design of optical components.

The wavelengths of optical data storage have been continuously reduced since the introduction of the first laser disk systems, which started at 830 nm. CD audio and early CD-ROM systems rely on infrared diodes at 780 nm while moderate power lasers at 680 nm are used in CD-R and MO products, and also in computer drives. The DVD-ROM standard relies on red-emitting diodes at 635 and 650 nm. The wavelength reduction has been one of the crucial factors for the sevenfold increase in storage capacity from CD to DVD (Figure 13.10). A shorter wavelength laser would support a higher storage density as shorter wavelength light can be focused to a smaller spot at the diffraction limit. As the diameter of a focused laser spot is proportional to its wavelength, the reduction of the wavelength will lead to the reduction of the spot size by the same factor, and consequently, to an increase of the storage density by the square of that factor.

Recent developments in the field of III–V semiconductor diode lasers allow for an almost revolutionary transition from the red to the blue-violet spectral range. Enormous advances have been achieved by incorporating short-wavelength blue-violet diode lasers in optical storage systems.

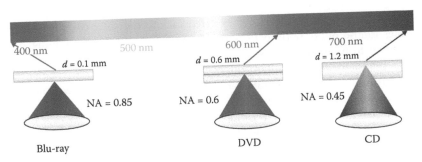

Figure 13.10 The increase in storage density from CD to DVD has been primarily achieved by decreasing the bit dimension but also by reducing the tracks spacing, by using shorter wavelength lasers and higher NA optics. At the same time, the substrate thickness has been reduced from 1.2 mm on a CD to 0.6 mm on a DVD, whereas a Blu-ray disk comprises a protective layer of 0.1 mm thickness only.

Blue-violet and ultraviolet diode lasers operating in the range of 370–430 nm have been first developed by Nichia Chemicals Corporation in Japan and introduced to the market in 1999. Announced as high-density DVD, the Blu-ray standard is based on a 50 mW blue-violet diode laser. By adopting a 405 nm laser, the "blue" optical disk technology minimizes its beam spot size by increasing the NA to 0.85. On the other hand, such an extremely sharply focused laser beam is characterized by a very small focus depth. Therefore, the substrate becomes also extremely thin with a thickness of 0.1 mm only so that the disk must be protected from outer influences and damages by a cartridge. Both issues, short wavelength and high NA, allow for reducing the pit size to approximately 0.2 μm, and the tracking pitch to 0.32 μm, almost half of that of a regular red DVD. All these refinements and improvements together push the DVD technology up to 27 GB high-density recording.

13.3.3 The servo system

Readout in optical storage relies on data reconstruction from a disk rotating with a several thousand rotations per minute. To provide a faithful retrieval of the stored data, the laser beam must be focused exactly on the disk track and then maintained accurately within it during the entire readout process. Typical tolerances in optical disk systems are 1 μm for positioning of the focus in the reflective data layer, and one-tenth of micron for focus positioning on the track center. The axial and radial runout of an optical disk are two or three order of magnitudes larger than these allowable focus positioning errors. The optical drive,

therefore, requires a servo system to compensate the radial and vertical runout of the disk as it spins and to provide submicron focus and track-locking schemes. The servo system is a closed-loop opto-electro-mechanical system, which couples optical position detectors to high bandwidth actuators to actively follow the disk rotation. Consequently, the servo control in an optical disk system involves accurate and continuous focus position error sensing and sophisticated feedback mechanisms that dynamically convert the detected error signals in corresponding actuator movements.

The task of actuators within the servo system is to correct and to control the position of the focusing optics, that is, objective lens. A typical objective lens has an NA of 0.45 or higher to create a focused beam spot smaller than 1 μm. In addition, the focused beam has a focus depth, which is only a fraction of a micron. On the other hand, a rapidly rotating disk has a tendency to wobble in and out of its ideal position in the optical drive. There is a variety of reasons that may cause such wobble effects, some of them are imperfections in the disk construction, substrates that are not ideally flat, other manufacturing errors in both disks and drives, disk tilt, and eccentricity, etc. In the ideal case, the disk mounted in a drive would be perfectly centered, and sufficiently flat to maintain an ideal perpendicular position with respect to the rotation axis at all times. Deviations up to ±100 μm in both vertical and radial directions usually occur during the disk rotation in an optical drive.

The task of the servo system is twofold: First of all, the laser beam spot focused into the reflective layer of the disk must remain within the depth of focus, and second, the focused spot must remain within a submicron-sized track while the disk rotates and wobbles in and out of both focus and track. The mechanism to maintain the laser spot on the disk within the focus depth is called automatic focusing and it is aimed to compensate vertical runout of the disk. In addition, automatic tracking servo is needed to maintain the position of the focused spot within a particular track—the mechanism is called automatic track following. In the ideal case, tracks are perfectly circular and concentric, and the disk is perfectly centered on the rotation axis. In practice, the eccentricity of both the tracks and disk creates demand for active track following mechanisms that compensate wobble effects and enable the laser beam spot to follow the track.

13.3.3.1 AUTOMATIC FOCUSING

Current optical drives rely on several different methods that provide the feedback mechanisms, that is, error signals that drive the focus servo system. The objective lens is mounted in a voice coil actuator with a bandwidth of several kilohertz, and the feedback mechanism is used to position the lens relative to the rotating disk in such a way as to maintain focus at all time. The basic premise is that an appropriate error signal is generated which is then fed back to the voice coil actuator for maintaining focus automatically. Depending on the detection scheme, various techniques have been proposed to generate the focus error signal (FES). The signal needed for the closed servo loop is usually derived from the light that is reflected from the disk. Several techniques use a field lens that creates a secondary focused spot; deviations from optimum focus are then analyzed by observing that secondary spot. The field lens, which might be a spherical, ring-toric, or astigmatic lens is placed after the objective lens in order to focus the light reflected back from the disk. The shape, size, and position of the focused spot depend on the position of the disk relative to the in-focus plane. Changes in the secondary spot caused by the off-focus status of the disk are detected via a photodetector and transformed into an electronic signal that contains the feedback information for the focus servo system.

A very popular mechanism for automatic focusing relies on the astigmatic lens detection method. The most of current optical disk systems use an astigmatic servo sensor that comprises an astigmatic lens having two different focal lengths along orthogonal axes, and a quad detector. Figure 13.11 shows a diagram of the astigmatic focus-error detection system used in many practical devices.

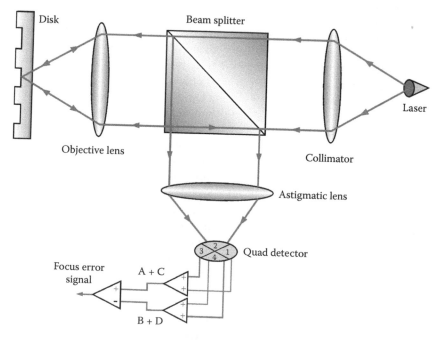

Figure 13.11 Astigmatic focus-error detection system.

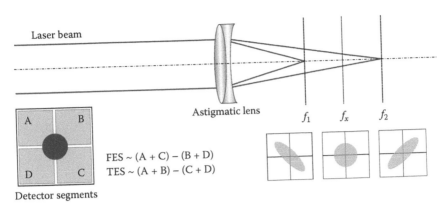

Figure 13.12 Four-quadrant-detector combined with an astigmatic lens is used in optical disk systems for both automatic focusing and track following. Control electronic signals are generated by an appropriate combination of the signals from four detector segments.

The light beam returned from the disk and collimated by the objective lens might be convergent or divergent, dependent on position of the disk relative to the plane of best focus. The reflected beam passes an astigmatic lens, which normally focuses the incident light beam to a circularly symmetric spot halfway between its focal planes. In the best, in-focus case, a quad detector placed at this plane receives equal amounts of light on its four quadrants. In contrast, when the disk is out of focus, the astigmatic lens will create an elongated spot on the detector so that individual quadrants will be illuminated differently and, consequently, they will create different electronic signals (Figure 13.12). Depending on the sign of defocus, this elongated spot may preferentially illuminate quadrants A and C or quadrants B and D of the detector. A bipolar FES can be then derived as the difference between diagonal quadrants, that is,

$$FES \propto \frac{(A+C)-(B+D)}{A+B+C+D}.$$

13.3.3.2 AUTOMATIC TRACK FOLLOWING

The information is recorded on an optical disk either around a series of concentric circular tracks or on a continuous spiral. Manufacturing errors and disk eccentricities caused by mounting errors or thermal expansion of the substrate, for example, will cause a given track to wobble in and out of position as the disk spins. Typically, a given track might be as much as ±100 μm away from its intended position at any given time. The focused

spot is only about 1 μm across and cannot be at the right place at all times. An automatic tracking scheme is, therefore, desired. The feedback signal for controlling the position of the objective lens within the tracking coil is again provided by the return beam itself. The four segments of the detector are combined in different ways for FES and track error signal (TES). Several mechanisms for automatic track following have been proposed and applied in commercial devices.

The push–pull tracking mechanism relies on the presence of either grooves or a trackful of data on the media. In the case of CD and CD-ROM, the data are prestamped along a spiral on the substrate, and the sequence of data marks along the spiral represents a sort of discontinuous groove structure. Writable media such as CD-R, MO, and PC require a tracking mechanism distinct from the data pattern, because prior to data recording, the write head must be able to follow the track before it can record anything. Once the data are recorded, the system will have a choice to follow either the original tracking mechanism or the recorded data pattern. Continuous grooves are the usual form of preexisting tracks on optical media. A typical groove is a fraction of a micron wide and one-eighth of the wavelength deep. As long as the focused beam is centered on a track, diffraction of light from the adjacent grooves will be symmetric. The symmetry of the reflected beam, as sensed by a quad detector in the return path, would produce a zero error signal (Figure 13.13). However, when the focused spot moves away from the center of the track, an asymmetry appears in the intensity

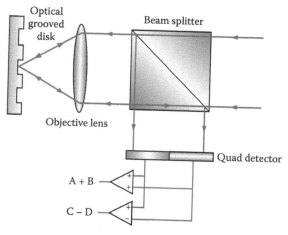

Figure 13.13 TES generated by push–pull method.

pattern at the detector. The difference signal is sufficient to return the focused spot to the center of the track. The TES is thus generated by the quad detector according to

$$TES \propto \frac{(A+B)-(C+D)}{A+B+C+D}.$$

In read-only media, where continuous grooves are not present, another method of tracking is applied, the so-called three-beam method. The laser beam is divided into three beams, one of which represents the main spot and follows the track under consideration, while the other two are focused on adjacent tracks, immediately before and after the addressed track. Consequently, three detectors are necessary for position error sensing. Any movement of the central beam away from its desired position will cause an increase in the signal from one of the outriggers and, simultaneously, a decrease in the signal from the other outrigger. A comparison of the two outrigger signals provides the information for the track-following servo.

In another possible tracking scheme, the so-called sampled servo scheme, a set of discrete pairs of marks is placed on the media at regular intervals. The wobble marks indicate the transversal boundaries of a track. Such marks might be embossed or written by a laser beam within the formatting procedure. These marks are slightly offset from the track center in opposite directions, so that the reflected light first indicates the arrival of one and then of the other wobble mark. The TES

is generated as the difference between the signals detected from each wobble mark. Depending on the spot position on the track, one of these two pulses of reflected light may be stronger than the other, thus indicating the direction of track error. The sample servo technique is often used in recordable media systems including both write-once and rewritable media.

In contrast, present DVD systems rely on the differential phase detection technique that is based on diffraction of the focused spot from the edges of data marks. If the spot is focused offtrack, the light reflected from the disk will show an asymmetric intensity distribution each time the spot strikes a mark edge. The intensity pattern rotates when the spot travels along the data mark and this rotation is then sensed by a quad detector that generates corresponding TESs.

13.3.4 Data coding and processing

An important step in storing digital data is encoding of the bit stream to be stored prior to recording, and, consequently, decoding of the readout signal after its conversion into digital form. Hereby, digital data are extracted from the analog signal obtained by the data detector. In a digital storage system, the input is typically a stream of binary data, that is, binary digits 1 and 0, which has to be recorded onto a storage medium in such a way as to provide reliable and error-free recovery. The storage system is requested to record the data, to store it, and to reconstruct it faithfully upon request.

The digital data are first converted into analog signals and then stored as a stream of data marks correspondingly modulated and embedded into the recording format. For retrieval, the original digital data are extracted from the analog signal that is collected by the optical head via the data detector. From the user input to the recovered output, the data will undergo several steps of electronic processing including analog/digital conversion, equalization, and filtering of the playback signal, error correction and modular coding, data synchronization, and organization within the recording channel, etc. [2].

Within the storage system, not only a recording unit is incorporated but also additional electronics units or subsystems that perform diverse steps in data processing from input to output after the stored data have been reconstructed by the optical read head. There are a variety of error sources that can cause errors in retrieval of information from a rapidly rotating optical disk. Some of them are related to the dynamic operation regime with the data structure reduced to the limit where the readout signal can only just be separated from the system noise. Also, media imperfections, defects, damages, etc., lead to errors in reconstruction of information sequences stored at the affected locations. To retrieve the data faithfully, all sorts of errors must be eliminated or compensated by appropriate error correction techniques. Therefore, the binary input data (user data) undergo several steps of encoding and modulation prior to being recorded on the storage medium.

The encoding process involves several measures against diverse error sources but it also entails other features that simplify the data processing and recovery. The flow of the data stream in an optical storage system is depicted in Figure 13.14. Binary user data are subjected to two encoding processes prior to recording on the storage medium: error-correction coding (ECC) and modulation or recording coding. The encoding process includes typically one or more ECC steps followed by a modulation coding step where appropriate features are incorporated into the bit pattern. The ECC step is designed to protect the data against random and burst errors, and the modulation step organizes the data to be stored so as to maximize the storage density and reliability. In the ECC step additional bits, the so-called check bits, are generated and added to the stream of user data in order to create an appropriate level of redundancy in the overall bit sequence. Both the source data and the data emerging from the first (ECC) encoder are unconstrained, that is, a randomly selected bit in the data stream may be either a "1" or a "0" with equal probability and arbitrarily long sequences of all "ones" and "zeros" may appear.

In addition to the error correction, binary sequences are encoded by a modulation coding whereby the bit-patterns to be recorded are expanded by certain additional features. These enable the generation of a clocking signal for the electronic waveform and also provide more efficient usage of the storage area available on the disk. The modulation step involves mapping of small blocks from the error-correction coded sequence into larger blocks known as modulation code words. The data emerging after the modulation encoding step are usually d, k constrained,

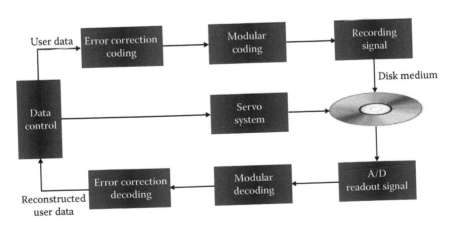

Figure 13.14 Channel data flow in an optical storage system.

that is, binary sequences are constrained in such a way that there must be at least d but no more that k zeros between two ones. Each binary segment consists therefore of a one that is followed by at least d but no more than k zeros. The data encoded in such a way is run-length limited (RLL) and it is referred to as channel data because it becomes then converted into electrical waveforms used to control the recording process. The CD optical disk technologies rely on a block modulation code known as *eight-to-fourteen* modulation (EFM) that expands blocks containing 8 data bits to 14-bit channel blocks. This modulation code is also applied in DVD systems in its advanced version, known as EFMPlus.

During retrieval, the RLL channel data are reconstructed by the detector and processed by a modulation decoder which gives the error-correction coded binary data as output. The modulation decoder must correctly determine the logical ECC code word boundaries within the stream of channel data to enable subsequent ECC decoding to take place. The ECC decoder processes the data leaving the demodulator to detect and correct any data error type for that it is designed to recognize and eliminate. Optical storage systems use typically Reed–Solomon (RS) codes for ECC. These are block codes, that is, encoded data consist of code words, or blocks, that contain a fixed number of bits. Encoding entails organizing the blocks of binary data into a succession of multibit symbols called information symbols, computing a number of additional symbols (of the same length) called parity checks and appending the parity checks to the information symbol to form a code word. RS codes employed in optical data storage systems use eight-bit information symbols, that is, they are designed to operate on bytes instead of bits. An RS code will correct up to a certain number of erroneous symbols in a given code word, and for correcting a specific number of erroneous bytes, it needs twice as many parity check bytes. For example, one of the two ECC used in CD audio systems is the RS code, which can correct up to 2 bytes of error in a 24-byte-long block of user data, with the addition of four parity bytes.

Information to be recorded on a disk is organized into uniformly sized blocks. Each of these blocks is written onto a portion of the storage medium that is referred to as a sector. Each track on the disk is then divided into a number of sectors

that contain the user data and any ECC parity information related to it, that is, calculated from it. Several different track formats find application in practical systems. As mentioned before, tracks can be realized either as concentric rings or a single continuous spiral whereby optical disks usually rely on the spiral format. For digital data storage, each track consists of sectors that are defined as small segments containing a single block of data. Spiral tracks are more suitable for writing of large data files without interruption while concentric rings better support multiple operation mode when different operations such as write, erase, verify, etc., are performed simultaneously in different tracks. The block of digital data stored within a sector has a fixed length, which is usually either 512 or 1024 bytes. Each sector has its own address, the so-called header that specifies the storage location of a given sector on the disk.

The storage location is given by the track number and azimuthal position in the track at which the sector will be written. The information that constitutes a sector is usually written onto the medium in two parts. The first part of the sector is the sector header, which consists of special patterns known as sector marks together with sector address data. The headers of all sectors are prerecorded, that is, they are placed on the disk either when it is manufactured or when it is formatted for use. The additional space used by the codes and by the header within each sector constitutes the overhead, which may take between 10% and 30% of disk's raw capacity depending on application.

Prior to retrieving the data from the recorded storage medium, the optical detection head must access the medium and find the data that are requested by the storage system controller. The optical head must move to a particular radial position on the disk and it must find the track and sector that contain the requested data. Especially, in the case of removable disk drives, the system checks other important information about the disk, for example, the sector size being used on the disk, the amplitude and polarity of signals obtained from prerecorded sectors and headers, etc. For reliable recording and readout, the design of a storage medium implements a defined recording format. The written data are embedded in this format, which entails certain system information to be prerecorded at specific locations on the disk. The recording format also provides that the user

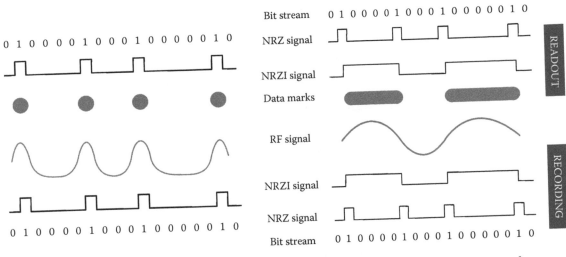

Figure 13.15 NRZ (left) and NRZI (right) conversion of binary sequences into electronic wave fronts.

data are written in its appropriate sector type at particular locations on the disk.

To conform to the recording format, the physical structure of the data to be stored on the disk must satisfy several requirements. In Figure 13.15, two possible schemes for conversion of binary sequences into electrical waveforms are illustrated. In the so-called non-return-to-zero (NRZ) scheme, each bit is allotted one unit of time during which the voltage is either high or low, depending on whether the bit is 1 or 0. In the ideal case, recording with an NRZ waveform will result in identical marks that have the same length proportional to one channel bit time. Neighboring data marks can have different center-to-center spacings, whereby the center of a mark represents a channel binary "one." The modified version of this scheme known as NRZI (non-return-to-zero-inverted) conforms much better to the EFM as here a 1 corresponds to a transition while a 0 is represented

by no transitions at all. An NRZI waveform will induce data marks and intervening spaces that have variable discrete lengths. Here, the appearance of a binary "one" corresponds to an edge of a recorded mark and both the mark length and the spacing between two successive marks are given by the number of binary "zeros" between two ones. Using such a scheme, a single data mark can contain more bits than only one as in a simple coding scheme where each pit is allotted one bit.

In summary, the optical resolution limit, known as the "Abbe barrier," has dictated progress and development among three generations of optical disk technology. The approach to higher data densities was straightforward—reducing the size of data marks by shortening the wavelength and by increasing the NA. Table 13.3 summarizes the optical specification, physical data structure, and storage performance of three optical disk standards. In its third generation, based on a 405 nm

Table 13.3 Three generations optical disk technology: optical specification, physical data structure, and storage performance

	CD	DVD	Blu-ray
Wavelength	780 nm	650 nm	405 nm
NA	0.45	0.6	0.85
Minimum pit size	0.83 μm	0.4 μm	200 nm
Minimum mark length	0.83 μm	0.4 μm	150 nm
Track pitch	1.6 μm	0.74 μm	320 nm
Storage density	1 bit/μm²	5 bits/μm²	20 bits/μm²
Data rate	1 Mbps	5 Mbps	50–200 Mbps
Storage capacity	700 MB	4.7 GB	27 GB

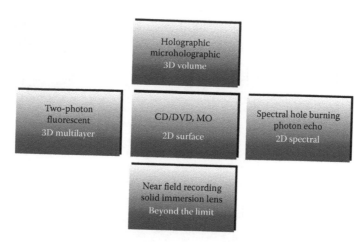

Figure 13.16 Novel approaches in optical storage to overcome the limitations of the surface storage by pits in CD/DVD as well as MO disk technology, including optical storage beyond the resolution limit by NFR and SIL, frequency-/time-domain optical storage by spectral hole burning and photon echo memories, multilayer storage within transparent materials—two-photon and fluorescent memories, page-oriented holographic memories, and bit-oriented microholographic disk.

laser, optical disk storage definitively encounters the physical limits imposed by diffraction of light. Meanwhile, new techniques that use more than the surface of a flat disk are under investigation.

13.4 NOVEL APPROACHES IN OPTICAL STORAGE

The acceleration of processor speeds and the evolution of new multimedia and Internet applications are creating an almost insatiable demand for high-performance data storage solutions. Storage requirements are growing at an exponential rate encouraged by immense technological advances, which have been achieved in recent decades. An ever-increasing amount of digital information is to be stored online, near-line, and off-line combining magnetic, MO, and optical systems. Rapidly growing demands are distributed through the data storage hierarchy, where diverse technologies are combined in complimentary way to satisfy specific requirements in different application environments. The usual computer applications are served mainly by hard disk drives but emerging consumer applications that combine audio, video, 3D image, and computer data files are creating an important category with substantially different requirements. Optical disk technology has established itself as a mainstream product provider for audio, video, and computer

storage. The extraordinary success of recordable and rewritable disk formats based on CD and DVD technology has opened new prospects but also new requirements. Advanced storage of digital contents requires both higher storage capacity and fast data transfer. Although the optical disk technology with its three generations satisfy storage demands of the entertainment and content distribution industry, novel application areas make essentially new technologies necessary.

In the future, optical data storage is expected to follow two directions to improve capacity and performance of disks that are currently available. The straightforward way predicts the further increase of the areal storage density by surpassing the limit imposed by the diffraction of light. Storage technologies that use only the surface of a medium for writing and reading are constrained to this direction. On the other hand, optical information storage uses laser–material interaction effects for recording and retrieval so that an entire spectrum of different optical phenomena can be applied to realize an optical memory. Developments in the field of nonlinear optical materials that exhibit strong laser-induced changes of their optical properties enable various novel approaches to become practically realizable. Using nonlinear optical effects, advanced technological solutions for optical storage may take advantage of new spatial and spectral dimensions (Figure 13.16).

13.4.1 Beyond the resolution limit

The traditional approach for increasing the areal density that has driven progress in data storage is to decrease the bit size. In optical storage, the attainable data density is largely determined by the size of the focused laser spot. A powerful way to surpass the density limit imposed by the diffraction of light in optical data storage is the usage of near-field optical recording. The creation and detection of pit marks smaller than predicted by the diffraction barrier can be realized by numerous near-field optical techniques. For conventional optical systems, the achievable spot size governed by the diffraction law is $\propto \lambda/2NA$. The resolution limit, known as the "Abbe barrier," was empirically discovered and named for the German physicist, Dr. Ernst Abbe, best known for his work in optics in the 1860s.

Recent progress in near-field optics has resulted in effective spot sizes smaller than 1/20 of the wavelength of light. To achieve such a high resolution, an aperture smaller than the resolution limit is placed between the light source and the medium. Light passing the aperture consists of propagating and evanescent waves; the smaller the aperture, the larger the fraction of evanescent field. The evanescent wave intensity decreases exponentially outside of the aperture; therefore, when the aperture-to-sample distance decreases, the evanescent power increases, and the resolution improves. If the aperture-to-medium distance is much less than the wavelength, the resolution will be determined by the aperture size rather than by the diffraction limit. Diverse techniques for near-field optical recording have been proposed making readout of subwavelength structures possible. However, a serious disadvantage for most of NFR techniques using small aperture is low optical efficiency. Although near-field optical recording can provide extraordinarily high areal densities, it is difficult to satisfy requirements on high data transfer while maintaining a working spacing of less than a wavelength.

Scanning near-field optical microscopy (SNOM) makes it possible to overcome the diffraction limit of conventional far-field optical systems by placing the pickup head very near (about 50 nm above) the media. In near-field microscopy, optical resolution beyond the diffraction limit is achieved by scanning a surface with the evanescent field behind a nanometer aperture. A small distance is necessary because the light field is confined only in the near field of the aperture. The technique can produce spots as small as 40 nm in diameter and conceptually can achieve areal densities in the order of 100 GB in.$^{-2}$. SNOM technology provides high areal densities but until now readout speed is low because the scanning process is very slow due to the small light power behind the aperture. Furthermore, the probe must be in near contact with the medium, making it difficult to prevent head crashes and support removable media. Another technique uses a metalized tapered optical fiber, in the end of which is a small aperture [3]. The tip of a fiber, which is smaller than the wavelength of the recording light, is positioned within 10 nm to the sample. This approach has been already used to write and detect 60 nm diameter MO domains. However, the tapered fiber approach also suffers from very low optical efficiency.

A technique that might solve the trade-off between an extremely high resolution and practical system implementation is based on the solid immersion lens (SIL). The principle of the SIL is that by focusing light inside a high refractive index glass where the propagation speed is slow, the spot size can be reduced below the minimum achievable spot size in air [4,5]. The SIL reduces the actual spot size by both refracting the light rays at the sphere surface and by having an increased index of refraction within the lens. The hemispherical glass of refractive index n receives the rays of light at normal incidence to its surface (Figure 13.17). These are focused at the center of the hemisphere to form a diffraction-limited spot that is smaller by a factor n compared to what would have been in the absence of the SIL. That becomes obvious if we consider the minimum achievable spot size which is given by

$$d_{min} = 0.61 \frac{\lambda}{n \sin \theta}$$

where θ is the aperture angle of the focusing lens. The SIL allows the aperture angle to be increased also. An increase of NA from 0.6 is typical for conventional optical disk systems to 0.95, and a refractive index change from 1.0 to 2.2 would result in a spot size of 0.2 μm at the flat surface of a SIL for light at 670 nm.

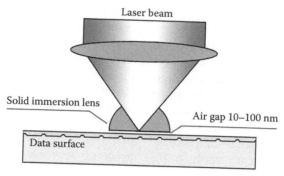

Figure 13.17 SIL for near-field optical storage. The light is focused internally in a semispherical lens onto the flat surface.

Although SILs cannot produce spot sizes as small as tip fiber apertures can, they have the advantage of a substantially higher optical throughput. Another advantage is that an SIL can be with ease integrated in any conventional system configuration as an addition to the objective lens. The application of the SIL also requires an extremely short working distance of the lens to the recording layer—about 100 nm—as well as a very thin protective layer. Flying heads as in hard disk storage can be used to provide such a distance, which is maintained over a disk that rotates rapidly. The SIL combined with short-wavelength lasers should enable the technique to reach areal densities of more than 10 GB in.$^{-2}$. To date, implementations of NFR with SIL are planned in MO storage. The first NFR products announced will have a storage capacity of 20 GB per disk.

13.4.2 Frequency-/time-domain optical storage

Looking for solutions to overcome the restrictions of two-dimensional optical storage systems, various approaches have been explored that use one further dimension in spatial, spectral, or time domain. Novel technologies, such as holographic storage, two-photon or fluorescent memories, persistent spectral hole burning (PSHB), photon echo memories, etc., are at various stages of development. Opening a new dimension in addition to the two-dimensional surface of a storage medium, they have the potential to improve tremendously both capacity and data transfer rates of optical storage systems.

Frequency- or time-domain optical storage techniques adhere to two-dimensional media but open one additional dimension in frequency or time domain. PSHB takes a step ahead of one-bit-per-spot memories, allowing multiple bits to be written, erased, and rewritten in a single location. High densities become possible because the diffraction limit does not limit PSHB memories to the recording of a single bit per spot, as it does in conventional optical data storage. In a PSHB material, it is possible to discriminate many different spectral addresses within a single λ^3 volume.

Frequency-domain optical storage based on PSHB involves burning "holes" in a material's absorption band [6,7]. The ideal material has many narrow, individual absorption lines that form a broad absorption band. Practical recording in PSHB media may be done as an extension of conventional holography with the difference that, instead of a single wavelength, a large number of independent spectral channels may be used. This number ranges from 10^4 to more than 10^7, depending on the material. For writing information, a frequency-tunable laser is focused on a single spot scanning down in wavelength to induce transition from one stable state to another in an absorbing center. As a result, there will be holes at certain frequencies that correspond to the presence of the written bits. PSHB using the frequency domain promises storage densities up to 10^3 bits μm^{-2}, which could be achieved with 10^3 absorbing centers, that is, spectral holes per diffraction limited laser spot.

Another PSHB storage method, the so-called photon echo optical memories, is based on time-domain storage [8,9]. Time-domain hole burning also uses spectral holes for data storage, but relies on coherent optical transient phenomena. Time-domain storage is realized by illuminating an inhomogeneously broadened material by two temporally separated resonant optical pulses. The first (reference) pulse creates an optical coherence in the material. The second pulse is temporally encoded with data and interferes with the optical coherence created by the reference pulse resulting in a frequency-dependent population grating. The recorded information is read out by illuminating the storage material with a read pulse (identical to the reference pulse) which generates a coherent optical signal having the same temporal profile as the data pulse.

The maximum number of bits that can be stored in a single spot using PSHB is given by the ratio of the inhomogeneous $\Delta\omega_i$, and homogenous $\Delta\omega_h$, absorption line width of the storage material. This ratio can range up to 10^7 in some materials. Because of the large number of frequency channels available, high storage densities may be possible with reasonable laser spot sizes. Even though both of these methods have advantage of increased storage density ($\geq$Mbit/laser spot), their application capability is significantly limited by the operating temperatures which should be kept extremely low around liquid helium temperature. In particular, present research efforts concentrate on achieving room temperature hole burning with novel materials, but there are still a number of technical challenges to overcome before PSHB becomes viable for data storage.

13.5 THREE-DIMENSIONAL OPTICAL INFORMATION STORAGE

In the present optical storage systems, one-dimensional serial information is stored in a two-dimensional medium. In three-dimensional optical memories, three independent coordinates specify the location of information. Multilayer [10–14], holographic [15–17], microholographic [18–23], and multidimensional [24–26] approaches are expected to exploit the entire volume of a storage medium. In addition to the spatial dimensions, optical multiplexing techniques allow superimposing multiple states that differ in wavelength, angle of incidence, phase, or polarization.

Optical storage in the form of holographic volume gratings has been investigated during the past three decades as a straightforward approach to realize three-dimensional high-density memories. Beside diverse holographic techniques, alternative solutions for three-dimensional optical memories are also under investigation. These include the extension of present disk systems to a layered format but also various new concepts of multilayer optical memories. In this case, the third dimension is introduced by recording the data in multiple layers through the thickness of a volumetric storage medium.

13.5.1 Multilayer optical information storage

The simplest way to use the third dimension of a storage medium is multilayer storage. Using multiple data layers instead of one, the overall storage capacity will grow linearly with the number of layers. Multilayer recording is a simple approach to higher density in optical data storage. Storing the data in multiple layers the third dimension of a disk, its depth, becomes useable for the optical disk technology and the overall data capacity linearly grows with the number of layers. A standard optical drive can easily address different depth positions while confocal filtering widely reduces interlayer cross talk. The success of dual-layer DVD-ROM has attracted interest but in conventional optical systems based on readout from the reflective layer, the multilayer approach has only a moderate potential to increase the storage capacity. It implies the costly production of multiple layer disks with rapidly decreasing tilt and flatness tolerances. Also there is a fundamental physical trade-off between the recording layer reflection (to generate the readout signal generation) and transmission (to optically access each data layer of a multilayer stack). The number of layers per side of a disk is limited strongly by higher optical power requirements, interlayer cross talk, and aberrations that appear while focusing on several layers at different depths simultaneously.

Optical recording in many layers primarily requires a homogeneous, low-absorption recording material while reflection mode is favored for confocal implementations of a standard optical pickup. Combined with other recording techniques, multilayer approach can become more attractive. In the case of two-photon or fluorescent memories that use transparent materials as storage media, the number of layers can become very large. Such quasi-three-dimensional optical memories use the volume of a storage medium by recording the data as binary planes stacked in three dimensions. The data are stored by discrete bits in the plane, but also through the volume (Figure 13.18).

Following the experimental advances in media and system concepts made in the last years, optical recording by two-photon excitation in photochromic as well as photorefractive media became very attractive as an alternative for three-dimensional optical memories. The modulation of the recording material is usually localized by using focused Gaussian beams. The laser beam induces a photochemical change in the focal range. Selectivity in addressing individual bit locations, that is, volume elements, is typically increased by nonlinearities of the laser–material interaction. Diverse nonlinear optical effects can be used for recording but the requirement for strong multiphoton absorption limits their practicability. Typical media sensitivity requires a strong laser-driven photon injection to initiate a chemical change in the medium. Upon excitation, a photochemical reaction leads to a permanent or reversible change in photochromic compounds of the medium. Chemical mechanisms such as photobleaching, photoisomerization, polymerization, and decomposition, have been

investigated for nonlinear optical recording. Depending on the photochemical reaction, the optically detectable change allow for permanent, reversible or "gray scale" recording where multiple gray levels can be addressed in a single data mark volume element.

Among nonlinear optical phenomena, two-photon absorption is particularly an attractive effect for multilayered optical recording. Two-photon excitation refers to the simultaneous absorption of two photons, whereby the excitation rate for this process is proportional to the square of the writing light intensity. Therefore, the excitation remains confined to the focal volume corresponding to the intensity distribution of a focused Gaussian beam. The basis of a two-photon recording system is the simultaneous absorption of two photons whose combined energy is equal to the energy difference between initial and final states of the recording material. This simultaneous absorption results in a structural phase transition that is reversible and detectable by measuring the fluorescence of the material. The read beam is unabsorbed and passes through the unwritten areas of the material while the recorded data marks will cause the absorption of the readout beam exciting the fluorescence at a longer wavelength.

A variety of materials have been proposed for two-photon recording. The most important material requirements concern the photochromism, that is, the ability to change the chemical

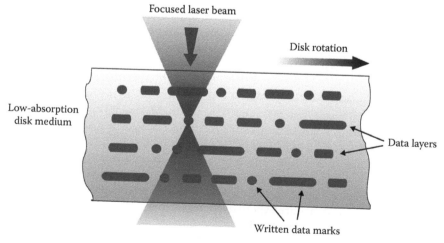

Figure 13.18 3D optical data storage disk. The disk medium is highly transparent so that the laser light can address many layers of information. Nonlinearity in the optical medium response localize the light–matter interaction to the high-intensity focal region.

structure under light excitation, fluorescence in one of two chemical states, stability of both states at room temperature, etc. Although two-photon excitation is the strongest multiphoton absorption mechanisms, it is still a very weak optical effect. Therefore, research efforts focus on chromophore components possessing high cross section for two-photon absorption.

A typical system involves two beams that are called data beam and address beam as shown in Figure 13.19. The data beam at 532 nm is modulated with a spatial light modulator (SLM) and focused at a particular plane within the medium. The addressing beam at 1064 nm provides the second photon required for the two-photon excitation. The data are written in the overlap region of the two beams and then read out by fluorescence when excited by single photons absorbed within the written bit volume. Hereby, the data beam is blocked and the read beam at 532 nm is focused to reconstruct the selected data page within the volume. The readout plane is then imaged onto a charge-coupled device (CCD). The spot size is limited by the recording wavelengths through diffraction. The approach promises not only ultra-high effective areal density but also parallel access to the stored data [27,28].

Besides the necessity for relatively high excitation energy, another critical issue in the realization of fluorescence-based multilayer memories is optical selectivity of readout. The read beam addressing an individual data mark simultaneously illuminates many other bits located at different layers. The read

beam light interaction with the medium can alternate or completely erase the written data. In addition, the mostly linear interaction involves a number of data marks and contaminates the readout. Optical resolution in 3D can be achieved by nonlinearity (e.g., two-photon microscopy) or by spatial filtering of the response (e.g., confocal laser microscopy). Detection of refractive index differences between the written marks and surrounding medium provides nondestructive readout but requires costly equipment such as phase-contrast microscope or confocal reflection microscope. Alternatively, nondestructivity can be achieved by using different absorption bands for recording and reading.

Advances in material science and technology allow the design and development of novel materials for optical recording. Tailored fabrication of nanostructured optical materials enables new recording and readout techniques. Based on multilayer memory concept, five-dimensional optical storage has been demonstrated in a recording material consisting of a plastic layer doped with plasmonic gold nanorods [26]. The longitudinal surface plasmon resonance (SPR) of gold nanorods is exploited to achieve a wavelength and polarization selectivity, whereas multilayer recording allows addressing three spatial dimensions. The readout mechanism is SPR mediated two-photon luminescence, which exhibits a high wavelength and angular selectivity necessary for crosstalk free readout.

Beside two-photon excited fluorescence, further optical techniques have successfully been used for

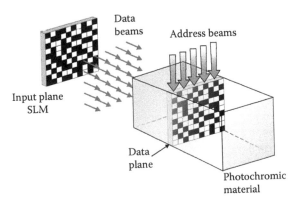

Figure 13.19 3D two-photon multilayer recording. 2D binary data planes recorded by two-photon absorption are stacked in the depth of a transparent storage medium. Readout relies on laser-excited fluorescence.

3D multilayer data storage. These include micro-holography and 3D submicron-resolved structuring of the medium by high-intensity laser pulses.

13.5.2 Holographic data storage

Three-dimensional optical information storage by volume holograms has been proposed first in 1963. One of the unique characteristics of optical volume storage is the very high bit packing density that can be attained. The ultimate upper limit of the storage density in three-dimensional storage media is of the order of $1/\lambda^3$, compared with $1/\lambda^2$ for surface of two-dimensional optical storage media. This results in 10^{12}–10^{13} bits cm^{-3}, although the practical limit set by other parameters of the optical system and by the constraints of the recording material may be lower than this.

Three-dimensional optical storage systems may generally be classified as bit oriented and page oriented. In holographic page-oriented memories, the information associated with stored bits is distributed throughout the whole volume of the storage medium. In bit-oriented memories, each bit occupies a specific location in three-dimensional space. Various approaches to realize a three-dimensional optical memory by bit-oriented storage have already been presented not only including holographic but also nonholographic. Such storage methods are based in general on a point-like or bit-by-bit three-dimensional recording by creating small data marks within the medium. Hereby, data marks represent single bits and are defined by large contrast in some optical property of the storage material. Using a nonlinear optical response of the material, the optical interaction can be confined to a focal volume. With a diffraction-limited, focused light beam, the physical size of bits can become as small as the wavelength of the laser beam in all three dimensions. Owing to the submicrometer or dimensions of a single bit, bit-oriented optical data storage requires very strict tolerances for the focusing optics and recording alignment.

In contrast to three-dimensional multilayered optical memories, in holographic storage, the information is recorded through volume. Recording is accomplished by interfering two coherent laser beams, the information-bearing signal beam and the reference or address beam. The resulting intensity pattern is then stored in a photosensitive material by inducing a grating-like modulation of its optical properties such as refractive index or absorption coefficient. The data are reconstructed by diffracting the address beam at the induced grating. A unique characteristics of thick holographic gratings is the Bragg selectivity, which allows many holograms to be stored overlapping by applying appropriate multiplexing methods. Compared to the conventional recording schemes, multiplexed optical recording provides an entirely new approach to higher information densities. Multiple individually addressable holographic gratings can be stored in the same volume, representing multiplexed data bits or pages.

13.5.3 Page-oriented holographic memories

Holographic memories usually store and recall the data in page format, that is, as two-dimensional bit arrays, which offers the way to realize high data rates and fast access. Combined with multiplexing, the inherent parallelism of holographic storage can provide a huge increase in both capacity and speed. For more than 30 years, holography has been considered as a storage approach that can change standards and prospects for optical storage media in a revolutionary manner. Depending on a number of supporting technologies, holographic memories became realizable with advances in photonics technology, particularly with improvements in liquid crystal modulators, CCDs, semiconductor detectors, and laser sources. Ongoing research efforts have led to impressive advances [17]. The first completed working platforms demonstrated high storage densities of more than 300 bits/μm^2, but they are still far from commercialization.

In contrast to the conventional optical recording where an individual data bit is stored as localized change in some optical property of two-dimensional storage media, holographic recording allows to store the data page-wise in the volume of the material. Instead of storing one single bit at each location, large data pages can be recorded and read out at once. The information to be stored is first digitized and then loaded onto an SLM as a two-dimensional pattern of binary ones and zeros. The SLM imprints that binary data page to the signal beam. The data are recorded by intersecting the signal beam with a reference beam inside the storage medium (Figure 13.20). The three-dimensional interference pattern induces a corresponding

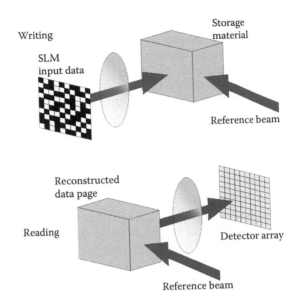

Figure 13.20 Page-oriented holographic storage. A two-dimensional data pattern created by SLM is stored by interfering signal and reference beam to record the hologram. One whole page is written at once, many pages overlap in the same volume. One page from many is read out with the corresponding reference beam.

spatial modulation of the refractive index of the recording material. Consequently, the data are stored as a refractive index grating representing a phase volume hologram.

As in page-oriented holographic storage, each data bit is distributed in three dimensions through the recording medium, there is no direct correlation between the data bit and a single volume element in the recorded structure. The stored data are retrieved by introducing the same reference beam used to record it, and read out by projecting the reconstructed signal beam onto the output detector array for optic-to-electronic conversion.

Due to the Bragg selectivity of holographic volume gratings, it is possible to store several holograms overlapping in the same volume element by either changing the angle or the wavelength of the reference beam. Such overlapping recording of multiple holograms in one single position is referred to as holographic multiplexing. In a multiplexing scheme, holographic structures are mixed together whereby the Bragg effect allows retrieval of an individual data page while minimizing cross talk from other pages stored in the same volume. The diffraction efficiency varies according to the mismatches in angle or wavelength between recording and readout. Deviations from the Bragg

condition lead to a rapid decrease in diffraction efficiency, which makes a selective reconstruction of multiple holographic gratings possible.

Various multiplexing methods have been proposed in diverse holographic storage systems. Wavelength and angle multiplexing result directly from the Bragg-selective character of thick volume holograms. The addressing mechanism here is the wavelength or angle of incidence of the reference beam. Both methods allow many holograms to be recorded in the same position but their practical impact is limited. Angle multiplexing requires complex optics, and the crucial component for efficient wavelength multiplexing is a laser light source that should be tunable in a sufficiently wide range. Such lasers are available but complex so that a significant increase of storage density by wavelength multiplexing only is difficult to realize in practical systems. New methods such as peristrophic or shift multiplexing have been developed for holographic storage systems which use a disk-shaped medium instead of a photorefractive crystal cube. Peristrophic multiplexing is based on the rotation of the plane wave reference beam around the optical axis; during the readout, the reconstructed holograms follow the motion of the reference beam. The address of an individual hologram is the relative rotational position of the storage medium. Shift multiplexing relies on a spherical wave as the reference beam while the signal beam is still a plane wave. Such spherical holograms can be shifted relative to the reference and stored next to another with a shift distance below 10 μm. The relative displacements are small enough that holograms in subsequent locations overlap significantly.

Volume holography is a powerful approach for digital storage systems with high densities (≥100 bits/μm²) and fast transfer rate (≥Gbitps). However, the practical realization of holographic memories suffers from the lack of suitable storage media. Both the performance and viability of systems under development are significantly limited by the characteristics of the available materials. To date, the requirement of adequate storage materials has been one of the most crucial aspects in the development of holographic memories. Indeed, there are very rigorous demands on storage materials, which should be satisfied to realize holographic memories as competitive, reliable optical storage systems of improved performance. The search for an optimum material to be used in holographic data storage does not appear to be finished yet. An ideal recording material should be of high optical quality, it should be able to hold the recorded data for a long time, and, for commercial applications and should be very reliable and not too expensive. Considering the physical processes, one can define a number of parameters to be controlled. The most important for the viability of holographic storage are high resolution (>3000 lines/mm), high photosensitivity, large dynamic range (i.e., diffraction efficiency of multiplexed data pages), archival storage time, low absorption, and low scatter. In particular, a large dynamic range of the storage material is necessary to allow multiplex recording of many holograms in the same volume.

Many kinds of materials have been investigated as holographic storage media. With sufficient material development efforts, the necessary optical quality has been achieved for both inorganic photorefractive crystals and organic photopolymer media. Photorefractive crystals such as lithium niobate, barium titanate, and strontium barium titanate were used previously in holographic systems. In the last years, a new class of photosensitive polymers has been introduced to satisfy the demand on adequate materials for holographic storage. Depending on the recording material,

different optical system architectures have been developed for holographic memories. Holographic storage media can be classified in two categories: thin photosensitive organic media and thick inorganic photorefractive crystals. Thin media (a few hundreds micrometers thick) are most sui for transmission type architecture using a variety of shift or phase multiplexing techniques, while angular multiplexing in various modifications is usually applied in thick media (about centimeter thick). Typical system architecture in this case is based on the 90° geometry, whereas thin phototable polymer layers are often used in a disk-based configuration. In this concept, digital holographic pages are stored on a disk-shaped medium and organized in tracks similar to those on conventional optical disks. The disk can rotate continuously and it can also move across tracks to allow the optical head to access the entire area of the medium. The storage medium is typically an organic photopolymer layer sandwiched between two glass plates.

13.5.4 Bit-oriented holographic storage

An alternative to page-oriented holographic memories is three-dimensional, bit-oriented optical storage on a holographic disk medium. The microholographic approach combines the bit-wise storage of CD/DVD and holographic volume recording, which makes it possible to advance the capabilities of conventional disk technologies by implementing spatial and wavelength multiplexing [18,19]. Microholography expands surface storage into three dimensions by storing the data as microscopic volume gratings instead of pits. Holographic recording is realized within a system that is in its main features very similar to the recordable CD or DVD systems. The technology provides volume storage of information while the areal structure of the stored data remains comparable with pits on a DVD. The microholographic storage concept benefits from both technologies: The bit-oriented storage allows for using many solutions of the highly developed CD/DVD technology. In addition, holographic recording offers a path to overcome the limitations of this technology, which are related to its two-dimensional nature.

In contrast to binary page-wise holographic data storage, the microholographic approach capitalizes on its fundamental compatibility with the established optical disk technology. The data are stored holographically in three dimensions but bit-wise in tracks and layers similar to those of a standard optical disk. The reflectivity of a photosensitive material, typically a holographic recording photopolymer, is locally varied by recording submicron-sized reflection gratings. Cross talk, typical in page-wise recording, is eliminated by the bit-wise nature of the recording and readout process: At any point in time, the focused light beam illuminates only one microholographic bit feature. Multiplexing techniques open additional paths to storage densities beyond the resolution limit imposed on 2D optical data storage.

For data recording, microscopic reflection gratings are holographically induced to vary the reflectivity of the disk by diffraction (Figure 13.21). The laser beam is focused into the photosensitive layer and reflected back with the mirror. The interference pattern of the incident and reflected beam induces a grating-like modulation of the refractive index of the storage medium. To retrieve the stored data, the original signal beam is reconstructed by reflection of the read beam at the induced gratings. Recording with sharply focused laser beams results in localized volume storage. The microgratings can be packed very densely and arranged in tracks similar to those on a CD. When the disk is rotating, microgratings of variable length are induced dynamically whereby grating fringes are extended in the motion direction.

The storage system (Figure 13.22) is very similar to the conventional optical drives: A focused laser beam is used for writing and reading, the data are stored bit-wise in tracks on the rotating disk, and similar systems for automatic focusing and track following are needed to control the position of the laser beam focus on the rotating disk. All these common aspects simplify the practical realization of the microholographic system as many components developed for the CD/DVD technology can be directly used or adapted for this purpose. The main difference here is the reflecting unit underneath the disk that is needed for writing. During readout, this unit, in the simplest realization an aspherical mirror, can be tilted or removed so that only reflection from the gratings will be detected.

The physical structure of the stored data is similar to the pit-land structure on a CD/DVD as stripe-shaped microgratings are written dynamically while the grating length corresponds to a coded bitstream. Multiplexing is integrated parallel into the coding scheme in order to maximize writing and reading speed. Each wavelength/layer channel corresponds to standard data channels in CD/DVD systems so that recording and error-correction code algorithms evolved in conventional technologies can be implemented. Being fundamentally a matching of the optical disk technology and holographic volume recording, the microholographic approach concomitantly offers reflective bit features and confocal optical pickup design, as well. In addition, photopolymer materials used as recording media allow a simple disk design with a monolithic photopolymer layer sandwiched between two substrates. Photopolymers developed for holographic storage possess the characteristics essential for multilayer recording including large thickness of up to 1 mm, low absorption for writing, sufficient refractive index change, and negligible absorption for reading.

Multilayer recording is a simple method of spatial multiplexing that relies on the depth localization and selectivity of holographic microgratings. Although multilayer recording in conventional optical disk technology requires a stack of physically distinct layers, the microholographic method allows a simple disk design consisting of a single homogeneous photopolymer layer. As photopolymers are substantially optically transparent, many microholographic layers can be stored through the depth of a disk. Starting from data densities comparable to DVD or Blu-ray, the microholographic multilayer approach targets the terabyte capacity range. At the same time, it allows a cost-effective and downward compatible technology implementation as the drive system has most optical and optoelectronic components in common with a standard optical drive. Multiple data layers are addressed by simple confocal movement to different depths within the photopolymer layer.

In addition to multilayer recording, the application of wavelength multiplexing would allow a linear increase of the storage density and also of the write/read rate with the number of wavelength used. Wavelength multiplexing can be realized by simultaneous recording of several gratings in the same position with write beams of different wavelengths. In this case, a complex periodical grating structure is induced that contains all single-color gratings. Due to the wavelength selectivity, each

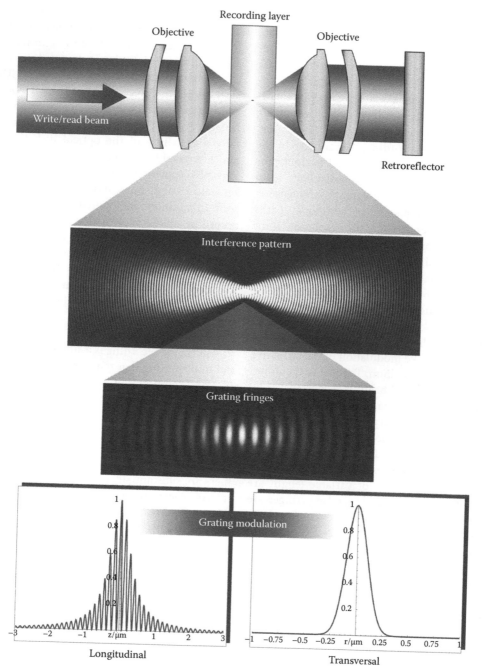

Figure 13.21 Microholographic recording in retroreflector configuration. (Top) The optical recording configuration is based on a single beam path: The "second" write beam is created by retroreflection to fully overlap with the incident write beam. Two identical objectives with high NA are adjusted to image the focal points of the incident and reflected beam at the same storage location. The grating formation takes place in the joint focal region of the two beams when a photosensitive polymer is exposed to their interference. (Middle) The interference pattern is plotted in the logarithmic scale to display the wave fronts. (Bottom) Driven by the linearity of the photoresponse, the index modulation mirrors the intensity distribution. With the given optical specification for high-density recording in the violet spectral range ($\lambda = 405$ nm, NA=0.75), microgratings effectively consist of less than 10 grating fringes.

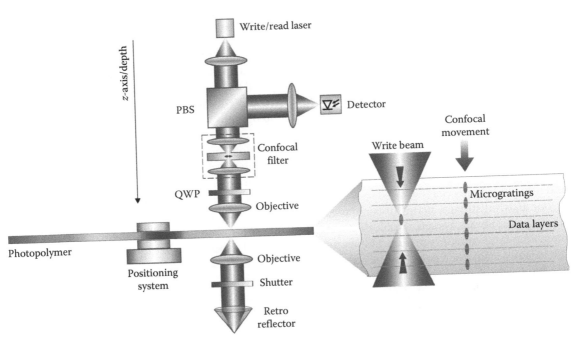

Figure 13.22 Optical write/read system for multilayer microholographic data storage. Essential features of the optical configuration primarily are single-beam path, diffraction-limited high-NA focusing, retroreflection, and confocal filtering.

laser beam with a certain wavelength detects only the corresponding grating during readout. The data rate is increased since all bits contained in one multiplex grating can be recorded and read in parallel.

The experimental results show that resolution-limited optical storage is also possible in 3D with submicron-sized volume gratings representing individual bit features [23]. The storage performance is governed by the interaction between the focused laser beam and photopolymer material. The micrometer-scaled depth localization of microgratings is the basis for multilayer recording. Depth multiplexing of microgratings recorded has been demonstrated by recording up to 75 layers spaced by $4\,\mu m$ [29]. These recordings achieved without any SA compensation evidence the potential of microholographic data storage to capitalize on multilayer recording for data capacities far beyond the state of the art. Furthermore, manufacturing technology can provide low-cost, removable media, and downward compatible systems. The key issues in the design and development of the optical system are 3D microlocalization of recording, selectivity, and sensitivity of readout.

Much research fails to recognize the importance and implications of the storage medium for system design. The microholographic disks may be successful as removable storage media that satisfy high-capacity demands in specific areas including data banks, archiving and security systems, image processing, and multimedia applications.

13.6 POTENTIAL IMPACT OF NOVEL TECHNOLOGIES

Optical systems for recording and retrieval of digital information represent a rapidly developing field with a huge potential to encompass entirely new applications and to provide solutions to problems arising from these applications. Continuous advances and discoveries in related technologies, devices, and materials have opened an entire spectrum of optical effects and materials that can be used to provide writing and reading mechanisms in an optical data storage system.

Trends toward rewritability and higher storage capacity have moved optical storage into competition with high-end magnetic storage. The key difference between these two technologies is the removability of optical media but also their excellent robustness, archival lifetime, and very low cost. An additional advantage of optical technology is the stability of written data, a feature that makes optical media suitable for archival lifetimes. An optical disk can be removed after recording and read out in any compatible drive which enables data storage separately from the main computer system or network. Typical applications range from archival storage, including software distribution, digital photographs and imaging, movies, and other video materials.

For archival storage, many disks are organized in an optical library system capable of storing and managing large amounts of data. Library systems are usually constructed as jukeboxes comprising hundreds of disks to provide high capacities for long-term storage. An important advantage of optical archiving systems is that the data are stored off-line, which releases computers or networks but also provides data security and retrieval even if the network is irreparably destroyed. Optical systems are already widely accepted in enterprise and institutional storage with applications ranging from extending existing server capacities to publishing and image storage. One of the most important applications for optical archival storage is document and image management—where documents and images such as receipts, X-rays, photographs, and other records are stored in the digital form on optical disks. Plurality of disks is then arranged in a database to facilitate rapid retrieval.

Optical disk technology across three generations has brought optical data storage to perfection at the physical limits of resolution and diffraction. CD, DVD, and Blu-ray rely on identical recording and readout processes while the differences originate from the format specifications. Despite impressive advances and continuous increases in storage density, the existing bit-wise, serial-access optical storage is far away from realizing the full potential of optical technology. With opening computer applications such as three-dimensional imaging, video mail and server applications including, for example, digital libraries, satellite imagery, medical document, and image archiving, optical storage should ensure capacities exceeding a terabyte per disk. Numerous techniques provide the ability to achieve high storage capacities by a more effective use of the volume of a storage medium and/or by taking advantage of additional degrees of freedom such as the recording/reading wavelength. Moreover, the inherent parallelism of optics offers the possibility to record and retrieve large data files with extremely high data rates compared to those achievable by electronics.

Three-dimensional optical storage by volume holography or two-photon recording hold promises for high-capacity, high-speed systems. In addition, microholographic disks or fluorescent multilayer disks that store the data bit-wise as length-coded marks can also satisfy the requirements for downward compatibility and low-cost media. Figure 13.23 gives a comparison of storage densities achievable in different technologies. A crucial aspect for the reliability of all these systems is the storage material itself. Many types of materials have been investigated in recent years as optical storage media including inorganic photorefractive crystals, organic photopolymers, and biological systems such as protein bacteriorhodopsin or DNA polymers. Progress in the last couple of years has been impressive, particularly in the field of photosensitive polymers that offer a wide variety of possible recording mechanisms including both write-once and rewritable media. In particular, a new class of photopolymer materials has been introduced for holographic storage. Optimization and further development of photosensitive polymeric media will be key to the success of this and other advanced optical storage technologies.

Page-oriented holographic memories hold the top of the table, promising terabyte devices and Gbit/s data rates, but it is questionable if they would be able to compete with the existing optical disks in daily life. Holographic storage can find important niche applications in professional storage, backup and archiving systems, and data banks, where large data files have to be stored and recalled with fast access. The possibility of associative retrieval enables holographic memories to be used as data search engines, that is, content-addressable database servers or large web servers.

In small end-user systems, the requirements will rather be governed by the convergence of entertainment and computing. From this point of view, bit-oriented optical storage offers more realistic solutions for the next generation. With the availability of adequate storage media, the microholographic disk technology might provide the successor to the blue DVD generation. The

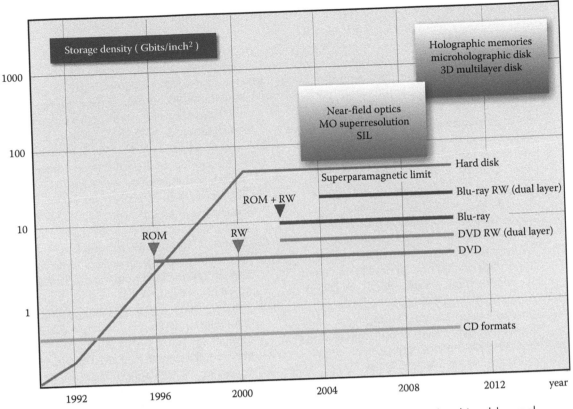

Figure 13.23 Potential impact of novel optical technologies in comparison with achievable areal densities in conventional optical storage systems and hard disk systems.

recording technique can improve performance significantly, reaching with ease the threshold of 100 GB per disk. Such advanced optical disks are asked for many applications and are particularly attractive for three-dimensional high-definition imaging, video and multimedia storage.

Although recent advances toward commercial devices are impressive, it remains to be seen which technology will be successful in providing the next generation optical storage media. A wide variety of materials as well as recording and readout techniques are under development. The requirements of foremost importance in optical information storage are high performance in terms of capacity and data rates, removability, and compatibility. The demand on downward compatibility is to ensure the use of universal drives, but also reliability and low-cost media. Meeting these requirements simultaneously is a major challenge to developers of novel optical technologies. Competition combined with an immense market potential and appetite for storage keeps the field exciting.

REFERENCES

1. Mansuripur, M. *The Physical Principles of Magneto-Optical Recording*, Cambridge, UK: Cambridge University Press (1995).
2. McDaniel, T. W. and Victora, R. H. *Handbook of Magneto-Optical Data Recording*, Westwood, NJ: Noyes Publications (1997).
3. Betzig, E., Trautman, J. K., Harris, T. D., Weiner, J. S., and Kostelak, R. L. *Science* 251, 1468 (1991).
4. Mansfield, S. M. and Kino, G. S. *Appl. Phys. Lett.* 57, 2615 (1990).
5. Terris, B. D., Mamin, H. J., and Rugar, D. *Appl. Phys. Lett.* 68, 141 (1996).
6. Parthenopolous, D. A. and Rentzepis, P. M. *Science* 245, 848 (1989).
7. McCormick, F. B., Zhang, H., Walker, E. P., Chapman, C., Kim, N., Costa, J. M., Esener, S., and Rentzepis P. M. *Proc. SPIE* 3802, 173 (1999).

8. Mossberg, T. W. *Opt. Lett.* 7, 7 (1982).

9. Bai, Y. S. and Kachru, R. *Opt. Lett.* 18, 1189 (1993).

10. Wilson, T., Kawata, Y., and Kawata, S. Readout of three dimensional optical memories. *Opt. Lett.* 21, 1003–1005 (1996).

11. Hunter, S., Kiamilev, F., Esener, S. C., Parthenopoulos D. A., and Rentzepis P. M. Potentials of two-photon based 3-D optical memories for high performance computing. *Appl. Opt.* 29, 2058–2066 (1990).

12. Cumpston, B. H., Ananthavel, S. P., Barlow, S., Dyer, D. L., Ehrlich, J. E., Erskine, L. L., Heikal, A. A., Kuebler, S. M., Lee, I. Y. S., McCord-Maughon, D., and Qin, J. Two-photon polymerization initiators for three-dimensional optical data storage and microfabrication. *Nature* 398, 51–54 (1999).

13. Wang, M. M. and Esener, S. C. Three-dimensional optical data storage in a fluorescent dye-doped photopolymer. *Appl. Opt.* 39, 1826–1834 (2000).

14. Kawata, S. and Kawata, Y. Three-dimensional optical data storage using photochromic materials. *Chem. Rev.* 100, 1777–1788 (2000).

15. Heanue, J. F., Bashaw, M. C., and Hesselink, L. Volume holographic storage and retrieval of digital data. *Science* 265, 749–752 (1994).

16. Homan, S. and Willner, A. E. High-capacity optical storage using multiple wavelengths, multiple layers and volume holograms. *Electron Lett.* 31, 621–623 (1995).

17. Coufal, H. J., Psaltis, D., and Sincerbox, G. T. (Eds) *Holographic Data Storage*, New York: Springer-Verlag (2000).

18. Eichler, H. J., Kuemmel, P., Orlic, S., and Wappelt, A. High density disk storage by multiplexed microholograms. *IEEE J. Sel. Top. Quantum Electron* 4, 840–848 (1998).

19. Orlic, S., Ulm, S., and Eichler, H. J. 3D bit-oriented optical storage in photopolymers. *J. Opt. A: Pure Appl. Opt.* 3, 72–81 (2001).

20. Day, D., Gu, M., and Smallridge, A. Rewritable 3D bit optical data storage in a PMMA-based photorefractive polymer. *Adv. Mater.* 13, 1005–1007 (2001).

21. Li, X. P., Chon, J. W. M., Wu, S. H., Evans, R. A., and Gu, M. Rewritable polarization-encoded multilayer data storage in 2, 5-dimethyl-4-(p-nitrophenylazo)anisole doped polymer. *Opt. Lett.* 32, 277–279 (2007).

22. Zijlstra, P., Chon, J. W. M., and Gu, M. Five-dimensional optical recording mediated by surface plasmons in gold nanorods. *Nature* 459, 410–413 (2009).

23. Orlic, S., Dietz, E., Frohmann, S., and Rass, J. Resolution-limited recording in 3D. *Opt. Express* 19, 16096–16105 (2011).

24. McLeod, R., Daiber, A. J., McDonald, M. E., Robertson, T. L., Slagle, T., Sochava, S. L., and Hesselink, L. Microholographic multi-layer optical disk data storage. *Appl. Opt.* 44(16), 3197–3207 (2005).

25. Dubois, M., Shi, X., Erben, C., Lawrence, B., Boden, E., and Longley, K. Microholograms recorded in a thermoplastic medium for three-dimensional data storage. *Jpn. J. Appl. Phys.* 45, 1239–1245 (2006).

26. Saito, K. and Kobayashi, S. Analysis of micro-reflector 3D optical disc recording. *Proc. SPIE* 6282, 628213 (2007).

27. Moerner, W. E. *J. Mol. Electron.* 1, 55 (1985).

28. Caro, C. D., Renn, A., and Wild, U. P. *Appl. Opt.* 30, 2890 (1991).

29. Orlic, S., Dietz, E., Frohmann, S., and Rass, J. Multilayer recording in microholographic data storage. *J. Opt.* 14, 072401 (2012).

14

Optical information processing

JOHN N. LEE
Naval Research Laboratory

14.1 INTRODUCTION AND HISTORICAL BACKGROUND

Optical technology has been developed for highly effective transport of information, either as very high speed temporal streams, e.g., in optical fibers or in free-space, or as in high-frame-rate two-dimensional (2D) image displays. There is, therefore, interest in performing routing, signal-processing and computing functions directly on such optical data streams. The development of various optical modulation, display, and storage techniques allows the investigation of processing concepts. The attraction of optical processing techniques is the promise for parallel routing and processing of data in the multiple dimensions of space, time, and wavelength at possible optical data rates. For example, in the temporal domain a 1 nm wide optical band at a wavelength of 1500 nm has a bandwidth of approximately 100 GHz, and temporal light modulators with 100 GHz bandwidth have also been demonstrated for optical fiber systems [1,2]. However, notional optical processing techniques can be envisioned that handle many such narrow-wavelength bands in parallel, and also operate in a combined spatiotemporal domain. Employing all domains simultaneously, it is theoretically possible to perform spatiotemporal routing and processing at an enormously high throughput in the four dimensions x, y, t, and λ. Throughput of 10^{14} samples/s would result from simple examples based on feasible modulation and display capabilities. In one case a 100 GHz temporal modulators can be combined with wavelength-selective devices to provide several hundred 1 nm wide channels at the operating wavelengths of existing photodetectors and light sources. A second example would consider that 2D spatial light modulator (SLM) devices can be constructed to have $>10^7$ pixels/frame (see Chapter 6) and that material developments allow optical frame update rates on the order of 1 MHz [3]. Unfortunately, although an optical processor operates on data in optical form, it is presently not possible to equate these maximal modulation and display rates to the expected information-processing throughput rates of such processors. There are penalties on the throughput due to necessary data pre-processing and post-processing in any information-processing system. These include the need to format and condition the input data to a processor, to compensate for shortcomings of any analogue signals (e.g., nonuniformities in space and time), and perhaps most importantly, to examine the processor's output data and extract the useful information. The latter is often an iterative process and requires fusion with other data processing results. An optical processor's speed advantage could be largely negated unless all processing operations can be performed at speeds commensurate with modulation and display rates. Thus, equally important considerations are the need to identify those operations that can be effectively performed optically, and the need to develop optical processing architectures that minimize the penalties on optical throughput. Because of these considerations optical information-processing approaches have covered a wide range of topics. Therefore, we first provide a brief review of the various paradigms that have been investigated in optical processing.

14.1.1 Analogue optical processing

In the oldest paradigm, optical data in analogue form can be manipulated to perform useful functions. The classic implementation of an analogue spatiotemporal processing function is the use of a simple lens to produce, at the back focal plane of a lens, the complex Fourier transform (both phase and amplitude) of optical data at the front focal plane [4]. In the most common and simplest configuration, shown in Figure 14.1 a one-dimensional (1D) object with complex transmission,

$$t(x_0) = a(x_0) \exp[-j2\pi b(x_0)] \qquad (14.1)$$

where a and b are the amplitude and phase value at pixel location x_0, is positioned at a distance d in front of a lens of focal length f_1. Illuminating the object with coherent light of wavelength λ, one obtains at one focal length distance behind the lens the amplitude distribution [4]

$$U_f(x_f) = \frac{A \exp[(j \pi/\lambda f_1)(1 - d/f_1)x_f^2]}{j\lambda f_1}$$

$$\int_0^L t(x_0) \exp[-j(2\pi/(\lambda f_1))(x_0 x_f)] \, dx \qquad (14.2)$$

where L is the spatial extent of $t(x_0)$, A is a constant, and the subscripts 0 and f are used to denote the object and output space, respectively. When

$$f_1 = d \qquad (14.3)$$

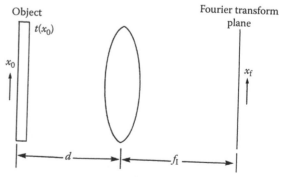

Figure 14.1 Optical arrangement for Fourier transformation with a single lens.

aside from the finite spatial limits of the integral, a Fourier transform of $t(x_0)$ results (to within a constant multiplicative factor). The 1D example given in Equation 14.2 employs a cylindrical lens. Use of a spherical lens results in a 2D Fourier transform, but since the x and y variables in the exponential term of the Fourier transform are separable, a spherical lens can be used even for the 1D data in Figure 14.1. Alternative lens configurations can produce the Fourier transform but require two instead of one lens, or produce the correct amplitude of the Fourier transform but with a curved phase front [4].

The core idea of Fourier transformation by a lens is the basis for many demonstrations of single-function optical processors. Electro-optic and acousto-optic modulation devices have been developed for optical input and output at the focal planes. To exploit this powerful concept, other mathematical integral transforms have often been re-cast as Fourier transformation problems, in particular the matched filter operation, or equivalently the correlation integral

$$R(\tau) = \int s_1(t)s_2^*(t+\tau)\mathrm{d}t. \qquad (14.4a)$$

According to the convolution theorem, Equation 14.4a is mathematically equivalent to the Fourier transform of the product of the instantaneous Fourier transforms of the two signals, $S_1(\omega)$ and $S_2^*(\omega)$, i.e.,

$$R(\tau) = \int S_1(\omega)S_2^*(\omega)\exp(j\omega\tau)\mathrm{d}\omega. \qquad (14.4b)$$

Alternative transform kernels to the Fourier Transform have been developed, and implemented using computer-generated holograms rather than lenses [5]. Processing applications involving massive amounts of linear processing have been successfully addressed with prototype analogue signal-processing systems that perform temporal spectral analysis [6] and correlation [7,8]. However, the optical processing performed in these systems implementations are linear operations involving only multiplication and addition. These analogue processors therefore address an application niche in signal processing, and because they are analogue, must contend with accuracy and dynamic

range concerns. Single-purpose analogue computation engines, described in the references, have been embedded into conventional electronic processing systems to accelerate specific signal-processing tasks, but not to perform general computations. It has often been the case that insertion of such engines leads to bottlenecks due to the optical-to-electronic and electronic-to-optical conversions, and the incompatibility of optical processor speeds with electronic computer limits such as memory access time. An approach that potentially relieves these bottlenecks involves "all-optical" systems whereby several analogue optical processing modules are cascaded, either via free-space or guided-wave optical paths or interconnections. At the end of the processing cascade the output data rate is presumably reduced to a manageable bandwidth for high-speed electronics. However, one problem with all-optical schemes is the need to store intermediate processor results and to access memory for subsequent processing. Advanced optical memories could be employed to enable all-optical schemes. Section 14.2.4.3.3 and Chapter 3.4 discuss advanced optical storage concepts such as page-oriented memories and holographic 3D memories, using SLMs and materials such as photorefractives, as described in Section 14.2.3.2.

14.1.2 Numerical optical processing

To obtain more general applicability of analogue optical processors, various encoding techniques have been investigated to improve the dynamic range and accuracy of such processors, and to overcome limitations of a processor that performs only multiplication and addition. Logarithmic encoding has been employed to compress dynamic range requirements into a smaller analogue voltage range, and to address the difficulty of performing the division operation optically, i.e., by converting division to an addition/subtraction problem. Numerical optical processing for potential implementation of an optical computer has also been considered. Approaches such as residue arithmetic [9] and digital multiplication by analogue convolution (DMAC) [10,11] have been explored extensively. However, all such approaches involve many nonlinear, logic operations that have to be performed optically to avoid numerous conversions between the optical and electrical domains. Consequently, many all-optical (optical

IN, optical OUT) switch devices have been investigated, for performing logic operations, such as nonlinear Fabry–Perot etalons and the self-electro-optic effect device (SEED) which is based on the quantum-confined Stark effect [12]. Significantly reduced power consumption in performing these nonlinear operations has been achieved with SEED devices [13] compared to early switches. However, no approach has, to date, resulted in an optical numerical computer competitive with electronic numerical computers. Fundamental arguments have been forwarded about the disadvantages of performing numerical operations optically, such as the large minimum-power requirement to perform nonlinear operations optically compared to electronically, even with SEED devices [14], the limits on integration density in such processors due to lower limits on the size of optical processors due to the wavelength of light [15], and the lower signal-to-noise of optical signals compared to electrical signals because the fundamental shot-noise power limit of optical signals is larger (for the same bandwidth) than the Johnson (or thermal) noise limit for electrical signals.

14.1.3 Optical interconnections/ optics in computers

An alternative paradigm for optical implementation of computational algorithms (vice the more familiar arithmetic formulation) is to recast particular signal and image processing algorithms in terms of routing of optical data among various nodes according to a particular interconnection scheme and with various weighting factors for data that recombine at nodes. This paradigm is similar to that used in neural-net formulations of processing problems [16]. The data nodes are "neurons". At these nodes only multiplication and addition are required, the former for applying weighting factors, and the latter at the recombination nodes, respectively. A simple threshold operation at the recombination node(s) residing at the output plane is often required. The advantage of using optics to perform the routing function is based on the free-space propagation characteristics of optical beams. Different optical channels (at reasonable power levels, so nonlinear effects can be ignored) can be routed in free space without interfering. Unlike physical wires, optical channels can overlap within the same space.

A data-routing paradigm can be used to describe the basic Fourier transform operation by a lens or a hologram, as has been described above and illustrated in Figure 14.1. The Fourier Transform is effected as a global optical "interconnection"; the interconnections are global since every input datum (or pixel) is connected to every output (transform) datum, and the input weighting factors are the amplitude and phase values at the pixel locations at front focal plane. A more general formulation involves both global and non-global interconnection, such as in neural-network formulations where layers of "neurons", or simple processors that perform addition and multiplication of data, are interconnected. The multiplicative weighting factors applied at various neurons can be adaptively adjusted to solve signal-and image-processing problems [16]. Optics has been explored for implementation of specific neural net algorithms [17,18]. The dynamic range and accuracy requirement for the additions and multiplications in neural net algorithms is generally not high, and therefore can be performed adequately with analogue data in an optical implementation. However, while optics has potential to implement massively parallel neural networks that might be difficult for electronic implementation, the maturity of neural network theory at this time has not yet progressed to the point where neural processor performance is limited by the size and interconnectivity possible with electronically implemented neural nets. Hence, the intrinsically large optical interconnection capability is not yet needed. Therefore, the alternate development of optical routing and processing capabilities has been towards enhancing the capabilities of computers and telecommunication networks (see Chapter 2). All-optical digital switching fabrics for interconnection of large fiber-optic networks have been addressed with arrays of SEED devices [19] that provide the logic operations needed for switching. While switching fabrics have been successfully demonstrated [20], the scale of fiber-optic telecommunication switching needs have not yet called for such ultra-high-bandwidth fabrics. Hence, optical interconnection has concentrated on use with within electronic computers, vice optical computing, and towards interconnecting large arrays of sensors and their associated electronic processors.

One avenue of research for using optics in computers has been to explore simply the interconnection among and within all-electronic computing elements in conventional and novel multiprocessor architectures. A second avenue has been to optically perform necessary switching, routing, and pre-processing operations for individual fiber links and networks of computing elements. Many investigations have addressed advanced concepts for interconnection and storage to implement novel multiprocessor and networked computing architectures [21]. Expectations have been driven by the fact that optical communications and optical storage have been the two most successful commercial optical technologies. However, these commercial technologies usually function exactly as in corresponding electronic subsystems. Therefore, significant research is required to adapt commercial technologies to fully exploit the interconnect potential of optics.

Optical interconnection schemes for computers range from computer-to-computer, box-to-box, card-to-card backplane, and finally chip-to-chip within a circuit card. The first of these already exists in several forms commercially; hence, research has concentrated on the latter applications. Guided-wave (also known as lightwave) interconnections are used for high throughput, low power and low bit error rate. The application is generally in novel multiprocessor computer configurations that focus on reducing or eliminating bottlenecks in conventional computers such as relating to memory access time. Free-space optical interconnections can be used advantageously as the number of processors increases, and they allow three-dimensional (3D) multiprocessor configurations [22]. For circuit-card and chip interconnections free-space optical beams do not have the limitations of physical wires, so circuit cards and elements may be laid out with more degrees of freedom.

14.1.4 Interconnection processing/ in-fiber processing

In addition to interconnecting network nodes for computer or communication networks, one can consider processing data residing within the interconnect path. Such processing is generally feasible only for fiber or guided-wave interconnections. The data may be either in digital or analogue form, the former for conventional digital computer and telecommunication systems, and the latter for arrays of fiber-optic sensors.

While the type of processing will vary depending on the nature of the optical fiber systems, the processing load can be expected to scale with the bandwidth of the data conveyed. With optical modulation rates now in excess of 100 GHz (see Section 4.2.1), commensurate processing speeds are required. Faster processing will allow more flexibility and capability in fiber systems, such as routing and switching a larger number of digital data channels. Use of conventional all-electronic approaches imposes a need to perform optical-to-electronic and electronic-to-optical conversions. Such conversions add complexity, and high-speed electronics will tend to be power hungry and may not provide the throughput required, e.g., due to electronic processor latencies. Hence, it is attractive to perform processing directly on the optical data stream at the transmission rate, e.g., in-fiber processing (see Section 14.2.4.3.3). The possibility of manipulating the information stream within an optical fiber has been made possible by (1) the development of optical fiber amplifiers, (2) the capability to build long fiber delay lines needed for short-duration buffer storage and for implementation of tapped delay lines, and (3) fiber couplers for tapping into delay lines and forming devices such as interferometers.

High-bandwidth digital and analogue data on fiber interconnects are often comprised of a large number of lower-bandwidth channels. There is need to multiplex and de-multiplex these channels. Digital systems may eventually require optical implementation of functions now performed electronically; these include clock recovery and reading of packet headers to enable packet switching and routing. However, if optical de-multiplexing is possible, electronic means for clock recovery will usually be available at the lower channel bandwidths. Routing of de-multiplexed channels requires reading of digital packet header information. As has been noted above, it has been difficult for optics to implement a digital computer; thus it is presently difficult to optically perform the logic needed to read packet headers and perform packet switching. However, it appears attractive to optically switch/rout high-bandwidth, fully-multiplexed, light streams, since these can usually be switched at rates much lower than the data bandwidth, and therefore avoids optical-to-electronic conversion at the signal bandwidth. Optical fiber networks (see Chapter 2) also can possess unique characteristics that require corresponding new processing functions. A major unique aspect is the wavelength-division multiplexing (WDM) capability of most optical networks, which then requires functions such as wavelength selection and amplitude equalization among all the wavelength channels in a WDM system.

Optical processing involving any of the various paradigms described above is practical only if materials and technology exist to construct devices that effectively implement the desired processing operations. The early literature generally showed architecture developments that required device performance exceeding capabilities at the time. Experience has shown that the former must work in conjunction with the development of the latter. Therefore, in the following we describe materials and devices that have been important to the development of optical processing techniques. Specific applications are described immediately after descriptions of device performance.

14.2 OPTICAL DEVICES AND PROCESSING APPLICATIONS

The needs in optical processing for optical modulation, display, storage and routing are shared with many applications, covered in other chapters in this Handbook (see Chapters 6, and 13). However, optical processors must not only efficiently modulate information onto optical beams, but must also rapidly manipulate the optical data to perform useful information processing. The needs for large temporal bandwidths and high-frame-rate 1D and 2D spatiotemporal optical modulation are clear, and are covered in Sections 14.2.1, 14.2.2 and 14.2.4 respectively, below. In addition, wavelength multiplexing can be performed in conjunction with both temporal and spatiotemporal processing, and various wavelengths can be selected or rejected out of either temporal or multidimensional data streams. Various wavelength-processing techniques are discussed in Section 14.2.3.

14.2.1 Temporal modulation and processing

Processing rate can be maximized via either increase in single-channel modulation rate or increase in number of modulated channels.

High-speed single-channel modulation can be done either by direct modulation of a semiconductor laser/light emitting diode (LED) or by using modulation devices external to the laser. The latter affords more flexibility to address a range of applications, although it is bulkier than the former. The former is attractive for their simplicity and high electrical-to-optical efficiency. However, modulated LED output power is limited, and there is a theoretical limit of approximately 30 GHz for high-bandwidth modulation with laser diodes, due to relaxation oscillations of charge carrier density in the laser cavity [23].

Temporal modulation with external devices achieve the highest speeds and lowest power in optically-guided lightwave structures such as fiber and channel waveguides. Either phase or intensity modulation devices can be employed.

The basic guided-wave structures for external phase modulation are the channel waveguide and the waveguide splitter where a channel waveguide branches out into two guides in the form of a "Y". Two waveguide splitters can be combined to form the guided-wave version Mach–Zehnder interferometer, as shown in Figure 14.2. Application of a voltage, V, across a channel waveguide fabricated in an electro-optic material such as $LiNbO_3$ will alter the optical path length, or equivalently, the phase of light passing through the waveguide due to the electro-optic effect [24]. If desired, the phase-modulated light can be converted to intensity-modulated light by using the guided-wave Mach–Zehnder and, most commonly, modulating the light in one of the two guides. High-speed modulation requires the voltage signal to be applied to a transmission line running parallel to the optical waveguide. The transmission line, which replaces the electrodes in Figure 14.2, is designed so the instantaneous voltage signal of the transmission line travels at close to the same velocity as the guided optical wave, greatly increasing modulation efficiency. This design approach has produced intensity-modulation devices with speeds and bandwidths of up to 100 GHz at less than 6 V drive signal [2].

The Mach–Zehnder modulator can be used for analogue intensity modulation. The output intensity as a function of V is a nonlinear but well-known function:

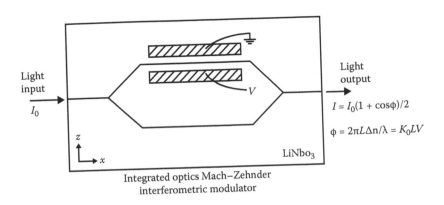

Integrated optics Mach–Zehnder
interferometric modulator

Figure 14.2 Guided-wave Mach-Zehnder interferometer.

$$I_1 / I_0 = 1 \pm \{\sin \pi(V / V\pi)\}, \qquad (14.5)$$

where $V\pi$ is the voltage to induce a π phase shift in the light beam, I_0 is the incident light beam intensity and I_1 is the output intensity. The sine-squared function can lead to spurious responses in a broadband signal. The production of spurious third-order signals can be seen from a Taylor-series expansion of the modulated quadrature-point signal

$$1 \pm \sin(\pi V(t) / V_\pi)$$

As

$$1 \pm (\pi V(t)/V_\pi)$$

$$\pm \{(1/6)(\pi V(t)/V_\pi)^3 \pm \text{higher order terms}\}. \quad (14.6)$$

Hence, if there are two frequency components f_1 and f_2 in the broad band signal, third-order intermodulation signals at $2f_1 - f_2$ and $2f_2 - f_1$ will appear within the band [25]. The effects of the sine-squared transfer function can be minimized by operating two Mach–Zehnder interferometers in parallel or back-to-back [26,27], but generally at an increase of optical insertion loss.

Fiber-optic versions of a Mach–Zehnder interferometer can be constructed, as shown in Figure 14.3a. Fused fiber couplers are at both ends of the interferometer legs to split and recombine the light beams from the single-mode fiber. Since silica fiber is not an electro-optic material, an additional component such as a piezoelectric cylinder is used to impose phase shifts onto the light in one leg

(Figure 14.3a) by increasing/decreasing the length of the fiber. Use of a piezoelectric element maximizes the possible phase change; however, it also limits the modulation speed of such devices.

Intensity modulation can be achieved using electroabsorption in III–V materials such as GaAs and its ternary and quaternary compounds with In, Al, and P [28]. In electroabsorption the optical density of materials whose bandgap is closely matched to a laser wavelength changes with applied voltage. In a device such as the electro-optic Mach–Zehnder device, the modulation voltage changes the optical density (imaginary part of the index of refraction), vice the optical phase (real part of the refractive index) as described above. Use of multiple quantum-well (MQW) materials, with an engineered exciton-line band edge, allows devices with faster operation at lower voltages and with higher signal-to-noise than the classical Franz–Keldysh effect in materials like bulk GaAs. MQW devices consist of a stack of ultra-thin, ~10-nm thick, layers of an alternating composition of III–V materials. Alternating layers of GaAs and AlGaAs are used for modulation at 850 nm wavelength. Alternating layers of quaternary materials such as InAlGaAs/InGaAs are used for operation at 1550 nm. Excitons produced by illumination of the MQW material have a much sharper and narrower exciton-absorption band than in bulk non-quantum-well material. Application of the electric field moves the exciton absorption line towards the red, causing change in the absorption at a fixed laser wavelength. Maximum optical contrast requires operation at the specific wavelength close to that of the bound exciton line in the MQW structure. The contrast change can also be greatly increased by placing the

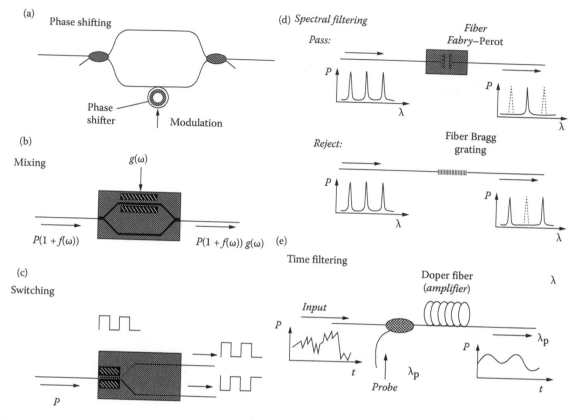

Figure 14.3 In-fiber optical processing functions: (a) Phase shifting in a Mach–Zehnder device constructed with fused fiber couplers, (b) frequency mixing of two signals $f(\omega)$ and $g(\omega)$, (c) switching of optical signal between two fiber channels, (d) spectral filtering using fiber Fabry–Perot or Bragg grating devices, (e) time filtering of a temporal signal using a probe beam at wavelength λ_p.

MQW stack within a Fabry–Perot cavity [29], so the incident light makes multiple passes through the stack. For maximum modulation speed a transmission line electrode must be used to maintain synchronism between the modulating signal and the optical signal, as described above for the Mach–Zehnder phase modulator. Since the modulation is due to shifting of a band edge, the modulation is highly nonlinear with respect to drive voltage. Hence, this type of modulator has been primarily used for digital modulation.

It is important to remember that speed, linearity, and efficiency are not the only parameters that need to be considered for temporal modulators. Other parameters that can be important include the capability to handle high optical power, and the device response to variables in the optical beam properties such as polarization state and wavelength. Hence, while the discussion above has concentrated on guided-wave devices, depending on

application one may need to consider slower and larger temporal modulators such as acousto-optic devices and micro-mechanical devices such as described in Sections 14.2.2.1 and 14.2.2.2 below.

14.2.1.1 TEMPORAL PROCESSING APPLICATIONS IN FIBER OPTIC SYSTEMS

14.2.1.1.1 Microwave optics and beamforming

Mach–Zehnder waveguide interferometer devices are crucial in the area of microwave optics. The most basic goal in microwave optics is transmission of high-bandwidth analogue signals. Microwave signals are modulated onto an optical carrier for transmission by a fiber-optic line; the advantage of optical transmission is the lower size and weight compared to the use of electrical transmission lines and microwave waveguides.

However, care must be taken in order to maximize analogue dynamic range, such as by minimizing third-order intermodulation signals, as discussed above for Equations 14.5 and 14.6. Areas of application for microwave-optic transmission include

- Antenna remoting for radar and communications systems, to minimize the weight of transmission lines when antennas must be at some distance from the receiver/transmitter, such as at the top of a mast
- In wireless networks, for feed lines from base stations to the cellular-network antennas

In addition to signal transmission, coherent optical devices can be constructed for antenna-array applications such as beamforming. In simplest terms, beamforming is the maximization of the transmit power of an array of antennas (for a transmitting array, or receive power for a receiving array) at particular look angles and frequencies. Minimization of receive sensitivity can also be an objective for rejection of spatially distributed interference. Maximization/minimization of transmit/receive power can be achieved by applying the correct phases to the antenna signals to produce constructive/destructive interference in the desired directions. For microwave-optic devices, the optical phases can be adjusted to produce the desired beams. The attraction of optical beamforming at microwave frequencies is the present difficulty in performing the necessary operations digitally at this high frequency range. To illustrate the beamforming operation we consider the simple case of a 1D linear array of N equally-spaced antenna elements. These elements can be either transmitters or receivers of a microwave signal at frequency f_0.

The phase difference of the signals from/by adjacent array elements at look angle a measured from the perpendicular to the line connecting the array elements is

$$\exp[j2\pi f_0(d\sin\alpha)/(v)]$$

where d is the array-element spacing, and v is the signal velocity. The phase shift at the nth element relative to the first is therefore

$$\exp[j2\pi f_0(nd\sin\alpha)/(v)]$$

If the signal strengths at the antenna elements x_n are sampled at a given instant of time and summed, one obtains (with wavelength $\lambda = f_0/v$)

$$b = \sum_{n=0}^{N-1} x_n \exp[j2\pi(nd\sin\alpha)/(\lambda)] \qquad (14.7)$$

To obtain signal gain, each term must be multiplied by a phase factor. For reasons soon to be apparent, the phase factor is chosen as

$$\exp[-j2\pi nk/N]$$

The new summation is

$$b_k = \sum_{n=0}^{N-1} x_n \exp[j2\pi(nd\sin\alpha)/(\lambda)]\exp[-j2\pi nk/N]$$

$$(14.8)$$

which is maximized when the exponential factors in every term cancel to produce unity, i.e., when

$$k = (Nd\sin\alpha)/v$$

The quantity k is thus a phase corresponding to a beam at look angle α. The choice of phase factor for producing the beam b_k is be seen to be equivalent to a spatial discrete Fourier transform (DFT) operation on the signal samples received from/by the transmit/receive array, respectively.

The above beamforming example applies best to a continuous-wave signal with small fractional bandwidth centered at microwave frequency f_0. If the microwave signal is short, e.g., an impulse function, as commonly for radar, the above beamforming algorithm is not optimum, since the signal has large fractional bandwidth $\Delta f/f_0$. In this large bandwidth case, time delays rather than phase shifts need to be applied to the array elements to produce a beam $B(t)$.

$$B(t) = \sum_{n=0}^{N-1} a_n x_n(t-\tau_n) \qquad (14.9)$$

where τ_n is the delay applied to the nth element in an array of size N, and a_n is a weighting function for shaping the angular characteristics of the beam, e.g., half-intensity mainlobe width and sidelobe levels [30]. The required time delays between adjacent elements is given by

$$\tau_{n+1} - \tau_n = (d \sin \alpha) / v. \qquad (14.10)$$

A fiber delay line for each array element can be used to implement the required delays, but this approach lacks the flexibility to easily change the beam direction. A novel alternate approach is to use different optical wavelengths for signal transmission to each array element. The wavelength-dispersion characteristics of the optical fiber, where different wavelengths will have different time delays over the same length of fiber [31], then contribute to the time delays between adjacent array elements.

The 1D example generalizes to unequal spacings between array elements and to 2D arrays in a straightforward manner.

14.2.1.1.2 Optical sensors

The Mach–Zehnder device, either fiber or waveguide version, is also an important class of device for sensors (see Chapter 11). For example, one leg of a fiber Mach–Zehnder is coated with a material that will change the length of the fiber in response to a specific external stimulus such as stress/strain, temperature, magnetic field, etc. Both the waveguide and fiber-optic Mach–Zehnder interferometers can perform a number of temporal operations on the optical signal: phase shifting (Figure 14.3a), mixing of a modulated signal with another (Figure 14.3b—note that the modulated input signal must be applied with a bias), switching of the polarity of signals (Figure 14.3c), spectral filtering using either a fiber Fabry–Perot or a Bragg grating (Figure 14.3d), and sampling or time filtering of high-bandwidth signals (Figure 14.3e). Note the necessity for optical amplification, and its implementation in the optical domain with fiber amplifiers, as shown in Figure 14.3e, where the fiber is doped with material such as erbium and the probe beam can also activate the amplification. When a similar fiber-optic Mach–Zehnder is used for demultiplexing and demodulation of sensor array data, higher accuracy is achieved because of compensation for the nonlinear $(1 + sine)$ transfer characteristic of a single Mach–Zehnder device, as already discussed above for Equations 14.5 and 14.6.

14.2.1.1.3 Optical interconnect for computers

Parallel high bandwidth optical channels, either guided-wave or fiber-optic, have been explored for optical interconnection applications [32], thereby multiplying the throughput rates that can be achieved on a single channel. However, simple aggregation of a multiplicity of single-channel optical-link hardware quickly becomes impractical, especially in applications where volume and power must be limited, e.g., within a computer. Fiber-optic "ribbon" cabling, consisting of a number of closely spaced fibers, is thus usually chosen to interconnect a number of high-speed processors or boards. Challenges in producing such optical interconnect hardware include the monolithic integration of arrays of lasers, modulators, and detectors. Achieving the requisite density for laser arrays that can be coupled to fibers has necessitated the development of vertical-cavity surface-emitting laser arrays (VCSELs) with low power threshold for lasing and high electrical-to-optical efficiency [33]. The required laser efficiency has been achieved through use of quantum well materials that reduce the number of allowed excited states that must be optically pumped [34].

14.2.1.1.4 Wavelength processing in fiber optic systems

WDM technology has been a major development in optical fiber telecommunications systems. Figure 14.3d illustrates how wavelength channels can be passed or rejected using, respectively, fiber devices such as the fiber Fabry–Perot filter, a mechanical device (Chapter 8.5), and fiber Bragg gratings, refractive index gratings written perpendicular to the long dimension of the fiber via the photorefractive effect. While WDM technology is primarily used in long-haul telecommunications to increase capacity without commensurate increase in physical plant, WDM technology can also be applied to short distance optical interconnects to provide an additional degree of freedom. For example, in ribbon fiber interconnects, each fiber can have a separate wavelength. The wavelength of a channel can be used as an identifier for data routing, reducing the need, or even avoiding the need to read the packet header (a difficulty for optics mentioned in Section 14.1.4). A key enabler of WDM optical interconnects is the monolithic multiple wavelength surface emitting laser array [33], which provides either a unique wavelength for each laser, or fewer but redundant wavelengths.

14.2.1.2 FREE-SPACE OPTICAL APPLICATION

Nonguided-wave temporal modulators can be applied to free-space communications, distributed computing architectures, laser radar, and laser designators. In the latter the use of high bandwidth coding is often needed for increased detection margin or range, using correlation methods, mathematically described in Section 14.2.2.1. For free-space optical communications bulk electro-optic modulators often require high drive powers and have low contrast. However, a novel quantum-well modulator device holds promise for both high speed and high contrast [35]. The modulator is positioned on the entry/exit facet of a corner cube retro-reflector. Electrical signals applied to the retro-reflector result in modulation of an incident laser beam. Data rates of up to 10 Mbps over several meters at bit error rates of 10−6 have been demonstrated; the approach has the potential for hundreds of megabits/second at power consumptions below 100 mW. Use of two retro-reflectors allows a two-way link to be established.

14.2.2 One-dimensional spatial light modulators and applications

14.2.2.1 ACOUSTO-OPTIC DEVICES

Optical modulation of data in a 1D spatial format allows parallel processing of blocks of temporal data. The most effective 1D light modulators have been acousto-optic, and this technology has played an important role in the demonstration of optical processing architectures. The basic construction of an acousto-optic device is shown in Figure 14.4. A high-bandwidth rf drive signal is applied to an acoustic, piezoelectric, transducer that has been bonded to one end of the acousto-optic cell, using acoustic impedance-matching materials. The resultant acoustic wave is a replica of the rf drive signal. The rarefactions and compressions of the acoustic wave produce corresponding refractive-index changes due to the elasto-optic effect. The cell thus contains a travelling phase grating corresponding to the acoustic wave. The main features of acousto-optic diffraction important to optical processing are now summarized.

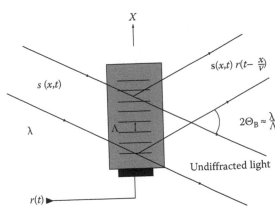

Figure 14.4 Construction of acousto-optic cell and Bragg cell geometry for data multiplication.

- The maximum diffraction efficiency, into a single order, occurs when the difference between the momentum vectors of the incident and diffracted light is equal to the acoustic-wave momentum vector. For an isotropic material momentum matching occurs when light is incident to the acoustic wavefront at the Bragg angle θ_B, defined by

$$\sin\theta_B = \lambda/(n\Lambda) = \lambda f/(2nv) \qquad (14.11)$$

where Λ, f, and v are the acoustic-wave wavelength, frequency, and velocity, respectively, λ is the optical wavelength of the monochromatic incident light, and n is the index of refraction of the medium at λ. To diffract light predominantly into only a single diffraction order one must examine the quantity Q,

$$Q = \frac{2\pi\lambda Z}{n\Lambda^2}.$$

For $Q > 7$, analysis shows that the first diffraction order contains >90% of the diffracted light [36].

- The diffraction efficiency is given by a nonlinear sine-squared relationship, that arises from considering the diffracted (I_1) and undiffracted (I_0) light as two modes of a coupled mode system [37].

$$I_1/I_0 = C_0 \sin^2\{M_2(\pi^2/2\lambda^2)(L/H)C_{rf}P_{rf}\}^{1/2}$$

$$(14.12)$$

where P_{rf} is the rf drive power, L/H is the ratio of width-to-height for the acoustic transducer, C_0 and C_{rf} are constants, and M_2 is a figure-of-merit for diffraction efficiency that depends only on material parameters, and is given by

$$M_2 = (n^6 p^2)/(\rho v^3) \qquad (14.13)$$

where n is the refractive index, p is the elasto-optic coefficient, p is the density and v is the acoustic velocity of the material. The sine-squared relationship, Equation 14.12, leads to similar considerations as for the Mach–Zehnder devices discussed earlier, Equation 14.5, such as limitation on wideband dynamic range due to third-order intermodulation products. Also, I_1/I_0 less than 3% in order for the relationship between I_1 and P_{rf} to be linear to better than 1%.

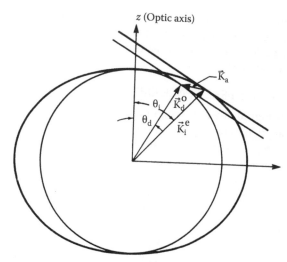

Wave-vector diagram for noncolliner acousto-optic filter.

Figure 14.5 Anisotropic Bragg diffraction geometry.

Extensive literature exists on materials and design of acousto-optic devices [38,39]. The figure-of-merit M_2, Equation 14.12, provides guidance on choice of optimum AO device material, especially dependence on refractive index n. Device development has converged on LiNbO$_3$ for operation at high frequencies and modest time-bandwidth (up to 4 GHz, ~100 TBW), TeO$_2$ for lower frequency application but large time-bandwidth (<100 MHz, several thousand TBW), and GaP for intermediate bandwidths and time-bandwidths (up to 1 GHz, several hundred TBW) [6].

The major aspects of acousto-optic devices are summarized below:

- For the isotropic case, the diffracted light is deflected by an angle $2\theta_B$. The more general nonisotropic diffraction case is shown in Figure 14.5, compared with the isotropic case. In this nonisotropic case, the incident light is polarized as an extraordinary ray with momentum vector $\boldsymbol{K}_i^e$ and the diffracted light is of opposite polarization from that of the incident light, $\boldsymbol{K}_d^0$. The angular relationship among the incident and diffracted light beams and the acoustic vector $\boldsymbol{k}_a$ is determined by the necessity for each beam to reside on its respective optical-index ellipsoid, and is shown in Figure 14.5.
- The information in the cell is constantly updated due to the travelling wave nature of

the acoustic signal. If the cell is driven by an electrical signal $r(t)$, the information displayed along the cell direction, x, is given by $r(t - x/v)$, as shown in Figure 14.4.
- Complex (both amplitude and phase) data on an incident light beam, $s(x, t)$, that spatially fills the cell aperture will be diffracted by the acoustic grating and therefore contain the product of $s(x, t)$ and the complex data on the acoustic wave, $r(t - x/v)$. The amount of spatial data contained on the modulated beam is equivalent to the number of resolvable deflection position for the optical beam and is the so-called time-bandwidth (TBW) product, equal to the product of the length of the cell, in units of time, and the temporal bandwidth of the drive signal [8]. This result is easily derivable from Equation 14.11 and the angular optical-diffraction limit due to the finite spatial extent of the incident light beam.
- If the continual temporal update of data in the cell, represented by $r(t - x/v)$, is not desired, a short pulse of light can be used to "freeze frame" the instantaneous contents of the cell. Alternatively, the incident light beam can be focused into a small diameter spot within the cell, resulting in a temporal modulator [40]; the bandwidth of such a temporal modulator is determined by the traversal time of the acoustic wave through the spot of light,

and initiation of modulation is determined by the delay caused by transit of the acoustic wave from the transducer to the optical spot location.

- The center frequency of the diffracted light is shifted relative to the incident light frequency by the carrier frequency of the acoustic signal. The frequency shift arises from energy and momentum conservation considerations in diffraction by a moving grating. Either frequency upshift or downshift is possible, depending on the momentum direction of the incident light relative to the acoustic beam momentum direction. Acousto-optic diffraction is an important means of producing small frequency shifts onto the ~100 THz optical carrier.

14.2.2.1.1 Multi-dimensional devices using acousto-optics

To better exploit the two spatial dimensions of optics, acousto-optic devices have been extended to 2D architectures either by, constructing an array of 1D devices in a single crystal, or by constructing a device with acoustic transducers on orthogonal edges of a crystal. In either of these approaches a large, high quality crystal is required. For multi-channel acousto-optic devices, a large number of channels is desirable, but in the absence of electrical crosstalk, the major fundamental limitation is acoustic diffraction. Acoustic diffraction depends not only on the dimensions of the transducer, but also on material. Ideally, the near-field acoustic wavefront is planar over the aperture of the transducer, and diffraction effects are observed in the far-field. Hence, it is desirable to maintain a near-field condition over as long a propagation distance

as possible. In an isotropic medium, the transition from near to far field occurs at a distance from the transducer of approximately

$$D = H^2/8L \qquad (14.14)$$

where H is the vertical dimension of the transducer and L is the acoustic wavelength. However, for an anisotropic medium, D can be increased by an additional factor of $c = (1-2b)^{-1}$. The quantity b is the coefficient of the q^2 term in a power series representation of the acoustic slowness surface [41],

$$K_a(q_a) = K_a(1 + bq_a^2 + dq_a^4 + L...) \qquad (14.15)$$

where q_a is the acoustic-wave direction relative to the normal to the transducer. For shear mode TeO$_2$ the quantity c is 0.02 so that the acoustic spreading is very rapid. Because of the severe acoustic spreading in shear mode TeO$_2$, multichannel devices have been constructed only of longitudinal mode TeO$_2$ where the spreading factor c is 2. A 32-channel TeO$_2$ device at a center frequency of 250 MHz has been demonstrated [42,43].

The second 2D approach uses orthogonally-propagating acoustic waves. Using two separate 1D cells orthogonal to each other, a set of anamorphic optics passes light from one cell to the other, where the data in the first cell are focused and passes through every point on the second cell. This scheme is illustrated in Figure 14.6. A much more compact alternative to individual 1D devices is to employ a single large square crystal with transducers along two edge facets and light propagating perpendicular to the square aperture. However,

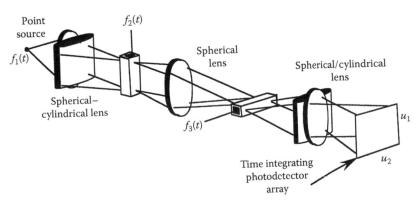

Figure 14.6 Triple-product processor using orthogonal acousto-optic Bragg cells.

this approach is feasible only in cubic materials. TeO$_2$ satisfies this criterion, and shear-mode TeO$_2$ devices have been demonstrated. However, if the two acoustic waves propagate within the same volume of crystal, nonlinear mixing of the signals will occur at lower power levels; this is particularly true for TeO$_2$, which exhibits acoustic nonlinearities at relatively low power levels. Thus, devices are made with the acoustic transducers offset from each other along the optical-beam direction. The transducer dimension along the optical path and the offset distance must be minimized to prevent loss of resolution or to avoid use of optics with large depth of focus.

14.2.2.1.2 Acousto-optic applications

Analogue processors

One-dimensional processing. The most basic application of the acousto-optic cell is to place information at the front focal plane of a Fourier Transform lens. The spatial spectrum of the travelling wave is then displayed at the back focal plane. Since acousto-optic devices have been constructed into the microwave frequency range, up to 4 GHz [6], an rf spectrum analyzer may be constructed as shown in Figure 14.7. A collimated optical beam with intensity profile $a(x)$ illuminates an acousto-optic cell P$_1$ that is driven by the signal $s(t)$. The signal $s(t)$ must first be mixed with a carrier signal $\cos(2\pi f_c t)$ to produce the drive signal $f(t)$ at the acoustic frequency f_c. The complex spectrum of $f(t)$ appears at the back focal plane P$_2$ of the Fourier transform lens. The spectral components with frequency upshift are represented by $F_+(\alpha, t)$, and with

frequency downshift by $F_-(\alpha, t)$. The Bragg diffraction geometry determines whether F_+ or F_- is displayed. The number of resolvable spectral positions is equal to the TBW product, as described above. The spectral update rate is approximately equal to the length of the illuminated portion of the Bragg cell, typically in the sub-microsecond to ten-microsecond range. The photodetector array at the back focal plane must therefore have corresponding output frame rate. The spectral information must also be measured with high fidelity and dynamic range, placing additional demands on the dynamic range performance of the photodetector array [44]. For a power spectrum analyzer only the amplitude $A(\alpha)$ is measured by square-law photodetector elements. If the phase information must be preserved, each photodetector element must be illuminated with a reference optical beam and the bandwidth of each element must support the difference frequency between signal and reference beam.

The back focal plane information of a spectrum analyzer can be cascaded into a second Fourier Transform lens, producing an optical image of the spatial information. Such an optical image can be multiplied with another information array by locating a second acousto-optic cell at the image plane, as shown in Figure 14.8. The travelling-wave nature of the acoustic-cell data has been advantageously used to implement the correlation integral, Equation 14.4. A final lens performs the Fourier Transform of the product of the functions shown in Figure 14.8, and the output of a small detector at the center of the back focal plane is the correlation integral

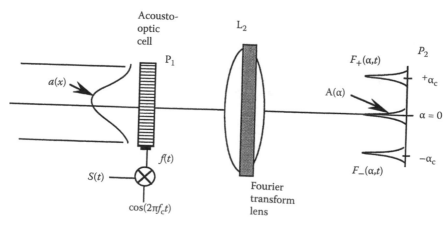

Figure 14.7 Acousto-optical spectrum analyzer.

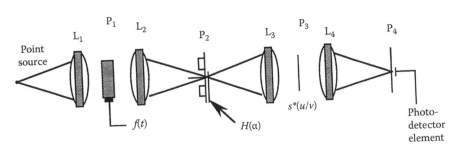

Figure 14.8 Space-integrating acousto-optical correlator.

$$R(t) = \int f(t + x/v)s(t - x/v)\,\mathrm{d}x = \int f(u)s(u - 2t)\,\mathrm{d}u.$$

$$(14.16)$$

The arrangement of Figure 14.8 has been used to perform high-speed correlation according to Equation 14.16 [7]. The integration can be thought of as performed by the focusing action of the lens, hence the correlation architecture is termed "space-integrating". Processing of a long-duration signal with the space-integrating architecture requires formatting of input data into frames of size, or length (when inserted into the cell), less than that of the illuminated portions of the cells. This formatting adds to the burden for high speed pre-processing in this processor.

An alternative to spatial integration using a lens, as shown in Figures 14.7 and 14.8, is to employ time-integration in the photodetection process [45]. The product terms within the kernel of integral transforms such as Equations 14.4 and 14.16 can be obtained by successive diffraction by acousto-optic cells of modulated light. The duration of signals is not limited by the size of the cells and photodetector array, but by the feasible integration time. The photodetector array does not need to be read until its charge capacity is reached, and digitization of the output allows further signal integration time. Output data rates in these "time-integrating" architectures are much lower than with the space-integrating architectures. However, only a small fraction of the resultant correlation function is produced, and some a priori knowledge of the location of the region of interest is required. Otherwise, the amount of output data does not decrease.

Multi-dimensional processing. A multiplicity of 1D spectrum analyzer and correlation processors can be implemented in parallel using the multichannel acousto-optic devices described in Section 14.2.2.1.1. The main application of these multi-channel processors is to process data from arrays of sensors. Array functions are performed in addition to the 1D analysis; these functions include

- Direction-finding in conjunction with spectral analysis [44] using an array of rf receive antennas. Each array element drives a separate channel of a multi-channel acousto-optic device where the acoustic transducers are relatively spaced exactly as for the array antennas. By coherently illuminating the multiplicity of channels, the phase differences among the various channels result in additional deflection orthogonal to the spectrum-analysis direction that provides information on the angle-of-arrival of the signal at the antenna array.
- Generalized multi-dimensional array beamforming [46]. The acousto-optical channels impose phase factors or time delays on array signals to produce directional transmit or receive beams, using a straightforward extension of the formalism described in Section 14.2.1.1 for a 1D array.

A more general extension of Fourier transform processing into two dimensions using acousto-optic input devices has been explored. With the orthogonal-cell arrangement shown in Figure 14.6, one can perform (1) 2D beam deflection in free space, for applications such as optical interconnection and raster addressing in displays, and (2) 2D processing where the processing kernel consists of two 1D factors. For the latter, the two orthogonal acousto-optic devices are combined with a temporally-modulated light source and a 2D integrating photodetector array to produce what is known as a triple-product processor (TPP) [47]. As illustrated in Figure 14.6, the time-integrating version of the TPP output has the form

$$g(u_1, u_2) = \int_0^{kT} f_1(t)f_2(t+u_1)f_3(t+u_2)\,dt \qquad (14.17)$$

where u_1 and u_2 are the coordinates of the output plane, T is the maximum integration time of the product signal onto the 2D photodetector array, and k is a constant ≤ 1 depending on the duration of the input signals. By employing various functions f_1, f_2, and f_3, the TPP, Equation 14.17, can implement a number of important signal-processing operations at extremely high speed:

- Time-frequency transformation, such as Ambiguity processing (also known as range-Doppler processing) in radar and sonar [48]. The product of the functions $f_1(t)$ and $f_2(t+u_1)$ constitute the cross-correlation kernel, as in Equation 14.4, to provide range determination. Function f_3 is the Fourier phase factor for compensating for any Doppler effects on the cross-correlation function, thereby resulting in the 2D ambiguity surface. Additional 2D time-frequency transforms that can be performed with the TPP include the Wigner function, instantaneous power spectrum, and cyclostationary function [46].
- Fourier transformation of very large 1D blocks of data in 2D format, also known as folded spectrum analysis [47]. Since each cell of the TPP can have TBW of about 1000, a signal can be spectrally analyzed into approximately one million narrow frequency bins.
- Matrix operations [49]. Matrix operations are employed, not for numerical processing, but to allow use of linear algebraic techniques to formulate an optical implementation of various signal-processing operations for the TPP. For example, analogue multiplication of three matrices has been implemented by cascading the TPP into a 2D SLM (covered in Section 14.2.4). In particular, the triple-matrix product allows the implementation of similarity transforms to be performed; these transforms, which diagonalize an input matrix, perform the important data-reduction function that

permits lower-bandwidth results to be passed to subsequent processing.

- Synthetic aperture radar (SAR) image formation [50]. SAR image formation is historically important as one of the first successful applications of optical signal processing. The original processors as input film onto which the radar return signals were written. For near-real time image formation the film can be replaced by a 2D SLM (discussed in Section 14.2.4 and Chapter 6). However, the use of acousto-optic cells allows more-rapid generation of SAR imagery over either film or present 2D SLMs. The original method where the radar return signals were written onto film and the developed film used as the input into an optical system. The ability to use 1D acousto-optic cells to produce 2D SAR imagery follows from recognition that the SAR image-formation equation is separable into factors involving integration over only the range variable η, followed by integration over only the azimuth variable ξ The equation governing the formation of an image point at coordinates (x, y) is given by

$$g(x, y) = \int_0^{A1} \int_0^{A2} \{f(\eta, \xi)\exp[j(\pi/\lambda D_1)(\eta+y)^2]d\eta\} \\ \exp[j(\pi/\lambda D_2(y))(\xi+x)^2]d\xi$$

$$(14.18)$$

where $f(\eta, \xi)$ represents the signal from point reflector on the ground, and D_1 and $D_2(y)$ are parameters or functions related to signal curvature (SAR systems transmit a train of short impulses; the resultant radar returns are chirp, or quadratic phase functions, but the chirp rates are different for the η and ξ directions). In the limits of integration A1 is proportional to the integrated number of radar pulses in the azimuth direction, and A2 determines the duration of the radar pulse train in the range dimension.

The analogue processors described in this section have found only niche application to date. However, a number of prototype acousto-optical processors have exhibited some advantageous features. The advantages generally lie in the size, weight and power consumption advantages over a comparable all-electronic implementation of equivalent processing power. For example, analogue data multiplication using acousto-optic technology at eight-bit precision can have a 350 × power consumption advantage over digital multipliers [51]. Specific examples of compact processors include a 6 in.^{3} rf spectrum analyzer [6], and a SAR image-formation processor in <1 ft^3 [52]. But the need for significant high speed pre-and post-processing electronics (as described in Sections 14.1 and 14.1.1) results in significant disincentive to use optical processing unless the size, weight and power advantages are paramount. Without such advantages even the early use of optics for SAR image formation has been largely replaced with digital electronic processing. An added measure of advantage could be obtained for optical processors if many of the electronic functions can be performed optically or if, at the output of an optical processor, the amount of data for subsequent electronic processing can be reduced. If a data-reduction algorithm can be implemented, such as the similarity transform mentioned above, one would not need to read out an entire 2D array or could use a linear array; however, this situation is clearly specific only to a special class of data processing. An alternative is to construct a "smart" photodetector array that selects only the region of interest for readout, such as a local peak [15]. Increasing the variety of operations that can be performed opto-electronically would also allow intelligent algorithms to be implemented. Many all-electronic approaches can be intelligently formulated to reduce the computation load required for a given processing task; use of the fast Fourier transform (FFT) algorithm to greatly reduce computation in digital electronic processing is an example that cannot yet be duplicated in optics.

14.2.2.2 ONE-DIMENSIONAL ELECTRO-OPTIC MODULATOR AND APPLICATIONS

An electro-optic modulator is a potential alternative to the acousto-optic device. A phase diffraction grating is produced by the electro-optic effect rather than a pressure wave. Such modulators constructed of LiNbO$_3$ have been demonstrated for high-speed printing applications [53,54]. However, in the most common configuration the electro-optic modulator requires many parallel addressing lines. Hence, there is significant difficulty in generating a multiplicity of phase gratings. Further, significant effort is required to format data for parallel addressing. However, careful design of the electrical addressing (the intrinsic speed of the electro-optic effect can be sub-picosecond) can result in a device that creates a grating much more rapidly than acousto-optic devices. An alternative approach to parallel electrodes is to use a high-speed shift register to address the electro-optic material. This approach is architecturally similar to that of acousto-optics. Use of a shift register has advantages such as capability to vary clocking rate.

14.2.2.3 ONE-DIMENSIONAL MECHANICAL MODULATORS AND APPLICATION

Spatial modulation can be performed using micro-electromechanical structures (MEMS). Devices are constructed entirely of silicon for both the addressing circuitry and the modulation structure. The micromechanical modulator design consist of support structures and cantilevered beams [55] or torsion beams extending from the structures; these are all fabricated out of silicon by an etching process [56]. A silicon circuit below the micromechanical structure provides addressing and activation voltages that deflect the structure. The deflection changes the angular position of the reflected light beam, producing intensity modulation in the readout system. The early devices were called deformable membrane devices (DMDs), and this terminology persists although no longer limited to the original membrane MEMS approach.

The importance of MEMS devices has been in their application to all-optical switching for fiber-optic telecommunication systems, where switch speed is not important. The advantage is that the data remain in optical form during the switching operation.

14.2.3 Wavelength processing

The bandwidth of optics can be exploited more fully by using a broad band of the optical spectrum and by subdividing an optical signal into many narrow-wavelength data channels and processing each channel individually. Processing in the wavelength dimension has already been introduced in Section 14.2.1.2 with respect to the importance of WDM fiber-optic systems. However, processing in the wavelength dimension can be applied to both temporal and multi-dimensional data. Wavelength-selective acousto-optic and electro-optic devices can be applied to spectroscopic systems, to channelize temporal data, and to perform 1D, 2D and 3D imagery analysis and display.

14.2.3.1 ACOUSTO-OPTICS

The separation of a broadband optical signal into narrow wavelength channels can be performed using acousto-optics. Many optical modulators are sensitive to the wavelength of operation. However, acousto-optic modulators are unique in that, without a large degree of modification, they can be used as flexible elements that can select a number of wavelengths simultaneously. The phase gratings produced by the acoustic wave will, as any grating, select out specific wavelengths at the appropriate angle of incidence for Bragg interaction. The grating period, corresponding to specific wavelength to be selected, is generated by driving the acousto-optic device with the appropriate rf frequency according to Equation 14.6. This spectral filtering action is notable in that by driving the device with a composite rf signal a multiplicity of wavelengths may be selected simultaneously, unlike optical glass and interference filters and resonant tunable structures such as Fabry–Perot cavities.

The resolving power of a grating,

$$R = \Delta\lambda / \lambda$$

is equal to the number of grating lines. Hence, for an acousto-optic tunable filter (AOTF), depending on whether the incident optical beam is parallel or perpendicular to the grating lines, the resolving power is then

$$R = D/\Lambda \text{ or } R = L/\Lambda, \text{ respectively} \quad (14.19)$$

where D is the extent of the illuminated region (parallel case, beam transverse to the acoustic wave) and L is the length over which the optical and acoustic wave overlap (perpendicular case, beam collinear with the acoustic wave).

AOTFs generally operate in the anisotropic diffraction mode, as shown in Figure 14.5, so that the filtered output light is orthogonally polarized to the input, i.e., the input beam must also be polarized to allow high contrast isolation between input and output optical beams by using crossed polarization. Good AOTF design for high spectral resolution must maximize

$$R = L(\Delta n)/\lambda$$

where Δn is the maximum difference in refractive indices for the birefringent material [57]. The rf frequencies required for specific wavelengths is governed by phase matching conditions for anisotropic diffraction [41,58]. Popular materials for AOTFs are $LiNbO_3$ and TeO_2 for the visible to mid-wave IR, Tl_3AsSe_3 for near-to long-wave IR, and quartz (SiO_2) and MgF for the UV. Both bulk and integrated-optic devices are possible. The latter are naturally compatible with optical fiber, and utilize the polarization anisotropy in thin-film optical waveguides in addition to material anisotropy [59].

The major opportunity for AOTF in information-processing applications is in their potential employment in WDM techniques and systems. For conventional fiber optic systems guided-wave AOTFs have been developed to filter the various WDM channels at moderate, microsecond, speeds [59]. However, significant interest has been generated in the use of AOTFs to flatten power across the WDM spectral band in long-distance communications networks. Since the erbium-doped fiber amplifiers used in long-distance networks do not amplify uniformly across the WDM band, and networks need to be dynamically reconfigurable, a rapid means is required to pre-condition the channel powers such that all channel power levels are always equal following the fiber amplifier, allowing proper functioning of the WDM network [60]. In the future, just as in fiber-optic communications, developments in optical interconnections are expected to encompass WDM techniques. AOTFs presently represent a viable wavelength switch for initial development.

The multiple-wavelength capability of AOTFs allows numerous other applications including

- Color generation and correction in multi-dimensional laser displays [61], such as for laser light shows. 3D displays systems have been developed using the rapid wavelength tunability and scanning capability of acousto-optic devices in conjunction with commensurate optical-beam scanning within a 3D medium such as with a rapidly spinning turbine blade.
- Scene imaging in many spectral bands, i.e., creation of image "cubes" for multi- and hyperspectral techniques for remote sensing in commercial, astronomical, and military applications.

- Wavelength-selective microscopic probing for research in biological processes, such as by selectively quenching fluorescence in biological samples with specific wavelengths [62].

Some of the present development issues in the use of AOTFs include, materials and devices for operation further into the UV and IR bands, power requirements due to the necessity for the rf drive, spectral sidelobe levels which limit channel selectivity in WDM networks, and imaging quality through AOTF devices. Finally, the Bragg diffraction requirement often limits the angular acceptance angle for imaging, and the use of rectilinear shapes for the AOTF can introduce astigmatism into the imaging system.

14.2.3.2 ELECTRO-OPTICS

Electro-optic devices can produce phase gratings analogous to those in acousto-optic devices, as previously mentioned in Section 14.2.2.2. Electro-optic devices have also been considered for wavelength filtering. By laying an array of electrodes onto a guided-wave structure, optical wavelength filtering can be performed analogous to the integrated-optic. However, it is desirable to have devices that are faster than either the electro-optic and acousto-optic devices discussed so far. One approach is to employ a bulk device with volume holograms as the wavelength-sensitive element.

Volume diffraction gratings can be produced optically in a number of materials using the photorefractive effect [63,64]. The photorefractive effect is a manifestation of the electro-optic effect on a microscopic scale. Internal space-charge electric fields, E_{sc}, arise from photocharge generation, followed by separation of the charges via photoconduction (transport of charge from illuminated to unilluminated regions) or via application of an external field, E_{dc}. Gratings are produced from an interference pattern obtained with two intersecting coherent optical beams. However, a rapid means is needed for either changing the grating period or changing to a different grating. Presently, sub-microsecond photorefractive processes are possible in quantum-well materials [3], but packet switching requires even faster speeds.

A recent development, entitled electroholography has potential to achieve nanosecond switching [65]. The quadratic electro-optic effect is used in paraelectric crystals such as $KTa_xNb_{1-x}O_3$ (KTN). In addition to a space-charge field E_{sc}, a dc field,

E_{dc}, is applied, leading to a cross term E_{sc} £dc. Until E_{dc} is applied, the information contained in the hologram is latent, the diffracted light containing only bias (E^2_{dc}) and unipolar (E^2_{sc}) terms. In principle, such switchable gratings would allow processing operations required in packet switches (e.g., multicasting, power management, data monitoring) to be performed at close to the data bit rate, entirely in the optical domain.

14.2.4 Two-dimensional spatial light modulators

2D SLMs are the most critical devices required for full and efficient exploitation of the inherent parallel-processing and interconnection capabilities of optics. As discussed above, and shown in Figure 14.6, 1D acousto-optic devices have successfully performed a number of important 2D signal-processing operations. However, that approach will work only for problems that can be separated into two 1D factors. Many image processing operations cannot be so factored.

Although Chapter 6 discusses SLM devices and technology, this section reviews necessary and desired performance characteristics for information processing. Devices will be cited without repeating descriptions already provided in Chapter 6. We first briefly review the major types of SLM.

14.2.4.1 TWO-DIMENSIONAL SLM DEVICES CATEGORIES

The first level of categorization is whether the device is electrically-addressed (E-SLM) or optically-addressed (O-SLM). The next level of categorization is by modulation mechanism or material. Within each category both O-SLMs and E-SLMs are possible.

- *Electro-optic.* These use traditional linear electro-optic materials such as $LiNbO_3$, KD*P, and bismuth silicon oxide, and quadratic electro-optic material such as PLZT. There is usually a strong trade-off between frame rate and resolution.
- *Liquid crystal.* There are two major classes.
  - Twisted nematics, as used in commercial displays. Large arrays are available (~10^7 pixels), but frame rate is limited by the natural relaxation time of the liquid crystal, of the order of 10 ms.

- Ferroelectric liquid crystals (FLCs) possess a permanent electric dipole. Both write and erase speed can be increased in proportion to drive signal level. Frame rate can therefore be high, limited by addressing circuitry. The most mature devices are binary contrast only. Analogue devices with large grey-scale capability are less mature

- *MQW structures.* These devices employ the quantum confined Stark effect in stacks of ultra-thin, ~ 10-nm thick, layers of III-V materials, generally of an alternating composition. Quantum well materials have intrinsically very fast response times, and have been described in Section 14.2.1. Large 2D arrays with high-speed readout have been produced. Monolithic integration with III-V, e.g., GaAs, electronic circuitry is a desired goal. However, integration with silicon circuitry is more common.

- *MEMS.* Previously introduced in Section 14.2.2.3, large 2D arrays have been developed, e.g., >750 × 500 pixels for display purposes, but the DMD has generally fewer pixels but higher frame rate. These are strictly E-SLMs and therefore the addressing circuitry limits the frame rate. However, MEMS arrays can be constructed of silicon, so the modulator and addressing circuitry can be monolithically integrated.

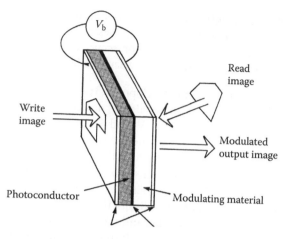

Figure 14.9 Generic sandwich construction for a 2D SLM.

14.2.4.2 SLM FUNCTIONS AND GOALS FOR OPTICAL PROCESSING

A large variety of SLM devices were categorized in Section 14.2.4.1. However, to understand the desired functioning of an SLM a general construct can be used, describing an SLM as a "sandwich" of structures (Figure 14.9): a modulating material between control and conversion structures. The conversion structures handle the input to and output from the SLM, the "write" beam and the "read" beam, respectively, and may consist of photosensitive material for an O-SLM, or an array of electrodes for an E-SLM. The control structure is the third port of this "three-port" device. Figure 14.9 can be used to illustrate the variants of SLMs and the basic functions they provide for optical processing:

- *Electrical-to-optical transduction.* The write image is in electrical form, e.g., the SLM input is via an electron beam, as in a cathode-ray

tube display, or via write lines on an electrical circuit. The input information is modulated onto an optical read beam (no read image data).

- *Optical-to-electrical transduction.* If the write image is in optical form, there is no read image, and the Modulated Output is electrical, then the SLM acts as a classical photodetector array.

- *Light conversion.* The write image may be on an optical beam of either incoherent light or coherent light of a particular wavelength. The read image is a readout beam of coherent light of another wavelength onto which the write image is modulated to produce an optical modulated output beam. For incoherent optical input, the image data is converted from incoherent to coherent light; in the coherent input case the image data can be converted from one wavelength to another. Additionally, the polarization state of the output can be changed relative to the input polarization.

- *Imagery projection.* The write image data may be either electrical or optical. If the read image is a readout beam, an optical beam of sufficient intensity then that beam may be used for projection display of the write image data modulated onto it.

- *Image processing.* An SLM can be used as an input and processing device for image-processing systems that perform operations such as transformations (e.g., Fourier transform), image correlation, nonlinear operations such as thresholding and level slicing, and linear re-mapping of images. The write image data may be either electrical or optical. The read image and the modulated output can then be either optical or electrical, respectively (depending on the write image). Some image processing may be performed first by the SLM, but the modulated output will be into either a further optical or electronic image-processing system.

- *Image amplification.* The read image is a readout beam that is modulated to produce an optical modulated output. The optical power in the read beam may be larger than in the write beam, or the SLM can amplify the write beam signal.
- *Analogue multiplication, addition and subtraction of data arrays.* Data on the write beam can interact with data on the read beam using any of a variety of mechanisms to perform these operations.
- *Storage of data arrays.* The modulating material may retain the information impressed onto it for some time. During the retention time a modulated output is obtained. If the SLM is optically read out, the modulating material is an optical memory.
- Input-to-output isolation and gain element for 2D data arrays, akin to a transistor for 1D temporal data. This three-port behavior is particularly important for optical interconnection and cascading optical elements in an all-optical architecture, since with a three-port device the modulated output levels are independent of the strength of the write image signal.
- *Analogue phase conjugator for adaptive optical systems.* Write and read images and modulated output are all optical; the SLM performs the phase conjugation operation.

For optical information processing systems utilizing 2D SLMs to be effective, SLM devices need to exhibit a minimum throughput (frame rate times pixels/frame) on par with that already demonstrated with acousto-optic devices, roughly 10^9 samples/s. However, this level of throughput can be partitioned between frame rate and frame size, as in the following two examples.

- A liquid crystal display of 10^7 pixels and 30 frames/s would approach the 10^9 samples/s goal.

- MQW modulator with 6.5×10^4 pixels (256×256 device) at 10^5 frames/s would exceed the goal. Such a modulator would likely be optically addressed, to avoid use of ultra-high speed electronic addressing circuits.

Depending on application it is likely that 10^9 pixels/frame would be desirable for processing of large images, and $>10^3$ framing rate would be needed for applications such as 3D displays and rapid search of volume holographic databases (see Section 14.2.4.3.3). In addition to these two parameters, additional requirements are

- *Large dynamic range and good image quality.* A uniform 1000:1 contrast ratio would satisfy most requirements.
- *Low power consumption.* A power consumption figure of merit would have units of mW cm^{-2} kHz^{-1}. This figure of merit would relate also to the size and resolution of the SLM. To keep power consumption low for a given number of pixels, the size of the pixels needs to be minimized. Pixellation corresponding to 50 line pairs/mm, easily within the state-of-the-art of most fabrication technologies, would minimize the focal length of the optics needed for relaying information to/from the SLM, and overall minimize optical system size and power.
- *Data storage time.* Flexibility to address a variety of applications (Section 14.2.4.3) is desirable, e.g., a fast frame rate SLM with storage times of seconds to almost indefinite. It may be necessary to keep the same data displayed in an SLM indefinitely as one rapidly processes imagery or searches databases. Another example, would be where the SLM responds to slowly-changing phenomena, such as environmental conditions, in an adaptive optics application.

At the extreme, the data are time invariant; except for write-time considerations, use of film as an SLM would be acceptable. Time-invariant applications include feature extractors and fixed filters for image processing.

14.2.4.3 APPLICATIONS OF 2D SLM DEVICES

14.2.4.3.1 Image processing

Next to the use of optical processing in synthetic aperture radar image formation, the 2D optical processor with the longest history is the

Fourier-transform based image correlator. This processor has arguably seen the most development over a number of years, with interest for high speed automatic target recognition systems. But performance has been determined by available SLM devices and photodetector arrays. Because of a lower requirement on space-bandwidth than alternate architectures, and ability to use direct phase encoding for the input and filter function, the preferred correlator architecture is a modification of the arrangement for temporal correlation, Figure 14.8. Known as the VanderLugt correlator, 2D SLMs are used at the input image plane $f(x, y)$ and the Fourier-filter plane, and a 2D photodetector is used to output the correlation function (at the location of the second acousto-optic cell in Figure 14.8, and the last lens and point photodetector are eliminated). Correlation of one image $f(x, y)$ with another $s(x, y)$ is performed as multiplication in Fourier space followed by inverse Fourier transformation. The key to use of this architecture is the scheme for encoding both phase and amplitude of the Fourier-plane filter onto an SLM device or film [66]. Variants of the VanderLugt filter have been explored to match the capabilities of existing SLMs and to improve throughput. These variants generally take advantage of the well-known fact in signal processing that clipping the amplitude of a signal does not greatly degrade the detection margin in the correlation process, only a 1 dB penalty typically. Hence, binary amplitude Fourier filters have been used. A further extension has been to additionally binarize the phase values in the filter, i.e., a binary phase only filter where the amplitude transmittances of ± 1 correspond to phase shifts of 0 and π. Various high-frame-rate SLMs with limited or no grey scale capability are therefore viable candidates for a high-speed correlation system such as the FLC SLM, the DMD SLM, and the electro-optic PLZT-on-silicon SLM. A compact correlator operating at 500 Hz using FLC technology has been packaged in a volume of $0.6' \times 1.0' \times 0.4'$ [15].

Many issues in producing viable optical processors have been addressed for optical correlators. The performance and size of an optical correlator is found to depend on SLM pixel pitch and size, and device flatness and uniformity. For example, the level of detail of the Fourier-plane filter will increase with number of pixels in the SLM and thus increase the ability to differentiate among various objects. Also, the correlator system size scales linearly with the number of pixels and quadratically with pixel pitch [15]. Thus, smaller pixel pitch in larger more-uniform frames are needed. However, equally important are the performance parameters of the photodetector array. Since it is often only necessary to detect the location of the peak of the correlation function, many proposals have been made for smart photodetector arrays that can report correlation results as peak/no peak at rates commensurate with the SLM rate. If peak detection is insufficient for the application, then faster image processing will be required to extract correlation plane features. Whether such processing can be done in another stage of processing remains a future challenge. Unless many of these issues are addressed, optical systems will remain noncompetitive with all-electronic implementations.

Adaptive techniques are a way to improve the performance of optical processing systems, and SLM technology can been applied to such techniques. One technique is to perform phase conjugation on an optical wavefront to restore the wavefront to its original state by removing aberrations and distortions introduced by the environment or imperfections in an optical system. Phase conjugation requires the storage of detailed interference fringe patterns produced by the aberrated wavefront and an unaberrated reference beam. A phase-conjugate mirror can be implemented with an SLM constructed of a photorefractive material [67]. Progress to date has centered on development of the required material parameters. High speed of response and efficiency are required from the photorefractive material. Photorefractive quantum-well material is capable of sub-microsecond response, and the efficiency has been increased by constructing pixels consisting of etalons [3]. Phase conjugate efficiency of 100% with high net gain has been achieved with polymer photorefractive materials [68].

14.2.4.3.2 Optical interconnection

Free-space: crossbar switch. Free-space optical interconnection can be described in matrix formulation that is easily related to capabilities in optical systems. Consider an array of optical input and output data channels, such as fiber-optic. If the array of input channels is represented as a vector u, any element or combination of elements can be routed to any element of the output array or vector

v using an appropriately constructed matrix M in the equation

$$v = Mu \qquad (14.20)$$

SLMs are needed in this concept because of several desirable characteristics they can provide. First, when interconnecting among a number of different communications channels an SLM provides isolation between input and outputs. Second, the routing process is naturally lossy, because each element of v must fan out onto a row of M. Optical-to-optical SLMs can then provide the amplification gain to compensate for these optical losses.

14.2.4.3.3 Optical storage

Effective processing systems need to have access to memory. It is natural to explore possible use of optical processing hardware with optical storage technology. One function of an SLM is storage of data, as noted in Section 14.2.4.2. An SLM can store the results from one stage of processing, and present those results to the next stage. However, the amount of information that can be stored on an SLM is limited to the number of pixels in the device. Optical devices that store large amounts of data are, however, of interest in optical information processing, e.g., large databases may need to be accessed rapidly and repeatedly, such as for cross-correlation. Conventional high-density optical storage technology has been reviewed in Chapter 13. However, conventional optical disk storage technology has not been found useful for optical processing without significant modification or additional electronics. One nonconventional approach stores a library of analogue holograms on a rotating photopolymer disk for subsequent correlation. Large numbers of holograms can be stored on a single disk, and rapid scanning of the library is possible with standard mechanical rotary drives. Another disk approach is to store conventional digital data in a 2D page format. Such holographic disk systems have been demonstrated that perform at 1 Gbit s^{-1} sustained and 10 Gbits s^{-1} peak transfer rates; along with the holography, is electronic processing that includes data modulation coding, interleaving, and error correction to a bit error rate of 10^{-3} [69].

Volume holographic storage has a more natural optical interface using 2D SLMs to transfer data from previous and to subsequent stages of 2D optical processing. Volume storage also would have a much higher storage capacity than disk, with a theoretical storage limit of ~10 Tbits cm^{-3} (~volume/λ^3 where λ is approximately the smallest linear dimension for storing one bit). In comparison, present single-layer optical disks have areal storage density of 15 Gbit s^{-2}, which can be extended with multilayer disks, the increase depending on the number of layers and the laser wavelength. Among the issues in cascaded optical SLM and storage elements is rapid addressing of and access to data arrays, and the proper registration of data arrays. The latter issue has been addressed with electronic postprocessing techniques [70].

14.3 SUMMARY

The area of optical information processing has explored many avenues of research. Areas of application are still developing, mainly as devices capabilities increase and new capabilities emerge. Many lessons have been learned on the proper application of optical capabilities.

- Using optics for general numeric computations is not presently desirable, because many of the digital operations are intrinsically nonlinear. Necessary nonlinear optical devices are still large compared to electronic integrated circuits, and also inefficient, i.e., power hungry. This does not mean the architectures that have been developed are without value. Such architectures were developed mainly to cleverly exploit the parallel-processing capabilities of optics, and may be applicable to building specialized processors [71].
- The advent of optical communications has led to a near-term processing application for fiber-optic interconnects between high-performance electronic digital processors and sensor arrays, whereby simple operations are performed on the optical data within fibers, such as multiplexing/demultiplexing and signal conditioning. Fiber interconnects are a straightforward insertion of optics, mainly to address input–output data-bandwidth bottlenecks that can occur with high-speed digital processing elements. But developments in fiber-optic technology have important implications for optical processing; these include the development of wavelength diversity techniques, and

of all-optical techniques, presently used to cascade fiber-optical subsystems in a network to avoid costly conversions between the optical and electrical domains. However, with fiber optics the full free-space 3D interconnection capability of optics is not fully exploited. Full exploitation of free-space 3D interconnection technology will require the use of wavelength diversity and all-optical techniques employing cascaded 2D SLMs. There will also be concurrent development of novel multi-processor computing architectures. Candidate architectures are likely to involve neural network parallel-distributed-processing concepts, where the interconnection scheme can be integral to the processing.

- Analogue optical signal processors, particularly for Fourier-transform-based operations such as spectral analysis and correlation, have been developed because they can perform linear operations, at a given rate, in implementations with smaller size-weight-power product than all-electronic implementations. Avoidance of power-hungry analogue-to-digital conversion by processing directly in the analogue domain and concentrating on applications in the microwave-frequency region make such processors a true complement to digital processors. Most development of compact systems has been done with 1D acousto-optic modulators, due to the maturity of acousto-optic technology. However, the highest performance should be obtained using 2D SLMs addressing intrinsically 2D problems such as pattern recognition and 3D optical interconnection. Use of wavelength-diversity techniques should further increase performance. Various 2D SLMs are capable of implementing functions needed for all-optical architectures, especially optical amplification, data storage, and data thresholding. All-optical systems are needed to address the fundamental problem with present analogue processors, by optically performing the present electronic pre-and post-processing of data into/out of optical processors. Such processing includes dynamic range re-scaling and data formatting, various nonlinear operations, and signal amplification and data encoding schemes. These needs can be met potentially with developments in optical-to-optical SLMs, adaptive

optics, and volume holography. Throughput improvements of 2D devices are still required, since present throughput rates are lower than for the acousto-optic devices. Hence, initial implementations may still be restricted to 1D and quasi-multi-dimensional architectures.

A feasible vision for the future for all application areas therefore involves all-optical architectures. Optical device technology exists for implementing subsystems that perform optical input, processing, and storage functions. The challenge is to develop efficient information and data flow between such subsystems in the optical implementations of pre- and post-processing functions that are done electronically for present optical systems.

REFERENCES

1. Gopalakrishnan, G. K., Burns, W. K., McElhanon, R. W., Bulmer, C. H. and Greenblatt, A. S. 1994. Performance and modeling of broadband LiNbO$_3$ traveling wave optical intensity modulators. *J. Lightwave Technol.* 12: 1807.
2. Noguchi, K., Mitomi, O. and Miyazawa, H. 1998. Millimeter-wave Ti: LiNbO$_3$ optical modulators. *J. Lightwave Technol.* 16: 615.
3. Bowman, S. R., Rabinovich, W. S., Beadie, G., Kirkpatrick, S. M., Katzer, D. S., Ikossi-Anastasiou, K. and Adler, C. L. 1998. Characterization of high performance integrated optically addressed spatial light modulators. *J. Opt. Soc. Am. B* 15: 640.
4. Goodman, J. W. 1968. *Introduction to Fourier Optics*. New York: McGraw-Hill.
5. Athale, R. A., Szu, H. H. and Lee, J. N. 1981. Three methods for performing Hankel transforms. *Optical Information Processing for Aerospace Applications*. NASA Conference Publ. 2207, NASA CP-2207 133.
6. Anderson, G. W., Webb, D., Spezio, A. E. and Lee, J. N. 1991. Advanced Channelization Devices for RF, Microwave, and Millimeterwave Applications (Invited Paper). *Proc. IEEE* 79, 355–388.
7. Griffin, R. D. and Lee, J. N. 1996. Acousto-optic wide band correlator system. *Acousto-Optic Signal Processing*, 2nd ed., chapter 11, ed. N. Berg and J. Pelligrino. New York: Dekker, pp. 367–400.

8. Berg, N. J. and Lee, J. N. (eds.) 1983. *Acousto-optic Signal Processing: Theory and Implementation.* New York: Dekker.

9. Huang, A. 1978. An optical residue arithmetic unit. *Proceedings of Fifth Annual Symposium on Computer Architecture.* New York: Association for Computing Machinery and IEEE Computer Society, pp. 17–23.

10. Whitehouse, H. J. and Speiser, J. M. 1977. *Aspects of Signal Processing with Emphasis on Underwater Acoustics,* ed. G. Tacconi. Hingham, MA: Reidel.

11. Athale, R. A., Hoang, H. Q. and Lee, J. N. 1983. High accuracy matrix multiplication with magnetooptic spatial light modulator. *Real-Time Signal Processing VI: Proc. SPIE,* vol. 431, ed. K. Bromley, p. 187.

12. Miller, D. A. B., Chemla, D. S., Damen, T. C., Gossard, A. C., Wiegmann, W., Wood, T. H. and Burrus, C. A. 1984. Band-edge electroabsorption in quantum well structures: The quantum confined Stark effect. *Phys. Rev. Lett.* 53: 2173.

13. Lentine, A. L., Hinton, H. S., Miller, D. A. mB., Henry, J. E., Cunningham, J. E. and Chirovsky, L. M. F. 1989. Symmetric self-electro-optic effect device: Optical set-reset latch, differential logic gate, and differential modulator/detector. *IEEE J. Quantum Electron.* 25: 1928.

14. Yu, S. and Forrest, S. 1993. Implementations of smart pixels for optoelectronic processors and interconnection systems: I. Optoelectronic gate technology, and II. SEED-based technology and comparison with optoelectronic gates. *J. Lightwave Technol.* 11: 1659.

15. Turner, R. M., Johnson, K. M. and Serati, S. 1995. High speed compact optical correlator design and implementation. *Design Issues in Optical Processing,* ed. J. N. Lee. Cambridge: Cambridge University Press, p. 169.

16. Denker, J. S., ed. 1986. *Neural Nets for Computing.* New York: American Institute of Physics.

17. Psaltis, D. and Farhat, N. 1985. Optical information processing models based on an associative-memory model of neural nets with thresholding and feedback. *Opt. Lett.* 10: 98.

18. Fisher, A. D., Lippincott, W. L. and Lee, J. N. 1987. Optical implementations of associative networks with versatile adaptive learning capabilities. *Appl. Opt.* 26: 5039.

19. McCormick, F. B. and Tooley, F. A. P. Optical and mechanical issues in free-space digital optical logic systems. *Design Issues in Optical Processing,* ed. J. N. Lee. Cambridge: Cambridge University Press, p. 220.

20. Lentine, A. L. et al. 1997. ATM distribution network using an optoelectronic VLSI switching chip. *Optics in Computing, OSA Technical Digest* 8: 2.

21. Husain, A., Crow, J. and Lee, J. N. (eds.) 1991. *Journal of Lightwave Technol: Special Issue on Optical Interconnects.* A. Husain and J. N. Lee (eds.) 1995. *Special Issue on Optical Interconnects.*

22. Feldman, M. R., Camp, J. L., Sharma, R. and Morris, J. E. 1995. Comparison between holographic and guided wave interconnects for VLSI multiprocessor systems. *Design Issues in Optical Processing,* ed. J. N. Lee. Cambridge: Cambridge University Press.

23. Uomi, K., Mishima, T. and Chinone, N. 1985. Ultra-high relaxation oscillation frequency (up to 30 GHz) of highly p-doped GaAlAs multiquantum well lasers. *Appl. Phys. Lett.* 51: 78.

24. Yariv, A. 1989. *Quantum Electronics,* 3rd ed., chapters 14 and 16. New York: John Wiley & Sons.

25. Hecht, D. L. 1977. Spectrum analysis using acousto-optic filters. *Opt. Eng.* 16: 461.

26. Korotsky, S. K. and DeRidder, R. M. 1990. Dual parallel modulation schemes for low-distortion analogue optical transmission. *IEEE J. Sel. Areas Commun.* 8: 1377.

27. Johnson, L. M. and Rousell, H. V. 1988. Reduction of intermodulation distortion in interferometric optical modulators. *Opt. Lett.* 13: 928.

28. Walker, R. G. 1991. High-speed III–V semiconductor intensity modulators. *IEEE J. Quantum Electron.* 27: 654.

29. Trezza, J. A., Pezeshki, B., Larson, M. C., Lord, S. M. and Harris, J. S. 1993. High contrast asymmetric Fabry–Perot electroabsorption modulator with zero phase change. *Appl. Phys. Lett.* 63: 452.

30. Farnett, E. C., Howard, T. B. and Stevens, G. H. 1970. Pulse compression radar. *Radar Handbook*, chapter 20, ed. M. I. Skolnik. New York: McGraw-Hill.

31. Frankel, M. Y., Matthews, P. J. and Esman, R. D. 1996. Two-dimensional fibre-optic control of a true time-steered array transmitter. *IEEE Trans. Microw. Theory Technol.* 44: 2696.

32. Wong, Y. M. et al. 1995. Technology development of a high-density 32-channel 16 Gb/s optical data link for optical interconnection applications for the optoelectronic technology consortium (OETC). *J. Lightwave Technol.* 13: 995.

33. Chang-Hasnain, C. J., Maeda, M. W., Harbison, J. P., Florez, L. T. and Lin, C. 1991. Monolithic multiple wavelength surface emitting laser arrays. *J. Lightwave Technol.* 9: 1665.

34. Lau, K. Y., Derry, P. L. and Yariv, A. 1988. Ultimate limit in low threshold quantum well GaAlAs semiconductor lasers. *Appl. Phys. Lett.* 52: 88.

35. Rabinovich, W. S. et al. 2001. InGasAs multiple quantum well modulating retro-reflector for free space optical communications. *Patent pending.* U.S. Naval Research Laboratory, 2000.

36. Klein, W. R. and Cook, B. D. 1967. Unified approach to ultrasonic light diffraction. *IEEE Trans. Sonics Ultrason.*, 14: 123–134.

37. Phariseau, P. 1956. On the diffraction of light by progressive ultrasonic waves. *Proc. Indian Acad. Sci.* 44A: 165–170.

38. Uchida, N. and Niizeki. 1973. Acousto-optic deflection materials and techniques. *Proc. IEEE* 61: 1073.

39. Lee, J. N. (ed.) 1995. *Design Issues in Optical Processing.* Cambridge: Cambridge University Press.

40. Young, E. H. and Yao, S. K. 1981. Design considerations for acousto-optic devices. *Proc. IEEE* 69: 54.

41. Cohen, M. G. 1967. Optical study of ultrasonic diffraction and focusing in anisotropic media. *J. Appl. Phys.* 38: 3821.

42. VanderLugt, A., Moore, G. S. and Mathe, S. S. 1983. Multichannel Bragg cell compensation for acoustic spreading. *Appl. Opt.* 22: 3906.

43. Amano, M. and Roos, E. 1987. 32-channel acousto-optic Bragg cell for optical computing. *Proc. SPIE*, 753: 37.

44. Anderson, G. W., Guenther, B. D., Hynecek, J. A., Keyes, R. J. and VanderLugt, A. 1988. Role of photodetectors in optical signal processing. *Appl. Opt.* 27: 2871.

45. Turpin, T. M. 1978. Time integrating optical processing. *Proc. SPIE*, 154: 196.

46. Lee, J. N. 1987. Architectures for temporal signal processing. *Optical Signal Processing*, ed. J. L. Horner. New York: Academic, p. 165.

47. Turpin, T. M. 1981. Spectrum analysis using optical processing. *Proc. IEEE* 69: 79.

48. Cohen, J. D. 1979. Ambiguity processor architectures using one-dimensional acoustooptic transducers. *Proc. SPIE* 180: 134.

49. Athale, R. A. and Lee, J. N. 1983. Optical systems for efficient triple-matrix-product processing. *Opt. Lett.* 8: 590.

50. Haney, M. and Psaltis, D. 1985. Acousto-optic techniques for real-time SAR imaging. *Proc. SPIE* 545: 108.

51. Lee, J. N. and VanderLugt, A. 1989. Acousto-optic signal processing and computing. *Proc. IEEE* 77: 1528.

52. Essex Corp. 2000. *ImSym Product Specification.*

53. Johnson, R. V., Hecht, D. L., Sprague, R. A., Flores, L. N., Steinmetz, D. L. and Turner, W. D. 1983. Characteristics of the linear array total internal reflection electro-optic spatial light modulator for optical information processing. *Opt. Eng.* 22: 665.

54. Hecht, D. L. 1993. Advanced optical information processing with total internal reflection electro-optic spatial light modulators. *Proc. SPIE* 2051: 306.

55. Brooks, R. E. 1985. Micromechanical light modulator on silicon. *Opt. Eng.* 24: 101.

56. Hornbeck, L. J. 1989. Deformable-mirror spatial light modulators. *Proc. SPIE* 1150: 86.

57. Chang, I. C. 1974. Noncollinear acousto-optic filter with large angular aperture. *Appl. Phys. Lett.* 25: 37.

58. Dixon, R. W. 1967. Acoustic diffraction of light in anisotropic media. *IEEE J. Quant. Electron.* 3: 85–93.

59. Smith, D. A., Baran, J. E., Cheung, K. W. and Johnson, J. J. 1990. Polarization-independent acoustically tunable filter. *Appl. Phys. Lett.* 60: 1538.

60. Willner, A. E. and Smith, D. A. 1996. Acousto-optic modulators flatten amplifier gain. *Laser Focus World*, June, p. 177.

61. Young, E. H. and Belfatto, R. V. 1993. Polychromatic acousto-optic modulators let users tailor output wavelengths. *Laser Focus World*, November, p. 179.

62. Spring, K. R. 2002. Acousto-optic tunable filters improve optical microscopy. *Laser Focus World*, January, p. 123.

63. Pepper, D. M., Feinberg, J. and Kukhtarev, N. V. 1990. The photorefractive effect. *Sci. Am.* 263: 34–40.

64. White, J. O. and Yariv, A. 1984. Photorefractive crystals as optical devices, elements, and processors. *Proc. SPIE* 464.

65. Agranat, A. J. 2002. Electroholographic switching devices and applications. *2002 CLEO Conference Technical Digest.* Washington, DC: Optical Society of America, p. 37, Paper CMH1.

66. VanderLugt, A. B. 1964. Signal detection by complex spatial filtering. *IEEE Trans. Inf. Theory* 10: 139.

67. Grunnet-Jepsen, A., Thompson, C. L. and Moerner, W. E. 1997. Spontaneous oscillation and self-pumped phase conjugation in a photorefractive polymer optical amplifier. *Science* 277: 549.

68. Peyghambarian, N., Meerholz, K., Volodin, B., Sandolphon and Kippelen, B. 1994. Near 100% diffraction efficiency and high net-gain in a photorefractive polymer. *Optics and Photonics News*, December, p. 13.

69. Orlov, S. S., Phillips, W., Bjornson, E., Hesselink, L. and Okas, R. 2002. Ultra-high transfer rate high capacity holographic disc digital data storage system. *Proceedings of 29th Applied Imagery Pattern Recognition Workshop.* New York: IEEE, p. 71.

70. Burr, G. W. and Weiss, T. 2001. Compensation of pixel misregistration in volume holographic data storage. *Opt. Lett.* 26: 542.

71. Tippett, J. T., Berkowitz, D. A., Clapp, L. C., Koester, C. J. and Vanderburgh, A. Jr. 1965. *Optical and Electro-optical Information Processing.* Cambridge, MA: MIT.

15

Spectroscopic analysis

GÜNTER GAUGLITZ
University of Tübingen

JOHN P. DAKIN
University of Southampton

15.1 INTRODUCTION

Optical spectroscopy is an invaluable tool for the characterization of many physical and chemical components and processes. It can provide complex and quantitative data and use it to precisely characterize a wide variety of physical objects or chemical analytes. Instruments can not only achieve much higher spectral resolution compared to eyes, but can also cover a much wider wavelength range and, if desired, even provide time- or distance-resolved results. A major advantage of optical methods of analysis is that they are usually nondestructive, noninvasive, and quick in operation. In optical spectroscopy, electromagnetic radiation in the range of 10^{12}–10^{15} Hz is most commonly used, covering infrared (IR), visible (VIS), and ultraviolet (UV) radiation. Optoelectronic devices detect the interactions of this electromagnetic radiation with matter.

This section will review the main methods of spectroscopy, concentrating mainly on covering general concepts of spectroscopy and introduce the technology behind some applications being covered in later chapters.

The passage of electromagnetic radiation can be influenced by matter in processes such as refraction, polarization change, specular reflection, scattering, or absorbance [1]. In most cases, the magnitude of these effects depends significantly on optical wavelength or photon energy.

The technique of absorbance spectroscopy, or spectrophotometry, monitors attenuation of a beam of light by matter and is one of the most commonly applied methods in spectrometry, particularly in the IR, VIS, and UV regions.

Measurement of light lost from a beam due to scattering (nephelometry or turbidity measurement) [2] is also often used. Apart from the more common light-scattering mechanisms, vis. Rayleigh [3], Rayleigh–Gans, Mie and geometric scattering, other elastic scattering (i.e., scattering with no change of wavelength) can arise due to the effects of diffraction from periodic structures (i.e., like X-ray scattering in crystals).

Inelastic scattering involves a change in optical wavelength (change in photon energy occurs inversely) and can arise, for example, in Raman [4] or Brillouin scattering, where a photon–phonon interaction process occurs.

After absorption of light, matter can dissipate the absorbed energy by several means, for example from an excited electronic state to lower vibrational or rotational states, i.e., to produce phonons or heat. The absorbed radiation can also be reradiated, resulting in luminescence behavior (this behavior can be subdivided into fluorescence and phosphorescence). In addition, after excitation with particularly energetic particles, matter can reemit strong radiation, a behavior made use of X-ray emission spectroscopy and atom emission spectroscopy.

In some cases (for example, in flames), thermally excited electrons in the hot gaseous materials can exhibit direct radiation of light by energy transitions from the excited states.

In addition to those that were discussed in Volume I, Chapter 1 (Introduction to Optoelectronics), processes relevant to spectroscopy will be briefly reviewed below [5–8]. The focus of this discussion will be particularly on the UV/VIS/NIR spectral region, especially when we describe quantitative aspects. The theoretical principles of atomic states and dispersion will be discussed, ending with some description of photophysics and aspects of polarization states in waveguides. Furthermore important spectroscopic methods and types of spectrometers will be mentioned together with a brief description of some interesting case studies of optical sensor systems.

15.2 THEORETICAL PRINCIPLES

15.2.1 Properties of electromagnetic radiation

Electromagnetic radiation is characterized by its amplitude, frequency, state-of-polarization, phase, and in the case of modulated light, the time dependence of its amplitude, phase, or frequency. Depending on the physical effects to be described, electromagnetic radiation can best be considered as either a particle or a wave, a behavior called de Broglie dualism. The choice of the wave description, however, is best used to understand interference and refraction effects [9,10]. The impact momentum, p, is proportional to the inverse of the wavelength, λ, of the radiation according to Equation 15.1:

$$p = \frac{h}{\lambda}. \qquad (15.1)$$

This can be given, in terms of effective mass (kg), m, and velocity (ms^{-1}), v, by

$$p = mv, \qquad (15.2)$$

where p is the momentum (kg m s^{-1}), and h is Planck's constant $= 6.63 \times 10^{-34}$ J s. Propagation of electromagnetic radiation can be explained by the well-known Maxwell's equations, which predict that, in a propagating E–M wave, the electric and the magnetic field vectors lie perpendicular to each other and are perpendicular to the direction of propagation (see Figure 15.1). Considering the particulate (or quantized) nature of the light, the radiation is composed of a stream of photons and the energy of each is proportional to the frequency in the wave model, so is therefore reciprocally dependent on the optical wavelength (see Equation 15.3).

$$E = h v = \frac{hc}{\lambda} = hc\tilde{v}. \qquad (15.3)$$

Here, $c = 3 \times 10^8$ m s^{-1} is the velocity of light in vacuum, and $\tilde{v}$ is the wave number in cm^{-1}.

There is a distinct lack of standardization in the units most spectroscopists use, depending on the field they work in. The unit of optical frequency that is most commonly used by scientists working in the IR region or using Raman scattering is the wave number. In the field of highly energetic radiation, eV (electron volts) is usually used. In the UV, VIS, and NIR wavelength range, most scientists use units of wavelength, λ, in nm, or, in the IR, units in μm. Yet again, many of these are not S.I. units, reflecting the fact that the reasons for the choice of the different units were historical in nature.

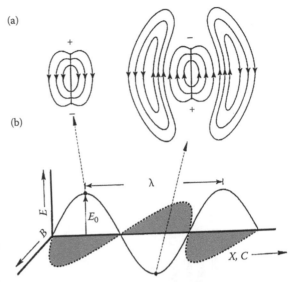

Figure 15.1 Electrical fields of a Hertz dipole. These are perpendicular to both the magnetic field and to the direction of propagation of the electromagnetic radiation (a). The time dependence of the electric field distribution is given in (b). The polar diagram of radiated intensity (power) is shown in (c).

The approximate order-of-magnitude ranges of wavelength, frequency, wave number, and photon energy that correspond to a number of different types of electromagnetic radiation are shown in Table 15.1 and Figure 15.2. The typical type of interaction that each type of electromagnetic radiation might have with matter is also shown, but this will be discussed later. As can be seen, the frequency and wavelength scales are logarithmic in the table, in order to cover a very wide range.

Electromagnetic radiation in the UV, visible, and IR wavelength range is especially interesting in spectroscopy. For this reason, this particular spectral range has been "zoomed out" in Figure 15.2 and we have indicated the colors associated with the different parts of the visible spectrum. In the

Table 15.1 Spectral regions of electromagnetic radiation, giving the approximate order of magnitude of relevant wavelengths, frequencies, wave numbers and energies, plus indications of typical interactions that particular type of radiation might have with matter

Spectroscopic technique	Approximate order of the wavelength (nm)	Approximate order of the frequency (s⁻¹)	Approximate order of the wave number (cm⁻¹)	Approximate magnitude of energy (in various units)			Type of interaction
				(kcal mol⁻¹)	(kJ mol⁻¹)	(eV)	
Nuclear magnetic resonance (NMR) 0.1–10 m	10^{10}	3×10^{7}	10^{3}	3×10^{-6}	1.2×10^{-5}	1.2×10^{-7}	Excitation of magnetic transitions of nuclei (spin > . 0)
Electron spin resonance (ESR) 0.1–10 cm	10^{8}	3×10^{9}	10^{-1}	3×10^{-4}	1.2×10^{-3}	1.2×10^{-5}	Excitation of unpaired electrons
Microwaves 0.1–10 cm	10^{6}	3×10^{11}	10	3×10^{-2}		1.2×10^{-3}	Excitation of rotation of molecules
IR 0.78–103 μm, 3000–106 nm	10^{4}	3×10^{13}	10^{3}	3	12	1.2×10^{-1}	Absorption by excitation of molecular vibrations. Raman scattering
VIS 380–780 nm, UV 200–400 nm, UVU 100–200 nm	10^{2}	3×10^{15}	10^{5}	3×10^{2}	1.2×10^{3}	12×10^{4}	Excitation of electronic transitions, emission and atomic absorption
X-rays 0.01–10 nm	1	3×10^{17}	10^{7}	3×10^{4}	1.2×10^{5}	1.2×10^{3}	Excitation and transitions of electrons involving inner-orbit energy levels. Raman scattering processes. Raman scattering
Moessbauer 100 keV, γ absorption	10^{-2}	3×10^{19}	10^{9}	3×10^{6}	1.2×10^{7}	1.2×10^{5}	Resonant absorption by nuclei
γ radiation, >1 MeV	10^{-4}	3×10^{21}	10^{11}	3×10^{8}	1.2×10^{9}	1.2×10^{7}	Nuclei transformation

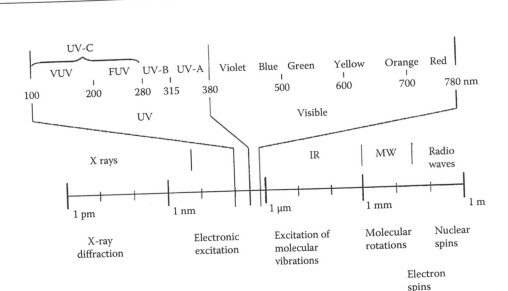

Figure 15.2 Depiction of the full electromagnetic spectrum of radiation.

case of simple absorption photometry, a transmission measurement is taken at only one wavelength (i.e., monochromatic light is used), but in more general spectrophotometry, a transmission spectrum is usually measured over a wider wavelength range, involving perhaps a superposition of many rotational, vibrational, and electronic interactions.

15.2.2 Interaction between radiation and matter

15.2.2.1 DISPERSION

The most usual and important interaction with matter for use in spectroscopy involves the electric field vector. The periodic change in the electric field vector of an incident E–M wave can cause either a new orientation of a molecule or a displacement of electrical charges in a dielectric medium (matter). Both of these result in electrical polarization of the matter, the so-called dielectric effect process. If there is any variation of this effect with optical wavelength or frequency, the effect is called dispersion [9], as it changes the velocity of the wave, depending on its wavelength.

Radiation induces a type of Hertz dipole in a dielectric material, and if this material is not optically homogenous with its surroundings, it can result in radiation, or scattering, into the surroundings. The charge displacement in a dielectric

by an electromagnetic wave can be explained in terms of a simple harmonic oscillator model, where an alternating dipole is induced in each molecule or microregion of the material. The total effect is proportional to the number of molecules or macromolecular particles affected, i.e., polarization is a volume-based (colligative) quantity.

This Hertz dipole oscillates at the excitation frequency of the radiation and behaves like a radiation source, emitting radiation most strongly in a direction perpendicular to the direction of polarization, as was shown earlier in Figure 15.1c. When propagating through matter, the radiation maintains its frequency, because of the conservation of photon energy $h\nu$, but decreases in wavelength and velocity. The velocity of propagation within matter is inversely proportional to the refractive index, n, given by the ratio between the electric field vectors of the radiation propagating in vacuum and matter:

$$n = \frac{\vec{E}(\text{vac})}{\vec{E}(\text{mat})}. \tag{15.4}$$

According to Maxwell's equation, the square of the optical refractive index (i.e., the value of refractive index in the optical range of the dispersion curve) is equivalent to the dielectric constant, which, in turn, can conveniently be described mathematically in terms of an equation with real

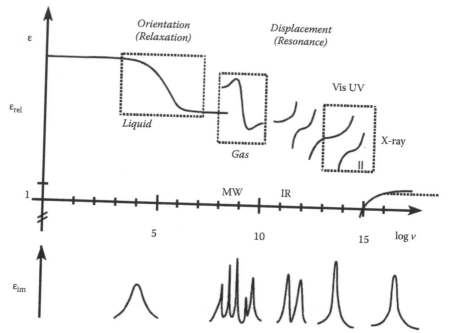

Figure 15.3 Depiction of optical dispersion, showing the real (refraction) and imaginary part (absorption) of dielectric constant versus frequency on a logarithmic scale. The lower curve shows various absorption bands.

and imaginary parts. This latter parameter can be graphically depicted, as shown in Figure 15.3. The imaginary part corresponds to absorption (or, more rarely, optical gain) in the medium.

In addition to polarization of matter, as occurs with scattering, absorption of light can also occur. Except in condensed matter (i.e., solids or liquids where spectral broadening occurs due to degeneracy of energy levels), absorption is usually not of a broadband nature, i.e., it often occurs at specific wavelengths. For many materials, e.g., liquids in the microwave region, a reorientation and a relaxation effect can be observed, which changes the effective dielectric constants and causes absorption.

A resonance (caused by displacement of charges) can typically be observed in the IR (vibration) and UV/VIS (electronic transition) spectral regions. This causes strong absorption bands (i.e., a larger imaginary part of the dielectric constant). In the case of gases, rotational resonance is also observable at microwave frequencies.

To summarize the above, two fundamental types of interaction between electric fields and dielectric matter can therefore occur: the first is scattering and the second is absorption (or, more rarely, optical gain). These will both be discussed

in more detail in the following sections and we shall start with the case of scattering where no absorption occurs and light energy is conserved. In contrast, inelastic scattering (Raman) will be discussed later.

15.2.2.2 SCATTERING

15.2.2.2.1 Scattering from single particles

Within molecules, charges may either be symmetrically (nonpolar) distributed or they may have an asymmetric distribution, making the molecule polar in nature. In nonpolar molecules, the nuclei and the electron density distribution have the same center of charge distribution. However, in an electrical field, positive and negative centers can be displaced relative to each other, forming a Hertz dipole acting as an electromagnetic transmitter.

The excited system behaves elastically, emitting energy directly from a nonresonant state (sometimes called a virtual state), with no energy absorbed. Even the smallest of molecules show such scattering effects, but, for such Rayleigh scattering (scattering by molecules or particles having a diameter of order of $\lambda/10$ or less), the effect is usually very small, as the intensity of scattered radiation increases in proportion to the sixth power of the length of the induced Hertz dipole [3,11]. For these small particles, the intensity of the radiation varies as the fourth power of the incident optical frequency (if the particle has low optical dispersion) and the light is scattered at essentially the same frequency as the incident beam. It is this strong dependence on wavelength that causes white light from the sun to be scattered mainly at short wavelengths, giving the blue appearance of sky on a clear day or the aurora/afterglow.

The radial distribution of scattered radiation depends, in general, on the state of polarization of incident radiation, on the shape of the particle or molecule, its dimensions and on the wavelength of scattered radiation. Thus, the following effects can be observed:

- An isotropic particle with a diameter smaller than $\lambda/10$, forms with polarized incident radiation a scattered intensity distribution (polar diagram) as shown in Figure 15.4a. Maximum energy is emitted in a direction perpendicular to the axis of the dipole, but zero energy is scattered in a direction along the axis of the dipole.
- Many very small particles ($<\lambda/10$) have some anisotropy (e.g., randomly oriented, needle-like, or egg-shaped particles) and so exhibit many induced Hertz dipoles at different angles (depending on the orientation). Then, even for incident polarized light, the polar diagram of the scattering is no longer as in Figure 15.4a, but has a new shape as in Figure 15.4b, and there is now a nonzero scattering intensity along the direction of polarization of the incident light. This phenomenon is known as depolarization of scattering.
- Randomly polarized (or unpolarized) radiation generates a set of random Hertz dipoles, whereby the intensity of radiation in the forward or backward direction is unchanged, but the radiation field at other angles (e.g., at 90°) is a superposition of the wavelets from all these different polarizations, with of course the contribution falling to zero whenever the incident polarization (or the induced Hertz dipole axis) is in the direction of observation. This gives a polar diagram with the greatest scattered intensity in the forward and backward

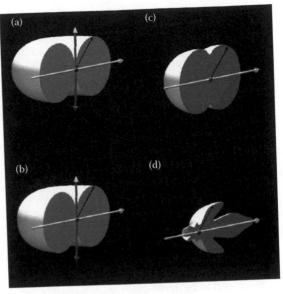

Figure 15.4 Polar diagrams for various cases of Rayleigh and Mie scattering. The horizontal arrow shows the direction of incident light. Where the incident light is plane polarized, the vertical arrow indicates the polarization direction. The inclined arrow shows a scattered light direction and the intersection with the outer surface indicates the scattered intensity at this angle. The top left figure (a) shows classical Rayleigh scattering, with small particles, where a Hertz dipole is induced along the polarization direction. The bottom left figure (b) shows the same when the particles are small, but have a small degree of anisotropy. The top right figure (c) shows the case for small particles with randomly polarized (or unpolarized) incident light. The bottom right figure (d) depicts a typical Mie scattering polar diagram for larger particles.

directions, but with half that intensity in a direction at right angles to the direction of incident radiation (see Figure 15.4c).

- If we now consider slightly larger particles (e.g., being of the same order as, or even larger than, the optical wavelength), but ones where the particle refractive index is very close to that of the surrounding medium, the behavior differs again. In this case, scattering occurs from many small points at different locations within each particle, yet the shape of the incident wave front remains essentially unchanged during passage through the particle. Then the excitation light has a phase that is dependent on the distance through the particle (Rayleigh Gans scattering condition). Interference between all the scattered wavelets from different positions will clearly influence the more complex radiation pattern, so, to obtain the far-field polar diagram, the Rayleigh scattering intensity from each has to be multiplied by a weighting function, which depends on the geometrical form of the particle (comparable to Mie scattering) [3]. If the scattering center or molecule is much larger and of significantly higher refractive index compared with its surroundings, then more than one center of scattering within the molecules will arise. However, the greater refractive index difference may now be enough to cause relative phase delays and even refraction of incident light, giving rise to significant wave front distortion. Thus, there is a phase shift, and in extreme cases, even a change in direction of the transmitted light, relative to light outside the particle. Under these conditions, the so-called Mie scattering [3,11] occurs. As with Rayleigh–Gans scattering, the scattered intensity distribution is determined by the interference of many scattered wavelets from different sites in the particle, but the situation is now far more complex as the phase of excitation of each point can no longer be so easily determined. The polar diagram is, therefore, generally complex and usually has many scattering lobes (see Figure 15.4d).

Thus, the intensity distribution and interference pattern contain information regarding the size and the shape of the molecules [3]. It is worth mentioning briefly that there is a further type of scattering, from very large, but optically homogeneous particles (raindrops are a typical example), several orders of magnitude greater than the wavelength of the light, where the scattering can be considered solely in terms of simple ray tracing, assuming Fresnel refraction and reflection at the optical interfaces. This form of scattering, which is observed in rainbows, is known as Fresnel or geometric scattering.

15.2.2.2.2 Scattering from multiple particles

Scattering from multiple particles depends strongly on whether the light is incoherent (broadband) or coherent (narrowband). If the light is incoherent, scattered power from each of the particles can simply be summed to give an overall polar far-field scattered intensity that is similar to that from one of the particles, but this is only if not too many strongly scattering particles are present. If the population density of scattering particles becomes very high (as can occur, for example, in milk or in white paints), then the light may suffer strong multiscattering paths, i.e., substantial secondary, tertiary, or even higher order scattering occurs before light from the first excitation (illumination) point can reach the observation point. A medium that causes significant light to be lost by scattering from a direct optical beam passing through it is called a turbid medium.

If an ensemble of particles is excited with highly coherent light (e.g., monochromatic laser light), the behavior is very different. Now the scattering from every small particle, or from every part of a larger particle, results in a monochromatic scattered wavelet being reemitted from each. Thus, at the point of remote observation, the electric fields of all these coherent wavelets will add coherently. This results in constructive, destructive, or intermediate levels of interference, depending on the relative phase of the light from the particles, which, in turn, depends on their relative positions relative to both the incident beam and to the direction or point of observation. These effects cause the familiar "speckle" effect when light from a visible laser is first scattered from a diffusely scattering object, such as a turbid liquid, or from a rough surface.

In a solid material, or particle in solid suspension, the positions of the particles may be reasonably stable, so then true elastic scattering occurs and a relatively fixed speckle pattern can be observed in the far-field scattering diagram, provided, of course, that no significant mechanical movement, vibrations, or thermal expansions occur. In fluid (gas or liquid) suspensions, however, the particle will generally undergo more significant and rapid movements due to Brownian motion, convection or turbulence, so the speckle pattern will move continuously with time. Because of the small Doppler shifts, the light scattering can be described as quasi-elastic. This time-dependent scattering behavior, which has come to prominence since the 1970s, can be used to deduce certain useful parameters, such as particle size information, using a method called photon correlation spectroscopy (PCS). The light scattered from a monochromatic laser beam interferes at the detector and gives rise to intensity changes (or to changes in photon-arrival-time statistics), which can, with knowledge of the temperature and viscosity of the fluid in which the particles are suspended, be related to the hydrodynamic radius of the particles using the Stokes–Einstein equation. The method works best with single-size (monodisperse) particles, but more complex correlation functions, from suspensions of two or even three particle types/sizes, can still be inverted using Laplace transformation to obtain useful information.

15.2.2.3 REFLECTION

Reflection usually describes the process where a light beam is reradiated back into the incident medium at a boundary. If only single particles (or random ensembles of such particles) are present in gases or liquids, they can be considered to be discontinuities that will cause diffuse light scattering, as we discussed earlier. However, if atoms and molecules are in very highly ordered arrays, as is the case in flat surface layers of solid matter (especially crystals and metals, but even in some amorphous materials, such as glasses, having only small-scale disorder), then the many Hertz dipoles can show constructive interference in preferred directions. This gives rise to directional or specular reflection at these solid surfaces, as seen, for example, on silvered-glass mirrors. This results in reflection at the same (yet opposite) angle to the normal to the surface as the initial or incident radiation [9,13]. If the reflection occurs due to refractive index differences (Fresnel reflection), then the amplitude of this reflectance depends on the refractive indices of both the media. For the simplest case of normal incidence and nonabsorbing media, the reflection is described by a simplified form of the Fresnel equation:

$$R = \left[\frac{(n_1 - n_2)}{(n_1 + n_2)} \right]^2. \tag{15.5}$$

If the reflection occurs at a polished metal surface, then the resulting more complete reflection can be predicted by Maxwell's equations, treating the metal as a conductive medium. If the metal is assumed to be a perfect electrical conductor, then the reflection is theoretically predicted both by Maxwell's equations and by simple conservation of energy considerations to be 100%. For nonperfect conductors, the reflection is <100%, and the

radiation actually extends within the material to a depth called the "skin depth." The reflection can therefore be used to determine the material properties of the metal.

15.2.2.4 REFRACTION

Refraction describes the effect of a boundary between two optical materials on a light beam passing through it. Radiation incident at an angle α on the interface between media with different refractive indices (n_1 and n_2) is refracted at an angle β, where, according to Snell's law:

$$n_1 \sin \alpha = n_2 \sin \beta. \tag{15.6}$$

Radiation, initially incident in a lower refractive index medium, is refracted toward the normal to the surface, so the angle of refraction becomes smaller, whereas in the opposite direction it is refracted away from the normal.

15.2.2.4.1 Total internal reflection and evanescent field

Equation 15.6 suggests there is a critical angle at which the angle of refraction is 90°. If the angle of incidence increases further, then total internal reflection occurs, a behavior only possible when light in an optically dense medium strikes an interface with a less dense medium. If the incident angle is greater than the critical angle, a simple explanation suggests radiation does not exit the denser medium, but is reflected back into this medium, as in Figure 15.5. This effect has become

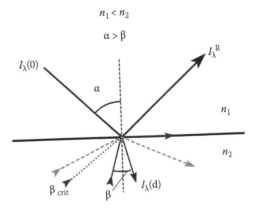

Figure 15.5 Behavior of light incident on a dielectric interface, demonstrating refraction and total internal reflection.

very important in recent years for optical or fiber-optical waveguides (see Chapters 1 and 7).

We shall now move on to consider inelastic processes, where either light energy is lost totally or where some light is reemitted with a change in photon energy.

15.2.2.5 ABSORPTION (THE LESS COMMON OPPOSITE OF THIS IS OPTICAL GAIN)

Absorption is a loss of light in a material, involving an internal energy exchange, resulting in the destruction of a photon. Optical gain is the opposite of this, where a light beam gains energy when traveling through a medium. Both mechanisms usually take place only at specific wavelengths, corresponding to a defined transition between two energy states. Such a transition occurs when the natural frequency of charge displacement in the material is in resonance with the frequency of the incident radiation. We shall mainly concentrate on absorption in the following discussion.

Absorption involves a process that leads to a new excited (either stable or metastable) energy state in the material. In the case of electronic transitions in an atom, this involves creation of a new electron density distribution. In the case of molecules, it is possible instead to excite a new resonant, vibrational or rotational, mode of the molecule (possibly in addition to electronic transition). An electronic transition, from the highest occupied molecular orbital (HOMO) to the lowest unoccupied molecular orbital (LUMO), as described mathematically by Equation 15.3, is demonstrated in Figure 15.6. Note in Figure 15.6, the phases of the orbitals are also given, in order to explain the symmetry of the orbitals, which causes specific transitions shown

by single, double, or triple arrows. These demonstrate the differences in intensity of the transition. According to transition rules, only transitions between even and odd states are permitted.

These HOMO and LUMO states correspond to bonding or antibonding orbitals [12]. The energy required for the transition is provided by the radiation and the process is known as (induced) absorption. Within this absorption band, anomalous dispersion is often observed (i.e., the refractive index increases with wavelength). Depending on the molecular environment and possible pathways of deactivation (see Table 15.2), the new excited state can exist for a time varying over the wide range of 10^{-13} to 10^{-3} s. From the Schrodinger equation, the corresponding energy states can be calculated as a set of eigenvalues, by using the electronic, vibrational, and rotational eigenfunctions and inserting the boundary conditions that are appropriate to the molecular structure. The relevant energy levels or states in the UV/VIS spectral region usually correspond to electronic levels, whereas in the IR area they correspond to the energies of molecular vibrational modes.

15.2.2.5.1 Energy level diagrams

In Figure 15.7, a Jablonski energy level diagram for a typical organic molecule is depicted. It shows the energy levels and allowable transitions. It shows both electronic and vibrational states, but rotational states have been omitted to keep the diagram simple. At room temperature, most of the molecules are resting in the lowest electronic and vibrational state, so resonance absorption usually takes place to various excited higher level vibrational levels or electronic states. To illustrate these, to the right of this energy level diagram, three types of resulting intensity spectra [absorbance (A), fluorescence (F), and phosphorescence (P)] are shown as a function of wave number.

The relative strengths of absorptions depend on the momentum of the transitions and are determined by a set of spectroscopic selection rules. Radiative transitions requiring a change of electron spin are not allowed and, in organic molecules, a normal ground state has a singlet property. Thus, a normal absorption spectrum is for a transition from S_0 to S_1. Such a transition occurs within a very short time, typically ~10^{-15} s. Usually, only the first excited electronic state is important in spectroscopy because all higher states have very

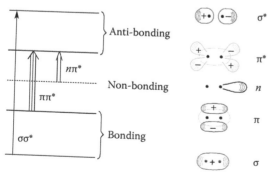

Figure 15.6 Bonding and antibonding orbitals (HOMO, LUMO) and their transition possibilities.

Table 15.2 Possible deactivation process from the vibrational ground state of the first excited singlet state

Process type	Abbreviation	Name of process	Process description
Radiationless transition	Te	Thermal equilibration	Relaxation from a high vibrational level within the present electronic state. Lifetime depends on matrix. Inner deactivation (transfer of energy in torsional vibrations).
	Ic	Internal conversion	Isoelectronic transition within the same energy level system from the vibrational ground state of a higher electronic state into the very high energy vibrational state of a lower electronic state.
	Isc	Intersystem crossing, intercombination	Isoelectronic transition into another energy level system (S ↔ T), usually from the vibrational ground state in the electronically excited state; respective radiative transition is forbidden because of the spin inversion prohibition (except for heavy nuclei). Therefore, phosphorescence is a "forbidden" process.
Spontaneous emission as a radiative process	F	Fluorescence	Without spin inversion from, e.g., S_1–S_0 (provided that lifetime of the electronically excited state is 10^{-8}s) within singlet system.
Photochemical reactions	P	Phosphorescence	Out of triplet into singlet system (provided that lifetime is 10^{-3}s) very low probability, only possible at low temperatures or in a matrix photo-induced reaction starting from the S_1 term leading to ionization, cleavage, or in a bimolecular step to a new compound or an isomer (trans–cis), provided that lifetime in excited state is relatively long.

short lifetimes (relaxation times). For organic molecules, usually paired spins are present (singlet states), whereas in inorganic transition metal complexes the ground state can be a triplet and other states with a large number of unpaired electrons may occur according to Hund's rule. An important example of a triplet ground state molecule is oxygen.

15.2.2.5.2 Electronic chromophores

Unlike the case for IR spectra, where vibrational bands are very useful for the analysis of components, or of functional groups, the occupancy of, and energy between, the electronic levels (i.e., those which usually give rise to UV/VIS spectra) does not depend so much on the molecular structure or on interactions with the matrix. However, changing the electronic ground or excited states can lead to a shift in the relative position of the electronic level or to a change in the degree of polarization. In general, UV/VIS spectra are less easily used to characterize chemical components. However, in some specific cases, for example, steroids, incremental rules can be determined that allow both determination of, and discrimination between, some of these molecules.

Various transitions between electronic and vibrational levels are shown in the energy level

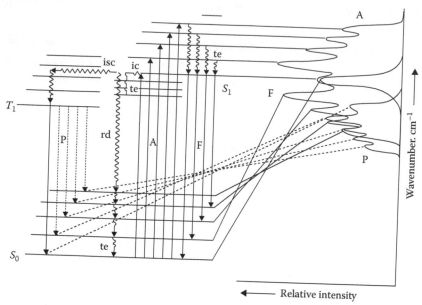

Figure 15.7 Jablonski energy level diagram, showing electronic and vibrational energy levels, and the resulting spectra for different transitions and pathways of deactivation.

diagram in Figure 15.7. The following abbreviations are used: rd (radiationless transition), IC (isoenergetic internal conversion within one electronic term scheme), ISC (isoelectronic intersystem crossing between term schemes). Depending on the electronic states involved, these transitions are called $\sigma\sigma^*$, $\pi\pi^*$, or $n\pi^*$ transitions (see Figure 15.6). In aromatic compounds, electron-attracting or electron-repelling components, such as functional groups in the ortho-, para-, or meta-positions of an aromatic ring, can affect these energy levels, and they may also be changed by solvent effects or with pH changes.

15.2.2.5.3 Linewidth and line broadening effects

Although the basic absorption process occurs at discrete wavelengths, there are many mechanisms that can effectively broaden the effective absorption line [7]:

- *Velocity or movement effects, causing translational Doppler shifts*: They are usually only important for gas absorption lines. Single-direction wavelength shifts of this nature will always occur in any fluids that are moving rapidly (e.g., fast-moving gas jets), but the most usual effect is due to the rapid movement of gas molecules, which is defined by well-known gas laws and of course depends strongly on the gas temperature.

- *Rotations of the molecule*: They can occur according to the number of degrees of freedom of the molecule and will cause a set of discrete energy (frequency) shifts, depending on the quantized rotational levels.

- *Interactions with other atoms in a molecule*: Such interactions increase, for example, the number of degrees of freedom of vibrational and rotational transitions (e.g., number of possible modes of vibration), so molecules containing more than about six atoms usually tend to lack a fine-line absorption band structure.

- *Interaction with other atoms or molecules by collisions*: A simple gas with atoms or molecules having a fine line absorption structure at low pressure will exhibit increasingly broader lines as pressure is increased. This is due to an effect called collision broadening or pressure broadening and is due to the additional energy levels that arise due to collisions. If the pressure is increased sufficiently, any closely spaced sets (manifolds) of narrow absorption lines will eventually merge into a continuous absorption band, just as occurs in condensed phases (liquids and solids).

- *Interaction with nearby atoms or molecules via electric field effects (ligand fields)*: Eventually, in the condensed state, the absorption is almost invariably of a broadband nature. This is because the large number of possible energy levels in absorbing bands can produce a near continuum. Electronic energy levels, for example, are affected by the electronic fields of nearby molecules, a phenomenon known as ligand field interaction.

15.2.2.5.4 Bandshifts of electronic levels

Apart from line broadening, line shifts can occur. As shown in Figure 15.8, the $\pi\pi^*$ and $n\pi^*$ transitions are influenced differently by polar solvents. If a cyclohexane solvent is replaced by methanol, the polarity increases considerably and, accordingly, the $\pi\pi^*$ band is red shifted by several nanometers, a so-called bathochromic effect, and the $n\pi^*$ band is shifted to the blue (hypsochromic). Such bandshifts can be used to obtain information regarding the properties of the energy levels and transitions. A change in intensity of the absorption band is described as either hyperchromic or hypochromic, respectively, depending on whether there is an increase or decrease in intensity.

15.2.2.5.5 Quantifying absorption levels in media

The power, $P(\lambda)$, transmitted through a sample in a small wavelength interval at a center wavelength λ, is given by Lambert's law:

$$P(\lambda) = P_0(\lambda)\exp[-\alpha(\lambda)\ell], \qquad (15.7)$$

where $P_0(\lambda)$ is the power entering the sample in this wavelength interval, $\alpha(\lambda)$ is the attenuation coefficient of the material at wavelength λ, and ℓ is the optical path length through the sample to the point at which $P(\lambda)$ is measured. This is only true for homogeneous nonturbid, nonfluorescent samples. The sample can be said to have a transmission $T(\lambda)$, at the wavelength λ, where

$$T(\lambda) = I_{out}/I_0. \qquad (15.8)$$

The transmission, T, is given by $0 \le T \le 1$; or $T*100\%$ in percent transmission. Alternatively, the sample can be said to have an absorbance $A(\lambda)$, where

$$A(\lambda) = \log_{10}[1/T(\lambda)] = \log_{10}(I_0/I_{out})$$
$$= \log_{10}[P_0(\lambda)/P(\lambda)] = 0.43\alpha(\lambda)\ell. \qquad (15.9)$$

The factor 0.43, which is the numerical value of $\log(e)$, has to be included to account for the use of $\log_{10}$ for decadic absorbance calculations, whereas natural (to base e) exponents are normally used for attenuation coefficients.

An absorbance of 1 therefore implies only 10% transmission. One should take care in the use of the term "absorbance," as defined above and as measured by most instruments, as this term seems to suggest the optical loss is only due to absorption,

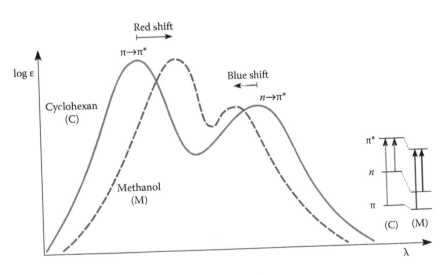

Figure 15.8 Illustration of band shifts caused by polar solvents.

whereas, in cases of turbid samples, it might actually arise from a combination of absorption and scattering (and reflection at boundaries of cuvettes). According to the Lambert–Beer–Bouguer law, the value of A for an absorbing analyte is given by

$$A = \log(I_0/I) = MCL, \qquad (15.10)$$

where M is the (decadic) molar extinction coefficient, C is the molar concentration, and L is the optical path length.

With merely measuring the intensity when using broadband radiation, there is a deviation from the above law whenever the absorption varies with wavelength. Thus, to avoid errors in calibration and/or absorption coefficient measurement, monochromatic radiation has to be used. Even at single wavelengths, however, the law has the wrong dependence on concentration if there is clustering or association of molecules or if a concentration-dependent or a pH-dependent chemical change occurs.

15.2.2.5.6 Frustrated total internal reflection, evanescent field absorption

In contrast to what is usually taught in classical optics, total internal reflection is not really "total" [12] at the interface between the media. A fuller examination, using either classical geometric optics or quantum optics, predicts that part of the radiation will penetrate the interface and then, for a certain distance, be guided outside the optically denser medium, leading to a lateral shift in the apparent point of reflection from the interface. This field that extends into the less dense medium is called the evanescent field. In the medium having the lower refractive index, it decreases exponentially away from the interface.

Any absorbing medium within an evanescent-field penetration depth, d, of this field, can absorb the radiation and result in an attenuation of the otherwise total reflectance. The value of d (measured from the interface) is given by Equation 15.11:

$$d = \frac{\lambda}{2\pi \sqrt{n_2^2 \sin^2 \Theta_2 - n_1^2}}. \qquad (15.11)$$

This penetration depth is typically of order of half the wavelength of guided radiation. Thus,

absorption of the evanescent wave will reduce the intensity of the otherwise "totally internally reflected" light via this electric field vector coupling.

This effect is called frustrated total internal reflection. It can be made practical use of for measurement of highly absorbing samples, which may absorb (or scatter) far too much light to be able to measure them by traditional methods. For example, complex samples such as paints, foodstuffs, or biological materials, which might absorb or scatter light very strongly, can be analyzed for their strongly absorbing components by placing them in contact with a high-index glass plate and measuring the internally reflected light.

15.2.2.6 FLUORESCENCE

Fluorescence is an inelastic energy relaxation process, which can follow after absorption of light in a material. In most materials, absorption of light (i.e., energy loss from a photon) merely causes heating of the absorbing material, with all the absorbed energy being converted to internal kinetic energy (e.g., to excite molecular vibrations or phonons). However, in some materials, only part of the energy of an optically excited state is dissipated as heat. Accordingly, one or more photons of lower energy than the incident one are radiated. The fluorescence therefore has lower photon energy (so is usually emitted at a longer wavelength) than that of the incident light (this is called a Stokes process). The emission of light by fluorescence has no preferred direction and is said to be omnidirectional. It is also randomly polarized. Of course, a single photon fluorescent event must, by its nature, have the direction and polarization of the single photon emitted, but over a period of time the average reradiated energy from many such events is omnidirectional and so fluorescent light is randomly polarized. This aspect of fluorescence can be used, apart from wavelength and temporal persistence (decay lifetime, after pulsed illumination) differences, to distinguish it from Rayleigh and/or Raman scattered light, which is usually strongly polarized.

Fluorescence detection is a valuable technique in chemical sensing where it can be used to directly monitor certain fluorescent chemicals (e.g., polyaromatic hydrocarbons, mineral oil, and fluorescein tracer dye). However, by deliberately introducing reactive fluorophores to act as a more-efficient chemical indicator, it can also be used to monitor reactions or reagents having no fluorescence. These are too numerous to cover here, as optical techniques have been used [14]. Many of the optical sensing applications of these are covered in detail in the proceedings of a series of international "Europt(r)ode" congresses. There is also an excellent textbook "Optical Sensors," edited by R. Narayanaswamy and O.S. Wolfbeis [15].

15.2.2.7 CHEMILUMINESCENCE AND BIOLUMINESCENCE

Fluorescent light can arise as the result of chemical reactions, an effect known as chemiluminescence. The reaction must be of a type to leave electrons in excited states that can then decay radiatively, usually with the emission of visible photons. Such reactions are now very commonly seen in the plastic-tube-enclosed chemical "light sticks" that children are often given, but which also have more serious uses for emergency lighting or as a distress signal at sea or in mountains. These lights operate by breaking chemical-filled ampules, enabling the reactants to mix and produce the light by a chemiluminescent reaction.

A compound commonly used to produce green light is called luminol ($C_8H_7N_3O_2$), which is known by several other chemical names, including 5-amino-2,3-dihydro-1,4-phthalazine-dione. When luminol is added to a basic solution of oxidizing compounds, such as perborate, permanganate, hyperchlorite, iodine, or hydrogen peroxide, in the presence of a metallic-ion catalyst, such as iron, manganese, copper, nickel, or cobalt, it undergoes an oxidation reaction to produce excited electronic states, which decay to give green light. The strongest response is usually seen with hydrogen peroxide, but very low concentrations of oxidizing agents such as ozone, chlorine, and nitrogen dioxide can also be measured. Numerous biochemicals can also cause a light-emitting reaction and hence be detected.

Nature has, however, used chemiluminescence long before man, involving a phenomenon known as bioluminescence, where the chemicals are generated within biological organisms. Well-known examples of this include the light from glowworms and fireflies and the dull glow arising from the bioluminescence of myriads of tiny organisms seen in agitated seawater, for example, in the wake of boats. Many deep-sea creatures use bioluminescence, either to make themselves visible or to confuse prey. The reactions are essentially of the same type as that with luminol, except that the origin of the chemicals is from biological reactions.

15.2.2.8 RAMAN SPECTROSCOPY

Raman spectroscopy [16] relies on a form of inelastic scattering, where the periodic electric field of the incident radiation again induces a dipole moment in a material, but, in this case, there is an interaction with material vibrations, or phonons, resulting in a change of photon energy. The effect depends on the polarizability α of the bond, at the equilibrium internuclear distance and on the variation of this distance, x according to Equation 15.12 below. This equation has three terms—the first represents the elastic Rayleigh scattering (scattering at the same frequency as the incident radiation: 00 transition).

$$\alpha = \alpha_0 + \left(\frac{\partial \alpha}{\partial x} \right) x. \tag{15.12}$$

In the equation below, there are polarizability terms that have the effect of reducing or increasing the reradiated frequency, compared with that of the incident light, by an amount equal to the molecular vibrational or/and rotational frequency. This causes so-called Stokes (downward) and anti-Stokes (upward) shifts in the reradiated optical frequency. This light is referred to as Raman scattered light.

$$p = \underbrace{\alpha_0 E_0 \cos 2\pi\, v_0 t}_{\text{Rayleigh scattering}} + \frac{1}{2} \left(\frac{\partial \alpha}{\partial x} \right)$$

$$x_0 E_0 [\cos 2\pi\, (v_0 - v_1)t + \cos 2\pi\, (v_0 + v_1)t]. \tag{15.13}$$

The coupling of vibrational eigenfrequencies (v_1) of the scattering molecule to the exciting frequency (v_0) is sometimes called a photon–phonon interaction process.

Raman spectroscopy occurs with selection rules that are different to those for normal vibrational-absorption IR spectroscopy. Because of this, many users regard analysis by Raman scattering and by conventional absorption-based IR spectroscopy to be complementary techniques. This can be very convenient, but unfortunately, the Raman scattering intensity is very weak, even when compared with Rayleigh scattering. Despite this, however, with the advent of many high-power lasers, it is in many cases now a preferred technique. Not only is it relatively free of problems that can arise due to turbidity or scattering from amorphous solids, it is now far more easy to achieve with low-cost visible or NIR semiconductor lasers, allowing compact instrumentation which has drastically improved in recent years [17].

15.2.2.9 MULTIPHOTON SPECTROSCOPIC PROCESSES

Apart from the simple processes described above, there are many light absorbing and/or re-emission processes that involve more than one photon. It is beyond the present text to discuss multiphoton spectroscopy, but it is worth making a few short comments. Examples include two-photon absorption, where it requires simultaneous involvement of two photons to excite an absorbing level (is very unlikely to occur, therefore a weak effect) or where the first photon excites the molecule to a metastable level and then the second photon excites it to a higher level from this intermediate level. Of course, many photons can be involved if an appropriate set of metastable levels is present.

We have also hardly discussed an optical gain process, which is the mathematical opposite of absorption. This occurs in media which have electronic levels which have already been excited, often by light injection. This is rarely used in spectroscopy, apart from an intercavity measurements within lasers, but the process is often outside the gain region anyway, so will not be discussed further here.

15.3 SPECTROSCOPIC METHODS AND INSTRUMENTATION

Instrumentation has always been a very fast developing field, so only the basic methods will be discussed. The interested reader should consult specialist texts for more details of instrument design and manufacturers data for the latest performance figures. The fastest developing component areas are those of (1) low-noise detector arrays, both 1-D and 2-D (two-dimensional) types, and (2) compact high-power semiconductor laser sources for Raman and fluorescence excitation,

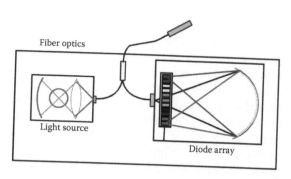

Figure 15.9 Spectrometer detection unit for reflectometric measurements, containing a white light source and a CCD camera. (A mirror is used in the light source to increase launch efficiency by imaging the filament on itself to "fill in" gaps in the filament.)

where performances are now achieved that were barely conceivable only 10 years ago.

Just as an example of these, to illustrate the type of components needed, a compact fiber-compatible spectrometer source and detection unit with a broadband source and concave diffraction grating is shown in Figure 15.9. In most cases, xenon or halogen lamps are a convenient source of broadband white light, which (often with the help of a rear reflector, to aid launch efficiency) is coupled into a fiber optic guide and then directed to the measurement probe. Here, transmitted or reflected light is collected and, via the coupler, guided back to a miniaturized spectrometer, which contains the diffraction grating and a photodiode array, such as a charge-coupled device (CCD) detector chip.

15.3.1 Spectrometer components

15.3.1.1 LIGHT SOURCES

Whenever an intense monochromatic light source or an intense light source that can be scanned over a small spectral range is required, the source of choice is nearly always a laser. This is clearly the optimum type of source for PCS or for Raman spectroscopy, as a monochromatic source is required. However, broadband sources, such as incandescent lamps, are desired for most full-spectrum spectrophotometry, as intense narrowband sources, such as lasers or low-pressure mercury lamps, either cannot be used for observing spectral variations over such a wide range or so many would be needed to cover the range that they would be prohibitively expensive.

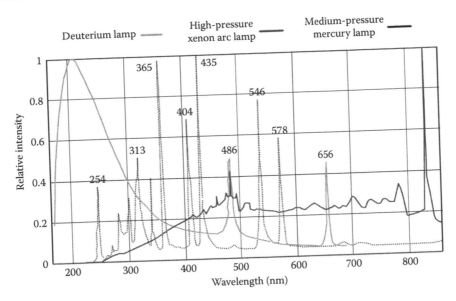

Figure 15.10 Comparison of emission spectra of a few discharge lamps and arc lamp sources.

Because most types of incandescent light sources were discussed in detail in Chapter 3, we shall cover only the most interesting spectral features of these broader band radiation sources, in particular types covering the important UV/VIS/NIR spectral regions. Tungsten lamps are low-cost and stable sources, which give a useful output over the VIS/NIR region (0.4–3.0 µm). Electrically heated thermal sources, such as Nernst emitters or modern variants, are usually used to cover the IR region beyond 3 µm.

High-pressure arc and discharge lamps are attractive sources in the visible and UV, as they can provide the desired broad spectral emission, but with significantly higher spectral intensity than tungsten lamps. In the visible and near IR region, the low cost and stable output of tungsten sources often means they are the preferred source. If operated at slightly reduced voltage, their lifetime is excellent, too. However, discharge lamps are far more effective sources than tungsten in the blue and violet regions and are far better in the UV region. Thus, the type of discharge lamp most often used to cover these regions in commercial spectrophotometers is the deuterium lamp, as this is the source that can provide the desired broadband performance at shorter optical wavelengths. Figure 15.10 provides a simple comparison of discharge lamps.

It is worth mentioning one more broadband source, the xenon flashlamp, that is very useful for spectroscopy. The output spectrum of it is, of course, very similar to that of the xenon arc lamp shown above. However, because it is only operated in short pulses and at a very low duty cycle (maybe a 100 µS flash every 0.01 s), it can be driven at very high current density without destroying the electrodes. This provides an even brighter source than the continuously operated arc-lamp version, albeit for very short "on" periods. These lamp sources also cover the UV, VIS, and part of the NIR spectrum. They are not only very compact and have low average power consumption, but can also provide a high peak energy output. This can give an optical signal at the detector that may be orders of magnitude higher than the thermal noise threshold of many optical receivers. They can, therefore, be used with optical systems, such as fluorimeters, where large energy losses often occur, yet still provide a good signal/noise ratio at the detector. Because of their pulsed output, they are also well suited for fluorescence lifetime measurements, provided the measured lifetimes are longer than the decay time of light output from the pulsed lamp.

15.3.1.2 COMPONENTS FOR OPTICAL DISPERSION (WAVELENGTH SEPARATION AND FILTERING)

15.3.1.2.1 Prisms

Many older spectrometers used prisms constructed of optical glass or vitreous silica. These exhibit

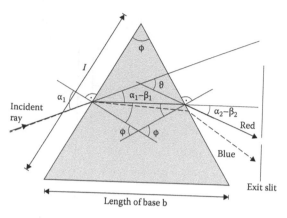

Figure 15.11 Dispersion using a prism: radiation incident on the prism is refracted and partially dispersed as it is deflected in φ direction toward the base of the prism. An exit slit selects the wavelength from the "rainbow" of radiation. Usually, the red wavelength is less refracted than the blue one.

a change of refractive index with wavelength, thereby giving wavelength dependent refraction, as shown in Figure 15.11. Collimating the input light and refocusing onto a slit behind the prism, it is possible to filter out "monochromatic" radiation.

In the UV, vitreous silica prisms are still occasionally used as wavelength dispersion elements, since they transmit well and and their dispersion is higher in this region than in the visible, but, even despite this, prisms are now rarely used in spectrometers. Because the prism material must be transparent and have a significant optical dispersion, greater problems with material selection arise in the mid-IR, so this was the historical reason why diffraction gratings were first used in this wavelength region.

15.3.1.2.2 Diffraction gratings

Diffraction gratings are based on coherent interference of light after reflection from, or refraction through, an optical plate with multiple parallel grooves. The dispersion of a grating depends on the spacing of the grooves, according to the well-known Bragg condition that describes the diffraction angle, i.e., the angle at which there is constructive interference of the diffracted light. In the more common reflective diffraction gratings, the surface has parallel grooves, of a "saw-tooth" cross-section, and these grooves have reflective surfaces. Apart from a high surface reflectivity,

the angle of the "saw-tooth" cross-section of the grooves to the plane of the surface is important to achieve good diffraction efficiency. The angle of the groove profile is called the blaze angle and this defines a blaze wavelength at which the diffracted power is maximized at the desired diffraction angle. The blaze angle can be chosen to maximize diffraction for the first order or for a higher order if desired (see Figure 15.12) [10].

Diffraction gratings were initially machined directly in a material that acted as the mirror reflector, and the greater ease of making gratings that had a greater intergroove spacing, meaning machining tolerances were less difficult, was another historical reason why gratings were first made for the IR region.

Most diffraction gratings are now made by one of two different processes. The first process, to make "replica" gratings, involves pressing (molding) of an epoxy-resin-coated glass substrate against a ruled (or "master") grating that has been previously coated with an aluminum or gold film, with pretreatment to ensure this film will "release" easily from the surface of the master when the epoxy has set and the two substrates are separated. This process duplicates the contours when the replica grating substrate, with the reflective metal film now attached to the outer surface of the resin, is pulled away from the master. The other production process, which is becoming ever more sophisticated, is to use photolithography to etch a grating in the material. The regular spacing is achieved by exposing photoresist using two converging parallel light beams to give a regular set of parallel and equally spaced interference fringes. These gratings can now be made by exposing the resist and etching the substrate in small steps, in a way that allows control of the blaze angle. These are called holographic gratings, because of the interference, or holography process that is used to expose the photoresist.

15.3.1.2.3 Optical detectors

Optical detectors and their associated amplifiers have been discussed in considerable detail in earlier sections, so there is no need to discuss them at length here. However, it is useful to comment that the most common single-element detectors used in wavelength-scanned spectrophotometers include photomultipliers (mainly for the UV range, below 400 nm), silicon photodiodes (VIS and NIR range

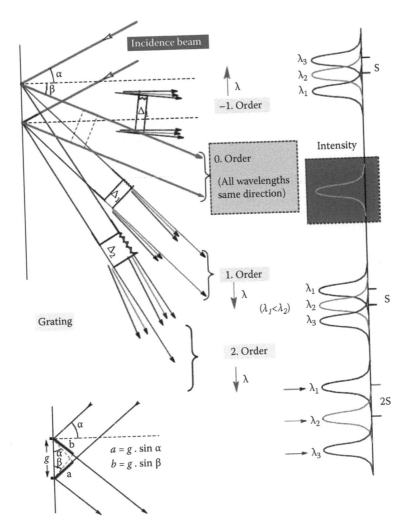

Figure 15.12 Dispersion at a diffraction grating: on the bottom left, the angle of incidence, a, cor-responding to the strongest diffraction, depends on the groove spacing constant, g. The groove profile can be varied to give more light in a preferred direction and the grating is usually specified as having a blaze angle for a desired wavelength. In the upper part of the diagram, it can be seen that dispersion by the grating depends on the interference order. Higher orders suffer more dispersion, so hence provide better wavelength resolution.

from 400 to 1000 nm), and photoconductive detectors (usually to cover above 1000 nm wavelength range). IR detectors are often cooled, using thermoelectric Peltier devices or mechanical thermal-cycle engines, to improve their inherently worse noise performance.

15.3.1.2.4 Optical detector arrays

Detector arrays are most commonly used in the visible and NIR region, primarily based on silicon (400–1050 nm) or GaInAs (500–1800 nm). One of the simplest approaches is to use a linear array of discrete photodiodes, each with its own separate preamplifier. However, it is more common in the silicon (400–1050 nm) wavelength range to use self-scanned arrays.

Advances in silicon CCD and other forms of self-scanned detector arrays have been tremendous in recent years, initially driven by developments for camera technology, but with better performances being achieved for low-light-level cameras or for scientific uses (e.g., spectroscopy and astronomy). Individual pixel noise levels are now getting down to levels of below 0.01

photoelectrons per second in the most advanced cooled scientific versions. All of these types of low-noise self-scanned detectors effectively store photoelectrons (in the capacitor formed by the individual detector diode elements) until they are electronically scanned out in the read-out process. Note that detectors and detector arrays were reviewed extensively in earlier sections.

15.3.2 Hardware instrumentation for spectroscopy

15.3.2.1 SPECTROMETERS: INSTRUMENTS FOR MEASURING OPTICAL INTENSITY, AS A FUNCTION OF WAVELENGTH

A spectrometer consists of a dispersive element, most commonly a diffraction grating, as described above, and a detection system. The latter is, most commonly, a silicon self-scanned CCD array (or a similar alternative), whenever visible or NIR (400–1050 nm) measurements are taken.

The signal to noise of the spectral analysis and detection part of spectrometers is often enhanced by various optoelectronic means. One of the most common methods for VIS/NIR use is to employ multielement self-scanned detector arrays with a fixed-grating light-dispersing system, as was shown in Figure 15.9 above.

A large proportion of these low-noise self-scanned arrays are 2-D types, with $n \times m$ detectors with up to several million elements, or pixels. Such 2-D arrays are clearly needed for cameras, for either low-light level terrestrial or astronomical use, and they offer an attractive advantage for use in spectroscopy, allowing use of a technique called spectral imaging. If light that passes through an input slit of a spectrometer is dispersed by, for example, a diffraction grating, and then is focused, or "imaged," onto such a 2-D detector, light from each part of the input slit will produce its own individual spectrum or "rainbow" image on this detector. This offers the possibility of measuring, independently, the spectrum of light that has entered at each point along the length of the input slit. The method therefore has the attractive property of measuring, simultaneously, the spectrum at many separate points on the input slit. There are many real-world applications of this. One, in

the field of fiber-remoted spectroscopy, is to allow multiplexing of light exiting from many optical fibers held in the focal plane of the spectrometer. Each of these can be positioned in line, just as the single fiber in Figure 15.9, but now, using a 2-D detector array in the output focal plane, a simultaneous measurement of the spectrum of light from each fiber can be performed. This is then called an imaging spectrometer.

15.3.2.1.1 Alternatives to detector arrays

A common alternative to array technology, particularly in the mid- and far-IR region, is to use some form of transform system, with a scanned optical filter, having a complex multiwavelength characteristic. Then, the temporal changes in the detected signal are electronically decoded to recover the spectrum. Such methods will be discussed in more detail in a later section.

15.3.2.2 THE SPECTROPHOTOMETER, AN INSTRUMENT TO MEASURE TRANSMISSION OR ABSORBANCE

Measurement of transmission loss usually involves the determination of light lost directly from a collimated light beam. This is normally performed in an instrument called a spectrophotometer. The result is usually required as a function of optical wavelength. The instrument normally includes a broadband light source, precise collimation optics to form the measurement beam or beams, some form of tunable optical filter to select the wavelengths measured, and finally a detector. This combination then allows transmission loss versus optical wavelength to be observed.

In the past, spectrophotometers were nearly always dual-beam instruments. In these, light from the source was split into two beams, which were alternately directed, first through a measurement sample path, where materials to be measured were placed, and then, via a reference optical path that contained no sample. This configuration allowed comparison (by signal division) of the transmission of each path, free of any intensity or spectral variations in the light source, or any spectral variations in the detector or in the collimation and filtering optics.

Many modern instruments are stable enough to first scan the wavelength range to take a wavelength-dependent intensity measurement without a sample in the beam, store the result on a PC and then rescan with the sample now inserted. Simple signal division, using the PC, then gives the transmission of the sample.

15.3.2.3 INDIRECT MEASUREMENT OF ABSORPTION USING FRUSTRATED INTERNAL REFLECTION METHODS

As mentioned above, frustrated internal reflection methods are useful for the measurement of highly absorbing samples, which may absorb (or scatter) far too much light to be able to measure them by traditional methods. For example, complex samples such as foodstuffs or biological materials, which might scatter light very strongly, can still be analyzed to detect strongly absorbing species, by placing them in contact with a high index glass plate and measuring the internally reflected light. Such a plate, with suitable coupling optics, is a common inclusion in the sets of optional attachments that can be purchased for use with commercial spectrophotometers. This evanescent field can be used for various measurements, as demonstrated in Figure 15.13.

Another characteristic of this evanescent field is that fluorescent materials or fluorophores close to the interface can absorb this evanescent field radiation and induce fluorescence. The evanescent field can be used to monitor effects very close to the interface because absorption clearly cannot take place beyond the penetration depth, so no fluorophores in the bulk are monitored.

15.3.2.4 MEASUREMENT OF INDIRECT EFFECTS OF ABSORPTION: PHOTOTHERMAL AND PHOTOACOUSTIC SPECTROSCOPY

Photothermal spectroscopy is a means of detecting or monitoring the absorption of materials, usually liquids or gases, by observing the resulting heat. The idea is that an intensity-modulated intense light source is used to illuminate the sample and this raises its temperature when light energy is lost from the beam. An intense laser beam is most commonly used. It first heats the fluid and then becomes deflected by the transverse variation of refractive index that occurs when the resulting heated region induces convection currents. Such currents give rise to thermal gradients. In addition, thermal-lensing effects can occur, which cause variation of beam diameter after passing through the fluid, because there is a higher temperature rise along the central axis of the beam.

Whichever method is used, the resulting deflections or beam diameter changes can be observed on optical detectors, often by enhancing the sensitivity to changes in the beam position or diameter by using lenses and spatial filters.

Photoacoustic spectroscopy is a special case of photothermal spectroscopy, where the light from an intense source, again usually a laser, is intensity modulated, for example, with a pulse, step or sinusoidal waveform, and the resulting changes in thermal expansion due to the heating are observed, usually with a traditional acoustic microphone. Gases have low specific heats and large expansion

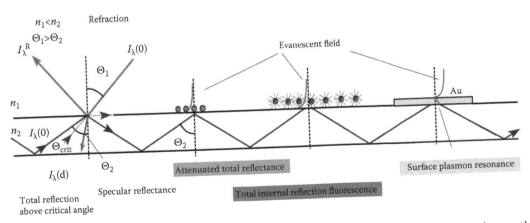

Figure 15.13 Illustration of the evanescent field caused by "total internal reflection." It can be partially absorbed (attenuated total reflection) or can excite fluorescence and surface plasmon resonance.

coefficients, so these are relatively easy to detect by this method. The light source can also be swept in optical frequency (or a broadband source can be passed through a swept optical filter) to allow spectral variations over an absorption line, or over a wider spectral region, to be observed.

A major advantage of the method for the measurement of liquids is that it can be used with turbid samples, for example, liquids containing small bubbles, or a suspension of nonabsorbing scattering particles, in a gas or liquid. In a conventional spectrometer, it is clearly not normally possible to distinguish between the light lost from the collimated beam due to elastic scattering (turbidity) and the light lost due to absorption.

15.3.2.4.1 Monitoring of chemiluminescence

In order to detect chemiluminescence and measure this weak luminol green light, it is best to use a photomultiplier, with some form of light-reflecting cavity (light-integrating chamber) to ensure most of the light hits the sensitive photocathode. Because photomultipliers have good green-light sensitivity and can even detect single photon events in the darkened state, very low concentrations of oxidizing agents can be measured, including hazardous oxidizing gases such as ozone, chlorine, and nitrogen dioxide. Numerous biochemicals can also cause a light-emitting reaction and hence be detected. Chemiluminescence is the basis of a number of commercial chemical sensors for important biochemicals.

A particularly useful reaction for law enforcement is the one that luminol has with blood, enabling crime scenes to be sprayed with this compound and then be viewed in the dark, when telltale glows appear wherever traces of blood are present, but here low-level camera detection is the simplest method.

15.3.2.5 INSTRUMENTATION FOR FLUORESCENCE SPECTROSCOPY

Fluorescence spectroscopy involves illumination of a sample with a monochromatic or filtered light source and observing the reradiated signal, which is almost invariably at a longer wavelength than the incident light. Light is either measured perpendicular to the incident optical axis, or for more sensitive detection, by collecting light emitted over wider angles with an integrating sphere.

It is common to observe either the fluorescence spectrum or the fluorescent decay curve, following pulsed excitation from a pulsed source. The latter is usually a pulsed laser or filtered light from a xenon flashlamp.

For fluorescence spectra, the most common method is to use a modified spectrophotometer where part of the omnidirectional fluorescent light is observed instead of the transmitted light. This can be done either with a dedicated instrument, or using a standard commercial spectrophotometer instrument fitted with a fluorescence attachment, which has appropriate optics to gather fluorescent light. There is no need to modulate the light source for this measurement, unless this is needed to allow lock-in (phase sensitive) detection methods to remove signals from unmodulated background light. When performing these measurements, considerable care has to be taken with the following filter-related aspects:

- Removal of any longer wavelength "side-band" light from the light source, which could be elastically scattered in the sample or the instrument and mistaken for fluorescence.
- Very careful filtering of the "fluorescent" light signal to remove any "cross talk" from elastically scattered source light.

The first aspect of source light filtering is less of a problem when using laser sources, although semiconductor lasers can have some residual spontaneous-emission component at longer wavelengths and gas or ion lasers can have residual light from the plasma discharge. Particular care is needed with incandescent, arc lamp, or xenon flashlamp sources, with their broadband emission. In a commercial spectrophotometer instrument or a dedicated instrument (spectrofluorimeter), the built-in grating spectrometer acts as a very good filter, provided "stray light" levels are low, and also has the advantage that the excitation wavelength can be tuned if desired. Additional rejection of long wavelength light is usually done with "short-pass" dichroic multilayer edge filters. The problem of removing elastically scattered source light from the "fluorescent" light signal can be solved in several ways. Narrowband dielectric laser mirrors make excellent rejection filters in transmission mode, as these can be designed with a reflectivity of 99.9% or higher. Dichroic long-pass edge filters are also now available with excellent performance. In addition, there is a wide range of long-pass optical glass filters, which have semiconductor-band-edge type transmission behavior, commonly having short-wavelength (i.e., shorter than the band edge) absorbance as high as 10^6 in a 2 mm thick filter. Care must be taken with these, however, as many of these fluoresce at longer wavelengths, so the first filtering stage is preferably done with a dielectric filter. Note that these filtering problems are even more acute for Raman scattering, so will be discussed further in the section below.

When it is desired to examine fluorescent lifetimes, the severity of the filtering problem is reduced by several orders of magnitude as the pulsed incident source will usually be no longer emitting by the time the intensity-decay curve of the consecutive fluorescence is now observed. However, strong light pulses can upset many sensitive detection systems or associated electronics for a short time after the pulse, so even here some degree of filtering is still desirable to remove light at the incident wavelength, and some source filtering may be necessary, too, if there is a spontaneous light decay occurring in the laser source itself.

When measuring fluorescent lifetime, a fast detector may be needed. Fluorescent decay is commonly in the form of an exponentially decaying curve, but lifetimes can typically vary from days, in the case of phosphorescence, to less than nanoseconds (important examples of samples having decay times in nanoseconds are the organic dyes often used in dye laser systems and some semiconductor samples having short excited state lifetimes). When measuring very fast fluorescent decays, it is common to use photon counting systems using photomultiplier (PMT) detectors. These have the advantage of high internal gain of the initial photoelectrons, so the input noise level of even a fast electronic preamplifier is easily exceeded. Also, as the desired detection time is reduced by using a fast PMT and amplifier, the effective peak current represented by a given size "bunch" of electrons, which will arrive at the anode from a single-photon event, actually increases as the time interval, over which it is measured, reduces (current=charge/time). Thus, using fast photon counting technology, where each photon count is fed into sets of digital registers according to its time of arrival, very fast fluorescent decay curves can be measured. It is now becoming more common

to design photon counting systems with avalanche photodiode detectors, which are operated in a so-called "Geiger" mode, where the incoming photon causes a full (but reversible, after the bias voltage is reapplied) conductive "breakdown" of the reverse-biased detector diode.

If there is more than one fluorophore or more than one fluorescent decay process occurring, the decay make take the form of a bi- or multiexponential decay curve, equivalent to linear addition of two or more exponentially decaying functions. In simple cases, these can be separated with software, but in some cases appropriate choice of excitation wavelength may help to isolate individual curves in mixtures of fluorophores.

Another common way of measuring fluorescence lifetime is to intensity modulate the source with a periodic waveform, usually a sinusoidal or square-wave signal, and observe the phase delay in the fluorescent signal intensity waveform, relative to that of the incident signal. This is commonly done using an electronic system, which can multiply the detected fluorescence signal by an electronic signal analog to the incident-intensity waveform, and then averages the product over a set time interval. If desired, more than one multiplication can be performed, using reference signals of different phases. Such an electronic detector system has several names, which include vector voltmeter, lock-in amplifier, or phase-sensitive detector. It not only allows for excellent signal averaging to be done to improve signal/noise, but also enables the phase difference between the two signals to be measured. The advantage is that cheaper, simpler detectors, such as silicon photodiodes, can now be used, as the illumination duty cycle is improved (to 50%, in the case of square-wave modulation), which helps to improve the signal-to-noise ratio, but the disadvantage is that the shape of the decay curve cannot be seen. Another significant disadvantage is that the system now requires much better optical filtering, as any residual optical cross talk, where elastically scattered light from the source might be detected, will alter the effective phase of the detected "fluorescent" signal.

An important feature is that fluorescence detection can be performed with highly scattering samples, such as roughly cut or even powdered materials, and can be used to analyze the surface of translucent materials, as a clear transparent sample is not required. Also, as the transmitted light level is not measured, very little sample preparation is needed. Another minor advantage is that the slightly shorter excitation wavelength means it can be focused on a slightly smaller diffraction-limited spot, enabling its use in fluorescence microscopes which excite the specimen via reasonably conventional optics. These advantages apply even more to Raman spectroscopy, which will be dealt with below, so they will be repeated again there for greater emphasis.

A particular problem with fluorescence detection is that many common materials will fluoresce, particularly if illuminated with UV light. These include, starting with particularly troublesome examples, organic dyes, compounds of rare-earth metals, chlorophyll, ruby, long-pass optical absorption filters, mineral oils, human sweat, many adhesives and even some optical glasses. Another problem can be that a high intensity excitation source may cause photodegradation.

15.3.2.6 INSTRUMENTATION FOR RAMAN SPECTROSCOPY

Raman spectroscopy can use conventional optical materials with low-cost semiconductor lasers and high-performance cooled-CCD detector arrays. A major advantage is that it can be used with highly scattering samples, such as roughly cut or even powdered and amorphous materials, and can even analyze the surface of apparently opaque materials. Obviously, a clear transparent sample is not required, as the transmitted light level is not measured, so very little sample preparation is needed. Another major advantage is the short excitation wavelength that means it can be focused on a smaller diffraction-limited spot, enabling its use in Raman microscopes, which excite the specimen via reasonably conventional optics and allow a spatial resolution that would be impossible with direct IR excitation. Confocal microscopy is also possible using Raman methods.

Raman spectroscopy requires a focused laser source to excite the very weak Raman scattering and a highly sensitive spectrometer to detect the very weak scattered light. The sample is illuminated with laser light, and the Raman scattered light is collected and often presented in terms of its power spectrum in wave numbers. Even in clear samples, such as optical glasses, which are only very weak Rayleigh scatterers, the Raman light intensity is usually several orders of magnitude even below

that from this elastic scattering. A major part of the problem was that highly scattering samples, such as cut materials or powders, can have elastic scattering levels 5 or more orders of magnitude higher than clear samples. Thus, it is necessary to separate out the very much stronger elastic scattering that can occur at the same wavelength as that of the incident laser light, and which might be between 5 and 9 orders of magnitude higher. The wavelength filtering to achieve separation usually requires at least two or maybe three stages of optical filtering to recover the desired Raman light. In the early days of Raman instrumentation, it was common practice to use rather large double or triple monochromators (i.e., two or three cascaded grating spectrometers). Second generation Raman systems reduced the problem by directing the light though a holographic Raman-notch filter, which is a compact optical filter designed just to reject a very narrowband of light, centered at the incident laser wavelength. It was easier to construct this compact component using low-fluorescence materials, thereby easing the problem of further separation in the dispersive spectrometers. With such a filter present in the detection system, a single spectrometer could be used, saving both space and cost. However, at the same time, another modern technology emerged in the form of greatly improved self-scanned silicon detector arrays, having very low noise and high quantum efficiency and being capable of providing even better low-light performance than photomultipliers on each tiny pixel in the array. This has allowed a compact state-of-the-art detector array to be placed in the output focal plane of a single monochromator, eliminating the need for a narrow output slit that rejected most of the light and now allowing all the wavelength components of the Raman spectrum wavelengths to be measured at the same time.

A more recent alternative to the Raman notch filter is the use of dichroic band edge filters. These interference filters, made in the usual way with a stack of dielectric layers on glass, have a long-wavelength pass characteristic that removes not only the undesirable elastically scattered light, but also the anti-Stokes Raman light. These new filters are highly effective, of relatively low cost, but far more stable than the previously used polymeric holographic ones used for notch filters. A minor disadvantage is the removal of the anti-Stokes signals, but as these are not used in over 90% of applications, this is not a major disadvantage.

The most troublesome residual problem with Raman systems is, as it always has been, that of undesirable background signals due to fluorescence. Fortunately, using NIR laser illumination, typically at 800 nm wavelength or above, goes a long way toward reducing the background fluorescence of most common materials and contaminants, so helps to make up for the weaker Raman signal when using longer wavelength excitation. There are also various signal subtraction methods that can be used. These can, for example, take advantage of the polarization dependence of Raman or of the temporal decay behavior of fluorescence, but these are beyond the scope of this introductory text.

All the above developments have given many orders of magnitude improvement in the spectrometer and detection system, to a situation where now Raman scattering is often a preferred technique to absorption spectroscopy. It is beyond the scope of this text to go into detail on other techniques for enhancing Raman signals, but it is useful to mention two briefly. The first is resonance Raman scattering, where the excitation wavelength is close to an absorption line of the material to be monitored. The second is surface-enhanced Raman spectroscopy (SERS), where the use of a metal surface, or a surface covered with metal particles, can greatly enhance the signal from Raman scattering. A few materials, such as silver or silver particles, are highly effective for this. Both of the above methods can enhance the Raman signal by between 4 and 8 orders of magnitude, depending on conditions. Being a surface effect, however, the SERS method is clearly very sensitive to surface preparation conditions and to subsequent treatment or contamination.

15.3.2.7 INSTRUMENTATION FOR PHOTON CORRELATION SPECTROSCOPY

The use of PCS for particle detection was discussed earlier. It is particularly useful for determination of particle size, over a range of a few nm to a few μm, simply by measuring scattered signals. Very small particles will undergo fast random movements as molecular collisions move them about (Brownian motion). In a conventional instrument, a polarized monochromatic TEM_{00} laser beam is usually used to illuminate the sample by focusing on a narrow beam waist, and a fixed detector is used to observe the scattered light from the beam-waist area. The

laser source must be stable better than a small fraction of 1% and usually has a power of a few mW. Originally gas lasers were used, but now compact semiconductor lasers are replacing them.

A single spatial mode (single "speckle") in the far field will exhibit more rapid intensity changes with small scattering particles present compared to larger particles at the same conditions. In order to detect the changes of phase, the scattered light is traditionally imaged through a very small hole acting as a spatial filter, in order to provide the greatest intensity modulation index (greatest fractional change in optical intensity as the optical phase changes). Clearly, a large detector without any pinhole aperture is unsuitable, as it would average the light from many interference points, and so would result in a smaller modulation index.

For successful operation, entire optics must be very clean, and additional opaque baffles are often used to reduce stray light. Samples must be prepared very carefully to avoid highly scattering large dust particles or bubbles, and clearly particle aggregation (clustering) must be avoided.

Following optical detection of the intensity changes, electrical spectral analysis (frequency analysis) of the signal scattered, for example, at 90°, could potentially yield valuable information on particle size, with smaller particles giving higher intensity modulation frequencies. Unfortunately, very small particles also scatter very weakly and, to act as an effective spatial filter, the receiving aperture has to be very small. As a result, the received photon flux is usually very low, sometimes only a few photons per second, making frequency analysis of a very noisy (random photon event) signal much more difficult. A preferred alternative method more suitable for use at low photon flux levels is the use of a method called PCS. Here, instead of analyzing the frequency of detected signals, the times of arrival of many individual photon pulses are correlated. The decay time of what, for monodisperse (single-size) particles, is usually an exponentially decaying correlation function, can be derived using digital correlator systems. As stated earlier, this can, with knowledge of the temperature and viscosity of the fluid in which the particles are suspended, be related to the hydrodynamic radius of the particles using the Stokes–Einstein equation. The method works best with single-size (monodisperse) particles, but more complex correlation functions from suspensions of two or even three particle types/sizes can be inverted using the Laplace transformation.

The photon correlation method works well at these very low light flux levels, but requires the use of single photon counting detectors, such as photomultipliers or silicon avalanche photodiodes (APDs). Significant developments have been made to actively quench APD photon counters that are operated in avalanche mode, at very high reverse bias, to allow fast recovery from their photon-induced breakdown and be ready to detect a subsequent photon. This allows instrumentation to take advantage of their superior quantum efficiency in the NIR [18].

Other developments have been to design fast correlators to process the photon count signals and recover particle size information. PCS can typically achieve an accuracy of order 0.5% in particle size on monodisperse samples. With more sophisticated signal processing, it is possible, provided conditions are suitable, to derive estimates of particle size distribution, polydispersity, molecular weight estimates (using Svedberg's equation), rotational diffusion behavior and particle shape, and many other parameters. The greatest practical problem is usually when large particles are present, either as an undesirable contaminant or as an inevitable feature of the sample to be monitored, as scattering from just a few large particles can often be intense enough to totally dominate the weak signals from many much smaller ones.

15.3.2.8 MEASUREMENT WITH OPTICAL FIBER OPTIC LEADS

Generally, if used efficiently, optical fiber leads offer tremendous advantages. First, expensive instrumentation can stay in a safe laboratory environment, free from risk of damage from chemicals, weather, or careless handling. Second, the remote measurement probe can be very small, robust, and immune to chemical attack. Third, there is no need to transport chemicals or other samples to the instrument, so real-time online measurements are possible with no risk to personnel.

15.3.2.8.1 In-fiber light delivery and collection for transmission measurements

Transmission (and hence absorption or turbidity) measurements can be most easily achieved over optical fiber paths by using a commercial spectrophotometer with specially designed extension leads. Typically, they have a unit that fits into the cell compartment of a standard instrument, with a first (focusing) lens that takes the collimated light that would normally pass through the sample chamber and focuses it instead into a large-core-diameter (usually >200 µm) optical fiber down-lead. A second lens that recollimates light, that has returned back from the sample area (via the return fiber lead), is used to reform a low-divergence beam suitable for passage back into the detection section of the instrument.

There is also a remote measurement cell in the sample probe, connected to the remote end of both these fiber leads. In this remote sample area, a first lens collimates light (coming from the spectrometer, via the down-lead) into a local interrogation beam. This beam passes through the remote measurement cell, after which a second lens collects the light and refocuses it into the fiber return lead going to the spectrometer instrument. Such optical transformations lead to inevitable losses of optical power (due to reflections, aberrations and misalignments) of typically 10–20 dB (equivalent to losing 1–2 units of absorbance in the measurement range). However, most modern spectrophotometer instruments have a typical dynamic range of >50 dB, so this optical loss is a price that many users are prepared to pay in order to achieve a useful remote measurement capability.

It should be noted that the optical power losses usually occur mainly due to misalignments and the imperfections of the focusing and recollimation optics, plus Fresnel reflection losses at interfaces, rather than arising from fiber transmission losses. If suitably collimated beams were to be available in the instrument, if large core diameter fibers could be used to connect to and from the probe, and if the entire optics, including fiber ends, could be antireflection coated, there should really be very little loss penalty. There are many other probe head designs that are possible. The simplest design, for use with measurement samples showing very strong absorption, is simply to have a probe that holds the ends of the down-lead and return fiber in axial alignment, facing each other across a small measurement gap, where the sample is then allowed to enter. Losses are low for fiber end spacing of the same order as the fiber diameter or less, but rapidly increase with larger gaps. The probe is far easier to miniaturize and to handle if the fiber down-lead and return lead are sheathed, in parallel alignment, in one cable. Use of such a cable can be accommodated using a right-angled prism or other retroreflecting device to deflect the beam in the probe tip through the desired 180° that allows it to first leave the outgoing fiber, pass through a sample and then enter the return fiber. Use of a directional fiber coupler at the instrument end allows use of a single fiber, but then any residual retroreflection from the fiber end will be present as a cross talk signal, adding light signal components that have not passed through the medium.

Clearly, there are many variants of such optical probes, some involving more complex optics (e.g., multipass probes), some constructed from more exotic materials to withstand corrosive chemicals. A very simple option that has often been used with such single fiber probes for monitoring the transmission of chemical indicators, is to dissolve the indicator in a polymer that is permeable to the chemical to be detected and also incorporate strongly scattering particles in the polymer. When a small piece of such a polymer is formed on the fiber end, the particles give rise to strong backscattered light, and the return fiber guides a portion of this to the detection system. This backscatter light had of course to pass through the indicator polymer in its path to and from each scattering particle, so the returning light is subject to spectral filtering by the indicator. Although this is a very lossy arrangement, it is extremely cheap and simple and has formed the basis of many chemical sensors, for example ones using pH indicators.

15.3.2.8.2 In-fiber light delivery and collection for Raman scattering and fluorescence

Raman scattering and fluorescence can also be measured via fiber leads, but the use of fibers causes far more loss of light due to the wide-angle reradiation patterns characteristic of both methods. However, the potential value of these methods, particularly of Raman scattering, for chemical sensing has encouraged workers to persevere to get useful performance, despite the low return light levels achieved with fiber-coupled systems. Both these mechanisms involve excitation of a sample with light, usually at a wavelength shorter than the scattered light to be observed; the reemitted light is then collected and narrowband filtered.

The loss due to launching of the excitation laser light into a fiber is usually negligible with Raman, as narrow-line laser sources are used, but ultimately the launch power limit may be set by nonlinear processes or, in the case of large-core multimode fibers, by optical damage thresholds. Similar excitation can be used for low-level fluorescence monitoring, provided no photobleaching or other photodegradation of the monitored substance can occur at high illumination intensity. The main potential loss is therefore that of light collection. The additional loss due to collection of Raman light via optical fibers is at least a factor of 90 greater than when using conventional optics, thus making the already low light levels about 2 orders of magnitude worse. Despite this, a number of fiber-remoted Raman chemical sensor probes are appearing as commercial items.

15.3.2.8.3 Evanescent field methods of in-fiber light delivery and collection: Frustrated total internal reflection

Frustrated internal reflection measurements are clearly attractive for use with multimode optical fibers, as the guidance in the fiber depends on such internal reflections. As stated already, total internal reflectance is often not total. One marked departure from traditional school teaching occurs during reflection at curved interfaces, where light can, under certain circumstances, be lost by radiation into the less dense medium, even when the angle of incidence is above the well-known critical angle. This occurs in multimode optical fibers if light is launched to excite skew rays, provided the light is launched at an angle to the fiber axis that is too large to expect guidance by total internal reflection, assuming rays were passing through the fiber axis. Thus, even if the actual angle of incidence on the fiber core–cladding interface for these skew rays is greater than the calculated critical angle, light will be lost into the cladding to form what are called leaky rays.

However, even with light incident on flat surfaces, so assuming no leaky rays are possible, there is always a small penetration of light into the less dense medium. Another aspect that is not normally taught in elementary optical texts is that the reflected light beam, although being reflected at the same angle to the normal as the incident light, undergoes a lateral shift in the direction of the interface plane between the two media. It is said to suffer a so-called Goos–Hänchen shift (this is like a displacement of the "point" of reflection at the interface). A fuller examination, using either classical geometric optics or quantum optics, predicts that part of the radiation, the evanescent field, penetrates the interface and then, for a certain distance, is guided outside the optically denser medium, leading, when it returns, to a lateral shift in the apparent point of reflection from the interface.

The concept of the evanescent field, which decreases exponentially away from the interface in the medium having the lower refractive index, was introduced earlier. This evanescent field can be used for various fiber-based experiments, as shown in Figure 15.13.

As stated earlier, material or molecules within a penetration depth, d, of this field can absorb the radiation and result in an attenuated total reflectance. The absorption can be enhanced further using thin metal layers on the fiber to cause evanescent field enhancement due to plasmon resonance. This effect will be discussed in more detail later. Another application of this evanescent field is that fluorescent materials or fluorophores close to the interface can absorb this evanescent field radiation and induce fluorescence. The evanescent field can be used to monitor effects very close to the interface because absorption clearly cannot take place beyond the penetration depth, so no fluorophores in the bulk are monitored.

15.3.2.8.4 In-fiber light delivery and collection for PCS

In PCS systems, monomode fibers are very well suited for both delivery and collection of light. A single-mode optical fiber not only makes an excellent delivery medium for a beam launched efficiently from a gas or semiconductor laser, but it can also form a near-ideal single-spatial-mode optical filter to replace the conventional pinhole used for traditional systems. The PCS technique is, therefore, easily adaptable to perform remote measurement with optical fibers. There is very little penalty in using optical fibers, as the lasers can be launched with high efficiency, and because a fiber-based single-spatial-mode receiving filter will not lose any more light energy than the alternative of a tiny hole in a metal plate. The fiber, in fact, makes a near-ideal spatial mode filter.

15.3.2.9 SPECIALIZED OPTOELECTRONICS AND SIGNAL PROCESSING METHODS FOR SPECTROSCOPY

This section will look at specialized spectroscopic methods such as scanned filters, modern fixed detector arrays, and use of Hadamard and Fourier transform signal processing methods. In simple dispersive spectrometer instruments, the desired selection of optical wavelength or frequency is achieved by monochromators using a prism or diffraction grating, with the necessary collimation and refocusing optics. The spectrum is then recorded sequentially in time as the frequency or wavelength of the light transmitted by the filter is scanned.

Another possibility, discussed in this section, is to use various parallel multiplexing techniques where all wavelengths are monitored at the same time. There are two generic lines of development that have gained success in recent years.

The first of these, which has already been mentioned above, is simply to use multiple detectors, either a discrete-multielement photodiode array with separate amplifiers, or a self-scanned CCD detector array. Both of these enable the parallel detection of all wavelengths, and thus a more efficient use of the available light. These components were discussed above.

The second generic approach is to pass the light through a more complex multiwavelength optical filter, which is capable of passing many wavelengths at once, but where the spectral transmission varies as it is scanned with time. Light then impinges on a single optical detector and finally the transmitted spectrum is decoded by applying mathematical algorithms to the observed temporal variations in the detected signal as the complex filter is scanned. Two common variations of this so-called transform approach are used, first the Hadamard, and second, the Fourier method. Both methods have the advantage of parallel multiplexing, thereby achieving a so-called Fellget advantage [19]. Both methods also have the additional advantage of high optical "throughput" [20], since instead of a single narrow slit a large area (normally a circular hole) is used, allowing far more light to enter.

In both of these transform methods, mathematical analysis is used to decode the temporal modulation pattern of optical signals arising from superposition of radiation at many different wavelengths. In the Hadamard spectrometer, an encoded mask with a transmissive (or reflective) pattern of markings is positioned in the focal exit plane of a normal dispersive spectrometer, and then a mathematical transformation called the Hadamard matrix is applied to the detected signal. The coded mask usually has a fine-detail binary orthogonal code on it, forming a pattern of elements like 0100101001101110010 such that the detected transmitted (or reflected) signal varies with the position of this mask [21]. Using the Hadamard matrix to perform the desired mathematical transform, the transmitted spectrum can be reconstructed from the detected modulation of light.

Fourier transform spectrometers require a filter having a sinusoidal variation of optical transmission, which can have its spectral period (wavelength or frequency between successive peaks in transmission) modulated with time in a defined manner. All two-path optical interferometers, as their path difference is varied with time, conveniently demonstrate the desired sinusoidal transmission changes which are a natural consequence of the interference process. The most convenient form of this to use in instruments is the Michelson interferometer configuration [22,23].

The simplest way to estimate how a Fourier transform spectrometer operates is to first consider what will happen with monochromatic light (e.g., single-frequency laser). Here, if the path difference of the interferometer is increased linearly with time, the intensity transmitted by the interferometer will vary in a purely sinusoidal manner with time, i.e., the output from the optical detector will be a single electronic frequency. Any more complex optical spectra can be considered to be composed of a superposition of many such pure optical frequencies, and each of these single-frequency components will generate a pure electronic frequency or tone, and so will give its own unique electronic signature. The temporal signal from the detector will be composed of a linear superposition of all such signals. Decoding of such an electronic signal to recover all such single-frequency sinusoidal signal components is a standard problem in electronic spectrum analyzers, and the well-known solution is that of Fourier analysis; hence, the use of the words Fourier transform spectrometer to describe its optical equivalent.

In all of these transform methods, the instantaneous condition of the complex optical filter is known from the position of the interferometer (Fourier method) or the coded mask (Hadamard method), thus allowing decoding of detected signals to produce the desired spectral output.

15.3.2.10 COMPARISON OF PROCESSING METHODS

Before finishing this section, it is instructive to discuss the relative strengths and weaknesses of each approach. In fundamental signal to noise ratio terms, it is preferable to use dispersive optics with a fully parallel array detector as, for example, used to great advantage in modern CCD array spectrometers. Unfortunately, such high-performance detector arrays are only available in the visible and NIR region, as detector technology for other regions is not yet as well commercialized [24]. Because of this, there is far greater use of Fourier transform methods in the mid- and far-IR regions, where it is only necessary to have one (often cooled) high-performance single detector. The Hadamard and Fourier methods gain advantage over using a single detector with a scanned narrowband filter, as more wavelengths are transmitted through the filter at any one time, and a greater optical throughput can be used in the optical system. The Fourier transform method has a further advantage, however, when high spectral resolution is desired, as it is difficult, particularly in the IR region, to make detector arrays with a small enough spacing to resolve very closely spaced wavelength, which would otherwise need very large dispersive spectrometers to achieve high resolution.

15.4 CASE STUDIES OF RESEARCH INTO OTHER SPECTROSCOPIC METHODS

In recent years, an increasing number of applications of optical sensors has been published. Many optical techniques using either fluorophore-labeled samples or applying direct optical detection have been investigated and well covered in the literature [1,25,26]. The direct optical methods can be classified according to whether they use refractometric or reflectometric principles. As discussed above, the first relies on the behavior of the evanescent field. Several types have been commercialized. The best known type is the BiaCore [27] using surface plasmon resonance. Several examples of refracto-metric methods are listed briefly below:

- An early method for refractive index monitoring, by Lukosz [28], was to embed a grating on the surface of a slab (rectangular cross section) waveguide, with the sensor taking advantage of the modification of the Bragg grating condition when the index of the external medium changed. The original concept has since been modified by many scientists. Read-out from the sensor can be conveniently performed using a CCD sensor array [29].

- Another approach used a prism coupler, which is a common means of coupling light into slab waveguides. This technique is also sometimes called a "resonant mirror" method. A layer of lower refractive index is arranged as a buffer layer between the prism and the waveguide. An incident beam (polarized at 45° to the normal in the plane of incidence) is reflected at the base of the prism in a manner dependent on the wavelength, the angle of incidence, and the optical properties of both the prism and the waveguide. The incident beam can excite TE (transversal electric) and/or TM (transversal magnetic) modes of the waveguide and the modes of the waveguide can recouple back into the prism, resulting in a phase shift relative to the directly reflected beam. Because both propagating modes (TE and TM) travel with different velocities within the waveguide and are differently influenced by the refractivity of the medium in the evanescent field region [30], the plane of polarization of the output light is in general elliptical, with a polarization state depending on the relative phase delay. The process is similar to the polarization changes that occur in a birefringent crystal.

- Direct interferometers, i.e., ones not making use of polarization changes, are also interesting and are useful arrangements for interrogation of refractive index. In a commonly used configuration, radiation is guided via two arms of a Mach–Zehnder interferometer [31], and one of the arms is covered by a sensing layer (polymer, biomolecular recognition layer). Guided radiation propagates within these two arms with different phase velocities, resulting in a phase shift that, after interferometric superposition at the coupling junction, can be measured as an intensity change, which is now dependent on the refractive index of the medium in contact with the recognition layer. Another possible configuration is the well-known Young interferometer [32] arrangement. Instead of recombining the two waveguides in the planar layer, the two beams are now focused to one plane only, using a cylindrical lens, and directed onto a CCD array where they form an interference pattern.

- One of the now most commonly used methods, surface plasmon resonance, involves enhancing the evanescent field intensity using thin metal films coated on the surface of a dielectric. The surface plasmons (electrons) in the metal film coating can be excited by the guided wave in the waveguide. The excitation depends on the refractive index of the medium on top of the metal film and varies dramatically near the resonance condition, as the intensity of the radiation propagating in the waveguide is substantially reduced by the stronger resonance coupling [33]. It can be used as a detection principle, either directly, using light propagating through the waveguide, or indirectly, using a prism coupled to the waveguide structure. Originally, the concept of surface plasmon resonance was introduced to the scientific community by Kretschmann [34] and Raether [35].

In addition to the above examples of interrogation principles, many other approaches have been published [36–38] and recently reviewed [1].

Apart from refractometry, reflectometry is frequently used in sensing. It often relies on multiple reflections and usually involves white light interferometry (similar to the Fabry–Perot) to give a wavelength-dependent or filtering action. For many decades, ellipsometry has been used to examine the properties of thin film layers. As mentioned above, reflection and refraction takes place at each interface between media of different refractive index. The partially reflected beams from the interfaces at each side of a layer will superimpose and exhibit interference. The intensity of reflection depends on the wavelength, the angle of incidence of radiation, the refractive index of the layer, and the physical thickness of this layer. In ellipsometry [39,40], polarized light is used and thus the refractive index and the physical thickness of the layer can be determined separately. In normal reflectometry, no polarization state is selected, resulting in a very simple, robust and easy-to-use method for monitoring effects in these layers, conveniently having negligible temperature dependence [37]. The reason is an opposite behavior of refractive index and of the thickness of the layer in dependence on temperature which compensate. Gases or liquids can be measured applying interference reflectometry. It has the combined advantages of small cell volume and the possibility of remotely monitoring interference effects in the cell via fiber optic leads.

15.5 CONCLUSIONS

We have now completed our introduction to the subject of spectroscopy, specifically to give a simple overview to nonspecialist readers. Of course, this area is so complex, it can fill large specialist textbooks, so it is hoped that the reader will forgive the inevitable omissions. We have given a short introduction to the basic theory, a brief practical overview of components and instruments, and finally we have introduced a few examples of recent research areas, mainly using fiber optics. Clearly, the interested reader can gain more insight into this fascinating subject from more-voluminous specialist textbooks, many of which have been mentioned in the references and further reading.

REFERENCES

1. Gauglitz, G. and Goddard, N. J. 2014. Direct optical detection in bioanalytics. In: Handbook of Spectroscopy, (eds) G. Gauglitz, and D. S. Moore (Weinheim: Wiley-VCH), Chapter 29, 1115–1158.
2. Strobel, H. and Heineman, W. 1989. Chemical Instrumentation: A Systematic Approach (New York: John Wiley & Sons).
3. van de Hulst, H. C. 1957. Light Scattering by Small Particles (New York: Wiley).
4. Schrader, B. (ed.) 1995. Infrared and Raman Spectroscopy. Methods and Applications (Weinheim: Wiley-VCH).
5. Gauglitz, G. 1994. Ultraviolet and visible spectroscopy. In: Ullmann's Encyclopedia of Industrial Chemistry, Vol. B5 (Weinheim: Wiley-VCH).
6. Ingle, J. D., Jr. and Crouch, S. R. 1988. Analytical Spectroscopy (Englewood Cliffs, NJ: Prentice-Hall).
7. Svehla, G. (ed.) 1986. Analytical visible and ultraviolet spectrometry. In: Comprehensive Analytical Chemistry, Vol. XIX (Amsterdam: Elsevier).
8. Skoog, D. A. and Leary, J. J. 1992. Principles of Instrumental Analysis (Fort Worth, TX: Saunders College Publishing).
9. Born, M. and Wolf, E. 1980. Principles of Optics (New York: Pergamon).
10. Hecht, E. and Zajac, A. 1974. Optics (Reading: Addison-Wesley).
11. Stacey, K. A. 1956. Light-Scattering in Physical Chemistry (London: Butterworths).
12. Harrick, N. J. 1979. Internal ReflectionSpectroscopy (New York: Harrick ScientificCorporation).
13. Klessinger, M. (ed.) 1995. Excited States and Photochemistry of Organic Molecules (New York: John Wiley & Sons).
14. Wolfbeis, O. S. (ed.) 1991. Fiber Optic Chemical Sensors and Biosensors (Boca Raton, FL: CRC Press).
15. Narayanaswamy, R. and Wolfbeis, O. S. 2004. Optical Sensors (Berlin, Heidelberg: Springer Verlag).
16. Colthup, N. B., Daly, L. H. and Wiberley, S. E. 1990. Introduction to Infrared and Raman Spectroscopy (New York: Academic Press).
17. Vandenabeele, P. 2013. *Practical Raman Spectroscopy an Introduction* (Weinheim: Wiley VCh).
18. Brown, R. G. W. and Smart, A. E. 1997. Appl. Opt., 36, 7480–7492.
19. Fellgett, P. B. 1958. J. Phys. Radium, 19, 187–191.
20. Jacquinot, E. 1954. J. Opt. Soc. Am., 44, 761–765.
21. Plankey, F. W., Glenn, T. H., Hart, L. P. and Winefordner J. D. 1974. Hadamard spectrometer for ultraviolet-visible spectrometry. Anal. Chem., 46(8), 1000–1005.
22. Griffiths, P. R. 1975. Chemical Infrared Fourier Transform Spectroscopy (New York: Wiley).
23. Griffiths, P. R. and De Haseth, J. A. 1986. Fourier Transform Infrared Spectrometry (New York: Wiley).

24. http://www.sensorsinc.com/products/area-cameras/.

25. Wang, X. D. and Wolfbeis, O. S. 2013. Fiber-optic chemical sensors and biosensors (2008–2012). Anal. Chem., 85(2), 487–508.

26. Conn, P. M. (ed.) 2013. *Fluorescence-Based Biosensors: From Concepts to Applications*, Vol. 113, (Series ed.) M. C. Morris, Progress in molecular biology translational science, (Academic Press).

27. http://www.gelifesciences.com/webapp/wcs/stores/servlet/catalog/en/GELifeSciences-de/products/AlternativeProductStructure_11932/.

28. Nellen, Ph. M. and Lukozs, W. 1993. Biosens. Bioelectron., 8, 129–147.

29. Brandenburg, A. and Gombert, A. 1993. Sens. Actuators, B, 17, 35–40.

30. Cush, R., Cronin, J. M., Stewart, W. J., Maule, C. H., Molloy, J. and Goddard, N. J. 1993. Biosens. Bioelectron., 8, 347–354.

31. Ingenhoff, J., Drapp, B. and Gauglitz, G. 1993. Fresenius J. Anal. Chem., 346, 580–583.

32. Brandenburg, A. and Henninger, R. 1994. Appl. Opt., 33, 5941–5947.

33. Piraud, C., Mwarania, E., Wylangowski, G., Wilkinson, J., O'Dwyer, K. and Schiffrin, D. J. 1992. Anal. Chem., 64, 651–655.

34. Kretschmann, E. 1971. Z. Phys., 241, 313–324.

35. Raether, H. 1977. Phys. Thin Films, 9, 145–261.

36. Gauglitz, G. 1996. Opto-chemical and opto-immuno sensors. In: Sensors Update, Vol I, (Weinheim: VCH Verlagsgesellschaft mbH).

37. Gauglitz, G. 2010. Direct optical detection in bioanalysis: An update. Anal. Bioanal. Chem., 398(6), 2363–2372.

38. Fan, X. D., White, I. M., Shopova, S. I., Zhu, H. Y., Suter, J. D. and Sun, Y. Z. 2009. Sensitive optical biosensors for unlabeled targets: A review. Anal. Chim. Acta, 620, 8–26.

39. Arwin, H. 2003. Ellipsometry in life sciences. In: Handbook of Ellipsometry (eds) G. E. Jellison and H. G. Tompkins (Park Ridge, NJ: Noyes Publications). Azzam, R. M. A. and Bashara, N. M. 1989. Ellipsometry and Polarized Light (Amsterdam: North Holland).

40. Azzam, R. M. A. and Bashara, N. M. 1989. Ellipsometry and Polarized Light (Amsterdam: North Holland).

FURTHER READING

Anton van der Merwe, P. 2001. Surface plasmon resonance. In: Protein–Ligand Interactions: Hydrodynamics and Calorimetry (Oxford: Oxford University Press).

Brecht, A., Gauglitz, G., Kraus, G., and Nahm, W. 1993. Sens. Actuators, B, 11, 21–27.

Dyer, S. A. et al. 1992. Hadamard methods in signal recovery. In: Computer-Enhanced Analytical Spectroscopy, Vol. III (ed.) P. C. Jurs (New York: Plenum Press) pp. 31–67.

Gauglitz, G. and Nahm, W. 1991. Fresenius' Z. Anal. Chem., 341, 279–283.

Gauglitz, G. and Proll, G. (2008). Strategies for label-free optical detection. In: Advances in Biochemical Engineering/Biotechnology, Vol. 109 (Biosensing for the 21st Century), (ed.) T. Scheper, pp. 395–432, Springer-Verlag, Berlin.

Kinning, T. and Edwards, P. 2002. Optical Biosensors (eds) F. S. Ligler, T. Rowe and A. Chris (Amsterdam: Elsevier).

Klotz, A., Brecht, A., Barzen, C., Gauglitz, G., Harris, R. D., Quigley, Q. R. and Wilkinson, J. S. 1998. Sens. Actuators, B, 51, 181.

Kuhlmeier, D., Rodda, E., Kolarik, L. O., Furlong, D. N. and Bilitewski, U. 2003. Biosens. Bioelectron., 18, 925.

Liebermann, T. and Knoll, W. 2000. Colloids Surf., A, 171, 115.

Mark, H. 1991. Principles and Practice of Spectroscopic Calibration (New York: John Wiley & Sons).

Robinson, J. W. (ed.) 1974. CRC Handbook of Spectroscopy, Vols. I–III (Cleveland, OH: CRC Press).

Smith, B. C. 1995. Fundamentals of Fourier Transform Infrared Spectroscopy (Boca Raton, FL: CRC Press).

Optical to electrical energy conversion: Solar cells

TOM MARKVART
University of Southampton

FERNANDO ARAUJO DE CASTRO
National Physical Laboratory

16.1 INTRODUCTION

Photovoltaics (PV) is about to celebrate 70 years of its existence in the modern era. Photochemical reactions that can produce energy from sunlight have been known for over 160 years [1] and photosensitive semiconductor devices were discovered in the second half of the nineteenth century [2,3]. However, it was not until the 1950s that the first practical solar cell was developed at Bell Laboratories in New Jersey [4].

Solar cells soon found applications in supplying electrical power to the first satellites. The first practical terrestrial systems followed in the 1970s. Their economic viability was restricted to what is now referred to as remote applications in isolated locations, including powering of radio repeater stations, control and measurement devices, and cathodic protection equipment for oil and gas lines. As costs fell, PV systems started to be economically feasible to provide power to the large number of people who did not have (or, in the foreseeable future, were unlikely to have) access to mains electricity.

Since the oil crises in the 1970s, solar electricity has been increasingly advocated as an alternative to conventional electricity supplies, which, in most countries, depended on fossil fuels. The first larger scale (utility) PV power plants started to be built in the United States and Europe during the 1980s.

In the last decade of the twentieth century, solar cell integration into roofs and facades of domestic and commercial buildings became widespread, enabling a new and attractive distributed form of power generation (see Table 16.1). In the early stages, economic viability often required government subsidies, but continued price reduction has led to a boom in PV production and increased attractiveness of this form of energy generation, with a market growth of more than 40% per year between 2000 and 2014.

Table 16.1 The growth of PV applications

1950s	First modern solar cells appeared
1960s	Solar cells become the main source of electric power for satellites
1970s	First use for remote industrial applications on the ground
1980s	Solar cells used in rural electrification and water pumping; first grid-connected systems
1990s	Major expansion in the use of building-integrated systems
2000s	PV installed capacity grows in average at over 40%/year
2014	PV installed capacity reaches over 1% of total global electricity demand

Solar cells capable of generating more than 38 GW were installed in 2014, providing electricity for applications of all scales, ranging from milliwatts to megawatts (Figure 16.1). This chapter aims to give an overview of this rapidly growing industry, with an outline of the device and system aspects of the field. The basic principles of solar cell operation are described in Section 16.2 along with an examination of the fundamental constraints on solar cell

efficiency. Section 16.3 reviews current PV technologies covering crystalline-silicon, thin-film, organic, and hybrid solar cell types. The integration of these into PV systems is briefly discussed in Section 16.4, with a short overview of the design and principal features of stand-alone and grid-connected systems.

Please note, however, that there is a special section on PV applications in Chapter 24 "Applications of electricity generation by solar panels" in Volume III.

16.2 PRINCIPLES OF SOLAR CELL OPERATION

From an engineering point of view, the solar cell can be considered as a semiconductor diode, connected in parallel with a current source, which is the photogenerated current I_l (Figure 16.2). This argument yields the Shockley ideal solar cell equation

$$I = I_l - I_0 \left(e^{\frac{qV}{kT}} - 1 \right) \tag{16.1}$$

where I_0 is the dark saturation current of the diode, q is the electron charge, V is the voltage at the

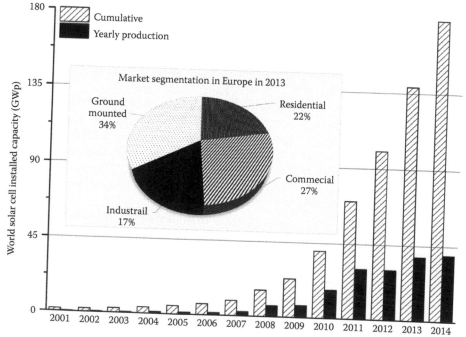

Figure 16.1 World annual production of solar cells (yearly and cumulative) and the market segmentation in Europe in 2013. (Data Adapted from Masson, G. et al., Global market outlook for photovoltaics 2014–2018, EPIA, p. 17, 2014.)

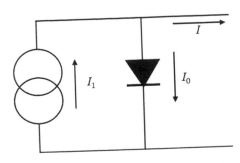

Figure 16.2 Simplified equivalent circuit of an *ideal* solar cell (internal resistances not shown).

terminals of the solar cell, k is the Boltzmann constant, and T is the absolute temperature. The solar cell characteristic is shown in Figure 16.3, depicting also the relevant parameters I_{sc}, V_{oc}, and P_{max}. The short-circuit current I_{sc} produced by this ideal solar cell is equal to the light generated current I_1.

The open-circuit voltage V_{oc} is given by

$$V_{oc} = \frac{kT}{q} \ln\left(1 + \frac{I_1}{I_0}\right) \qquad (16.2)$$

It is customary to define a "fill factor" that is used to calculate the power produced by the cell at the maximum power point P_{max}:

$$FF = \frac{P_{max}}{I_{SC}V_{OC}} \qquad (16.3)$$

To describe practical devices, a diode "ideality" factor n is sometimes introduced as a coefficient of the thermal voltage kT in the exponent of Equation 16.1. A second diode with $n=2$ can also be added, as well as series and parallel resistances to improve the agreement of equations with measured data. More complex circuits can also be used, mainly during the R&D phase to help optimize device structure aimed at achieving higher efficiency and stability [6].

Constraints to solar cell efficiency imposed by fundamental physical laws have been reviewed on a number of occasions [7–12]. The best known of these, the "Shockley–Queisser detailed balance limit" is briefly reviewed in this section. This limit reflects the fact that the solar cell is a quantum energy converter, which is not subject to some of the usual constraints normally applicable to the conversion of heat to mechanical work. Notwithstanding this, we shall see that the thermodynamic Carnot factor does appear in a subtle way.

Before the efficiency of a solar cell can be discussed, a common standard of solar radiation has to be agreed upon. Workers in the area of terrestrial solar cells and systems use the most commonly accepted standard, which represents a good compromise between convenience and observation: solar spectrum at air mass (AM) 1.5, normalized to a total energy flux density of $1\,kWm^{-2}$ [13]. The value of $1\,kWm^{-2}$ for the total irradiance is

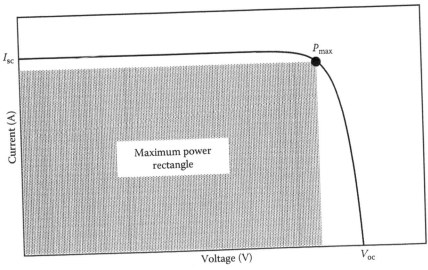

Figure 16.3 *I–V* characteristic of a solar cell. The power generated at the maximum power point P_{max} is equal to the area of the maximum power rectangle.

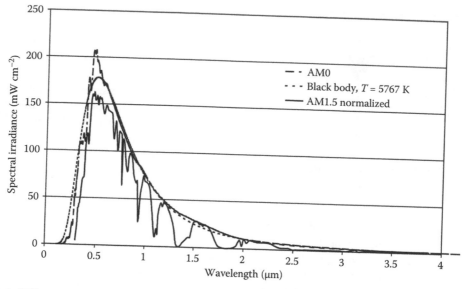

Figure 16.4 Different types of solar radiation spectra that are commonly used as a basis for solar cell theory and characterization.

particularly convenient for the system design (see Section 16.4) and, together with the cell temperature of 25°C, corresponds to the usual conditions for the calibration of terrestrial solar cells and modules. The rated solar cell output at the maximum power point under these standard test conditions is usually known as "watt peak" (W_p).

Space solar cells and systems operate under broadband extraterrestrial (AM0) radiation, and this spectrum is therefore used for their calibration [14]. The total irradiance, equal to the average solar irradiance immediately outside the earth's atmosphere, is called the solar constant S. The commonly accepted value of S is now 1.367 kWm^{-2}.

Theoretical calculations are often based on a third definition of solar radiation: the radiation of a black body that agrees with the observed AM0 irradiance, giving the appropriate temperature of solar radiation T_s= 5767 K [15]. To complicate matters further, some calculations use a less accurate but more convenient value of T_s= 6000 K. A geometrical factor f_ω is introduced to allow for the size of the solar disk as perceived from the earth:

$$f_\omega = \left(\frac{R_S}{R_{SE}}\right)^2 = \frac{\omega_S}{\pi} \qquad (16.4)$$

where R_S is the radius of the sun, R_{SE} is the mean distance between the sun and the earth, and ωS= 6.85×10^{-5} sr is the solid angle subtended

by the sun. The spectra that correspond to these different types of solar radiation are shown in Figure 16.4.

In the calculation of their detailed balance limit, Shockley and Queisser [11] used the black body photon flux ($\delta\phi$) on earth received by a planar surface of unit area in a frequency interval $\delta\nu$, equal to

$$\delta\phi = \frac{2\pi}{c^2} f_\omega \frac{\nu^2 \delta\nu}{e^{\frac{h\nu}{kT_S}} - 1} \qquad (16.5)$$

where h is the Planck constant and c is the speed of light. An ideal solar cell made from a semiconductor with bandgap E_g absorbs all photons with frequency $\nu > \nu_g = E_g/h$ and each such photon gives rise to an electron in the external circuit. The short-circuit current density is then

$$J_{SC} = q\int\delta\phi = \frac{2\pi}{c^2} q f_\omega \int_{\nu_g}^{\infty} \frac{\nu^2 \delta\nu}{e^{\frac{h\nu}{kT_S}} - 1} \qquad (16.6)$$

A similar approach can be used to obtain the dark saturation current density J_0, but now the black body radiation (which completely surrounds the cell) is at the ambient temperature T:

$$J_0 = \frac{4\pi}{c^2} q \int_{\nu_g}^{\infty} \frac{\nu^2 \delta\nu}{e^{\frac{h\nu}{kT_S}} - 1} \qquad (16.7)$$

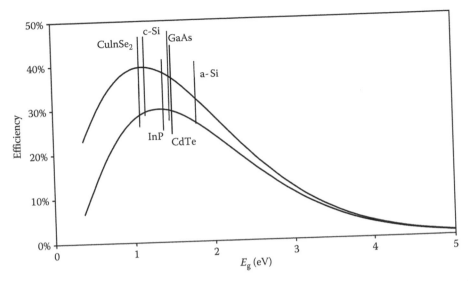

Figure 16.5 Shockley–Queisser detailed balance limit on the solar cell efficiency. The lower efficiency corresponds to ordinary sunlight with geometrical factor f_v given by Equation 16.4. The higher value corresponds to the case where the radiation source completely surrounds the solar cell obtained by setting $f_\omega = 1$. Also shown are the bandgaps of the most common PV materials (c-Si and a-Si denote crystalline and amorphous silicon, respectively).

Note that there is an extra factor of 2 included to account for the fact that the black body radiation at ambient temperature is emitted from twice the cell area (i.e., from the front and back of the cell). The open-circuit voltage can now be determined from Equations 16.6 and 16.7 with the aid of Equation 16.2. The maximum solar cell efficiency thus obtained depends only on the bandgap of the semiconductor from which the solar cell is made (Figure 16.5).

Higher power generation for a given semiconductor area can be obtained if solar radiation is "concentrated" by means of lenses or mirrors, and higher cell conversion efficiency can be obtained if several semiconductors with different bandgaps are used to convert different parts of the spectrum. The former situation is shown in Figure 16.5 by assuming that the cell is completely surrounded by the source of the black body radiation (i.e., by setting $f_\omega = 1$). The combination of different semiconductors to make "tandem cells" is discussed later in this chapter.

Other methods to calculate the solar efficiency for a single-gap cell include arguments based on thermodynamic principles, which underline the observation that, near the open circuit, the solar cell behaves as an ideal thermodynamic engine with Carnot efficiency [16]. Using a simplified

two-level model, Baruch et al. [10] showed that the maximum open-circuit voltage produced by a solar cell made from a semiconductor with bandgap E_g can be approximated by the expression:

$$V_{OC} = \frac{E_g}{q}\left(1 - \frac{T}{T_S}\right) + \frac{kT}{q}\ln\left(\frac{f_\omega}{2}\right) \quad (16.8)$$

The first term in Equation 16.8 is the bandgap energy in eV multiplied by the Carnot efficiency, and the second term gives the reduction in V_{oc} on account of the spread of the solar black body radiation on earth. A factor of two has been added in the denominator of the second term, in keeping with a similar term in Equation 16.7. A closer agreement between the methods of Baruch et al. [10] and Shockley and Queisser [11] can be obtained by adding a term $kT/q \ln(T_S/T)$ to the right-hand side of Equation 16.8 that takes into account the fact that conversion occurs in semiconductor bands rather than in a two-level system [17]. Figures 16.6 and 16.7 compare the theoretical values of the short-circuit current and open-circuit voltage with data for the best solar cells to date from different materials.

The ideal solar cell efficiencies discussed earlier refer to single-junction semiconductor devices

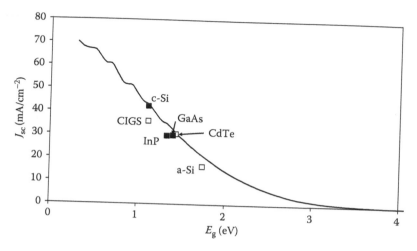

Figure 16.6 Maximum theoretical short-circuit current density under the standard AM1.5 illumination (full line) and the best measured values for solar cells from different materials. (CIGS stands for copper indium–gallium diselenide.

whose principal efficiency limitations are due to the inability of the semiconductor to absorb photons of energy lower than the bandgap, and to the fact that a part of the energy of above-bandgap photons is inevitably lost as heat. To overcome these limitations, one can form a tandem cell by stacking two or more cells on top of each other, each converting its own part of the spectrum. Devices of this form, which are now available commercially, include high-efficiency solar cells for satellites from Spectrolab or SolAero Technologies in the United States and thin-film amorphous silicon or silicon/germanium double- or triple junction tandem cells. The tandem cell technology may, in principle, increase efficiency to 42% (55% under concentrated sunlight) for a double-tandem structure, or

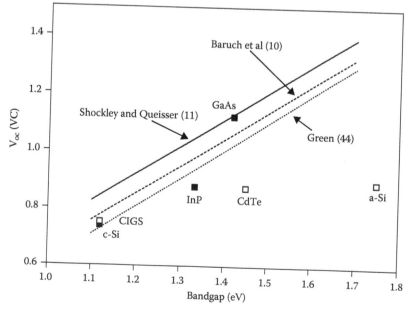

Figure 16.7 Best measured values of open-circuit voltage for crystalline materials and thin films (filled and open squares, respectively) compared with theoretical predictions. The theories of Shockley and Queisser and Baruch et al. are discussed in the text. The theory of Green [44] gives a semiempirical estimate of the maximum achievable voltage.

to 86.8% in the limit of an infinite number of cells [8,18]. In 2015, the certified efficiency of an experimental four-junction cell reached 46% under concentrated sunlight.

Further improvements are possible by a variety of means that has recently been referred to as "third generation" technology [19]. These ideas involve, for example, the creation of more than one electron-hole pair from the higher energy incident photons—in other words, achieving quantum efficiencies in excess of unity. This use of "impact ionization" to improve the cell efficiency was first proposed by Landsberg et al. [20].

Kolodinski et al. [21] demonstrated the feasibility of this approach but a working solar cell has yet to be developed. Other theoretical third-generation concepts invoke hot electrons [22,23], or the use of impurities [24,25], or intermediate bands [26,27] to attempt to utilize the below-bandgap light. Quantum wells have also been suggested as a means of improving efficiency [28,29]. (Nelson [30] gives a review of this field.) A thermodynamic efficiency limit for this general form of solar energy converter is known as the Landsberg efficiency [31,32].

16.3 CURRENT SOLAR CELL TECHNOLOGIES

At the time of publication, the PV market is dominated by crystalline silicon, in the single crystal or multicrystalline form (Figure 16.8). Thin-film

technologies still accounted for only about 10% of the global market in 2013, despite significant growth in volume. There is also a fast growing market, albeit from a very small base, for terrestrial applications of solar cells using concentrator systems that can make use of expensive high-efficiency solar cells, of a type previously only used extensively in the space technology domain.

Space solar cells are not included in Figure 16.8 as their production is much smaller in generated energy, even though, because of high cost, they are not insignificant in terms of financial turnover. Although similar in structure, the main requirements on solar cells used to power satellites in space differ substantially from their terrestrial counterparts. The primary drivers for space cells are the power output for a given size and weight (but, most importantly, the power-to-weight ratio) and, for many applications, the requirement for good radiation resistance. The latter factor is the main reason for the predominant n-on-p configuration of current silicon solar cells, since the p-type base, where most of the power is generated, is more radiation resistant than the n-type material.

The principal effect of particle radiation on the solar cell arises from the change in minority-carrier lifetime and the resultant reduction in the diffusion length. Over recent years, whenever there is a requirement for high efficiency and radiation tolerance, the balance has swayed in favor of GaAs cells, sometimes in the form of a double-junction structure based on GaInP/GaAs. Indium phosphide

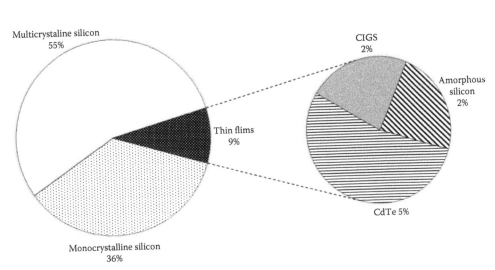

Figure 16.8 Market share of different PV technologies (Kost, C., Mayer, J. N., Thomsen, J. et al., Levelized Cost of Electricity Renewable Energy Technologies, Fraunhofer ISE, Freiburg, Germany, 2013.)

cells have been suggested for use in harsh radiation environments as their radiation resistance is particularly high, due to partial annealing of radiation damage under illumination [33].

16.3.1 Crystalline silicon

At the time of publication, a typical monocrystalline silicon solar cell module in production has an efficiency between 15% and 18%, usually slightly less for cells made from multicrystalline material. Cells are typically made by phosphorus diffusion of the top layer (emitter) into a p-type wafer, with screen-printed metal contacts and a thin layer of antireflection coating (arc; Figure 16.9) [34]. Photons, absorbed mainly in the base of the cell, create electron–hole pairs. These are separated by the electric field of the junction (Figure 16.10a). The principal efficiency losses incurred in such a cell are shown in Figure 16.11.

Improvements to the basic structure can be made by texturing the top surface to reduce optical reflection. In combination with an optically reflecting back surface, this can be used to produce a significant degree of light trapping to offset the poor optical absorption of silicon. Surfaces can be passivated to reduce surface recombination, and p^+ diffusion can be used to create a back-surface field (a barrier to minority-carrier transport) and

reduce recombination near the back surface. Laser-grooved buried-contact technology, invented by Martin Green at the University of New South Wales in Australia and utilized by the BP Solar "Saturn" cells, reduces shading by top surface metallization and pushes up the efficiency of production cells close to 18%. In the laboratory, the top silicon solar cell efficiency has now exceeded 25%—about 80% of the theoretical maximum for single-junction devices. Two high-efficiency structures that have achieved record conversion efficiencies are shown in Figure 16.12. The Stanford cell, with both sets of contacts at the rear to avoid shading, is intended for operation under concentrated sunlight. The highest efficiency silicon-based solar cell (Sanyo/Panasonic HIT cell, with efficiency 25.6%) uses a back contacted heterojunction of crystalline and amorphous silicon.

Silicon is an indirect-gap material, and the low absorption coefficient necessitates relatively thick devices to make good use of the available sunlight (Figure 16.13). This is most easily achieved by a wafer-based technology. The major part of crystalline silicon cell production involves cutting wafers from monocrystalline or multicrystalline ingots. To reduce costs, there has been a push toward thinner and thinner silicon wafers, from over 300 μm in the late 1990s to around 200 μm a decade later. The trend is expected to continue as reliable production

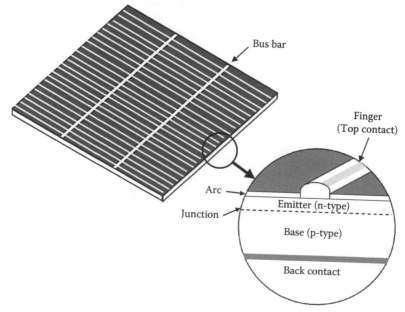

Figure 16.9 Structure of a typical crystalline silicon solar cell in production today.

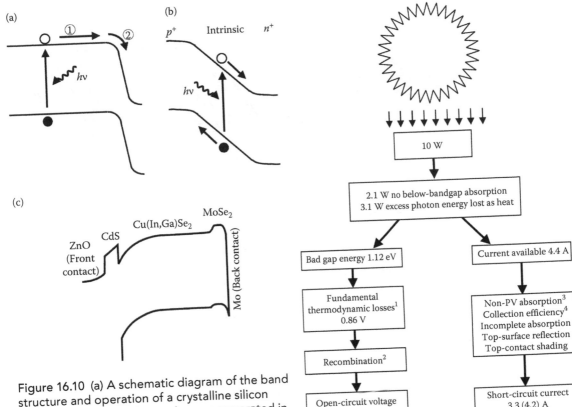

(a)

(b)

p^+ Intrinsic n^+

(c)

ZnO
(Front
contact)

CdS   Cu(In,Ga)Se$_2$ MoSe$_2$

Mo (Back contact)

Figure 16.10 (a) A schematic diagram of the band structure and operation of a crystalline silicon solar cell. Most minority carriers are generated in the p-type base, and must diffuse to the junction (arrow 1) before charge separation (arrow 2). (b) In a typical amorphous silicon p–i–n structure, electron–hole pairs are generated in the space charge region where the built-in field aids carrier collection. (c) The band structure of a Cu(InGa) Se$_2$ solar cell (After Rau, U. and Schock, H. W., *Cu(In, Ga)Se$_2$ Solar Cells Clean Electricity from Photovoltaics*, World Scientific, Singapore, 2001.)

10 W

2.1 W no below-bandgap absorption
3.1 W excess photon energy lost as heat

Bad gap energy 1.12 eV

Current available 4.4 A

Fundamental
thermodynamic losses[1]
0.86 V

Non-PV absorption[3]
Collection efficiency[4]
Incomplete absorption
Top-surface reflection
Top-contact shading

Recombination[2]

Open-circuit voltage
0.58 (0.7) V

Short-circuit currect
3.3 (4.2) A

Series resistance losses
Fill factor 0.75 (0.8)

Cell output 1.5 W (2.5W)
Efficiency 15% (25%)

and handling technologies for even thinner wafers become available.

To avoid the *kerf loss* (amount of material lost during the cutting process), a number of companies (including ASE, Evergreen Solar, and Ebarra) are turning to "non-ingot" wafer technologies such as dendritic or edge-defined film growth [37]. Other fabrication methods that use less silicon include the growth of thin crystalline silicon solar cells deposited on a ceramic substrate, as demonstrated by Astropower. Since the early days of PV, however, there has been a substantial effort directed toward finding a cheap replacement material for crystalline silicon—for a semiconductor with good optical properties that can be deposited as a thin film to reduce the material requirements.

Figure 16.11 Principal efficiency losses in a commercial 100×100 mm crystalline silicon solar cell, compared with those for the passivated emitter rear, locally diffused solar cell (PERL) (in brackets, see Figure 16.12). *Notes:* 1. See Equation 16.8. Losses due to dilution of solar radiation (the second term) are absent under maximum concentration of sunlight. 2. Surface and volume recombination in the base, emitter, and junction regions. 3. Principally free-carrier absorption. 4. Collection efficiency is limited mainly by recombination in the base and emitter (After Boes, E. C. and Luque, A., *Photovoltaic Concentrator Technology Renewable Energy: Sources for Fuels and Electricity*, Earthscan, London, 1993.)

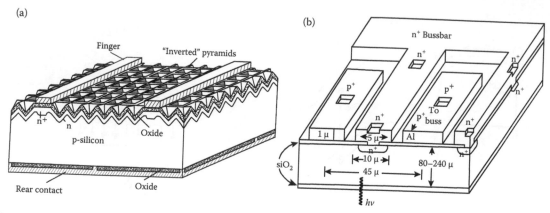

Figure 16.12 The UNSW PERL cell [45] and the Stanford point-contact cell [46].

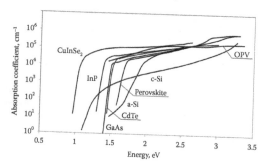

Figure 16.13 Absorption coefficients of selected PV materials. Data for organic solar cell poly[N-9'-heptadecanyl-2, 7-carbazole-alt-5, 5-(4', 7'-di-2-thienyl-2', 1', -benzothiadiazole)] 1:4 wt/wt blend with [6,6]-Phenyl-C71-butyric acid methyl ester, PCDTBT:PC71BM, Adapted from [35]. Data for methylammonium lead triiodide, $CH_3NH_3PbI_3$, (perovskite solar cell) (Adapted from Löper, P. et al., *J. Phys. Chem. Lett.*, 6, 66–71, 2015.)

16.3.2 Thin-film solar cells

The origin of thin-film solar cells dates back to the same year as crystalline silicon solar cells [38]. Commercial production of the first thin-film devices had to await the development of hydrogenated amorphous silicon (a-Si) in the second half of the 1970s. Unfortunately, the discovery of an amorphous Si solar cell degradation mechanism associated with light-induced metastable changes in the properties of hydrogenated amorphous silicon (the Staebler–Wronski effect [49]), represented a major setback to the new technology, and most a-Si solar cells produced today require some measures to reduce its impact.

The present single junction a-Si cells (used principally to power commercial products such as watches and calculators) are invariably based on p–i–n or n–i–p technology. The incorporation of the intrinsic layer serves to enhance carrier collection since most electron–hole pairs are generated in a region where electric field assists their separation (Figure 16.10b). The manufacture involves first the deposition of a layer of tin oxide on glass to act as a transparent front contact. The p–i–n cell is then formed by depositing a p-layer of boron-doped silicon (or silicon/silicon carbide alloy), intrinsic silicon, the phosphorus-doped n-layer, and finally a metal back contact. Tandem a-Si cells, which are increasingly more common, sometimes also incorporate a-Si/a-Ge layers to optimize the bandgap. Other structures involve a combination of microcrystalline/amorphous silicon solar cells, as marketed by Sharp and Kaneka Solar Energy.

Other thin-film solar cell technologies based on compound semiconductors—cadmium telluride (CdTe) and copper indium gallium disellenide (CIGS) or its derivatives—have been overtaking a-Si in terms of commercial importance and, in 2013, CIGS production was more than twice that of a-Si.

CdTe cells are usually manufactured in the form of CdTe/CdS heterojunctions. They present a direct transition bandgap of 1.51 eV, which allows the use of very thin films, and have achieved certified module efficiency of >18% in 2015, with best research cells reaching 21.5%. At the time of writing, the major company developing CdTe products is First Solar who has installed millions of CdTe modules in one of the largest PV power plants in the World, Topaz, in the United States.

The other main compound semiconductor that is being used in commercial production of solar cells is CIGS. It forms a chalcopyrite crystal

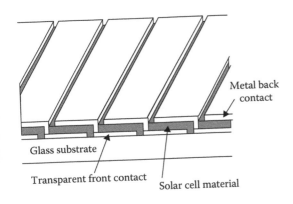

Figure 16.14 A schematic diagram of an integrally interconnected module [47].

structure and the bandgap can be optimized by controlling the amount of gallium in the structure, $CuIn_xGa_{(1-x)}Se_2$. Optimized structures have achieved research cell efficiencies of 21.7%. The band structure of a typical CIGS solar cell, based on a $CdS/Cu(InGa)Se_2$ heterostructure, is shown in Figure 16.10c. There is also much interest in other types of compound semiconductors that do not use scarce elements (i.e., In), such as compounds that form kesterite structure (e.g., Cu_2ZnSnS_4), that have shown laboratory cell efficiencies of 12%.

An advantage of thin-film solar cells over the wafer-based technologies lies not only in the lower material requirements but also in the possibility of integrated module manufacture. Alternating layers can be deposited and laser scribed with an offset in such a way that a series connection of cells is produced without mechanical handling of individual cells (Figure 16.14). The major drawback of thin-film technologies in the past has been lower efficiencies combined often with stability problems (e.g., due to moisture). The recent increase in production and the growing efficiency seem to indicate that thin films are ready for large-scale installations, as demonstrated by its use in the Topaz solar farm. It remains to be seen if this will be sufficient to increase market share that has stagnated in the last years, despite large overall market growth.

16.3.3 Organic and hybrid solar cells

The search for new cost-effective materials includes the development of devices where the PV energy conversion is carried out by molecules instead of traditional semiconductor structures. In the early 1990s, Michael Grätzel and colleagues at the Ecole

Polytechnique Fédérale de Lausanne demonstrated dye-sensitized solar cell (DSSC) architecture where a Ru-based molecular dye both absorbs the light and participates in charge separation. In effect, the monomolecular dye layer "pumps" electrons from one electrode (liquid electrolyte) to the other (solid titanium dioxide). The use of mesostructured nanocrystalline titanium oxide particles coated by the dye ensures high optical absorption: virtually all light within the spectral range of the dye is absorbed by a coating not more than few angstroms thin. Laboratory cell efficiencies of over 10% were quickly achieved; however, the presence of a liquid electrolyte combined with serious stability issues severely affected industrialization of this technology. Several attempts were made to substitute the liquid electrolyte with a solid-state material but often lead to lower efficiencies. Recently, many companies and research groups have been changing focus from DSSCs to the recently discovered perovskite solar cells.

The light absorbing element in hybrid organic–inorganic Perovskite solar cells are crystals with an ABC_3 structure, where A is an organic cation, B a metal cation, and C a halide anion. They were introduced as sensitizers in solid-state DSSC configurations but soon it became clear that high efficiencies can be achieved in a thin-film configuration. It has shown the fastest growth in device efficiency of any PV technology to date and in 2014, certified lab efficiencies of >20% were demonstrated. It is still too early to determine if such technology will ever become commercially viable. Advantages include the potential for low temperature processing and even for solution-based manufacturing, which could mean low production costs and fast energy payback time. However, early results indicate significant stability issues (e.g., due to moisture). Additionally, the most efficient perovskites use Pb in the structure, which could lead to regulatory challenges when dealing with end-of-life and disposal of the products. Alternative lead-free combinations are being investigated and, with further research, it is likely that improvements in stability should be achieved.

Another area of interest is that of truly molecular solar cells. Both polymer and small organic molecules have been used as active ultrathin layers in solar cells. These materials are direct bandgap semiconductors with extremely high absorption coefficients, therefore, only a few hundred nanometer

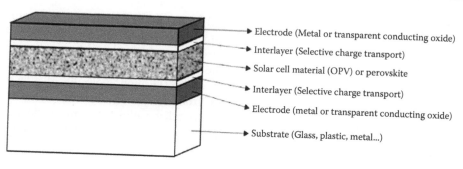

Figure 16.15 A schematic generic diagram of the typical organic solar cell or perovskite planar heterojunction solar cell today.

thick layers are needed to absorb all light. A typical organic photovoltaic (OPV) is formed by two materials, one electron donor and one electron acceptor, somewhat mimicking a p–n junction. However, in OPV, the pair of charges (exciton) generated upon illumination is strongly bound and will not generate free charges unless charge transfer takes place at the heterojunction. That, combined with a very short exciton lifetime and low charge-carrier mobility, puts restrictions on device architecture [50]. At the time of writing, the preferred device configuration uses a blend of donor and acceptor materials to form an ultrathin film (~100–200 nm) with an interpenetrating network of phase domains (in order to maximize donor–acceptor interface area and, therefore, maximize charge generation) (Figure 16.15). This active layer is sandwiched between two electrodes, one of which is typically a transparent oxide, such as indium tin oxide. Between the active layer and the electrode, typically an interlayer is used to improve charge collection and avoid interdiffusion of species. Modules are monolithically integrated, similarly to thin-film PV modules. Note however, in contrast to silicon PV, OPV is not one single technology and device structure. Thousands of active molecules, interlayer materials, and combinations of these have been under investigation and each reported breakthrough in efficiency is typically achieved with a new material.

Significant industrial development has taken place since 2010 and, in 2014, Toshiba announced the first certified OPV module efficiency of 8.7%, while the cell efficiencies are reported between 11% and 12%. Despite the efficiency being significantly lower than that of c-Si, OPV provides a series of potential advantages that could lead to market entry in niche applications. The first advantage is related to low temperature and potential for fast roll-to-roll manufacturing via printing or coating, which enables direct coating onto flexible substrates such as plastic, thin metal sheets, and even paper. This should allow very low production costs and very short energy payback time. Other characteristics are lightweight and easy customizability, which should allow easy integration into consumer products. For example, in 2015, Heliatek and RECKLI announced an OPV solar concrete wall installation in their headquarters in Germany. One major drawback of OPV technology has been the limited lifetime that has hindered the use of flexible products. Recent advances in intrinsic stability and improved barrier layers have led to significant improvements in demonstrated stability in field trials. There is much to look forward to if knowledge from the booming organic light-emitting diodes industry can be used to bring OPV closer to the market.

16.4 ISSUES IN SYSTEM DESIGN

For practical use, solar cells are laminated and encapsulated to form PV modules. These are then combined into arrays, and interconnected with other electrical and electronic components—for example, batteries, charge regulators, and inverters—to form a PV system. A number of issues need to be resolved before an optimum system design is achieved. These issues include the choice between a flat plate or concentrating system, and the required array configuration (fixed tilt or tracking). Answers to these questions will vary depending on the solar radiation at the site of the installation. Specific issues also relate to whether the system is to be connected to the utility supply (the "grid") or is intended for stand-alone operation.

The available solar radiation, of course, varies from day to day, season to season, and depends on the geographical location. One can, however, obtain an estimate for the mean daily irradiation (solar energy received by a unit area in one day), averaged over the surface of the globe. To this end, we note that the total energy flux incident on the earth is equal to the solar constant S multiplied by the area of the disk presented to the sun's radiation by the earth. The average flux incident on a unit surface area is then obtained by dividing this number by the total surface area of the earth. Making allowance for 30% of the incident radiation being scattered and reflected into space, the average daily solar radiation G_d on the ground is equal to

$$G_d = 24 \times 0.7 \times \frac{\pi R_E^2}{4 \pi R_E^2} \times S = 24 \times 0.7 \times \frac{1}{4} S$$

$$= 5.74 \text{ kWh m}^{-2} \tag{16.9}$$

where R_E is the radius of the earth, which was introduced in Section 16.2. The value should be compared with the observed values. Figure 16.16 shows the daily solar radiation on a horizontal plane for four locations ranging from the humid equatorial regions to northern Europe. The solar radiation is highest in the continental desert areas around latitude 25–30°N and 25–30°S, and falls off toward the equator because of the clouds, and toward the poles because of low solar elevation. Equatorial regions experience little seasonal variation, in contrast with higher latitudes where the summer/winter ratios are large.

Most PV arrays are installed at fixed tilt and, wherever possible, oriented toward the equator. The optimum tilt angle is usually determined by the nature of the application. Arrays that are to provide maximum generation over the year (e.g., many grid-connected systems) should be inclined at an angle equal to the latitude of the site. Stand-alone systems that are to operate during the winter months have arrays inclined at a steeper angle of latitude +15°. If power is required mainly in summer (e.g., for water pumping and irrigation), the guide inclination is latitude −15°.

The amount of solar energy captured can be increased if the modules track the sun. Full two-axis tracking, for example, will increase the annual energy available by over 30% over a nontracking array fixed at the angle of latitude (Figure 16.17)— at the expense, however, of increased complexity. Single-axis tracking is simpler but yields a smaller gain. Tracking is particularly important in systems that use concentrated sunlight. These systems can partially offset the high cost of solar cells by the use of inexpensive optical elements (mirrors or lenses). However, the cells then usually need to be cooled and it should also be borne in mind that only direct (beam) solar radiation can be concentrated to a significant degree, thus reducing the available energy input. This effectively restricts the application of concentrator systems to regions with clear skies.

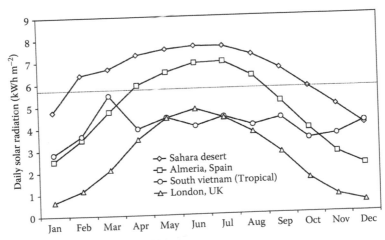

Figure 16.16 Mean daily solar radiation and its monthly variation in different regions of the world. The dashed line indicates the average value (Equation 16.9).

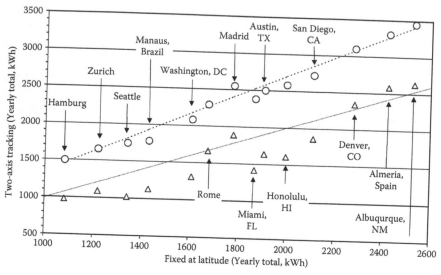

Figure 16.17 The energy available to a PV array with two-axis tracking, in comparison with a flat plate system inclined at an angle equal to the latitude of the site. Global radiation (circle) and direct beam (triangle) are shown. The full and dotted lines correspond to the global radiation available to the fixed array and a 34% increase over this value, respectively (After Boes, E. C. and Luque, A., *Photovoltaic Concentrator Technology Renewable Energy: Sources for Fuels and Electricity*, Earthscan, London, 1993.)

There is a considerable difference between the designs of stand-alone and grid-connected systems. Much of the difference stems from the fact that the design of stand-alone systems endeavors to make the most of the available solar radiation. This consideration is less important when utility supply is available, but the grid connection imposes its own particular constraints that must be allowed for in the system design.

16.4.1 Stand-alone systems

The various applications of PV systems in isolated locations have been mentioned briefly in Section 16.1 and more examples are given in Chapter 24 "Applications of electricity generation by solar panels" in Volume III. The diversity of uses leads to different requirements on the system and different system specifications. An important parameter that reflects the nature of the application is the required security of supply. Telecommunication and systems used for marine signals, for example, need to operate at a very high level of reliability. In other applications, the user may be able to tolerate lower reliability in return for a lower system cost [40]. These considerations have an important bearing on how large PV

array and how large energy storage (battery) need to be installed, in other words, on system sizing.

Among the variety of sizing techniques, sizing based on energy balance provides a simple and popular technique that is often used in practice. It gives a simple estimate of the PV array necessary to supply a required load, based on an average daily solar irradiation at the site of the installation, available now for many locations in the world. Choosing the month with the lowest irradiation (e.g., December in northerly latitudes) and the value appropriate to the inclined panel, the energy balance equation can then be found with the use of data in Figure 16.16:

$$\text{Array size (in } W_p) = \frac{\text{Daily energy consumption}}{\text{Daily solar raditaion}}$$

(16.10)

Equation 16.10 specifies how many PV modules need to be installed to supply the load under average conditions of solar irradiation. The battery size is then estimated "from experience"—the rule of thumb recommends, for example, installing three days of storage in tropical locations, five days in southern Europe and ten days or more in the United Kingdom. As an illustration, Table 16.2 shows the

Table 16.2 Area and nominal power of a PV array, facing south and inclined at latitude angle, needed to produce 1 MWh of electricity per year at different locations, for two-module efficiencies h

Location	Nominal array power (kW$_p$)	Module area (m²)	
		η= 15%	η= 10%
Sahara desert	0.41	2.7	4.1
Almeria, Spain	0.53	3.5	5.3
London	1.04	6.9	10.4
Vietnam	0.65	4.4	6.5

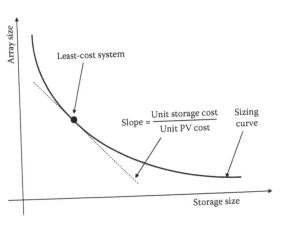

Figure 16.18 The sizing curve, representing the locus of PV system configurations with the same reliability of energy supply. The least-cost system can be found by a simple geometrical construction.

size of the PV array needed to generate 1 MWh per year in different locations around the world.

Although easy to use, the sizing method based on energy balance does not give an indication of how the PV system will operate under fluctuating solar radiation. In particular, it does not predict the reliability of energy supply. The relationship between the reliability of supply and sizing is illustrated well by the random walk method [41]. The method consists of treating the possible states of charge of the battery as discrete numbers, which are then identified as sites for a random walk. Each day, the system makes a step in the random walk depending on solar radiation: one step up if it is "sunny" and one step down if it is "cloudy." The magnitudes of these steps and probabilities of weather being "sunny" or "cloudy" are determined from the solar radiation data and the daily load. When the random walker resides in the top state, the battery is fully charged; when it is in the bottom state, the battery is completely discharged and the load is disconnected. The calculations are carried out by assuming that, after a certain time, the random walker reaches a steady state. The loss of load probability (LLP) is then equated to the probability of the random walker residing in the lowest state. Bucciarelli [42] subsequently extended this method to allow for correlation between solar radiation on different days. For a given LLP, the result can be expressed in the form of a sizing curve: a functional dependence of the array size on the battery size from which the least-cost system can easily be determined (Figure 16.18). These and other more complex sizing techniques have been summarized in more detail by Gordon [43].

16.4.2 Grid-connected systems

Grid-connected systems have grown considerably in number since the early 1990s spurred by government "feed-in tariffs" in a number of countries, led initially by Switzerland and followed by more substantial programs in Germany, Japan, California, and Spain. One feature that affects the system design is the need for compliance with the relevant technical guidelines to ensure that the grid connection is safe; the exported power must also be of sufficient quality and without adverse effects on other users of the network. Although a common set of international standards is still some way off, it is probably fair to say that the fundamental issues have been identified—partly through the work of the Task 5 of the International Energy Agency. In a number of countries, the required statutory guidelines have now been produced: in the United Kingdom, for example, the relevant Engineering Recommendation G77 was published in 2000 and some updated recommendations in 2012 (Engineering Recommendation G83/2). An example of the requirements imposed on the grid interface of a PV system by the utility is shown in Table 16.3.

Many of the grid connection issues are not unique to PV. They arise from the difficulties in trying to accommodate "embedded" or distributed generators in an electricity supply system designed around large central power stations. It is likely that many of these features of grid connection

Table 16.3 A summary of principal requirements on the grid interface of PV generators, covered in the UK engineering recommendation G77

Function	Reference
Protection	
General	IEC 255
Under/over voltage	UK guidelines[a]
Under/over frequency	UK guidelines[a]
Loss of mains	Specific for PV
Supply quality	
Harmonics	BS and EN[b]
Voltage flicker	BS and EN[b]
Electromagnetic compatibility	BS and EN[b]
DC injection	UK guidelines[a]
Safety	
Earthing	BS[c]

[a] Covered by existing UK guidelines for parallel connection of embedded generators.
[b] Covered by existing British standard and European norm.
[c] Covered by existing British standard.

will undergo a review as electricity distribution networks evolve to absorb a higher proportion of embedded generation: wind farms, cogeneration (or CHP) units, or other local energy sources. The electricity supply system in 20 or 30 years' time might be quite different from now, and new and innovative integration schemes will be needed to ensure optimum integration. PV generators are likely to benefit from these changes, particularly from the recent advances in the technology of small domestic size cogeneration units (micro-CHP) that have a good seasonal synergy with the energy supply from solar sources, and can share the cost of the grid interface.

An example of how elegant architecture can be combined with forward looking engineering is offered by the Mont Cenis Energy Academy at Herne-Sodingen (Figure 16.19). This solar-cell clad glass envelope at the site of a former coal mine provides a controlled Mediterranean microclimate that is powered partly by $1\,MW_p$ PV array and partly by two cogeneration units fueled by methane released from the disused mine. To ensure a good integration into the local electricity supply, the generators are complemented by a 1.2 MWh battery bank. In addition to the Academy, the scheme also exports electricity and heat to 250 units in a nearby housing estate and a local hospital. The Mont Cenis Academy is a fine flag carrier for PV and new energy engineering—without a doubt, similar schemes will become more prolific as PV and energy efficient solutions become the accepted norm over the next few decades.

16.5 CONCLUSIONS

PV technology has come a long way since the first solar-powered satellites in the late 1950s. Based today mainly on crystalline silicon, new approaches that utilize thin-film technologies and molecular materials are beginning to make their mark in commercial production. If the current rate of growth continues, a combination of distributed solar power systems with large PV farms will supply a significant part of our energy needs during the early part of this new millennium. One can foresee a bright future for this new, clean energy source.

ACKNOWLEDGMENTS

The authors would like to thank Dr. Neil Ross and Dr. James Blakesley for their careful reading of the manuscript and a number of useful comments.

Figure 16.19 The Mont Cenis Academy at Herne-Sodingen.

REFERENCES

1. Becquerel, A. E. 1839. Recherches sur les effets de la radiation chimique de la lumière solair au moyen des courants électriques. *C. R. Acad. Sci.* 9, 561–567.

2. Smith, W. 1873. The action of light on selenium. *J. Soc. Telegraph Eng.* 2, 31–33.

3. Adams, W. G. and Day, R. E. 1877. The action of light on selenium. *Proc. R. Soc. A* 25, 113–117.

4. Chapin, D. M., Fuller, C. S., and Pearson, G. O. 1954. A new silicon p–n junction photocell for converting solar radiation into electrical power. *J. Appl. Phys.* 25, 676–677.

5. Masson, G., Orlandi, S., and Rekinger, M. 2014. Global market outlook for photovoltaics 2014–2018, ed Tom Rowe (Brussels: EPIA) p. 17.

6. Castro, F. A., Heier, J., Nüesch, F., and Hany, R. 2010. Origin of the kink in current-density versus voltage curves and efficiency enhancement of polymer-C60 heterojunction solar cells. *IEEE J. Sel. Top. Quantum Electron.* 16(6), 1690–1699.

7. Landsberg, P. T. and Badescu, V. 1998. Solar energy conversion: List of efficiencies and some theoretical considerations. *Prog. Quantum Electron.* 22, 211–230.

8. de Vos, A. 1992. *Endoreversible Thermodynamics of Solar Energy Conversion* (Oxford: Oxford University Press).

9. Würfel, P. 1982. The chemical potential of radiation. *J. Phys. C* 15, 3967–3985.

10. Baruch, P., Picard, C., and Swanson, R. M. 1980. Thermodynamical limits to photovoltaic solar energy conversion efficiency. *Proceedings of 3rd European Photovoltaic Energy Conference.* Springer, Netherlands, p. 927.

11. Shockley, W. and Queisser, H. J. 1961. Detailed balance limit of efficiency of p-n junction solar cells. *J. Appl. Phys.* 32, 510–519.

12. Müser, H. A. 1957. Thermodynamische behandlung von elektronenprozessen in halbleiter-randschiten. *Z. Physik* 148, 380–390.

13. Hulstrom, R., Bird, R., and Riordan, C. 1985. Spectral solar irradiance data sets for selected terrestrial conditions. *Sol.Cells* 15, 365–391.

14. Thekaekara, M. P. 1977. *Solar Energy Engineering*, ed A. A. M. Sayigh (New York: Academic) p. 37.

15. Parrott, J. E. 1993. Choice of an equivalent black body solar temperature. *Sol. Energy* 51, 195.

16. Baruch, P. and Parrott, J. E. 1990. A thermodynamic cycle for photovoltaic energy conversion. *J. Phys. D: Appl. Phys.* 23, 739.

17. Markvart, T. 2007. Thermodynamics of losses in photovoltaic conversion. *Appl. Phys. Lett.* 91, 064102.

18. de Vos, A. 1980. Detailed balance limit of the efficiency of tandem solar cells. *J. Phys. D: Appl. Phys.* 13, 839.

19. Green, M. A. 2001. Third generation photovoltaics: Ultra high conversion efficiency at low cost. *Prog. Photovoltaics Res. Appl.* 9, 123–135.

20. Landsberg, P. T., Nussbaumer, H. and Willeke, G. 1993. Band-band impact ionization and solar energy efficiency. *J. Appl. Phys.* 74, 1451–1452.

21. Kolodinski, S., Werner, J. H., Wittchen, T., and Queisser, H. J. 1993. Quantum efficiencies exceeding unity due to impact ionisation in silicon solar cells. *Appl. Phys.* 63, 2405.

22. Ross, R. T. and Nozik, A. J. 1982. Hot-carrier solar energy converters. *J. Appl. Phys.* 53, 3813–3818.

23. Würfel, P. 1997. Solar energy conversion with hot electrons from impact ionisation. *Sol. Energy Mater. Sol. Cells* 46, 43.

24. Würfel, P. 1993. Limiting efficiency for solar cells with defects from three-level model. *Sol. Energy Mater. Sol. Cells* 29, 403–413.

25. Keevers, M. and Green, M. A. 1994. Efficiency improvements of silicon solar cells by the impurity photovoltaic effect. *J. Appl. Phys.* 75, 4022–4031.

26. Luque, A. and Martí, A. 1997. Increasing the efficiency of ideal solar cells by photon induced transitions at intermediate levels. *Phys. Rev. Lett.* 78, 5014–5017.

27. Luque, A. and Martí, A. 2001. A metallic intermediate band high efficiency solar cell. *Prog. Photovoltaics Res. Appl.* 9, 73–86.

28. Barnham, K. W. J. and Duggan, G. 1990. A new approach to high-efficiency multi-band gap solar cells. *J. Appl. Phys.* 67, 3490–3493.

29. Luque, A. and Martí, A. 1997. Entropy production in photovoltaic conversion. *Phys. Rev. B.* 55, 6994–6999.

30. Nelson, J. 2001. *Quantum Well Solar Cells Clean Electricity from Photovoltaics*, ed M. A. Archer and R. Hill (Singapore: World Scientific) pp. 447–480.

31. Petela, R. 1964. Energy of heat radiation. *Trans. ASME Heat Transfer* 36, 187.

32. Landsberg, P. T. and Mallinson, J. R. 1976. Thermodynamic constraints, effective temperatures and solar cells. *International Colloquium on Solar Electricity* (CNES, Toulouse) p. 46.

33. Coutts, T. J. and Yamaguchi, M. 1988. Indium phosphide based solar cells: A critical review of their fabrication, performance and operation. *Curr. Top. Photovoltaics* 3, ed T. J. Coutts and J. D. Meakin pp. 79–234.

34. Cuevas, A. 2000. Crystalline silicon technology, in *Solar Electricity*, ed T. Markvart (Chichester: Wiley) pp. 46–62.

35. Schmiedova, V., Heinrichova, P., Zmeskal, O. and Weiter, M. 2015. Characterization of polymeric thin films for photovoltaic applications by spectroscopic ellipsometry. *Appl. Surf. Sci.* 349, 582–588.

36. Löper, P., Stuckelberger, M., Niesen, B., Werner, J., Filipic, M., Moon, S-J., Yum, J-H., Topic, M., De Wolf, S., and Ballif, C. 2015. Complex refractive index spectra of $CH_3NH_3PbI_3$ perovskite thin films determined by spectroscopic ellipsometry and spectrophotometry. *J. Phys. Chem. Lett.* 6, 66–71.

37. Bruton, T. M. 2002. General trends about photovoltaics based on crystalline silicon. *Sol. Energy Mater. Sol. Cells* 72, 3–10.

38. Reynolds, D. C. 1954. Photovoltaic effect in cadmium sulfide. *Phys. Rev.* 2, 31–33.

39. Rau, U. and Schock, H. W. 2001. *Cu(In, Ga)Se₂ Solar Cells, in Clean Electricity from Photovoltaics*, ed M. A. Archer and R. Hill (Singapore: World Scientific) pp. 277–346.

40. Lorenzo, E. 1997. Photovoltaic rural electrification. *Prog. Photovoltaics Res. Appl.* 5, 3–27.

41. Bucciarelli, L. L. 1984. Estimating loss-of-power probabilities of stand-alone photovoltaic solar-energy systems. *Sol. Energy* 32, 205–209.

42. Bucciarelli, L. L. 1986. The effect of day-to-day correlation in solar-radiation on the probability of loss-of-power in a standalone photovoltaic energy system. *Sol. Energy* 36, 11–14.

43. Gordon, J. M. 1987. Optimal sizing of stand-alone photovoltaic solar power systems. *Sol. Cells* 20, 295–313.

44. Green, M. A. 1982. *Solar Cells* (New York: Prentice Hall).

45. Zhao, J., Wang, A., Altermatt, P., and Green, M. A. 1995. 24% efficient silicon solar cells with double layer antireflection coatings and reduced resistance loss. *Appl. Phys. Lett.* 66, 3636–3638.

46. Sinton, R. A., Kwark, Y., Gan, J. Y. and Swanson, R. M. 1986. 27.5-percent silicon concentrator solar cells. *IEEE Electron. Device Lett.* 7, 567–569.

47. Hill, R. 2000. Thin film solar cells, in *Solar Electricity* 2nd edn, ed T. Markvart (Chichester: Wiley) pp. 73–74.

48. Boes, E. C. and Luque, A. 1993. *Photovoltaic Concentrator Technology Renewable Energy: Sources for Fuels and Electricity*, ed T. B. Johansson, H. Kelly, A. K. N. Reddy and R. H. Williams (London: Earthscan) pp. 361–401.

49. Staebler, L. D. and Wronski, C. R. 1980. Reversible conductivity change in discharge produced amorphous silicon. *J. Appl. Phys.* 51, 3262.

50. Nicholson, P. G. and Castro, F. A. 2010. Organic photovoltaics: Principles and techniques for nanometre scale characterization. *Nanotechnology* 21, 492001.

Optical nano- and microactuation

GEORGE K. KNOPF
University of Western Ontario

17.1 INTRODUCTION

Nano- and microactuators are very small material structures and mechanisms that perform mechanical work in response to external stimuli. The size of these devices can range from a few molecules to several hundred microns. The mechanical action produced by the actuating structure is able to generate tiny displacements or induce microforces on the surrounding medium (Knopf 2006, 2012; Tabib-Azar 1998). Examples of microscale actuators include cantilever beams, microbridges, flexible diaphragms, torsional mirrors, shape memory wires and strips, and expansive polymer gels. These miniature devices may exploit mechanical principles similar to much larger analogous systems or merely involve the subtle expansive and contractive characteristics of environmentally sensitive metals and polymers. In contrast, nanoactuators are often assembled from small groups of interconnected molecules that move in unison under an external energy source. Molecular motor proteins found in living cells are an example of nanoactuators capable of producing piconewton (pN) forces (Kang et al. 2009; Li and Tan 2001; Setou et al. 2000; Vale 2003).

Different nano- and microactuation methods will take advantage of mechanical, electrostatic, piezoelectric, magnetic, thermal, fluidic, acoustic, chemical, biological, or optical principles. Although optical actuators may be one of the least studied, in the world of the *very very small* (Feynman 1960) optically driven transducers provide a number of interesting features and design opportunities. These actuators can be created to either *directly* or *indirectly* transform the light energy into structural movement (Knopf 2006, 2012; Tabib-Azar 1998). Direct optical methods use the photons to interact with the photosensitive properties of the actuator material in order to initiate mechanical displacement. One example of direct optical microactuation occurs when photoresponsive shape memory polymers (SMPs) (Jiang et al. 2006) are physically deformed by exposing the material to ultraviolet (UV) radiation. On the other hand, indirect optical methods exploit the ability of light to generate heat when the focused beam strikes the material surface and influences the thermal properties of nearby gases, fluids, and solids. One application of this concept is to use a focused light source to heat a liquid (e.g., Freon 113) that transforms into a gas and expands sufficiently to deform a very thin mechanically flexible diaphragm (Mizoguchi et al. 1992). The remote optical heating of liquids in Lab-on-a-Chip (LOC) devices also represents a novel approach to move fluid along a microchannel (Weinert and Braun 2009; Weinert et al. 2009) or mix adjacent streams of liquids (Shiu et al. 2010) prior to chemical analysis.

At the nano- and microscale optically driven devices have important advantages over conventional microelectromechanical systems because streams of photons provide both the energy into the system and the control signal used to regulate the actuator response. Optical actuators can be created to operate under different properties of light such as intensity, wavelength, phase, and polarization characteristics. Many of these light-driven mechanisms are also free from small electrical current leakage, resistive heat dissipation, and mechanical friction forces that may significantly reduce performance and efficiency.

The fundamental operating principles and unique attributes of a wide variety of light-activated nano- and microactuators are explored in this chapter. The discussion begins with a brief look at how light can exert pressure on mechanical structures and even levitate artificial micro-objects. Section 17.3 introduces light-driven nanomotors comprised of single protein or DNA molecules that are able to switch between states causing the mechanism to shrink and expand like an inchworm. Several direct methods of optical actuation that take advantage of newly engineered photoresponsive materials are examined in Section 17.4. Some of these materials are able to alter their shape, bend, or expand in volume when exposed to specific wavelengths of light. Other direct optical methods are also introduced using *photon-generated electrons* to alter the electrostatic pressure on a microcantilever beam and that induce mechanical deformation. Indirect optical actuation methods, caused by the photothermal interactions arising from light striking a material surface, is described

in Section 17.5. The principles of light material interactions are discussed and applications involving diaphragm actuators and light-driven micro-flows are introduced. In addition, examples of temperature-induced phase transformation of solids and photothermal vibrations of optical wave-guides for mechanical actuation are presented. In all cases, light is used to initiate movement and control the actuating mechanism that performs work. Finally, a summary of light-driven actuation and future opportunities for exploiting this technology is provided in Section 17.6.

17.2 FORCES GENERATED BY LIGHT

In the late 1800s, the Scottish mathematical physicist James Maxwell (1873) predicted that it could be theoretically possible to generate measurable forces using a source of light. However, it was not until the dawn of the 20th century when Russian physicist Pyotr Lebedev (1901) experimentally confirmed Maxwell's prediction. For many years, Lebedev's observations were treated as minor scientific curiosities that would have little influence or impact on practical engineering applications. Although perceived to be impractical, the possibilities of using light to drive machines have continued to inspire the imagination of numerous science fiction writers for more than a century.

Recent advances in precision instrumentation and a deeper understanding of material science have opened doors not envisioned by Maxwell or Lebedev. The modern tools for scientific investigation have enabled a broad spectrum of scientists and engineers from around the globe to explore the micron and submicron worlds where electromagnetic radiation plays a dominant factor in changing the behavior of certain materials. In other words, in the world of the *very very small* light does matter. Through advanced material engineering and clever design, the properties of light (radiation pressure, intensity, wavelength, phase) can be transformed into small, yet meaningful, mechanical pressure, force, and displacement.

In general, electromagnetic radiation from a focused light source will exert a small amount of pressure upon any exposed surface. The *radiation pressure* absorbed by that surface can be described as the power flux density divided by the speed of light. The driving force generated by light radiation pressure is based on the transfer of photons.

The photons have no mass but carry energy at the speed of light. The momentum, p_{ph}, of each photon in a light beam is the result of this energy (Ashkin 2000; Steen 1998; Steen and Mazumder 2010) and can be given by

$$p_{ph} = \frac{h\upsilon}{c} \qquad (17.1)$$

where h is the Planck's constant (6.63×10^{-34} Js), c is the speed of light in a vacuum ($\sim 2.99 \times 10^{8}$ m/s), and υ is the optical frequency related to the wavelength (λ) of the light source by

$$\upsilon = \frac{c}{\lambda} \qquad (17.2)$$

The optical forces arising from this light–material interaction are the result of the exchange of momentum between the incident photons and irradiated object (Jonáš and Zemánek 2008). A focused light beam with power of P will generate ($P/h\upsilon$) photons per second. The force acting on the structure is equal to the change-in-momentum per unit time, (dp_{Total}/dt). If a light beam strikes a surface that absorbs 100% of the photons per second then corresponding force on the structure is

$$F_{absorber} = \left(\frac{dp_{Total}}{dt}\right) = \left(\frac{P}{h\upsilon}\right)\left(\frac{h\upsilon}{c}\right) = \frac{P}{c} \qquad (17.3a)$$

Alternatively, the principle of conservation of momentum (Ashkin 2000) states that if the same stream of photons strikes a highly polished mirror surface then all the photons are reflected straight back. As a consequence, the total change-in-momentum per second doubles the force acting on the structure, that is

$$F_{mirror} = \left(\frac{dp_{Total}}{dt}\right) = \left(\frac{2P}{h\upsilon}\right)\left(\frac{h\upsilon}{c}\right) = 2\frac{P}{c} \qquad (17.3b)$$

A single-focused laser beam will, however, only produce a *very weak* force. For example, a single 1 mW diode laser pointer pen will generate a force in the range of a pN. Therefore, to suspend a small solid object, such as a coin, in the air it would be necessary to use 10^9 similar pen lasers all pointed at the same spot on the object's surface (Tang 2009). Increasing the power of the light source 10^9 times will result in an equivalent kW laser with sufficient energy to ablate or vaporize the coin surface (depending on beam spot size). Although such tiny forces are not able to suspend or move large objects, this perceived weakness can be turned into an engineering strength if the goal is to manipulate solid objects that weigh only few pictograms or move microliters of liquid. At this scale, optics can be used to shape and redirect light beams to strike the object surface at precise locations (Knopf 2012).

The *optical gradient forces* necessary to hold and induce micro-object movement were first explored more than four decades ago. In the early 1970s, Ashkin (1970) developed a single-beam gradient force optical trap based on the principle that a laser beam brought into sharp focus will generate a restoring force that can pull particles into that focus. The generated force is the result of the elastic scattering of laser photons by the particle such that the object alters the direction of the photon momentum without absorbing any of the beam energy. The basic operating principle of an optical trap is that light carries momentum in the direction of propagation that is proportional to its energy. Any change in the direction of the light rays arising from either reflection or refraction will produce a change in the momentum. If the object in the trap bends the light rays and changes the momentum of the light, then the object undergoes an equal and opposite change in momentum. The momentum transfer gives rise to a force acting on the object. In a typical setup, Figure 17.1, the incoming laser light has a Gaussian intensity profile that is brighter in the center of the beam than at the edges. When the light interacts with the object, the rays are bent according to the laws of reflection and refraction. The sum of the forces from all light rays in the beam can be divided into scattering (f_s) and gradient (f_g) components. The scattering force (f_s) points in the direction of the incident light, whereas the gradient force (f_g) arises from the gradient intensity profile pointing toward the center of the beam. The gradient force is the restoring force that pulls the object into the beam center. If the contribution to the scattering force of the refracted rays is larger than that of the reflected rays, then a restoring force is also created along the incident light direction and a stable optical trap is formed.

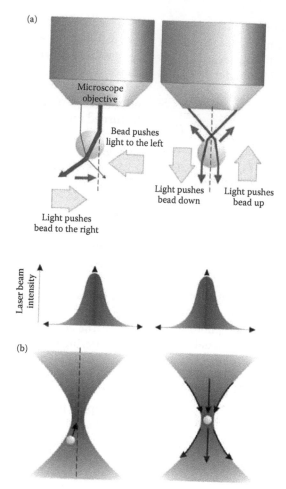

(a)

Microscope objective

Bead pushes light to the left

Light pushes bead to the right

Light pushes bead down

Light pushes bead up

Laser beam intensity

(b)

Figure 17.1 The light–particle interactions that occur in a focused Gaussian laser beam during transverse and axial optical trapping. (a) The light refracted through a transparent object with a high refractive index produces an optical force that is balanced when the object reaches the beam center. (b) The electromagnetic field of the light causes the object to act as an induced dipole, which is drawn into the beam focus where its energy is minimized. (Reprinted from *Nanotoday*, 1, Dholakia, K. and Reece, P., Optical micromanipulation takes hold, 18–27, Copyright 2006, with permission from Elsevier.)

In his seminal work, Ashkin was able to experimentally observe the forced acceleration of freely suspended particles by radiation pressure from a continuous wave (CW) visible laser beam. Based on this observation, he was also able to lift and levitate a very small glass sphere (Ashkin 1970). A later experiment with his colleague Dziedzic (Ashkin and Dziedzic 1971, 1980) showed that it was possible to trap a glass sphere in a vertical laser. The concept of optical trapping and tweezing (Dholakia and Reece 2006; Dienerowitz et al. 2008, Kröner et al 2007; Nieminen et al. 2006) has more recently been applied to a variety of scientific areas including colloidal dynamics, statistical mechanics, cell

biology, and nanomanipulation. In the late 1990s, Higurashi et al. (1997) demonstrated how carefully shaped fluorinated polyimide micro-objects with a 6–7.5 μm cross-sectional radius can be rotated using Ashkin's method of optical trapping and tweezing. In this series of experiments, applying radiation forces near the focal point were used to reposition and rotate the micro-object about the laser beam axis.

Over the past decade, Shoji Maruo and his colleagues at the Yokohama National University (Maruo and Inoue 2007; Maruo 2008, 2012) developed a large variety of light-driven micromachines based on the concept of an optical trap. Polymeric microstructures such as microgears (Figure 17.2) were fabricated using two-photon microstereolithography techniques and driven by a circular scanning laser beam. During operation, the gear is optically trapped when the laser beam is focused on the center of the tooth. However, if the focus is moved slightly to one side of the gear tooth center then the net radiation pressure will produce an external force on the tooth moving it toward the focus. In this manner, a circular scanning laser beam was used to rotate the microgear and various motion patterns were generated by modifying the beam trajectory. In addition to driving gears, it is also possible to use optical tweezing to control multiple micromanipulators. For example, the polymeric three-hand micromanipulator shown in Figure 17.3 was developed by the research team to grasp and manipulate tiny objects. The manipulator arms were driven sequentially using a single laser beam and a time-divided laser scanning method (Maruo 2008). By optimizing the repetition rate (~100 Hz) between trajectories, it was possible to handle the 5 μm diameter glass microbead.

The 3D microfabrication technology was also used by the Yokohama researchers to create an optically driven microfluidic pump. The micropump was comprised of two interlocked spinning 9 μm diameter rotors that could regulate the liquid flow in a microfluidic channel (Maruo 2008, 2012). Each individual rotor had two lobes and was held within the microchannel by a stationary shaft. A tightly focused laser beam was then used to create forces necessary to drive the micropump. Once more, the pumping mechanism was controlled by changing the trajectory of the scanning beam. The velocity of tracer particles added to the fluid was observed and determined to be 0.2–0.7 μm/s, which was directly proportional to the rotation speed of the rotors. An optically driven micropump that used only a single-disk microrotor was later introduced by Maruo and his colleagues. In this particular design, the single 10 μm diameter disk rotor had three columns as targets for the optical trap. In addition, the shaft had been eliminated by confining the disk to a U-shaped microchannel. The laser beam was focused on the center

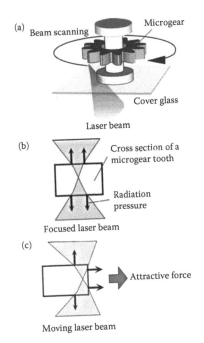

Figure 17.2 A circular scanning laser beam is used to optically drive a polymeric microgear as shown in (a). The gear is trapped (b) when the beam is focused on the center of the tooth causing the radiation pressure exerted on the microstructure to be balanced. If the focus is moved slightly to the side of the gear tooth (c), then the net radiation pressure is directed to the focus creating an external force. (With kind permission from Springer Science+Business Media: Nano-and Micromaterials, Optically driven micromachines for biochip application, 2008, 291–309, Maruo, S., Figure 12.9.)

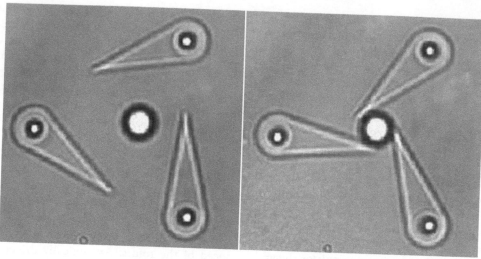

Figure 17.3 Optical microscopic images of three-hand micromanipulators grasping a 5 μm diameter microglass bead. By optimizing the repetition rate (~100 Hz) between trajectories, it is possible to handle the microbead. (With kind permission from Springer Science+Business Media: *Nano-and Micromaterials,* Optically driven micromachines for biochip application, 2008, 291–309, Maruo, S., Figure 12.16.)

column in an effort to hold the rotating disk without the need for a mechanical shaft. Experiments by the Yokohama researchers showed that the flow velocity is proportional to the rotational speed of the microrotor.

Expanding on these concepts, researchers at MIT (Tang 2009) were able to perform nanomanipulation on a microchip using gradient optical forces. Specifically, the researchers showed it was possible to generate a gradient force in the pN range that was sufficient to activate a nanoscale oscillator. The investigated device was constructed from two parallel optical waveguides and required a light source at a known frequency (Li et al. 2008, 2009; Tang 2009). Hong X. Tang (2009) and his team at Yale continued this work and produced oscillations using only one single-mode waveguide. The net force was generated by creating an asymmetrical optical field around the free-standing single-mode waveguide. A measureable transverse force exists on the waveguide because the guided light is evanescently coupled to the dielectric substrate. In this configuration, both the *effective refractive index* and the force on the waveguide depend on the separation distance between the waveguide and substrate. The calculations by Li et al. (2008) show that as the separation is reduced from 500 to 50 nm, the magnitude of the optical force increases from 0.1 to 8 pN μm^{-1} mW^{-1}. Although these transverse gradient optical forces are very small, this basic principle can be used to throw switches in silicon optical circuits (Tang 2009) and to develop nanomechanical beam resonators embedded in a photonic circuit with an on-chip interferometer for displacement sensing (Li et al. 2008, 2009).

17.3 PHOTON-DRIVEN NANOACTUATORS

Nanoactuators are pN force-generating structures constructed from several independent molecules to several thousand interconnect molecules functioning as a single system under an external energy source. An example commonly found in biology is a molecular nanomotor comprised of DNA molecules or a single protein. The DNA nanomotors are able to switch between intramolecular tetraplex and intermolecular duplex states by alternating through DNA hybridization and strand exchange reactions causing the motor to shrink and expand like an inchworm (Kang et al. 2009; Li and Tan 2001). From an engineering perspective, these molecular devices can be exploited for developing novel methods in drug delivery, biochip design, and even nanoscale manufacturing.

Kang et al. (2009) describes a light-driven single-molecule DNA hairpin-structured nanomotor that utilizes a photosensitive azobenzene molecule to induce movement (Figure 17.4). Azobenzene undergoes a reversible *cis–trans* isomerism when

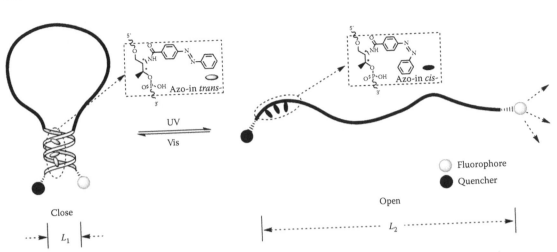

Figure 17.4 An illustration of the photoswitchable single-DNA nanomotor. The basic nanomotor components are a hairpin backbone, photosensitive azobenzene molecules, and fluorophore/quencher pair for signaling motor movement. The average size of hairpin structure is L_1, and the average size of extended molecules based on persistence length of a single DNA strand is L_2. The single-DNA nanomotor exhibits a CLOSE state when exposed to visible irradiation and an OPEN state with irradiated by UV light. (Reprinted with permission from Kang, H. et al., *Nano Letters*, 9(7): 2690–2696. Copyright 2009, American Chemical Society.)

exposed to alternating UV and visible radiation. Note that the term *cis* means "on the same side" and *trans* means "on the other side." When azobenzene is linked to other molecules then the switching mechanism can cause a relatively large dimensional change in the functionalized material. By controlling the azobenzene moieties integrated on the DNA bases in the hairpin's duplex stem segment, it is possible to modify the dehybridization (open) and hybridization (closed) state of a hairpin structure and control movement. Since the open–close cycle of the hairpin molecule exhibits reversible extension and contraction behavior, it can be identified as a nanomotor. The photoregulation of this simple system is concentration-independent and, therefore, suitable for fabricating high-density molecular motors.

The notion of a molecular photoswitch has also been used to control the movement of living microorganisms. Neil Branda and colleagues at Simon Fraser University (Al-Atar et al. 2009; Boyer et al. 2010; Carling et al. 2009) have demonstrated that the light-induced reactions of photoresponsive dithienylethene can be used to trigger paralysis in the transparent nemotode worm, *Caenorhabditis elegans*. The *C. elegans* neotodes are fed a "ring-open" mixture of bipyridinium dithienylethene and a buffer containing 10% dimethlsulfoxide and exposed to a UV light (365 nm) for 2 min. The shorter UV wavelength radiation causes the dithienylethene photoswitch to undergo a ring-closing reaction that immobilizes the organism and turns it blue. These researchers further demonstrated that the paralyzed *C. elegans* worms could be reanimated and returned to their original colorless state by exposing the organism to visible light (> 490 nm) for 20 min. This was possible because the visible light triggered the reverse reaction in the photoswitch by producing the ring-open isomer. Although the intent of Branda and his fellow researchers was not to create an optically driven nanoactuator, they were exploiting light to regulate the movement of a biological organism to perform a simple directed task. This study and its possible extension to *in vivo* drug delivery are of particular interest to researchers who are developing new actuation technologies for biomedical applications (Al-Atar et al. 2009; Carling et al. 2009).

17.4 PHOTOMECHANICAL RESPONSE OF MATERIALS

Since the early 1990s, a large variety of materials have been investigated that exhibit optical-to-mechanical energy conversion properties. These materials include light-induced shape-changing polymers (Jiang et al. 2006; Mitzutani et al. 2008), shape-changing liquid crystal elastomers (LCEs; Finkelmann et al. 2001; Tabiryan et al. 2005; van Oosten et al. 2007, 2009; Warner and Terentjev 1996, 2003; Yu et al. 2003), photostrictive materials that take advantage of photovoltaic and inverse piezoelectric effects (Morikawa and Nakada 1997; Poosanass et al. 2000; Uchino 1990, 2012), photomechanical actuators with charge-induced surfaces (Datskos et al. 1998; Lagowski et al. 1975; Suski et al. 1990), chalcogenide glasses (ChGs) influenced by mechanical polarization effects (Krecmer et al. 1997; Stuchlik et al. 2001), and photomechanical actuation of carbon nanotubes (CNTs) (Kroerner et al. 2004; Liu et al. 2009; Piegari et al. 2002; Verissimo-Alves et al. 2001; Zhang and Iijima 1999). The functional behavior of several types of light-driven materials is summarized below. These categories are based on the underlying photomechanical response behavior and include shape-changing polymers, ferroelectric materials exhibiting photorestrictive effects, electrostatic structures mediated by photogenerated electrons, and light-induced elastic CNTs.

17.4.1 Shape-changing polymers

A variety of polymeric gels and polymer films have been developed that exhibit relatively large displacements when exposed to light radiation. For example, hydrogels are fluid-filled polymeric networks that can undergo dramatic volume changes when exposed to an environmental stimulus such as a change in surrounding temperature, pH, electric field, specific ion concentration, and light. This unique ability to generate relatively large actuation dynamics and specificity of the triggering mechanisms has made smart hydrogels the preferable candidate for the core sensing and actuation transducers in a variety of microsystem technologies (Al-Aribe et al. 2012; Knopf 2012; Knopf and Al-Aribe 2012). Poly(N-iso-propyl acrylamide) (PNIPAM) with triphenylmethane leuco derivatives is a UV light responsive hydrogel that swells when exposed to UV radiation and shrinks back to the original shape when the light is removed (Hirasa 1993). The swelling of the hydrogel is caused by an increase in osmotic pressure within the gels due to ionization reactions and ion-pair formation initiated by the UV irradiation. It is also possible to create hydrogels that respond to

visible light by introducing trisodium salt of copper chlorophyllin into the PNIPAM gel. The swelling behavior of these thermosensitive gels are linked to the increase in temperature arising from the absorption of light energy. Some PNIPAM gels have also been observed to undergo very slow volume changes under radiation exposure even when no photosensitive molecules are attached to the macromolecules (Hirasa 1993).

An alternative approach is to create hydrogel microstructures that are responsive to specific wavelengths of light. To accomplish this Sershen et al. (2005) introduced an optomechanically responsive nanocomposite hydrogel that undergoes pronounced and reversible changes in shape when exposed to different wavelengths of light. The materials are composites of a thermally responsive polymer (PNIPAM with a 95:5 comonomer ratio) and nanoparticles that have distinct optical absorption profiles. Duff and Baiker (1993) used gold colloids, while Oldenburg et al. (1998) used gold nanoshells, with an 110 nm diameter silica core and 10 nm thick gold shell, for similar purposes. The composites were formed by mixing the nanoparticles with the monomer solution thereby trapping the particles within the hydrogel matrix after polymerization. These nanocomposite materials responded to different wavelengths of light. Sershen et al. (2005) further demonstrated the independent control of two valves along the T-junction as shown in Figure 17.5. One valve is constructed from gold–colloid nanocomposite hydrogel and the other is gold–nanoshell nanocomposite hydrogel. The channels are 100 μm wide. The entire device is illuminated with 1.6 W/cm², 532 nm (green) light source. The gold–colloid valve opened while the nanoshell valve remained closed. However, when the device is illuminated with 2.7 W/cm², 832 nm (near infrared) light then the opposite response was observed. In both cases, the valves opened within 5 s.

In order to respond directly to light, shape-transforming solid polymer films must often contain photosensitive fillers (Jiang et al. 2006) such as triphenylmethane leuco derivatives that undergo photoinduced ionic dissociation, photoreactive molecules (e.g., cinnamates), and photoisomerizable molecules (e.g., azobenzenes). For example, the triphenylmethane leuco derivatives will dissociated into ion pairs under UV illumination and recombine under dark conditions when the light is

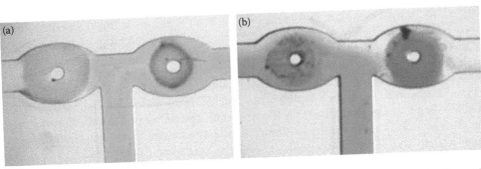

Figure 17.5 Two light-activated nanocomposite hydrogel valves inserted at a T-junction in a micro-fluidic chip where one valve is made of an Au–nanoshell composite (left) and the other Au–colloid composite (right). The channels are 100 μm wide. (a) When the T-junction is illuminated with green light (532 nm, 1.6 W/cm²), the Au–colloid valve opened while the nanoshell valve remained closed. (b) However, when the device was illuminated with near-infrared light (832 nm, 2.7 W/cm²), the opposite response was observed. (Sershen, S. R. et al. Independent optical control of microfluidic valves formed from optomechanically responsive nanocomposite hydrogels. *Advanced Materials*, 2005, 17, 1366–1368. Copyright Wiley-VCH Verlag GmbH & Co. KGaA. Reproduced with permission.)

extinguished. If the triphenylmethane leuco derivatives are embedded in a nonrigid solid-state polymer or gel then the reversible variation of electrostatic repulsion between the photogenerated charges can induce expansion and contraction in the polymer. Cinnamate-type groups can also be used to create light responsive SMPs because these photochemically reactive molecules form photoreversible covalent cross-links in the material. The most common photosensitive molecules are azobenzenes because these molecules switch reversibly from the *cis* or *trans* conformation states when exposed to light of a predetermined wavelength (Finkelmann et al. 2001; Jiang et al. 2006; Warner and Terentjev 1996). If the azobenzene groups are linked to macromolecules then the light-driven switch can cause relatively large changes in the polymeric material.

Some researchers have explored how nematic LCEs can be used to create photoresponsive SMPs. The term nematic refers to molecules that tend to align in loose parallel lines. For nematic LCEs containing azobenzene groups, the light-induced reaction from *trans → cis* isomerization of the azobenzene units produces the movement of the liquid crystal (LC) domains and subsequently causes the collapse of the alignment order producing a significant contraction. Studies have also shown that the bending direction of certain LCE films can be controlled by applying linearly polarized light (Tabiryan et al. 2005; Yu et al. 2003). In these cases, the incident light is largely absorbed by the surface layer of the polymer film

because of the strong absorption by the azobenzene moieties.

Light intensity, wavelength, and the length of time that a LCE sample is irradiated will all impact the extent and speed of shape recovery. The response times of nematic elastomers is slow but can be improved by either decreasing the thickness of the polymer film or increasing the light intensity (Yu et al. 2003). LCEs obtain their unique shape-changing properties from the interrelationship between the elastic properties of the polymer networks and the ordering of the mesogenic liquid crystalline moieties. As a result, nematic LCEs are able to change their shape by up to 400% over a relatively narrow temperature interval straddling their nematic–isotropic (N–I) transition temperature. Finkelmann et al. (2001) found that at a temperature of 60°C, the fractional contraction of nematic elastomers was as large as 22%. Yu et al. (2003) demonstrated that the large bending of a single film of LC network containing an azobenzene chromophore can be produced by UV light. Tabiryan et al. (2005) reported a reversible bidirectional bending of the azo-LC polymer by switching the polarization of the light beam between orthogonal directions. For these samples, both the magnitude and sign of photoinduced deformation were controlled by the polarization state of the light beam.

Yamada et al. (2008) performed a series of tests on a laminated structure composed of a LCE layer and a flexible plastic sheet in an effort to study the mechanical forces generated on the composite film when exposed to UV light radiation. In these tests, both ends of the film were rigidly clamped and the film was then loaded with a force of 44 mN at 30°C. Although a single-layer LCE layer was experimentally shown to be brittle and crack at high intensities due to low mechanical strength, the composite film was able to generate significant forces without breakage. The researchers also used the laminated LCE film as a plastic belt attached to a pulley system (Figure 17.6) to create a light-driven motor that directly converted light energy into rotation. The pulleys of the system rotate by exposing the top of the belt to UV light while irradiating the bottom side with visible light. The simultaneous contraction and expansion forces at different regions along the long axis of the belt induce rotation in the pulleys and enabled them to move continuously.

Van Oosten et al. (2008) investigated the reversible nonlinear response of photostimulated LC polymer by looking at the bending action of a planar uniaxially aligned film. When exposed to laser light (351 nm) from one direction, the LC film would rapidly bend toward the light source (Figure 17.7). After a period of prolonged exposure, the film initially relaxes and then curves in the opposite direction over time. The speed of the transformation was experimentally found to increase as the LC film uncurls because of the unbending sample will be exposed to more direct light. The artificial cilia were similar to the natural cilia found in paramecia in that it may produce a flapping, asymmetrical motion causing the surrounding liquid to flow. The asymmetric motion was the result of a backward stroke different from the forward stroke, which had

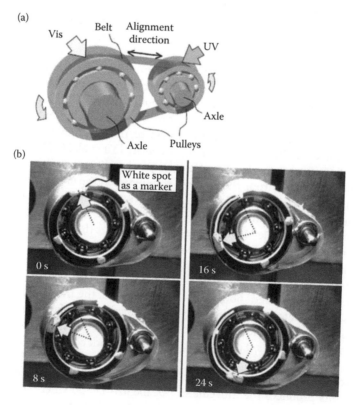

Figure 17.6 A light-driven motor with a plastic belt constructed from LCE-laminated film. The basic motor design and relationship between the rotation direction and UV irradiation positions are illustrated in (a). The diameter of the larger left pulley is 10 mm and the smaller right is 3 mm. The LCE plastic belt is 36 mm × 5.5 mm. The time series of images in (b) show the rotation profiles of the LCE-laminated film induced by 366 nm, 240 mW/cm² UV irradiation and >500 nm, 120 mW/cm² visible light. (Yamada, M. et al. Photomobile polymer materials: Towards light-driven plastic motors. *Angewandte Chemie International Edition.* 2008, 47, 4986–4988. Copyright Wiley-VCH Verlag GmbH & Co. KGaA. Reproduced with permission.)

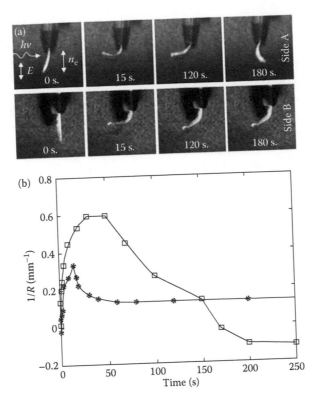

Figure 17.7 Photostimulated bending of light-driven LC actuators driven by the isomerization of azobenzene is illustrated in (a). The top row of images show the response of the LC film oriented with Side A toward the 351 nm, 150 mW/cm² laser light. The bottom row is the same LC film actuator with Side B oriented toward the laser light. (b) The bending radius (1/R) of the sample as it is exposed to light for both Side A (ⓒ) and Side B (*). (Reprinted with permission from Van Oosten, C. L. et al. *Macromolecules* 41: 8592–8596. Copyright 2008, American Chemical Society.)

been introduced into the LC azobenzene artificial cilia by temporally varying the intensity and wavelength of the light over the microactuator surface. The researchers envisioned that this simple bending principle could be used to create light-driven artificial cilia that function as a microfluidic mixer or pump (van Oosten et al. 2009).

17.4.2 Photostrictive ferroelectric materials

When irradiated by light, various ferroelectric materials (Uchino 1990) will generate mechanical strain that can produce observable movements and measureable microforces. The photostrictive properties of these materials are the result of combined photovoltaic and the converse piezoelectric effects (Poosanaas et al. 2000). The photovoltaic property causes the ferroelectric material to generate voltages in the kV/cm range when irradiated, while

the converse piezoelectric effect causes the same material structure to simultaneously expand or contract. The most commonly used piezoelectric materials are the lanthanum-doped lead zirconate titanate (PLZT) ceramics. The photostriction properties of these PLZT ceramics are almost instantaneous when exposed to light and, therefore, have been used to develop a number of rapid response optical microactuators, relays, and photon-driven micromachines (Uchino 1990, 2012).

A bimorph-type optical actuator created by Morikawa and Nakada (1997) was able to produce displacements of several hundred microns. The light-driven microcantilever beam was constructed from a pair of adhesively bonded PLZT ceramic elements oriented in opposite polarized directions. When the upper PLZT element in the cantilever device was illuminated by a UV light source, the element stretched in the polarized direction, while the lower nonradiated PLZT element did not expand. Since the two PLZT elements were connected to a common electrode and the lower nonexpanding element was oriented in the opposite polarized direction of the top, the lower element would experience a negative voltage and correspondingly contract due to the same piezoelectric phenomena. The combined effect of the two PLZT ceramic elements was a significant downward displacement of the microcantilever beam. By switching the illumination to the bottom element, it was possible to drive the cantilever beam upward.

A similar photomechanical effect has also been used by Suski et al. (1990) to produce *photostimulated vibrations* in silicon (Si) microcantilever beams that were covered in a thin polycrystalline ZnO film. Earlier studies (Gatos and Lagowski 1973; Lagowski and Gatos 1972; Lagowski et al. 1975) had shown that thin crystals of polar (noncentrosymmetric) semiconductors can exhibit piezoelectric properties by light-induced electronic transitions. The photomechanical response arises from the depopulation and population of surface states by sub-bandgap illumination while the overall number of bulk free carriers remains the same (Suski et al. 1990). The barrier height of the depleted layer can also change under these illumination conditions resulting in surface stress variations. When the light source is modulated, the barrier height will vary according to the fluctuations in photon intensity producing surface stresses with the same frequency. Consequently, it is possible to create resonant vibrations in the silicon (Si) microcantilever beam by optically altering the frequency of the surface barrier so that it closely matches the natural frequency of the beam (Lagowski et al. 1975; Suski et al. 1990). The observed photomechanical effect is consistent with the surface piezoelectric phenomenon where the external stress applied to polar semiconductors leads to a modification of the surface barrier height and causes pronounced changes in contact potential difference (Lagowski et al. 1975).

Suski et al. (1990) demonstrated the photomechanical effect in the early 1990s by fabricating a 10 mm×1.5 mm×50 μm Si/SiO$_2$/ZnO microcantilever beam and activating it using an argon laser (520 nm). The microcantilever resonated at a frequency of ~350 Hz with a maximum deflection of 160 nm under 130 μW of light power. The photostimulated effect was also observed in bulk semiconductor materials where the photon energy irradiating on the Si microcantilever was above the bandgap. As the photons become absorbed in the semiconductor material the free electrons are excited and move from the valance band into conduction band (CB) leaving holes in the lattice. The movement of electrons creates local mechanical strain in the material.

Another photomechanical effect has been observed in ChG when the material absorbs polarized light (Stuchlik et al. 2001, 2004). Reversible photoinduced anisotropy was the first reported by Krecmer et al. (1997) who showed that when a thin amorphous film of As$_{50}$Se$_{50}$ deposited on a clamped atomic force microscope cantilever was exposed to polarized irradiation it would exhibit reversible nanocontraction movement. To demonstrate the effect, the research team performed measurements on a 200 μm cantilever beam with a thickness of 0.6 μm. The surface of the beam was covered with a thin 250 nm As$_{50}$Se$_{50}$ film. When exposed to polarized light, the beam bent approximately ±1 μm (Stuchlik et al. 2001). Upon irradiation with polarized light, a very small movement in the ChG was measured parallel to the direction of the electric field of the light and a very small expansion was also observed along the axis orthogonal to the electric field.

Poly(vinylidene difluoride) (PVDF) is another ferroelectric polymer with both pyroelectric and piezoelectric properties that can be optically driven (Mizutani et al. 2008). When the temperature of the PVDF material is increased through light irradiation, it causes a piezoelectric effect to occur and the material mechanically deforms. Mizutani et al. (2008) further demonstrated how a strip of PVDF could be used to create a small leg for a light-driven microrobot (Figure 17.8). In this case, one surface of the PVDF film was coated with a thin silver (Ag) electrode. When irradiated with a He–Ne laser, the PVDF cantilever generated an electric field by means of the pyroelectric effect. The thickness of the PVDF cantilever was 28 μm and the Ag electrode was 6 μm thick. The pyroelectric effect produced by the PVDF-generated conduction electrons that were dispersed on the Ag electrode. The electric field in the cross section of the PVDF film induced the inverse piezoelectric effect that caused the microcantilever leg to bend toward the light. In these experiments, a 10 mW irradiation was used to move the microcantilever

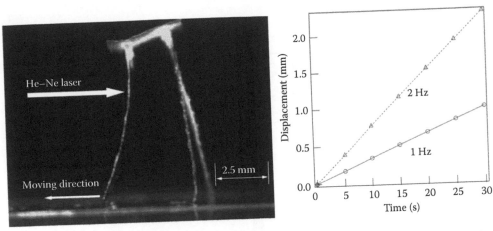

Figure 17.8 Photograph of the optically driven PVDF leg actuator (a) and the displacement of the leg for two different frequencies (1 Hz and 2 Hz) (b). (With kind permission from Springer Science+Business Media: *Optical Review*, Optically driven actuators using poly (vinylidene difluoride), 15, 2008, 162–165. Mizutani, Y. et al., Figures 6 and 7.)

250 µm in 0.5 ms. The velocities were 33.3 µm/s at 1 Hz and 76.7 µm/s at 2 Hz.

17.4.3 Electrostatic pressure mediated by photogenerated electrons

A "warped capacitor" microactuator that uses photogenerated electrons to change the electrostatic pressure on a thin silicon (Si) microcantilever beam was introduced by Tabib-Azar and his colleagues in 1990 (Tabib-Azar 1990; Tabib-Azar and Leane 1990; Tabib-Azar et al. 1992; Tabib-Azar 1998). The parallel plate capacitor consists of the thin P+ silicon microcantilever beam attached to a glass substrate insulator and suspended over a gold (Au) ground plate. Photogenerated electrons occur when the Au ground plate is exposed to a light source with sufficient energy. The photoelectrons travel through the air gap to the deformable microcantilever beam, reducing the charge on the capacitor, and causing the cantilever to bend. The capacitance (C) of the flat parallel plates is given by the expression

$$C = \frac{k\varepsilon_0 A}{d} = \frac{\varepsilon_0 Lb}{d}, \qquad (17.4)$$

where k is relative permittivity of the dielectric material between the plates ($k \approx 1$ for air), ε_0 is the free-space permittivity (8.854×10^{-12} F/m), b and L are the width and length of the cantilever beam, respectively, A is the surface area of the parallel plate capacitor, and d is the distance between the cantilever beam and the ground plane. From the definition of capacitance, the unit Farad (F) is equal to C/V. A bias current must be used in order to make the microbeam capacitor function effectively as a microactuator. A separate voltage source, V_0, through a resistor is connected across the capacitor to form the bias current (Figure 17.9).

If the change in capacitance as the cantilever beam deforms is neglected, the steady-state deflection at the end of the microbeam can be described in terms of the charge on the capacitor. However, the cantilever beam will bend in a stable manner up to the threshold of spontaneous collapse (Tabib-Azar 1998) given by

$$\delta_{\text{threshold}} \approx \frac{6\varepsilon V^2}{Yb^3}\frac{L^4}{8d^2}, \qquad (17.5)$$

which is a function of the Young's modulus (Y) and the thickness of the cantilever beam (b). The fastest time that it can smoothly traverse this distance is approximately equal to the period of the fundamental mode of free vibration.

Tabib-Azar (1998) constructed a microactuator that uses a $600 \times 50 \times 1$ µm^3 cantilever beam with a gap of 12 µm to demonstrate this concept. A bias voltage of 6 V and optical power less than 0.1 mW/cm^2 was used to move the cantilever 4 µm in approximately 0.1 ms. The light-controlled actuation is possible by charging the capacitor

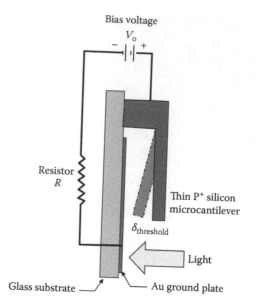

Figure 17.9 The basic design of the electrostatic "warped capacitor" microactuator originally investigated by Tabib-Azar (1998). Photogenerated electrons occur when the gold ground plate is exposed to light with sufficient energy. The photoelectrons across the air gap and reduction in the charge on the parallel plate capacitor cause the microcantilever to bend.

with a current from the battery circuit. A continuous photon flux, $\Phi < i/\eta$, where η is the quantum efficiency, short circuits the capacitor more slowly than the battery charges it, causing a charge buildup which closes the plates. A photon flux $\Phi > i/\eta$ causes an opposing photocurrent greater than the charging current. The net charge then decreases, and the capacitor plates relax open. In this manner, the light source is not the primary source of energy to the actuating system rather it is sufficient to enable the action by pushing the current level above a threshold.

17.4.4 Light-induced elastic CNTs

CNTs are a relatively new class of one-dimensional carbon nanomaterials that have extraordinary mechanical, electrical, and thermal properties (Iijima 1991; Sun and Li 2007). The Young's modulus of a CNT is over 1 TPa and the tensile strength is approximately 200 GPa. The thermal conductivity can be as high as 3000 W/mK. Cylindrical single-walled carbon nanotubes (SWCNTs) have relatively small diameters (~0.4 nm) and based on their structure can be either metallic or semiconducting. Studies in the late 1990s suggest that the structure distortion caused by the van der Waals

forces can modify the electronic structure of the CNT and influence both the mechanical and optical properties of SWCNT bundles.

Light-induced elastic responses from SWCNT bundles and fibrous networks were first reported by Zhang and Iijima (1999). These researchers observed the elastic movement of bundles of ~20–50 μm long SWCNT filaments when exposed to visible light. The electrostatic interaction of the SWCNT bundles was believed to be the cause of the elastic filament behavior. The authors concluded that the effect was the result of photovoltaic or light-induced thermoelectric effect related to the modification of electronic structure during the bundle formation. The orientation and alignment of highly anisotropic CNTs has also been shown to be an important factor in determining the photomechanical properties. Liu et al. (2009) studied the effects of CNT alignment on the photoconductivity of thin films and showed that a high degree of nanotube orientation can improve the power conversion efficiency ~10%. In addition, films containing partially oriented CNTs produced faster response times and attained a higher internal photon to electron power conversion efficiency than the film made up of nonoriented nanotubes.

Polymer CNT and nanocomposites have been receiving a great deal of attention in recent years because of enhanced mechanical properties and their unique electrical and thermal properties. One of the earliest material-based systems was a CNT-filled thermoplastic elastomer (Morthane) nanocomposite introduced by Kroerner et al. (2004). Morthane is a linear, hydroxyl terminated polyester polyurethane that exhibits a low glass transition temperature ($T_g = -45°C$), near-ambient melting temperature of the soft-segment crystallites ($T_{m,s} = 48°C$), and exhibits huge strain-induced deformations (~700%). Significant deformations at room temperature ($T_g < T_{room} < T_{m,s}$) causes the flexible polymer segments to crystallize. The crystallization process forms physical cross-links, which prevent the polymer from undergoing strain recovery when the applied stress is removed. Subsequent heating and melting of the strain-induced soft-segment crystallites releases the constrained polymer chains forcing the material to revert back to its stress-free shape. The researchers were able to deform the composite Morthane material by 300%, exerting ~19 J to lift 60 g weight more than 3 cm with a force of approximately 588 N.

More recently, Panchapakesan and his colleagues (Lu and Panchapakesan 2006, 2007) exploited CNTs to create strain-induced crystallization for SMPs. Shape memory is the ability of the polymeric material to reversibly recover inelastic strain energy when exposed to specific environmental stimuli. The strain energy is captured in the material through a reversible morphology change induced by shape deformation or by the suppression of molecular relaxation. Molecular relaxation occurs in SMPs when the material is quenched through the glass transition or crystallization temperature for the SMP. For most cases, the material is able to recover the original shape when the temperature is raised above the critical thermal transition.

17.5 PHOTOTHERMAL INTERACTION WITH MATERIALS

A number of optical actuators have been designed that exploit the ability of light to generate heat and, thereby, directly influence the thermal properties of surrounding gases, liquids, and solids. A focused light source can be used to produce sufficient heat energy that can introduce thermo viscous effects (Hockaday and Waters 1990), transform a confined liquid into a pressurized gas (Hale et al. 1988, 1990; Hockaday and Waters 1990; McKenzie and Clark 1992; McKenzie et al. 1995; Mizoguchi et al. 1992; Tabib-Azar 1998), drive microflows (Shiu et al. 2007; Weinert and Braun 2008), initiate temperature induced phase transformation of solids (Okamura et al. 2009; Okamura 2012; Yoshizawa et al. 2001), or create photothermal vibrations in fiber optic waveguides (Inaba et al. 1995; Jankovic et al. 2004; Otani et al. 2001).

Most engineering applications utilize lasers as the light source because the light rays from a CW or pulsed lasers travel in the same direction (unidirectional), are monochromatic (single wavelength), and are coherent (all rays are in-phase). When a stream of photons from a laser is projected onto a solid surface, the interaction may result in illumination, light reflection, energy absorption, thermal and thermodynamic changes, melting, vaporization, or plasma effects (Steen 1998; Steen and Mazumder 2010). The precise nature of the interaction depends on the optical power of the laser, duration of exposure, reflective and absorption properties of the target surface, thermal properties of the exposed material, and local environmental conditions. The absorption properties of the target surface determine how efficiently the photon energy is transferred into the material. This is dependent on the material's absorption coefficient (relates to the amount of optical energy transferred per unit depth), reflectivity (if 100% reflective then no light is absorbed), and material surface finish (smooth or rough).

The laser light striking a nonmirrored surface will produce a measureable localized thermal change in the material. Irradiance (E) is a measure of the incident laser power per unit area (W/cm^2 or W/m^2) and can be given as

$$E = \frac{P}{A},$$ (17.6)

where P is the power of the laser source in watts (W) and A is the area of the beam spot in (m^2). The amount of energy delivered to an object surface is, therefore, a function of the beam diameter where a smaller beam diameter produces a higher irradiance and, thereby, greater amount of energy striking the material. For example, a 2 kW laser focused to a 0.2 mm diameter beam produces an irradiance of 6.3×10^{10} W/m^2 (Steen 1998). In other words, a kW laser with a small beam size will vaporize an element in a fraction of a millisecond and, therefore, a concentrated (focused) light beam can be used as a *powerful "heat source"* in the microworld.

Applying laser light to the material heating process also requires a basic understanding of the thermal properties of the target material (Steen 1998; Steen and Mazumder 2010). Thermally conductive, insulating, and semiconducting materials all behave differently when exposed to a focused, highly concentrated light beam. Since no energy bandgap exists between the valence and conduction bands of a thermally conductive material, a large amount of the electrons in the CB can easily absorb the photon energy. These CB electrons transfer their energy to the material through electron–lattice collisions. With *very low energy* from a laser, the photons are easily absorbed by conductors and the acquired energy is turned to heat. In contrast, the large bandgaps cause the insulator materials to have essentially no CB electrons and, therefore, exhibit no thermal conduction. A large energy bandgap in the material will require a significant amount of energy from the laser for photon absorption. The thermal behavior of semiconductors is between conductors (e.g., metals) and insulators. A very small bandgap exists between the conduction and valence bands for semiconductor materials and the energy can be transferred fairly easily between the CB and the lattice structure. However, the amount of energy necessary from the laser for photons to be absorbed by the material is greater than conductive metals.

Thermal properties also determine how the heat energy delivered by the laser will flow into the target material. Important engineering properties include *thermal conductivity* and *thermal diffusivity*. Thermal conductivity describes how fast the heat flows through the material whereas thermal diffusivity reflects how fast the material will conduct the thermal energy. The *heat flow* through a material depends on the thermal conductivity (k) and on the specific heat (c_h) of a material. Thermal diffusivity (α) tells us how fast materials will accept/conduct thermal energy and can be used to approximate the depth that heat will travel per pulse (with time t) through the material. Materials with a low value for thermal diffusivity, such as stainless steel and some nickel alloys, will limit the penetration depth into the material. Note that the heat flow is dependent on the *specific heat* of a material as it is used to determine the rate of change of temperature.

Laser intensity and pulse duration also influence heat penetration in a material. For some metals like stainless steel that have *low thermal diffusivity*, a lower-powered laser with a long pulse is used. In contrast, for metals such as copper that have a *high thermal diffusivity*, a higher power laser with shorter pulses to overcome the losses can be used. In addition to conductivity and diffusivity, the *thermodynamic properties* relate to the amount of energy required to heat, melt, or vaporize the material. This depends upon the target material's density, heat capacity, melting and vaporization temperatures, and the latent heats of fusion and vaporization (Steen 1998). Heat capacity is a measureable physical quantity related to heat, mass, and change in temperature. A basic understanding and appreciation of light material interaction is necessary for designing viable optically driven microactuators based on the photothermal effects.

17.5.1 Light-driven diaphragms

Many indirect optical microactuators take advantage of the heat generated by the light source to create the desired force or pressure to move the actuator structure (Hale et al. 1988, 1990; Hockaday and Waters 1990; McKenzie and Clark 1992; McKenzie et al. 1995). When a simple gas is heated, it expands according to the ideal gas law

$$\rho G_v = nRT, \qquad (17.7)$$

where ρ is the gas pressure, G_v is the gas volume, n is the number of moles, R is the gas constant (0.0821 L atm/mol K), and T is the absolute temperature. Equation 17.7 states that a fluid in a confined space will expand and increase the pressure exerted on the walls of the enclosed structure as the temperature increases. The pressure in the enclosure may be sufficient to mechanically deform one of the walls if it is constructed from a flexible membrane. As the membrane, or diaphragm, expands under pressure, it produces the desired deflection, δ, and performs the desired mechanical work. The displacement produced by a diaphragm

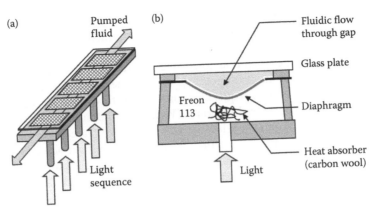

Figure 17.10 An illustration of a simple light-driven micropump composed of five microcells (Mizoguchi 1992). Each microcell has a carbon wool absorber sealed in a chamber filled with liquid Freon 113 (boiling temperature ~47.5°C). The light delivered through the optic fiber is used to heat the absorber and, thereby, raise the temperature of Freon 113. The phase change from liquid to an expanding gas forces the diaphragm to deflect upward and push the working fluid that resides in the gap between the flexible membrane and rigid glass plate.

microactuator at the center from its equilibrium position, δ, can be described as (Tabib-Azar 1998),

$$\rho = \frac{4a_1b}{L^2}\sigma_0\delta + \frac{16a_2f(\gamma)b}{L^4}\left(\frac{Y}{1-\gamma}\right)\delta^3, \quad (17.8)$$

where L is the length, σ_0 is the residual stress, $(Y/1-\gamma)$ is the biaxial modulus, and b is the thickness of the diaphragm. The dimensionless parameters a_1, a_2, and $f(\gamma)$ depend on the geometry of the diaphragm. Tabib-Azar (1998) describes a square diaphragm given as $a_1 = 3.04$, $a_2 = 1.37$, and $f(\gamma) = 1.075 - 0.292\gamma$. These microfabricated flow controllers have speeds of 21 ms in air flow and 67 ms in oil flow, with sensitivities of 304 Pa/mW and 75 Pa/mW, respectively.

Mizoguchi et al. (1992) used this same simple concept to create a micropump that included an array of five closed diaphragm-actuated devices called microcells as shown in Figure 17.10. Each microcell consisted of a predeflected 800×800 μm square membrane that was micromachined in 0.25 mm³ of silicon and filled with Freon 113, a liquid with a boiling point of approximately 47.5°C. A carbon wool absorber was placed inside the cell to convert the incident light from the optic fiber into heat. The microcell exhibited a relatively large deflection, approximately 35 μm, when the cell's content was heated and the Freon 113 undergone a phase change from liquid to a gas. The fluid being transported by the pump is fed into a flow channel between the glass plate and deflecting membrane using very small harmonic movements. The harmonic order of the membrane's deflection determines the fluid flow rate and direction. The small quantities of Freon in each cell allowed relatively low optical powers to be used to change liquid to gas, giving the large membrane deflections needed to operate the pump. The microcell was fabricated and operated by a laser source with no more than 10 mW. The micropump achieved a head pressure of approximately 30 mmag and flow rate of 30 nL/cycle.

17.5.2 Light-driven microflows

The miniaturization and integration of various analytical laboratory processes on a single platform is a key design requirement for microfluidic and LOC devices used for medicine and environmental monitoring (Dittrich and Manz 2006; Eijkel and van den Berg 2005). The reduction in physical size of these functional components has increased the speed of analysis, lowered operating costs due to the consumption of small quantities of reagents, and enabled novel system designs that avoid biological sample cross-contamination.

One of the most critical components on the microsystem platform is the microfluidic mixer. Efficient mixing of chemicals and biological substances in order to create the desired reactions is an essential step in preparing samples for analysis. The typical mechanism for mixing fluids in the larger macroworld is to create turbulent flows at high Reynold's numbers. In the microdomain, however, mixing liquids becomes more difficult because laminar flow dominates. As a consequence, the much slower process of molecular diffusion becomes the primary mixing mechanism (Hessel et al. 2005; Nguyen and Wu 2005). To improve the mixing rate over shorter length microchannels, researchers have developed a variety of micromixer designs (Nguyen and Wereley 2002). Passive mixers do not require external power sources and, typically, use specially designed microfluidic channel microstructures to improve the overall mixing rate. T-and Y-micromixers are the simplest passive micromixer designs for combining two adjacent streams of liquid (Kamholz et al 1999; Kamholz and Yagar 2002). Unfortunately, a relatively long microchannel is required because the mixing mechanism is completely depended on the molecular diffusion rate. In contrast, active micromixers use an external energy source to create disturbances in liquid streams to improve mixing rates.

Weinert and Braun (2008) developed a light pump approach for active mixing that moves liquid through LOC devices. The authors demonstrate how small volumes of fluid could be optically pumped along a predefined path of a moving warm spot that was created using an infrared laser. The repetitive motion of the laser beam is used to remotely drive 2D microflows of liquid with a resolution of 2 μm. The experiments produced pumping speeds of 150 μm/s with a maximum 10°C temperature increase at the localized light spot. The study also confirmed that the fluid motion was the result of the dynamic thermal expansion in the gradient of liquid viscosity. The viscosity of the liquid at the light spot is reduced by an increase in local temperature, resulting in a broken symmetry between the thermal expansion and thermal contraction in the front and wake of the spot. Consequently, the fluid is observed to move in an opposite direction to the movement of the heated spot because of this asymmetric thermal expansion at the spot front with respect to the corresponding asymmetric thermal contraction in the wake.

In comparison, pressure-driven microfluidics will operate at significantly faster flow speeds. However, to achieve these flow speeds, it is necessary to incorporate millimeter-sized interfacial connections to the outside of the pumps. The light pump approach enables highly localized fluid actuation. Since the pump pattern is programmable in real time, it can be adapted to a variety of geometric patterns. Interestingly, the light pump based on thermoviscous expansion can be used to move fluids through an unstructured environment (Weinert and Braun 2009; Weinert et al. 2009). In this work, an infrared laser was used to melt liquid channels into a sheet of ice. Since the entire channel was not melted at once, the liquid would rapidly refreeze behind the laser spot. The repetition of the laser spot motion with high frequencies enabled the water to undergo a series of melting and freezing cycles, thereby increasing the velocity of the pumping action. As the warm spot formed by the laser was moved through a thin ice sheet, the ice was observed to thaw at the front of the spot while simultaneously freezing in the wake. The solid ice boundaries confine the liquid flow such that the movement is from the back of the molten spot to the front. The pumping action was performed with a repetition rate of $f=650\,\text{Hz}$ and a chamber temperature of $T_o=-10°C$ (Weinert et al. 2009). With densities $\varphi_{water}=1000\,\text{kg/m}^3$ and $\varphi_{ice}=917\,\text{kg/m}^3$, the pump velocity is $v_{pump}=9.5\,\text{mm/s}$. Experimentally, the authors were able to measure the pump velocity at 11 mm/s. The length of the molten spot depends on the temperature of the ice sheet. At low ice temperatures, only a short molten spot is formed. At higher ice temperatures, the molten spot can reach lengths beyond 500 μm with the pump velocity exceeding 50 mm/s.

From a design perspective, the light pump approach implemented on "ice sheets" as introduced by Weinert and Braun (2009) does not require separate valves to switch between pump paths, thereby significantly reducing the hardware necessary for the LOC devices. Further, the approach to fluid transport does not require prefabricated channel walls to define the fluid motion path or permit external pressure control to drive flow. In other words, it is possible to locally drive fluids without structural changes to the substrate, eliminating some of the steps in microfabrication.

17.5.3 Temperature-induced shape-changing metal alloys

The photothermal effect arising from light striking a solid material can also be used generate measurable forces (Okamura et al. 2009; Okamura 2012;

Yoshizawa et al. 2001). Shape memory alloys (SMAs) such as 50/50 nickel–titanium (Ni–Ti) and 50/50 gold–cadmium (Au–Cd) are a group of metal alloys that experience a discontinuous change in their physical structure near their crystalline phase transformation temperature. The dimensional change arising from this phenomenon is significantly greater than the linear volume change that occurs under the normal thermal expansion. Further, the phase transformation temperature can be often modified by varying the alloy composition.

The SMA Ni–Ti is commonly used to transform thermal energy into mechanical work (Gilbertson 1993). The mechanism for microactuation is the forward and reverse martensite-to-austenite phase transformations that occur as Ni–Ti material is heated and cooled. The transformation produces a hysteresis effect where the temperature at which the material undergoes a phase change during heating is different from the temperature that causes the same material to return to the martensite state during cooling. The alloy can be formed into a wire or strip at high temperatures when it resides in an austenite condition. Increasing the temperature applied to a preloaded Ni–Ti wire, originally at ambient temperature, will cause the material to undergo a phase transformation and move the position of the attached load a distance of approximately 4% of the overall wire length. In other words, the small force created during the contraction period can be used to perform mechanical work. The reduction in the wire length can be recovered by cooling the material back to ambient temperature. The number of times the Ni–Ti material can exhibit the shape memory effect is dependent upon the amount of strain, and consequently, the total distance through which the wire is displaced. The amount of wire deflection is also a function of the initial force applied. Thicker wires will generate greater forces but require more heat and longer cooling time. For example, a 200 μm Ni–Ti wire produces 4× the force (~ 5.8 N) than a 100 μm wire but takes 5.5× as long (~2.2 s) to cool down once heating has ceased (Gilbertson 1993).

Although SMA materials exhibit unique and useful design characteristics such as large power/weight ratio, small size, cleanness, and silent actuation, the successful application of the material has been limited to a small number of actuation devices that produce small linear displacements (Okamura et al. 2009; Okamura 2012; Yoshizawa et al. 2001). Okamura et al. (2009) investigated a number of different pretensioned light-driven SMA actuators for grasping and manipulating macroscopic-sized objects. One effective design was a simple torsion spring with the arm extensions formed into a tweezer as shown in Figure 17.11. The 50 μm Ni–Ti wire wrapped around the arms provided the compression to close the end tip and the spring provided the necessary bias force to return the SMA back to its original length after cooling. In this work, a 1.0 W Argon-ion laser was used to heat the actuating wire. However, the design challenge is to find an efficient method to optically heat a very thin wire.

17.5.4 Photothermal vibration of optical fibers

Inaba et al. (1995) investigated how the photothermal vibration of the quartz core of an optical fiber by laser light can be used to construct a vibration-type transducer. The microcantilever beam in this design was the quartz core of the fiber and fabricated by etching the clad layer from the optical fiber tip. The resonance frequency is dependent upon the physical qualities of the cantilever such as size, material density, and Young's modulus. The effect is also partially dependent upon the density of the gas or liquid, which surrounds the cantilever because the resonance shapeness of the beam is a function of the viscosity coefficient for the gas or liquid. The resonance frequency for the microcantilever was observed to decrease from 16.69 to 16.59 kHz with an increase in pressure from 1 to 100 Pa, and a reduction in the resonance sharpness with an increase in pressure from 100 Pa to 10 kPa (Inaba et al. 1995).

Based on the concept of photothermal vibration, Otani et al. (2001) proposed a dynamic optical actuator that was driven solely by light. The device is a walking miniature robot constructed from three optical fibers, which represent legs, attached to a base. Each fiber was cut for a bevel and the surface was painted black so that it could absorb light and convert it to heat. The photothermal effect occurred in response to a flashed incident beam onto one side of the optical fiber leg. The flashing light source with a constant cycle time produced a stretch vibration on the tip of the fiber that enabled

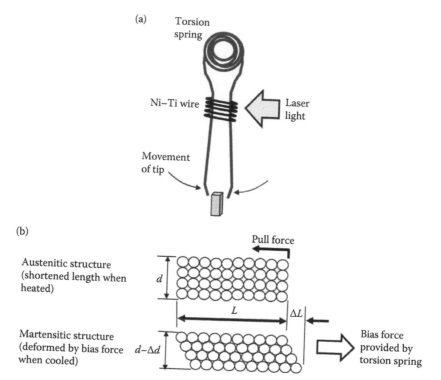

Figure 17.11 A schematic drawing of a light-driven tweezer originally proposed by Okamura et al. (2009) for grasping small objects (a). The SMA actuator is based on the crystalline structure transformation of the Ni–Ti alloy as it is heated laser light and cooled (b). The torsion spring provides the bias force that extends the Ni–Ti wire when cooled.

it to operate like a flat spring. The authors also experimentally demonstrated that the diameter of the fiber has a direct influence on the amount of deformation. It was observed that a 10 mm long fiber with a diameter of 250 μm would deform by 30 μm, whereas a 1000 μm diameter fiber of the same length would deform by as much as 50 μm. Furthermore, Otani et al. (2001) studied the effect of fiber length on the amount of displacement generated. A 1-mm-long fiber with a diameter of 250 μm was found to deform 10 μm, whereas a 15 mm fiber of the same diameter deformed 90 μm.

In the microrobot design, the optical fiber became bent due to the thermal expansion that occurred on the beveled side of the fiber when the light was turned on. If the light is turned off, it returns to its original shape. The switching frequency of the 12.1 mW HeCd laser (442 nm) was 4 Hz. The light was delivered to an individual optical fiber by a mirror and object lens mounted on a moving stage that followed the movement of the photothermal actuator. The size of the optical

actuated walking robot was 3 mm × 10 mm and moved 2.3 mm at 25 μm per second using a pulsating light source.

A micromanipulator based on the photothermal bending effect experienced by a beveled optical fiber was described in a paper by Jankovic et al. (2004). The micromanipulator design incorporated four fingers, two bendable fibers for actively grasping small objects and two stationary fibers to provide structural support while holding the object as shown in Figure 17.12. Each finger was a 1 mm diameter acrylic optic fiber with a 25 mm beveled edge near the tip. The beveled edge as coated with a thin layer of black paint where the thickness has a measurable impact on the amount of tip deflection. A light beam, from a 150 W halogen illuminator, was directed into the fixed end of the sculpted optic fiber causing the tip at the free end to deflect by approximately 50 μm. Several experiments were conducted to demonstrate that this simple microgripper is able to grasp, hold, and release a variety of small metal screws and ball bearings.

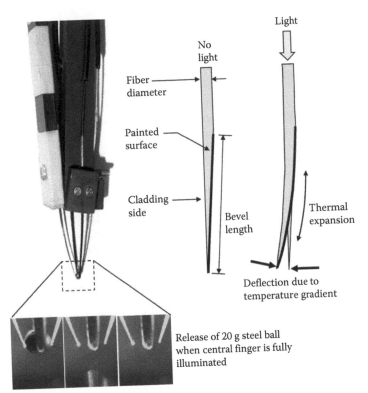

Figure 17.12 A micromanipulator based on the photothermal bending effect of optical fibers as described by Jankovic et al. (2004). The device consists of two bendable fingers and two stationary fingers.

17.6 SUMMARY

The fundamental aspects and unique characteristics of light-driven actuators were discussed in this chapter. In general, nano- and microscale actuator technologies evolved rapidly in the past decade because of unprecedented advances in innovative materials, new fabrication processes, and a multidisciplinary approach to product and process design. These optically powered devices are typically small remotely activated transducers that transform the spectral, intensity, or phase properties of light into very small structural displacements and forces. Radiation pressure and optical gradient forces have been used to manipulate minuscule objects, whereas the spectral properties of light have changed the mechanical behavior of various stimulus responsive polymers. Alternatively, optically driven photothermal effects have been used to heat liquids and gases in an effort to increase the fluid pressure acting on a flexible diaphragm. Many of these induced effects are not readily observable with the human eye and may not appear to be significant in the larger more familiar world, but within the micro- and nanodomain of the *very very small* these light-driven mechanisms become viable solutions.

As predicated by Richard Feynman in his 1959 lecture to the American Physical Society at Caltech, "There's Plenty of Room at the Bottom" (Feynman 1960), the world of the very small will play a significant role in future technology innovations. The future, he believed, lies in "manipulating and controlling things on a small scale" (Feynman 1960). With respect to light, a tightly focused beam provides a highly versatile precision tool for reaching out and manipulating tiny cells, or providing a feathered touch to rearrange molecules, or creating the gentle forces necessary to drive a submicron motor. Specific wavelengths of electromagnetic radiation can also induce physical changes in the shape of preformed photoresponsive polymers and gels. The smaller the polymer structure, the faster and more pronounced the observed reaction to the incident light. These light-induced shape-changing materials can be exploited by creative engineers to develop embedded mechanisms for regulating microfluidic flow or manipulating solid objects in Feynman's world of the *very small*. An interesting illustration of exploiting the unique characteristics of photoresponsive materials is the light-driven single-molecule DNA hairpin-structured nanomotor introduced by Kang et al. (2009). The nanomotor incorporates photoisomerizable azobenzene molecules that enable the DNA structure to undergo a reversible light-controlled switching operation.

REFERENCES

Al-Aribe, K., Knopf, G.K. and Bassi, A.S. 2012. Fabrication of an optically driven pH-gradient generator based on self-assembled proton pumps. *Microfluidics and Nanofluidics* 12: 325–335.

Al-Atar, U., Fernandes, R., Johnsen, B., Baillie, D. and Branda, N.R. 2009. A photocontrolled molecular switch regulates paralysis in a living organism. *Journal of American Chemical Society* F91: 15966–15967.

Ashkin, A. 1970. Acceleration and trapping of particles by radiation pressure. *Physical Review Letters* 24: 156–159.

Ashkin, A. and Dziedzic, J.M. 1971. Optical levitation by radiation pressure. *Applied Physics Letters* 19: 283–285.

Ashkin, A. and Dziedzic, J.M. 1980. Observation of light scattering using optical levitation. *Applied Physics* 19: 660–668.

Ashkin, A. 2000. History of optical trapping and manipulation of small-neutral particle, atoms and molecules. *IEEE Journal of Selected Topics in Quantum Electronics* 6(6): 841–856.

Boyer, J.C., Carling, C.J., Gates, B.D. and Branda, N.R. 2010. Two-way photoswitching using one type of near-infrared light, upconverting nanoparticles, and chaining only light intensity. *Journal of American Chemical Society* F92: 15766–15772.

Carling, C.J., Boyer, J.C. and Branda, N.R. 2009. Remote-control photoswitching using NIR light. *Journal of American Chemical Society* F91: 10838–10839.

Datskos, P.G., Rajic, S. and Datskou, I. 1998. Photoinduced and thermal stress in silicon microcantilevers. *Applied Physics Letters* 73(16): 2319–2321.

Dholakia, K. and Reece, P. 2006. Optical micro-manipulation takes hold. *Nanotoday* 1(1): 18–27.

Dienerowitz, M., Mazilu, M. and Dholakia, K. 2008. Optical manipulation of nanoparticles: A review. *Journal of Nanophotonics* 2: 021875 (32 pp).

Dittrich, P.S. and Manz, A. 2006. Lab-on-a-chip: Microfluidics in drug discovery. *Nature Reviews* 5: 210–218.

Duff, D.G. and Baiker, A. 1993. A new hydrosol of gold clusters. 1. Formation and particle size variation. *Langmuir* 9: 2301–2309.

Eijkel, J.C.T. and van den Berg, A. 2005. Nanofluidics: What is it and what can we expect from it. *Microfluid Nanofluid* 1: 249–267.

Feynman, R.P. 1960. There's plenty of room at the bottom. *California Institute of Technology Journal of Engineering and Science* 4: 23–36.

Finkelmann, H., Nishikawa, E., Pereira, G.G. and Warner, M. 2001. A new opto-mechanical effect in solids. *Physical Review Letters* 87(1): 015501.

Gilbertson, R.G. 1993. *Muscle Wires Project Book*. San Rafael, CA: Mondo-Tronics.

Gatos, H.C. and Lagowski, J. 1973. Surface photovoltage spectroscopy—A new approach to the study of high-gap semiconductor surfaces. *Journal Vaccum Science and Technology* 10(1): F90–F95.

Hale, K.F., Clark, C., Duggan, R.F. and Jones, B.E. 1988. High-sensitivity optopneumatic converter. *IEE Proceedings* F95(5): 348–352.

Hale, K.F., Clark, C., Duggan, R.F. and Jones, B.E. 1990. Incremental control of a valve actuator employing optopneumatic conversion. *Sensors and Actuators* (A21–A23): 207–210.

Hessel, V., Lowe, H. and Schonfeld, F. 2005. Micromixers—A review on passive and active mixing principles. *Chemical Engineering Science* 60: 2479–2501.

Higurashi, E., Ohguchi, O., Tamamura, T., Ukita, H. and Sawada, R. 1997. Optically induced rotation of dissymmetrically shaped fluorinated polyimide micro-objects in optical traps. *Journal of Applied Physics* 82(6): 2773–2779.

Hirasa, O. 1993. Research trends of stimuli-responsive polymer hydrogels in Japan.

Journal of Intelligent Material Systems and Structures 4: 538–542.

Hockaday, B.D. and Waters, J.P. 1990. Direct optical-to-mechanical actuation. *Applied Optics* 29(31): 4629–4632.

Iijima, S. 1991. Helical microtubules of graphitic carbon. *Nature* 354: 56–58.

Inaba, S., Kumazaki, H. and Hane, K. 1995. Photothermal vibration of fiber core for vibration-type sensor. *Japanese Journal of Applied Physics* 34: 2018–2021.

Jankovic, N., Zeman, M., Cai, N., Igwe, P. and Knopf, G.K. 2004. Light actuated micromanipulator based on photothermal effect of optical fibers. In *Optomechatronic Sensors, Actuators and Control*, K. Moon (Ed), Proceedings of SPIE. Vol. 5602, 63–72.

Jiang, H., Kelch, S. and Lendlein, A. 2006. Polymers move in response to light. *Advanced Material* 18: 1471–1475.

Jonáš, A. and Zemánek, P. 2008. Light at work: The use of optical forces for particle manipulation, sorting and analysis. *Elecrophoresis* 29: 48F9–4851.

Kamholz, A.E., Weigl, B.H., Finlayson, B.A. and Yager, P. 1999. Quantitative analysis of molecular interactive in microfluidic channel: The T-sensor. *Analytical Chemistry* 71: 5340–5347.

Kamholz, A.E. and Yager, P. 2002. Molecular diffusive scaling laws in pressure-driven microfluidic channels: deviation from one-dimensional Einstein approximations. *Sensor Actuators B* 82: 117–121.

Kang, H., Liu, H., Phillips, J.A., Cao, Z., Kim, Y., Chen, Y., Yang, Z., Li, J. and Tan, W. 2009. Single-DNA molecule nanomotor regulated by photons. *Nano Letters* 9(7): 2690–2696.

Knopf, G.K. 2006. Optical actuation and control. In *Handbook of Optoelectronics*, J. Dakin and R.G.W. Brown (Eds), Boca Raton, FL: CRC Press, 1453–1479.

Knopf, G.K. 2012. Light driven and optically actuated technologies. In *Optical Nano and Micro Actuator Technology*, G. Knopf and Y. Otani (Eds), Boca Raton, FL: CRC Press, 3–46.

Knopf, G.K. and Al-Aribe, K. 2012. Light driven micro and nanofluidic systems. In *Optical Nano and Micro Actuator Technology*, G. Knopf and Y. Otani (Eds), Boca Raton, FL: CRC Press, 375–403.

Krecmer, P., Moulin, A.M., Stephenson, R.J., Rayment, T., Well and, M.E. and Elliott, S.R. 1997. Reversible nanocontraction and dilatation in a solid induced by polarized light. *Science* 277: 1799–1802.

Kroerner, H., Price, G., Pearce, N.A., Alexander, M. and Vaia, R.A. 2004. Remotely actuated polymer nanocomposites-stress recovery of carbon nanotube filled thermoplastic elastomers. *Nature Materials* 3: 115–120.Kröner, G., Parkin, S., Nieminen, T.A., Loke, V.L.Y., Hechenberg, N.R. and Rubinsztein-Dunlop, H. 2007. Integrated optomechanical microelements. *Optics Express* 15(9): 5521–5530.

Lagowski, J. and Gatos, H.C. 1972. Photomechanical effect in noncentrosymmetric semiconductors-CdS. *Applied Physics Letters* 20(1): 14–16.

Lagowski, J., Gatos, H.C. and Sproles, E.S. 1975. Surface stress and normal mode of vibration of thin crystals: GaAs. *Applied Physics Letters* 26(9): 493–495.

Lebedev, P. 1901. *Untersuchungen über die druckkräfte des lichtes. Annals Physics* 311: 433–458.

Li, J.J. and Tan, W. 2001. A single DNA molecule nanomotor. *Nano Letters* 2(4): 315–318.

Li, M., Pernice, W.H., Xiong, C., Baehr-Jones, T., Hochberg, M. and Tang, H.X. 2008. Harnessing optical forces in integrated photonic circuits. *Nature* 456(27): 480–484.

Li, M., Pernice, W.H. and Tang, H.X. 2009. Broadband all-photonic transduction of nanocantilevers. *Nature Nanotechnology* 4: 377–382.

Liu, Y., Lu, S. and Panchapakesan, B. 2009. Alignment enhanced photoconductivity in single walled carbon nanotube films. *Nanotechnology* 20: 1–7.

Lu, S. and Panchapakesan, B. 2006. Nanotube micro-optomechanical actuators. *Applied Physics Letters* 88: 253107.

Lu, S. and Panchapakesan, B. 2007. Photomechanical responses of carbon nanotube/polymer actuators. *Nanotechnology* 18: 305502 (8 pp).

Maruo, S. and Inoue, H. 2007. Optically driven viscous micropump using a rotating microdisk. *Applied Physics Letters* 91: 084101 (3 pp).

Maruo, S. 2008. Optically driven microma-chines for biochip application. In *Nano-and Micromaterials*, K. Ohno, M. Tanaka, J. Takeda and Y. Kawazoe (Eds), Berlin, Heidelberg: Springer: 291–309.

Maruo, S., Takaura, A. and Inoue, H. 2009. Optically driven micropump with a twin spiral microrotor. *Optics Express* 17(21): 18525–18532.

Maruo, S. 2012. Optically driven microfluidic devices produced by multiphoton microfabrication. In *Optical Nano and Micro Actuator Technology*, G. Knopf and Y. Otani (Eds), Boca Raton, FL: CRC Press, 307–331.

Maxwell, J.C. 1873. *A Treatise on Electricity and Magnetism*. Oxford, UK: Clarendon Press.

McKenzie, J.S. and Clark, C. 1992. Highly sensitive micromachined optical-to-fluid pressure converter for use in an optical actuation scheme. *Journal of Micromechanics Microengineering* 2: 245–249.

McKenzie, J.S., Hale, K.F. and Jones, B.E. 1995. Optical Actuators. In *Advances in Actuators*, A.P. Dorey and J.H. Moore (Eds), IOP Publishing Ltd: 82–111.

Mizoguchi, H., Ando, M., Mizuno, T., Takagi, T. and Nakajima, N. 1992. Design and fabrication of light driven pump. *Micro Electro Mechanical Systems'* 92: 31–36.

Mizutani, Y., Otani, Y. and Umeda, N. 2008. Optically driven actuators using poly (vinylidene difluoride). *Optical Review* 15(3): 162–165.

Morikawa, Y. and Nakada, T. 1997. Position control of PLZT bimorph-type optical actuator by on-off control. *IECON Proceedings* 3: 1403–1408.

Nguyen, N.T. and Wereley, S.T. 2002. *Fundamentals and Applications of Microfluidics Fabrication Techniques for Microfluidics*. Artech House.

Nguyen, N.T. and Wu, Z. 2005. Micromixers—A review. *Journal of Micromechanics Microengineering* 15: R1–R16.

Nieminen, T.A., Higuet, J., Kröner, G., Loke, V.L.Y., Parkin, S., Singer, W., Hechenberg, N.R. and Rubinsztein-Dunlop, H. 2006. Optically driven micromachines: Progress and prospects. *Proc. SPIE* 6038: 237–245.

Okamura, H., Yamaguchi, K. and Ono, R. 2009. Light-driven actuator with shape memory alloy for manipulation of macroscopic objects. *International Journal of Optomechatronics* 3: 277–288.

Okamura, H. 2012. Light activated and powered SMA actuators. In *Optical Nano and Micro Actuator Technology*, G. Knopf and Y. Otani (Eds), Boca Raton, FL: CRC Press, 553–568.

Oldenburg, S.J., Averitt, R.D., Westcott, S.L. and Halas, N.J. 1998. Nanoengineering of optical resonances. *Chemical Physics Letters* 288: 243–247.

Otani, Y., Matsuba, Y. and Yoshizawa, T. 2001. Photothermal actuator composed of optical fibers. In *Optomechatronic Systems II*, H.-Y. Cho (Eds), Proceedings of SPIE, 4564: 216–219.

Piegari, E., Cataudella, V., Marigliano Ramaglia, V. and Iadonisi, G. 2002. Comment on "Polarons in Carbon nanotubes." *Physical Review Letters* 89(4): 049701.

Poosanaas, P., Tonooka, K. and Uchino, K. 2000. Photostrictive actuators. *Mechatronics* 10: 467–487.

Sershen, S.R., Mensing, G.A., Ng, M., Halas, N.J., Beebe, D.J. and West, J.L. 2005. Independent optical control of microfluidic valves formed from optomechanically responsive nanocomposite hydrogels. *Advanced Materials*, 17: 1366–1368.

Setou, M., Nakagawa, T., Seog D.-H. and Hirokawa, N. 2000. Kinesin superfamily motor protein KIF17 and mLin-10 in NMDA receptor-containing vesicle transport. *Science* 288: 1796–1802.

Shiu, P.P., Knopf, G.K., Ostojic, M. and Nikumb, S. 2007. Rapid fabrication of micromolds for polymeric microfluidic devices. In *IEEE 20th Canadian Conference Electrical Computer Engineering (CCECE) (Vancouver, CA, April 07)*, 8–11.

Shiu, P.P., Knopf, G.K. and Ostojic, M. 2010. Laser-assisted active microfluidic mixer. In 2010 International Symposium on Optomechatronic Technologies (ISOT), Toronto, ON, October 25–27, 1–5.

Steen, W.M. 1998. *Laser Material Processing*. 2nd Ed. London: Springer-Verlag.

Steen, W.M. and Mazumder, J. 2010. *Laser Material Processing*. 4th Ed. London: Springer-Verlag.

Stuchlik, M., Krecmer, P. and Elliott, S.R. 2001. Opto-mechanical effect in chalcogenide glasses. *Journal of Optoelectronics and Advanced Materials* 3(2): 361–366.

Stuchlik, M., Krecmer, P. and Elliott, S.R. 2004. Micro-nano actuators driven by polarised light. *IEE Proceedings Science Measurement and Technology* 151(2): F91–F96.

Sun, X. and Li, C.-Z. 2007. Fundamental aspects and applications of nanotubes and nanowires for biosensors. In *Smart Biosensor Technology*, G.K. Knopf and A.S. Bassi (Eds), Boca Raton, FL: CRC Press, 291–330.

Suski, J., Largeau, D., Steyer, A., Van de Pol, F.C.M. and Blom, F.R. 1990. Optically activated ZnO/SiO$_2$/Si cantilever beams. *Sensors and Actuators A* 24: 221–225.

Tabib-Azar, M. 1990. Optically controlled silicon microactuators. *Nanotechnology* 1: 81–92.

Tabib-Azar, M. and Leane, J.S. 1990. Direct optical control for a silicon microactuator. *Sensors and Actuators A* 21 (1–3): 229–235.

Tabib-Azar, M., Wong, K. and Ko, W. 1992. Aging phenomena in heavily doped (p$^+$) micromachined silicon cantilever beams. *Sensors and Actuators A* 33: 199–206.

Tabib-Azar, M. (1998). *Microactuators: Electrical, Magnetic, Thermal, Optical, Mechanical, Chemical, and Smart Structures*. Norwell, MA: Kluwer Academic.

Tabiryan, N., Serak, S., Dai X.-M. and Bunning, T. 2005. Polymer film with optically controlled form and actuation. *Optics Express* F9(9): 7442–7448.

Tang, H.X. 2009. May the force of light be with you. *IEEE Spectrum* 46(10): 46–51.

Uchino, K. 1990. Photostrictive actuator. *IEEE Ultrasonics Symposium*: 721–723.

Uchino, K. 2012. Photostrictive microctuators. In *Optical Nano and Micro Actuator Technology*, G. Knopf and Y. Otani (Eds), Boca Raton, FL: CRC Press, 153–175.

Vale, R.D. 2003. The molecular motor toolbox for intracellular transport. *Cell* 112: 467–480.

Van Oosten, C.L., Harris, K.D., Bastiaansen, C.W.M. and Broer, D.J. 2007. Glassy photomechanical liquid-crystal network actuators for microscale devices. *European Physical Journal E*. 23: 329–336.

Van Oosten, C.L., Corbett, D., Davies, D., Warner, M., Bastiaansen, C.W.M. and Broer, D.J. 2008. Bending dynamics and directionality reversal in liquid crystal network photoactuators. *Macromolecules* 41: 8592–8596.

Van Oosten, C.L., Bastiaansen, C.W.M. and Broer, D.J. 2009. Printed artificial cilia from liquid-crystal network actuators modularly driven by light. *Nature Materials*, 8: 677–682.

Verissimo-Alves, M., Capaz, R.B., Koiller, B., Artacho, E. and Chacham, H. 2001. Polarons in carbon nanotubes. *Physical Review Letters* 86(15): 3372–3375.

Warner, M. and Terentjev, E.M. 1996. Nematic elastomers-a new state of matter? *Progress Polymer Science* 21: 853–891.

Warner, M. and Terentjev, E.M. 2003. *Nematic Liquid Crystal Elastomers*. Oxford: Clarendon Press.

Weinert, F.M. and Braun, D. 2008. Optically driven fluid flow along arbitrary microscale patterns using thermoviscous expansion. *Journal of Applied Physics* 104: 104701.

Weinert, F. and Braun, D. 2009. Light driven microfluidics-pumping water optically by thermoviscous expansion. In *Optomechatronic Technologies*, Istanbul Turkey (September 21–23, 2009): 383–386.

Weinert, F.M., Wühr, M. and Braun, D. 2009. Light driven microflow in ice. *Applied Physics Letters* 94: 1F9901.

Yamada, M., Kondo, M., Miyasato, R., Mamiya, J., Yu, Y., Kinoshita, M., Barrett, C.J. and Ikeda, T. 2008. Photomobile polymer materials: Towards light-driven plastic motors. *Angewandte Chemie International Edition* 47: 4986–4988.

Yoshizawa, T., Hayashi, D., Yamamoto, M. and Otani, Y. 2001. A walking machine driven by a light beam. In *Optomechatronic Systems II*, H.-Y. Cho (Eds), Proceedings of SPIE, 4564: 229–236.

Yu, Y., Nakano, M. and Ikeda, T. 2003. Directed bending of a polymer film by light. *Nature* 425: 145.

Zhang, Y. and Iijima, S. 1999. Elastic response of carbon nanotube bundles to visible light. *Physical Review Letters* 82(17): 3472–3475.

FURTHER READING

Cho, H. 2002. *Opto-mechatronic Systems Handbook: Techniques and Applications.* Boca Raton: CRC Press.

Cho, H. 2005. *Optomechatronics: Fusion of Optical and Mechatronic Engineering.* Boca Raton: CRC Press.

Hosaka, H., Yoshitada, K., Terunao, H. and Kiyoshi, I. 2004. *Micro-Optomechatronics.* Boca Raton: CRC Press.

Knopf, G.K. and Otani, Y. 2012. *Optical Nano and Micro Actuator Technology.* Boca Raton: CRC Press.

Saleh, B.E.A. and Teich, M.C. 1991. *Fundamentals of Photonics.* New York: Wiley.

Tabib-Azar, M. 1998. *Microactuators: Electrical, Magnetic, Thermal, Optical, Mechanical, Chemical, and Smart Structures.* Norwell, MA: Kluwer Academic.

PART IV

Optoelectronic systems

The art of practical optoelectronic systems

ANTHONY E. SMART
Scattering Solutions, Inc.

18.1 INTRODUCTION

To create great paintings, an artist not only needs a magnificent vision but must also know how to stretch canvas and mix paint. This chapter discusses practical skills for the design and implementation of optoelectronic systems. Because art has no agreed boundaries, technical depth may appear inconsistent, with items perhaps appearing trivial to one reader being useful insights for another and vice versa. A specific optoelectronic system is chosen as an example to provide continuity while illustrating many techniques.

Departing from the more formal style of others, this chapter stresses practicalities, examples from painfully acquired experience, and methods of reducing risk. Ideas both familiar and less common are offered to permit the reader, at whatever level, to explore beyond the obvious, and to anticipate problems that may otherwise compromise success. While earlier chapters give depth and essential numerical specificity, practical arts are predominantly hints, recipes, and general guidelines. This chapter attempts to offer useful ideas to the beginner, while including sufficient subtleties to interest those with extensive experience.

The need for specifications and definitions of success (Section 18.2), understanding requirements (Section 18.3), and modeling (Section 18.4) apply generally. Practicalities of design, planning, and organizational discipline (Section 18.5) are important to the success of any complex optical and electronic endeavor. Rather than repeating quantitative details readily available elsewhere, emphasis is given to properties and things that go wrong at system design and component levels, with some ideas of how to avoid or recover from them (Section 18.6). Later sections introduce testing (Section 18.7), some effects of hostile environments (Section 18.8), checklists (Section 18.9), aphorisms (Section 18.10), and a further reading list selected for its practicality.

18.2 SPECIFICATIONS

Optoelectronic programs can be put at risk by insufficient definition of the problem, the environment, the constraints, and/or the available tools. A dominant precursor to success in the art of optoelectronics is therefore to get a comprehensive written specification. Introducing our example, we quote as

Table 18.1 the abbreviated target specification for an optical air data system (OADS), intended to replace the pneumatic system currently in the control loop of intrinsically unstable high-performance aircraft, to allow operation closer to the limits of the flight envelope than otherwise possible. Outputs are aircraft speed and attitude with respect to the ambient air, whose temperature and pressure were also required but provided by an independent optical system.

The following sections exemplify methods used to meet these specifications and how success may be defined. Table 18.2 shows a hierarchy of success, enhanced by proper specifications, top-down conceptual design, and control of risk. Available components and circumstances may occasionally force bottom-up modifications.

18.3 REQUIREMENTS

The art of practical optoelectronics is about combining components, whose specifications we believe are plausible, into instruments and systems to solve problems for which we will be rewarded. At least four main factors are necessary: motivation, specifications, aptitude, and components. First, because optoelectronics is a challenging discipline, motivation must be sufficient, usually personal satisfaction or even financial gain. Second, we must demand explicit specifications and accept implicit boundary conditions, over both of which we have only limited control. Third, aptitude enhances the ability to create and configure an effective system. Frequently appearing as experience, it guides the designer and prevents unnecessary innovation. Fourth, the choice of components is based upon optimization of complex interrelationships of what is available, and what quoted claims are trustworthy. Possible consequences of the omission of one or more of these four factors are shown in Figure 18.1. The practicality of any optoelectronic system also depends strongly on the control of risk, and the optimization of many complex interactions, not all of which may be fully understood or can be wholly controlled.

Supplementary to these requirements, but at least as important for success, is time, money, and a champion. Even with ample resources, things take time, but neither time nor money will be enough unless someone has made the personal commitment to succeed. Occasional digressions into areas

Table 18.1 Optical air data system specifications

Performance

Speed (true air)	15–1500 ms^{-1}	± 0.1 ms^{-1}
Pitch (angle of attack)	+25 to −10°arc	±0.1°arc
Yaw (angle of sideslip)	±22.5°arc	± 0.1°arc
Update rate	40 Hz	64 Hz preferred
Pressure (altitude)	−500 m to 16 km	±8 m (higher preferred)
Temperature (amb.)	−70°C to 50°C	Unspecified (best)

Environment

Altitude	Ground to 25 km	37 km preferred
Temperature	−70°C to 140°C	Local ambient
Acceleration	−3 to +12 g	Peak
Ambient light	Darkness to sun stare	Darkness to 1.5 kW m^{-2}
Sensed particle radius	0.2 μm and larger	<0.12 μm preferred
Particle concentration	<1 m^{-1} to dense cloud	

Physical

Throw	1.2 m	More preferred
Weight	<50 kg	Less preferred
Lifetime	100 000 h	MTBF
Final cost	<$100k each	>1000 units
Power consumption	<1 kW	28V ± 50% (noisy)

Other

Performance must be acceptable with naturally occurring particles anywhere in the global atmosphere in all weathers.
The system must be 'eye-safe' and 'stealthy'.

Table 18.2 Hierarchy of success

Ideal	All specifications are met
Adequate	Specifications are met and performance confirmed after acceptable modification
Incomplete	Not all specifications are met, but useful information and diagnostic data are recovered
Malfunction	The system produces no useful scientific results, but diagnostic data are recovered
Failure	No science is recovered and the system diagnostics do not indicate what went wrong
Catastrophe	The system destroys itself; we have no idea why
Disaster	The system destroys itself, causing collateral damage to the facilities and possibly personnel

less susceptible to scientific analysis are intended to help distinguish what is important from what is not, because much risk lies here.

Earlier chapters of this book give precise numerical facts about optical, electronic, and mechanical components, subsystems, and behaviors. Many are complex, interactive, and frequently counterintuitive, typically more so in optics than in electronics. Familiarity with these subtleties is necessary, as is studying what others have done, and why, and whether these led to success or ingeniously redefined failure.

Almost all measurements and systems for which optoelectronics is conventionally used may be

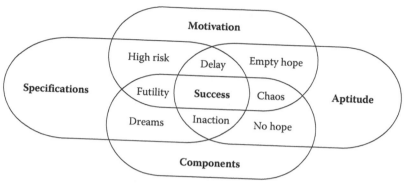

Figure 18.1 Venn diagram of success factors.

described as instrumentally assisted observation of a phenomenon. Ideally, the phenomenon can be controlled, where it essentially becomes part of the system. It may however only be probed with negligible perturbation, as with the OADS example, or merely observed passively as in astronomy. The challenge is to create a specified instrument or experiment to enhance quantitative knowledge of the phenomenon, confirm existing wisdom, and perhaps suggest new ideas.

18.4 MODELING

Making an optoelectronic system to the customer's satisfaction demands that specifications and requirements are understood. An initial approach may be to construct a mental model of a possible system, from concept through design, assembly, testing, operation, and interpretation of the output. Table 18.3 lists eight major functions of the initial thought experiment. At each design level new tasks are spun off, although questions are asked and answered to permit continuation of the planning and design sequence.

Eliminating or solving problems uncovered while addressing these items frequently redirects attention. For example, maintaining adequate temperature stability for OADS originally seemed impossible. Combining air circulation and heat pipes offered a possible if inelegant solution permitting design progress to continue, and although this might have been acceptable, the final choice to circulate a temperature stabilized liquid through a sealed labyrinth held temperatures with 0.5°C, maintaining optical alignment and electronic performance. Eliminating such obstacles as early as possible reduces redesigns, which get more difficult and expensive the later their necessity is accepted. The thinking process implicit in Table 18.3 depends strongly upon feedback to achieve performance satisfactory to the customer and consistent with available resources.

Figure 18.2 maps a possible design sequence, where arrows represent potential information flow upon which changes may be contingent. The starting point is the customer's specification. If this appears infeasible or even unreasonably demanding, for example by violating the laws of

Table 18.3 Functions of a model

1. Quantification of the physics and phenomenology
2. Verification of compatibility of specifications and constraints
3. Understanding quantitative interactions between parameters
4. Optimization of architecture, design configuration, and components
5. Prediction of ideal performance possible with the planned design
6. Prediction of performance deterioration when implementation is not ideal
7. Diagnosis of anomalies anticipated during construction, testing, and use
8. Reduction of risk and enhancement of confidence

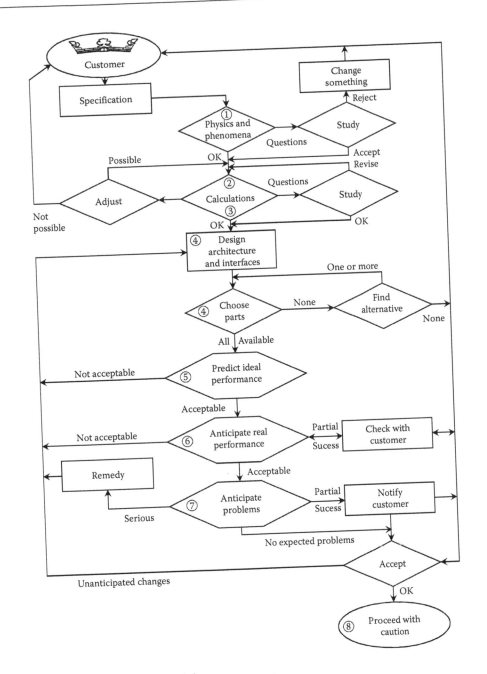

Figure 18.2 Operation of design model.

physics or similarly extreme constraints, early and amicable negotiations can yield great advantages in simplicity, cost, and timeliness—less commonly in performance. Quantitative examination of the proposed physics and phenomenology ① may confirm acceptability or demand changes. For OADS, the 3-D velocity was calculated from the measured transit times of naturally occurring particles crossing each of three inclined pairs of parallel light sheets projected 1.2 m from the optical unit looking sideways from the aircraft. Pressure, altitude, and temperature were obtained from simultaneous pulsed measurements of total elastic scattering and fluorescence of molecular oxygen over a common volume of air outside the boundary layer of the vehicle. Simple calculations ② with tests for scientific plausibility and technical capability can confirm that nothing silly is being planned ③. Plausible and self-consistent numerical criteria, which for OADS were difficult but not infeasible, may then allow a preliminary ④ design. The projection of 8 mm light sheets with a near diffraction limited full width to e^{-2} intensity of 70 μm, separated by 11.2 mm, looked reasonable, but required transit time measurements accurate to 10^{-4}, challenging but not impossible—see below. External and internal interfaces with a coherent logical and structural architecture (Figure 18.3) iterate to a self-consistent set of available parts ④, again. Clear and unambiguous names for each module help to avoid confusion among participants. Defined hardware and software modules must be isomorphic, with each other and also with the necessary functions and interfaces to allow testing and possible remediation. Within this structure an ideal performance can be predicted ⑤. Independent knowledge, measurements, and/or testing allow consideration of effects of ageing, deterioration during use, or even failure of parts or subsystems to anticipate likely real performance ⑥. In critical applications where single point failure cannot be tolerated, initial choices for system architecture may require adaptation. Expanding the rigor of preliminary thoughts and testing at this level may show detrimental problems ⑦, requiring more serious remediation. For OADS, the pressure and temperature system (PATS) technology was

changed from the originally planned differential absorption to Schumann–Runge fluorescence of O_2 because of installation constraints. However, tedious and expensive preliminary modeling might appear, it is incomparably cheaper than later fixes, even should these prove possible ⑧. Design errors are better corrected on paper or the more modern electronic equivalent documents than in hardware components or computer code. Three outcomes from this exercise may be (1) no obvious problems, (2) problems with reasonable remedies, or (3) problems requiring negotiation with, or at least reporting to, the customer. Such a sequence of increasingly demanding analytical processes for OADS identified five major risks: (1) Would there be enough particles everywhere in the atmosphere? (2) Is the velocity measurement geometry acceptable? (3) Can the system be calibrated? (4) Is everything stable enough? (5) Would PATS work? In the event enough particles were found everywhere flown, the velocity measurements were sufficiently accurate—more so than systems used to calibrate them, stability was just acceptable, and PATS performance was marginal. Ironically the major risk was not technical, and the program, of which OADS was but a tiny fraction, was cancelled by the customer's customer for wholly unconnected reasons.

However, rather than insisting that everything be modeled it is often more sensible to evaluate

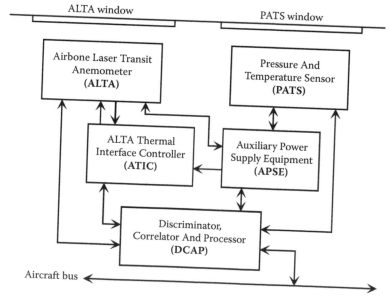

Figure 18.3 OADS system components.

where effort is most wisely spent in terms of the overall requirements of the customer. Run-away costs, or late delivery, can be just as serious as an inadequate instrument.

18.5 DESIGN

Design, like art, is a creative process involving flashes of insight. However, the first practical step in building a cathedral is pounding rocks into a ditch, and in this sense the preparation of foundations is an important, if a less apparently rewarding, precursor of more detailed design work. Using appropriate tools for design planning amply repays the effort; examples follow. A more colloquial style is adopted here to lighten the overwhelming complexity that can be engendered by rigorously performing the following tasks for which a fairly logical sequence is established below, and followed later in Section 18.6.

A top-down design methodology helps to assure that specifications can and will be met. Important aspects of the system, although initially existing only in imagination, can be solidified into pathways along which the physical construction can later proceed. Nine ideas introduced in Table 18.4 are expanded in the following paragraphs. General program management and design activities must often be augmented for specific systems by special software packages for optical design, signal analysis optimization, computer aided design (CAD), and finite element analysis (FEA) for mechanical and thermal optimization.

18.5.1 Riding-the-ray

Riding-the-ray tracks information-bearing photons, later transformed into electrons, through successive components. Examining the effects of each component, including the phenomenon, from generation of photons to the recovery of the numerical and graphical representations of the final measurement, can explore the properties of many devices in the system. This approach is particularly useful for examining the effects of aberrations, polarization, spatial and temporal coherence, optical power budget, pointing stability, alignment criteria, and various other component-related properties. Where there is more than one optical path, each may be analyzed independently or in various combinations. This approach is the essence of ray tracing, estimating aberrations and ghosts. It may also show unanticipated effects.

18.5.2 Assembly methods

In addition to describing the structure, design must specify how the system may be constructed, how it will be aligned, and the alignment maintained. For example, mechanical support may be (1) fixed by design and manufacture, (2) initially adjustable over a specified range and then locked, or (3) dynamically compensated during operation. In mass production, where adjustments are neither necessary nor permitted, functionality may be compromised to accommodate manufacturing tolerances. A more versatile design avoids such performance reduction at the expense of increased complexity. The extreme is to have everything adjustable, with ranges and sensitivities sufficient to accommodate almost any manufacturing imperfection. This can be a dangerous and unwise choice. A better solution optimizes reasonably tight manufacturing tolerances with a few well-chosen adjustments, whose ranges are best limited to the smallest necessary. Each adjustment should be independent

Table 18.4 Tools and purposes

1. Riding-the-ray	Analyze all optical components
2. Assembly methods	Provide adequately sensitive adjustments and criteria
3. Calibration	Build trust in the planned measurements
4. Operation	Use and experience the virtual system
5. Risk reduction	Consider the unknowns, and the unknown unknowns
6. Signal estimation	Predict signal, noise, accuracy, precision, and errors
7. Information retrieval	Simulate signal processing and information recovery
8. Interfaces	Estimate power, environment, and communication needs
9. Management	Track status, changes, costs, and consequences

Table 18.5 Adjustments and alignments

Available adjustments
    Provide independent control for each variable
    Arrange suitable sensitivity of adjustment
    Assure that every adjustment may be rigidly locked
Alignment criteria
    Agree a definition of acceptable alignment
    Specify a quantitative criterion for its achievement
Completion
    Confirm that the alignment criterion is finally met
    Mark the final position and document all activities and observations

and appropriately sensitive. After meeting a clear criterion of acceptability, each should be locked, and not be susceptible to residual creep. If such stability is impossible, perhaps because of changing external conditions, then adjustments may be driven by a closed-loop feedback system controlled by a prespecified algorithm to an established alignment or performance criterion. The most likely candidate for real time control is temperature stabilization, but other adverse effects may be similarly mitigated.

Table 18.5 summarizes guidelines for mounting and aligning the chosen components, with specific techniques given later (Section 18.6.3). For any particular system, some line items may appear either obvious or too detailed. It is important to know which, and to assign effort intentionally, rather than by crisis intervention.

Every intentional, or even accidental, variable should be reviewed in the light of Table 18.5. Listing all variables allows the definition of acceptability criteria for each alignment. Early attention to surprises that often emerge from this exercise greatly enhances assurance of success.

18.5.3 Calibration

Although calibration is often necessary to assure the desired accuracy of measurements, it may be simplified by careful manufacture, by ingenious choice of parameters, or by specifying measurements of more stable output quantities, such as time and frequency, rather than intensity or voltage. If possible, the reported measurement should be independent of quantities that are difficult to measure or guarantee constant, such as luminance, wave front flatness, stray light, noise level, and others. Table 18.6 suggests a sequence of informal questions to which proper answers can improve confidence.

Calibration implies that the system is both stable and well characterized and that its initial behavior is understood and close to what is intended. Residual disparities are assigned numerical values to correct or compensate measurements, possibly even while the instrument is operating. Implicit in calibration is the planning of methods to test the system for efficacy, performance, reliability, constancy, and other properties discussed in Section 18.7.

Table 18.6 Calibration and characterization

Do the numerical results mean what they seem to mean?
What precision can be expected, or achieved?
What accuracy can be expected, or achieved?
What evidence links the current results to previously known results?
Can errors be bounded using comparisons of known and measured data?
Do the measurements depend exclusively on the phenomenon of interest?
If not, what other effects must be allowed for?
Can these effects be measured and the data corrected?
If not, can the errors be otherwise bounded or some compensation performed?

18.5.4 Operation

Anticipating what instructions may be necessary and what responses are expected from an operator is desirable. A proper balance between automatic operation and user intervention can accommodate unforeseen situations, without making routine tasks so complex that they invite errors. Routine operation should be automatic, perhaps run by prewritten scripts, requiring user intervention only if something unexpected arises. If results or monitored quantities differ from expectations, the operator can request additional diagnostic displays, preempt a script in favor of a new one, make changes to existing commands, or take over progressively more complete control of the system. This hierarchy of increasingly operator-intensive actions must be fairly robust to operator errors induced by fatigue or stress caused, for example, by something unexpected. Even if the operator must assume complete control, with increased risk of mistakes, online error checking can prevent incompatible instructions.

18.5.5 Risk reduction

Table 18.7 lists five main areas of risk, although optoelectronics carries special types of risk with each element potentially leading to further questions. Experience helps identify critical areas, and often a barely remembered thought about some device or approach can intimate something that has perhaps been neglected, at future peril.

One great risk may be not staying aware of what is wanted, perhaps by misunderstanding the customer's original desire or intention, or even a change of priorities during design and/or development. Sharing risk with the customer by frequent timely exchanges of current status information, thoughts, feelings, concerns, intuitions, and expectations enhances the probability of a mutually successful outcome. Rarely, will a customer become impatient with such attention, but neglect of personal relationships and communication is a sure way to fail. Of particular hazard in optoelectronics is suboptimization, achieving flawless performance in areas with which one is most familiar, while less well understood areas receive less, or perhaps even inadequate, attention. The following paragraphs address areas of risk not always stressed in optoelectronics texts, but critical nevertheless.

Ideal and real component behaviors may differ. Buying or borrowing critical parts for early testing increases confidence. Design reviews are vital and once a design is accepted, changes should be evolutionary rather than revolutionary, with ripple-through effects reduced by the early definition of modularity and interfaces.

Testing to limit functional risk can only be complete when physical hardware exists, too late for conceptual and design changes without great inconvenience. However, components, software and hardware modules, and supporting structures can be independently tested as soon as they are available. To be effective, testing requires a clear definition of what is expected, a valid criterion for its manifestation, and quantitative documentation of how and to what extent this is observed.

Robustness to external conditions, misuse, and misunderstanding is necessary once control of the system passes to a user who may typically have different attitudes and capabilities from those of the designer. The term foolproof is probably optimistic—fools are more creative than can be imagined.

Complexity increases the technical risk of not performing as wished, or of doing inadvertently something neither expected nor understood. Having fewer parts increases the chance of success. The least trustworthy components are complex or

Table 18.7 Risk areas

Customer	Are there enough timely, frequent, and honest exchanges?
Resources	Are planned schedule, funding, personnel, and expertise available?
Technical	Is the physics right to show the desired effects?
	Will the experiment design allow proper measurements?
	Is the system implementation acceptable in all areas?
Implementation	Can what is designed be built?
	Can adequacy be confirmed by well-specified testing?
Deployment	Will it stay that way during all conditions of use?

nonstandard parts for which little prior experience exists. Using common commercially available parts where possible, even with some slight performance compromise, is often recommended. Although this might apply particularly to optics and/or mounts, commercial mounting assemblies for optics may be intended for environments different from the current need, and some commitment to modification, such as adding locking screws, must be accepted. An almost trivial example of simplification is the specification of screw sizes. As far as practicable, the screws should be the same type and size; that way the ability to find the right wrench is much enhanced, especially if you buy many of that size. The ideal of toolless assembly and maintenance is rarely achievable in optoelectronics.

In optics, common sense is often corrected by experience, and optical designs that seem acceptable on paper, or even in the laboratory, can fail in use in various subtle ways. Time and effort devoted to anticipating failure modes is better directed to the more likely ones, identified seemingly only by painfully acquired experience. Here, design for graceful degradation can permit remedial action before complete failure. Time invested with suppliers before committing to final design, parts, or assembly methods is never wasted.

Although durability is not always a direct concern, many optoelectronic systems are so diverse and complex that it is well to build them as though they must last forever, or at least for much longer than the specification demands. Three approaches may be useful: (1) design the system to tolerate all anticipated conditions, (2) preserve the system from unplanned conditions that could cause deterioration, and (3) arrange continuous or frequent monitoring to assure that no relevant parameter is, or has been, outside its permissible range.

An optoelectronic system should be robust to any change in conditions within its specification. This has sometimes been called hardening, which is commonly used with three meanings, the robustness of (1) the design, (2) the resultant instrument, and (3) the ability to operate without serious deterioration in radiation environments. Even allegedly benign conditions can have hazards. From experience, the author once expected to lose one or two digital camera charge coupled device (CCD) pixels to radiation damage on every long-haul commercial airline flight until the technology of substituting less vulnerable complementary metal–oxide–semiconductor (CMOS) sensors became almost universally adopted.

Although rarely in the specifications, a robust design not only establishes initial operation, but also accommodates inevitable later changes. Design quality is enhanced by having experienced people review it for potential weaknesses before commitment to procurement and manufacture. Reviewers are best selected from those who are skeptical that it can be made to work at all, and are looking for justification of their prejudice. Heeding their observations and improving the objects of their concern improves the robustness of the design to many unquantifiable properties of real life. Four aspects merit review: (1) general concept, (2) interface and layout drawings, (3) design frozen prior to commitment to hardware, and (4) software responsible for the operational experience of the user. Once testing is begun, the only permitted design changes should be those demanded by problems uncovered by or during that testing.

18.5.6 Signal estimation

Quantitative estimates of expected system properties can be obtained by representing each component by a cell on a spreadsheet, entering a plausible estimate of the power or signal properties at that point, from source to detection, allowing for all the plausible or possible effects. Following columns contain best-case and worst-case approximations that may raise confidence, suggest remedial action, or demand redesign. This technique can identify potential problems arising from many sources, from stray light to amplifiers or subsequent processing. Commonly, modern optoelectronic systems make at least two sequential transitions: optical radiation to analog electrical signals, and digitization of the analog signal. Unless the system is based upon individual photon detections that are in some sense already digital, converting from analog to digital transmission as close to the detector as practicable minimizes further increase of noise from parasitic and transfer effects.

18.5.6.1 OPTICAL NOISE SOURCES

However good, detectors and electronics can never recover properties of the optical signal lost or diluted by unintended effects. Examples vary from a mismatched aperture to windows that deteriorate or get dirty, but can be improved by care and attention to component choice, environment, and operating methods. Stray light from various internal and/or external sources can dilute the optical signal. Most commercially available lens and optical design programs have the facility to analyze ghosts, which although reducible by multilayer antireflection coatings, may still be detrimental in laser-based systems. With high power pulsed lasers even ghost reflections can focus enough of the beam to cause ionization along the optical path or permanent damage if the focus is near or within a component. Cleanliness is always essential. Although superpolishing is sometimes recommended, it is expensive and parts require special handling.

Internal flare can be improved by machining fine annular grooves with a matt black coating on all enclosing barrels. Conventional optical stops may be augmented by a few well-placed discs with internal knife-edges to reduce stray light. Despite potential external attractiveness, shiny or polished surfaces should be avoided internally, substituting surface finishes such as black anodizing for matt aluminum, and various other oxidation processes for materials such as brass or steel. Black plastic shrouds also help reduce reflections and flare, and enclosing the light path often suppresses convection, an occasional source of image degradation where density gradients occur. Carbon-based matt black paints can be useful and cobalt oxide pigments survive and stay black at high temperatures.

The advent of lasers as illumination sources has introduced the phenomenon of cavity feedback, which is less rare than its cause might suggest. Even a tiny percentage of light fed back coherently into the laser cavity can destabilize cavity gain and hence the output intensity or mode shape. Even a minutely fluctuating feedback from optical components can be significantly amplified in the resonator to become a serious instability. Sensitivity to mechanical or thermal instabilities becomes more serious as the instrument is better aligned, or contains surfaces conjugate with the laser output face. For example, when launching the laser beam into a single-mode fiber, the best match of numerical aperture, focus, and centering is also the most likely condition for laser destabilization. Tilting the input face sometime helps, but wedging is better. The opposite end of the single-mode fiber is also conjugate, and if the fiber is shorter than the laser coherence length, or a near multiple of the beat length between close modes if more than one is stimulated, then this also will make the laser flicker.

18.5.6.2 DETECTION AND NOISE

Currently the only directly detectable property of an electromagnetic field at optical frequencies is its intensity, although techniques such as coherent mixing can visualize phase, and spectral characteristics can be derived via various dispersive or non-dispersive optical processes. Square-law detectors measure light intensity in one of three regimes: (1) low intensity, where quantum properties prevail, (2) high light levels, with classically continuous intensity levels, and (3) a "gray" area between, where observed behavior depends on the power level and detector type. Noise effects differ significantly between regimes, easily estimated by noting that a green photon has energy of about $4e^{-19}$ J, giving about $2.5e^{18}$ photons s^{-1} in a 1 W beam.

Detectors capable of resolving individual photons often show defects such as after-pulsing, fluctuating sensitivity, and photoelectron pulse pile-up, which can distort the statistics upon which information depends. If these effects are reduced to be negligible, the detector output at sufficiently low light levels, and/or high bandwidths, consists of a series of amplified photoelectron pulses. Even if many photons are missed because of reduced quantum efficiency, the statistics can remain an unbiased estimate of incident intensity, usually as modifications to the Poisson distributed arrival times from an unmodulated coherent source. Noise sources in the detector itself, the following amplifier(s), and resistive loads, usually become significant only in the classical regime. Certain types of detector, such as photomultipliers (PMTs) and avalanche photodiodes (APDs) in the Geiger mode, have enough internal gain to produce a discriminable electrical pulse from each released photoelectron. The statistical fluctuation, commonly called shot noise, which increases as the square root of the optical signal power, is an effect of quantization of the optical field, and thus is not really noise at all. For detectors of lesser capability internal noise forces approach to the gray area.

For sufficiently high light levels, particularly for restricted bandwidth systems, the dominant noise in a uniform level signal is the shot noise, typically approaching Gaussian statistics, and although the signal-to-noise ratio (SNR) improves, the actual noise increases with the square root of the power and the system bandwidth.

The optical power incident upon a detector is proportional to the square of the electric field, whereas photon noise typically depends on the square root of optical power or the electric field itself. In later circuitry, the electrical power is typically described by the product of the current generated by the incident light and the voltage across a load, becoming the fourth power of the electric field. A transimpedance amplifier, presenting a high input impedance to the detector with a low impedance to following devices, can provide wide-bandwidth high-gain linear amplification without seriously worsening the noise inherent in the signal.

After optical detection, electronic conditioning using pulse-height or time discriminators as photon counting circuits, or box-car analyzers and lock-in amplifiers to average repetitive signals, can extract a signal seemingly buried in noise. Further improvements may be possible using auxiliary timing or other information that is either independently known or can be derived from the signal. Techniques such as photon correlation can make good use of the photon arrival statistics, using information a light beam carries in the intervals between successive detected photons. For OADS, this permitted accurate speed measurements from correlated pairs of pulses as particles crossed two parallel sheets, whether the pulses were from individual or clusters of scattered photons, from particles of <0.2 μm radius up to snowflakes.

Many real experiments fall between photon resolved and classical, and measurements are required where noise is a combination of statistical effects, whose relative importance depends on properties of the optics and detectors, as well as the phenomenon of interest. In this regime it is essential to consider all sources of noise, the relative importance of each, and their effect on the accuracy of the measurements. Simple calculations supplementing answers to carefully worded questions to device manufacturers improve confidence.

For single-photon detection, the signal is, in some sense, already digital, allowing various ingenious time dependent pulse processing techniques to examine the information encoded as variable intervals between one-bit level changes. Processing methods available here can be complex, as intimated by the information in the further reading list. Where detection yields an analog signal, transformation to a digital form is preferable before further signal manipulation. An analog-to-digital converter (ADC) produces a sequence of multi-bit words, immediately exploitable by digital processors. Transformation of an optical measurement to a usable digital signal usually demands low-noise linear amplification, DC offset control, and sufficient resolution for acceptable quantization noise, arising from the number of discrete levels, 256 with 8 bits, 4096 with 12 bits. Only for an extremely noise-free input or a large dynamic range is a 16-bit ADC or higher occasionally justified. Where suitable averaging is possible, quantization noise can be reduced by adding known noise to provide random jitter between discrete levels, a technique originally used in RADAR. Most ADCs tend to be power hungry, and careful choice is necessary for applications where power is scarce, such as from batteries or in spacecraft. Noise specifications in 2-D arrays, such as CCDs or addressable CMOS arrays, can be less optimistic than at first apparent. Cooling can reduce the readout noise that dominates for CCDs, but excessive cooling can reduce sensitivity. All relevant properties of any detector must be well understood before committing to its use.

18.5.7 Information retrieval

Two important questions are "What is to be inferred from the measured optical signal?" and "How is the inference to be drawn?" Relating the value of the quantity to be measured to the detected light is not always obvious, as the subject and study of inverse problems attests.

Care is taken to optimize the signal in terms of intrinsic or additive noise, nonlinearity, overranging, etc. It is convenient to distinguish 'signal and data processing' from 'information retrieval'. The former is typically intended to manipulate, rearrange, and/or often selectively reduce raw numerical data. The latter, is weighted toward extracting the values of the required quantities, and their boundaries, accuracy, precision, constraints, and even significance. Its final aspects are to present these values to the user in an easily understood form, and to archive all pertinent information and provenance in a format that may be accessed for as long as it may be required.

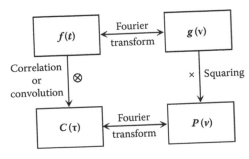

Figure 18.4 Weiner–Khintchine relationship.

18.5.7.1 SIGNAL AND DATA PROCESSING

Signal and data processing may include precursor analog manipulation, and the later rearrangement of one or more rich 1-D raw data streams into a more useful form. Whether from a single-point detector, or serially accessed pixelated sensor array, the data stream must typically be processed to emphasize the required measurement over irrelevant factors. Figure 18.4 is a symbolic diagram of the Weiner–Khintchine relationship linking the conservative Fourier transform (horizontal) with the selective discard of data (vertically down). Although both routes from upper left to lower right are mathematically identical, practically the route via correlation can be preferable because unwanted data are discarded sooner.

Conservative transformation may make data more intelligible, whereas selective discard of the irrelevant may improve recovery of useful information. For the illustrative example of OADS, the analog APD pulse from each particle transit was processed by automatic gain control and dynamically matched filtering. Successive subsamples of each pulse were digitized and buffered by a first-in–first-out (FIFO) memory with conditional exclusion of digital samples not part of any useful pulse, obviating the need to process information-free noise between pulses. The FIFO was drained and processed to produce an amplitude and interpolated time for each probable particle transit, as a discriminated event. Infeasible sequences were removed by a conditional editor, again reducing processing load. A table of transit time pair probabilities, analogous to a correlogram, was accumulated in random access memory (RAM). Particles not crossing both sheets contribute a noise floor, while useful transits give a clear peak, whose centroid is parabolically interpolated to better than the 5 ns resolution available from the time tagging of pulses. This sequence of processes, combined with the parallel sheet geometry and its control, achieved

the 1 part in 10^4 accuracy demanded here, but not usually accepted as possible with conventional optical anemometry. Accuracy was also aided by the absence of residual turbulence since the instrument was moved through relatively undisturbed air in which are embedded near stationary particles.

Several processes must often be performed sequentially, and choices between real-time and off-line processing can be the subtle. If either the data input rate or the processing overhead vary significantly, then asynchronous interfacing with local buffering or even storage can be appropriate (Section 18.6.6). OADS used the FIFO to tolerate wildly fluctuating input pulse rates, dual port RAM to accumulate correlograms, and conditional peak finding so that interrogation by the aircraft bus always accessed the most current measurement. Unless the measurements are required in real time, or with as little latency as possible, more accurate or comprehensive information may often be recovered by complete recording of the raw data streams, with sufficient digitization rate and bit depth, for later off-line processing, perhaps by several different techniques or even including information from independent sources.

Estimating computing resources predicts bottlenecks to be alleviated by appropriate hardware. Machine architecture is tailored to provide sequential, pipelined, or parallel task execution, and assignment of the various synchronous or asynchronous tasks. However, the cost, risk, and suitability of available software can force hardware choices.

Many systems rapidly generate enormous numbers of data as one or more signal streams too rich to be recorded in real time as suggested above. Examples are a laser-based velocity sensor applied to a wind tunnel, an industrial flame, a ship's wake at sea, transonic flow, or inside an aeroengine. While waiting for hardware performance capabilities to catch up with the ever more demanding ideal of raw data archiving, techniques such as correlation can compress the incoming event stream, selectively discarding only that which is less interesting, in this case, relative phase information. Where tens of thousands of correlograms may be acquired, often at great expense, a balance must be established between archiving all that could later be necessary and a reasonable storage expense, in time, money, or hardware resources. As a minimum, data as raw as possible, together with all system health and status monitors, should be stored and backed up as comprehensively as can be arranged.

Built and flown between 1989 and 1993, OADS acquired six incoming data streams each at 20 Mbytes s^{-1} to be processed by various dynamic and conditional discriminators to give three simultaneous channels, each of forty 32-bit correlograms per second with sub nanosecond delay time discrimination, for each typically 2-h flight costing several hundred thousand US$ (1993). This system used all of the techniques outlined above, and indeed many more. Data processing choices of how to display updates to the experimenter for real-time inspection and system control during research and development were also extensively different from those necessary for a final deployable system.

18.5.7.2 INFORMATION RETRIEVAL

Although distinguishing between data and information may seem merely philosophical, it can help to establish hardware requirements, and both usability and intelligibility of the reported numbers. Although some systems are passive, perhaps even with data only available after the experiment, others may yield a real-time indication permitting user intervention, or possible redirection of the experiment.

By information retrieval is implied the recovery of the measurement in a format for archiving, with experiment provenance and monitors, and which is also immediately accessible to the observer. This implies that the presentation is meaningful to a human, important when real-time control or intervention may be necessary, or where proper experiment progress must be monitored or confirmed.

18.5.7.3 SOFTWARE

For any instrument or system involving optoelectronic disciplines, an organizational methodology is necessary to make it all play together. Functional and structural design must be augmented by a software specification as early and as complete as possible. Without such discipline, a monster is created, a hostage to fortune, later to move out of control into cost overruns, chaos, and disillusionment.

Typically with optoelectronic instrumentation, signal and data processing and information retrieval will require software, as will the user interface, with its control, monitoring, and archiving needs. Software must be modular, testable, and appropriate for the planned hardware, although more usually nowadays the hardware is chosen on the basis of computing capabilities and resources demanded by software, which becomes more expensive and less flexible than hardware.

18.5.8 Interfaces

One critical area in any optoelectronic activity is the establishment and comprehensive definition of external and internal interfaces. Each interface must be compatible with both the elements that it is intended to connect. Isomorphic mapping of software and hardware modules (Section 18.4 and Figure 18.3) introduced the need for ideas expanded in Table 18.8, without which testing and diagnostic analysis are made more difficult. An initial top-down design process is attractive, although it may often be necessary to accommodate bottom-up changes driven by available components, modules, subsystems or blocks of reusable software or code.

Accurate definition of all interfaces is imperative. Even, or perhaps especially, when other criteria may superficially appear to be more important in the establishment of system architecture, it is essential to resolve priorities to create consistent structure. Interface definitions start with a connectivity map of the proposed system, including its interaction with the outside world (e.g., Figure 18.3), proceed via required functionality, ultimately to details of actual wires or numerical quantities to be exchanged, and the formats to which they are constrained.

The importance of these documents centers upon their practical use. Interface drawings are best as portable files in a format that can be read by everyone interested, and printed at least once as hard copy on a paper that will accept bold red marks, for peer or final review. The existence of interface diagrams as working documents, spread upon a conference table or a videoconference screen for examination and critical discussion, can greatly enhance the effectiveness of the final system, instrument, measurement, or scientific understanding obtained from its use.

Commitment of definitions of interfaces and connectivity to a formal document improves confidence that the system can perform its intended task, often suggesting simplifications and more logical ways of doing things, but more importantly defining modular boundaries. The specification of every module allows each part of the system to be tested independently of others, an ability whose importance cannot be stressed too highly.

Table 18.8 Interface maps

1. *Operator interface*
   Command, control, access, reports, displays, archiving

2. *Modularity*
   Block diagram and naming of functions, hardware, software structures

3. *Information flow*
   Control, data, health and status monitoring, communication, formats, timing
   protocols, archive, retrieval

4. *Power and wiring*
   Input power availability, conditioning, connections
   Internal power supplies: average and peak voltages and currents, acceptable noise,
   cross-modulation limits
   Wires: type, gauge, insulation, color
   Connectors: type, gender, shells, cable harnesses; runs, and makeup

5. *Optical train*
   Power, wavelength, spectrum and bandwidth, polarization, acceptable aberrations,
   spatial and temporal coherence, noise, sources of stray light, ghosts

6. *Physical structure*
   Mechanical: hardware systems and mounting structure, detail drawing index,
   assembly methods and sequence, available adjustments, alignment criteria
   Thermal: sources, sinks, pathways, effects, stability

A second advantage is that critical components may be tested as soon as they become available, without waiting for everything else to arrive, not only for functionality, which is mostly reasonable for commercially purchased parts, but more importantly for fitness and agreement with expectations.

Within the larger context of the initially specified environment of the system, Table 18.8 lists six general types of interface or definable structure. During design, each must be described down to a level sufficiently detailed to prevent inevitable later changes from rippling through the system with unanticipated effects. A hierarchy of mapped and documented interfaces reduces later problems, although some interference is to be expected between the tabulated items, expanded somewhat below.

18.5.8.1 OPERATOR INTERFACE

The highest level, and in many cases the only one visible to the operator, includes the appearance of the system, the user friendliness of the available commands, the methods of reporting and displaying the required measurements, either in real time or recalled from archive, the confidence given by suitable values of health and status monitors, and a method of deciding what to do if something unexpected occurs. Despite the obviousness of this, many user interfaces could be better designed. Past examples were once exemplified by the arcane programming of video recorders. Even the continuing quest to make user interfaces more intuitive, especially as devices have an increasingly extended functionality, is threatened by lack of standardization and a distressing need to learn a new "operating system" for every new device, once the prerogative of the engineering world but now as universal as telephones, refrigerators, transport and entertainment. Typical optoelectronic instruments are complicated and are required to make difficult measurements. As far as possible the operator should be protected from this complexity, but nevertheless have immediate access to more detailed information should anything unexpected occur. Successful operation should need minimal or preferably no intervention, presenting measurements succinctly and intelligibly, with a comforting status indication that everything is within its design range. If more control is desired, then it is available immediately on interrupt, which need not pause data acquisition, but only make more information or control available to the operator. This includes the ability to request more detailed monitoring of any function, to modify the existing control scripts, to recall archived data for comparison, and above all to have sufficient feedback to know what is going on either as a result of operator instruction or unanticipated factors. It can be expensive and embarrassing to have an experiment yield a record that is either unintelligible or has uncertain provenance.

18.5.8.2 MODULARITY

Most purchased optoelectronic parts or subsystems already have well defined interfaces. Even if they are not always the most convenient, it may be better to accommodate something that already exists. This can save time and reduce risk, but conversely much design effort may be consumed exploiting standard components and interfaces in a system being designed to satisfy unique needs.

The system overview document should be a connected block diagram where each block is labeled by function, hardware name, and software requirement, as an expansion of Figure 18.3. On a yet more detailed view of this same diagram will be labeled the types of interconnectivity representing module control, information flow, health and status monitors, and supplied power. Even at the system level not all blocks need have all connections, and as the same discipline is applied to lower levels of modularity down to individual components, the connections per component will become fewer, simpler, and more specific. At the lowest levels of the functional hierarchy, many components will be passive. For example a lens, even though passive and appearing first in item 5 in Table 18.8, must be included in item 6, so that a later drawing may be created to specify its physical mounting structure and alignment requirements.

At each level the specifications of interfaces with the real world environment are again reviewed, understood, confirmed, and fully documented.

18.5.8.3 INFORMATION FLOW

Interface drawings for control, information flow, and health and status monitors are powerful tools. The thought devoted to creating these diagrams becomes the design work necessary to assure the complete and proper functionality of the final system. This is also a valuable method of checking internal consistency, and raising concerns that may

even require customer choices. The optical part of this can specify everything about the radiation, the optical train, and optical components. For the electronic parts, except for standard communication interfaces, the actual wires that carry different types of information should be clearly identified to avoid confusion. Three separable overlays of connectivity are useful, (1) the command and control information passed to modules and components, (2) the health and status information returned from modules and components, and (3) signals, which also include the light, pertaining to the desired parametric measurements. In addition to their separate functions, the first two types allow closed loop control where this is desirable.

Active components have command and control capabilities with responses and properties monitored as often as necessary. Optical signals typically end at the detector, transforming to electrical signals, and thence to numerical signals for interpretation and archive, with inspection possible at various intermediate positions. Wherever possible, information exchange should be via standardized interfaces of which many are available to be selected according to the application. The five standardized communication layers of physical, data link, network, transport, and application should be understood, with special care necessary for real-time instruments.

18.5.8.4 POWER AND WIRING

Power supplies deteriorate or fail in various ways, devolving from improper specification or performance all the way to poor mounting or inadequate cooling. A connector can be the wrong gender, sometimes the wrong connector, occasionally even with the wrong number of pins, and the actual voltages delivered in service may be anybody's guess. Careful specification of input power availability, instrument requirements, and bounding conditions can mitigate such mishaps. Auxiliary aspects must not be taken for granted—for example the occasional need for a trickle current to sustain settings—and are better discovered early to prevent the interconnect wiring from declining from a traceable color coded minimal set into a rat's nest of post-assembly fixes.

Voltages and currents to various parts of the system may be quantified on the power interface diagram. Conditioning the supplied voltage and minimizing noise is the main function of a power supply, of which there will frequently be more than one. Each must accommodate maximum currents under all operational combinations and must be stable, with adequate freedom from noise, either intrinsic or introduced by switching or operation of some other part of the system. Where several power supplies are involved, they may interfere with each other to introduce mysterious noise properties whose spurious cross-modulation signals can masquerade as legitimate measured effects until properly diagnosed. Problems traceable to power supplies are not rare (Section 18.6.7).

Specification of wire gauge, type, insulation, color, and especially connectors should be consistent throughout the system, but will probably be compromised in purchased subassemblies. Heatshrink insulation of each termination may be desirable, but does prevent electrical probing. Choosing the appropriate gauge, type, and insulation is obvious, and specifying an appropriately consistent color coding is extremely useful. The author once had the opportunity to diagnose a cross-modulation problem within a customer's power supply where well over 230 wires supplied more than 54 different subassemblies from seven different types of allegedly independent power supply connected in parallel to the same aircraft bus. The compact wiring looms were tightly laced. Every wire was unlabeled, 16-gauge, Teflon coated, and white.

Specifying connectors at the top level of modularity, subsequently percolating down to every component for which electrical connectivity is required, as early as possible, prevents much pain later. Many connectors have long delivery times, and early definition of the type and gender of connectors, and the length and construction of cable harnesses, is valuable.

18.5.8.5 OPTICAL TRAIN

The experiment in Section 18.5.1 implicitly creates the interface diagram for all optical paths, and this evolves into formal documentation of power budget, wavelength spectrum and bandwidth, polarization, spatial and temporal coherence, noise, acceptable aberrations, sources of stray light, ghosts, and many other properties. In fact these may not all appear explicitly on the interface diagram, but are included here since a rethinking of these properties in the context of interfaces and compatibilities almost always identifies previously unconsidered items.

While poorly understood or excessive aberrations can seriously compromise performance, effort can be wasted reducing aberrations more than necessary. Monochromatic wave front or ray aberrations can be generally represented by a polynomial expression of increasing order in aperture, field angle and azimuth and their products. The five third-order Seidel aberrations not correctable by a focal shift are primary spherical aberration, coma, astigmatism, field curvature, and distortion. Occasionally higher orders must be considered. Chromatic aberrations must also be considered in multiwavelength systems. Unless the system demands special devices with exotic optimizations, good-quality components optimized for general use should be examined for suitability.

With many modern components, tolerancing is of more interest for performance prediction than correction, although as with ghost analyses it merits sufficient consideration to justify its subsequent dismissal. Where optical or physical path length is at a premium, substantial gains may be realized by folding the path. Although this may also improve rigidity and general robustness, there may be hidden penalties in design costs, performance, and flare.

Because in recent years, smaller, brighter, and more coherent optical sources have become commonplace, calculations based upon Gaussian beam behavior are often necessary, and the appropriate diagram and formulae are repeated here. Figure 18.5 and the inset box show the relationships of the beam radius and wave front radius of curvature at a distance z from a Gaussian source of e^{-2} intensity radius w_0, with the asymptotic approximations in the far field. For a converging lens of focal length f at distance z_0 from the effective waist of the source, the remaining two equations give the aberration-free waist position and radius.

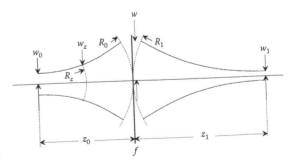

Figure 18.5 Gaussian beam parameters.

Beam radius w_z at distance z is

$$w_z = \sqrt{w_0^2 \left[1 + \left(\frac{\lambda z}{\pi w_0^2} \right)^2 \right]} \approx \frac{\lambda z}{\pi w_0}$$

Radius of curvature of wavefront R_z is

$$R_z = z \left[1 + \left(\frac{\pi w_0^2}{\lambda z} \right)^2 \right] \approx z$$

Axial position of waist w_1

$$z_1 = f + \frac{(z_0 - f)f^2}{(z_0 - f)^2 + \left(\frac{\pi w_0^2}{\lambda} \right)^2}$$

Radius of waist w_1

$$w_1 = \frac{w_0 f}{\sqrt{(z_0 - f)^2 + \left(\frac{\pi w_0^2}{\lambda} \right)^2}}$$

18.5.8.6 PHYSICAL STRUCTURE

Mechanical layout drawings are necessary for optical component mounting, later detailed by a draughtsman for a machinist to make. Mostly now these processes may be automated by CAD systems and optical design software easing the more tedious tasks of the designer and allowing semiautomated extensions to thermo-physical behavior, mechanical rigidity, strength, alignment retention and many other necessary features of any optical system. Beyond these capabilities and specifically for optoelectronic systems they can quantify ranges of necessary adjustments, confirmation of adjustment orthogonality, with appropriate locking techniques, establish a thermal interface map, including sources and sinks, effects of conduction, convection, and radiation on the heat budget and temperature distribution. Thermal effects can compromise success, and show different behavior in service from that expected during design or noted during assembly, testing, calibration, and evaluation. In the

absence of gravity, for example, convection is suppressed, and under different operating conditions very different thermal distributions can be expected, perhaps changing calibration or causing irreversible effects. Gradual drift from thermal cycling is common but not always suspected in its early stages. Beyond these automated capabilities lies the necessity to apply personal experience to assure that the design has not included some ridiculous feature as a result of an unsuspected error or omission. If experience suggest that "it does not *feel* right, then it probably is not," whatever the beautifully produced automated design might suggest.

Occasionally an optoelectronic system is so complex that a FEA of the mechanical structure is a good investment, permitting mechanical and thermal loads to be simulated throughout any environmental range. Although design programs are increasingly able to interface with each other, the inclusion of all aspects necessary to guarantee a successful system remains far from trivial. Occasionally, creating a supposedly effective design can consume too many of the available resources, and may be better assured by modularization, testing and frequent empirical closure.

18.5.9 Management and documentation

An individual scientist or engineer responsible for creating a simple optoelectronic device can occasionally track everything well enough to retain control using only a notebook and memory. However, this is rarely sufficient. The notebook is essential, and will likely evolve into a design document in electronic format. Four things are emphasized: that documentation should (1) be subject to configuration control from the first keystroke, (2) be backed up frequently and at more than one site, (3) be in a portable format that can be read by everyone who needs it, and (4) not become unreadable with time, either because it is stored on obsolete media, or because the generating software is no longer available. Few can now read 5¼ in. floppy disks, once a useful standard. 3½ in. disks rapidly followed them, and CDs and DVDs became as quaint as vinyl records. Whatever medium is chosen for archive one can guarantee that it will have a finite life, sometimes quite short. Even in comprehensive corporate networks, programs used to create and store information, particularly drawings, become obsolete surprisingly quickly, and translators disappear. Fortunately, server farms and the distributed "cloud" claim to overcome these problems, but be suspiciously cautious for at least two reasons, (1) there is so much growth of stored information that it may be impossible to find the source, and (2) decoding the information may become infeasible.

Documentation is key to success, whether by allowing design reviews to preempt problems, or by providing the necessary audit trail to diagnose and fix them. Table 18.9 suggests documents that support the success of the program and the measurements.

The minimum program plan is that which, in addition to its technical merit, will satisfy the customer about cost and schedule. Full-service project management software is usual for general programs, but for optoelectronic activities and systems the match is rarely perfect, representing overkill in some areas and under-specification in others. A simple spreadsheet can be a valuable tool for keeping track of optoelectronic system development. For example, the first column can contain the nonorthogonal set of all components, activities, and concerns, followed by exemplary column headings such as specifications, source, supplier, address, universal resource locator (URL), phone, name of personal contact, cost, order date, promised delivery date, status, actual delivery date, tests to be performed, test results, assembly status, problems, comments, contingency plans, second sources, alternatives, mysteries, and more as design and implementation progress. More still will be added during building, testing and evaluation. This spreadsheet is a convenient place to keep current information, which might otherwise be spread among many people, documents, mailrooms, desks, file drawers, memories, and wastebaskets.

Configuration control assures that information is current. Editing an obsolete file, repeating a task already completed, misplacing an unlabeled lens, or machining a part from a drawing that was changed in review but not sent to the machine shop, wastes resources that may not easily be replaced. Of particular concern is that purchased parts are often not exactly what the catalogue offered, and ripple-through effects can only be minimized by good configuration control. Everyone involved needs the most current information.

Comprehensive and powerful documents are wasted unless people have access to them, can read them, can understand them, and will act upon them. File portability and file backups are mostly handled by modern computer networks, but a poor choice of commercial software or unfamiliar modeling program can inhibit dissemination of, and access to, necessary information.

The primary responsibility for a design document is that of the technical champion who has assumed personal responsibility for satisfying the customer. Although in covering all levels from overview to the smallest detail, it can seem difficult to create and maintain, the power in comprehensive understanding of the final system justifies the investment. A convenient sequence of chapters might be (1) introduction and physical principles, (2) specifications, (3) system architecture, (4) interfaces, (5) performance modeling and prediction, (6) module function and physical descriptions (several chapters), (7) software, (8) testing and calibration, (9) power supplies, (10) table of all relevant and known numerical values, (11) method of contacting suppliers or other expert sources,

Table 18.9 Documentation

Program plan	Cost and schedule plan, tracking of all activities
Configuration control	What documentation is current, and where to find it
Interface specifications	Modularity, information flow, monitoring, power
File backup	All documents immune to loss from single-point failure
File portability	All documents readable by all, now and in the future
Design document	A single source for all current data

(12) reasons for design choices, and why alternatives were rejected, (13) log of evolutionary changes, (14) unresolved mysteries, and (15) several specific technical appendices. If this document is brought into existence at the beginning of the optoelectronic project and maintained merely by ordered accretion of all pertinent information, then it will prove invaluable. Abstracted sections can become the final report, comprehensive provenance for a scientific publication, the operator's handbook, the service manual, the diagnostic methodology, the platform for evolutionary improvements, and a delight to the customer.

18.6 COMPONENTS

Planning and design prepare the way for the choice and acquisition of components. In optoelectronics, the design process naturally iterates with what is available, and the loop between design and component choice is always tight. Components improve almost daily and innovations can sometimes solve problems considered challenging at the beginning. This is often magnified by a steep learning curve, where the uncertainty of initial unfamiliarity gives way to confidence as knowledge increases.

While Section 18.5 is too brief to include all design concerns, this section addressing physical components must be even less comprehensive and current. Deterioration of relevance with time is exacerbated because quantities, qualities, nature, and even functional principles of the components available to the scientist or engineer practicing optoelectronics evolve almost daily. To avoid becoming even more outdated than is unfortunately inevitable, rather than current device specifications, this section emphasizes questions and classes of properties that may perhaps continue to be relevant for the future.

Component choices are often guided by familiarity, cost, and/or common sense, any of which can be disastrous. Typically in any optoelectronics-based activity the cost of the components, however high it may appear, is largely irrelevant compared with the total program cost. Buy the best and cry only once. In optoelectronics, and often optics in general, the common sense approach is often derailed by the intrinsically counterintuitive nature of the subject. Even after many years of familiarity, what should be a naturally obvious improvement can often makes things worse. Conversely a less familiar component may seem attractive, if expensive. Rather than basing a choice on the manufacturer's recommendation, a better technique is to talk with someone who has used that component, ask the vendor for a demonstration, or borrow one and try it. Of course in some cases final choices may depend upon what we already have, or can get, or think we understand. Pragmatic examination of every aspect, reaffirming the validity of even well-formalized beliefs, carries high dividends. Check every assumption; check it again.

18.6.1 Light sources

In a passive experiment, the phenomenon provides photons from which information is to be extracted. In active experiments using light to probe or intentionally stimulate a phenomenon of interest, the light source must have known, constant, and well-understood properties. The performance of the source may depend on properties of its final environment not necessarily suspected when the equipment is characterized in the laboratory. Where many types of light source are available, the simplest device satisfying the requirements is usually the best. Lasers have high intrinsic luminance, a desirable property for most applications, but where coherence is either not required or even undesirable, a superluminescent diode may be better. Light emitting diodes (LEDs), cheap and available over a wide range of wavelengths, may be adequate. Increasing source power alone improves the signal less often than might be wished, and accidental modification of the probed phenomenon by a poor choice of light source is not impossible.

18.6.1.1 TYPE OF SOURCE

The light source must have sufficient power and be stable under all likely conditions. Its properties must be matched to the system. Spatial coherence, implying the capability for good optical beam control, is usually desirable, whereas temporal coherence, essential for applications such as holography and interferometry, in others can be a serious disadvantage, giving rise to unwanted fringes or signal fluctuations unrelated to the measurement. Often not all the desirable properties may be available from a specific light source, and a trade-off may be necessary between, power, wavelength, coherence, polarization, modes, and their distribution and stabilities. The availability of a commercial product may even dominate the trade-off. Occasions where a thermal source is best are becoming less common with the increasing availability of many other ways to generate light, from innovative discharge lamps such as microwave excited sodium, through room temperature solid-state sources whose properties may be well specified over an increasingly wide range, to ingeniously contrived quantum-well emitters, and many more available both now and in the unpredictable future. Although much significant optical knowledge was gained with sunlight or wax candles, new principles of light production and photonic materials continue to drive the optical arts at an ever-increasing pace.

For OADS each of six thin sheets of light was imaged from a 1W GaAlAs laser at 812 nm with 20 adjacent $1\,\mu m \times 3\,\mu m$ emitters on the same die, with weak coupling to give constant output phase with a longitudinal intensity uniformity better than 20%. This was in the 1980s—the world moves on at an ever increasing pace. Keeping up is difficult but essential.

18.6.1.2 WAVELENGTH

Optimum wavelength is often an early consideration. The choice is rarely trivial since both the phenomenon and the optical system can force complex trade-offs. The spreadsheet technique introduced earlier to predict signals and noise levels can also be used to optimize wavelength, or indeed, any aspect of the system, by setting up a figure-of-merit involving all the parameters considered important and exploring the effects of various values, available components, and plausible changes.

18.6.1.3 POWER AND STABILITY

Since optical power must be both sufficient and stable, it is highly desirable to either monitor the intensity (Section 18.7.1) or better still use the monitor signal to stabilize the power with a suitable closed-loop controller. Despite manufacturers' claims, it is essential to characterize the source under the conditions of anticipated use, including ambient temperature and changes, power supply modulations, drifts and transients, mechanical vibrations and shocks, contamination, and ageing of construction materials. Although visual confirmation of spatial properties of the source, enhanced by special techniques if it is outside the visible spectrum, is often adequate, it is sometimes

desirable to quantify the spatial distribution and transverse mode structure by a beam profiler or by image capture and analysis, requiring an additional 2-D sensor. If direction of polarization or longitudinal mode structure, particularly their effects on coherence, for example, can change the measurement, then these too must be monitored and/or controlled.

Beam-pointing stability is particularly important in optical systems with high magnification or with strict directional requirements. A sufficient change of laser beam pointing direction can seriously compromise launching into a fiber, spatial uniformity of illumination, position of a measurement volume, or receiver alignment. Experimental quantification of effects increases confidence. Diode lasers typically must be temperature stabilized to retain power, wavelength, mode structure and pointing stability, with residual stabilization of polarization purity and orientation particularly important when launching into single-mode and/ or polarization-maintaining fibers. The OADS lasers exhibited a negative wavelength dependence on temperature of about $0.3\,nm\ °C^{-1}$ and all six were kept near the center of the 3.5 nm full-width half-maximum (FWHM) narrowband filter, also temperature stabilized. Because each laser was slightly different the base temperature of each was individually controlled differently to give the same wavelength. A better technique would have been to use six slightly different wavelengths and six separate filters to reduce signals from particles crossing other channels. The potential advantage was considered but outweighed by the necessary increase of size and physical complexity.

18.6.2 Optics

The term "optics" conventionally describes the components whereby light is transmitted between source and detector via the phenomenon of interest. Wavelength is conserved in the most common cases, but inelastic processes such as fluorescence, or other nonlinear effects, change wavelength in complex ways. Optical components include lenses, beam splitters, polarizers, fiber-optics, graded-index devices, coatings, stops, filters, attenuators, acousto-optic, Faraday or Kerr effect devices, windows, transparent adhesives, holographic elements, photonic devices and structures, and many others. The choice from so many components should always be the minimum set of the simplest parts that will meet the specification. Avoid moving parts; avoid unusual or nonstandard parts; where possible, avoid parts. While the usual goal is to transmit as much light as possible that carries useful information, the reduction of flare can be as or more important, because excessive flare can mask desired effects and also consume bandwidth or processing resources. The optical properties of the system are crucial, for once information is lost it can never be recovered. Even if it is merely swamped recovery may be impracticable.

18.6.2.1 LENSES AND OTHER COMPONENTS

Competing criteria affect the choice of optical components. These include maximum transmission at the wavelength of interest, control of aberrations, implying not only reduction but also exploitation such as apodization, surface placement and curvature, and treatments designed to reduce unwanted scattering and reflections such as ghosts and flare. However good a standard design, it must be evaluated for each application, and any sufficiently innovative requirement may demand special consideration. A particular example of this is to think that zoom lenses, whose compromises seem adequate for photography, might be applicable to optoelectronic instruments. Despite the tempting cost and apparent versatility the results are often disappointing and sometimes disastrous, because of the complexity and unsatisfactory compromises introduced to solve a problem not present in the system of interest.

Sometimes an unconventional component may promise, or even deliver, great rewards. However, excess attention to the improvement may hide shortcomings not apparent with a more conventional approach. All that may be said about lenses applies to flatware such as mirrors and prisms, or a variety of more complex and unusual components such as diffractive or holographic optics, nonimaging concentrators, or other less common devices and techniques that may yet solve a specific problem effectively. Smaller components have better mechanical integrity and are less prone to distortion with mounting and conditions of use, but they can be more expensive and difficult to handle, and are more prone to impaired properties close to the edges. Even with well-made optics it is good to have the component at least 10%–20%

bigger than the active dimension. For Gaussian beams an operational aperture radius of at least twice the e^{-2} intensity contour of the beam is highly recommended to avoid intensity profile or wave front structure changes. For conventional optical systems the presence of implicit apertures forced by component clear dimensions or edge imperfections may introduce effects in addition to, and different from, the intentionally designed stops. Vignetting, intensity roll-off with radius, can impair imaging systems. Attempt at compensation in software by radial gain scaling can reduce the usable dynamic range of the sensor and increase the effective noise.

18.6.2.2 STRAY LIGHT

Reducing stray light is a black art. For monostatic systems, where common optics illuminate the phenomenon and receive the returned light, control of stray light is always a challenge. Finding all sources of flare demands a full design understanding of the optical geometry and component properties, which can still yield surprises. For bistatic systems the problem is alleviated by physical separation of illuminating and receiving optical trains, but even here high multiple-order reflections and scattering in any common parts, such as windows, can contribute unwanted scattered light comparable to the often extremely low signal from which information must be extracted.

Stray light from sources external to the instrument may often be reduced by baffles, optical labyrinths, stops, and/or filters. Spectral control using narrowband or sharp-edged filters usually requires a collimated beam, even for such as the temperature-tunable and no-longer fashionable Christiansen filter, where a cell contains transparent liquid and powder with different dispersion curves intersecting at the transmission wavelength. OADS used a temperature-tuned dielectric stack in a well collimated beam. If necessary, spatial and spectral methods may be augmented by temporal methods, such as a modulated source with phase-sensitive detection. This can vary from simple ac coupling with a tuned or lock-in amplifier, with phase-locked or frequency-locked loop capabilities, to driving the source with a pseudorandom code generator and correlating it with the detector output. Where flare does not have the same time characteristics as the signal, many orders of magnitude of rejection become possible, and in the last case range-gating is an additional bonus. Modern solid-state sources, for whose physical properties a nanosecond is a long time, especially permit these techniques and the associated improvement of SNR.

Internal flare is often more difficult to suppress, but is amenable to the usual techniques of high-quality components, properly chosen and applied coatings, cleanliness, stops, and the choice of surface shapes. A tunnel diagram can help understand systems with reflective components, where the optical system is unfolded about each reflection, and all possible stray light paths become straight lines. This is good for minimizing ghost reflections and finding where best to put stops and baffles, and quantitatively augments the powerful technique of actually looking down the system from the detector end. This visual assessment of the system often finds sources of stray light, particularly in systems designed for visible wavelengths, where the dark-adapted human eye is almost as sensitive as typical detectors. Even looking through systems not designed for the visible wavelengths can still warn of unexpected consequences.

Ray-tracing programs extend the tunnel diagram concept, allowing for surface curvatures distorting the tunnel. Almost all offer ghost analyses, and in many cases help to identify sources of stray light and allow its reduction by proper choice and placement of stops, for example. Optimization is less obvious than software designers might suggest and there is no substitute for thorough optical knowledge and experience. Using optical design programs without having a realistic expectation for what the result should look like can be dangerous. It is easy to mis-set a single variable, and produce a design that may be a poor optimization or even nonfunctional. One aspect of the art is that if something 'feels' wrong, then it probably is. For the 150 mm diameter receiver common to all six channels of OADS, quite elaborate spatial and spectral restriction was essential to discriminate against direct sunlight.

18.6.2.3 COATINGS

Coatings are applied to most optical components to minimize surface reflection, to confer spectral selectivity, or even to protect. Using multiple layers improves efficiency and antireflection effectiveness but can increase scatter. The trade-off depends upon a calculated figure-of-merit for the chosen system. Well-chosen coatings can increase the nongeometrical aspects of light gathering power, reduce ghosts and increase SNR, particularly in monostatic systems. Conversely, they can increase scatter, reduce SNR(,) and/or be sensitive to angle. Optical coatings are sometimes more fragile than the base material, preventing cleaning, but conversely can sometimes be chemically or physically protective. Less commonly considered coatings, for example, electrically heated indium tin oxide, decrease transmission but may prevent worse damage from condensation, as with a cool window in a condensing environment. Where surfaces are superpolished, coatings require extra quality control and handling to avoid worsening the expensively achieved improvement.

18.6.2.4 WINDOWS

Preservation of optical access with clean, undamaged, birefringence-free windows is often mandatory, but is rarely trivial. Windows and optical access in general can be ruined by local melting, stress cracking, surface crazing, ablation, thermophoretic deposition, condensation damage, defects induced by the phenomenon, improper mounting, or even by physical accidents. Occasional manual or automatic cleaning may be essential, as may cooling to avoid deterioration, or heating to avoid condensation. Table 18.10 suggests some factors of interest when specifying optical access, but other techniques are sometimes necessary, for example a high-pressure gas curtain or liquid film may sometimes be superior to a solid material as a window to protect the instrument from the phenomenon or its environment. Sacrificial membranes can be occasionally or continually replaced, such as those to prevent spalling damage in laser machining. Transmission losses arise both from bulk properties and from surface effects. Temperature dependence of bulk absorption effects, stress birefringence or mechanical variability are determined exclusively by the material and specified accordingly. Surface properties including adsorption are more variable and less well controlled. Good material, polishing, coating, and the assurance of their constancy are the best ways to reduce attenuation, scattering, polarization effects, and changes with time. Superpolishing under liquid reduces scattering from surface defects, but later coating or improper handling may compromise this. A trade-off must be

Table 18.10 Window properties

Material	Transmissivity, suitability, availability, preparation, cost
Physical	Strength, hardness, damage resistance, distortion
Optical	Refractive index, homogeneity, aberrations, surface finish, birefringence
Coatings	Effectiveness, durability, scatter, cleaning capability
Mechanical	Thickness, width, shape, stress, strain, retention, sealing
Thermal	Expansion, conductivity, ablation, cooling, heating, embrittlement
Stability	Change of state, etching, chemical or biological attack, crazing, ageing

made between scatter, reflectance, and vulnerability to damage, as it is often impractical to achieve the best of all three simultaneously.

Transmission through one or more windows must not unacceptably change intensity, wave front curvature, polarization, wavelength distribution, nor introduce scattering or detrimental reflections during operation. Note, for example, that in an uncollimated beam even a perfectly flat plate has aberrations.

If requirements are either mutually exclusive or demand an expensive material, then double windows with an intervening balancing medium sometimes offer a cheaper and more practical solution. For example, simultaneous exposure to extreme pressure and temperature is possible using a thin magnesium oxide window exposed to the heat, followed by a cooled gas at high pressure, and a subsequent thick float glass window. Edge polishing and annealing reduce mechanical stress-raising weak points. Support and sealing are best distributed over one or more faces with a suitable material that retains its flexibility. Elastic materials are good for repeated cycling over temperatures tolerated by the material. Annealed copper is good for one time sealing, and reasonable temperature cycling. For higher temperatures and cycling ranges, internally pressurized stainless steel bellows work well, as do thermally compensated materials. For many applications an adhesive with slight residual elasticity is good, but a rigid ledge may be better to support mechanical load, provided that stress is acceptably distributed. Thermal as well as physical shock can break materials.

More expensive and exotic materials are not necessarily better. For example, specific conditions in a flame chamber test rig broke all its sapphire windows immediately. Short of spares, time and money I had a local spectacle lens maker cut replacement windows from a piece of float glass.

Surprisingly, a simple set of these cheap windows survived through the several month test program.

18.6.2.5 OPTICAL FIBERS

Optical fibers are commonly used both to transmit information, and to sense various physical properties as a technique for measurement. Because these capabilities are mutually exclusive, optical fibers are less easy to optimize for any given application than might be expected. Most investment has been funded by telecommunication requirements in the 1.3–1.5 μm wavelength region. Although this knowledge and experience has wide utility in optoelectronics, the properties of interest, either required or accidentally exhibited, are not always well controlled nor characterized, since they are not necessarily those of most concern to the manufacturers. Great care is essential in examining the properties affecting the specific application. Table 18.11 lists a few questions best answered before commitment to a specific optical fiber.

Optical fibers essentially transmit intensity alone, but extremely complex effects can alter the relationship between input and output so that the fibers act like intricate optical systems, behaving both as integrated optics, and also capable of applications, such as amplification, formerly occupied by electronics. The existence of photonic bandgaps and special optical properties, derived from physical structure on the order of the wavelength or even much less, has an increasingly wide field of potential applications.

Transmission of information is possible in many ways, including simple intensity, polarization, wavelength or frequency division multiplexing, solitons, and individual photons. Two applications of fibers are common, as a sensor, requiring that the output shall depend uniquely on the sensed quantity, and as a channel, requiring that the output shall indicate only the required properties of

Table 18.11 Optical fiber questions

Is single- or multimode fiber necessary?
Which of many fiber types is optimal, or even adequate?
What materials are appropriate and available?
Are transmission losses acceptable? And will they remain so?
Does transmitted or external hard radiation cause deterioration?
Should the fiber maintain, select, or control polarization? How?
How is injection mode-matching to be established and maintained?
Is the fiber sheath opaque to ambient light? Does it matter?
Must cladding modes be suppressed? If so, how?
Can properties of the fiber introduce signal fluctuations?
Does high light intensity induce nonlinear effects or cause damage?
Will the fiber be properly sensitive to the measured properties?
Will the fiber be properly insensitive to ambient properties?
What adhesives are usable or necessary for construction?
How shall terminations be arranged, aligned, and controlled?
How do vibration, bending stress, temperature, and ageing affect performance?

the input. In the second case, filtering using prior knowledge can improve the signal. For example, a single-mode fiber of sufficient length to attenuate cladding modes can operate as an excellent spatial filter. Since a typical single-mode fiber remains single mode over a factor of two in wavelength, injecting the output from a superluminescent diode gives an excellent temporally incoherent point source, an important device once only available as an inferior approximation. Replacing the source with a laser permits retention of temporal coherence and again gives a versatile point source with small mass to align accurately. In the latter case, however, earlier comments must be augmented, and the fiber may have to be terminated with a wedged face to prevent coherent reflections from generating instability in the laser cavity or elsewhere, since the fiber can act as a sensitive Fabry–Perot interferometer. An inclined face changes fiber emission geometry from circular to slightly elliptical, with a pointing direction no longer parallel to the fiber axis. Dome polishing the fiber end can also mitigate detrimental effects of back-reflections, and facilitates compression coupling, but sometimes has other limitations.

Launching into a single-mode optical fiber requires matching convergence angle or numerical aperture and waist size to preserve the Lagrange invariant, the product of aperture, field angle and refractive index, similar to the Abbé sine condition or the brightness theorem, which determines the highest possible freedom from loss of light gathering power (étendue). It is rarely possible to meet this condition totally, even with accurate and stable alignment, and birefringence and polarization properties of fiber and other components may worsen it further. Once light is launched into an optical fiber, it is not desirable to let it out and then try to relaunch it—the power loss is always worse than expected. Contamination of end-faces exposed to an open atmosphere can massively reduce the amount and quality of transmitted power, or even destroy the component. Emergent power density near the fiber face for one transmitted watt of green light can be 10^{11} W m^{-2}, eight orders of magnitude greater than sunlight. This may be sufficient to attract dust by thermophoresis, causing enough local heating to craze or ablate the fiber material. Such problems are easily recognized. Output damage is identified by the appearance of Airy-type rings or other intensity structure within the ideally Gaussian output beam in the far field. Input damage causes only loss of power. Optical fibers are typically fragile, requiring one or more tough concentric sheaths outside the cladding for protection. Their use also demands high-precision mechanical designs and mounts, typically to submicron and microradian tolerances. Even standard connectors for single-mode optical fibers have complex and exacting requirements, and transmission efficiency can fall with each reconnection.

Transmitting high optical power, especially through single-mode fibers, requires careful specification of fiber material, such as a germanium-free core that will not form color centers, and special optical adhesives that do not denature. Transmission may reduce with exposure to high transmitted power or cosmic or other radiation, micro-cracking with vibration or under sustained strain from too small a bend radius, or inadequate physical support. Unless intentionally exploited, sensitivity to external effects is detrimental. Microphonics may sometimes be reduced by special care with alignment, and sometimes not, depending on fiber and environment, either of which may be insufficiently well specified. The only way to assure adequate performance is by testing a sample of the specific optical fiber from the batch available.

The conjugacy of end faces of coherently bundled fibers allows imaging of relatively inaccessible situations from body cavities to turbine blades. However, beware of the potentially ambiguous meaning of 'coherent' as applied to fibers. A coherent bundle is a loom of fibers whose ends map spatially so that an image may be transmitted—like the naturally occurring mineral ulexite, whereas a coherent fiber typically conserves spatial and/or temporal coherence of the light. Optical fibers based upon coherent constraint of the propagating modes, by step or graded refractive index control in regions of a few microns diameter, are distinguished from incoherent optical light guides, typically with step index confinement within a few tens of microns up to millimeters in solid material or liquid-filled tubes.

A special case is the GRaded INdex (GRIN) lens exploiting the radial control of refractive index to perform the operation of miniature lenses, using the techniques of optical fibers. Control of refractive index gradient in fibers may also inhibit dispersion and increase bandwidth. In a step-index single-mode fiber an evanescent wave extends into the cladding material. Ambient conditions may be sensed by their effect on this evanescent wave, and hence on the propagated intensity or polarization. This same sensitivity manifests as detrimental effects for pure transmission, where bending radius, temperature changes, stress microcracking, vestigial inhomogeneity from irregular sleeking near the core–cladding interface, and other manufacturing imperfections can give rise

to a range of curious effects only fully appreciated by exhaustive empirical testing. Polarization rotation, temporal fluctuations over a wide range of frequencies from seconds to months, and cladding mode coupling are all possible. Used with GRIN lenses that exhibit residual birefringence, the fiber can behave as a temperature dependent wave plate. Sometimes fluctuating light leaks are visible where the sheath is not opaque. Even with polarization-maintaining fiber, using bow-tie stress birefringence or elliptical cores, a slight error in launch orientation can yield large and unexpected sensitivity to fiber environment, and not necessarily in a predictable way. Polarization preservation where one of the two orthogonal axes is attenuated gives a purer polarization output, but occasionally at the expense of intensity instability. As fiber manufacturing control and methods of characterization improve, many of these effects may be reduced, or at least their sources better understood and avoided. Meanwhile specification for a given application must allow for experience specific to the individual fiber type, manufacturer, and indeed the production batch. Six 1.2 mm diameter clad quartz fibers were used in OADS as light pipes following a single chrome etched field stop to direct each of the six sheet images to its own APD.

18.6.3 Physical mounting

Physical mounting of components must maintain correct position without stress or distortion. To avoid stress concentration, support must spread the load either by flexible media or sprung clamping. The hole, slot and plane mount, with gravity or light spring loading, provides kinematic location but cannot be locked without over-constraint. It is thus useful for exact relocation, for example of a hologram for reconstruction, or a component that must be removed and then replaced exactly, but is not robust and is typically confined to the vibration-isolated stable table of a temperature controlled laboratory. Even in that environment, where optics are mounted on posts, the posts should be as short as possible with the largest possible diameter. Where a mount is fixed to a supporting metering structure with screws, the component face should be dished to provide the largest footprint, and controlled locking torque specified to constrain distortion. While the metering structure may be as simple as a flat table with tapped holes or magnetic clamps for greater versatility, for specialized instruments it may be a complex casting, machined frame, or composite assembly of diverse elements. The retention of its physical properties, especially dimension and rigidity, is important. In OADS the necessary beam pointing stability of $<5\,\mu\text{rad}$ was achieved by a metering structure of low-expansion stainless steel, with weight and rigidity optimized by FEA of the complete structure, isothermalized by ethylene glycol circulating in labyrinthine manifolds. Each laser was precisely located in one of the six circularly symmetric transmitter lens assemblies with five lockable adjustments, two tilt, two transverse, and one focus. Each completed transmitter assembly was directed to the correct point in space by rotation in a hollow spherical bearing near the output nodal plane, using one rotation and two more tilt adjustments. Alignments were made in a set sequence to meet specified criteria. The adjustments were locked by screws to a specified torque and all critical sliding or pressure points encapsulated in rigid adhesive. Most applications require neither this precision nor its associated complexity.

Whether in benign or more demanding environments, there is a trade-off between the rigidity of accurate placement and available adjustment to accommodate changing conditions. Active control is complex, to be avoided if possible, but sometimes essential. Large or fragile components may be held in a sine wave mount, a perforated annulus of elastomer. To seal and/or distribute stress, O-ring seals or well-chosen flexible adhesives often suffice. Metal clamps are hazardous unless appropriately sprung. Where no movement of the component can be tolerated, rigid sealing may exploit materials with graduated properties to distribute the stress.

Ideally, optical mounts must be robust and rigid, without affecting the optical properties. For materials with optical activity or birefringence induced by stress, this can be confirmed by viewing the component between crossed polarizers, often an instructive test, with surprising results if applied, for example, to spectacle lenses. For high-quality components even small mounting stress can introduce aberrations, and rigid retention for high performance competes with more flexible location for field survival.

Adhesives are common in optical packaging, from component retention to optical function. Several popular transparent epoxy-based materials can be cured by ultraviolet exposure; others require two components, a catalyst, heating or time to reach the required state. When used for retention, the final mechanical properties dominate, but when adhesives are in the optical path more careful choice is advisable. Transparency is usually necessary, and this may be compromised in use by ageing, nonoptical radiation, stress cracking, change of polymer properties, crystallization, or high light level. The last is especially relevant where, for example, an illuminator has a GRIN lens glued to a single-mode fiber.

18.6.4 Detectors

Ideally, any square-law detector yields an electrical signal corresponding to incident intensity, manifest in the semiclassical approximation as successive quantizations, which may not be resolvable. For sufficiently low light levels, an appropriate detector reports a given fraction, based on quantum efficiency, of individual photon arrivals, missing some but ideally not changing the arrival statistics. As the light level increases, this becomes a semicontinuous signal whose shot noise increases with the intensity, approximately as the square root of the equivalent number of photon arrivals, improving SNR. An audible analogy of this might be hail on the metal roof of a noisy workshop. This is true for a single-point detector such as a PMT or APD or for one pixel of a CCD, CMOS or other array. Criteria for optimizing detector type assume different relative importance depending on the application. For photon correlation, the ideal is a single-point detector free from dead time, afterpulsing, pulse pile-up, internal correlations, and with a sensitivity and quantum efficiency as high as practicable, to produce single resolvable pulses from every field quantization, with no triggering events from any other source. This may be approached quite closely for high-energy ultraviolet quanta with a rubidium telluride photocathode, as with OADS pressure and temperature sensing. For red photons, both energy and cathode sensitivity fall, allowing more false triggers and impairing the correspondence of pulse statistics with those of the optical field. Single-photon performance is approached with reverse biased silicon APDs in the Geiger mode. To obtain highly accurate measurements of high intensity at low bandwidth, the criteria may shift from shot noise intrinsic to the signal to other sources such as Johnson noise from the load resistor or preamplifier, and keeping the noise low moves from optical to electronics design expertise. Cooling the detector and its electronics below a certain temperature often helps with noise, more so where the detector relies upon absorbed energy as with a thermopile or bolometer.

For single-channel detectors, PMTs are fast and can have huge sensitive areas, with the penalty of being physically large and requiring high voltages for the accelerating grids. However they offer amplification of many orders of magnitude up to thousands of amperes per watt, without adding excess noise to the original quantum realization variance. Wavelength sensitivity typically falls from mediocre (~25%) in the blue and green to rather poor (~1%–4%) in the red, although newer implementations claim significant improvement. APDs have much better quantum efficiency (~90%) in the red, and are physically small. Although their noise figure can be as low as 10^{-14} W $Hz^{-1/2}$, when stably cooled to around 0°C, they also require a moderately high (~ 200+ V) and stable voltage to achieve a gain from a few to a few tens of amperes per watt. The sensitive area is often inconveniently small (<1 mm^2) with a longer dead time than PMTs. Lower-performance bulk detectors of various materials are often cheaper, and may be adequate for less demanding applications. Operating in either photovoltaic or photoconductive modes without intrinsic gain, their noise typically increases with surface area and temperature. Most detectors are fairly robust to optical overload, but take varying times to recover. For example, a PMT will typically recover its sensitivity lost from transient optical overload as soon as the dynode decoupling capacitor chain recharges, but to recover its noise characteristics may take several days in darkness with full voltage applied. APDs flip into an avalanche overload state if the optical input exceeds a level not much larger than the average, but are not damaged if the current is externally limited. Recovery after the removal of excess light can be quite rapid, but must sometimes be encouraged by the temporary complete removal of the reverse bias voltage. Thus both PMT and APD detectors can be protected by self-limiting mechanisms, and are not necessarily permanently damaged until the optical input melts the sensing or amplifying material. This is also true for conventional bulk material sensors, although, without limiting the electrical current, thermal runaway can cause permanent damage.

Array detectors operate on similar photoelectric mechanisms but store charge locally for an equivalent exposure time. CCDs read the charge out serially via a "bucket brigade," which is a major source of noise at low light level. Individual pixels rarely have uniform sensitivity and a calibration table is desirable for quantitative work. Ageing and radiation damage also change the characteristics of individual pixels, demanding calibration and update of lookup tables. Types of array detector are available with different characteristics such as pixel size and shape, dead space between pixels (fill-factor), well depth (how much charge can be stored at an individual pixel site without nonlinear effects), readout speed and method, and even manufacturing technology. For example, CMOS requires less power and can be constructed so that each pixel may be separately addressed, clocked, and amplified. This conditional interrogation can greatly increase the dynamic range of the array, because those pixels that are more brightly illuminated may be read out more rapidly, accumulating the total from each pixel in external memory. Pixels in the darker areas are merely read less frequently or even at the end of the image acquisition time. In this mode, low pixel crosstalk is essential. Although excessive cooling can cause a CCD to stop functioning, the effect on a CMOS detector is desirably to reduce the noise. Typically, the output is read as a filtered analog signal to an ADC, either on or close to the detector chip. Although the digitization noise might appear to be no more than the inverse of the number of bits, real devices rarely perform this well and a bit or two of additional noise is common. Claims for the relative merits of each type of device change frequently and for any application it is important to understand how the manufacturer's claims relate to the proposed experiment.

The art here is to choose a detector that seems appropriate and become as familiar with it as possible, including how best to drive and control it, how to compensate for its shortcomings, and what its limitations dictate for the accuracy of the experiment and the quantitative validity of the conclusions.

18.6.5 Miscellaneous observations

Light gathering power can easily fall below the upper limit set by the Lagrange invariant unless the numerical aperture is matched everywhere. Errors can occur readily when coupling single- to multimode fibers, or at the entrance to a detector, under the impression that an alignment criterion may be relaxed. Typically the receiving optics, collecting light scattered from the probed phenomenon, is merely a photon-bucket, but this does not necessarily mean that aberrations can be tolerated; for example, the proper performance of a field stop may rely upon its high-definition image at some other location. Aperture stops may be similarly critical. The performance of dielectric stack narrowband filters is typically compromised in a poorly collimated beam.

A given specification may be met and implemented by many different candidate components, devices and the phenomena upon which they rely. New techniques and innovative devices are constantly becoming available. Naturally occurring materials are now being augmented by artificial structures, including sequences of differently doped layers, physically and/or chemically controlled photonic bandgaps, and many other ways of spatially or temporally manipulating light from classical to quantum domains. The formerly popular division of experiments was into classical, where the signal is assumed to be continuous analog with additive noise, and "photon resolved," where all the information is contained in the intervals between photon arrivals, assuming monochromatic light, but with device limitations such as dead time, pulse pile-up—endemic with Poisson statistics—and other limitations. This separation represents rare extremes, and increasingly the interesting typically low-light-level experiments are in the "gray" area between the two, demanding a more carefully consideration of the nature of the signal and how best to use it to measure the phenomenon of interest. New manufacturing methods, extending beyond chemical vapor deposition and molecular beam epitaxy, are being introduced. Although fashion, implying ready availability and perhaps low cost, favors the new, many older ideas also have merit. Having at least heard of as many obscure or archaic devices and techniques as possible can be very useful, where old ideas may find renewed application because of other perhaps unrelated advances. For example, etching diffraction gratings from laser created fringes exceeds the capability of the once necessary complex and precise mechanical ruling engines. A broad knowledge of natural history is beneficial, since many subtle optical and electronic ideas have been common in nature for millions of years. For example, the Bayer mask used to obtain color information from sensor arrays with monochromatic pixels, without too serious a resolution penalty is similar to the implementation of color vision in the pigeon by color masking of retinal cones. The eye of Anableps has bifocal lenses to give clear images above and below the water surfaces it favors. Contraction of a cat's pupil to a slit permits a larger dynamic intensity range than a circle would, without impairing the motion sensitivity of an already fovea-deprived organ. However, other than as examples, such details are beyond the present discussion of arts and ideas to enhance success.

18.6.6 Electronics

The transition from detected photons to sensible information is mostly performed by electronic devices, starting with amplification, via digitization, signal processing, and data reduction, to retrieval of information for archive and/or presentation as a human-readable display. The hardware implementation of ideas introduced in Section 18.5.7 has expanded to once unimaginable capabilities. Since this trend seems likely to continue, a discussion of specific devices is less useful than brief comments about general techniques, although even those become more comprehensive with each passing publication. An important trend is that optical techniques and devices are increasingly able to perform tasks once exclusive to electronics, from the extreme example of optical fiber replacing copper wire, to the integration of optical components with architectures formerly used only for electronic integrated circuits. The discipline of photonics parallels that of electronics, exploiting nanostructures with capabilities as yet barely imagined.

Once the detector has responded in whatever terms are available, the transition to a digital format should follow as soon as possible, usually via a pulse discriminator or ADC. Although in principle the same bandwidth product of data throughput may be represented as a rapidly varying 1-bit signal or as a less frequently updated multi-bit word, available hardware often suggests an optimal operation between these two extremes.

For real time systems where the input is synchronized with an independent activity, bottlenecks impose undesirable limits, and techniques such as dedicated hardware with a real-time operating system may be necessary, perhaps demanding low-level coding and testing, even with integer arithmetic and/or bit-sliced architecture. However, hardware devices change so rapidly that a once clever way to overcome a specific limitation can become an irrelevant complication. It seems that advances in hardware may shift risk areas toward increasingly complex software. However, that has not always been so and situations change quickly. All recent choices made to achieve system function are susceptible to critical review in the light of innovations. Sadly, past experience may be a detriment here.

Where delayed throughput is allowed, asynchronization can be achieved by a FIFO buffer, through single- or dual-port RAM, to hard disk files in a suitable format for later access, depending upon the required buffering and accumulation rates. More than one asynchronous interface may be needed to retain clean modularity in some complex systems, and to complete specified tasks within the time available. Hardware choices vary widely, from application specific or very high-speed integrated circuits (ASIC or VHSIC), through digital signal processors (DSPs), via field programmable gate arrays (FPGAs), to generalized multicore serial microprocessors. Each has its pros and cons. ASICs and VHSICs are fast, expensive and inflexible. DSPs are fast, versatile, and typically externally preprogrammed. FPGAs can be slower, but are dynamically, even conditionally, reconfigurable. Graphics processors, developed for online gaming, can off-load many of the more calculation-intensive tasks to enhance performance. Microprocessors are completely versatile, but with the most common operating systems may not be able to process signals in real time because of intrinsic and uncontrolled task timing, precluding interrupt driven actions. Real-time operating systems mitigate this, but application software is less readily available. At each stage of the processing one wishes to have fewer data to handle, but can perhaps tolerate more complex processing

algorithms. Hardware and software tools evolve rapidly. They become faster, more powerful, and more versatile, but also more complex, more difficult to use, and often have reduced backward compatibility.

Processing algorithms and hardware must be optimized as a system, retaining modularity and testability. Analysis is based on information intrinsic to the signal, but can be augmented by supplementary data to extend the limits of what information can be extracted, and the boundaries of attainable resolution and accuracy. Exploitation of independent auxiliary data, or any of a wide range of mathematical and/or statistical techniques, such as maximum entropy, analytic continuation, and super-resolution, may improve results. Understanding the output is not always trivial and iteration between the design concepts in Section 18.5.7, and the hardware choices here, may be necessary and can be challenging, particularly where the physics and phenomenology of the available measurement make extracting the required information an "inverse problem" with all its pitfalls of multiple values, measure theory errors, nonlinearities and numerical instabilities.

For OADS, the 200 MHz raw signal was dynamically scaled and filtered before the ADC, followed by discrimination and conditional editing implemented in discrete components prior to loading into the FIFO, subsequently accessed by a TMS320C25 DSP, using scaled integer arithmetic with a real-time operating system. Its program was downloaded on initialization, permitting easy upgrades during development, and later evolution of function. For example, velocity, pressure and temperature were reported during twenty-four 1–2 hour flights of the F-16B aircraft in 1990. Velocity and particle properties were measured during later research flights on F-104 and SR-71 aircraft, with final emphasis changed towards particle size and concentration measurements on some final DC-8 flights.

18.6.7 Power supplies

Without sufficient attention, power supplies can be a high risk item. Not only can they fail, but they may also induce into the signal instabilities, correlated fluctuations, noise in one or more frequency bands, and inconstancy with input voltage, temperature, load, and/or ambient conditions.

Although these are mitigated by careful design, specifications must be confirmed by testing, calibration, characterization and monitoring of any artifacts in the light source or signal. For thermal sources, inertia prevents significant fluctuations faster than a few tens of kilohertz, but solid-state sources can have sub-nanosecond fluctuations. Accordingly, great care must be given to the power supplies driving the light sources, detectors, and sensitive electronics. Although nontrivial, feedback stabilization should always be considered and may often be necessary. Note also that while a single power supply may perform acceptably, simultaneous operation of several even apparently unrelated supplies in the field environment may be susceptible to greatly worsened noise performance.

18.7 TESTING AND CALIBRATION

Much can be learned from just looking at the system and phenomenon, and even incidental observations may indicate unexpected behavior, whose rectification builds confidence in subsequent measurements and the validity of conclusions. Components should be rigorously tested as soon as available. Curious effects are often noticed near the limit of awareness, and although they may be neither repeatable nor completely understood, their dismissal as insignificant may invite later catastrophe. Table 18.12 augments Table 18.6 suggesting further questions whose proper answers may improve confidence.

No amount of testing can ever prove that software is infallible. The likelihood of unknown error is high. Good designs allow upload of new code modules to correct anomalies found in testing, permitting analysis and rectification before damage occurs. Power supplies must not only be tested as individual units in isolation, but also in their final operational configuration. It is well to impose test conditions outside those expected, and for longer periods, logging all relevant parameters with higher accuracy than seems necessary. It is important to examine these test results in various graphical formats and actually think about what trends might mean. Surprisingly many failures have origins in data measurable, or even measured, long before, but neither understood nor even properly examined.

Tests should simulate operation as closely as possible. In OADS as a subset of system testing, the lasers could be independently modulated and the response of each APD to its own and other channels examined for various targets. Complete test documentation allows later insights about why the results are not identical with what was expected—they never are—and can allow compensation or correction to improve the measurements or performance, or at least to predict the boundaries of accuracy.

Calibration is based on knowledge gained by testing. Quantitative characterization may permit adjustment of output values to reflect their input origins during operation, or to correct measurements later. The two aspects of calibration are the quantitative aptitude of the apparatus for its design purpose, and the traceability of the recovered information to required properties of the examined phenomenon.

For OADS, the only important physical calibration was the separation and parallelism of the sheet pairs, performed by exposing a dimensionally stable sensitive film to the illuminated pattern at several axial stations and using a traveling microscope to measure the geometry.

18.7.1 General health and status monitoring

Extensive monitoring is desirable. Power on system (or self) test (POST), and routines to exercise the equipment as it approaches operational status, are conventionally provided, but should be augmented by health and status monitors for all quantities that could affect performance or the values of retrieved measurements. Monitored values should be

Table 18.12 Testing and calibration

Initial function	Does everything work as planned? Does it meet design expectations?
Anomalies	What do unexpected observations mean?
Deterioration	In a simulated environment, does it get worse at an acceptable rate?
Contingencies	How is less-than-ideal operation to be accommodated? How is the unexpected to be handled?

recorded as a time-logged data stream at whatever rate is appropriate for the parameter concerned. Table 18.13 lists examples of usefully monitored quantities with supplementary notes following.

For temperature, pressure, and thermal controls, a readout rate of 1 Hz is mostly adequate. Power supplies require either a more rapid readout, or a separate assessment of root mean square (rms) fluctuation, noise, crest factor, or other warning of unexpected operational noise or transient spikes. If the illumination intensity can fluctuate on a timescale shorter than the inverse monitoring rate, then additional properties should be assessed, as with power supplies, above. Most serious may be fluctuations and drifts on a timescale similar to that of the detector or camera exposure or readout, but not necessarily in phase. This can lead to quite subtle effects in later processing or, for 2-D sensing, picture-to-picture variations, and moiré; or other patterns in single pictures. If such obvious effects persist despite design approaches intended to remove them, then the bad data may be rejected later only at the cost of efficiency, if they may be rejected at all. However, small attendant changes in SNR in the detector signal or picture are not so easily discovered nor compensated. Indeed, these effects may be almost impossible to compensate, so it is important during design and testing to eliminate or at least reduce problems associated with the interactive effects of power supplies and the devices they drive. For example, calibration may change if only parts of a multifunction apparatus are currently activated, and this must be reported in the health and status archive attached to the provenance file for the specific experiment. Monitoring helps support claims for the validity of measurements.

Many internal systems merit stabilization by negative feedback, with the loop signal monitored. It is relatively cheap and easy to characterize such behavior. Experimental data may be only poorly understood without at least a time record of rms and peak-to-peak fluctuations, together with the amplitude and phase, or even power spectrum of the light source noise and drift.

18.8 HOSTILE ENVIRONMENTS

Optoelectronics can be challenging enough even in benign conditions. Hostile environments that threaten success are of two kinds, hostile to the instrumentation, and hostile to the observer. These include all aspects of the measuring process during operation, and the program that brings the measuring device to the current situation.

In the first class, conditions such as extremes of temperature, pressure, radiation, acceleration, vibration, shock, chemical attack, electromagnetic fields, noise, power source fluctuations, cosmic radiation and perhaps other specific conditions lead to potentially less than satisfactory results. Even if the system is designed to undergo graceful degradation, successive levels of deterioration may include (1) loss of calibration leading to poor or untrustworthy measurements, (2) loss of function, preventing any measurements, (3) destruction of the instrument, or finally (4) collateral damage, implying that the instrument is not only destroyed but also causes associated damage that may be much greater. The loss of a $1.5b Mars mission because the same type of vulnerable component was used in all three of the triply redundant timing circuits was a sad example of instantaneous and unexpected failure.

Table 18.13 Candidate health and status monitors

ISO date, time, and all experimental provenance
Power supply input and output voltages and currents
Illumination level and stability of the light source(s)
A sampling of temperatures throughout the apparatus
Internal pressure, humidity, and other vapor pressures
Electrical and thermal control voltages and currents
Fluid pressure, temperatures, flow rates, leaks
Loop signal in negative feedback stabilization wherever used
Independent views of the phenomenon and instrument (video cameras)

Table 18.14 Ideal properties

Simple, low technology	Easy to understand
Rigid, robust, small	Resistant to damage
Modular, reliable, cheap	Easy to diagnose and fix
Automated by default	Reduces operator mistakes
Low power demand	Stays cool, widely applicable
Calibrated, verified	Constant and trustworthy
Tested, characterized	Known and understood

Table 18.15 Warnings

Chemical attack ruins	Noise reduces capability
Gravity changes disturb	Hard radiation damages
Vibration changes things	Power supplies fail
Electromagnetic radiation gives errors	Windows get dirty
Stress reduces performance	Software crashes
Temperature drifts misalign	Pressure distorts
Changes affect things unexpectedly	Fluids leak

Three aspects of environments hostile to the designer of the system are (1) not enough money, (2) not enough time, and (3) not having a champion. Although the first two are common and obvious, the last is the most serious. To make anything happen takes someone dedicated to its successful completion. During instrument deployment the observer may also be exposed to hostile environments, such as extremes of temperature, pressure, vibration, acceleration, radiation, noise, antagonism, unreasonable expectations, or stress.

Four classes of hostility merit consideration during the design of the instrument or methods for its deployment and use. These are where (1) conditions in the phenomenon are extreme, (2) access may be constrained, (3) the instrument is exposed to an extreme environment, and (4) the observer is under difficult or stressful conditions. Table 18.14 introduces a few of the ideal properties whose consideration may improve success.

Obvious though they may appear, warnings in Table 18.15 can apply in benign as well as hostile environments and may apply to the instrument or to the operator, or both.

Outdoor or field installations are often exposed to diurnal temperature cycling or steep temperature gradients, whether in use or not. Direct or reflected sunlight, rainwater, and unsuspected fixed or pulsed magnetic fields can introduce problems. Exposure to moisture or high humidity can age optical surfaces and dielectric coatings, also possibly causing fungus buildup or other biologically driven damage surprisingly rapidly. Any one of these is easily averted or accommodated: avoiding them all, and also the unanticipated ones, can be challenging.

In summary, optoelectronic measurements in hostile environments are difficult. Analysis and thoughtful design mitigate risk. Planning graceful degradation, anticipating the unknown unknowns, talking honestly and often with the customer, and having plenty of money, time, support, and good luck are all beneficial.

18.9 CHECKLISTS

Table 18.16 lists some component properties deserving examination, followed by Table 18.17 stressing environmental aspects.

18.9.1 Analyzing unwanted effects

Any part of a system or its environmental exposure may have unanticipated effects on any other part, and to minimize later surprises a useful technique is to create a spreadsheet whose first column contains all possible affective properties and whose first row contains all aspects that could suffer effects. Examples of both properties

Table 18.16 Component considerations

Optical radiation source

Type	Thermal, laser, diode, flash, arc, discharge, synchrotron, fluorescence, phosphorescence, explosion, biological, others
Geometry	Size, 2- and 3-D beam shape, divergence, pointing, spatial coherence
Wavelength	X-ray, far and near ultraviolet, visible, near and far infrared
Spectrum	Bandwidth, distribution, temporal coherence, tuning
Intensity	Intrinsic luminance, brightness, spatial and temporal stability, noise
Power	Efficiency, electrical and thermal stabilization, cooling
Polarization	Type, orientation, degree
Stability	Frequency, pointing, intensity, beam shape

Optical components

Lenses	Type, material, design, specification, compromises, speed, resolution, surface finish, tolerances, handling and mounting
Beam splitters	Flatness, thickness, multiple reflection, parallelism
Filters	Bandwidth, tuning, edge roll-off, efficiency, reflections, absorption
Aberrations	Seidel, chromatic, higher order
Geometry	Complexity, aperture, stops, vignetting
Refraction	Refractive index, angular dependence of reflection, dispersion
Materials	Birefringence, transmissivity, homogeneity, strength, chemical stability
Coatings	Complexity, efficiency, materials, durability, scatter

Optical system

Mechanical	Design, rigidity, robustness, beam control, adjustments (and ranges)
Alignment	Methods, sequence, sensitivity, criteria, stability, locking, interactions
Polarization	Necessity, preservation, type ("S," "P," circular or elliptical)

Detectors

Type	Array, line, single point, static (e.g., film), dynamic (e.g., electronic)
Principle	Photoelectric effect, permanent or reversible chemical change, thermal or mechanical effect
Mechanical	Sensitive area, packaging, thermal stabilization, protection
Detection	Capabilities, quantum efficiency, sensitivity, gain, noise
Speed	Dead time, read rate, fatigue, recovery time
Noise	Shot, dark, amplifier, statistical, after-pulsing

Table 18.17 Environmental considerations

Mechanical	Vibration, shock, *g*-loads, rigidity, static and dynamic stresses, strength, materials, construction, pressure, extraneous damage, weight, size, cost
Thermal	Heat, cold, static/dynamic gradients, changes, distortion, misalignment, insulation, control
Radiation	Internal and external stray light, stops, filters, surface treatments
Aero-optics	Variable refractive distortions of wave front
Noise	Electromagnetic or radio frequency interference (EMI, RFI), grounding, hardening, screening, isolation, spurious signals, power supplies

and components are given in Table 18.18. In the body of the spreadsheet, a manual entry is made into each and every cell to verify that any mutual effects between row and column heading have been considered. Effects here include any possible interaction or sensitivity, which may then be quantified, or less tangible warnings, such as excessive complexity. The exercise may trigger a memory of a similar or analogous situation with good or bad former outcome. The cell entry creates an audit trail later to become a powerful diagnostic tool. If you think this is a tedious and unnecessary exercise, try fixing things later. Most entries will indicate that there is no cross effect. Other possible entries are that the effect has been considered and is acceptable or that the effect has been analyzed and remedial action taken, as indicated by a traceable reference. Perhaps most seriously, and a major justification for this exercise, is that an effect is found to be likely but unknown, requiring further effort directed either towards its understanding and rectification, or its quantitative inclusion in the risk analysis.

Table 18.18 is not intended to be exhaustive, but merely contains an exemplary set of possible subject items to be reduced, refined and/or augmented according to the specific optoelectronic system.

As an example, a reduction in ambient pressure once seriously compromised the behavior of an argon ion laser. With the reduced refractive index of the lower-pressure air, the emission angle from the Brewster windows inside the cavity changed sufficiently to detune the resonator. Unexpectedly, retuning did not restore the power since the new direction in the cavity passed through a slightly thicker section of the wavelength selection prism, also intracavity. Even though the increase in absorptive attenuation was slight, the many passes within the resonator substantially reduced the cavity gain. The first lesson is that reduced ambient pressure changed the pointing direction and power in a way that was not immediately recoverable by realigning the cavity—the prism also had to be physically displaced. The second lesson is that since such ion lasers are no longer in common use, the first lesson is largely irrelevant. The real lesson therefore is that a situation must be examined in greater depth and with more imagination than might at first seem necessary.

In a working career thousands of such examples appear, but using the cross-reference effects matrix postulated above can improve the probability of creating a working design with minimal evolutionary iterations.

Table 18.18 Possible effects matrix

Examples of properties	Aberrations, absorption, adjustments, ageing, alignment, blooming, breaking, calibration, changes, charging, chemistry, collimation, constancy, corrosion, cost, damage, design, detectivity, dimensions, dispersion, distortions, durability, efficiency, electromagnetic interference, errors, external stray light, field, flare, friction, fungal growth, ghosts, g-loads, ground loops, hard radiation, heat sources and sinks, insulation, internal stray light, isolation, Lagrange invariant, mounting methods, noise, nonlinear effects, optical feedback, outgassing, packaging, pointing, polarization, positioning, pressure, quality, quantum efficiency, quantum statistics, radio-frequency interference, reflections, refractions, rigidity, robustness, screening, sensitivity to everything, shock, signal and data formats, size, specifications, spectrum, stability, static and dynamic stresses, stiction, strength, temperature, testing, thermal gradients, transmission, vibration, vignetting, weight
Examples of components	Adhesives, apertures, baffles, beam splitters, bearings, coatings, construction, detectors, electronics, Faraday and Kerr effect devices, fibers, filters, GRINs, health monitors, lenses, light sources, materials, mounts, locking, moving parts, polarizers, power supplies, software, stops, structures, surface treatments, thermal control, windows

Table 18.19 Design homilies

- Software and power supplies give more trouble and consume more resources than can possibly be imagined; plan not to be disappointed by this
- Performance and ruggedization are better achieved by good initial design than by any number of later attempts at remediation
- Underspecified science leads to impossible engineering
- Communication, error checking, and conflict resolution are equally applicable to systems, hardware, software, scientists, engineers, managers, and customers who provide the funds
- Knowledge alone is not sufficient ... choosing people with proper experience is essential
- Configuration control from the start, where design evolves most rapidly, may prevent errors; unconscious neglect makes later disasters almost inevitable
- A top-down straw-man design, starting at system level, and percolating down to isolate any area of uncertainty, is desirable as soon as possible
- Hardware and software must be modular, and should map isomorphically
- Optical, mechanical, interconnection, and functional layouts must be available early
- Everything is made of India rubber—nothing is truly rigid
- The Internet is an extremely valuable source of ideas and information; however, since content is not typically peer reviewed, it must be independently corroborated
- Fifty hours in the lab will easily save 1 h in the library
- Simplicity is a worthy goal, but oversimplification can invite problems
- Innovation is only acceptable where no possible method currently exists
- Quality is meeting the specifications
- If in doubt, make it stout, out of things you know about

Table 18.20 Attitude homilies

- Talk to the customer
- Intend the excellence of your design
- Be honest, pragmatic, and courageous
- Study how others have approached and solved similar problems
- Test potential solutions and guidelines for relevance to your purposes
- Understand why something was done the way it was done, before committing to doing it differently
- Seek advice, comments, and recommendations; detractors are especially creative, and will be most helpful, if somewhat indirectly
- Consider all guidance carefully: suspect that which does not fit with intuition
- Never trust opinion, however informed, without supporting evidence
- If you are being brilliantly innovative, think about it harder, and insist others check it
- Question everything; analyze and understand the answers
- Check every assumption; check it again
- Historical reasons can vary from wise cautionary guidelines to unexamined prejudice
- Distinguish between intelligent compromises and short cuts; there are no short cuts
- Avoid moving parts; avoid nonstandard parts; if possible, avoid parts
- Buy the best and cry only once
- Hope is not a strategy

18.10 HOMILIES FROM EXPERIENCE

Q. How do you succeed?	A. Good choices
Q. How do you make good choices?	A. Experience
Q. How do you get experience?	A. Bad choices

Many lessons may be encapsulated in aphorisms, whose illusion of triviality belies the pain associated with truly apprehending their significance. As light relief, two tables give examples of maxims encapsulating arts acquired by bitter experience. Table 18.19 passively addresses system design issues, whereas Table 18.20 advances less scientifically tractable ideas and attitudes whose comprehension helps to reduce risk. These more psychologically based concepts are cautiously offered to mitigate the anguish incurred by those with the temerity to embark upon optoelectronic projects.

FURTHER READING*

The following bibliography merely skims a vast collection of interesting and useful works. The criterion for their choice is that each represents an in-depth study of one or more aspects of optoelectronics, which is either useful of itself or presents a method or analogy of more general application. Although access to the Internet is now necessary, its content is not peer reviewed and may not be trusted without the corroborative evidence of which the bibliography includes examples.

Abbiss, J. B. and Smart, A. E. (eds) 1988. *Photon Correlation Techniques and Applications: OSA Conference Proceedings Series* vol. 1 (Washington, DC: Optical Society of America).

Bass, M. (ed.) 1995. *Handbook of Optics*, vols 1–4 (New York: McGraw-Hill).

* Author's Note: Many ideas and occasional short sections of modified text are reproduced from Brown, R. G. W. and Smart, A. E. 1997 Practical considerations in photon correlation experiments. *Appl. Opt.* 36: 7477–7479, with kind permission of the Optical Society of America.

Berne, B. J. and Pecora, R. 1976. *Dynamic Light Scattering* (New York: John Wiley & Sons).

Born, M. and Wolf, E. 1980. *Principles of Optics*, 6th edn (Oxford, UK: Pergamon).

Brown, N. J. 1986. Preparation of ultra-smooth surfaces. *Annu. Rev. Mater. Sci.* 16: 371–388.

Brown, R. G. W. 1987. Dynamic light scattering using monomode optical fibers. *Appl. Opt.* 26: 4846–4851.

Brown, R. G. W. and Daniels, M. 1989. Characterization of silicon avalanche photodiodes for photon correlation measurements. 3: Sub-Geiger operation. *Appl. Opt.* 28: 4616–4621.

Brown, R. G. W., Jones, R., Ridley, K. D. and Rarity, J. G. 1987. Characterization of silicon avalanche photodiodes for photon correlation measurements. 2: Active quenching. *Appl. Opt.* 26: 2383–2389.

Brown, R. G. W., Ridley, K. D. and Rarity, J. G. 1986. Characterization of silicon avalanche photodiodes for photon correlation measurements. 1: Passive quenching. *Appl. Opt.* 25: 4122–4126.

Brown, W. (ed.) 1993. *Dynamic Light Scattering* (Oxford, UK: Clarendon).

Chu, B. 1991. *Laser Light Scattering*, 2nd edn (San Diego, CA: Academic Press).

Dainty, J. C. (ed.) 1984. *Laser Speckle and Related Phenomena* (Berlin: Springer).

Danielsson, L. and Pike, E. R. 1983. Long-range anemometry—A comparative review. *J. Phys. E* 16: 107–118.

Dautet, H., Deschamps, P., Dion, B., MacGregor, A. D., MacSween, D., McIntyre, R. J., Trottier, C., and Webb, P. P. 1993. Photon counting techniques with silicon avalanche photodiodes. *Appl. Opt.* 32: 3894–3900.

Driscoll, W. G. and Vaughan, W. (eds) 1978. *Handbook of Optics* (New York: McGraw-Hill).

Durst, F., Melling, A., and Whitelaw, J. H. 1976. *Principles and Practice of Laser Doppler Anemometry* (London: Academic).

Hecht, E. 1987. *Optics* (Boston, MA: Addison Wesley).

Johnson, C. S. and Gabriel, D. A. 1994. *Laser Light Scattering* (New York: Dover)

Kingslake, R. 1983. *Optical System Design* (New York: Academic).

Lasers, Photonics and Environmental Optics 20 August 2000 *Special Issue of Applied Optics* 40.

Lasers, Photonics and Environmental Optics 20 October 1997 *Special Issue of Applied Optics* 36.

Luxmoore, A. (ed.) 1983. *Optical Transducers and Techniques in Engineering Measurement* (London: Applied Science).

Macleod, H. A. 1986. *Thin Film Optical Filters*, 2nd edn (London: Adam Hilger).

Mandel, L. and Wolf, E. 1995. *Optical Coherence and Quantum Optics* (Cambridge, UK: Cambridge University Press).

Mayo, W. T. and Smart, A. E. (eds) 1980. *Photon Correlation Techniques in Fluid Mechanics* (Stanford, CA: Stanford University).

MIL-HDBK-141. 1962. *Military Standardization Handbook on Optical Design* (Washington, DC: US Government Printing Office).

Minnaert, M. 1954. *Light and Color in the Open Air* (New York: Dover).

Photon Correlation and Scattering: Theory and Applications 20 July 1993 *Special Issue of Applied Optics* 32.

Saleh, B. E. A. and Teich, M. 1991. *Fundamentals of Photonics* (New York: Wiley-InterScience).

Schätzel, K. 1987. Correlation techniques in dynamic light scattering. *Appl. Phys. B* 42: 193–213.

Siegman, A. 1986. *Lasers* (Mill Valley, CA: University Science Books).

Smart, A. E. 1991. Velocity sensor for an airborne optical air data system *AIAA: J. Aircraft* 28 163–164.

Smart, A. E. 1992. Optical velocity sensor for air data applications. *Opt. Eng.* 31: 166–173.

Smart, A. E. 1994. Folk wisdom in optical design. *Optics and Photonics News, Engineering and Laboratory Notes* 5.

Smith, W. 1972. *Modern Optical Engineering* (New York: McGraw-Hill).

Suparno, D. K., Stamatelopolous, P., Srivastra, R., and Thomas, J. C. 1994. Light scattering with single mode fiber collimators. *Appl. Opt.* 33: 7200–7205.

van de Hulst, H. C. 1981. *Light Scattering by Small Particles* (New York: Dover).

Vaughan, J. M. 1989. *The Fabry–Perot Interferometer* (Bristol: Hilger).

Weast, R. C. *Handbook of Chemistry and Physics*, current edn (Boca Raton, FL: CRC Press).

Welford, W. T. and Winston, R. 1978. *The Optics of Nonimaging Concentrators* (New York: Academic).

Wyatt, C. L. 1991. *Electro-Optical System Design for Information Processing* (New York: McGraw-Hill).

Yariv, A. 1995. *Optical Electronics*, 5th edn (Oxford, UK: Oxford University Press).

Index

Note: Page numbers followed by "f" indicate figures; those followed by "t" indicate tables.